Emerging Non-destructive Technologies for Food Quality Analysis

C. K. Sunil • Ashish Rawson • D. V. Chidanand
S. Shanmugasundaram
Editors

Emerging Non-destructive Technologies for Food Quality Analysis

Springer

Editors
C. K. Sunil
National Institute of Food Technology
Entrepreneurship and Management
(NIFTEM-T)
Thanjavur, Tamil Nadu, India

Ashish Rawson
National Institute of Food Technology
Entrepreneurship and Management
(NIFTEM-T)
Thanjavur, Tamil Nadu, India

D. V. Chidanand
National Institute of Food Technology
Entrepreneurship and Management
(NIFTEM-T)
Thanjavur, Tamil Nadu, India

S. Shanmugasundaram
National Institute of Food Technology
Entrepreneurship and Management
(NIFTEM-T)
Thanjavur, Tamil Nadu, India

ISBN 978-3-031-95047-6 ISBN 978-3-031-95048-3 (eBook)
https://doi.org/10.1007/978-3-031-95048-3

This Springer imprint is published by the registered company Springer Nature Switzerland AG
The registered company address is: Gewerbestrasse 11, 6330 Cham, Switzerland

Contents

1 Food Quality and Analysis—An Overview 1
P. Krishnakumar, Panoth Abhirami, C. K. Sunil, Ashish Rawson, D. V. Chidanand, and S. Shanmugasundaram

2 Spectroscopy—Optical Methods: Visible, NIR, FTIR 39
A. Shivani Indumathi, S. Charan Adithya, J. Deepa, G. Jeevarathinam, and R. Pandiselvam

3 Nuclear Magnetic Resonance 65
Muhammed Navaf, Monica R. Nemtanu, Amin Mousavi Khaneghah, and Kappat Valiyapeediyekkal Sunooj

4 Computer Vision Systems 99
Shubham Chandrakar, M. Suji, and R. Mahendran

5 X-Ray, CT, and MRI 135
Mathangi Raja Sekhar, Sruthileena Bommi, B. N. Vinutha, M. A. Yogesh, C. K. Sunil, Ashish Rawson, S. Shanmugasundaram, and D. V. Chidanand

6 Hyperspectral Imaging 159
Princess Tiffany D. Mendoza, Paul R. Armstrong, and Kaliramesh Siliveru

7 Multispectral Imaging Technique: Principles and Recent Applications in Food Processing 187
Tej Bhan Singh, Yuvaraj Ghosh, Riya Kundu, and Kambhampati Vivek

8 Backscattering Imaging 215
Abhishek Dasore and Norhashila Hashim

9 Biospeckle Imaging 237
T. Seeranguraya, T. Senthilkumar, and V. Chelladurai

10 Thermal Imaging Applications in Food Processing 259
D. Ethaya, H. Cubarani, R. Himavanth, and S. Anandakumar

11 Terahertz Spectroscopy 287
N. Sharath Kumar, C. K. Sunil, Ashish Rawson, D. V. Chidanand, and S. Shanmugasundaram

12 Ultrasonics 325
S. Guberavishnu, S. Surender, Pranav Nanda Kishore, Ashish Rawson, C. K. Sunil, D. V. Chidanand, and S. Shanmugasundaram

13 Electronic Nose and Electronic Tongue 351
Veerapandian Chandrasekar, Vijayaram Eyarkai Nambi, and Saravanan Shanmuga Sundaram

14 Biosensors 379
Anantika Sharma, B. O. Madhu, Ashok Suraj, C. Indumathi, C. K. Sunil, Ashish Rawson, D. V. Chidanand, and S. Shanmugasundaram

15 Techniques Based on Electrical Properties of Food 417
Ashok Kumar Yadav and Sandeep Janghu

16 Color and Texture Measurement Methods 451
Ravi Harini, K. Sandhya, Panoth Abhirami, M. Sivasathiya, and N. Venkatachalapathy

17 AI and Machine Learning Applications 501
Vinass Jamali, Vismaya K. Sachithanandhan, Eyarkai Nambi, Loganathan M., S. Shanmugasundaram, and V. Chandrasekar

Index 557

Chapter 1
Food Quality and Analysis—An Overview

P. Krishnakumar, Panoth Abhirami, C. K. Sunil, Ashish Rawson, D. V. Chidanand, and S. Shanmugasundaram

1.1 Introduction

Food Quality is the degree of excellence for factors such as taste, appearance, and nutritional content. It refers to differentiation among a product's properties and influencing users as a result by improved acceptability. It is classified into organoleptic characteristics (e.g., taste, aroma, appearance) and nutritional value of food. Food safety standards are also essential for determining food quality and are a marketing tool for consumer acceptance (Nollet et al., 2007; Vizzarri, 2021). The acceptable level of food quality also varies depending on the consumer region, culture, and economic level. Food quality is determined by confirmation of product specifications in the food industry. Food quality measurement and control has gained much importance considering growing consumer demand and stricter regulatory scrutiny (Chhetri, 2024).

1.1.1 The Importance of Food Quality Control

Food quality control is critical in eliminating hazards and safeguarding consumers' safe food consumption. It also denotes that food products should fulfill the Food Safety Management standards. Further, it reduces buyers' threats from duplicate traders and avoids poor quality goods through food safety standards and regulations. In the food industry, keeping high-quality standards for food is vital to deliver user

P. Krishnakumar · P. Abhirami · C. K. Sunil (✉) · A. Rawson · D. V. Chidanand · S. Shanmugasundaram
National Institute of Food Technology, Entrepreneurship and Management (NIFTEM-T), Thanjavur, Tamil Nadu, India
e-mail: sunil.ck@iifpt.edu.in

C. K. Sunil et al. (eds.), *Emerging Non-destructive Technologies for Food Quality Analysis*, https://doi.org/10.1007/978-3-031-95048-3_1

safety, fulfillment, and healthiness (Abass et al., 2024). Food quality and safety are regularly discussed in the media because of food calamities. Until now, food quality and safety have not been distinguished because they vary depending on user opinion (Vizzarri, 2021). In the food supply chain, buyers are safeguarded from foodborne illness by food quality control, and it also gives self-assurance to buyers through proper labeling to fulfill their specific nutritional needs. Food quality control assures buyer safety through Hazard Identification, Compliance with regulations, Inspections and testing, Traceability, and Recall procedures. It gives buyers confidence to purchase food products available in the market. Food quality control aids the product: Benchmark setting of food quality assures supplier to maintain standard quality, monitors and controls various processes in food production, frequently inspects and tests food products, and implements corrective actions in the areas that need improvement.

Food products are developed based on taste, texture, appearance, and nutritional value through the effects of quality control. It builds buyer trust and loyalty, and through these, sales and revenue can be increased. Benefits of food quality control actions include buyer satisfaction, brand reputation improvement, food hygiene, business profitability, and waste reduction. Assuring food quality, especially within the contemporary global fast-paced food industry environment, remains one of the key priorities for business players. Technological advancements, including artificial intelligence (AI) and machine learning advances, are critical to improving quality control in food quality analysis through prediction (Kumar et al., 2021). These technologies improve data analysis, outcome prediction, and enhancement of processes in the food industry (Mohd Ali et al., 2021).

1.1.2 Food Quality Standards

World Trade Organization's (WTO) definition of a standard is a technical trade barriers (TBT) agreement, as general and repeated use, guidelines, or characteristics for products or related processes and production methods, with which compliance is not mandatory provided in the approved document of recognized body. A guideline for food companies can develop their policies and execute higher level food safety and food law compliance, food quality and customer requirements, and the protection of people, animals, and environmental welfare through a quality management system by customer or third-party standards. Globally, food safety and quality standards developed and executed by public and private organizations have increased in terms of growing customers, buyer expectations, interest in food safety, and legal and official requirements (Nollet et al., 2007; Vizzarri, 2021). A comprehensive customer or third-party standard will comprise training employees, specifications of ingredients, traceability, quality manual, standard operating procedures, critical control points, sampling and analytical schedules, reporting systems, and review processes. Significant hurdles can be overcome in the food business by ensuring food safety and upholding high standards (Chhetri, 2024).

1.2 Food Quality and Factors Affecting It

From the viewpoint of end user food quality may have varied definition. However, from a consumer's perspective intrinsic and extrinsic factors together defines the quality of the food product.

1.2.1 Intrinsic Factors

These characteristics connect to the product's physical appearance and provide it with usefulness. Each product has unique intrinsic qualities that vanish after consumption and cannot be changed without altering the product's fundamental characteristics. It is the result of product properties like water activity (aw), pH/total acidity, microbiological counts of the product, oxygen availability, reduction potential (Eh), chemistry of the product, added preservatives (e.g., salt, spices, antioxidants), and product formulation (Nollet et al., 2007; Vizzarri, 2021). The influence of intrinsic properties (e.g., oils and fats, carbohydrates, proteins, pH, physical structure, water, and salt) will affect the quality of essential oils (Zheng et al., 2024).

1.2.2 Extrinsic Factors

Extrinsic factors, covers those features which can be related to the product but not incorporated into it. These include but are not limited to particulars related to price, brand name, shop name, or sometimes as general as the nation of origin. Extrinsic factors also include product distribution and storage constraints.

1.2.3 Other Factors Which Affect the Food Quality

Food quality is affected by several factors, which are classified into two categories: technical standing and customer fondness. Based on the technical opinion, the factors affecting the quality of food are structure deterioration, essences of color and flavor, nutritional value, functional ingredients, and adulteration. Based on the customer's perspective, the factors affecting the quality of food are directly related to human sensory characteristics like visual (color, moisture, and overall appearance), touch (elasticity, sliminess, softness, and hardness), taste and smell (chemical composition), and mouthfeel (texture, tenderness, softness, and chewy sensation) (Nollet et al., 2007; Vizzarri, 2021).

1.2.3.1 Nutrition

Among all the factors, the nutritional quality of the food is essential. From an everyday perspective, food quality means food with a high nutritional rate. Due to increased awareness towards health promoting properties of food components consumers select food quality based on its nutritional content, like high protein, high fiber, rich in vitamins and minerals, and crucial nutrients, *i.e.,* omega-3 fatty acids, low cholesterol, zero trans-fat, sugar-free, gluten-free, *etc.* Familiarity is created among consumers via nutritional labeling based which contains information about nutrients and is compulsory for packaged and supplementary foods, which should be mentioned in the front-of-pack labeling (FOPL). It gives consumers the ability to select the quality of food based on nutritional, healthier, and inspiring composition (Nollet et al., 2007; Vizzarri, 2021).

1.2.3.2 Organoleptic

The stimulus, which is an objective evaluation of the effector of the sensory perception of color, smell, taste, and palatability, and in the broader sense, the physiological effect of a physio chemistry unit are the two determinants of the organoleptic or gastronomic worth of food (Nollet et al., 2007; Vizzarri, 2021), which give the receptor an interpretable meaning out of the inputs in terms of a signal. The objective stimuli initiate these cascade signals, leading to the receptor behaving in a particular manner (Tremolieres, 1970).

1.2.3.2.1 Flavor and Aroma

Among the sensory properties, flavor and aroma are significant properties of consumers' desired food quality. In the food, more than one thousand flavor compounds were found based on literature. During food processing and storage, changes in flavor and aroma are among the most critical factors in determining the quality of the desired food. Each food product and livestock species' typical flavor characteristics vary because of specific flavor originators. Generally, Foods that contain a small number of volatile compounds are odorless (For example, Fresh saltwater fish). In contrast, foods that contain large amounts of volatile compounds have some odor (For example, Freshwater fish contain more pyrrolidine due to this earthy odor). Flavor and aroma are among the quality indexes of food. For example, trimethylamine or dimethylamine, among many presences, or putrescine and cadaverine absence indication of best and worst index, respectively (Nollet et al., 2007; Vizzarri, 2021).

1.2.3.2.2 Color

Buyer attraction to food products by color, from surface characteristics. Among these colors is food quality. Food products can be selected or rejected by color, and aroma, texture, and taste can be evaluated earlier than aroma. In the meat and poultry product processing industry, color is most important and influences the indication of the quality of the final product. In food processing, the color of the product is important because of several factors, namely, uniform color to maintain the entire process, to avoid any discoloration during processing by external and internal parameters, to enhance the color of the final product after processing, and to establish quality standards for acceptability by the buyer through the color of the final product (Nollet et al., 2007; Vizzarri, 2021).

1.2.3.3 Microbiology and Safety

Food products contain beneficial as well as harmful microorganisms to humans. In fermented foods, microorganisms are beneficial, but some potential pathogenic microorganisms cause illness in consumers. Hence, government agencies should develop safety standards to ensure quality and safe food for consumers (Nollet et al., 2007; Vizzarri, 2021).

1.2.3.4 Processing Methods

Living organisms make foods, and they deteriorate over time. Food spoilage accords by physical, chemical, and biological factors. The food must be processed to avoid spoilage. The quality of food is affected by processing. Two types of processing methods: (1) Thermal Processing and (2) Nonthermal processing.

1.2.3.4.1 Thermal Processing Methods

Thermal process technology is one of the most common methods used in the food industry, contributing to the microbiological safety of products. These methods depend mainly on the production of heat except when heating is done on the outside of the product through burning fuel or a heater resistive and its conversion into the product through conduction and convection methods (Pereira & Vicente, 2010).

1.2.3.4.2 Nonthermal Processing Methods

The term 'nonthermal processing' is commonly used to describe technologies effective at ambient or sublethal temperatures. Nonthermal technologies include high-pressure processing, pulsed-field electric treatment, ultrasonic waves, UV light, light pulses, irradiation, etc. Both direct and alternating magnetic fields can inactivate microorganisms to a certain extent (Butz & Tauscher, 2002). Some of these treatments may include heat because of the generation of internal energy such as Adiabatic heating and resistive heating in HHP or PEF but regarded as nonthermal on contrast to thermal processing technologies, they are capable of eradicating reliance on high temperatures to neutralize the microorganism to prevent the undesirable effects of heat on the taste, appearance, and nutritional quality of foods.

1.3 Global Strategy on Food Safety

Safety and Quality goes hand in hand, and many a times they affect each other. If the safety is compromised the quality of food may also be low. Ensuring access to safe food is fundamental to human well-being. To uphold consumer's right to food safety, governmental authorities must oversee and regulate food production to ensure compliance with established safety standards. In an increasingly interconnected world, strengthening food safety measures is imperative, necessitating the enhancement of national food safety management systems to facilitate greater collaboration both domestically and internationally. Governments must proactively implement comprehensive food safety policies; otherwise, ensuring reliable access to safe food will remain a challenging and unattainable objective. Government organizations, food scientists, and industry have one of the aims of ensuring food safety for consumers. National and International organizations are eliminating food risks through food safety (Mu et al., 2024).

Since its establishment in 1948, the World Health Organization (WHO) has been dedicated to reducing the global burden of foodborne diseases. It has played a crucial role in assessing food borne diseases impacts, strengthening national surveillance systems, and advocating for prevention strategies through education and awareness initiatives. WHO has also ensured that Codex Alimentarius food safety standards prioritize public health while supporting resource-limited countries through capacity-building programs. By fostering global cooperation, WHO has shaped the food safety agenda and managed crises under its Global Strategy for Food Safety in the year 2002 and Strategic Plan for Food Safety during 2013 to 2022. However, many Member States still face challenges in regulatory frameworks, enforcement, surveillance, laboratory capacity, crisis management, and food safety education. Strengthening these systems and integrating food safety into national and regional policies on health, agriculture, trade, and development remains essential.

WHO formulated the Global Strategy for Food Safety with guidance from the Technical Advisory Group for Food Safety, regional advisors, international partners, NGOs, and WHO Collaborating Centres. The strategy integrates key regional frameworks, including those for the Western Pacific, South-East Asia, and the Eastern Mediterranean, along with food safety initiatives in Africa, Europe, and the America. Additionally, it aligns with Codex Alimentarius standards and the FAO

Food Safety Strategy (2022–2031) to ensure a comprehensive, globally coordinated approach to food safety.

To uphold food safety standards in manufacturing, a comprehensive quality evaluation is essential, encompassing systematic testing, continuous monitoring, and strict adherence to regulatory requirements to mitigate contamination risks and safeguard public health. Establishing robust quality assessment frameworks—integrating microbiological analysis, chemical testing, and the principles of Hazard Analysis and Critical Control Points (HACCP)—is imperative for ensuring food integrity and consumer protection.

1.3.1 Food Safety: A Fundamental Pillar of Sustainable Development Goals (SDGs)

The Sustainable Development Goals represent a global commitment by 193 nations to foster prosperity while safeguarding the environment. Serving as a framework for a more equitable and sustainable future, the 17 interlinked goals are collectively targeted for achievement by 2030. Access to sufficient, safe, and nutritious food is fundamental to all SDGs, highlighting the intrinsic connection between health, nutrition, food safety, and food security.

Integrating food safety into the SDGs is essential, particularly in advancing SDG 2 -Zero Hunger, SDG 3 -Good Health and Well-being, and SDG 8 Decent Work and Economic Growth. Additionally, it plays a critical role in achieving SDG 1-No Poverty, SDG 12- Responsible Consumption and Production, and SDG 17 -Partnerships for the Goals. Moreover, SDG 6 -Clean Water and Sanitation serves as a foundational element of food safety. Without ensuring the availability of safe and nutritious food, especially in low- and middle-income countries, the successful realization of these goals remains uncertain.

1.3.2 Safety as Important Component in Food Trade

Food safety is fundamental not only to public health but also to national economic stability and the integrity of global trade. The repercussions of consuming unsafe food extend beyond health complications manifesting as illness, disability, and mortality to considerable economic losses. Failures in food safety undermine trade, diminish consumer confidence, and obstruct sustainable development. Consequently, the establishment of robust national food safety systems is imperative for advancing the Sustainable Development Goals, minimizing economic losses, and fostering resilient food systems.

The Codex Alimentarius Commission formulates science-based international food standards to safeguard consumers and promote equitable trade. To enhance food safety and align with global standards, Member States should prioritize the following strategic objectives such as

- Reinforcing food control mechanisms and regulatory frameworks
- Enhancing co-ordination among national agencies
- Harmonizing national food safety systems with codex standards
- Strengthening engagement with international regulatory networks

1.4 Management of Food Quality

Managing food quality through developing schemes such as HACCP enhances providing safe foods to consumers. In this context, a food system contains several diverse aspects; a primary concern is moving food from the production center to the food market and court (Forster, 2013). The principles of food quality management include: raw material control, process control and final product examination.

Raw material control The development of food products hinges on incorporating raw materials such as post harvested fruits and vegetables, meat, eggs, and dairy products. The selection and control of these raw materials are paramount, as they are the key ingredients that set the stage for successful processing. This underscores the crucial role of food industry professionals, quality assurance managers, and food safety regulators in ensuring the quality and safety of the final product (Liboreiro, 2013; Kahindi, 2016; Agarwal et al., 2024).

Process control This principle is the cornerstone of food processing. It requires constant vigilance and the implementation of both remedial and anticipatory measures. The importance of this principle cannot be overstated, as it is the key to resolving quality issues that may arise due to external conditions. This underscores the crucial role of food industry professionals, quality assurance managers, and food safety regulators in maintaining the quality and safety of the final product (Jacxsens et al., 2009).

Final product examination The third principle focuses on the final and polished product, the packaged food that goes through quality assurance for vouching to suppliers and consumers. Visual observation and Physicochemical examinations are examples of different stages. Besides, microbiological examination is instrumental in defining health and hygiene controls and relentlessly monitoring progress in data collection, records, and pattern analysis. A dynamic activity of quality administration can be significantly identified in the food industry across the numerous sourcing stages, processing, and packaging (Nordenskjöld, 2012).

1.4.1 Good Manufacturing Practices (GMP)

From a quality assurance point of view, Good Manufacturing Practices ensure food quality during production and manufacturing. They define quality measures in production and testing and have a legal constituent covering delivery, processing, testing, and finished goods faults and consumer grievances. The World Health Organization implemented Good Manufacturing Practices to monitor food production and manufacturing quality. Other advantages of the system include minimizing waste and improving profits. Initially, the guidelines for the GMP were set for the pharmaceutical industry only but later generalized to embrace the food industry. GMP entails procedures for managing production premises, raw materials, hygiene, and equipment and should be bureaucratically written down and implemented in the organizational culture. Properly running GMPs are required when operating under systems like HACCP or norms like 22,000 or BRC. Compared to GMP, HACCP is a system that moves further and delves deeper into triaging the risks and how it can be prevented. It improves the quality of smoked fish (*Scombercolias*) (Misra, 2024). After adoption, without any change in the physical quality of fish, there is a reduction in aerobic and fungal load, and an additional advantage is the value of peroxide, and the level of histamine reduced to an acceptable range (Geraldo et al., 2024). GMP and SSOP are adopted in the production process to reduce contamination (Mouludi & Sipahutar, 2024).

1.4.2 Good Hygiene Practices (GHP)

All food hygiene systems make food production safe and meet consumers' requirements based on good hygiene practices. By governing food safety hazards, these systems will provide the foundation for absolute food safety and implementation for food business operators. Food is contaminated by physical, chemical, and biological hazards in the food chain. It is valid when the GHP complies with Codex Alimentarius, especially for companies operating in or exporting their products to other countries (Insfran-Rivarola et al., 2020). In the online food safety toolbox, GHP should be the reference resource for users, and it enhances food quality management (Hasani et al., 2024).

1.4.3 Hazard Analysis Critical Control Points (HACCP)

As defined by the Codex Alimentarius Commission, HACCP categorizes, assesses, and regulates risks relevant to food safety. It is globally recognized as the method of reassurance of food safety. The HACCP system was initially applied to minimize the contamination of food products with microorganisms (Misra, 2024). The application scope was expanded to all food safety hazards, including chemical, physical, and allergenic hazards. The HACCP approach can also be implemented in other supporting systems, such as the water system (Motarjemi, 2014). The Food and Drug Administration (FDA) and the United States Department of Agriculture (USDA) say that their mandatory HACCP programs for juice and meat are a practical approach to food safety and protecting public health. The USDA regulates meat HACCP systems, while the FDA regulates seafood and juice. The use of HACCP is currently voluntary in other food industries. In the online food safety toolbox, HACCP will be the reference resource for the users, and it improves food safety management (Hasani et al., 2024).

1.5 Global Standards for Food Safety and Quality

Every nation must establish its own food safety regulations to guarantee that consumers have access to safe, high-quality food that aligns with local dietary practices, agricultural conditions, and legal frameworks. Implementing national food safety measures is essential for minimizing the risk of foodborne illnesses, protecting public health, and enhancing consumer trust. Governments play a crucial role in this process by developing, enforcing, and continuously improving these standards through specialized regulatory bodies. For instance, India regulates food safety through the Food Safety and Standards Authority of India (FSSAI), established under the Food Safety and Standards Act, 2006. In the United States, the Food and Drug Administration (FDA) and the United States Department of Agriculture (USDA) oversee food safety regulations. The European Union (EU) relies on the European Food Safety Authority (EFSA) and the General Food Law (Regulation (EC) 178/2002) to ensure compliance. In China, food safety is managed by the National Health Commission and the State Administration for Market Regulation (Madilo et al., 2024). These regulatory frameworks are vital in ensuring that only safe and compliant food reaches consumers, thereby promoting public health and facilitating international trade. Some of the globally recognized standards are described below in detail.

1.5.1 Codex Alimentarius

Codex Alimentarius provides food quality and safety standards and international food trading guidelines. The Food and Agriculture Organization and the World Health Organization frame these standards. The Codex Alimentarius is a collection of internationally adopted food standards and related texts presented uniformly. It provides global standards for all raw, semi-processed, processed, and distributed food concerning food hygiene, additives, adult tents, contaminants, labeling requirements, analysis, and sampling methods during the importing and exporting of food commodities (Hasani et al., 2024). The Codex Alimentarius is divided into general, specific, and regional areas. One of the critical characteristics of the committees is the involvement of specialists from nearly all countries that developed science-based guidelines and support the risk communication process. An example is the Joint Expert Committee of Microbial Risk Assessment (JEMRA), formed in 2000 and has authored over 40 publications on risk assessment of commodities such as fresh produce or pathogens (foodborne viruses). Other joint committees work on pesticide residues, additives, nutrition, and veterinary drugs. As one could infer, the Codex has operated for more than 60 years since its formation. It has accumulated vast scientific knowledge and information about food safety and quality issues. Using this in developing new Guidelines on Front-of-Pack Nutrition Labelling reduces consumer confusion in Public health and prevents inappropriate marketing (Boatwright et al., 2024).

1.5.2 International Organization for Standardization (ISO)

ISO is an independent, non-governmental international organization. It brings global experts together to agree on the best ways of doing things. From climate change and healthcare to quality management and artificial intelligence, Their mission is to make lives easier, safer, and better—for everyone, everywhere. ISO 22000: ISO 22000 is the international standard that provides the specifications for a comprehensive food safety management system across all sizes and all types of organizations within the food chain (Misra, 2024). Key potential benefits of using the standard include the capacity for routine development and delivery of food-related products and services that are safe and compliant with regulations and increased efficiency in managing risk factors involved in food safety production chains. They are closely associated with the 'United Nations Codex Alimentarius,' an international reference for formulating government food safety standards. The objective is to bring together the plethora of current national and private standards and integrate the management systems approach of the ISO 900115 standard specifically designed for food safety management (Neves et al., 2014; Panpatte, 2019).

1.6 Introduction to Food Quality Analysis

Food quality analysis includes the preparation of samples, analysis, and determining major and associated components. Complexity and diversity of food components by food quality analysis through various analytical techniques (Gupta et al., 2003). Food researchers and analysts regularly determine the chemical composition and physical characteristics of foods for food quality. The importance of food analysis has increased for ensuring food quality and safety among buyers, government organizations, and food industries. Three major steps involved in analysis are

Table 1.1 List of destructive and non-destructive techniques in food quality analysis

Destructive techniques	Non-destructive techniques
Gas chromatography	Image analysis
High-performance thin-layer chromatography (HPTLC)	Reflectance, transmittance and absorbance spectroscopy
Ultra-high-pressure liquid chromatography (UHPLC	Laser spectroscopy
High-performance liquid chromatography (HPLC)	Ultrasonic
Mass spectroscopy, Proton transfer reaction mass spectrometry (PTR-MS)	X-ray image and CT
Matrix-assisted MALDI-TOF/MS	MR/MRI
Ultraviolet-visible spectrophotometry (UV–vis)	Vibrated excitation
Polymerase chain reaction (PCR)	Sonic

selecting and preparing samples, performing the assay, results calculations, and interpretation (Nielsen, 2024). Food analysis techniques can be classified into two categories and are given in Table 1.1.

1.6.1 Destructive Technologies for Food Quality Analysis

Destructive testing requires using an optical device to examine a specimen when it reaches failure or optically analyze an unbroken specimen from an interrupted experiment by taking a cross-section of the specimen and analyzing the cracks/defects (Kong et al., 2020). These techniques include identifying the foreign components in the commodity based on the product's chemical composition. They should follow standard protocols that are stringent and devoid of any possibility of failure. However, they are usually detrimental, time-consuming, costly, and necessitate certain information about the analytical procedures, so they cannot be easily employed at the laboratory level. These drawbacks of destructive techniques often make them unsuitable for fast processing, and in-line quality evaluation and decision-making in scale-up and industry settings (Modupalli et al., 2021). Various Destructive Technologies for Food Quality Analysis is shown in Fig. 1.1 and their principles, and applications are presented in Table 1.2.

1.6.2 Non-Destructive Technologies for Food Quality Analysis

Non-destructive methods (NDMs) are one of the elements of high-quality control functions and contribute to other established techniques. The classification of Non-Destructive Technologies for Food Quality Analysis is shown in Fig. 1.2 and their principles and applications are given in Table 1.3. The non destructive analysis

Fig. 1.1 Destructive Technologies for Food Quality Analysis

is defined as the exterior examination of fruits and vegetables in a manner that introduces no undue influence on the food dimension or quality. The non-destructive assessment methods provide information on food attributes, including structural, mechanical, physical, and chemical properties. In other words, non-destructive measurement should be employed as the best strategy for food processing (Saldaña et al., 2013). The most widely used non-destructive techniques include machine vision, near-infrared spectroscopy, hyperspectral imaging, electronic noses, ultrasound measurement, and acoustic emission measurements (Nelson et al., 2006; Lorente et al., 2012; Alander et al., 2013).

1.6.2.1 Spectroscopy—Optical Methods—Visible, NIR, FTIR

Spectroscopy is an effective, non-destructive method to analyze the quality as well as quantity with the advantage of being pollution-free (Qi et al., 2024) In the electromagnetic spectrum, wavelengths from 50,000 to 4000 cm^{-1} (200–2500 nm) used in spectroscopy are called ultraviolet-visible-near–infrared (UV-vis-NIR) spectroscopy. This will again be divided into the UV, visible, and NIR regions of the electromagnetic spectrum. In spectroscopy, electronic transitions of molecules observed in UV region probes, implications of molecules observed in NIR region

Table 1.2 Destructive Technologies for Food Quality Analysis, Principle and their Applications

Sl. No.	Name of non-destructive technologies	Principle	Application	Reference
1.	Moisture Content (Gravimetric)	Based on the Change in weight of food concerning the initial weight before drying.	Determination of the shelf life of food	Bonner (1981), Wrolstad et al. (2005)
2.	Biochemical analysis	Use of Chemicals to determine the nutritional contents of foods	Determination of carbohydrates, proteins, fats, antioxidants, etc.,	Wang et al. (2022)
3.	Instrumentation Methods—HPLC	Separation of compounds (Nonvolatile) based on their affinity to stationary phase and mobile phase.	Quantitative and Qualitative analysis of vitamins, amino acids, etc.,	Ismail and Reuhs (2024)
4.	GC	Separating compounds (volatile) based on their affinity to stationary and mobile phases.	Quantitative and Qualitative analysis of veterinary drugs, pesticides, contaminants, etc.,	Samakradhamrongthai (2024)
5.	Texture profile analyzer	Based on the amount of force and the time required for the deformation of the food to a certain extent.	Determination of hardness, chewiness, springiness, cohesiveness, compressibility, hardness.	Ahmad et al. (2024)
6.	Viscosity analysis (Viscometer)	Based on the torque required to rotate an object in the liquid.	Rheological study of foods, flowability, pasting properties/ etc.,	Nashima and Fujitha (2009)

probes, and combination vibrations of specific vibration modes observed in region 100–4000 cm^{-1} called Raman and IR spectroscopy (Wachs & Keturakis, 2013). In the electromagnetic spectrum, wavelengths from 4000 to 400 cm^{-1} are called FTIR spectroscopy, and it is used as a non-destructive technique in solids, liquids, and gases (Fadlelmoula et al., 2022). It is a current alternative to analysis by destructive methods (Saji et al., 2024). In the food industry, the quality and quantity of ingredients and complete foods are analyzed by infrared spectroscopy. Compared with conventional wet chemical or chromatographic techniques, quantity analysis takes less time than mid-IR, near-IR, and Raman spectroscopy. Molecules in food or other substances are measured by radiation absorption in near or mid-infrared regions of the spectroscopy (Rodriguez-Saona & Ayvaz, 2024). Water quality parameters like chemical oxygen demand, heavy metals, pollution-causing micro-organisms, and organic matter are analyzed using UV-Vis spectroscopy and near-infrared (NIR) spectroscopy (Qi et al., 2024). Agrifood products are analyzed by combining near infrared spectroscopy and Fourier transform infrared spectroscopy

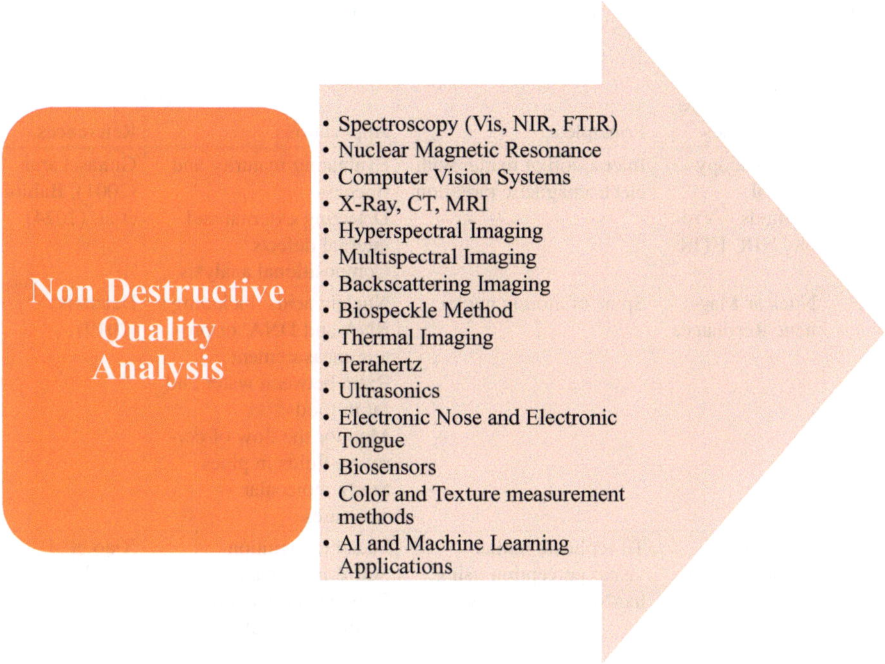

Fig. 1.2 Non-Destructive Technologies for Food Quality Analysis

with chemometric patterns and multivariate calibrations (Abdul et al., 2024). In dairy products like milk, ghee adulteration is detected and characterized by a combination of Fourier transform infrared spectroscopy (FTIR) and chemometrics, as well as contaminants identification, compositional change prediction, and ensuring food quality and safety by this method (Saji et al., 2024).

1.6.2.1.1 Fluorescence Spectroscopy

Fluorescence spectroscopy is a non-destructive technique for evaluating the fluorescence characteristics of an object (Lenhardt et al., 2015). It delivers information concerning fluorophores in simple and complex biological systems (Ahmad et al., 2016). In general, the fluorescence process involves three main steps: excitation where light is taken up by molecules and promoted to the electronically excited state, vibrational inward conversion through which particles move from the higher electronically excited state to the lower one without radiation and emit longer wavelengths of light and return the molecules to the ground state (Keramat et al., 2011; Zhang et al., 2012; Gabor et al., 2014).

Table 1.3 Non-Destructive Technologies for Food Quality Analysis principle and their Applications

Sl. No.	Name of non-destructive technologies	Principle	Applications	References
1.	Spectroscopy—Optical methods—Visible, NIR, FTIR	Interaction of matter with electromagnetic radiation	Evaluating maturity and ripeness Detecting external and internal defects Compositional analysis	Gunasekaran (2001), Bahlol et al. (2024)
2.	Nuclear Magnetic Resonance	Spins of atomic nuclei	Nucleic acids, including RNA and DNA, or protein measurement Ratio between water and fat in foods Monitor the flow of corrosive fluids in pipes Study molecular structures	Hatzakis (2019)
3.	Computer vision systems	To replicate human visual perception using hardware and software.	Facial recognition Self-driving cars Robotic automation Medical anomaly detection Sports performance analysis Manufacturing fault detection Agricultural monitoring Plant species classification Text parsing	Xiao et al. (2022)
4.	X-ray	Reduction of energy is caused by absorption	Contaminants detection Chemical changes in baked food products Interferometry in food systems	Olakanmi et al. (2023)
5.	CT	Most notably radiation attenuation, Lambert-Beer's Law	Non-destructive characterization of food microstructure	Schoeman et al. (2016)
6.	MRI	Nuclear magnetic resonance (NMR), is a spectroscopic technique used to obtain microscopic chemical and physical information about molecules. MRI is based on the absorption and emission of energy in the electromagnetic spectrum's radiofrequency (RF) range.	Evaluation of qualitative and quantitative properties of biological materials	Cleary and Guimarães (2014)

(continued)

Table 1.3 (continued)

Sl. No.	Name of non-destructive technologies	Principle	Applications	References
7.	Hyperspectral Imaging	Acquisition of spatial images in many spectrally contiguous bands Capturing a series of closely spaced images across various narrow wavebands within the visible and infrared regions of the electromagnetic spectrum	Capturing morphological and chemical information from food and food products	ElMasry and Sun (2010), Yuen and Richardson (2010)
8.	Multispectral Imaging	Different materials have unique spectral signatures, which are patterns of how they interact with light across different wavelengths Captures different wavelengths of the electromagnetic spectrum. This can be achieved with multiple sensors or using instruments with multi-band pass optical filters providing the same results as a single but complex sensor	Fruits and vegetables, dairy goods, meat products, seafood, grains, and legumes are all evaluated for quality and safety Plant disease detection Skin disease detection Retinal disease detection Military target tracking Weather forecasting	Stamford et al. (2023)
9.	Bio speckle Method	A beam of light with a consistent wavelength interacts with uneven surfaces, it creates patterns of light and dark spots on the equipment used to capture the image	Identifying ripening stages Grading and selection system	Al-Ogaidi and Al-Temeemy (2024), Pavarin et al. (2024)
10.	Thermal Imaging	All objects emit infrared (IR) radiation, or heat, as a function of their temperature. Convert the energy in the infrared wavelength into a visible light display. To record the infrared radiation emitted from an object by an infrared detector and convert these records into two-dimensional temperature data	Detection of diseases Insect infestation in the whole plant Estimating soil water status Evaluating crop water stress Identifying irrigation schedules Estimate crop yield Assess the maturity status of fruits and vegetables Functional or structural parameters like leaf	ElMasry et al. (2020)

(continued)

Table 1.3 (continued)

Sl. No.	Name of non-destructive technologies	Principle	Applications	References
			orientation, heat capacity, surface properties, infrared absorption, and transpiration rate Seed quality evaluation	
11.	Terahertz	Terahertz is a non-destructive spectroscopic technique. The terahertz radiation frequency falls between microwave and infrared radiation (0.1 to 10 THz) in the electromagnetic spectrum.	Biological detection Medical imaging Security inspections, safety, aviation and automation, cultural heritage, and food	Wang et al. (2024), Magaway and Navarro-Cía (2024)
12.	Ultrasonication	Cheese and poultry	Air instability compensation transducers to acquire non-contact ultrasound images to detect metal and glass fragments in boneless chicken breasts and to identify holes in cheese	Cho and Irudayaraj (2003)
13.	Electronic nose and Electronic Tongue	E-nose and E-tongue are an array of gas/chemical sensors that detect and identify food's volatile compounds and sensory properties.	Detects pesticide presence Evaluate flavor and authenticity Controls quality Assesses freshness Taste without destroying the commodity	Tibaduiza et al. (2024)
14.	Biosensors	Combining biological elements such as enzymes and antibiotics with physicochemical detectors like electrochemical, optical, and mass-based sensors enables the specific detection and quantification of analytes.	Detects pathogens and allergens Measures specific nutrients and additives Verifies the quality and authenticity of the product	Aggarwal et al. (2021)
15.	Techniques based on the Electrical properties of food	Electrical properties play a significant role in processing food by ohmic heating, microwave heating, pulsed electric field, and radiofrequency. Based on	Minimizes nutrient loss Enables rapid heating Extends shelf life Improves texture Enhances extraction efficiency	Bogale Teseme and Weldemichael Weldeselassie (2020)

(continued)

Table 1.3 (continued)

Sl. No.	Name of non-destructive technologies	Principle	Applications	References
		the food's conductivity and dielectric properties, the food's quality characteristics can be analyzed.		
16.	Color and Texture measurement methods	Machine vision (X-ray, hyperspectral imaging) helps ensure food quality and safety by capturing and processing high-resolution images.	Grading Removing foreign matter Detecting internal defects Determining the ripeness and maturity of the commodity	Zhu et al. (2021)
17.	AI and Machine Learning Applications	Analyses large data sets from various sources and forecasts the presence of pathogenic organisms in the food.	Enhance real-time monitoring Traceability Transparency in the food supply chain.	Onyeaka et al. (2024)

1.6.2.1.2 Near-infrared Spectroscopy

Near-infrared reflectance spectroscopy (NIRs) is another widely used method for food detection; (He & Sun, 2015; Wang et al., 2017a, b) which proves spectral extending and bending feedback where the chemical bonds C-H, N-H, S-H, and O-H are involved. Only a part of the near-infrared range is divided in short wave near-infrared (SW-NIR) the spectral region of 780–1100 nm, which is where the estimation of transmission for fabrics is usually done with weak uptake and long wave near-infrared (LW-NIR) in the broad spectral range of 1100–2500 nm, in which most of the radiation is produced is diffuse reflection (Manley, 2014; Lapchareonsuk & Sirisomboon, 2015). A NIR spectroscopy instrument typically comprises a radiant energy source, an instrument for measuring the wavelength and showing the samples, a detector for converting energy into an electrical signal, and an individual measuring unit, signal processor, and reader. Quartz halogen bulbs and light-emitting diode bulbs are two of the many types of lighting devices that are used to provide light diodes (LEDs) and are often employed as a broadband near-infrared repository for analytical utilization (Hussain et al., 2019).

1.6.2.1.3 Fourier Transform Infrared Spectroscopy

Fourier Transform Infrared (FTIR) spectroscopy is a rapid, cost-effective, and non-destructive analytical technique utilized to assess the chemical properties of biological specimens. By analyzing the interaction between infrared radiation and molecular bonds, FTIR spectroscopy provides detailed insights into the biochemical composition of a sample, which aids in detecting adulterants and ensure food safety (Holse et al., 2011). It is used to determine the power of infrared radiation in the range of 700 nm to 1 mm with the function of frequency or wavelength (Ferreira et al., 2014). This technique is preferred for food applications because it is useful in wet and dry analysis and can yield high-quality spectra from a small sample (Reder et al., 2014). FTIR imaging can measure the absorbed infrared radiation and revealing the functional groups present, indirectly finding the presence of particular compound. Fourier Transform Infrared (FTIR) spectroscopy has been effectively employed in evaluating fish quality, particularly in monitoring microbiological spoilage in gilthead sea bream fillets. This technique facilitates the rapid detection of spoilage indicators, thereby ensuring product safety and potentially extending shelf life (Candoğan et al., 2021). Additionally, FTIR spectroscopy has been utilized to detect adulteration in edible oils, such as identifying the presence of rapeseed and soybean oils in sunflower oil. This capability is crucial for maintaining product authenticity and safeguarding consumer health. The versatility of FTIR imaging extends to analyzing the structural and functional relationships of food components, including proteins, fats, and carbohydrates. By providing detailed information on molecular composition, FTIR aids in quality control and ensures compliance with industry standards.

1.6.2.2 Hyperspectral Imaging

Hyperspectral imaging (HSI) essentially combines spectral and digital imaging or computer vision (Wang & Sun, 2003; Du & Sun, 2005; Jackman et al., 2011) into a single system to acquire spatial, spectral, and multi-component information for any object. Technically, the hyperspectral imaging method can produce hundreds of pictures of a similar object in different spectral bands (Lorente et al., 2012; Wu & Sun, 2013). HSI system mainly comprises three fundamental parts: a light source, a light scattering element, and an image capturing means, which is an eye and the brain to be able to make the decision (Huang et al., 2014). The light emitted from the light source and the food are affected by that light source, which contains the physical and chemical information about the sample and is then spread and expected on the two-dimensional detector array on the imaging spectrograph. The signal is then transmitted to the computer for other processes, including digitization, modeling, data storage, and decision-making (Qin et al., 2013; Huang et al., 2014; Baiano, 2017).

1.6.2.3 Multispectral Imaging

A multispectral image is composed of grey-scale images. Each one is associated with a particular wavelength or a range of electromagnetic spectrum wavelengths (Bandara et al., 2018). MSI is a technique for obtaining pictures of various spectral bands by acquiring spatial and spectral data. Imagers based on MSI technology can feature the wavelength at near-UV, visible, near-IR, mid-IR, and far-IR bands (Jayasundara et al., 2020). Thus, MSI can provide more information than RGB images. The acquired wavelength channels can be used directly for real-time applications in some industry categories (for example, fruit packing plants and food processing plants). The system uses an adjustable focusing lens for acquiring a high resolution of 1290 × 960 pixels. This has six bands, each sweeping a rather broad range of wavelengths, for which it is strong and fast. MSI (430–1010 nm) combined with linear and non-linear regression methods has been used to evaluate fish spoilage. Indicators include Total-Volatile Basic Nitrogen (TVB-N) Psychotropic Plate Count (PPC) and sensory score of the fillets. Based on nine optimal bands selected by the genetic algorithm, the non-linear models showed higher performance (Fan & Su, 2022).

1.6.2.4 Magnetic Resonance Imaging

MRI is a real-time tool for internal process control systems in the food industry (Divya et al., 2012). MRI works the same way as nuclear magnetic resonance (NMR) spectroscopy. The images of the food are produced using MRI. When one or more, other than constant field gradients, are applied on the magnetic field, the frequency at which the spins can resonate depends on the location of the detection element concerning the axis. Subsequently, the occurrence of their synchronicity can develop a position-dependent spectrum, producing the image and its characteristics. MRI systems consist of imaging probes, magnets, shim, radio frequency transmitters, gradient power supplies, radio frequency receivers and Pumps, transmitter coils, and computer workstations (Schmidt et al., 1996). Research focuses on classification purposes with hams. These authors wanted to distinguish between hams and pigs fed on different diets and the level of salt intake (Pérez-Palacios et al., 2011; Caballero et al., 2016). As in the two previous works, images were obtained from HF-H1 MRI scanners using a quadrature whole-body coil with T1-weighted SE sequences. As presented in the former study, findings revealed that visual MRI of hams from pigs fattened with an acorn label showed the difference between grass and high oleic acid accumulations.

1.6.2.5 X-ray

The X-ray imaging technique is already popular in different areas of agriculture and food quality assessment. It is an inexpensive, non-destructive technique of determining the quality of the food substance by using real-time image processing and can easily be used for online food monitoring engaged in the food industries (Herremans et al., 2014). X-rays are passed via food, where the finished products are located at the end of the processing line. The detectors develop a visional picture of the products line by line for vision analysis. These are then stored in the computer system, and that is analyzed (Magwaza & Opara, 2014). In a study of Bakery products, the X-ray technique is used to check the Bread crumb structure, crispness, foam setting in bread, and starch crystallization (Vidhya et al.,2017). The micro-structure of banana slices and banana slice crispiness in bananas were checked (Léonard et al., 2008).

1.6.2.6 Computed Tomography

CT or CAT scan is a non-invasive diagnostic procedure for obtaining cross-sectional body views. In conventional CT, an X-ray source is rotated around the sample; the attenuation is measured on a detector. The ability to define an object's location is restricted by the focal spot size of the X-ray tube, while the true resolution depends on the sample and the magnification used. In Grain Sorghum Caryopses: Caryopses of sorghum grain, Phenotypic variability was substantial for all the traits, including the volume of the embryo, hardness of endosperm, the texture of endosperm, the volume of endosperm, volume of the pericarp, volume of the kernel (Crozier et al., 2022).

1.6.2.7 Bio Speckle Method

Bio speckle is an optical technique employed to analyze biological materials without causing any damage to the surface. Here, the object of interest is illuminated by coherent laser light. Because of the interference from backscattered light, a speckle pattern is formed in an observation plane. The speckle pattern of biological samples generally possesses two components: a static pattern due to the stationary particles of the tissue and a variable pattern emanating from the tissues that the movements of the tissues have caused. Among these, the variable speckle pattern is a unique biological characteristic known as speckle (Ansari et al., 2012), whereas the inorganic parts are linked with more structural and less dynamic structures. Research has shown that damaged or infected tissues have been found to have low speckle activity. Contamination, damage, maturation, etc., cause changes in its biological activity. Hence, these factors could be non-destructively analyzed by the bio speckle method along with lesser numerical processing techniques (Samuel et al., 2016).

1.6.2.8 Thermal Imaging

Every object receives and transfers thermal radiation proportional to its surface temperature. The technology of assessing this emitted temperature and portraying it as a visual image is known as Thermography. Thermal imaging (TI) can be used to monitor the variability of temperature at the core of the foods during different processing conditions (Feng & Sun, 2012; Takacs et al., 2020). TI creates a digital copy of any food material according to the heat released or spread out on the surface and can measure high-resolution temperature by non-contact method (Rahkonen & Jokela, 2003; ElMasry et al., 2020). It gives information concerning the stability and texture of food, size, shape, and abnormal heat conditions (Giorleo & Meola, 2002). In a study, the Shelf life of mushrooms in different polymer packings was observed (Taghizadeh et al., 2010). Detection of husk and prediction of moisture content, maturity, and foreign materials in paddy (Bejo-Khairunniza et al., 2016).

1.6.2.9 Terahertz

Terahertz offers special characteristics for studying molecular vibration in the spectroscopy type compared to other spectroscopies. This band ranges between 0.1 and 10 THz and is referred as terahertz: the space between the infrared and microwave bands of the electromagnetic spectrum (Zhang et al., 2023). Due to its high permission to inspect and deliver inside data of strong examples, THz spectroscopy is viewed as a decent substitution for risky procedures like X-ray. As a result, it has been applied to a broad range of fields, including foreign body detection (Lee et al., 2012; Wang et al., 2019; Sun et al., 2022), toxic detection (Chen & Xie, 2014; Liu et al., 2019b), microbial detection (Park et al., 2014; Fawole et al., 2015; Pu et al., 2023, 2024), adulteration detection (Liu et al., 2016, 2019a; Zhang et al., 2024).

1.6.2.10 Ultrasonication

Ultrasonic imaging stands as a robust non-destructive technique for evaluating food quality, thereby enhancing food safety standards. By employing high-frequency sound waves, this method facilitates the internal examination of food products without inducing physical deformation or damage. This capability is particularly advantageous for assessing internal attributes such as composition, structure, and physical condition, all of which are critical indicators of quality and safety. Ultrasonic imaging serves as a potent non-destructive technique for identifying internal defects, voids, or foreign materials within food products that are not externally visible. For instance, this method has been effectively utilized to detect skin cracks in tomatoes and defects in husked sweet corn, ensuring that compromised products are identified and removed from the distribution chain, thereby upholding quality standards and consumer safety. This method enables the internal assessment of produce without causing damage, facilitating the determination of optimal harvest and consumption periods, thereby reducing waste and enhancing consumer satisfaction (Sarkar & Wolfe, 1983). Studies have demonstrated a correlation between ultrasonic propagation and the ripening quality of fresh produce, indicating its efficacy in monitoring textural changes during storage (Mohd Shah 2016). Ultrasonic imaging facilitates real-time, online measurements during food processing, enabling precise monitoring and control of various operations. For example, it can assess the stability of emulsions and suspensions, detect foreign objects in canned products, and analyze crystallization profiles across diverse food matrices. This continuous oversight ensures consistency and safety in processed food items. Furthermore, ultrasonic techniques have been successfully applied in the degassing and anti-foaming treatment of liquids, enhancing the quality of carbonated beverages and fermentation systems. Additionally, the integration of ultrasonic technology with machine learning techniques presents promising prospects for real-time process monitoring in the food industry (Gallo et al., 2018; Yan et al., 2024).

1.6.2.11 Computer Vision Systems

However, unlike AI techniques such as neural nets, which have been used significantly since the 1940s, computer vision techniques only existed in the late 1960s and early 70s. Initially, they work with notions like emulation of human vision based on different understandings of how camera schematics, projection, or photogrammetry functions. Within a few decades, computer vision could find its combined existence in all possible aspects, like pattern recognition, machine learning, computer graphics, 3D reconstructions, virtual reality, and augmented reality (Kakani et al., 2020). Applying a computer vision system allows many characteristics, such as texture, shape, color, size, and defects, to be graded or inspected. However, some defects due to texture and color are like skin, which is challenging to identify. A hyperspectral computer vision system closely resembles a spectroscopic imaging system, as it provides spectral data for each individual pixel within a spatial image (Zahra et al., 2024).

1.6.2.12 Nuclear Magnetic Resonance

Nuclear magnetic resonance (NMR) spectroscopy is a technique free from sample degradation. It utilizes the ability of the nucleus to feel the surrounding chemical environment in a molecular structure. NMR can work in liquids and solids in 1D and 2D, and nD experiments with information regarding structure, content, purity, molecular weight, dynamics, and diffusion of food material (Cheng, 2021). When a sample is introduced in the external magnetic field, then the nuclear spins of the sample align themselves either parallel or antiparallel to the external magnetic field, leading to net magnetization that an NMR instrument can pick. When the sample is subjected to a radio frequency pulse, nuclei, spin depolarises from the equilibrium plane, which gives a signal upon bringing it back to the equilibrium plane when it releases the signal, measured by the NMR spectrometer (Fu & Ying, 2016). The use of NMR spectroscopy in determining the conjugated-LA content in the lipid fraction of dairy products is relatively unexplored but more attractive compared to conventional methods such as GC due to its sample derivatization drawback. The NMR analysis approach can be applied in the dairy industry for the primary screening of total CLA concentration in many cheese samples and for the screening of conjugated LA content in several dairy products (Prema et al., 2013).

1.6.2.13 Electronic Nose

A set of chemical or gas sensors that simulate a human tongue or nose is called an e-nose or e-tongue. Chemical sensor arrays are called "electronic tongue (e-tongue)," whereas gas sensor arrays are called "electronic nose (e-nose)" (Orlandi et al., 2019). Surface acoustic wave (SAW) sensors, conducting polymers (CP), metal-oxide semiconductors (MOS), and quartz crystal microbalance (QCM) are among the several types of gas sensors that may be categorized based on the sensing materials (Wilson, 2012). Reversible electrical properties, like conductivity, are produced when the target gases react with the sensing materials. Conductivity is usually measured by monitoring the sensor's output voltage and characterizing the output voltage pattern using parameters like voltage peak, response time, and recovery time (Tan & Xu, 2020). The Process of sensing of Electronic Nose is shown in Fig. 1.3.

1.6.2.14 Electronic Tongue

The five essential tastes with sour, salty, umami, bitter, and sweet flavor compounds create a sense of taste. For each type of flavor that a person can taste food or drink, they use special organs on the tongue known as taste buds (Trivedi, 2012). Every taste information is transmitted to the taste nerves and passes to the brain's gustatory area as a mediator tissue due to the release of neurotransmitters (Margolskee et al., 2007). Regarding the background, the primary method of food quality assessment in the food industry is sensory analysis, which is when actual samples are tasted by professional assessors known as sensory assessors or sensory panelists. However, these tests have limitations: low objectivity and reproducibility and high stress on panelists during their work. The electronic tongue, a sensor device for scientifically distinguishing and quantifying food taste, was developed in response to this problem (Tahara & Toko, 2013).

Electronic tongues based on sensor arrays have two characteristics:

1. They can statistically analyze the outputs from several sensors
2. They have poor and high cross-selectivity rather than high selectivity.

These characteristics—low selectivity, strong cross-selectivity, and statistical analysis—have led to investigating sensing technologies connected to electronic noses. Some applications are a technique for employing a taste-sensing technology to determine when food expires (Kobayashi et al., 2010), Black tea taste (Hayashi

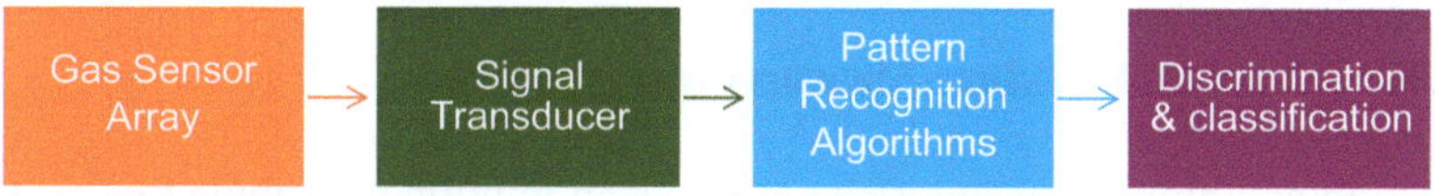

Fig. 1.3 Process of Sensing Electronic Nose

et al., 2007), Green tea tasting (Hayashi et al., 2008), Rice (Uyen et al., 2004) Prosciutto ham (Kobayashi et al., 2010) and Pork (Uyen et al., 2004).

1.6.2.15 Biosensors

To decide the existence of one or more analytes together with their focus and their kinetics in a test example, a biosensor is a specialist instrument that comprises various biologically sensitive detection systems, including antibodies, nucleic acids, enzymes, organelles, whole cells, aptamers, etc. fixed on a physicochemical transducer with a detector (Perumal & Hashim, 2014). Thus, recognizing biomolecules in the biosensor system is an important index of biosensor function and efficiency, as these biomolecules incorporate a higher specificity and binding affinity with the target molecules. Such binding, also known as specific binding or coupling, determines whether a contact occurs and what gives an electrical signal that is being registered and amplified (Sharma et al., 2017). In any case, very high selectivity is achieved because the recognition element of the matrix is selective towards the analyte and thus generates signals from just these specific interactions (Idili et al., 2014). There are examples of glucose meters that have a working electrode coated with glucose oxidase to get a response only with glucose. The three main areas of food analysis expectations that biosensors address are authenticity, quality, and safety (Ye et al., 2017). It was used in the detection of pesticide and antibiotic residues (Ribeiro et al., 2014), allergens detection (Andjelković et al., 2017), detection of biological toxins (Malhotra et al., 2014), to determine pathogenic microbes (Singh et al., 2014).

1.6.2.16 Techniques Based on Electrical Properties

Electrical characteristics are important when applying radio frequency, microwave, ohmic, induction, and pulsed electric fields in food processing. These characteristics are also helpful in assessing food quality or processing conditions (Mouludi & Sipahutar, 2024). Electrical conductivity- Ohmic heating can be described as passing an a.c through food, turning electricity to heat, where the chief parameter is conductivity. The scientific realization of the effective conductivity or the total resistance of liquid-particle mixtures is important because its practical use was indicated in pasteurizing fluids (Halden et al., 1990). Dielectric properties are measured characteristics of food items that describe the interaction of food with electromagnetic fields (Tıraş et al., 2019).

1.6.2.17 Color and Texture Measurement Method

In the food and bioprocess industries, color is an important attribute that customers use to make their choices. These changes in the chemical, metabolic, microbiological, and physical properties during growth, maturation, post-harvest handling, and processing affect the color of foods. Because it is easy, fast, and has a high positive relationship with some physicochemical attributes, color measurement of food commodities has been used to index other quality attributes such as taste and pigment content in the foods. Several color coordinate systems may be used to characterize an object's color (Clydesdale & Ahmed, 1978; Francis, 1980; Hunter & Harold, 1987). Among the most often used systems are RGB (red, green, and blue), which is utilized in colour video displays; they include Hunter Lab, CIE's L*a*b*, CIE XYZ, CIE L*u*v*, CIE Yxy, and CIE LCH. Some studies done include banana Peel (Boudhrioua et al., 2003), Tomato (Maharaj et al., 1999), Navel oranges, Potato chips and Salmon fillet (Quevedo et al., 2010).

1.6.2.18 AI and Machine Learning Application

The present global food industry faces on-going challenges in fulfilling the consumer demands for safe and wholesome food products. They include issues related to adulteration, contamination, and maintaining product quality, given that they are produced in large quantities and transported to different destinations. This desire for new solutions has led to the development of Artificial Intelligence (AI), which offers solutions to problems in food safety and quality control in detection, monitoring, and management. The current applications of AI in various food industries, including computer vision, industrial robots, machine learning techniques, and the Internet of Things. With the help of AI, the food sector can gather more relevant and timely data for analysis, maintenance, and risk identification, as well as enhance the protection and quality of foods (Palakurti, 2022). The performance of Convolutional Neural Networks (CNNs) is very effective in analyzing the images of the food samples to identify contamination. To detect contamination, they operate through several layers to go through the stepwise features as well as the patterns of the representation as shown in Fig. 1.4.

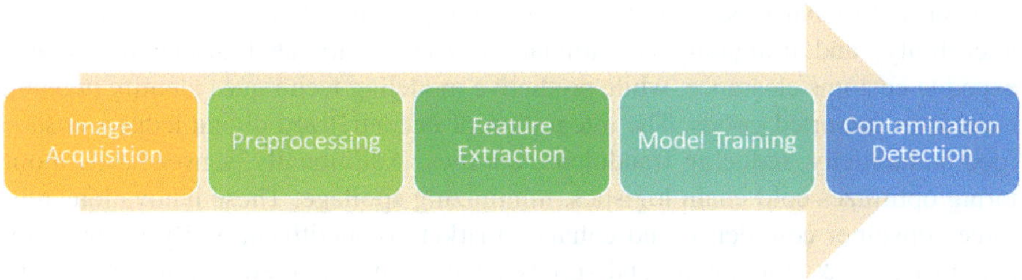

Fig. 1.4 Process of Contamination Detection by Convolutional Neural Networks

1.7 Recent Trends in Food Safety and Quality Evaluation

The integration of artificial intelligence and machine learning in food integrity encompasses multiple facets. Generally, food safety and quality assessments relied heavily on manual inspections, necessitating labor-intensive, repetitive, and time-consuming procedures prone to human error and bias. The growing sophistication of artificial intelligence and machine learning and related technologies highlights their critical role in enhancing food integrity across the supply chain. The incorporation of advanced computational methodologies such as machine learning and artificial intelligence into food safety and quality assessment is transforming the industry by strengthening traceability, mitigating health hazards and ensuring food authenticity. Although certain challenges persist, continuous technological progress is anticipated to further refine safety protocols and enhance adherence to regulatory frameworks. The growth of computational modelling approaches paved the way for predicting shelf life of various foods. By systematic collection and analysing of various factors such as temperature, humidity and microbial dynamics during storage, these methodologies support the construction of predictive models capable of accurately estimating product stability across different environmental conditions. These innovations assist food manufactures in optimizing production strategies and implementing effective storage and transportation measures, ultimately enhancing shelf life and reducing food wastes. Advanced predictive modeling facilitates the evaluation of food safety hazards, the identification of contaminants, and the mitigation of fraudulent practices. Intelligent sensors, integrated with the Internet of Things (IoT), enable real-time monitoring of storage environments, thereby optimizing shelf stability and minimizing waste. Computational methodologies for pathogen detection expedite the identification of foodborne illnesses, while predictive modeling techniques assess antimicrobial resistance patterns. Moreover, technological innovations contribute to the development of tailored nutritional solutions and the formulation of food products that adhere to regulatory standards. Although challenges persist, including data reliability and compliance with evolving regulatory frameworks, continuous progress in computational analytics, IoT, and large-scale data processing is anticipated to reinforce food safety measures. The convergence of these technologies enhances quality assurance, reduces public health risks, and fosters greater consumer confidence in the integrity of the global food supply chain. The advancement of computational technologies has significantly transformed global food trade by strengthening regulatory compliance, enhancing traceability, and mitigating contamination risks. Automated inspection systems expedite customs clearance, while predictive modeling techniques identify potential hazards in imported goods. The integration of decentralized digital ledgers ensures food authenticity, reducing fraudulent activities. Additionally, sensor-based monitoring optimizes cold chain logistics, minimizing spoilage. These innovations reinforce consumer confidence and enhance market competitiveness. By streamlining compliance, risk assessment, and supply chain monitoring, emerging technologies contribute to a more efficient, secure, and transparent international food trade system (Shi et al., 2024).

1.8 Conclusion

Ensuring food quality is paramount to providing consumers with safe, nutritious, and palatable products. The management of food quality and safety entails a multifaceted interaction of various elements, encompassing both sensory characteristics and nutritional composition. These determinants can be classified into intrinsic factors inherent to the product and extrinsic factors related to storage and processing conditions. Rigorous food quality control is essential for mitigating risks, ensuring regulatory compliance, and fostering consumer trust in food products. This process necessitates systematic monitoring, prompt corrective measures, and adherence to established food safety frameworks such as HACCP, ISO, and GFSI, which offer comprehensive standards for maintaining both safety and quality. Technological advancements, including artificial intelligence and non-destructive analytical techniques such as hyperspectral imaging and spectroscopy, have revolutionized food quality assessment by enabling rapid and precise identification of defects and contaminants. These innovations are particularly crucial in the modern, fast-paced global food supply chain. A robust commitment to food quality control not only safeguards consumer health and satisfaction but also strengthens brand reputation, minimizes waste, and enhances profitability within the food production sector. By refining quality management strategies and leveraging cutting-edge analytical methodologies, the food industry can effectively address the increasing consumer demand for safer, healthier, and superior-quality products.

References

Abass, A. S., Abdalla, I. F., & Ahmed, A. E. (2024). Contribution of higher education institutions to food security: The case of King Faisal University. In *Food and Nutrition Security in the Kingdom of Saudi Arabia, Vol 1: National Analysis of Agricultural and Food Security* (pp. 425–461). Springer Nature. https://doi.org/10.1007/978-3-031-46716-5_18

Abdul, R., Riana, L., & Fajar Aji, L. (2024). Near infrared (NIR) and fourier transform infrared (FTIR) spectroscopy combined with chemometrics for agrifood products analysis. In *Chemometrics* (pp. 125–146). Elsevier.

Agarwal, R., Yadav, R., Singh, P., et al. (2024). Applications of management practices in food processing industries: An overview. *EC Nutrtion, 19*, 01–05.

Aggarwal, S., Sehgal, S., & Tandon, A. W. (2021). Biosensors-types and application in food processing industry. *Journal of Postharvest Technology, 2021*, 1–19.

Ahmad, M. H., Nache, M., Hinrichs, J., & Hitzmann, B. (2016). Estimation of the nutritional parameters of various types of wheat flours using fluorescence spectroscopy and chemometrics. *International Journal of Food Science and Technology, 51*, 1186–1194. https://doi.org/10.1111/ijfs.13080

Ahmad, K., Khan, S., Chen, P., et al. (2024). Enhancing mucoadhesion: Exploring rheological parameters and texture profile in starch solutions, with emphasis on micro-nanofiber influence. *International Journal of Biological Macromolecules, 275*, 133392. https://doi.org/10.1016/j.ijbiomac.2024.133392

Alander, J. T., Bochko, V., Martinkauppi, B., et al. (2013). A review of optical nondestructive visual and near-infrared methods for food quality and safety. *International Journal of Spectroscopy, 2013*, 1–36. https://doi.org/10.1155/2013/341402

Al-Ogaidi, M., & Al-Temeemy, A. A. (2024). Bio-Speckle laser technology in revealing microbial radial growth: A review. *Pakistan Journal of Life and Social Science, 22*, 1890–1901.

Andjelković, U., Gavrović-Jankulović, M., Martinović, T., & Josić, D. (2017). Omics methods as a tool for investigation of food allergies. *TrAC - Trends in Analytical Chemistry, 96*, 107–115. https://doi.org/10.1016/j.trac.2017.07.011

Ansari, M. Z., Minz, P. D., & Nirala, A. K. (2012). Fruit quality evaluation using biospeckle techniques. In *2012 1st International Conference on Recent Advances in Information Technology, RAIT-2012* (pp. 873–876). https://doi.org/10.1109/RAIT.2012.6194540

Bahlol, H. S., Li, J., Deng, J., et al. (2024). Recent progress in nanomaterial-based surface-enhanced Raman spectroscopy for food safety detection. *Nanomaterials, 14*, 1750. https://doi.org/10.3390/nano14211750

Baiano, A. (2017). Applications of hyperspectral imaging for quality assessment of liquid based and semi-liquid food products: A review. *Journal of Food Engineering, 214*, 10–15. https://doi.org/10.1016/j.jfoodeng.2017.06.012

Bandara, W. G. C., Prabhath, G. W. K., Dissanayake, D. W. S. C. B., et al. (2018). A multispectral imaging system to assess meat quality. In *IEEE Region 10 Humanitarian Technology Conference, R10-HTC*. https://doi.org/10.1109/R10-HTC.2018.8629858

Bejo-Khairunniza, S., Azman, N., & Jamil, N. (2016). Paddy grading using thermal imaging technology. *International Food Research Journal, 23*, S245–S248.

Boatwright, M., Lawrence, M., Carriedo, A., et al. (2024). Understanding the politics of food regulation and public health: An analysis of codex standard-setting processes on food labelling. *International Journal of Health Policy and Management, 13*, 8310. https://doi.org/10.34172/ijhpm.8310

Bogale Teseme, W., & Weldemichael Weldeselassie, H. (2020). Review on the study of dielectric properties of food materials. *American Journal of Engineering and Technology Management, 5*, 76. https://doi.org/10.11648/j.ajetm.20200505.11

Bonner, F. T. (1981). *Measurement and management of tree seed moisture* (p. 10). US Department of Agriculture, Forest Service, Southern Forest Experiment Station. https://doi.org/10.2737/SO-RP-177

Boudhrioua, N., Giampaoli, P., & Bonazzi, C. (2003). Changes in aromatic components of banana during ripening and air-drying. *LWT - Food Science and Technology, 36*, 633–642. https://doi.org/10.1016/S0023-6438(03)00083-5

Butz, P., & Tauscher, B. (2002). Emerging technologies: Chemical aspects. *Food Research International, 35*, 279–284. https://doi.org/10.1016/S0963-9969(01)00197-1

Caballero, D., Caro, A., Rodríguez, P. G., et al. (2016). Modeling salt diffusion in Iberian ham by applying MRI and data mining. *Journal of Food Engineering, 189*, 115–122. https://doi.org/10.1016/j.jfoodeng.2016.06.003

Candoğan, K., Altuntas, E. G., & İğci, N. (2021). Authentication and quality assessment of meat products by fourier-transform infrared (FTIR) spectroscopy. *Food Engineering Reviews, 13*(1), 66–91.

Chen, M., & Xie, L. (2014). A preliminary study of aflatoxin b1 detection in peanut oil by terahertz time-domain spectroscopy. *Transactions of the ASABE, 57*, 1793–1799. https://doi.org/10.13031/trans.57.10725

Cheng, K. (2021). Lignin characterization. In *Natural polyphenols from wood* (pp. 147–182). Elsevier.

Chhetri, K. B. (2024). Applications of artificial intelligence and machine learning in food quality control and safety assessment. *Food Engineering Reviews, 16*, 1–21. https://doi.org/10.1007/s12393-023-09363-1

Cho, B. K., & Irudayaraj, J. M. K. (2003). Foreign object and internal disorder detection in food materials using noncontact ultrasound imaging. *Journal of Food Science, 68*(3), 967–974.

Cleary, J. O. S. H., & Guimarães, A. R. (2014). Magnetic resonance imaging. In *Pathobiology of human disease* (pp. 3987–4004). Elsevier.

Clydesdale, F. M., & Ahmed, E. M. (1978). Colorimetry — methodology and applications∗. C R C Critical Reviews in Food Science and Nutrition, 10(3), 243–301. https://doi.org/10.1080/10408397809527252

Crozier, D., Riera-Lizarazu, O., & Rooney, W. L. (2022). Application of X-ray computed tomography to analyze the structure of sorghum grain. *Plant Methods, 18*, 3. https://doi.org/10.1186/s13007-022-00837-7

Divya, S., Thyagarajan, D., & Sujatha, G. (2012). Magnetic resonance imaging technology for process control and quality maintenance in food quality operation. *International Journal of Engineering and Technology, 4*, 441–449.

Du, C. J., & Sun, D. W. (2005). Pizza sauce spread classification using colour vision and support vector machines. *Journal of Food Engineering, 66*, 137–145. https://doi.org/10.1016/j.jfoodeng.2004.03.011

ElMasry, G., & Sun, D.-W. (2010). Principles of hyperspectral imaging technology. In *Hyperspectral imaging for food quality analysis and control* (pp. 3–43). Elsevier.

ElMasry, G., ElGamal, R., Mandour, N., et al. (2020). Emerging thermal imaging techniques for seed quality evaluation: Principles and applications. *Food Research International, 131*, 109025. https://doi.org/10.1016/j.foodres.2020.109025

Fadlelmoula, A., Pinho, D., Carvalho, V. H., et al. (2022). Fourier Transform Infrared (FTIR) spectroscopy to analyse human blood over the last 20 years: A review towards Lab-on-a-Chip devices. *Micromachines (Basel), 13*, 187. https://doi.org/10.3390/mi13020187

Fan, K.-J., & Su, W.-H. (2022). Applications of fluorescence spectroscopy, RGB- and multispectral imaging for quality determinations of white meat: A review. *Biosensors (Basel), 12*, 76. https://doi.org/10.3390/bios12020076

Fawole, O., Sinha, K., & Tabib-Azar, M. (2015). Monitoring yeast activation with sugar and zero-calorie sweetener using terahertz waves. *IEEE SENSORS - Proceedings, 2015*, 19–22. https://doi.org/10.1109/ICSENS.2015.7370322

Feng, Y. Z., & Sun, D. W. (2012). Application of hyperspectral imaging in food safety inspection and control: A review. *Critical Reviews in Food Science and Nutrition, 52*, 1039–1058. https://doi.org/10.1080/10408398.2011.651542

Ferreira, D. S., Galão, O. F., Pallone, J. A. L., & Poppi, R. J. (2014). Comparison and application of near-infrared (NIR) and mid-infrared (MIR) spectroscopy for determination of quality parameters in soybean samples. *Food Control, 35*, 227–232. https://doi.org/10.1016/j.foodcont.2013.07.010

Forster, A. (2013). Sustainability: Best practices in the food industry. *UW-L Journal of Undergraduate Research, XVI*, 1–9.

Francis, F. J. (1980). Color quality evaluation of horticultural crops. *HortScience, 15*, 58–59.

Fu, X., & Ying, Y. (2016). Food safety evaluation based on near infrared spectroscopy and imaging: A review. *Critical Reviews in Food Science and Nutrition, 56*, 1913–1924. https://doi.org/10.1080/10408398.2013.807418

Gabor, R. S., Baker, A., McKnight, D. M., & Miller, M. P. (2014). Fluorescence indices and their interpretation. In *Aquatic organic matter fluorescence* (pp. 303–338). Cambridge University Press. https://doi.org/10.1017/cbo9781139045452.015

Gallo, M., Ferrara, L., & Naviglio, D. (2018). Application of ultrasound in food science and technology: A perspective. *Food, 7*(10), 164.

Geraldo, V., Ampofo-Asiama, J., Adinortey, C. A., et al. (2024). Implementation of good manufacturing practices (GMP) to improve the quality of smoked fish (Scomber colias). *Heliyon, 10*, e27401. https://doi.org/10.1016/j.heliyon.2024.e27401

Giorleo, G., & Meola, C. (2002). Comparison between pulsed and modulated thermography in glass-epoxy laminates. *NDT and E International, 35*, 287–292. https://doi.org/10.1016/S0963-8695(01)00062-7

Gunasekaran, S. (2001). *Nondestructive food evaluation techniques to analyze properties and quality* (1st ed.). CRC Press.

Gupta, K., Jain, V., Jain, S., et al. (2003). Analysis of food. In *Encyclopedia of food sciences and nutrition* (pp. 206–215). Elsevier.

Halden, K., De Alwis, A. A. P., & Fryer, P. J. (1990). Changes in the electrical conductivity of foods during ohmic heating. *International Journal of Food Science and Technology, 25*, 9–25. https://doi.org/10.1111/j.1365-2621.1990.tb01055.x

Hasani, M., Camacho-Martinez, S. V., Zai, B., et al. (2024). Development of an online food safety toolbox based on the Codex Alimentarius General Principles of Food Hygiene: Engaging users through mapping, chunking, and learning-by-asking. *Food Quality and Safety, 8*, fyae030. https://doi.org/10.1093/fqsafe/fyae030

Hatzakis, E. (2019). Nuclear Magnetic Resonance (NMR) spectroscopy in food science: A comprehensive review. *Comprehensive Reviews in Food Science and Food Safety, 18*, 189–220. https://doi.org/10.1111/1541-4337.12408

Hayashi, N., Chen, R., Ikezaki, H., et al. (2007). Evaluation of the astringency of black tea by a taste sensor system: Scope and limitation. *Bioscience, Biotechnology, and Biochemistry, 71*, 587–589. https://doi.org/10.1271/bbb.60514

Hayashi, N., Chen, R., Ikezaki, H., & Ujihara, T. (2008). Evaluation of the umami taste intensity of green tea by a taste sensor. *Journal of Agricultural and Food Chemistry, 56*, 7384–7387. https://doi.org/10.1021/jf800933x

He, H. J., & Sun, D. W. (2015). Microbial evaluation of raw and processed food products by Visible/Infrared, Raman and Fluorescence spectroscopy. *Trends in Food Science and Technology, 46*, 199–210. https://doi.org/10.1016/j.tifs.2015.10.004

Herremans, E., Verboven, P., Defraeye, T., et al. (2014). X-ray CT for quantitative food microstructure engineering: The apple case. *Nuclear Instruments and Methods in Physics Research B, 324*, 88–94. https://doi.org/10.1016/j.nimb.2013.07.035

Holse, M., Larsen, F. H., Hansen, Å., & Engelsen, S. B. (2011). Characterization of marama bean (Tylosema esculentum) by comparative spectroscopy: NMR, FT-Raman, FT-IR and NIR. *Food Research International, 44*, 373–384. https://doi.org/10.1016/j.foodres.2010.10.003

Huang, H., Liu, L., & Ngadi, M. O. (2014). Recent developments in hyperspectral imaging for assessment of food quality and safety. *Sensors (Switzerland), 14*, 7248–7276. https://doi.org/10.3390/s140407248

Hunter, R. S., & Harold, R. W. (1987). The measurement of appearance. John wiley & sons.

Hussain, N., Sun, D.-W., & Pu, H. (2019). Classical and emerging non-destructive technologies for safety and quality evaluation of cereals: A review of recent applications. *Trends in Food Science and Technology, 91*, 598–608. https://doi.org/10.1016/j.tifs.2019.07.018

Idili, A., Amodio, A., Vidonis, M., et al. (2014). Folding-upon-binding and signal-on electrochemical DNA sensor with high affinity and specificity. *Analytical Chemistry, 86*, 9013–9019. https://doi.org/10.1021/ac501418g

Insfran-Rivarola, A., Tlapa, D., Limon-Romero, J., et al. (2020). A systematic review and meta-analysis of the effects of food safety and hygiene training on food handlers. *Food, 9*, 1169. https://doi.org/10.3390/foods9091169

Ismail, B. P., & Reuhs, B. L. (2024). High-performance liquid chromatography. In *Nielsen's food analysis* (pp. 193–205). Springer.

Jackman, P., Sun, D. W., & Allen, P. (2011). Recent advances in the use of computer vision technology in the quality assessment of fresh meats. *Trends in Food Science and Technology, 22*, 185–197. https://doi.org/10.1016/j.tifs.2011.01.008

Jacxsens, L., Kussaga, J., Luning, P. A., et al. (2009). A microbial assessment scheme to measure microbial performance of food safety management systems. *International Journal of Food Microbiology, 134*, 113–125. https://doi.org/10.1016/j.ijfoodmicro.2009.02.018

Jayasundara, D., Ramanayake, L., Senarath, N., et al. (2020). Multispectral imaging for automated fish quality grading. In *2020 IEEE 15th International Conference on Industrial and Information Systems, ICIIS 2020 - Proceedings* (pp. 321–326). https://doi.org/10.1109/ICIIS51140.2020.9342726

Kahindi, B. B. (2016). *Food safety management practices of small and medium sized food industry enterprizes in Tanzania* [Masters Theses & Specialist Projects].

Kakani, V., Nguyen, V. H., Kumar, B. P., et al. (2020). A critical review on computer vision and artificial intelligence in food industry. *Journal of Agriculture and Food Research, 2*, 100033. https://doi.org/10.1016/j.jafr.2020.100033

Keramat, J., LeBail, A., Prost, C., & Soltanizadeh, N. (2011). Acrylamide in foods: chemistry and analysis. A review. *Food and Bioprocess Technology, 4*, 340–363. https://doi.org/10.1007/s11947-010-0470-x

Kobayashi, Y., Habara, M., Ikezazki, H., et al. (2010). Advanced taste sensors based on artificial lipids with global selectivity to basic taste qualities and high correlation to sensory scores. *Sensors, 10*, 3411–3443. https://doi.org/10.3390/s100403411

Kong, Y., Bennett, C. J., & Hyde, C. J. (2020). A review of non-destructive testing techniques for the in-situ investigation of fretting fatigue cracks. *Materials and Design, 196*. https://doi.org/10.1016/j.matdes.2020.109093

Kumar, I., Rawat, J., Mohd, N., & Husain, S. (2021). Opportunities of artificial intelligence and machine learning in the food industry. *Journal of Food Quality, 2021*. https://doi.org/10.1155/2021/4535567

Lapchareonsuk, R., & Sirisomboon, P. (2015). Sensory quality evaluation of rice using visible and shortwave near-infrared spectroscopy. *International Journal of Food Properties, 18*, 1128–1138. https://doi.org/10.1080/10942912.2013.870572

Lee, Y. K., Choi, S. W., Han, S. T., et al. (2012). Detection of foreign bodies in foods using continuous wave terahertz imaging. *Journal of Food Protection, 75*, 179–183. https://doi.org/10.4315/0362-028X.JFP-11-181

Lenhardt, L., Bro, R., Zeković, I., et al. (2015). Fluorescence spectroscopy coupled with PARAFAC and PLS DA for characterization and classification of honey. *Food Chemistry, 175*, 284–291. https://doi.org/10.1016/j.foodchem.2014.11.162

Léonard, A., Blacher, S., Nimmol, C., & Devahastin, S. (2008). Effect of far-infrared radiation assisted drying on microstructure of banana slices: An illustrative use of X-ray microtomography in microstructural evaluation of a food product. *Journal of Food Engineering, 85*, 154–162. https://doi.org/10.1016/j.jfoodeng.2007.07.017

Liboreiro, C. (2013). Quality management: Important aspects for the food industry. In *Food industry*. InTech.

Liu, W., Zhang, Y., & Han, D. (2016). Feasibility study of determination of high-fructose syrup content of Acacia honey by terahertz technique. *Infrared, Millimeter-Wave, and Terahertz Technologies IV, 10030*, 100300J. https://doi.org/10.1117/12.2245966

Liu, J., Fan, L., Liu, Y., et al. (2019a). Application of terahertz spectroscopy and chemometrics for discrimination of transgenic camellia oil. *Spectrochimica Acta. Part A, Molecular and Biomolecular Spectroscopy, 206*, 165–169. https://doi.org/10.1016/j.saa.2018.08.005

Liu, W., Zhao, P., Wu, C., et al. (2019b). Rapid determination of aflatoxin B1 concentration in soybean oil using terahertz spectroscopy with chemometric methods. *Food Chemistry, 293*, 213–219. https://doi.org/10.1016/j.foodchem.2019.04.081

Lorente, D., Aleixos, N., Gómez-Sanchis, J., et al. (2012). Recent advances and applications of hyperspectral imaging for fruit and vegetable quality assessment. *Food and Bioprocess Technology, 5*, 1121–1142. https://doi.org/10.1007/s11947-011-0725-1

Madilo, F. K., Kunadu, A. P. H., & Tano-Debrah, K. (2024). Challenges with food safety adoption: A review. *Journal of Food Safety, 44*(1), e13099.

Magaway, E. J. Y., & Navarro-Cía, M. (2024). The potential of THz microscopy for non-destructive evaluation applications. *Journal of Non Destructive Testing and Evaluation (JNDE), 21*, 23–32. https://jnde.isnt.in/index.php/JNDE/article/view/83

Magwaza, L. S., & Opara, U. L. (2014). Investigating non-destructive quantification and characterization of pomegranate fruit internal structure using X-ray computed tomography. *Postharvest Biology and Technology, 95*, 1–6. https://doi.org/10.1016/j.postharvbio.2014.03.014

Maharaj, R., Arul, J., & Nadeau, P. (1999). Effect of photochemical treatment in the preservation of fresh tomato (Lycopersicon esculentum cv. Capello) by delaying senescence. *Postharvest Biology and Technology, 15*, 13–23.

Malhotra, B. D., Srivastava, S., Ali, M. A., & Singh, C. (2014). Nanomaterial-based biosensors for food toxin detection. *Applied Biochemistry and Biotechnology, 174*, 880–896. https://doi.org/10.1007/s12010-014-0993-0

Manley, M. (2014). Near-infrared spectroscopy and hyperspectral imaging: Non-destructive analysis of biological materials. *Chemical Society Reviews, 43*, 8200–8214. https://doi.org/10.1039/c4cs00062e

Margolskee, R. F., Dyer, J., Kokrashvili, Z., et al. (2007). T1R3 and gustducin in gut sense sugars to regulate expression of Na +-glucose cotransporter 1. *Proceedings of the National Academy of Sciences of the United States of America, 104*, 15075–15080. https://doi.org/10.1073/pnas.0706678104

Misra, G. (2024). *Quality control and regulatory aspects for biologicals*. CRC Press.

Modupalli, N., Naik, M., Sunil, C. K., & Natarajan, V. (2021). Emerging non-destructive methods for quality and safety monitoring of spices. *Trends in Food Science and Technology, 108*, 133–147. https://doi.org/10.1016/j.tifs.2020.12.021

Mohd Ali, M., Hashim, N., Abd Aziz, S., & Lasekan, O. (2021). Quality inspection of food and agricultural products using artificial intelligence. *Advances in Agricultural and Food Research Journal*. https://doi.org/10.36877/aafrj.a0000237

Mohd Shah, L. B. (2016). *Instrumental and ultrasonic techniques in quality evaluation of fresh fruit and vegetables* [Doctoral dissertation, University of Leeds].

Motarjemi, Y. (2014). Hazard Analysis and Critical Control Point System (HACCP). In *Food Safety Management* (pp. 845–872). Elsevier.

Mouludi, I., & Sipahutar, E. (2024). Tea packaging design using integration of Kansei engineering and eye tracking for enhancing consumer appeal. *Journal of Engineering and Management in Industrial System, 12*, 78–86. https://doi.org/10.21776/ub.jemis.2024.012.02.2

Mu, W., Kleter, G. A., Bouzembrak, Y., et al. (2024). Making food systems more resilient to food safety risks by including artificial intelligence, big data, and internet of things into food safety early warning and emerging risk identification tools. *Comprehensive Reviews in Food Science and Food Safety, 23*, 1–18.

Nashima, T., & Fujitha, Y. (2009). Accurate viscosity measurement by a rotational viscometer. *Netsu Bussei, 23*, 148–153. https://doi.org/10.2963/jjtp.23.148

Nelson, S. O., Trabelsi, S., & Kays, S. J. (2006). Dielectric spectroscopy of honeydew melons from 10 MHz to 1.8 GHz for quality sensing. *Transactions of the ASABE, 49*, 1977–1981. https://doi.org/10.13031/2013.22278

Neves, T., Drohomeretski, E., Gouvea Da Costa, S. E., & De Lima, E. P. (2014). Sustainable operations management: Practices and measures in the food industry. *International Journal of Data Mining, Modelling and Management, 6*, 335–352. https://doi.org/10.1504/IJAOM.2014.066827

Nielsen, S. S. (2024). Introduction to food analysis. In *Nielsen's food analysis* (pp. 3–14). Springer.

Nollet, L., Boylston, T., Chen, F., et al. (2007). *Handbook of meat, poultry and seafood quality*. Wiley.

Nordenskjöld, J. (2012). *Implementation of a quality management system in food production*. Swedish University of Agricultural Sciences.

Olakanmi, S. J., Jayas, D. S., & Paliwal, J. (2023). Applications of imaging systems for the assessment of quality characteristics of bread and other baked goods: A review. *Comprehensive Reviews in Food Science and Food Safety, 22*, 1817–1838. https://doi.org/10.1111/1541-4337.13131

Onyeaka, H., Akinsemolu, A., Miri, T., et al. (2024). Advancing food security: The role of machine learning in pathogen detection. *Applied Food Research, 4*, 100532. https://doi.org/10.1016/j.afres.2024.100532

Orlandi, G., Calvini, R., Foca, G., et al. (2019). Data fusion of electronic eye and electronic tongue signals to monitor grape ripening. *Talanta, 195*, 181–189. https://doi.org/10.1016/j.talanta.2018.11.046

Palakurti, N. R. (2022). AI applications in food safety and quality control. *ESP Journal of Engineering & Technology Advancements, 2*, 48–61. https://doi.org/10.56472/25832646/JETA-V2I3P111

Panpatte, S. (2019). Management practices in food processing industries. *International Journal of Research in Engineering, Science and Management, 2*, 154–157.

Park, S. J., Son, B. H., Choi, S. J., et al. (2014). Sensitive detection of yeast using terahertz slot antennas. *Optics Express, 22*, 30467. https://doi.org/10.1364/oe.22.030467

Pavarin, F. F. A., Fracarolli, J. A., & Falcão, A. X. (2024). Biospeckle laser for assessing tomatoes ripeness indexes. *Revista Ciencia Agronomica, 55*, e20207677. https://doi.org/10.5935/1806-6690.20240022

Pereira, R. N., & Vicente, A. A. (2010). Environmental impact of novel thermal and non-thermal technologies in food processing. *Food Research International, 43*, 1936–1943. https://doi.org/10.1016/j.foodres.2009.09.013

Pérez-Palacios, T., Antequera, T., Durán, M. L., et al. (2011). MRI-based analysis of feeding background effect on fresh Iberian ham. *Food Chemistry, 126*, 1366–1372. https://doi.org/10.1016/j.foodchem.2010.11.101

Perumal, V., & Hashim, U. (2014). Advances in biosensors: Principle, architecture and applications. *Journal of Applied Biomedicine, 12*, 1–15. https://doi.org/10.1016/j.jab.2013.02.001

Prema, D., Pilfold, J. L., Krauchi, J., et al. (2013). Rapid determination of total conjugated linoleic acid content in select Canadian cheeses by 1H NMR spectroscopy. *Journal of Agricultural and Food Chemistry, 61*, 9915–9921. https://doi.org/10.1021/jf402627q

Pu, H., Yu, J., Sun, D. W., et al. (2023). Distinguishing pericarpium citri reticulatae of different origins using terahertz time-domain spectroscopy combined with convolutional neural networks. *Spectrochimica Acta. Part A, Molecular and Biomolecular Spectroscopy, 299*, 122771. https://doi.org/10.1016/j.saa.2023.122771

Pu, H., Yu, J., Luo, J., et al. (2024). Terahertz spectra reconstructed using convolutional denoising autoencoder for identification of rice grains infested with Sitophilus oryzae at different growth stages. *Spectrochimica Acta. Part A, Molecular and Biomolecular Spectroscopy, 311*, 124015. https://doi.org/10.1016/j.saa.2024.124015

Qi, X., Lian, Y., Xie, L., et al. (2024). Water quality detection based on UV-Vis and NIR spectroscopy: A review. *Applied Spectroscopy Reviews, 59*, 1036–1060. https://doi.org/10.1080/05704928.2023.2294458

Qin, J., Chao, K., Kim, M. S., et al. (2013). Hyperspectral and multispectral imaging for evaluating food safety and quality. *Journal of Food Engineering, 118*, 157–171. https://doi.org/10.1016/j.jfoodeng.2013.04.001

Quevedo, R. A., Aguilera, J. M., & Pedreschi, F. (2010). Color of salmon fillets by computer vision and sensory panel. *Food and Bioprocess Technology, 3*, 637–643. https://doi.org/10.1007/s11947-008-0106-6

Rahkonen, J., & Jokela, H. (2003). Infrared radiometry for measuring plant leaf temperature during thermal weed control treatment. *Biosystems Engineering, 86*, 257–266. https://doi.org/10.1016/S1537-5110(03)00138-7

Reder, M., Koczoń, P., Wirkowska, M., et al. (2014). The application of FT-MIR spectroscopy for the evaluation of energy value, fat content, and fatty acid composition in selected organic oat products. *Food Analytical Methods, 7*, 547–554. https://doi.org/10.1007/s12161-013-9652-2

Ribeiro, F. W. P., Barroso, M. F., Morais, S., et al. (2014). Simple laccase-based biosensor for formetanate hydrochloride quantification in fruits. *Bioelectrochemistry, 95*, 7–14. https://doi.org/10.1016/j.bioelechem.2013.09.005

Rodriguez-Saona, L., & Ayvaz, H. (2024). Infrared and Raman spectroscopy. In *Nielsen's food analysis* (pp. 95–116). Springer.

Saji, R., Ramani, A., Gandhi, K., et al. (2024). Application of FTIR spectroscopy in dairy products: A systematic review. *Food and Humanity, 2*, 100239. https://doi.org/10.1016/j.foohum.2024.100239

Saldaña, E., Siche, R., Luján, M., & Quevedo, R. (2013). Review: Computer vision applied to the inspection and quality control of fruits and vegetables. Revisão : visão computacional aplicada à inspeção e ao controle da qualidade de frutas e verduras. *Brazilian Journal of Food Technology, Campinas, 16*, 254–272.

Samakradhamrongthai, R. S. (2024). The extraction and isolation. In *Aroma and flavor in product development: Characterization, perception, and application* (pp. 107–138). Springer Nature Switzerland.

Samuel, B., Retheesh, R., Nampoori, V. P. N., & Mujeeb, A. (2016). Nondestructive evaluation of fruits using cross correlation and time history of biospeckle pattern. In *Indian National Seminar & Exhibition on Non-Destructive Evaluation NDE 2016* (pp. 470–473). NDT India.

Sarkar, N., & Wolfe, R. R. (1983). Potential of ultrasonic measurements in food quality evaluation. *Transactions of the ASAE, 26*(2), 624–629.

Schmidt, S. J., Sun, X., & Litchfield, J. B. (1996). Applications of magnetic resonance imaging in food science. *Critical Reviews in Food Science and Nutrition, 36*, 357–385. https://doi.org/10.1080/10408399609527730

Schoeman, L., Williams, P., du Plessis, A., & Manley, M. (2016). X-ray micro-computed tomography (μCT) for non-destructive characterisation of food microstructure. *Trends in Food Science and Technology, 47*, 10–24. https://doi.org/10.1016/j.tifs.2015.10.016

Sharma, A., Goud, K. Y., Hayat, A., et al. (2017). Recent advances in electrochemical-based sensing platforms for aflatoxins detection. *Chemosensors, 5*, 1. https://doi.org/10.3390/chemosensors5010001

Shi, C., Zhao, Z., Jia, Z., Hou, M., Yang, X., Ying, X., & Ji, Z. (2024). Artificial neural network-based shelf life prediction approach in the food storage process: A review. *Critical Reviews in Food Science and Nutrition, 64*(32), 12009–12024. https://doi.org/10.1080/10408398.2023.2245899

Singh, R., Das, M., Sumana, G., et al. (2014). Sensors and Actuators B: Chemical biosensors for pathogen detection: A smart approach towards clinical diagnosis. *Sensors and Actuators B Chemical, 197*, 385–404. https://doi.org/10.1016/j.snb.2014.03.005

Stamford, J. D., Vialet-Chabrand, S., Cameron, I., & Lawson, T. (2023). Development of an accurate low cost NDVI imaging system for assessing plant health. *Plant Methods, 19*, 9. https://doi.org/10.1186/s13007-023-00981-8

Sun, X., Cui, D., Shen, Y., et al. (2022). Non-destructive detection for foreign bodies of tea stalks in finished tea products using terahertz spectroscopy and imaging. *Infrared Physics & Technology, 121*. https://doi.org/10.1016/j.infrared.2021.104018

Taghizadeh, M., Gowen, A., Ward, P., & O'Donnell, C. P. (2010). Use of hyperspectral imaging for evaluation of the shelf-life of fresh white button mushrooms (Agaricus bisporus) stored in different packaging films. *Innovative Food Science and Emerging Technologies, 11*, 423–431. https://doi.org/10.1016/j.ifset.2010.01.016

Tahara, Y., & Toko, K. (2013). Electronic tongues-a review. *IEEE Sensors Journal, 13*, 3001–3011. https://doi.org/10.1109/JSEN.2013.2263125

Takacs, R., Jovicic, V., Zbogar-Rasic, A., et al. (2020). Evaluation of baking performance by means of mid-infrared imaging. *Innovative Food Science & Emerging Technologies, 61*, 102327. https://doi.org/10.1016/j.ifset.2020.102327

Tan, J., & Xu, J. (2020). Applications of electronic nose (e-nose) and electronic tongue (e-tongue) in food quality-related properties determination: A review. *Artificial Intelligence in Agriculture, 4*, 104–115. https://doi.org/10.1016/j.aiia.2020.06.003

Tibaduiza, D., Anaya, M., Gómez, J., et al. (2024). Electronic tongues and noses: A general overview. *Biosensors (Basel), 14*, 190. https://doi.org/10.3390/bios14040190

Tıraş, B., Dede, S., & Altay, F. (2019). Dielectric properties of foods. *Turkish Journal of Agriculture - Food Science and Technology, 7*, 1805–1816. https://doi.org/10.24925/turjaf.v7i11.1805-1816.2650

Tremolieres, R. J. (1970). Organolepsis and nutrition: The sensory perception of food. In *A behavioural approach to organoleptic properties of food* (pp. 285–292). Cambridge University Press.

Trivedi, B. P. (2012). Gustatory system: The finer points of taste. *Nature, 486*, S2–S3. https://doi.org/10.1038/486S2a

Uyen Tran, T., Suzuki, K., Okadome, H., et al. (2004). Analysis of the tastes of brown rice and milled rice with different milling yields using a taste sensing system. *Food Chemistry, 88*, 557–566. https://doi.org/10.1016/j.foodchem.2004.02.007

Vidhya, M., Varadharaju, N., Kennedy, Z. J., et al. (2017). Applications of X-Ray computed tomography in food processing. *Journal of Food Processing & Technology, 08*, 10–12. https://doi.org/10.4172/2157-7110.1000673

Vizzarri, F. (2021). Factors affecting food quality-technologies. In Conference Proceedings-*FOODIE - Quality soil as a pathway to healthy food in the EU*. At: Nitra (Slovak Republic)

Wachs, I. E., & Keturakis, C. J. (2013). Monolayer systems. In R. Schlogl (Ed.), *Comprehensive inorganic chemistry II* (2nd ed., pp. 131–151). Elsevier.

Wang, H. H., & Sun, D. W. (2003). Assessment of cheese browning affected by baking conditions using computer vision. *Journal of Food Engineering, 56*, 339–345. https://doi.org/10.1016/S0260-8774(02)00159-0

Wang, K., Sun, D. W., Pu, H., & Wei, Q. (2017a). Principles and applications of spectroscopic techniques for evaluating food protein conformational changes: A review. *Trends in Food Science and Technology, 67*, 207–219. https://doi.org/10.1016/j.tifs.2017.06.015

Wang, L., Sun, D. W., Pu, H., & Cheng, J. H. (2017b). Quality analysis, classification, and authentication of liquid foods by near-infrared spectroscopy: A review of recent research developments. *Critical Reviews in Food Science and Nutrition, 57*, 1524–1538. https://doi.org/10.1080/10408398.2015.1115954

Wang, C., Zhou, R., Huang, Y., et al. (2019). Terahertz spectroscopic imaging with discriminant analysis for detecting foreign materials among sausages. *Food Control, 97*, 100–104. https://doi.org/10.1016/j.foodcont.2018.10.024

Wang, P., Huang, J., Sun, J., et al. (2022). Evaluating the nutritional properties of food: A scoping review. *Nutrients, 14*, 2352. https://doi.org/10.3390/nu14112352

Wang, W., Sun, K., Xue, Y., et al. (2024). A review of terahertz metamaterial sensors and their applications. *Optics Communication, 556*, 130266. https://doi.org/10.1016/j.optcom.2024.130266

Wilson, A. D. (2012). Review of electronic-nose technologies and algorithms to detect hazardous chemicals in the environment. *Procedia Technology, 1*, 453–463. https://doi.org/10.1016/j.protcy.2012.02.101

Wrolstad, R. E., Acree, T. E., Decker, E. A., et al. (2005). *Handbook of food analytical chemistry*. Wiley.

Wu, D., & Sun, D. W. (2013). Advanced applications of hyperspectral imaging technology for food quality and safety analysis and assessment: A review - Part I: Fundamentals. *Innovative Food Science and Emerging Technologies, 19*, 1–14. https://doi.org/10.1016/j.ifset.2013.04.014

Xiao, Z., Wang, J., Han, L., et al. (2022). Application of machine vision system in food detection. *Frontiers in Nutrition, 9*. https://doi.org/10.3389/fnut.2022.888245

Yan, J., Zhang, Y., Jiao, Z., Song, L., Wang, Z., Zhang, Q., et al. (2024). Opportunities and challenges of ultrasonic diagnostic techniques for plant-based food monitoring: Principle, machine system, and application strategies. *Critical Reviews in Food Science and Nutrition*, 1–20. https://doi.org/10.1080/10408398.2024.2418891

Ye, W., Guo, J., Bao, X., et al. (2017). Rapid and sensitive detection of bacteria response to antibiotics using nanoporous membrane and graphene quantum dot (GQDs)-based electrochemical biosensors. *Materials, 10*. https://doi.org/10.3390/ma10060603

Yuen, P. W., & Richardson, M. (2010). An introduction to hyperspectral imaging and its application for security, surveillance and target acquisition. *The Imaging Science Journal, 58*, 241–253. https://doi.org/10.1179/174313110X12771950995716

Zahra, A., Qureshi, R., Sajjad, M., et al. (2024). Current advances in imaging spectroscopy and its state-of-the-art applications. *Expert Systems with Applications, 238*, 122172. https://doi.org/10.1016/j.eswa.2023.122172

Zhang, R., Ying, Y., Rao, X., & Li, J. (2012). Quality and safety assessment of food and agricultural products by hyperspectral fluorescence imaging. *Journal of the Science of Food and Agriculture, 92*, 2397–2408. https://doi.org/10.1002/jsfa.5702

Zhang, L., Zhang, M., & Mujumdar, A. S. (2023). Terahertz spectroscopy: A powerful technique for food drying research. *Food Reviews International, 39*, 1733–1750. https://doi.org/10.1080/87559129.2021.1936004

Zhang, G. M., Hu, S. Y., Chen, G., et al. (2024). Age identification of the root of Huanren mountain cultivated ginseng and differentiation with cultivated ginseng using terahertz spectroscopy. *Journal of Food Composition and Analysis, 125*. https://doi.org/10.1016/j.jfca.2023.105790

Zheng, L., Guo, H., Zhu, M., et al. (2024). Intrinsic properties and extrinsic factors of food matrix system affecting the effectiveness of essential oils in foods: A comprehensive review. *Critical Reviews in Food Science and Nutrition, 64*, 7363–7396. https://doi.org/10.1080/10408398.2023.2184767

Zhu, L., Spachos, P., Pensini, E., & Plataniotis, K. N. (2021). Deep learning and machine vision for food processing: A survey. *Current Research in Food Science, 4*, 233–249. https://doi.org/10.1016/j.crfs.2021.03.009

Chapter 2
Spectroscopy—Optical Methods: Visible, NIR, FTIR

A. Shivani Indumathi, S. Charan Adithya, J. Deepa, G. Jeevarathinam, and R. Pandiselvam

Abbreviations

AFM1	Aflatoxin M1
AI	Artificial Intelligence
AV	Acid Value
BC	Butter Cheeses
CCD	Charge-Coupled Device
cm	centimeter
DRIFT	Diffuse Reflectance Spectroscopy
ELISA	Enzyme Linked Immunosorbent Assay
EMD	Empirical Mode Decomposition
EVOO	Extra Virgin Olive Oils
FIR	Far Infrared
FTIR	Fourier-Transform Infrared
FTIR-ATR	Fourier-Transform Infrared Attenuated Total Reflectance Spectroscopy
FTIR-PAS	Fourier-Transform Infrared Photoacoustic Spectroscopy
FT-MIR	Fourier-Transform Mid-Infrared
FT-NIR	Fourier-Transform Near Infrared
GCMS	Gas Chromatography and Mass Spectrometry
HFCS	High Fructose Corn Syrup
HPLC	High Performance Liquid Chromatography

A. S. Indumathi · S. C. Adithya · J. Deepa · G. Jeevarathinam
Department of Food Technology, Hindusthan College of Engineering and Technology, Coimbatore, TN, India

R. Pandiselvam (✉)
Chemical & Biochemical Processing Division (CBPD), ICAR-Central Institute for Research on Cotton Technology (CIRCOT), Mumbai, Maharashtra, India

C. K. Sunil et al. (eds.), *Emerging Non-destructive Technologies for Food Quality Analysis*, https://doi.org/10.1007/978-3-031-95048-3_2

HSI	Hyperspectral Imaging
InGaAs	Indium Gallium Arsenide
IoT	Internet of Things
IR	Infrared
kJ mol^{-1}	Kilojoules Per Mole
LED	Light-Emitting Diodes
LVF	Linear Variable Filter
LW-NIR	Long-Wave NIR
MEMS	Micro-Electromechanical Systems
MOEMS	Micro-Opto-Electromechanical Systems
NIR	Near-Infrared
nm	nanometers
PCA	Principal Component Analysis
PCR	Principal Component Regression
PDA	Photodiode Array Detector
PET	Polyethene Terephthalate
PICS	Partly Inverted Cane Syrup
PLS	Partial Least Square
PMT	Photomultiplier Tubes
QIM	Quality Index Method
S/N	Of Signal-To-Noise Ratio
SO	Soybean Oil
UHT	Ultra-High Temperature
UV	Ultraviolet
VIS	Visible
κ-casein	Kappa Casein
λ	wavelengths

2.1 Introduction

In recent era of food consumption, the consumer's expectations of food quality and safety have widely increased. There are numerous causes for food adulteration which have created greater attention to preserving the product authentication. To preserve the quality of food products, analysis must be performed in the areas of crop production, processing, microbial contaminant levels, sensory evaluations, authenticity, traceability, and storage conditions (Kharbach et al., 2023). Various analytical techniques such as Liquid Chromatography, Gas Chromatography and spectroscopy techniques like Gas Chromatography and Mass Spectrometry (GCMS) were used to analyse the quality of food samples. However, these techniques are not portable for quick field research because sample preparation consumes a lot of time and costs a lot of money in terms of equipment. Spectroscopy holds significant promise for both qualitative and quantitative analysis in food testing. Among them, Raman spectrometers, Near-infrared (NIR) and Fourier-transform infrared (FTIR) are superior substitutes to analytical instruments that are chosen over other competing analytical methods. These methods are non-invasive, non-destructive, low-cost, portable and offer high-speed measurements with the absence of sample preparation (Fakayode et al., 2020).

The study of spectroscopy deals with how electromagnetic radiation interacts with the material which results in the production, measurement, and interpretation of spectra. Numerous spectroscopic techniques are available to address a broad spectrum of analytical issues. Different methodologies are used depending on the type of radiation-molecule interaction to be tracked (diffraction, emission, or absorption), the molecules to be examined (molecular spectroscopy), and the region of the electromagnetic spectrum being studied. Because they are very precise and informative, spectroscopic techniques are widely used for both quantitative and qualitative analysis (Penner, 2017).

2.2 Different Types of Spectroscopy Methods

The most often utilized spectroscopic methods in traditional food analysis labs are those that rely on radiation absorption or emission in the visible (Vis), ultraviolet (UV), infrared (IR), and nuclear magnetic resonance (NMR) frequency ranges. The infrared region is extremely broad and its energy ranges from 0.12 to 150 kJ mol^{-1} (10–12,500 cm^{-1}). In IR, two ends energy difference was found to be more than 1000 times greater than the NIR area and lower than the far-IR (FIR) region. The infrared area is shown in Fig. 2.1 as being divided into three regions: the IR region NIR region (12,500–4000 cm^{-1}; 800–2500 nm), the mid-IR (4000–400 cm^{-1}; 2500–25,000 nm) and the far-IR region (FIR; 400–10 cm^{-1}; 25 μm^{-1} mm) (Ozaki et al., 2021).

2.2.1 Fourier Transform Infrared (FTIR) Spectroscopy

Fourier Transform Infrared (FTIR) spectroscopy is a widespread and efficient method for identifying and analysing the structure of chemical substances in food. The vibrational modes of the molecules present in a sample are represented by the peaks in its infrared spectrum, which provides information about the functional groups and chemical bonds in the food complex structure. Consequently, the infrared spectrum functions as a unique "fingerprint" that allows a molecule to be identified by its unique characteristics. An intensity spectrum versus frequency is created from an interferogram using the Fourier transformation which is a mathematical method. It makes the process easier by converting a signal into a frequency representation from a time domain. When two infrared beams interfere, an interferogram is generated in a Fourier Transform Infrared device that accounts for changes

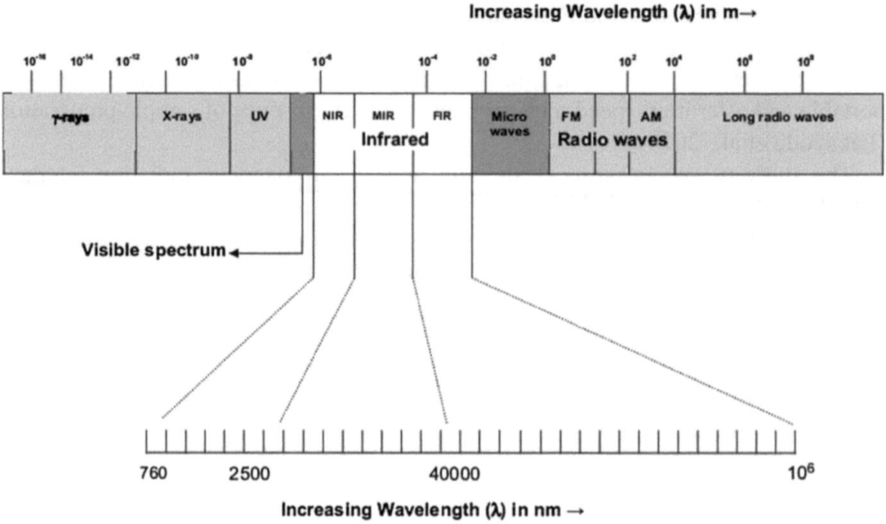

Fig. 2.1 Schematic representation of the electromagnetic spectrum. (Source: Cheng et al., 2013)

in the beam wavelengths. For this, an interferometer called Michelson is frequently employed. The interferogram results can be quantitatively analyzed to determine the frequencies that make up the signal using Fourier transform techniques (Upadhyay et al., 2021). The advantages of FTIR spectrometers include higher signal-to-noise ratios, more precise wavenumber measurements, quicker scan durations for getting spectra at all frequencies, greater resolution, a larger scan range, and a reduction in stray light interference. The mid-infrared portion of the infrared spectrum is known as FT-MIR spectroscopy. This spectroscopy allows to tracking of the basic molecular stretching and rotational modes, providing a comprehensive chemical profile for the material. FTMIR spectroscopy offers more thorough chemical information than FT-NIR (He et al., 2021).

2.2.1.1 Principle

The principle of FTIR spectroscopy is that molecules absorb infrared light at frequencies that correspond to their functional groups and chemical bonds. A portion of the infrared light that travels through a food sample is absorbed by the food molecules and the remaining light passes through the sample. The absorption spectrum is then ascertained by analysing the transmitted light. Different food samples have different chemical compositions that produce different infrared absorption spectra, which makes them suitable targets for Fourier transform infrared (FTIR) analysis. Different chemical bonds and functional groups can be recognized and distinguished by comparing the absorption spectra of known food samples to unknown samples (Chen et al., 2016).

2.2.1.2 Equipment and Working

A spectrophotometer is a device that measures a compound's absorbance spectrum. The Fourier transform spectrophotometer offers faster-infrared spectrum analysis compared to conventional spectrophotometer. Figure 2.2 depicts the major parts of an FTIR spectrophotometer. The major components of FTIR spectrometers are a light source, a sample holder, a detector, an interferometer, and a converter (analog-to-digital). The black body resource emits a beam of IR irradiation. When the beam travels through the interferometer, there the spectral encoding takes place. Beams with varying in-path lengths recombining in the interferometer produce an interferogram, which is a mixture of constructive and destructive interference. The sample absorbs the precise frequency of energy from the interferogram that is unique to the sample when the beam enters the sample compartment. The unique interferogram signal is then concurrently measured by the detector for all frequencies in energy against time. In the interim, a beam is superimposed to act as a background or point of reference for the instrument's operation. In the end, the required range was acquired by automatically extracting the background spectrum from the sample spectrum in the interferogram using computer software that carried out Fourier transformation. The appearance of an absorption band in the mid-IR range is used (4000–400 cm^{-1}) to determine if a particular functional group is present in the molecule. Transition energies associated with modifications in the vibrational energy levels of different functional groups are included in this region. An FTIR spectrometer's output of the infrared spectrum is mostly in the mid-IR range of 2.5–15 mm, or between 4000 and 666 cm^{-1}. Higher wavenumbers (2500–4000 cm^{-1}) allow for the detection of single bonds (N–H, O–H, and C–H). The intermediate wavenumber regions of 2000–2500 cm^{-1} and 1500–2000 cm^{-1}, respectively, are where the triple and double bonds are observed. Moreover, the molecule's vibration generates a complex pattern of vibrations at the low wavenumber region of 650–1500 cm^{-1}, which is indicative of the molecule as a whole and can be used for identification (Mohamed et al., 2017).

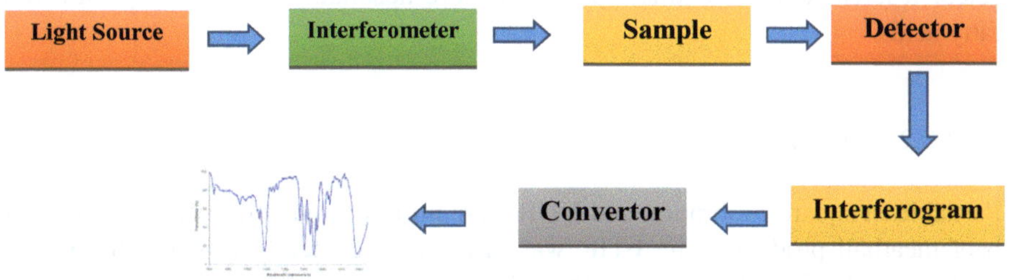

Fig. 2.2 Fundamental component of a Fourier transform infrared spectrometer. (Source: Mohamed et al., 2017)

2.2.1.3 Application of FTIR

2.2.1.3.1 Dairy Products

Dvorak et al. (2016) analysed the adulteration in cheeses prepared from goat and cow milk using two different analytical methods: FT-NIR and the immunological method ELISA kit. The authors discovered that goat cheese adulterated with cow's milk could be detected using FT-NIR techniques down to 1%. This result was found to be better than the commercial ELISA kit since the test was not suitable for a quantitative study of cheese because the process of producing fresh cheese severely damaged the κ-casein in cow's milk.

Spina et al. (2022) conducted an experiment using completely automated equipment for the determination of the milk composition. The adulteration of buffalo milk with cow milk was quantitatively detected by the author through the use of FTIR in conjunction with multivariate analysis by PLS regression.

The author developed a calibration model with an R^2 value of less than 0.99 for routine examination of buffalo milk composition. FTIR spectroscopy in conjunction with artificial intelligence was used to identify and quantify lactose as well as contaminated sugars like glucose and galactose in raw, pasteurized, and ultra-high temperature (UHT) milk. According to Ribeiro et al. (2023), the artificial intelligence algorithms' measurement of lactose, galactose, and glucose showed 95% accuracy, with coefficients of determination (R^2) values of 81%, 92%, and 86%, respectively. ATR-FTIR and chemometrics were employed by Harshitha et al. (2023) to investigate the connection between aflatoxin M1 (AFM1) and milk proteins. The 1700–1600 cm^{-1} range of the second-order derivative spectrum for AFM1 indicates an affinity for milk proteins. The conformational changes in the secondary structure of milk proteins caused by AFM1 are reflected in the PCA data from the regions between 1700 and 1600 cm^{-1}. Yaman et al. (2023) examined the infrared spectra of the soluble fractions in a variety of solvents (water, methanol, and ethanol) to determine the composition of Turkish white cheese throughout ripening. The author created regression and classification predictive models using chemometrics and FTIR spectra. These models exhibited outstanding accuracy in predicting target compounds throughout the cheese-ripening process. The water extracts showed more accurate in estimating organic acids and total free amino acids, whereas ethanol extracts provided more accurate forecasts of specific free fatty acids.

2.2.1.3.2 Meat Products

Partial least square calibration and FTIR were utilized to identify and detect pork in beef meatball products. The FTIR was efficient in detecting and quantifying the adulteration by the calibration model in the characteristic region (1000–1200 cm^{-1}) using PLS regression (Rohman et al., 2011). Alamprese et al. (2016) also found that minced beef samples that were cooked, frozen-thawed, and fresh could all be accurately measured and adulterated with turkey meat using FT-NIR. The efficiency of a handheld device and a benchtop FT-NIR was studied by Grassi et al. (2018) to differentiate between two distinct kinds of fish fillets. The author discovered that there was no discernible difference in the two devices' sensitivity or specificity ($P > 0.05$).

2.2.1.3.3 Oils and Fats

Adulteration in fats and oils can also be detected using FTIR. Jiang et al. (2016) created a Fourier transform infrared-based method to calculate the acid values (AVs) of edible oils using the hydroxyl stretching (3535 cm^{-1}) band of the COOH in free fatty acids, peak (3508 cm^{-1}), and spectral data from 3340 to 3390 cm^{-1}. ATR-FTIR was explored to characterize and detect adulteration of soybean oil (SO) in butter cheeses (BC) in combination with multivariate techniques. The scientists observed a drop in band intensities associated with proteins after adding adulterants. But when soybean oil (SO) was substituted for butter oil (BO), the majority of lipid-associated bands showed higher intensities, with adulteration levels rising to 70%. A particular band at 3007 cm^{-1} was also found exclusively in cheese that had been tampered with (Leite et al., 2019). Noman, 2024 reported that FTIR has shown higher accuracy and reduced the overall analysis time and it is also feasible to conduct iodine value, saponification number and peroxide value of fat and oil.

2.2.1.3.4 Other Applications

FTIR was used to evaluate the physical properties and composition of wine, including its pH, individual organic acids, reducing sugars, glucose, glycerol, total phenols, alcohol content, sucrose, relative density, fructose, total acidity, extract, and sugar-free extract (Patz et al., 2004). FT-MIR was utilized by Kelly and Downey (2005) to detect adulterants in apple juice, including sugar, beet sucrose (BS), high fructose corn syrup (HFCS), partially inverted cane syrup (PICS), and artificial fructose: glucose: sucrose (FGS) solutions. Additionally, FTIR was also employed as a non-destructive technique to forecast mango sweetness. Amir et al. (2013) employed Fourier transform infrared spectroscopy (FTIR) to examine various wheat quality characteristics, including hardness of grain, carbohydrate, fat, moisture, protein, and ash based on the identified points. The OH functional group allowed for the observation of the aqueous peaks in the 3300 cm^{-1} area. Amide band I (stretching of the C=O) and Amide II band (NH_2 deformation) which were within the spectral region of 1600 cm^{-1} to 1700 cm^{-1} and 1550 to 1570 cm^{-1} respectively were used to identify the protein, while the C–H bond was used to identify fat. The author observed that the starch was recognized in the spectral region of 2800–3000 cm^{-1} and 3000–3600 cm^{-1} based on C–H and O–H stretch areas.

Based on provenance, FTIR was also successful in differentiating between the different types of green coffee. This was accomplished by running an algorithm for principal component analysis on infrared spectra that were acquired for each sample in the 600–4000 cm^{-1} range (Obeidat et al., 2018). Thangavel and Dhivya (2019) used FTIR to determine the quality parameters of the turmeric. The author reported that curcumin content ranged from 2.78% to 4.81% using FT-NIR whereas, curcumin ranged between 2.82% and 4.91% using a conventional method which helped in the online grading of spices. Wei et al. (2021) compared the effectiveness of the three FTIR techniques by analysing the quality of rice. In contrast to FTIR attenuated total reflectance spectroscopy (FTIR-ATR) and diffuse reflectance spectroscopy (DRIFT), the FTIR technology employed in the study is FTIR photoacoustic spectroscopy (FTIR-PAS). The author concluded that FTIR-PAS was the best technique for quantitative analysis of rice samples. Jenner et al. (2022) reported that FTIR is also used to detect microplastics in human lung tissues. The three most common microplastics found in human lungs were resin (15%), polypropylene (PP; 23%), and polyethene terephthalate (PET; 18%). The author stated that this might be due to inhalation and exposure to environmental microplastics.

2.2.2 *Near Infrared Spectroscopy*

The identification of organic matter constituents is a typical use of near-infrared spectroscopy. It depends on the unique absorption of the sample by electromagnetic radiation at wavelengths ranging from 780 to 2500 nanometers. The NIR absorption frequencies wavelengths (λ) are expressed in nanometers. NIR spectra are made up of broad bands with overlapping absorption. These spectra are mostly representative of mixtures of vibrational modes of chemical bonds with the carboxyl, amino and hydroxyl group vibrational modes (Osborne, 2000).

2.2.2.1 Principle

The instrument NIR operates on the principle of light reflection, absorption, transmission or scattering in or through a food material by the Beer-Lambert law. In NIR spectral regions combinations of fundamental vibration and the overtones occur from the responses obtained from the chemical bonds in which chemical bonds provide the feature information of organic molecules. Since many food material quality parameters are measured and correlated with light reflectance or absorption in an identified range of wavelengths, NIR spectroscopy has become a valuable analytical tool in the food industry, especially when combined with chemometric methods for data mining and processing (Tan et al., 2012).

2.2.2.2 Equipment

The foremost components of NIR are a wavelength selector, a detector, and a light source.

2.2.2.2.1 Light Source

The light source used in NIR is a tungsten halogen bulb. It is a source of thermal radiation in which an electric current runs through a filament, causing resistive heating. Through a halogen cycle, tungsten travels between the filament and the halogen gas that occupies the bulb's volume. It emits a dazzling amount of light, the intensity and emission profile of which depend on the filament temperature and inner wall temperature of the lamp (Krzysztof et al., 2021).

2.2.2.2.2 Wavelength Selection Technique

The wavelength selector is a spectrometer's most crucial component. There are instruments on the market that are based on the concepts of a Hadamard mask, digital micromirror array, linear variable filter (LVF) or Fabry Perot interferometer. Moreover, a Michelson interferometer is implemented on a microscale in certain miniature designs (Beć et al., 2020). The wavelength selection methods determine whether the spectrometer should utilize an intricate array detector or a more affordable detector which is a single pixel (i.e., a single spectral resolution element). Several wavelength selectors could be made smaller by using micro-electromechanical systems (MEMS) or if micro-optics are also incorporated, micro-opto-electromechanical systems (MOEMS) (Antila et al., 2013). These optomechanical devices can process digital light (DLP). Microfabrication in silicon, or MEMS, is produced using a method akin to that of integrated circuits. A Michelson interferometer in a tiny spectrometer primarily encountered execution challenges, even though it eventually became the standard in tabletop spectrometers. However, there are successful commercial instruments based on this constituent (Neo Spectra sensors, for example). Furthermore, low light-throughput efficiency is a challenge for MEMS-based FTNIR spectrometers; nonetheless, steps are taken to address this aspect. Miniaturization of a Hadamard-transform (HT) spectrometer appears increasingly realistic. Its design is like that of a grating tool (Dyer, 1991).

2.2.2.2.3 Detectors

Photovoltaic Si diodes are used in small, low-cost spectrometers which are widely operated in visible spectrometers and NIR areas because they retain an acceptable sensitivity over the 700–1100 nm region. Moreover, the level of signal-to-noise ratio

(S/N) obtained from this approach is rather low. Wavelength cutoff filters are necessary when using photodiodes in portable spectrometers to reduce the possibility of the detector reacting to sunlight. If the 1050–2500 nm mid- and long-wave regions of the near-infrared spectrum are taken into consideration, then additional detectors are necessary. These devices primarily use high-performance Indium gallium arsenide (InGaAs) photodetectors since a small spectrometer needs to maintain a suitable S/N. The sensitivity typically ranges from 900 to 1700 nm. Quick scanning with good S/N is made possible by InGaAs because it has rapid reaction time, better quantum efficacy and minimal dark current at a specific sensor zone when compared to other detectors. An extended InGaAs version is available that can detect wavelengths up to 1700 nm, however, it has a lower signal-to-noise ratio and needs to be cooled (Crocombe, 2018).

2.2.2.2.4 Optical Materials

The majority of optical materials are usually transparent over a large amount of the near-infrared (NIR) region and do not absorb in the visible (Vis) range. This characteristic is crucial, as it makes glass optics suitable for use, but for optimal performance in the long-wave NIR (LW-NIR) region, superior quality bonded silica (bonded quartz) optics, free from impurities containing hydroxyl groups, may be necessary. A major advantage of convenient near-infrared spectrometers is the absence of alkali halides as photosensitive materials, making these spectrometers less expensive and better suited for use in humid environments. These features are particularly beneficial for process monitoring and on-site analysis.

A scratch-resistant optical window at the sample interface is favored for practical contact-mode operation in order to guarantee durability. Sapphire is an ideal material for such applications, as seen in devices like the MicroNIR spectrometer. Since reflection causes optical loss, its high refractive index (over 1.7 in the Vis and NIR bands) makes it more appropriate for devices with high optical throughput, like multiplexed or multichannel spectrometers. Notably, mirrors are the only reflective component required in certain miniature spectrometers. Gold-plated surfaces excel in this role, reflecting at least 96% of light between 700 and 2500 nm, covering the entire near-infrared spectrum.

2.2.2.3 Imaging and Processing

Spectral distribution information is obtained by combining digital pictures and spectroscopy in near-infrared chemical imaging. A hyperspectral data cube is produced by capturing the spectrum in every pixel of the sample image. The resulting chemical pictures display the distribution of each analyte in the sample matrix. An FTIR interferometer, a tunable filter, and a hyperspectral camera are examples of NIR instrumentation. Light sources for imaging include a halogen lamp for laboratory testing and a source for outside scans. A hyperspectral data cube, with a spectrum assigned to each X and Y element, is the representation of the measurement findings. Chemometric methods can be used to derive information about the sample's chemical composition from its data cube (Gendrin et al., 2008).

2.2.2.4 Applications of Near Infrared Spectroscopy

2.2.2.4.1 Dairy Products

Domingo et al. (2014) examined melamine detection in milk using MIR in conjunction with NIR spectroscopy. But the extremely low concentrations of melamine (parts per million) were only present in milk which made it difficult to detect and measure using near-infrared spectroscopy. Melamine in milk powder was detected using chemometrics and NIR hyperspectral imaging. Fu et al. (2014) used spectral similarity analysis and NIR hyperspectral imaging (990–1700 nm) to detect different levels of melamine adulteration in milk powders, ranging from 0.025% to 1%. NIR hyperspectral imaging has the advantage of analyzing a wider surface area.

2.2.2.4.2 Meat Products

During commercial beef carcass grading, visible and NIR spectroscopy was employed not only to predict chemical parameters but also to analyze the tenderness of longissimus steaks (Shackelford et al., 2004). Schmutzler et al. (2015) used three distinct NIR devices with differing degrees of precision to study the adulteration of pork sausages. Interestingly, they were also able to accurately evaluate the samples through polymer containers. Qu et al. (2018) used NIR spectroscopy to assess the pH and nitrogen content of pork to determine its freshness.

2.2.2.4.3 Fats and Oils

The chemometric prediction model combined with NIR to detect the presence of α-tocopherol in edible oil (Szlyk et al., 2005). The NIR approach was used to determine the amount of glycosinolate present in rapeseed (Kumar et al., 2010). Li et al. (2016) conducted a quantitative investigation of the relationship between spectral accumulation and temperature. The author discovered that it is feasible to precisely ascertain the composition of an oil mixture at different temperatures by constructing a temperature-induced spectral variation coefficient, with R^2 values as high as 0.9992. Using NIR and HPLC techniques, the authors noted accuracy and recoveries. The recoveries were 96.8–103.2% and 97.2–102.4%, and the relative standard deviations (RSD) were 0.68–2.80% and 0.79–3.06%. These results demonstrated that NIR could accurately predict the amount of α-tocopherol in an unknown edible oil. NIR was used to determine the total tocopherol and α-tocopherol content of olive oil. Due to the model's excellent performance, the residual predictive deviations (RPD) were 2.01 and 2.37 (Cayuela & Carcia, 2017). Meanwhile, R^2 values were greater than 0.7 when phenols and glucosinolates in rapeseed were measured using NIR (Sen et al., 2017). Cheevitsopon and Sirisomboon (2018) employed NIR spectroscopy to track the CH_2 vibration band to estimate the fat content in curry soup. The amount of carotenoid and tocopherol in crude oil was rapidly predicted quickly using NIR. The findings demonstrated the high predictive capability of the combined tocopherol and carotenoid contents. To identify the phytosterol and tocopherol contents in sunflower seeds, the critical wavelengths must be fixed (Kahrıman et al., 2019).

2.2.2.4.4 Fruits and Vegetables

Vegetables and fruits are special types of food items since they differ in size, colour, form, and chemical makeup even when they are picked at the same location and time. Sorting them according to quality is therefore crucial. Using traditional analytical methods requires a lot of labour and time. According to Kays (1999), NIR spectroscopy is a visually appealing non-destructive technique that works well for measuring the moisture content of fruits and vegetables. According to Flores et al. (2008), the NIR reflectance spectrum can reach a depth of 4–20 mm depending on the type of fruit and the intensity of the light. This can be enough to assess the soluble solid contents for small and thinly peeled fruits since sufficient spectral information can be obtained. Omar (2013) revealed that anthocyanin concentration can also be assessed using NIR in the spectral region 640–1100 nm using a PLS–based model. Spectral data in the 600–830 nm wavelength region is the most useful for identifying internal browning in 'Cripps Pink apples during cold storage (Mogollon et al., 2020). A spectral range analysis of 550–1650 nm was utilized to identify the internal fault in apples, and a spectral range analysis of 700–1000 nm was used to detect color changes (Huang et al., 2020). NIR was predicted to have great accuracy in detecting internal defects in Apple.

2.2.2.4.5 Other Applications

NIR spectroscopy was utilized by Aculey et al. (2010) and Teye et al. (2020) to distinguish and describe cocoa beans cultivated in Ghana's several cocoa-growing regions. Bazar et al. (2016) examined changes in honey assortments using a combination of aqua photometric and NIR spectroscopy by identifying minute variations in the water contained in the honey. They made use of the absorption bands in the spectral range of 1300–1800 nm that are linked to both water and carbohydrates. Lapcharoensuk and Nakawajana (2018) distinguished 30 distinct kinds of syrup by using NIR methods to find a vibrational band at 6940 cm that matched the combined band of O–H stretching in sucrose. In a more recent study, Guelpa et al. (2017) reported that NIR spectroscopy, when combined with PLS-DA, is a most dependable technique to differentiate South African honey from honey that has been purposefully contaminated (with sugar solutions, such as glucose and fructose) or imported. Overall categorization accuracy was found to range from 93.3% to 99.9%. The authors also tested various instruments and showed that classification accuracy was not affected using a portable device.

2.2.3 *UV-Visible Spectroscopy*

Measurements of light absorption, transmission, or reflection within the visible region of the electromagnetic spectrum (about 400–700 nanometers) can be made with the strong analytical tool known as visible spectroscopy (Tarhan, 2020). It is predicated on the idea that distinct materials absorb light at particular wavelengths, which can be applied to the identification and measurement of a wide range of molecules, including those found in food science and technology.

2.2.3.1 Theory

VIS spectroscopy measures the absorbance, transmittance, and reflectance of visible light (400–700 nm) by a sample, predicated on how light and matter interact (Barbin & ElMasry, 2020). While visible light interacts with a sample, it can be absorbed, reflected, transmitted, or scattered, with absorption being the primary focus. Absorption occurs due to electronic transitions within the molecules, where electrons move from lower to higher energy levels upon absorbing specific wavelengths of light. The Beer-Lambert Law connects absorbance to the sample's concentration and path length which is stated as follows, $A = \epsilon \cdot c \cdot l$, where ε is the molar absorptivity, also known as the extinction coefficient, c is the absorbing species' concentration, and l is the sample cell's path length. This principle underpins the quantitative analysis capabilities of visible spectroscopy, making it essential for identifying and measuring substances based on their unique absorption spectra. The technique involves a light source, monochromator or filter, sample holder, detector, and data processing unit to generate and analyse the spectrum (Heia et al., 2003).

2.2.3.2 Equipment and Instrumentation

Visible spectroscopy instrumentation comprises a light source, a monochromator or filter, a detector, a sample holder, and a data processing unit (Cavdaroglu & Ozen, 2022).

2.2.3.2.1 Light Source

The choice of light source depends on the desired wavelength range and application.

Tungsten Halogen Lamps These lamps offer a broad spectrum of light across the visible and near-infrared (NIR) regions. They are cost-effective and widely used, but their intensity can decrease over time.

Xenon Arc Lamps These lamps provide a more intense and stable light output across the visible and ultraviolet (UV) regions. They are ideal for applications requiring high sensitivity, but they are generally more expensive than tungsten halogen lamps.

LEDs (Light-Emitting Diodes) LEDs offer a narrowband light source with high intensity and stability. They can be tailored to emit specific wavelengths, making them suitable for targeted analysis of specific food components. However, they may require multiple LEDs to cover the entire visible range (Suhandy & Yulia, 2017).

2.2.3.2.2 Sample Holder

The sample holder is essential to maintaining accuracy and reproducibility measurements. It should be made from a material that does not absorb light in the desired wavelength range (Cavdaroglu & Ozen, 2022) (e.g., quartz cuvettes for liquid samples).

Adjustable Allow for proper positioning of the sample within the light path to minimize scattering effects. Leak-proof: For liquid samples, to prevent contamination and ensure consistent path length. Sample holders come in various designs depending on the sample type such as Cuvettes, Reflectance Probes and Flow Cells.

2.2.3.2.3 Monochromator

The monochromator acts as the heart of a spectrometer, separating the broad spectrum of light from the source into its wavelengths. Common types include Diffraction Gratings: These gratings are made up of a reflecting surface etched with a sequence of parallel lines that are tightly spaced apart. Light diffracts from the grating, separating the different wavelengths based on their properties. The rotation of the grating allows for the selection of specific wavelengths to pass through the system. Interference Filters: these filters selectively transmit specific wavelength ranges while blocking others. They offer simpler designs but may have limitations in terms of wavelength resolution compared to diffraction gratings (Ríos-Reina et al., 2020).

2.2.3.2.4 The Detector

The detector converts the light intensity at each selected wavelength into an electrical signal that can be analyzed and quantified (Huck et al., 2017). Common detector types include photodiodes and photomultiplier tubes.

Photodiodes These solid-state devices generate an electric current proportional to the intensity of light they receive. They are reliable, possess a broad spectrum of sensitivity, and offer good response times.

Photomultiplier Tubes (PMTs) These highly sensitive detectors amplify the incoming light signal before conversion to an electrical signal. PMTs are ideal for low-light applications but may be more expensive and require higher operating voltages compared to photodiodes. These also include multiplex detectors, which may measure an entire spectrum in a matter of seconds or minutes, contingent upon the apparatus, the experimental configuration, and the sample under examination. Arrays of charge-coupled devices (CCDs) or linear photodiodes are used to construct these detectors. The common detectors and light sources were showed in Table 2.1 (Sharma, 2019).

2.2.3.2.5 Instrument Designs for Molecular UV/Vis Absorption

Single-Beam Spectrophotometer

This type of spectrophotometer is a device that selects wavelengths using a monochromator. Figure 2.3 illustrates the simplest form of a spectrophotometer, a monochromator having a fixed wavelength used in a single beam device (Lu, 2014). The calibration and operation of single-beam spectrophotometers are like those of photometers. The Thermo Scientific Spectronic 20D+ is one type of single-beam spectrophotometer. The wavelength range of the Spectronic 20D+ is 340–625 nm, with an extension to 950 nm when used a red-sensitive detector. Its fixed effective bandwidth is 20 nm. Portable, battery-operated single-beam spectrophotometers are readily available for the field use (Sadecka & Jakubikova, 2022). Additionally, there are single-beam spectrophotometers with effective bandwidths ranging from 2 to 8 nm. Unfortunately, the laborious and time-consuming procedure of manually

Table 2.1 The common detectors and light sources

Detectors	Sensitivity	Radiation source	Emission range
Photomultiplier tube	<190–850 nm	Lamps with tungsten-halogen filaments A greater filament temperature is achieved by adding iodine.	320–3200 nm
Silicon photodiodes	200–1100 nm	Deuterium lamp	185–400 nm
Silicon charge-coupled	200–1100 nm	Xenon arc lamp	270–1000 nm

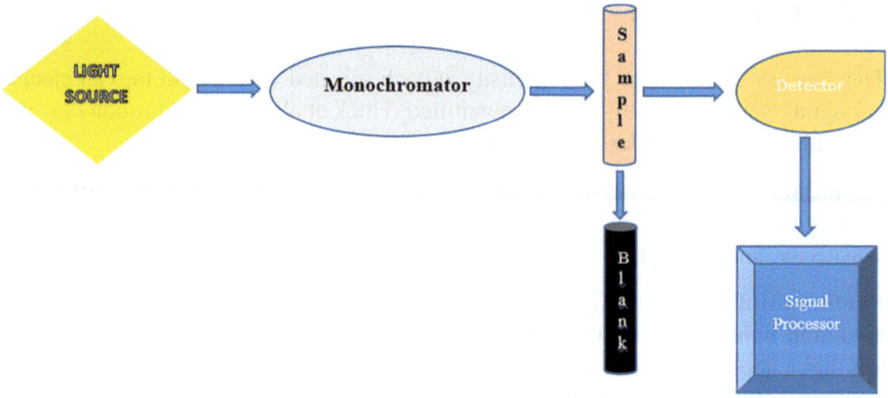

Fig. 2.3 Single-Beam Spectrophotometer

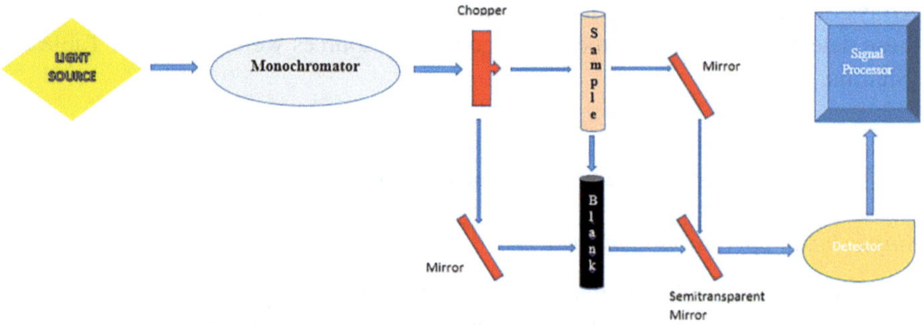

Fig. 2.4 Cary UV-Vis Double-Beam Spectrophotometer

altering the wavelength and recalibrating the device makes fixed-wavelength devices unsuitable for capturing spectra. The precision of a monochromatic spectrophotometer depends on the source's and detector's long-term stability.

Double-Beam Spectrophotometer

The disadvantages of a fixed-wavelength, single-beam spectrophotometer are lessened by using a double-beam spectrophotometer. Figure 2.4 illustrates how a chopper controls the radiation's path by switching the beam between a shutter, the sample, and the blank. The signal processor divides the signals that arrive at the detector into the transmission of the blank (P_O) and the sample (P_T) using the chopper's spinning speed. It is also possible to continuously control 0% transmission by using a shutter with an opaque surface. A double-beam spectrophotometer's effective bandwidth can be changed by altering the monochromator's entry and exit slits, which are normally between 0.2 and 3.0 nm in length (Smith & Johnson, 2018). Spectra can be automatically recorded with a scanning monochromator. Double-beam instruments are more valuable and versatile for both quantitative and qualitative studies, despite being less portable and more expensive than single-beam instruments. In the UV-vis module of this course, a double-beam spectrophotometer was used, specifically the Cary UV-vis double-beam spectrometer, as shown in Fig. 2.4.

Diode Array Spectrometer

One wavelength can only be monitored at a time by an instrument with a single detector. An array of photodiodes, on the other hand, can be used to replace a single photomultiplier to build a detector that can record the full spectrum in as little as 0.1 seconds. As seen in Fig. 2.5, the source radiation in a diode array spectrometer travels through the sample before being scattered by a grating. Every diode in the photodiode array detector records radiant power within a particular wavelength range, and it is positioned at the focal plane of the grating. This configuration makes the diode array spectrometer lightweight and compact by doing away with the requirement for a full monochromator. This type of spectrometer, known as an in-line detector or photodiode array detector (PDA), will be used in the HPLC experiment, and is incorporated into the HPLC machine.

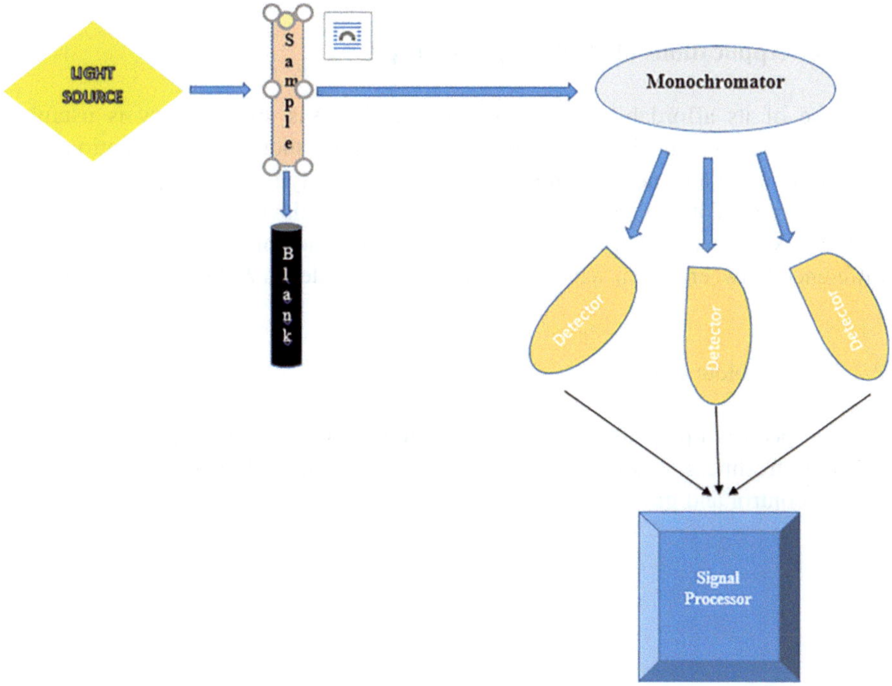

Fig. 2.5 Double-Beam Spectrophotometer

2.2.3.3 Imaging and Processing

Visible spectroscopy imaging and processing, covering the visible spectrum (400–700 nm), involves capturing and analyzing spectral information to study the composition, structure, and properties of materials. Hyperspectral imaging (HSI) combines spectroscopy and imaging to create detailed datasets where each pixel contains full spectral information. This technique is used in medical diagnostics, agriculture, environmental monitoring, and industrial quality control. Data processing includes calibration, noise reduction, spectral unmixing, chemometric analysis, and pattern recognition, utilizing software and custom algorithms to visualize, quantify, and identify substances.

UV-Vis spectroscopy systems focus electromagnetic radiation from a light source onto a sample, with light either transmitted through or reflected off the sample and collected for analysis (Kucharska et al., 2017). The monochromator uses dispersing elements to separate light by wavelength, which is then measured by a charge-coupled device (CCD) and displayed on a computer, producing a spectrum showing light intensity versus wavelength. These systems, which cover ultraviolet to infrared radiation, offer comprehensive tools for precise identification and quantification of various compounds and properties in complex samples, despite challenges like managing large data volumes and high equipment costs. Figure 2.6 shows the working of UV-Vis Spectrophotometer.

2.2.3.4 Applications of Visible Spectroscopy

Because of its affordability and ease of use, UV-Vis spectrometry is useful in screening various suspicious samples before a more thorough examination using any of the more advanced methods. This valuable analytical technique was utilized for classifying Tokaj wines according to varieties and flavours. The author reported that UV-Vis spectrometry when combined with chemometry discriminated the difference between Tokaj wines and other wines (Sadecka & Jakubikova, 2022).

2.2.3.4.1 Colour Analysis

Visible spectroscopy is used to assess the colour attributes of food products, such as ripeness, freshness, and colour uniformity. Quantifying colour parameters, aids in quality control and grading processes.

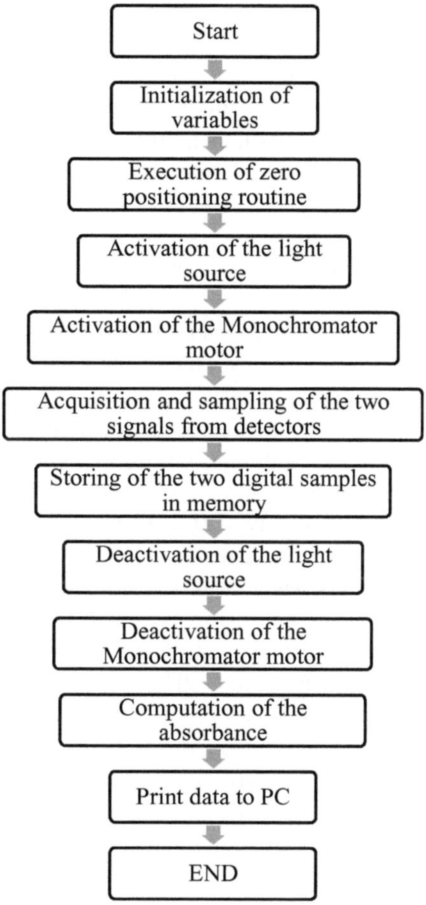

Fig. 2.6 Working of UV-Vis Spectrophotometer

2.2.3.4.2 Nutritional Analysis

Spectroscopic techniques can determine the concentration of nutrients in food, including vitamins, minerals, and pigments. This information is vital for assessing nutritional value and ensuring compliance with regulatory standards.

2.2.3.4.3 Detection of Contaminants

Visible spectroscopy facilitates the detection of contaminants in food, such as pesticides, toxins, and microbial contaminants. Rapid screening methods based on spectroscopic analysis help to ensure food safety.

2.2.3.4.4 Authentication and Traceability

Spectroscopy enables the authentication of food products by identifying characteristic spectral signatures associated with specific ingredients or geographical origins. It also supports traceability efforts by verifying the authenticity of food labels and certifications

2.2.3.4.5 Fish Quality

VIS spectroscopy can be used to determine the freshness of food as it is related to storage conditions. One of the most often used applications of VIS spectroscopy is the evaluation of seafood quality. A fish's freshness is affected by its storage method since different fish storage methods cause the fish's quality to gradually decline. In their 2002 study, Nilsen et al. (2002) used multivariate statistics to describe the link between spectral data and storage duration using VIS spectroscopy to evaluate the freshness quality of cod. The visible wavelength range was used to identify the best-fit model, which produced a 0.97 correlation of prediction with an error value of 1.04 days. Heia et al. (2003) evaluated the state of freshness of the hake seafood variety and cod using VIS spectroscopy. This study showed how the fish section's demands for the creation of industrial methods may be satisfied using VIS spectroscopy. Costa et al. (2011) utilized VIS spectroscopy to distinguish the grade of European sea bass raised in various environments. The analysis's findings showed that, in comparison to 96 hours after death, spectral measurements could more accurately identify individual animals 48 hours after death.

VIS spectroscopy was used to estimate the quality index method (QIM) score for cod, and it was shown that a small band in the visible spectral area was sufficient for identifying fish freshness. Fish freshness was monitored and frozen-thawed and fresh fish fillets were distinguished using VIS spectroscopy. The findings demonstrated that a narrow range of visible wavelengths might be used to completely discriminate between frozen, thawed fish fillets and fresh fish fillets.

2.2.3.4.6 Other Foods

One of the best methods for identifying less concentrated chemicals in food samples is UV-visible spectroscopy. This technique uses electromagnetic radiation with a wavelength range of 200–800 nm to identify two separate aspects of fat oxidation and colour (Cavdaroglu & Ozen, 2021). Recent studies used a mix of analytical techniques, such as UV-Vis and PLS regression, to measure the exact amount of squalene present in extra virgin olive oils (EVOO) (Tarhan, 2020). Wu et al. (2021) used empirical mode decomposition in conjunction with SVR (EMD-SVR) to assess the quality of culinary blend oil samples. They came to the conclusion that EMD-SVR produced superior findings for the quantitative analysis of ternary edible blend oil. Zhang et al. (2015) built models for the evaluation of acid value in different oils using UV-Vis in conjunction with principal component regression (PCR) and partial least squares regression (PLS). PLS models performed well in comparison to PCR models.

UV-Vis spectroscopy was found to be a more effective method when combined with PLS for quantifying grape-must caramel in Balsamic vinegar of different wine varieties than univariate methods. This was on top of earlier research on oil. UV-Vis spectroscopy, when combined with PLS, was found to be a more effective way than monovariate instruments for assessing grape-must caramel in Balsamic vinegars of various wine vinegar varietals. In addition to olive oil specimens, two different species of mint have been distinguished from one another using multivariate discriminating tools that use UV-Vis spectroscopy, such as PLS-DA and SVM (Kucharska-Ambrozej et al., 2021). Similarly, coffee has been precisely categorized and differentiated between ordinary coffee and Peaberry coffee by UV-Vis spectroscopy analysis (Suhandy & Yulia, 2017). In a different study, UV-Vis spectroscopy was used to identify between vinegar contaminated with spirit vinegar or acetic acid (Cavdaroglu & Ozen, 2022).

When used in conjunction with multivariate techniques, UV-visible spectroscopy may effectively examine food adulteration and improve discrimination. As such, it is frequently a useful tool for testing and managing food quality. Also, offers several benefits like cost-effectiveness, simplicity, non-destructiveness, and adaptability. However, when utilizing this technique, one should be aware of its limits on sensitivity, interference, and surface analysis.

2.3 Future Direction and Innovations

The future of visible spectroscopy in non-destructive food analysis is characterized by several emerging trends (Jeevarathinam et al., 2025).

Miniaturization and Portable Devices Advances in miniaturized spectroscopic instruments enable on-site and real-time analysis, enhancing convenience and accessibility in food quality assessment.

Integration with IoT and Big Data Analytics Visible spectroscopy is being integrated with Internet of Things (IoT) platforms and big data analytics to enable continuous monitoring of food quality throughout the supply chain. This integration enhances traceability, quality control, and predictive maintenance capabilities.

Multimodal Imaging and Hybrid Techniques Future developments may focus on combining visible spectroscopy with other imaging modalities, such as infrared spectroscopy and ultrasound imaging, to obtain comprehensive information about food composition and structure.

Machine Learning and AI-Assisted Analysis Machine learning algorithms and artificial intelligence (AI) techniques are being utilized for data analysis and interpretation in spectroscopy. These tools enhance the accuracy and efficiency of food quality assessment by automating complex pattern recognition tasks.

Visible spectroscopy plays a pivotal role in non-destructive food quality analysis, offering rapid, sensitive, and cost-effective solutions for assessing various attributes of food products. With ongoing technological advancements and emerging trends, the future holds promising opportunities for the continued evolution and application of visible spectroscopy in ensuring food safety, quality, and authenticity.

2.4 Conclusion

Ultimately, this chapter has offered a thorough examination of the concepts, procedures, and uses of UV-visible, near-infrared (NIR), and Fourier-transform infrared (FTIR) spectroscopy. These optical methods have proven indispensable across a wide range of fields, from medical diagnostics to environmental monitoring, owing to their ability to provide detailed insights into the composition, structure, and properties of materials. The integration of advanced technologies, particularly machine learning and artificial intelligence, has significantly enhanced data analysis and interpretation, making these techniques more powerful and versatile. Despite challenges such as managing large data volumes and the high costs associated with sophisticated equipment, the benefits and applications of these spectroscopic methods far outweigh these limitations.

Future advancements in technology and computational methods promise to further improve the accuracy, efficiency, and accessibility of spectroscopic analyses.

References

Aculey, P. C., Snitkjaer, P., Owusu, M., Bassompiere, M., Takrama, J., Nørgaard, L., Petersen, M. A., & Nielsen, D. S. (2010). Ghanaian cocoa bean fermentation characterized by spectroscopic and chromatographic methods and chemometrics. *Journal of Food Science, 75*(6), S300–S307. https://doi.org/10.1111/j.1750-3841.2010.01710.x

Alamprese, C., Amigo, J. M., Casiraghi, E., & Engelsen, S. B. (2016). Identification and quantification of Turkey meat adulteration in fresh, frozen-thawed and cooked minced beef by FT-NIR spectroscopy and chemometrics. *Meat Science, 121*, 175–181.

Amir, R., Anjum, M., Khan, F. M., Khan, M. I., Khan, M. R., Pasha, I., & Nadeem, M. (2013). Application of Fourier transform infrared (FTIR) spectroscopy for the identification of wheat varieties. *Journal of Food Science and Technology, 50*, 1018–1023.

Antila, J., Tuohiniemi, M., Rissanen, A., Kantojarvi, U., Lahti, M., Viherkanto, K., Kaarre, M., & Malinen, J. (2013). *MEMS- and MOEMS-Based near-infrared spectrometers in encyclopedia of analytical chemistery*. Wiley.

Barbin, D. F., & ElMasry, G. (Eds.). (2020). *Hyperspectral imaging analysis and applications for food quality*. Woodhead Publishing.

Bazar, G., Romvari, R., Szabo, A., Somogyi, T., Eles, V., & Tsenkova, R. (2016). NIR detection of honey adulteration reveals differences in water spectral pattern. *Food Chemistry, 194*, 873–880.

Beć, K. B., Grabska, J., Siesler, H. W., & Huck, C. W. (2020). Handheld near-infrared spectrometers: Where are we heading? *NIR News, 31*, 28–35.
Cavdaroglu, C., & Ozen, B. (2021). Prediction of vinegar processing parameters with chemometric modelling of spectroscopic data. *Microchemical Journal, 171*, 106886.
Cavdaroglu, C., & Ozen, B. (2022). Detection of vinegar adulteration with spirit vinegar and acetic acid using UV–Visible and Fourier transform infrared spectroscopy. *Food Chemistry, 379*, 132150.
Cayuela, J. A., & Carcia, J. F. (2017). Sorting olive oil based on alpha-tocopherol and total tocopherol content using 403 near-infra-red spectroscopy (NIRS) analysis. *Journal of Food Engineering, 202*, 79–88.
Cheevitsopon, E., & Sirisomboon, P. (2018). Rapid evaluation of fat content in curry soup containing coconut milk by using near infrared spectroscopy. *Journal of Near Infrared Spectroscopy, 26*(1), 16–25.
Chen, B., Han, X., & Jiang, X. (2016). In situ FTIR analysis of the evolution of functional groups of oil shale during pyrolysis. *Energy & Fuels, 30*(7), 5611–5616.
Cheng, J. H., Dai, Q., Sun, D. W., Zeng, X. A., Liu, D., & Pu, H. B. (2013). Applications of non-destructive spectroscopic techniques for fish quality and safety evaluation and inspection. Trends in food science & technology, 34(1), 18–31.
Costa, C., D'Andrea, S., Russo, R., Antonucci, F., Pallottino, F., & Menesatti, P. (2011). Application of non-invasive techniques to differentiate sea bass (Dicentrarchus labrax, L. 1758) quality cultured under different conditions. *Aquaculture International, 19*(4), 765–778.
Crocombe, R. A. (2018). Portable spectroscopy. *Applied Spectroscopy*. https://doi.org/10.1177/0003702818809719
Domingo, E., Tirelli, A. A., Nunes, C. A., Guerreiro, M. C., & Pinto, S. M. (2014). Melamine detection in milk using vibrational spectroscopy and chemometrics analysis: A review. *Food Research International, 60*, 131–139.
Dvorak, L., Mlcek, J., & Sustova, K. (2016). Comparison of FT-NIR spectroscopy and ELISA for detection of adulteration of goat cheeses with cow's milk. *Journal of AOAC International, 99*(1), 180–186.
Dyer, S. A. (1991). Hadamard transform spectrometry. *Chemometrics and Intelligent Laboratory Systems, 12*, 101–115.
Fakayode, S. O., Baker, G. A., Bwambok, D. K., Bhawawet, N., Elzey, B., Siraj, N., et al. (2020). Molecular (Raman, NIR, and FTIR) spectroscopy and multivariate analysis in consumable products analysis1. *Applied Spectroscopy Reviews, 55*(8), 647–723.
Flores, K., Sánchez, M. T., Pérez-Marín, D. C., López, M. D., Guerrero, J. E., & Garrido-Varo, A. (2008). Prediction of total soluble solid content in intact and cut melons and watermelons using near infrared spectroscopy. *Journal of Near Infrared Spectroscopy, 16*, 91–98.
Fu, X., Kim, M. S., Chao, K., Qin, J., Lim, J., Lee, H., Garrido-Varo, A., Perez-Marín, D., & Ying, Y. (2014). Detection of melamine in milk powders based on NIR hyperspectral imaging and spectral similarity analyses. *Journal of Food Engineering, 124*, 97–104.
Gendrin, C., Roggo, Y., & Collet, C. (2008). Pharmaceutical applications of vibrational chemical imaging and chemometrics: A review. *Journal of Pharmaceutical and Biomedical Analysis, 48*(3), 533–553.
Grassi, S., Casiraghi, E., & Alamprese, C. (2018). Handheld NIR device: A non-targeted approach to assess authenticity of fish fillets and patties. *Food Chemistry, 243*, 382–388.
Guelpa, A., Marini, F., du Plessis, A., Slabbert, R., & Manley, M. (2017). Verification of authenticity and fraud detection in South African honey using NIR spectroscopy. *Food Control, 73*, 1388–1396.
Harshitha, C. G., Sharma, N., Singh, R., Sharma, R., Gandhi, K., & Mann, B. (2023). Interaction study of aflatoxin M1 with milk proteins using ATR-FTIR. *Journal of Food Science and Technology, 60*(1), 64–72.

He, Y., Bai, X., Xiao, Q., Liu, F., Zhou, L., & Zhang, C. (2021). Detection of adulteration in food based on nondestructive analysis techniques: A review. *Critical Reviews in Food Science and Nutrition, 61*(14), 2351–2371.

Heia, K., Esaiassen, M., Nilsen, H., Sigernes, F., Luten, J., Oehlenschläger, J., et al. (2003). Visible spectroscopy-evaluation of storage time of ice stored cod and frozen stored hake. In *Quality of fish from catch to consumer: Labelling, monitoring and traceability*. Wageningen Academic Publishers.

Huang, P., Zheng, N., Zhou, H., & Huang, J. (2020). Curcumin inhibits BACE1 expression through the interaction between ERβ and NFκB signaling pathway in SH-SY5Y cells. *Molecular and Cellular Biochemistry, 463*(1), 161–173.

Huck, C. W., Bonn, G. K., & Huck-Pezzei, V. A. (2017). *Handbook of food analysis instruments*. CRC Press.

Jenner, L. C., Rotchell, J. M., Bennett, R. T., Cowen, M., Tentzeris, V., & Sadofsky, L. R. (2022). Detection of microplastics in human lung tissue using μFTIR spectroscopy. *Science of the Total Environment, 831*, 154907.

Jeevarathinam, G., Deepa, J., Nishevidha, P. B., Janika, A. S., Nandhakumar, S., Neethu, C., Kishore, S. G., Monicka, A. A., & Pandiselvam, R. (2025). Spectroscopy food functionality and safety. Advances in Food and Nutrition Research https://doi.org/10.1016/bs.afnr.2025.01.003

Jiang, X., Li, S., Xiang, G., Li, Q., Fan, L., He, L., & Gu, K. (2016). Determination of the acid values of edible oils via FTIR spectroscopy based on the OH stretching band. *Food Chemistry, 212*, 585–589.

Kahrıman, F., Onaç, I., Turk, F. M., Oner, F., & Egesel, C. O. (2019). Determination of carotenoid and tocopherol 508 content in maize flour and oil samples using near-infrared spectroscopy. *Spectroscopy Letters, 52*, 1–9.

Kays, S. J. (1999). Nondestructive quality evaluation of intact, high moisture products. *NIR News, 10*, 12–15.

Kelly, J. D., & Downey, G. (2005). Detection of sugar adulterants in apple juice using Fourier transform infrared spectroscopy and chemometrics. *Journal of Agricultural and Food Chemistry, 53*(9), 3281–3286.

Kharbach, M., Alaoui Mansouri, M., Taabouz, M., & Yu, H. (2023). Current application of advancing spectroscopy techniques in food analysis: Data handling with chemometric approaches. *Food, 12*(14), 2753.

Krzysztof, B. B., Grabska, J., & Huck, C. W. (2021). The new avenue - theoretical simulation of NIR spectra and its potential in analytical applications. In *The International Conference on NIR* (pp. 32–46). Springer Nature Singapore.

Kucharska AZ, Sokół-Łętowska A, Oszmiański J, Piórecki N, Fecka I. (2017) Iridoids, Phenolic Compounds and Antioxidant Activity of Edible Honeysuckle Berries (Lonicera caerulea var. kamtschatica Sevast.). *Molecules, 22*(3):405 https://doi.org/10.3390/molecules22030405

Kucharska-Ambrozej, K., Martyna, A., Karpińska, J., Kiełtyka-Dadasiewicz, A., & Kubat-Sikorska, A. (2021). Quality control of mint species based on UV-VIS and FTIR spectral data supported by chemometric tools. *Food Control, 129*, 108228.

Kumar, S., Chauhan, J. S., & Kumar, A. (2010). Screening for erucic acid and glucosinolate content in rapeseed-mustard seeds using near infrared reflectance spectroscopy. *Journal of Food Science and Technology, 47*(6), 690–692.

Lapcharoensuk, R., & Nakawajana, N. (2018). Identification of syrup type using fourier transform-near infrared spectroscopy with multivariate classification methods. *Journal of Innovative Optical Health Science, 11*(02), 1750019.

Leite, A. I. N., Pereira, C. G., Andrade, J., Vicentini, N. M., Bell, M. J. V., & Anjos, V. (2019). FTIR-ATR spectroscopy as a tool for the rapid detection of adulterations in butter cheeses. *LWT, 109*, 63–69.

Li, Z., Zhou, M., Luo, Y., Li, G., & Lin, L. (2016). Quantitative determination based on the differences between spectra-temperature relationships. *Talanta, 155*, 47–52.

Lu, R. (2014). *Hyperspectral imaging for food quality analysis and control*. Academic Press.

Mogollon, M. R., Jara, A. F., Contreras, C., & Zoffoli, J. P. (2020). Quantitative and qualitative VIS-NIR models for early determination of internal browning in 'Cripps Pink' apples during cold storage. *Postharvest Biology and Technology, 161*, 111060.

Mohamed, M. A., Jaafar, J., Ismail, A. F., Othman, M. H. D., & Rahman, M. A. (2017). Fourier transform infrared (FTIR) spectroscopy. In *Membrane characterization* (pp. 3–29). Elsevier.

Nilsen, H., Esaiassen, M., Heia, K., & Sigernes, F. (2002). Visible/near infrared spectroscopy: A new tool for the evaluation of fish freshness? *Journal of Food Science, 67*(5), 1821–1826.

Obeidat, S. M., Hammoudeh, A. Y., & Alomary, A. A. (2018). Application of FTIR spectroscopy for assessment of green coffee beans according to their origin. *Journal of Applied Spectroscopy, 84*, 1051–1055.

Omar, A. F. (2013). Spectroscopic profiling of soluble solids content and acidity of intact grape, lime, and star fruit. *Sensor Review, 33*, 238–245.

Osborne, B. G. (2000). *Near infrared spectroscopy in food analysis* (pp. 1–14). Wiley. (Chapter 1).

Ozaki, Y., Huck, C., Tsuchikawa, S., & Engelsen, S. B. (Eds.). (2021). *Near-infrared spectroscopy: Theory, spectral analysis, instrumentation, and applications* (pp. 978–9811586477). Springer.

Patz, C. D., Blieke, A., Ristow, R., & Dietrich, H. (2004). Application of FT-MIR spectrometry in wine analysis. *Analytica Chimica Acta, 513*(1), 81–89.

Penner, M. H. (2017). Basic principles of spectroscopy. In *Food analysis* (pp. 79–88). Springer.

Qu, F., Ren, D., He, Y., Nie, P., Lin, L., Cai, C., & Dong, T. (2018). Predicting pork freshness using multi-index statistical information fusion method based on near infrared spectroscopy. *Meat Science, 146*, 59–67.

Ribeiro, D. C., Neto, H. A., Lima, J. S., de Assis, D. C., Keller, K. M., Campos, S. V., & Fonseca, L. M. (2023). Determination of the lactose content in low-lactose milk using Fourier-transform infrared spectroscopy (FTIR) and convolutional neural network. *Heliyon, 9*(1), e12898. https://doi.org/10.1016/j.heliyon.2023.e12898

Ríos-Reina, R., Azcarate, S. M., Camiña, J., & Callejón, R. M. (2020). Assessment of UV–visible spectroscopy as a useful tool for determining grape-must caramel in high-quality wine and balsamic vinegars. *Food Chemistry, 323*, 126792.

Rohman, A., Erwanto, Y., & Man, Y. B. C. (2011). Analysis of pork adulteration in beef meatball using Fourier transform infrared (FTIR) spectroscopy. *Meat Science, 88*(1), 91–95.

Sadecka, J., & Jakubikova, M. (2022). Classification of Tokaj wines by ultraviolet–visible spectroscopy. *Food Analytical Methods, 15*(1), 56–66.

Schmutzler, M., Beganovic, A., Bohler, G., & Huck, C. W. (2015). Methods for detection of pork adulteration in veal product based on FT-NIR spectroscopy for laboratory, industrial and on-site analysis. *Food Control, 57*, 258–267.

Sen, R., Sharma, S., Kaur, G., & Banga, S. S. (2017). Near-infrared reflectance spectroscopy calibrations for assessment of oil, phenols, glucosinolates and fatty acid content in the intact seeds of oilseed Brassica species. *Journal of the Science of Food and Agriculture, 98*, 4050–4057.

Shackelford, S. D., Wheeler, T. L., & Koohmaraie, M. (2004). Development of optimal protocol for visible and near-infrared reflectance spectroscopic evaluation of meat quality. *Meat Science, 68*, 371–381.

Sharma, S. K. (2019). *Artificial intelligence in food quality monitoring and control*. CRC Press.

Smith, A. B., & Johnson, C. D. (2018). *UV-visible spectrophotometry in the laboratory*. CRC Press.

Spina, A. A., Ceniti, C., Piras, C., Tilocca, B., Britti, D., & Morittu, V. M. (2022). Midinfrared (MIR) spectroscopy for the detection of cow's milk in buffalo milk. *Journal of Animal Science and Technology, 64*(3), 531.

Suhandy, D., & Yulia, M. (2017). Peaberry coffee discrimination using UV-Visible spectroscopy combined with SIMCA and PLS-DA. *International Journal of Food Properties, 20*, S331–S339.

Szlyk, E., Szydlowska-Czerniak, A., & Kowalczyk-Marzec, A. (2005). NIR spectroscopy and partial 614 least-squares regression for determination of natural r-Tocopherol in vegetable oils. *Journal of 615 Agricultural and Food Chemistry, 53*, 6980–6987.

Tan, C., Chen, H., Xu, Z., Wu, T., Wang, L., & Zhu, W. (2012). Improvement of spectral calibration for food analysis through multi-model fusion. *Spectrochimica Acta Part A: Molecular and Biomolecular Spectroscopy, 96*(10), 526e531.

Tarhan, I. (2020). A comparative study of ATR-FTIR, UV–visible and Fluorescence spectroscopy combined with chemometrics for quantification of squalene in extra virgin olive oils. *Spectrochimica Acta. Part A, Molecular and Biomolecular Spectroscopy, 241*, 118714.

Teye, E., Anyidoho, E., Agbemafle, R., Sam-Amoah, L. K., & Elliott, C. (2020). Cocoa bean and cocoa bean products quality evaluation by NIR spectroscopy and chemometrics: A review. *Infrared Physics & Technology, 104*, 103127.

Thangavel, K., & Dhivya, K. (2019). Determination of curcumin, starch and moisture content in turmeric by Fourier Transform Near Infrared Spectroscopy (FT-NIR). *Engineering in Agriculture, Environment and Food, 12*, 264–269. https://doi.org/10.1016/j.eaef.2019.02.003

Upadhyay, N., Harshitha, C. G., Pathak, N. K., & Sharma, R. (2021). Fourier Transform Infrared (FTIR) spectroscopy with chemometrics: Evaluation of food quality and safety. In *Handbook of research on food processing and preservation technologies* (pp. 271–310). Apple Academic Press.

Wei, L., Ma, F., & Du, C. (2021). Application of FTIR-PAS in rapid assessment of rice quality under climate change conditions. *Foods, 10*(1), 159.

Wu, X., Bian, X., Lin, E., Wang, H., Guo, Y., & Tan, X. (2021). Weighted multiscale support vector regression for fast quantification of vegetable oils in edible blend oil by ultraviolet-visible spectroscopy. *Food Chemistry, 342*, 128245.

Yaman, H., Aykas, D. P., & Rodriguez-Saona, L. E. (2023). Monitoring Turkish white cheese ripening by portable FT-IR spectroscopy. *Frontiers in Nutrition, 10*, 126.

Zhang, W., Li, N., Feng, Y., Su, S., Li, T., & Liang, B. (2015). A unique quantitative method of acid value of edible oils and studying the impact of heating on edible oils by UV–Vis spectrometry. *Food Chemistry, 185*, 326–332.

Chapter 3
Nuclear Magnetic Resonance

Muhammed Navaf, Monica R. Nemtanu, Amin Mousavi Khaneghah, and Kappat Valiyapeediyekkal Sunooj

3.1 Introduction

Nuclear Magnetic Resonance (NMR) can be applied to a wide spectrum of applications, including the characterization of solids, liquids, biomaterials, etc., without affecting the sample properties or producing any hazardous materials. NMR-based methods have numerous benefits over alternative analytical techniques like HPLC, GC MS, etc. NMR represents a rapid technique requiring less than 5 minutes for the acquisition of a ^{1}H NMR spectrum. Unlike many other analytical methods, NMR does not necessitate calibration. Its exceptional selectivity enables the determination of unknown compounds with molecular precision. A key advantage lies in its capacity to provide both structural and quantitative insights across a broad spectrum of chemical groups in a single analysis and also has outstanding repeatability and reproducibility. Furthermore, in the realm of food analysis, where products are inherently prone to biochemical and chemical transformations, NMR serves as a formidable tool for monitoring such changes. A high-resolution NMR is particularly useful in analyzing liquid food samples like juice, puree, beverages, wines, and oils. Nevertheless, NMR can also be applied effortlessly to samples with minimal extraction or preparation procedures (Bertelli et al., 2010).

In the late 1940s, NMR technology found its initial application in revealing the molecular structures within organic chemistry. Despite this, the varied uses of NMR spectroscopy in food science didn't emerge till the 1980s. This delay was mainly due

M. Navaf · K. V. Sunooj (✉)
Department of Food Science and Technology, Pondicherry University, Puducherry, India

M. R. Nemtanu
National Institute for Laser, Plasma and Radiation Physics, Bucharest-Magurele, Romania

A. M. Khaneghah
Department of Fruit and Vegetable Product Technology, Prof. Wacław Dąbrowski Institute of Agricultural and Food Biotechnology – State Research Institute, Warsaw, Poland

C. K. Sunil et al. (eds.), *Emerging Non-destructive Technologies for Food Quality Analysis*, https://doi.org/10.1007/978-3-031-95048-3_3

to a scarcity of scientific proficiency, the steep equipment costs, and the unavailability of NMR components tailored for food-related studies. Nonetheless, it's important to mention that pulsed NMR techniques had already been explored in the analysis of foods and their derivatives before this timeframe (Marcone et al., 2013).

3.2 Basic Principles of NMR Spectroscopy

The atomic nucleus is composed of protons and neutrons providing quantum mechanical characteristics to the nucleus. These quantum mechanical characteristics, called spin, have magnitude as well as dimensions. Likewise, the molecules in the nucleus possess a certain level of energy. The nucleus, in its spinning state, exhibits a circulating current capable of producing magnetic fields (MF) and creating magnetic moments (MM). This phenomenon is distinguished by a nuclear spin quantum number denoted as "I". Each nucleus possesses a unique MM. When a spinning nucleus with a non-zero magnetic moment ($I \neq 0$) encounters an external MF, B0, it behaves akin to a gyroscope's motion, termed Lamohr precession. The following equation governs the frequency of this nuclear spin within the field.

$$\omega = \gamma B_0 = 2\Pi \upsilon$$

where ω is the angular frequency, υ is *Lamohr precession* frequency.

When the frequency of a radio-frequency (RF) electromagnetic field matches the Larmor frequency of the nucleus, resonance occurs. The resonance between the frequency of the radio-frequency field and the Lamohr frequency of the nucleus leads to the absorption of radio-frequency energy by the nucleus. Consequently, the nucleus transitions from a low-energy to a high-energy state, initiating NMR. Nuclei with zero spin ($I = 0$) lack a MM, rendering them incapable of producing observable NMR signals.

NMR is a process where atomic nuclei, possessing a non-zero nuclear spin, absorb and re-emit electromagnetic waves when placed in a magnetic field. This absorption and emission occur as the nuclei transition between different energy levels associated with their spin states. This absorption process provides insight into the structure of the nucleus and its environment, which is then examined through chemical shifts (δ, ppm) and spin couplings.

NMR facilitates the exploration of nuclear spins possessing magnetic features, rendering it pivotal in food analysis. Primary nuclei of focus encompass ^{1}H and ^{13}C, with ancillary interest in ^{15}N, ^{19}F, and ^{31}P. The ^{1}H nucleus, possessing a high gyromagnetic ratio and abundant presence in nature (99.99%), stands as the nucleus of choice for food analysis (Van et al., 2008). It has the ability to give valuable insights into the minor elements within the food samples. However, the ^{1}H NMR spectra of most foodstuffs exhibit a high degree of complexity. The presence of strong signal overlaps and scalar couplings within a limited resonance frequency range (around 15 ppm) renders interpretation quite challenging.

In the realm of NMR analysis for foods, the ^{13}C nucleus stands as the second most frequently employed nucleus. It is possible to gain a deeper understanding of molecules through ^{13}C NMR spectroscopy than can be achieved through ^{1}H NMR. The ^{13}C nucleus has a broad range of resonance frequencies (250 ppm). The ^{13}C NMR experiments are conducted with proton decoupling, which results in simplified spectra even for complex food matrices. Further, the much wider range of resonance (~1000) reported for the ^{31}P nucleus ensures good separation of signals obtained under proton decoupling (Apostolos Spyros, 2013).

3.3 NMR Parameters

NMR is broadly explored as a tool for quantitative and qualitative analysis. The significant parameter obtained from the NMR spectra includes chemical shift, the scalar spin-spin coupling constant, and the signal intensity. The nuclear overhauser effect and the spin-lattice (T_1) and spin-spin (T_2) relaxation times, etc., could be obtained from a specific type of NMR analysis (e.g. 2D NMR).

The chemical shift in NMR refers to the displacement of resonance frequencies of atomic nuclei within a molecule relative to a reference standard, typically tetramethylsilane (TMS) for proton NMR (^{1}H NMR) and carbon tetrafluoride ($CFCl_3$) for carbon NMR (^{13}C NMR). The chemical shift denoted in ppm offers vital insights into the local electronic environment surrounding the nuclei and the chemical constituents of nuclei. It is determined by factors such as electron density, neighboring atoms, hybridization, and molecular geometry. A chemical shift in NMR spectroscopy serves as a useful instrument for the structural elucidation of organic compounds, allowing chemists to analyze molecular environments and infer structural features based on the observed resonance frequencies.

Scalar spin-spin coupling is a property of the nuclear spins that are manifested in the NMR spectrum by the signal splitting into several components. In NMR spectroscopy, scalar spin-spin coupling refers to the interactivity among the MM of nuclei due to their inherent magnetic properties. This interaction arises from the coupling of the nuclear spins through the chemical bonds that connect them. Scalar coupling manifests as the splitting of NMR signals into multiple peaks in the spectrum, known as multiplets. Scalar coupling in NMR spectroscopy is essential for elucidating molecular connectivity, determining stereochemistry, and characterizing chemical environments within the molecules.

Signal intensity in NMR spectra refers to the magnitude of the signal detected in the spectrum. It represents the amplitude or strength of the signal generated by the resonating nuclei in the sample. The intensity of an NMR peak correlates straightaway with both the quantity of nuclei involved in the specific resonance and their relaxation characteristics. In 1D NMR spectra, signal intensity is often depicted by the height or area under the peaks. A higher signal intensity indicates a greater abundance of nuclei with the corresponding chemical environment. In 2D NMR spectra, signal intensity can be visualized in contour plots or cross-peaks, reflecting the correlations between different nuclei in the sample (Apostolos Spyros, 2013).

3.4 1D and 2D NMR Analysis

The 1D and 2D NMR analyses were carried out for the quantitative measurement of food samples. 1D spectra could be obtained by using a single radiofrequency pulse, causing nuclei to absorb energy and resonate at specific frequencies. The 2D NMR spectroscopy involves the correlation of two frequency domains. The key advantage of using 2D NMR is the spread of chemical shifts in two dimensions that help the separation of overlapping signals that hamper the reliable quantification with 1D NMR. Correlation spectroscopy (COSY), total correlation spectroscopy (TOCSY), hetero-nuclear single quantum correlation (HSQC), and heteronuclear multiple bond correlation (HMBC), etc. are the few types of 2D spectroscopy (Van et al., 2008). Figure 3.1 shows the image of the NMR spectrometer.

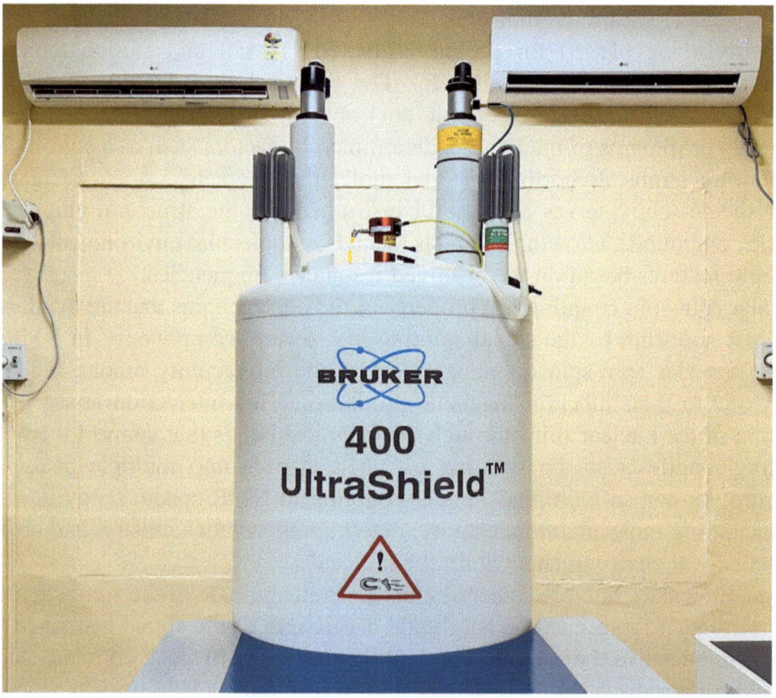

Fig. 3.1 Image of NMR spectroscope. (Photo courtesy Pondicherry University-Central instrumentation facility)

3.5 NMR Relaxometry

NMR relaxometry, also recognized as time domain NMR (TD NMR), stands as a formidable approach for scrutinizing the dynamic characteristics of condensed matter and solids. This encompasses a broad spectrum, ranging from elementary, diminutive molecules and ions to intricate, large-scale macromolecular systems. In the realm of NMR relaxometry, a specimen is introduced into a potent, consistent magnetic field. Here, radiofrequency (RF) pulses are administered to control the nuclear spins. Following the cessation of the RF pulse, the spins revert to their balanced orientations, emitting RF signals discerned by the NMR apparatus. This duration, during which the spins ease back into equilibrium, is termed relaxation time. NMR relaxometry offers distinctive benefits, including its capability to investigate dynamic processes spanning a wide spectrum of correlation times (ranging from milliseconds to nanoseconds) within a single experiment. Additionally, it provides the capacity to discern the mechanisms underlying molecular and ionic motions (Kruk & Florek-Wojciechowska, 2020). The application of NMR relaxometry, especially fast field-cycling NMR, has shown its effectiveness in examining various liquid food items like dairy products, fats, oils, honey, and others. The NMR relaxometry application in dairy includes the evaluation of curd syneresis, phase transition of milk during cooling, changes in cheese during brining, etc. (Ates et al., 2021).

3.6 NMR Sample Preparation and Analysis

Sampling plays a vital role in the analytical techniques. A sample is a small representative of a large group, and the information obtained from the small group could be used to judge a large group. In general, the botanical and geographical origin of the product could be known for analyzing food samples. The subsequent consideration pertains to the distribution of samples, it has a significant role in classification and multivariate calibration studies. Likewise, the food composition analysis of plant origin is affected by the growth or ripening stage, processing conditions, and metabolic changes like enzymatic and oxidative reactions. Figure 3.2 gives an overview of the NMR analysis of food samples

Preparing a food sample for the analysis typically includes several preliminary steps.

Conservation For a successful analysis, rapid sample pre-treatment followed by immediate analysis is essential. However, this approach is often impractical, necessitating sample conservation for a brief period before analysis as a crucial initial step to prevent potential alterations in food composition. For example, rapid cooling in liquid nitrogen could arrest the enzymatic and metabolic changes.

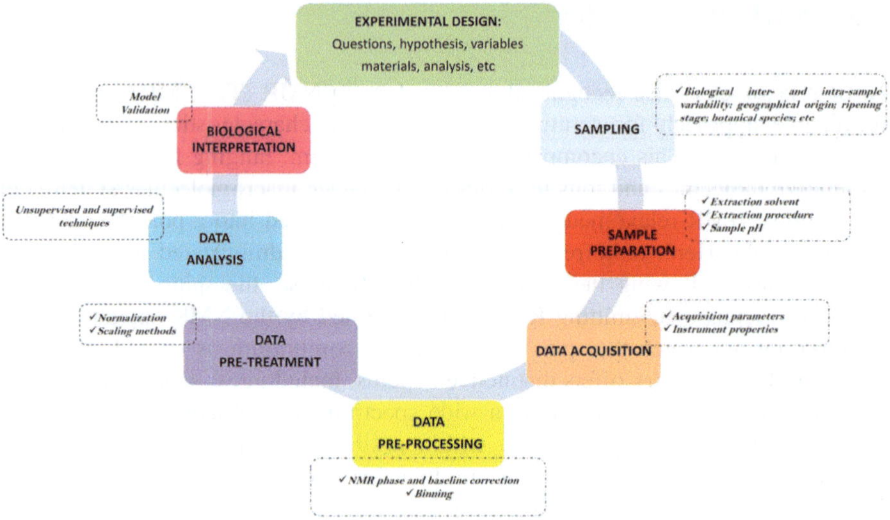

Fig. 3.2 The workflow of an NMR analysis of a food sample. (Sobolev et al., 2022)

Homogenization Based on the characteristics of the food specimen, it may need to be homogenized to ensure uniformity. This step is crucial for solid or heterogeneous samples to obtain consistent results. Techniques such as grinding, blending, or chopping may be employed for this purpose.

Extraction In some cases, extraction of specific components from the food matrix may be necessary to narrow down the analysis to a particular compound of interest. Various extraction methods, like solvent, solid-phase, or liquid-liquid extraction, etc., may be utilized depending on the type of analytes.

Dehydration If the food sample contains a high water content, dehydration may be required to prevent interference with NMR signals. This can be achieved through methods such as freeze-drying, oven-drying, or vacuum-drying, ensuring that the process does not induce chemical changes in the sample.

Solvent selection Deuterated solvents are frequently employed to establish a neutral baseline for NMR assessments, effectively reducing potential interference arising from the solvent. Protonated solvents should be avoided because their strong signals may overlap with those of the sample signals. Deuterated chloroform is the most explored solvent for the proton NMR analysis and its chemical shift is typically observed around 7.26 ppm when referenced to tetramethylsilane (TMS) as the internal standard.

Data analysis Once the NMR analysis is performed, the data obtained from the metabolomic study is transferred to a suitable format to perform multivariate data analysis (MvDA) like principal component analysis (PCA). The pre-processing steps in an NMR-based metabolomic analysis consist of baseline correction, alignment, binning, normalization, and scaling.

Baseline correction Baseline correction is a crucial pre-processing step in NMR analysis aimed at removing unwanted signal distortions or artifacts originating from various sources, such as baseline drift, solvent impurities, instrument noise, or imperfect shimming. These distortions can mask the genuine spectral characteristics aim to observe, thereby affecting the precision of quantitative study and the interpretation of spectra. There are different methods to perform the baseline correction, for example, polynomial fitting, wavelet transform, Spline fitting, etc.

Alignment Alignment involves rectifying deviations in the chemical shifts of peaks across a spectrum, usually stemming from disparities in instrumental settings, sample conditions, or experimental setups. Initially, spectra are adjusted relative to an internal standard within each spectrum, utilizing TMS at 0.0 ppm for this task. Although this initial alignment aids in eliminating global shifts from the dataset. Consequently, fine-tuning or local alignment becomes essential. Local alignment can be achieved through diverse alignment algorithms, such as interval correlated shifting, Correlation Optimised Warping, etc.

Binning Binning is conducted to aid in MvDA, where the spectra are segmented into distinct groups referred to as bins, and the intensity of the spectrum within each bin is computed.

Normalization In the realm of NMR-based metabolomics, normalization entails a mathematical process aimed to adjust the overall concentration of the sample. The aim is to ensure comparability among profiles by accounting for metabolite dilution.

Scaling For the purpose of MvDA, scaling is crucial. It is the process of adjusting the intensity values of NMR spectra to facilitate comparison or interpretation. As a result of scaling, the multivariate data models are less likely to be dominated by the most prominent peaks. This adjustment is crucial owing to variations in the sensitivity of NMR instruments, fluctuations in sample concentration, or differences in experimental setups. Different methods are employed for the scaling process, for example, auto-scaling, Pareto scaling, etc.

3.7 NMR in Food Component Analysis

3.7.1 Water

Water is one of the major constituents in most food products, and hence, the analysis of measurement of water has a prime role. Supercritical water has gained much attention in the recent past since it has great biological importance. Water's exceptional traits in typical environmental conditions stem from hydrogen bonding. Analysis at the molecular level often revolves around the intermolecular correlation function of the O–H pair, elucidating the bonding between two water molecules. Meanwhile, the intermolecular correlation function of the O–O pair unveils the intricate network of hydrogen bonds involving multiple water molecules. The proton NMR is one of the techniques that probe the density effect on the structure of water molecules. The H-NMR spectra of water are temperature dependent, and they exhibit peak shifts up the field as the temperature is raised. At 30 °C (1.00 g/cm), water exhibits a chemical shift at 2.7. Similarly at 300 °C (0.71 g/cm^3), 400 °C (0.19 g/cm^3), and 400 °C (0.41 g/cm^3) water exhibits chemical shifts at 2.12, 1.26, and 0.65, respectively (Matubayasi et al., 1997).

Several effective methods are employed to investigate water behavior in food systems, notably NMR spectroscopy alongside pulsed-field gradient NMR and MRI, which have gained prominence. Instruments designated for moisture determination through NMR are termed low-field NMR (Schmidt, 1991). Recent remarkable developments in NMR technology have significantly broadened the spectrum of methods used for water study. These techniques now encompass assessing water quantity via proton (^{1}H) pulsed NMR, scrutinizing water's detailed structure and dynamic attributes through deuterium (^{2}H) and oxygen-17 (^{17}O) pulsed Fourier transform NMR, as well as employing ^{1}H and ^{2}H pulsed-field gradient NMR and MRI (Schmidt & Lai, 1991).

In the analysis of food samples using NMR spectra, water signals take precedence, serving as indicators for gauging water content and its dispersion within the food system. MRI, on the other hand, unveils the intricate molecular layout of water within the sample, providing a multidimensional insight into its internal structure. In NMR spectra, peak intensity mirrors the water quantity present in the food, aligning with MRI's ability to quantitatively determine water content and distribution via magnetic resonance parameters. Moreover, it provides information regarding the physicochemical properties of water molecules in the food system. By practicing repetitive MRI, the changes in moisture content could be determined, which could be associated with maturity, ripening, shelf life, etc. The NMR results revealed noticeable changes in the moisture content of fruits among different seasons and different geographical regions (Marcone et al., 2013).

3.7.2 *Protein and Amino Acid*

The protein structure and amino acid profile are essential traits that affect the characteristics of protein-rich foods. NMR spectroscopy stands as a prominent technique pivotal for scrutinizing protein structure, interactions, and their dynamics at the atomic level. It furnishes insights into the conformational and interactional dynamics occurring within fractions of time spans extending over days across diverse sample conditions like solution, solid, and membrane states. Analysis entails gathering and dissecting data gleaned from the chemical shifts of key atoms, notably ^{1}H, ^{15}N, and ^{13}C, within the protein molecule (Hu et al., 2021).

The utilization of NMR spectroscopy enables the extraction of atomic-resolution data concerning the intermolecular attraction and binding-induced configurational alteration within protein complexes. Such complexes occur in solution, within the cell membrane, and in bulky macromolecular complexes (Purslow et al., 2020). The assessment of protein structure relies on various experimental constraints, such as local distance, coming out from the Nuclear Overhauser Effect (NOE) (≤6 Å), hydrogen bonds, and dihedral angle restraints. Additionally, long-range distance restraints can be inferred by paramagnetic relaxation enhancement techniques (Hu et al., 2021).

The low-field NMR spectral analysis of mustard and rape seed showed an average protein content of 28.02% and 20.80%, respectively (Pedersen et al., 2000). The amino acid profiling of Amarone wine protein using NMR spectra revealed that amino acids exhibit distinct NMR signals, like leucine at 0.88, 1.44, and 3.37 ppm, threonine at 1.42 and 4.38 ppm, alanine at 1.45 and 3.65 ppm, proline at 2.01, 2.08, 2.35, 3.38, and 4.11 ppm and choline at 3.20 ppm (Fig. 3.3) (Consonni et al., 2011). Similarly, significant variations in amino acid content were observed in grana Padano cheese at varying ripening levels through NMR analysis (de Angelis Curtis et al., 2000). Figure 3.4 illustrates the NMR spectra of Padano cheese at 12 months of ripening.

3.7.3 *Fat and Fatty Acids*

NMR spectroscopy is capable of examining fat content in different types of foods, enabling not just the measurement of overall fat content but also the identification of fatty acid composition. Over time, NMR has been utilized explicitly for quantifying

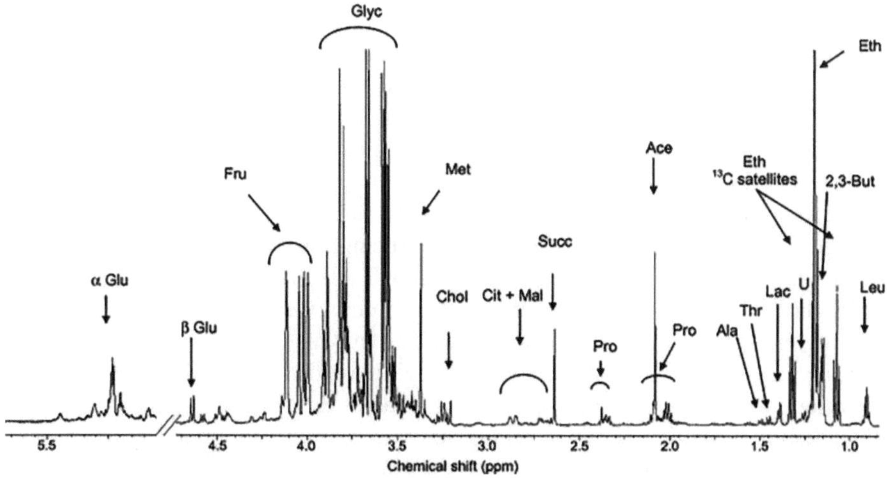

Fig. 3.3 ^{1}H NMR spectrum of Amarone wine. (Consonni et al., 2011)

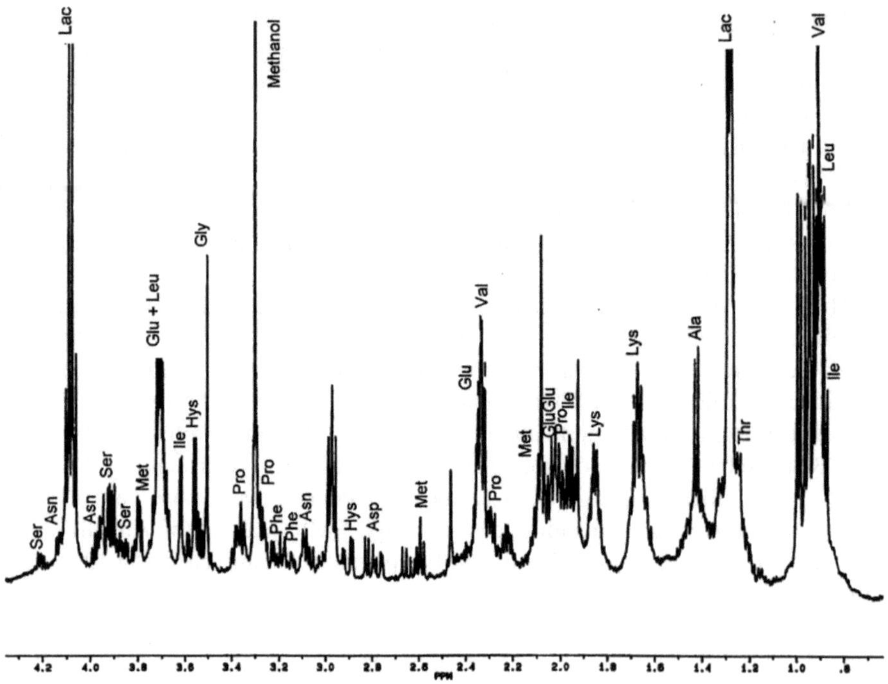

Fig. 3.4 NMR spectra of free amino acids of Grana Padano cheese. (de Angelis Curtis et al., 2000)

fats and oils across various agricultural and food products, encompassing both animal and vegetable sources (Marcone et al., 2013). In the NMR analysis of fats, each hydrogen proton present in the fat will contribute to NMR signals. The protons situated at various positions within a molecule encounter subtly distinct magnetic fields due to the varying degrees of shielding by the electronic clouds enveloping the nuclei from the external field. Hence, the molecular position of different protons can be identified from the signal intensity of NMR spectra (Fig. 3.5) (Peterson et al., 2021). The observable signals in ^{1}H-NMR that serve for quantification of unsaturated fatty acids include those originating from olefinic protons (5.3–5.4 ppm), protons bonded to bis-allylic ‘C’ (2.7–2.8 ppm), protons bonded to allylic ‘C’ (2.0–2.1 ppm), and terminal CH_3 protons (0.8–0.9 ppm). Additionally, saturated fatty acid quantities can be assessed by referencing the peaks from CH_2 protons at 1.2–1.4 ppm (Knothe & Kenar, 2004) Fig. 3.6, Table 3.1.

^{1}H and ^{13}C NMR spectroscopy is extensively employed for analyzing various oils and fats, encompassing vegetable oils as well as animal and milk fats. Glycerolipids, typically abundant in food, are effectively quantified via ^{13}C NMR analysis, allowing for precise determination of fatty acids in buttermilk triacylglycerols. Remarkable distinctions in fatty acid profiles were observed among milk fat, lard, and margarine through their respective ^{13}C NMR spectra. Particularly, signals for butyric acid, predominantly present in milk fat, were absent in both lard and

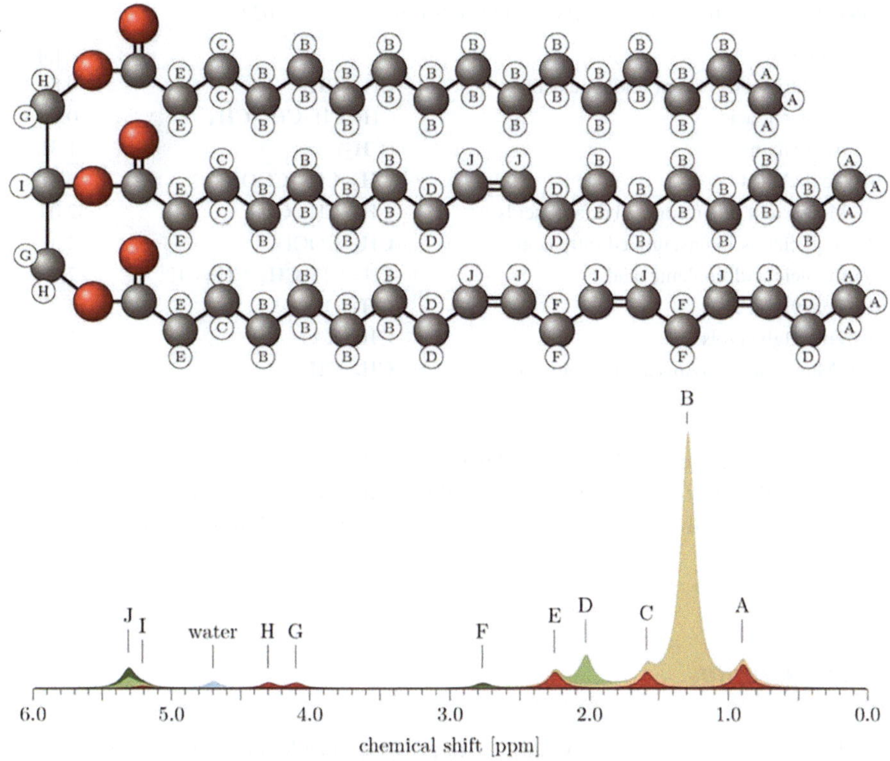

Fig. 3.5 Schematic representation of fat and corresponding magnetic resonance spectra. (Peterson et al., 2021)

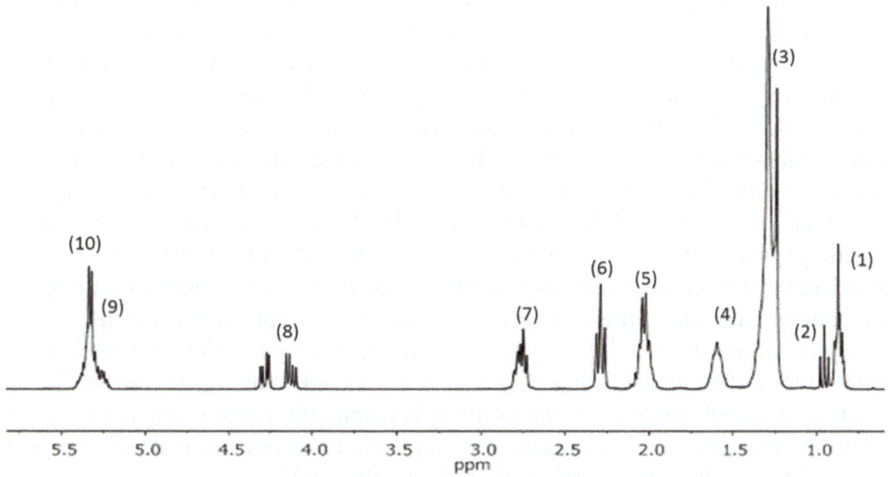

Fig. 3.6 ^{1}H NMR spectrum of vegetable oils. (Siudem et al., 2022)

Table 3.1 ^{1}H NMR chemical shifts of edible oils (Siudem et al., 2022)

Compound		δ [ppm]
(1) All fatty acids except linolenic acid	$–CH_2–\mathbf{CH_3}$	0.89
(2) Linolenic acid	$–CH{=}CH–CH_2–\mathbf{CH_3}$	0.97
(3) Acyl chains	$–(\mathbf{CH_2})_n$-	1.31
(4) Acyl chains	$–\mathbf{CH_2}–CH_2–COOH$	1.61
(5) Mono- and polyunsaturated fatty acids	$–\mathbf{CH_2}–CH{=}CH–$	2.04
(6) Acyl chains in unsaturated fatty acids	$–\mathbf{CH_2}–COOH$	2.31
(7) Linoleic and linolenic acid	$–CH{=}CH–\mathbf{CH_2}–CH{=}CH–$	2.77
(8) Triacylglycerols	$–\mathbf{CH_2}–OCO–$	4.27
(9) Triacylglycerols	$–\mathbf{CH}–OCO–$	5.28
(10) Mono- and polyunsaturated fatty acids	$–\mathbf{CH{=}CH}–$	5.35

margarine (Sacchi et al., 2018). Similarly, the NMR spectral examination of cow and buffalo milk specimens revealed a notable contrast in their fatty acid profiling, encompassing both saturated and unsaturated varieties (Antonietta Brescia et al., 2004)

3.7.4 Glycans

NMR spectroscopy plays a pivotal role in polysaccharide analysis, offering detailed insights into their structural composition and conformational properties. By examining the chemical shifts and coupling patterns of nuclei within polysaccharides, NMR enables the identification of monosaccharide residues, linkages between sugar units, branching patterns, and anomeric configurations. Additionally, NMR can elucidate the dynamic behavior of polysaccharides in solution, providing valuable information about their flexibility, hydration, and interactions with other molecules.

NMR investigations on glycans typically utilize different nuclei, including ^{1}H, ^{13}C, ^{15}N, and ^{31}P, to probe resonances and inter-spin correlations in multidimensional analyses. In ^{1}H NMR spectra, non-exchangeable protons typically manifest within the δH range of approximately 1.0 to 6.0 ppm. However, in aqueous solutions, $–NH_2$ and $–CONH_2$ protons may be discerned around δH ~ 8.0, while hydroxyl protons typically appear within the range of δH ~ 6.0 to 8.0. The relatively restricted spread of proton chemical shifts often results in signal overlap, particularly in complex regions, where strong coupling effects can further complicate the analysis. Alternatively, ^{13}C resonances exhibit a broader chemical shift range (δC ~ 15 to 180 ppm), mitigating issues associated with signal overlap. In the context of carbohydrates, ^{31}P chemical shifts commonly range from approximately 10 to −5 ppm, while ^{15}N resonances are typically observed within the δN range of approximately 30–125 ppm (Fontana & Widmalm, 2023).

In the ^{1}H NMR profiles of carbohydrates, diverse signals are discernible: for instance, anomeric protons of aldoses resonate approximately between δH 4.4 and

6.0. Protons linked to ‘C’ bearing OH or $-CONH_2$ groups typically manifest around δH 3.2 to 4.2, while CH_2 groups tend to appear within the range of δH 1.6 to 2.8. CH_3 protons from N- or O-acetyl typically exhibit a single peak at around δH 2.0 to 2.2, whereas CH_3 protons from 6-deoxy-hexoses often present a double peak at δH 1.2. Protons situated at O-acylated locations experience a downfield shift of approximately 0.5–1.7 ppm in comparison to non-substituted protons lacking ester groups. Consequently, they might overlap with the anomeric proton resonances within the same spectral region (Fontana & Widmalm, 2023).

In ^{13}C NMR spectra of carbohydrates, the anomeric ‘C’ of reducing pyranoses typically manifests at δC around 90 to 100. Pyranoses adopting the β-anomeric configuration tend to exhibit less shielding compared to those with the α-anomeric configuration. Secondary hydroxyl group carbons are discernible within the spectral range of approximately δC 65 to 85. Hydroxymethyl carbons resonate within the vicinity of δC 60 to 65, while nitrogen-bearing carbons tend to appear at δC around 50 to 60. The ‘C=O’ carbons originating from carboxylic acids, esters, or amide groups are identifiable at δC approximately 170 to 180. Methyl carbons stemming from CH_3CO- manifest at δC around 20 to 25, whereas those from 6-deoxy-sugars typically appear in the range of δC 15 to 20 (Fontana & Widmalm, 2023).

3.8 NMR in Food Authentication

Ensuring the authenticity of food products remains a critical issue in maintaining their quality standards. NMR technology holds a pivotal role in this realm, primarily employed for product authentication. Validating the genuineness of food items is paramount for fostering a transparent market landscape, given its substantial economic and health ramifications, thereby engaging both producers and consumers. Given that food composition involves a multitude of variables, authentication predominantly relies on integrating NMR with complementary tools such as MvDA (Hatzakis, 2019). NMR finds extensive utility across a spectrum of food items, including honey, beverages, dairy products, meat, and seafood, as well as a variety of fruits and vegetables (Marcone et al., 2013). NMR analysis is mainly explored to identify the variety, geographical origin, postharvest changes, adulteration, etc., of the products.

The proton NMR is typically used for food analysis. However, there are some exceptions. The utilization of ^{13}C NMR analysis in coffee authentication aims to address issues pertaining to chemical shift alignment caused by interactions between caffeine and chlorogenic acids (Wang et al., 2011). In the domain of olive oil verification, ^{13}C NMR is paired with discriminant analysis for accurate assessment. Similarly, when examining the patterns inherent in fish oil capsules, ^{13}C NMR is complemented by both supervised and unsupervised statistical methodologies. Likewise, ^{31}P NMR spectroscopy finds numerous applications in the realm of food analysis, notably in the analysis of olive oil (Hatzakis, 2019).

3.8.1 Honey

Honey stands out as a highly regarded food item owing to its valuable nutritional profile. Abundant in sugars, proteins, organic acids, flavonoids, vitamins, minerals, and various volatile constituents, it offers a diverse range of health benefits. Various monofloral and polyfloral honey variants exist, each distinguished by unique physical, chemical, and sensory characteristics (Consonni & Cagliani, 2019). In the recent past, limited production led to a hike in the price of honey, and honey adulteration became a common practice. The adulteration of honey could be performed by diluting it with water, incorporating cheaper sugar, or wrong information regarding both botanical or geographical sources (Marcone et al., 2013). Research has been directed towards evaluating the purity and genuineness of honey through metabolomic analysis using NMR techniques. Investigations have revealed that 1D NMR proved to be superior in terms of efficiency, simplicity, and speed as compared to 2D NMR for identifying honey adulteration, achieving a 92.5% accuracy rate surpassing FTIR methods (Bertelli et al., 2010). Numerous research endeavors concentrated on characterizing and distinguishing honey based on its botanical and geographical attributes across different locations.

Natural honey compositions and flavors are intricately pertinent to their maturity, as well as their botanical and geographic source, resulting in great variations in popularity and pricing on the market (Bertelli et al., 2010). NMR analysis can play a prevalent role in giving information assisting the traceability of the botanical and geographical source of honey (Sobolev et al., 2016). The geographical or botanical discrimination mainly depends on the variation in the concentration of compounds within a particular group of constituents like sugar, amino acids, and organic acids (Spiteri et al., 2015). Traditionally, the NMR analysis on honey is performed by dissolving honey in water or DMSO. However, nowadays the organic extract of honey is explored for the NMR analysis. In NMR analysis of the organic extract of honey, differentiation occurs through the identification of particular marker compounds rather than variations in compound concentrations. Consequently, the chloroform extracts from honey yield distinct fingerprints closely linked to their source (Schievano et al., 2019). The proton NMR analysis of citrus honey diluted in water exhibited spectra characterized by particular compound resonance like the aliphatic (δ 0.00–3.00), carbohydrates (δ 3.00–6.00), and aromatic (δ 6.00–10.00) regions. In the aliphatic region, it exhibits distinct NMR signals of valine (*δ* 0.98 and 1.03 ppm), alanine (*δ* 1.47 ppm), proline (*δ* 2.30–2.37 and 1.96–2.08 ppm), lactic acid (*δ* 1.33 ppm), acetic acid (*δ* 2.02 ppm), and ethanol (*δ* 1.16 ppm). In the carbohydrate spectrum, it exhibits distinct NMR signals of sucrose (*δ* 5.40 and 4.20 ppm), fructose (*δ* 4.10 ppm), and glucose (*δ* 5.23, 4.64, and 3.24 ppm). In the aromatic region, characteristic NMR signals of formic acid (*δ* 8.42 ppm), aromatic peaks of phenylalanine (*δ* 7.29–7.42 ppm), and tyrosine (*δ* 7.18, 6.89 ppm) also could be noticed (Fig. 3.7) (Zhang et al., 2020).

In the proton NMR analysis of citrus honey, sucrose content outweighed other components, whereas eucalyptus honey exhibited elevated levels of lactic acid. The

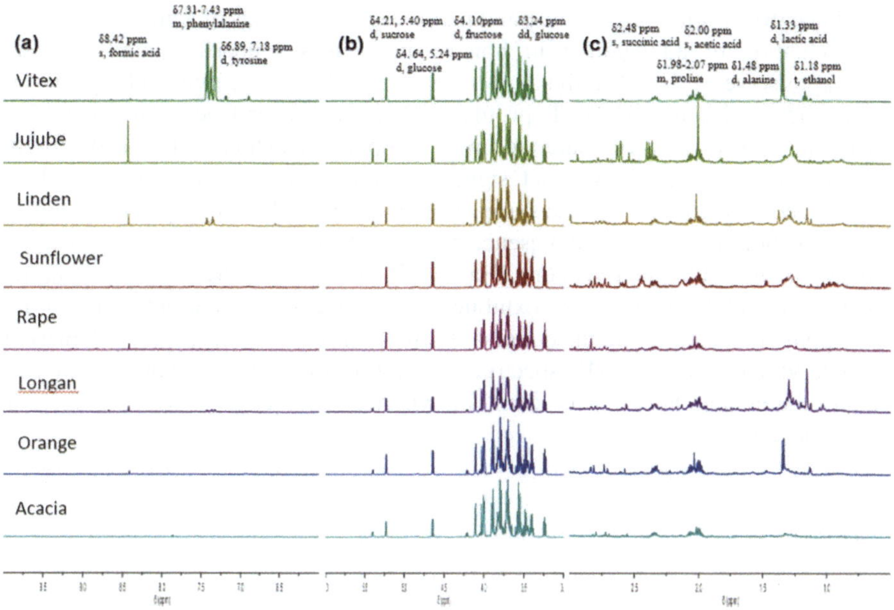

Fig. 3.7 NMR spectra of different mono-floral honey (**a**) Aromatic region, (**b**) carbonhydrate region, (**c**) aliphatic region. (Zhang et al., 2020)

spectra of Assa-peixe honey closely resembled those of eucalyptus and citrus varieties. Sugar cane honey exhibited certain resemblances to both eucalyptus and citrus honey varieties, yet it also displayed distinctive signals for tyrosine and phenylalanine (Boffo et al., 2012). An analysis of 218 monofloral honey samples sourced from China using ^{1}H NMR testing indicated that vitex honey possessed the highest concentration of phenylalanine, succeeded by linden and longan honey. A minimal amount of phenylalanine was detected in rape honey, while within the prescribed detection threshold, phenylalanine was absent in the remaining honey varieties (Zhang et al., 2020).

Chloroform extracts of honey yield detailed NMR fingerprints that provide detailed information regarding its botanical and entomological source. The NMR spectra of chloroform extract of acacia honey exhibited an unidentified signal at 5.25 ppm, which was not observed in any other type of honey. Moreover, Italian and Eastern European acacia honey samples did not exhibit significant changes in five principal marker concentrations (Schievano et al., 2019).

In recent times, customers have shown a growing concern for the quality of honey due to the prevalent issue of adulteration. Adulteration, which alters both the quality and safety of honey, has grown as a notable obstacle in the honey industry. Various substances like lower-grade honey or alternative syrups that include high-fructose corn syrup, maltose syrup, etc., and even adulterated rape honey are commonly used for this purpose. These adulterations in honey could be determined by using NMR analysis. The presence of various sugars in adulterated honey became evident in the carbohydrate spectrum (3–6 ppm) of NMR analysis. Specific peaks within this range exhibited discernible shifts corresponding to heightened levels of adulteration. In the determination of adulteration in honey, the signals primarily found near the 5.40 ppm serve as a marker signal (Biswas et al., 2023). Identifying adulteration in acacia honey through ^{1}H NMR profile analysis is simpler when it involves corn syrup or maltose syrup compared to invert syrup and rape honey. This is due to the presence of at least four distinct signals between pure acacia honey and those adulterated with corn or maltose syrup. Alternatively, in the carbohydrate spectrum, the distinction lies in the mere presence of two to three unique peaks between acacia honey and its adulterated versions (Hao et al., 2022). The NMR spectra of barley malt syrup adulterated honey exhibited an augmented signal intensity in the 5.40 ppm compared to the unadulterated sample (Biswas et al., 2023). Likewise, a notable variance in the NMR spectrum ranging from 5.3 to 5.5 ppm was evident upon adulterating honey with corn, brown rice, and jaggery syrups (Fig. 3.8) (Rachineni et al., 2022).

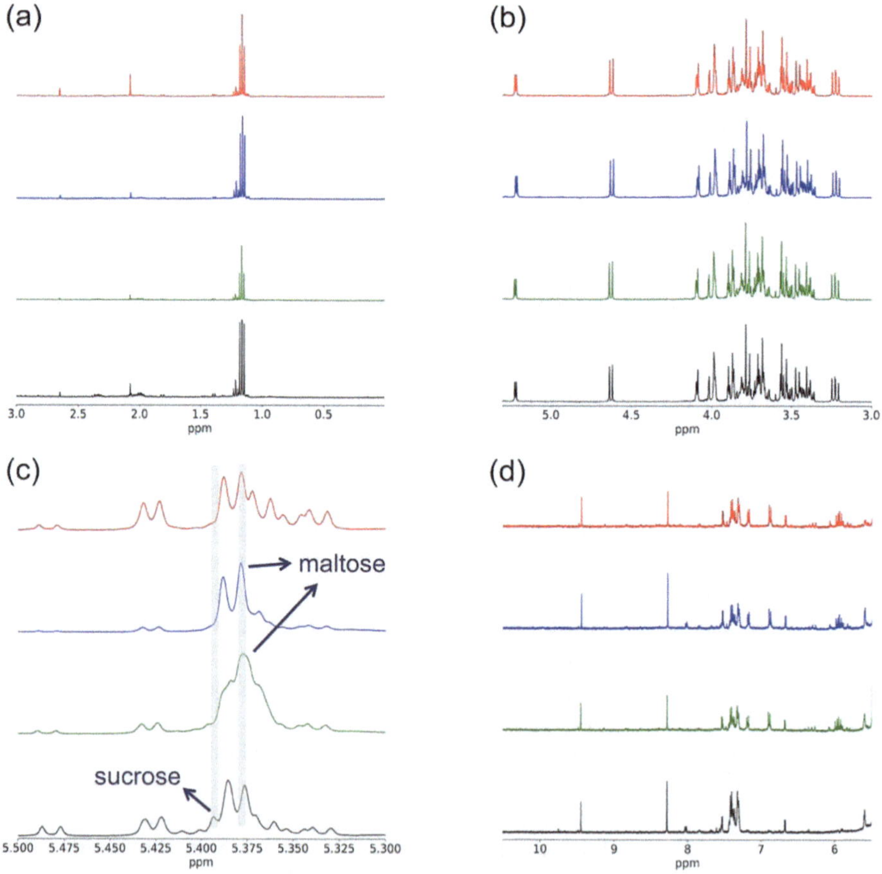

Fig. 3.8 NMR spectra of Indian rape seed honey (red) adulterated with brown rice syrup (blue), corn syrup (green), and gaggery (black) (**a**) 0-3 ppm, (**b**) 3-5.3 ppm, (**c**) 5.3-5.5, (**d**) 5.5-10.5 ppm. (Rachineni et al., 2022)

3.8.2 Meat

NMR spectroscopy finds extensive application across various types of meat. Several studies have explored low and high-frequency NMR to assess meat quality. The ^{1}H NMR spectra of meat display a multitude of signals, presenting challenges in analysis due to extensive overlapping (Brescia et al., 2002). In meat, NMR spectroscopy is widely explored to study muscle metabolism. The feasibility of utilizing ^{1}H NMR spectroscopy to measure lactate content in post-mortem muscles was first demonstrated by Lundberg et al. (1986), who observed the lactate methyl resonance at 1.3 ppm. The consequential rise in post-mortem lactate plays a critical role in determining meat quality, largely due to its effect on pH. ^{1}H NMR spectroscopy can be used to monitor this process dynamically (Bertram et al., 2004). ^{13}C NMR is utilized to study muscle glycogen since it has a potential role in post-mortem anaerobic glycolysis and lactate formation. The changes in the pH of meat can be analyzed by using ^{31}P NMR analysis since the magnetic resonance of different constituents, such as inorganic phosphate and ATP, etc., are influenced by changes in pH (Bertram & Ersen, 2004).

Utilizing a 60 MHz ^{1}H NMR approach, Jakes et al. (2015) identified variance among beef and horse meat by examining the discrepancies in their triglyceride constituents. The analytical procedure involved the use of a basic chloroform-based extract. It was reported that when spectra from the two species are magnified, a systematic difference becomes evident, which is only evident at certain portions of the spectrum. Likewise, in their study, Bischof et al. (2024) utilized NMR analysis for the authentication and classification of aged beef. NMR spectra was used to determine the aging time and cattle aging type.

A research endeavor utilized proton NMR analysis to assess meat quality across distinct pork breeds. Through high-resolution proton NMR analysis of freeze exudate and meat extracts, discernible distinctions emerged in the metabolite composition among the studied breeds. Notably, breed variation appeared to influence post-mortem lactate production, shedding light on observed differences in water-protein interactions and water-holding ability between breeds (Straadt et al., 2011). The utilization of the ^{1}H NMR technique in conjunction with chemometrics proved effective in discerning and quantifying the adulteration of ground beef with pork and duck meat (Leng et al., 2023). Notably, despite the high similarity in the ^{1}H NMR spectra of beef, pork, and duck, there were slight variations in their signal intensities. These distinctions in signal intensities between beef and pork/duck could potentially be attributed to the elevated concentrations of methionine and glutamine found in beef. Akhtar et al. (2021) explored ^{1}H-NMR spectroscopy conjugated with MvDA to dicern meat from chicken, chevon, beef, and donkey. Different metabolites like amino acids, dipeptides, organic acids, vitamins, sugars, and nucleic acids were analyzed. It was noticed that all the samples exhibited a similar NMR spectrum, and chicken meat exhibited high-intensity signals, indicating its higher content of metabolites (Fig. 3.9).

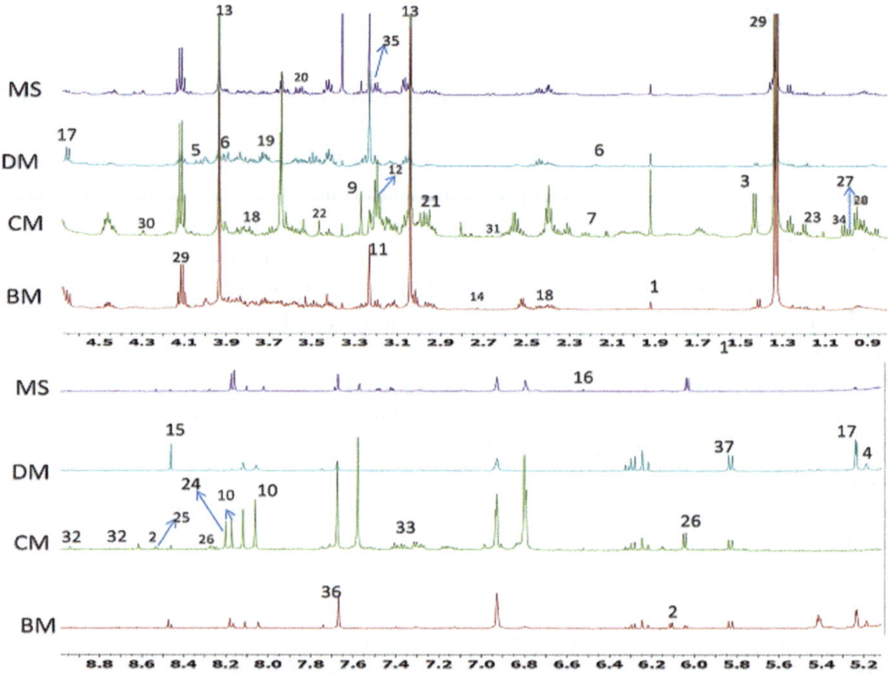

Fig. 3.9 H-NMR spectra of Chevon (MS); Donkey (DM); Beef (BM); Chicken (CM) meat. (Akhtar et al. 2021)

The fatty acid constituents of meat products can be determined through the utilization of high-resolution ^{1}H NMR spectroscopy. In their study, Siciliano et al. (2013) utilized a high-resolution ^{1}H NMR technique to examine the acyl chain constituents of lipid extracts extracted from two traditional Calabrian salamis. Their quantitative NMR analysis unveiled the predominant presence of oleyl chains and noteworthy levels of polyunsaturated fatty acid chains in both salami variants.

3.8.3 Milk and Milk Products

Milk is one of the very complex food materials that contains nutritionally rich components like protein, lipids, etc. The study of milk and milk products using NMR spectroscopy helps with the authentication and determination of adulteration. In dairy analysis, one-dimensional ^{1}H-NMR spectroscopy has a significant place as the favored method for milk analysis. This preference resulted from the high isotopic natural abundance (~99%) and minimal relaxation time of ^{1}H nuclei, resulting in higher sensitivity and quick data extraction. Consequently, it furnishes comprehensive quantitative and intricate structural insights into milk composition. The

^{1}H-NMR spectrum of milk encompasses approximately 296 metabolites and metabolic species. The exploration of ^{1}H-NMR in milk research spans various areas, including analyzing chemical composition, determining nutritional quality, distinguishing between milk from different feeding methods and species, assessing the impact of processing on milk quality, and detecting adulteration. In the examination of milk lipid constituents, ^{13}C-NMR spectroscopy is commonly preferred due to its superior resolution over ^{1}H-NMR. ^{31}P-NMR spectroscopy stands out as a valuable technique in milk analysis, shedding light on the phospholipid composition of milk and finding successful applications in studies aimed at authenticating bovine milk (Eltemur et al., 2023).

Milk is commonly adulterated with whey, urea, H_2O_2, synthetic urine, etc. This adulteration in milk can quickly be determined by using ^{1}H TD-NMR. Examination of ^{1}H TD-NMR relaxation decay patterns indicated that the milk samples were comprised of a singular water component. Furthermore, discernible differences in T2 relaxation times were noted, depending on the extent of adulteration detected (Santos et al., 2016). Milk could also adulterated by adding cheap bovine milk to goat milk and by adding soy milk. The ^{1}H and ^{13}C NMR technique with PCA of these milks shows 11 metabolites, which are different from each other (Li et al., 2017). The NMR chemometric approach combined with multivariate and univariate analysis identified *N*-acetylglucosamine and *N*-acetylgalactosamine as biomarkers for uncovering fraud with caprine milk. It was found that a high positive correlation exists between the *N*-acetyl carbohydrates and the percentage of bovine milk (Rysova et al., 2021).

NMR studies are also performed to understand the physical state of milk and milk product compounds and for the identification of constituents from different dairy origins. NMR elucidates the physical traits of milk fat, particularly through ^{1}H-NMR analysis, which has emerged as an important tool for assessing the distribution of liquid and solid fats. The contrasting mobilities of these two fractions serve as the basis for the identification. Solid fat, with its reduced mobility, exhibits swift relaxation and a broad peak, contrasting sharply with the sharper signal of liquid fat. Mathematical techniques can effectively segregate and quantify both signals. Mass spectrometry coupled with NMR is frequently employed to evaluate the carbohydrate content within milk. Likewise, 2D ^{1}H-NMR spectra could give information regarding the assessment of peptides in whey proteins (Belloque, 1999).

NMR has been extensively documented as a powerful tool for delving into the world of cheese. It focuses on diverse aspects, including pinpointing its geographical roots, analyzing the molecular dynamics of water within, and evaluating its capacity to retain water. NMR analysis was also explored to identify the fatty acid content of cheese. A simple and rapid TD-NMR technology could be performed to quantify the fat and moisture present in cheese (Castell-Palou et al., 2013). The ^{1}H-NMR spectra of hard and white fermented cheese fats revealed 18 different fatty acids and helped the development of a chemometric equation. Consequently, the proportion of long-chain saturated fatty acids to saturated fatty acids can be rapidly determined. This will help to identify the adulteration of cowmilk cheese with vegetable oils (Tociu et al., 2018). The proton NMR is also explored to characterize the lipid constituents and metabolic profiling of Italian fossa (pit) cheese. It was reported that significant changes in the peak intensity of NMR spectra of cheeses at different levels of ripening (Scano et al., 2019).

The regional discrimination of cheeses can be determined by analyzing the difference in the constituents in the sample. The distinction between Emmenthal cheese from distinct regions has been made possible in recent times through the application of MAS-NMR techniques coupled with chemometric methods (Shintu & Caldarelli, 2006). Similarly, Employing a high-resolution NMR technique coupled with multivariate statistical methods, geographical disparities between Parmigiano Reggiano and Eastern European cheese samples were successfully delineated (Consonni & Cagliani, 2008). The ripening and maturation of Cheddar cheese can also be characterized by using NMR techniques. In evaluating the ripening stages of Cheddar cheese, the citrulline-to-arginine ratios within aqueous extracts emerged as paramount metrics. Throughout the maturation process, there was a discernible reduction observed in the levels of these metabolites within the extracts (Chen et al., 2021).

The ^{1}H NMR technique, along with MvDA, is commonly employed to identify the adulteration in butter with lard, margarine, etc. The NMR signals of butter and margarine exhibited significant differences in the range of 2.5 and 1.0 ppm. A notable variance in the spectral readings of butter and margarine becomes evident around 2.17 ppm. This NMR signal arises only in butter and is an indication of a CH_3 hydrogen bonded to a carbonyl group. The detection of numerous peaks centered near 2.0 ppm in margarine suggests the prevalence of extended CH_2 chains. Further, a clear distinction was observed between 1.68 and 1.57 ppm, revealing a higher number of alkane chains in margarine (Obeidat, 2020). In a study on ^{1}H NMR spectra of lard and butter, an obvious difference in spectra was noticed in signal intensities at 2.60–2.84 ppm. These are special peaks found only in lard, with their own unique characteristic, in which only PUFA gives a signal at 2.63, revealing the presence of double allylic methylene protons (Fadzillah et al., 2017).

3.8.4 Beverages

The application of NMR is extended to beverages as well for authentication and discrimination. Its application lies in distinguishing green tea on the basis of its country of origin and quality. In a single green tea extract, ^{1}H NMR successfully identified catechins, amino acids, organic compounds, phenolic compounds, fatty acids, and sugars simultaneously. It is also employed for the detection of catechins, caffeine, 5-galloyl quinic acid, and 2-O- (â-L-arabinopyranosyl)-myo-inositol. Likewise, it also explored the authentication of alcoholic beverages like beer, wine, etc. (Marcone et al., 2013). Quantitative high-resolution NMR technique has proven invaluable in analyzing both natural and synthetic compounds, enabling precise determination of their quantitative characteristics. This technique capitalizes on the quantitative relationship between the ^{1}H-NMR signal area and the number of protons of identical types within a molecule. The quantitative study of sucrose content in beverages becomes feasible through the utilization of diffusion ordered-quantitative ^{1}H-NMR (Cao et al., 2015).

The main components in tea extract such as theanine (δ 1.10, 2.13, 2.37, 3.19, and 7.97 ppm), caffeine (δ 3.22, 3.38, 3.77, and 7.63 ppm), theaflavins (δ 2.74, 2.75, 2.86, 2.87, 4.21, 4.29, 5.41, 6.016.13, 6.86, 7.02, and 7.15 ppm), theaflavin 3,3′-digallate (δ 2.93, 2.94, 3.00, 3.01, 5.03, 5.42, 5.52, 6.046.13, 6.58, 6.61, 6.86, 6.92, and 6.94 ppm), and thearubigin 3,3 -digallate (δ 1.88, 1.98, 2.05, 2.08, 3.54, 3.62, 4.02, 4.14, 6.046.13, 6.85, 7.03, and 7.16 ppm) were determined in the ^{1}H NMR spectra (Ohno et al., 2011). The ^{1}H NMR spectra of green tea extract of different teas growing in three different geographical areas exhibited noticeable discrimination in their amino acid and polyphenolic components constituents (Lee et al., 2010). Likewise, the NMR spectra of tea grown at various altitudes also exhibited significant differences in their peak intensities (Ohno et al., 2011). Further, the NMR analysis is used for the metabolic profiling of tea during the fermentation process (Lee et al., 2011). The NMR technique conjugated with multi-variate analysis is also employed for the quality characterization of the tea (Tarachiwin et al., 2007).

Fruit juice analysis via NMR has emerged as a technique for verifying and authenticating fruit juice quality. In addition to quantifying various compounds, this automated screening method employs statistical models to estimate fruit content or determine the type and source of the juice (Spraul et al., 2009). NMR analysis, when paired with chemometric techniques, is a prevalent method for profiling the metabolites in fruit beverages. Additionally, it serves to ascertain the proportional concentrations of distinct juices within blends (Marchetti et al., 2019).

The ^{1}H NMR spectra of citrus fruit juice exhibited frequencies of the organic acids and free amino acids in the region of 0 to 3, region 3 to 6 composed of different sugars present in the juice, which is a highly complex superimposed region, and the region 6 to 8 exhibits weak peaks, representing the phenolic metabolites. Using 1D ^{1}H and ^{13}C-{^{1}H} NMR techniques, along with 2D homo and heteronuclear NMR analysis, a comprehensive analysis identified 23 low molecular weight compounds within orange juice (Salvino et al., 2021). Likewise, NMR techniques were utilized for the qualitative and quantitative determination of components of mango juice. More than 50 components with varying concentrations were reported. As the ripening process advances, numerous compositional alterations take place. Through signal integration, precise measurements are obtained for the absolute concentrations of organic acids, amino acids, and sugars (Duarte et al., 2005). The quality of grape juice was determined by using NMR techniques during the storage. The presence of ethanol in the grape juice could act as an indicator of the fermentation, and the quantitative measurement of ethanol could be performed by combined application of NMR and the chemometric analysis (da Grandizoli et al., 2014).

NMR analysis combined with PCA is also employed for the qualitative and quantitative analysis of carbonated beverages. The ^{1}H NMR spectra of cola beverages can offer insights into their composition, including brand identity, as well as quantitative data on key ingredients (Maes et al., 2012).

The abundance of proton, ^{1}H NMR spectroscopy in alcoholic beverages serves as a common tool for profiling metabolites like amino acids, sugars, and polyphenols. It also explored to determine the alcoholic content, which helps for their authentication. NMR spectroscopy reveals the alcohol beverage metabolome, highlighting a nuanced chemical composition subject to fluctuations attributed to factors like variety, origin, maturation, fermentation conditions, etc. (Tabago et al., 2021). The principal components contributing to the proton content of wine are water, ethanol, and acetic acid. Therefore, measuring the intensity of their signals in the ^{1}H NMR spectrum should enable the assessment of wine spoilage levels (Marcone et al., 2013).

The NMR spectra of beer dominate the peaks of water (δ 4.86 ppm) and ethanol (δ 1.18 and 3.65 ppm). The non-volatile carbohydrate is exhibited mainly in the region of δ 3.2–4.2 and 4.5–5.5 ppm, and organic acids were observed in the range of δ 1.4–3.2 ppm. The aromatic compounds were characterized by weak peaks in the range of δ 6.5–8.8 ppm. The quantification of components in beer from NMR analysis can be done by the integration of peaks along with the use of references and by the use of partial least-squares regression (Nord et al., 2004).

3.8.5 Edible Oils

The NMR techniques are commonly used for the characterization and adulterant determination of edible oils since the sample can be directly used for the analysis without any pre-treatment, and it has the ability to provide extensive information in a single spectrum. Numerous studies have utilized NMR analysis to explore the composition of edible oils. The ^{1}H-NMR spectra of oils have the advantage of the ability to retrieve significant information about the oil by identifying the minor constituent present in it. In this concept, NMR analysis could provide information about the adulteration, freshness, or oxidation of oils from the spectral peaks corresponding to β-sitosterol, α-linolenic, squalene, etc. (Castejón et al., 2014). ^{1}H-NMR and ^{13}C NMR are also explored for the determination of geographical origin and classification of oils (Fauhl et al., 2000). ^{1}H NMR can also provide information regarding the relative quantities of PUFAs, MUFAs, and SFAs, with distinct proton signatures for PUFAs identifiable at 2.75 ppm. Resonance at 2.00 ppm is indicative of both MUFAs and PUFAs (Hama & Fitzsimmons-Thoss, 2022).

NMR spectroscopy, along with MvDA, could be explored for the metabolomic profiling of oils and their linkage with the major cultivar constituents and the source of oil. An NMR study conducted on the extra virgin olive oils collected over 4 years from different parts of the world exhibited significant differences between Italian and other origins. In addition, utilizing NMR-based metabolic profiling to forecast the geographic origins of olive oils and identify potential correlations with climate changes (Rongai et al., 2017). Similarly, in the aspects of authentication, the CH_3 group of linolenic acyl chains exhibits a tiny variation from the CH_3 groups of the other fatty acid chains. Hence, vegetable oils have less linolenic acyl, which can be easily distinguished from other oils using an NMR spectrum (Almoselhy et al., 2014).

In the ^{1}H-NMR spectra of olive oil, the ^{1}H nuclei bonded to 'C' located at the 1 and 3 sites of the glycerol unit exhibit signals at 4.2 ppm. This is a beneficial group of peaks, as it gives information pertaining to the overall glyceride content in the oil and could be utilized as a reference peak. The ^{1}H nuclei bonded to an olefinic carbon exhibit peaks at 5.2 ppm. These peaks are indicative of the number of unsaturated linkages in a triglyceride, irrespective of their position, like mono or polyunsaturated chains. The quantitative measure of polyunsaturated bonds is given by the minor peak at 2.7 ppm, coming out from the bis-allylic protons from the $–CH_2–$ group present between the unsaturated bonds.

The addition of hazelnut oil to olive oil can be easily decerned by analyzing the spectral region at region 3.6–5.8 ppm, and it represents olefinic and glyceride groups. The ratio of the olefinic-to-glyceride peaks area is utilized for the determination of the hazelnut oil adulteration (Parker et al., 2014). Similarly, the oxidative stability of various oils can be determined by using NMR techniques. The oxidation of oils produces hydroperoxides and aldehydes, which could be detected using NMR analysis. The presence of hydroperoxides peak at the range of 8.09–8.19 ppm and aldehydes at 9.30–9.90 ppm denotes the oxidation of oils (Almoselhy et al., 2014).

The ^{13}C NMR could be used to differentiate between fatty acid isomers like oleic and vaccenic acids. Additionally, ^{13}C NMR spectroscopy enabled the discrimination and characterization of fatty acids with distinct chain lengths but identical unsaturated carbon configurations, e.g. the ^{13}C NMR spectra of oil are classified into four groups: carbonyl carbons (173.3 to 172.8 ppm); unsaturated carbons (132.1 to 126.8 ppm); glycerol carbons (69.1 to 61.6 ppm); and aliphatic carbons (34.5 to 13.9 ppm). In the ^{13}C NMR spectrum, the carbonyl carbons of oil triglycerides give two sets of resonances: a high frequency set at 1(3)-glycerol locations and a low frequency set at 2-glycerol location (Hama & Fitzsimmons-Thoss, 2022). Generally, oils exhibit variations primarily in the amount of chain constituents instead of qualitative acid profiles. As a result, they yield similar spectra in the ^{13}C NMR study, even though with differing intensities, which are unique to each oil. The specific location and intensities of these peaks are attributed to the fatty acid profile of the oils and the arrangement of acyl chains within the triacylglycerols (Hidalgo & Zamora, 2003).

3.8.6 Fruits and Vegetable

The NMR technique is broadly utilized for the characterization and authentication of fruits and vegetables. Methods utilizing magnetic resonance, such as NMR spectroscopy, MRI, and NMR relaxometry, are suitable for analyzing plants. The study of plant tissue microstructure greatly benefits from the utilization of NMR characteristics like spin-spin relaxation time and spin-lattice relaxation. Furthermore, a notable benefit of MRI lies in its capacity for conducting multiple measurements on individual fruits over consecutive time intervals. This capability enables researchers to effectively explore microscopic and macroscopic structural changes (Musse et al., 2009).

In fruits, the NMR study is commonly used to analyze the composition of a variety of fruits and to measure the ripening stages. The ^{1}H NMR spectra exhibit aliphatic amino acids like leucine, valine, etc., in the chemical shift region from 0.50 to 3.00 ppm. The aromatic amino acids like phenylalanine exhibited a chemical shift range from 7.32 to 7.36–7.38 ppm, and the organic acid exhibited in the range of 2.5 to 4.5 ppm (Md Nor et al., 2022). Similarly, the ^{1}H NMR spectra of blueberry were characterized by solid peaks in the region of 3.0–5.5 ppm, denoting the sugars present in it (Capitani et al., 2014).

The ^{1}H NMR was successfully utilized to understand the variation in the composition of melon during ripening and discriminate between different varieties. The NMR spectra of its aqueous extract were characterized by the presence of 42 metabolites. Likewise, the spectra of lipophilic extract indicated the presence of fatty acids, tri-acylglycerols, carotenoids, and sterols (Tristán et al., 2022). Similarly, a high-resolution ^{1} H-NMR study of banana extract was used for the determination of its ripening stage by measuring the difference in the amount of various soluble sugars in the cellular sap at various ripening stages. Additionally, using low-resolution ^{1}H-NMR, water transverse relaxation times, and self-diffusion coefficients have been employed for assessing sub-cellular water distribution and tracking alterations in water dynamics linked to the chemical composition shifts observed during banana ripening (Antonio Raffo et al., 2005). The 1D and 2D ^{1}H-NMR spectroscopic analysis of bananas during postharvest senescence showed a similar NMR profile, whereas the concentration of components changed noticeably (Yuan et al., 2017).

The quantitative MRI imaging study observed a reduction in postharvest changes of tomatoes, with the transverse and longitudinal relaxation times decreasing by approximately 25% and 25–30%, respectively, as ripening progressed. Furthermore, NMR relaxometry is applied to examine the key compartments and the shifting distribution of water within these compartments as ripening progresses (Musse et al., 2009). Similarly, in the NMR analysis of peaches during cold storage, metabolites like alanine, choline, ethanol, fumarate, lactate, and methanol tend to increase gradually but steadily. The cold storage of plums decreased metabolites like asparagine, flavonoids, and polyphenols (Santucci et al., 2015). As mango ripens, its ^{1}H-NMR spectrum reveals a notable surge in sucrose levels compared to fructose and glucose. Simultaneously, there is a decline in citric acid content and a rise in alanine content (Gil et al., 2000).

Like fruits, the NMR techniques were also explored for the compositional analysis and postharvest quality determination of vegetables. For example, the ^{1}H NMR spectroscopy, along with PCA analysis, exhibited significant differences in the metabolic profile of cabbage grown in Korea and China. Acetate, asparagine, etc., are a few primary metabolites responsible for the discrimination in the metabolic profile of various cabbage cultivars (Kim et al., 2013). The postharvest quality analysis of cucumber using ^{1}H NMR spectroscopy noticed that during the storage, the changes in moisture content are mainly observed in the pulp of the cucumber (Feng et al., 2018). The ^{1}H NMR spectra of the lipidic fraction of lettuce leaves exhibited signals of almost all glycerolipids, sterols, and pheophytins in the region of the 3–5 ppm range. In addition, the carotenoids exhibited a range of 6.7–6.0 ppm (Sobolev et al., 2005). Similarly, the ^{1}H—NMR 2D spectra, together with PCA analysis, observed that metabolite-like flavonoids and glucosinolates played a pivotal role in delineating varietal disparities and developmental stages within *Brassica rapa* (Abdel-Farid et al., 2007).

3.9 NMR in Sensory Analysis

The NMR and MRI techniques could be explored for the sensory analysis of food items, which performs as the magnetic tongue. However, for effective sensory analysis using magnetic resonance, a relation should be established between the sensory property of interest and the NMR and MRI observations. The magnetic resonance technique coupled with other techniques, was explored for the examination of food texture. NMR can quantify the water and its mobility in food samples, and the MRI used helps to determine the distribution of water molecules. The salty taste of the food sample can be determined by using NMR. The NMR helps to determine the binding of Na^{+} in both ionic and nonionic solutions. Similarly, in the perception of sweetness, NMR is commonly utilized for the structural determination of sugar and its interaction with water molecules (Lee et al., 2008). Low-resolution NMR spectroscopy was successfully employed for the sensory analysis of soup powder (Cheigh et al., 2011). Similarly, the sensory attributes of roasted coffee beans were successfully established using NMR profiling (Wei et al., 2014). The NMR profile of canned tomatoes can correlate to sensory perceptions like bitterness, sweetness, saltiness, etc.(Malmendal et al., 2011).

3.10 NMR in Nutritional Research

In nutritional metabolomics, the NMR enables the exploration of constituents, physicochemical attributes, and the use of food metrics for individualized dietary diagnosis and prognosis. By integrating NMR with MS and chromatography methodologies, researchers can unearth metabolic insights vital for nutritional studies. Renowned for its reliability and robustness, NMR excels in pinpointing both known and unidentified metabolites, offering precise quantitative outcomes consistently and requiring minimal sample preparation. NMR technology enables the concurrent analysis of numerous components, thereby expanding the scope of dietary interventions and enhancing the quality of nutritional care provided to patients (Chowdhury et al., 2023). The utilization of NMR techniques in nutritional studies is classified into three primary divisions: (1) dietary intervention analysis, (2) dietary biomarker discovery analysis, and (3) diet-related disease analysis.

Utilizing metabolomics in dietary intervention analysis empowers researchers to delve into the mechanistic ramifications of interventions, elucidating the influence of diets or specific food samples on metabolic pathways (Brennan, 2014). The influence of chocolate consumption on metabolism was investigated using ^{1}H-NMR and MS-based metabonomics techniques. The NMR-MS-based analysis of blood plasma exhibited a large number of amino and organic acids, ketone bodies, sugars, osmolytes, and saturated and unsaturated fatty acids. Likewise, the ^{1}H-NMR profile of urine exhibited low molecular weight molecules like ketone bodies, organic acids, amino acids, and aromatic metabolites. Ingesting dark chocolate led to a hindrance in the excretion of stress hormones cortisol and catecholamines in urine, alongside a partial restoration of stress-induced variation in energy metabolism and gut microbial functions (Martin et al., 2009).

Another application of NMR in nutrition research is the identification of biomarkers of dietary intake. There is broad consensus that incorporating dietary biomarkers is crucial for enhancing dietary assessment techniques, with metabolomics emerging as a promising avenue in this regard (Brennan, 2014). The ^{1}H-NMR study of urine from citrus fruit-consuming individuals exhibited an apparent metabolic change in urine profiles between pre-fruit and post-fruit meal intake. Proline betaine was identified as a putative biomarker of citrus fruit intake (Heinzmann et al., 2010).

Utilizing metabolomics in the study of diet-related ailments such as Type 2 diabetes and cardiovascular disease shows potential for unveiling the roots of these disorders and identifying innovative biomarkers (Brennan, 2014). Substantial variances were observed in the ^{1}H-NMR profiles of urine collected from the northern and southern Chinese populations. The urinary metabolites like dimethylglycine, alanine, lactate, etc., were found to be higher for the northern population. Meanwhile, metabolites like gut microbial metabolites such as succinate, creatine, etc., were higher in the southern population. The dissimilarities in metabolic composition noted between northern and southern Chinese populations could potentially account for the differing susceptibility to cardiovascular diseases observed in these regions (Yap et al., 2010).

3.11 Future Trends

The NMR techniques are widely explored for the characterization and quality assessment of different food products. The process of analysis involves several steps, and it does not have a uniform or standardized protocol for NMR analysis. A future aspect of great interest to the NMR community is the unification of the analytical protocols used to investigate foods. Many foods have been analyzed using different experimental methods (extraction procedures, solvents, magnetic fields, etc.). This lack of uniformity often yields results that cannot be compared and, in any case, cannot be proposed as official methods. A great effort expected from the NMR scientific community is, therefore, to standardize specific analytical procedures for each food. Finally, the scientific community is apt to create a food NMR database that is shareable not only among the scientific world but also with companies to provide in-depth knowledge of the products (Sobolev et al., 2022).

Further, in food science, the challenges include the need for expertise, and the safety-related challenges to maintain the magnetic field. Low-field NMR techniques are relatively highly explored for food applications, but their utility is limited. The extension of the application of NMR beyond the research level to industrial application and quality control is still in its infancy and needs to be explored. Thus, they required instruments with economically viable prices (Marcone et al., 2013). The development of a comprehensive database for NMR spectra of foodstuff could be beneficial. The development of Food Screener™ platforms will rely on the setting up of NMR food databases, and widely shared experimental protocols and will benefit from the interface with other analytical platforms (Sobolev et al., 2019).

3.12 Conclusions

The NMR technique has developed as an emerging tool in the realm of food analysis. It performs as a non-destructive, high-resolution technique. It helps with the composition, quality, and safety analysis of the food products. Different NMR techniques, like NMR spectroscopy, MRI, etc., are explored for the analysis of food. The utilization of NMR spectroscopy gives valuable insights into the constituents, structure, and molecular arrangement, and the dynamic attributes of several food components can be elucidated. Through the NMR profiling of food components like protein, oil, carbohydrate, etc., it could be explored for the determination of adulterants and contaminants in food items. It also explores the geographical identification and classification of the products. Further, the advancement of NMR technology like high-resolution NMR, TD-NMR, etc., and the application of NMR techniques coupled with other techniques and MvDA to expand the capabilities and application of NMR in the food sector.

References

Abdel-Farid, I. B., Kim, H. K., Choi, Y. H., & Verpoorte, R. (2007). Metabolic characterization of Brassica rapa leaves by NMR spectroscopy. *Journal of Agricultural and Food Chemistry, 55*(19), 7936–7943. https://doi.org/10.1021/jf071294b

Akhtar, M. T., Samar, M., Shami, A. A., Mumtaz, M. W., Mukhtar, H., Tahir, A., Shahzad-ul-Hussan, S., Chaudhary, S. U., & Kaka, U. (2021). 1H-NMR-Based metabolomics: An integrated approach for the detection of the adulteration in Chicken, Chevon, Beef and Donkey Meat. *Molecules, 26*(15), 4643. https://doi.org/10.3390/molecules26154643

Almoselhy, R. I. M., Allam, M. H., El-Kalyoubi, M. H., & El-Sharkawy, A. A. (2014). 1H NMR spectral analysis as a new aspect to evaluate the stability of some edible oils. *Annals of Agricultural Sciences, 59*(2), 201–206. https://doi.org/10.1016/j.aoas.2014.11.006

Antonietta Brescia, M., Mazzilli, V., Sgaramella, A., Ghelli, S., Paolo Fanizzi, F., & Sacco, A. (2004). 1H NMR characterization of milk lipids: A comparison between cow and buffalo milk. *Journal of the American Oil Chemists' Society, 81*(5), 431–436. https://doi.org/10.1007/s11746-004-0918-3

Apostolos Spyros, P. D. (2013). In P. S. Belton (Ed.), *NMR spectroscopy in food analysis* (Vol. 1, 1st ed.). Royal Society of Chemistry.

Ates, E. G., Domenici, V., Florek-Wojciechowska, M., Gradišek, A., Kruk, D., Maltar-Strmečki, N., Oztop, M., Ozvural, E. B., & Rollet, A.-L. (2021). Field-dependent NMR relaxometry for food science: Applications and perspectives. *Trends in Food Science & Technology, 110*, 513–524. https://doi.org/10.1016/j.tifs.2021.02.026

Belloque, J. (1999). Application of NMR spectroscopy to milk and dairy products. *Trends in Food Science & Technology, 10*(10), 313–320. https://doi.org/10.1016/S0924-2244(00)00012-1

Bertelli, D., Lolli, M., Papotti, G., Bortolotti, L., Serra, G., & Plessi, M. (2010). Detection of honey adulteration by sugar syrups using one-dimensional and two-dimensional high-resolution nuclear magnetic resonance. *Journal of Agricultural and Food Chemistry, 58*(15), 8495–8501. https://doi.org/10.1021/jf101460t

Bertram, H. C., & Ersen, H. (2004). Applications of NMR in meat science. In *Annual reports on NMR spectroscopy* (pp. 157–202). https://doi.org/10.1016/S0066-4103(04)53003-X

Bertram, H. C., Schäfer, A., Rosenvold, K., & Andersen, H. J. (2004). Physical changes of significance for early post mortem water distribution in porcine M. longissimus. *Meat Science, 66*(4), 915–924. https://doi.org/10.1016/S0309-1740(03)00188-8

Bischof, G., Witte, F., Januschewski, E., Schilling, F., Terjung, N., Heinz, V., Juadjur, A., & Gibis, M. (2024). Authentication of aged beef in terms of aging time and aging type by 1H NMR spectroscopy. *Food Chemistry, 435*, 137531. https://doi.org/10.1016/j.foodchem.2023.137531

Biswas, A., Hazra, S. K., & Chaudhari, S. R. (2023). Detection of barley malt syrup as an adulterant in honey by 1H NMR profile. *Food Chemistry, 429*, 136842. https://doi.org/10.1016/j.foodchem.2023.136842

Boffo, E. F., Tavares, L. A., Tobias, A. C. T., Ferreira, M. M. C., & Ferreira, A. G. (2012). Identification of components of Brazilian honey by 1H NMR and classification of its botanical origin by chemometric methods. *LWT, 49*(1), 55–63. https://doi.org/10.1016/j.lwt.2012.04.024

Brennan, L. (2014). NMR-based metabolomics: From sample preparation to applications in nutrition research. *Progress in Nuclear Magnetic Resonance Spectroscopy, 83*, 42–49. https://doi.org/10.1016/j.pnmrs.2014.09.001

Brescia, M. A., Jambrenghi, A. C., di Martino, V., Sacco, D., Giannico, F., Vonghia, G., & Sacco, A. (2002). High resolution nuclear magnetic resonance spectroscopy (NMR) studies on meat components: Potentialities and prospects. *Italian Journal of Animal Science, 1*(2), 151–158. https://doi.org/10.4081/ijas.2002.151

Cao, R., Nonaka, A., Komura, F., & Matsui, T. (2015). Application of diffusion ordered-1H-nuclear magnetic resonance spectroscopy to quantify sucrose in beverages. *Food Chemistry, 171*, 8–12. https://doi.org/10.1016/j.foodchem.2014.08.105

Capitani, D., Sobolev, A. P., Delfini, M., Vista, S., Antiochia, R., Proietti, N., Bubici, S., Ferrante, G., Carradori, S., De Salvador, F. R., & Mannina, L. (2014). NMR methodologies in the analysis of blueberries. *Electrophoresis, 35*(11), 1615–1626. https://doi.org/10.1002/elps.201300629

Castejón, D., Mateos-Aparicio, I., Molero, M. D., Cambero, M. I., & Herrera, A. (2014). Evaluation and optimization of the analysis of fatty acid types in edible oils by 1H-NMR. *Food Analytical Methods, 7*(6), 1285–1297. https://doi.org/10.1007/s12161-013-9747-9

Castell-Palou, A., Rosselló, C., Femenia, A., & Simal, S. (2013). Simultaneous quantification of fat and water content in cheese by TD-NMR. *Food and Bioprocess Technology, 6*(10), 2685–2694. https://doi.org/10.1007/s11947-012-0912-8

Cheigh, C.-I., Wee, H.-W., & Chung, M.-S. (2011). Caking characteristics and sensory attributes of ramen soup powder evaluated using a low-resolution proton NMR technique. *Food Research International, 44*(4), 1102–1107. https://doi.org/10.1016/j.foodres.2011.03.026

Chen, Y., MacNaughtan, W., Jones, P., Yang, Q., Williams, H., & Foster, T. (2021). Selection of potential molecular markers for cheese ripening and quality prediction by NMR spectroscopy. *LWT, 136*, 110306. https://doi.org/10.1016/j.lwt.2020.110306

Chowdhury, C. R., Kavitake, D., Jaiswal, K. K., Jaiswal, K. S., Reddy, G. B., Agarwal, V., & Shetty, P. H. (2023). NMR-based metabolomics as a significant tool for human nutritional research and health applications. *Food Bioscience, 53*, 102538. https://doi.org/10.1016/j.fbio.2023.102538

Consonni, R., & Cagliani, L. R. (2008). Ripening and geographical characterization of Parmigiano Reggiano cheese by 1H NMR spectroscopy. *Talanta, 76*(1), 200–205. https://doi.org/10.1016/j.talanta.2008.02.022

Consonni, R., & Cagliani, L. R. (2019). The potentiality of NMR-based metabolomics in food science and food authentication assessment. *Magnetic Resonance in Chemistry, 57*(9), 558–578. https://doi.org/10.1002/mrc.4807

Consonni, R., Cagliani, L. R., Guantieri, V., & Simonato, B. (2011). Identification of metabolic content of selected Amarone wine. *Food Chemistry, 129*(2), 693–699. https://doi.org/10.1016/j.foodchem.2011.05.008

da Grandizoli, S. C. W. P., Campos, F. R., Simonelli, F., & Barison, A. (2014). Grape juice quality control by means of 1 H NMR spectroscopy and chemometric analyses. *Química Nova, 37*(7). https://doi.org/10.5935/0100-4042.20140208

de Angelis Curtis, S., Curini, R., Delfini, M., Brosio, E., D'Ascenzo, F., & Bocca, B. (2000). Amino acid profile in the ripening of Grana Padano cheese: A NMR study. *Food Chemistry, 71*(4), 495–502. https://doi.org/10.1016/S0308-8146(00)00192-8

Duarte, I. F., Goodfellow, B. J., Gil, A. M., & Delgadillo, I. (2005). Characterization of mango juice by high-resolution NMR, hyphenated NMR, and diffusion-ordered spectroscopy. *Spectroscopy Letters, 38*(3), 319–342. https://doi.org/10.1081/SL-200058713

Eltemur, D., Robatscher, P., Oberhuber, M., Scampicchio, M., & Ceccon, A. (2023). Applications of solution NMR spectroscopy in quality assessment and authentication of bovine milk. *Food, 12*(17), 3240. https://doi.org/10.3390/foods12173240

Fadzillah, N. A., Rohman, A., Salleh, R. A., Amin, I., Shuhaimi, M., Farahwahida, M. Y., Rashidi, O., Aizat, J. M., & Khatib, A. (2017). Authentication of butter from lard adulteration using high-resolution of nuclear magnetic resonance spectroscopy and high-performance liquid chromatography. *International Journal of Food Properties, 20*(9), 2147–2156. https://doi.org/10.1080/10942912.2016.1233428

Fauhl, C., Reniero, F., & Guillou, C. (2000). 1H NMR as a tool for the analysis of mixtures of virgin olive oil with oils of different botanical origin. *Magnetic Resonance in Chemistry, 38*(6), 436–443. https://doi.org/10.1002/1097-458X(200006)38:6<436::AID-MRC672>3.0.CO;2-X

Feng, L., Zhang, M., Bhandari, B., & Guo, Z. (2018). Determination of postharvest quality of cucumbers using nuclear magnetic resonance and electronic nose combined with chemometric methods. *Food and Bioprocess Technology, 11*(12), 2142–2152. https://doi.org/10.1007/s11947-018-2171-9

Fontana, C., & Widmalm, G. (2023). Primary structure of glycans by NMR spectroscopy. *Chemical Reviews, 123*(3), 1040–1102. https://doi.org/10.1021/acs.chemrev.2c00580

Gil, A. M., Duarte, I. F., Delgadillo, I., Colquhoun, I. J., Casuscelli, F., Humpfer, E., & Spraul, M. (2000). Study of the compositional changes of mango during ripening by use of nuclear magnetic resonance spectroscopy. *Journal of Agricultural and Food Chemistry, 48*(5), 1524–1536. https://doi.org/10.1021/jf9911287

Hama, J. R., & Fitzsimmons-Thoss, V. (2022). Determination of unsaturated fatty acids composition in Walnut (Juglans regia L.) oil using NMR spectroscopy. *Food Analytical Methods, 15*(5), 1226–1236. https://doi.org/10.1007/s12161-021-02203-0

Hao, S., Yuan, J., Cui, J., Yuan, W., Zhang, H., & Xuan, H. (2022). The rapid detection of acacia honey adulteration by alternating current impedance spectroscopy combined with 1H NMR profile. *LWT, 161*, 113377. https://doi.org/10.1016/j.lwt.2022.113377

Hatzakis, E. (2019). Nuclear Magnetic Resonance (NMR) spectroscopy in food science: A comprehensive review. *Comprehensive Reviews in Food Science and Food Safety, 18*(1), 189–220. https://doi.org/10.1111/1541-4337.12408

Heinzmann, S. S., Brown, I. J., Chan, Q., Bictash, M., Dumas, M.-E., Kochhar, S., Stamler, J., Holmes, E., Elliott, P., & Nicholson, J. K. (2010). Metabolic profiling strategy for discovery of nutritional biomarkers: Proline betaine as a marker of citrus consumption. *The American Journal of Clinical Nutrition, 92*(2), 436–443. https://doi.org/10.3945/ajcn.2010.29672

Hidalgo, F. J., & Zamora, R. (2003). Edible oil analysis by high-resolution nuclear magnetic resonance spectroscopy: Recent advances and future perspectives. *Trends in Food Science & Technology, 14*(12), 499–506. https://doi.org/10.1016/j.tifs.2003.07.001

Hu, Y., Cheng, K., He, L., Zhang, X., Jiang, B., Jiang, L., Li, C., Wang, G., Yang, Y., & Liu, M. (2021). NMR-Based methods for protein analysis. *Analytical Chemistry, 93*(4), 1866–1879. https://doi.org/10.1021/acs.analchem.0c03830

Jakes, W., Gerdova, A., Defernez, M., Watson, A. D., McCallum, C., Limer, E., Colquhoun, I. J., Williamson, D. C., & Kemsley, E. K. (2015). Authentication of beef versus horse meat using 60MHz 1H NMR spectroscopy. *Food Chemistry, 175*, 1–9. https://doi.org/10.1016/j.foodchem.2014.11.110

Kim, J., Jung, Y., Song, B., Bong, Y.-S., Ryu, D. H., Lee, K.-S., & Hwang, G.-S. (2013). Discrimination of cabbage (Brassica rapa ssp. pekinensis) cultivars grown in different geographical areas using 1H NMR-based metabolomics. *Food Chemistry, 137*(1–4), 68–75. https://doi.org/10.1016/j.foodchem.2012.10.012

Knothe, G., & Kenar, J. A. (2004). Determination of the fatty acid profile by 1 H-NMR spectroscopy. *European Journal of Lipid Science and Technology, 106*(2), 88–96. https://doi.org/10.1002/ejlt.200300880

Kruk, D., & Florek-Wojciechowska, M. (2020). Recent development in 1H NMR relaxometry. In *Annual reports on NMR spectroscopy* (pp. 119–184). https://doi.org/10.1016/bs.arnmr.2019.10.001

Lee, Y., Lee, S.-Y., & Schmidt, S. J. (2008). Probing the sensory properties of food materials with nuclear magnetic resonance spectroscopy and imaging. In *Modern magnetic resonance* (pp. 1889–1894). Springer. https://doi.org/10.1007/1-4020-3910-7_215

Lee, J. E., Lee, B.-J., Chung, J.-O., Hwang, J.-A., Lee, S.-J., Lee, C.-H., & Hong, Y.-S. (2010). Geographical and climatic dependencies of Green Tea (Camellia sinensis) metabolites: A 1H NMR-Based metabolomics study. *Journal of Agricultural and Food Chemistry, 58*(19), 10582–10589. https://doi.org/10.1021/jf102415m

Lee, J. E., Lee, B.-J., Chung, J.-O., Shin, H.-J., Lee, S.-J., Lee, C.-H., & Hong, Y.-S. (2011). 1H NMR-based metabolomic characterization during green tea (Camellia sinensis) fermentation. *Food Research International, 44*(2), 597–604. https://doi.org/10.1016/j.foodres.2010.12.004

Leng, T., Hu, X., Chen, Y., Gan, B., Xie, J., & Yu, Q. (2023). Rapid identification and quantitation of pork and duck meat of binary and ternary adulteration in minced beef by 1H NMR combined with multivariate data fusion. *Food Control, 154*, 110018. https://doi.org/10.1016/j.foodcont.2023.110018

Li, Q., Yu, Z., Zhu, D., Meng, X., Pang, X., Liu, Y., Frew, R., Chen, H., & Chen, G. (2017). The application of NMR-based milk metabolite analysis in milk authenticity identification. *Journal of the Science of Food and Agriculture, 97*(9), 2875–2882. https://doi.org/10.1002/jsfa.8118

Lundberg, P., Vogel, H. J., & Rudérus, H. (1986). Carbon-13 and proton NMR studies of postmortem metabolism in bovine muscles. *Meat Science, 18*(2), 133–160. https://doi.org/10.1016/0309-1740(86)90089-6

Maes, P., Monakhova, Y. B., Kuballa, T., Reusch, H., & Lachenmeier, D. W. (2012). Qualitative and quantitative control of carbonated cola beverages using 1 H NMR spectroscopy. *Journal of Agricultural and Food Chemistry, 60*(11), 2778–2784. https://doi.org/10.1021/jf204777m

Malmendal, A., Amoresano, C., Trotta, R., Lauri, I., De Tito, S., Novellino, E., & Randazzo, A. (2011). NMR spectrometers as "Magnetic Tongues": Prediction of sensory descriptors in canned tomatoes. *Journal of Agricultural and Food Chemistry, 59*(20), 10831–10838. https://doi.org/10.1021/jf203803q

Marchetti, L., Pellati, F., Benvenuti, S., & Bertelli, D. (2019). Use of 1H NMR to detect the percentage of pure fruit juices in blends. *Molecules, 24*(14), 2592. https://doi.org/10.3390/molecules24142592

Marcone, M. F., Wang, S., Albabish, W., Nie, S., Somnarain, D., & Hill, A. (2013). Diverse food-based applications of nuclear magnetic resonance (NMR) technology. *Food Research International, 51*(2), 729–747. https://doi.org/10.1016/j.foodres.2012.12.046

Martin, F.-P. J., Rezzi, S., Peré-Trepat, E., Kamlage, B., Collino, S., Leibold, E., Kastler, J., Rein, D., Fay, L. B., & Kochhar, S. (2009). Metabolic effects of dark chocolate consumption on energy, gut microbiota, and stress-related metabolism in free-living subjects. *Journal of Proteome Research, 8*(12), 5568–5579. https://doi.org/10.1021/pr900607v

Matubayasi, N., Wakai, C., & Nakahara, M. (1997). Structural study of supercritical water. I. Nuclear magnetic resonance spectroscopy. *The Journal of Chemical Physics, 107*(21), 9133–9140. https://doi.org/10.1063/1.475205

Md Nor, S., Ding, P., Abas, F., & Mediani, A. (2022). 1H NMR reveals dynamic changes of primary metabolites in purple passion fruit (Passiflora edulis Sims) juice during maturation and ripening. *Agriculture, 12*(2), 156. https://doi.org/10.3390/agriculture12020156

Musse, M., Quellec, S., Cambert, M., Devaux, M.-F., Lahaye, M., & Mariette, F. (2009). Monitoring the postharvest ripening of tomato fruit using quantitative MRI and NMR relaxometry. *Postharvest Biology and Technology, 53*(1–2), 22–35. https://doi.org/10.1016/j.postharvbio.2009.02.004

Nord, L. I., Vaag, P., & Duus, J. Ø. (2004). Quantification of organic and amino acids in beer by 1H NMR spectroscopy. *Analytical Chemistry, 76*(16), 4790–4798. https://doi.org/10.1021/ac0496852

Obeidat, S. M. (2020). Investigation of butter adulteration with margarine using 1H NMR and PCA. *Applied Magnetic Resonance, 51*(8), 691–701. https://doi.org/10.1007/s00723-020-01211-3

Ohno, A., Oka, K., Sakuma, C., Okuda, H., & Fukuhara, K. (2011). Characterization of tea cultivated at four different altitudes using 1 H NMR analysis coupled with multivariate statistics. *Journal of Agricultural and Food Chemistry, 59*(10), 5181–5187. https://doi.org/10.1021/jf200204y

Parker, T., Limer, E., Watson, A. D., Defernez, M., Williamson, D., & Kemsley, E. K. (2014). 60MHz 1H NMR spectroscopy for the analysis of edible oils. *TrAC Trends in Analytical Chemistry, 57*, 147–158. https://doi.org/10.1016/j.trac.2014.02.006

Pedersen, H. T., Munck, L., & Engelsen, S. B. (2000). Low-field 1H nuclear magnetic resonance and chemometrics combined for simultaneous determination of water, oil, and protein contents in oilseeds. *Journal of the American Oil Chemists' Society, 77*(10), 1069–1077. https://doi.org/10.1007/s11746-000-0168-4

Peterson, P., Trinh, L., & Månsson, S. (2021). Quantitative 1 H MRI and MRS of fatty acid composition. *Magnetic Resonance in Medicine, 85*(1), 49–67. https://doi.org/10.1002/mrm.28471

Purslow, J. A., Khatiwada, B., Bayro, M. J., & Venditti, V. (2020). NMR methods for structural characterization of protein-protein complexes. *Frontiers in Molecular Biosciences, 7*, 9. https://doi.org/10.3389/fmolb.2020.00009

Rachineni, K., Rao Kakita, V. M., Awasthi, N. P., Shirke, V. S., Hosur, R. V., & Chandra Shukla, S. (2022). Identifying type of sugar adulterants in honey: Combined application of NMR spectroscopy and supervised machine learning classification. *Current Research in Food Science, 5*, 272–277. https://doi.org/10.1016/j.crfs.2022.01.008

Raffo, A., Gianferri, R., Barbieri, R., & Brosio, E. (2005). Ripening of banana fruit monitored by water relaxation and diffusion 1H-NMR measurements. *Food Chemistry, 89*(1), 149–158. https://doi.org/10.1016/j.foodchem.2004.02.024

Rongai, D., Sabatini, N., Del Coco, L., Perri, E., Del Re, P., Simone, N., Marchegiani, D., & Fanizzi, F. (2017). 1H NMR and multivariate analysis for geographic characterization of commercial extra virgin olive oil: A possible correlation with climate data. *Food, 6*(11), 96. https://doi.org/10.3390/foods6110096

Rysova, L., Legarova, V., Pacakova, Z., Hanus, O., Nemeckova, I., Klimesova, M., & Havlik, J. (2021). Detection of bovine milk adulteration in caprine milk with N-acetyl carbohydrate biomarkers by using 1H nuclear magnetic resonance spectroscopy. *Journal of Dairy Science, 104*(9), 9583–9595. https://doi.org/10.3168/jds.2020-20077

Sacchi, R., Paduano, A., Caporaso, N., Picariello, G., Romano, R., & Addeo, F. (2018). Assessment of milk fat content in fat blends by 13 C NMR spectroscopy analysis of butyrate. *Food Control, 91*, 231–236. https://doi.org/10.1016/j.foodcont.2018.04.011

Salvino, R. A., Colella, M. F., & De Luca, G. (2021). NMR-based metabolomics analysis of Calabrian citrus fruit juices and its application to industrial process quality control. *Food Control, 121*, 107619. https://doi.org/10.1016/j.foodcont.2020.107619

Santos, P. M., Pereira-Filho, E. R., & Colnago, L. A. (2016). Detection and quantification of milk adulteration using time domain nuclear magnetic resonance (TD-NMR). *Microchemical Journal, 124*, 15–19. https://doi.org/10.1016/j.microc.2015.07.013

Santucci, C., Tenori, L., & Luchinat, C. (2015). NMR fingerprinting as a tool to evaluate post-harvest time-related changes of peaches, tomatoes and plums. *Food Research International, 75*, 106–114. https://doi.org/10.1016/j.foodres.2015.05.046

Scano, P., Cagliani, L. R., & Consonni, R. (2019). 1H NMR characterisation of the lipid fraction and the metabolite profiles of Fossa (pit) cheese. *International Dairy Journal, 90*, 39–44. https://doi.org/10.1016/j.idairyj.2018.10.007

Schievano, E., Stocchero, M., Zuccato, V., Conti, I., & Piana, L. (2019). NMR assessment of European acacia honey origin and composition of EU-blend based on geographical floral markers. *Food Chemistry, 288*, 96–101. https://doi.org/10.1016/j.foodchem.2019.02.062

Schmidt, S. J. (1991). Determination of moisture content by pulsed nuclear magnetic resonance spectroscopy. *Advances in Experimental Medicine and Biology, 302*, 599–613. https://doi.org/10.1007/978-1-4899-0664-9_32

Schmidt, S. J., & Lai, H.-M. (1991). Use of NMR and MRI to study water relations in foods. *Advances in Experimental Medicine and Biology, 302*, 405–452. https://doi.org/10.1007/978-1-4899-0664-9_23

Shintu, L., & Caldarelli, S. (2006). Toward the determination of the geographical origin of Emmental(er) cheese via high resolution MAS NMR: A preliminary investigation. *Journal of Agricultural and Food Chemistry, 54*(12), 4148–4154. https://doi.org/10.1021/jf060532k

Siciliano, C., Belsito, E., De Marco, R., Di Gioia, M. L., Leggio, A., & Liguori, A. (2013). Quantitative determination of fatty acid chain composition in pork meat products by high resolution 1H NMR spectroscopy. *Food Chemistry, 136*(2), 546–554. https://doi.org/10.1016/j.foodchem.2012.08.058

Siudem, P., Zielińska, A., & Paradowska, K. (2022). Application of 1H NMR in the study of fatty acids composition of vegetable oils. *Journal of Pharmaceutical and Biomedical Analysis, 212*, 114658. https://doi.org/10.1016/j.jpba.2022.114658

Sobolev, A. P., Brosio, E., Gianferri, R., & Segre, A. L. (2005). Metabolic profile of lettuce leaves by high-field NMR spectra. *Magnetic Resonance in Chemistry, 43*(8), 625–638. https://doi.org/10.1002/mrc.1618

Sobolev, A. P., Circi, S., & Mannina, L. (2016). Advances in nuclear magnetic resonance spectroscopy for food authenticity testing. In *Advances in food authenticity testing* (pp. 147–170). Elsevier. https://doi.org/10.1016/B978-0-08-100220-9.00006-0

Sobolev, A. P., Thomas, F., Donarski, J., Ingallina, C., Circi, S., Cesare Marincola, F., Capitani, D., & Mannina, L. (2019). Use of NMR applications to tackle future food fraud issues. *Trends in Food Science & Technology, 91*, 347–353. https://doi.org/10.1016/j.tifs.2019.07.035

Sobolev, A. P., Ingallina, C., Spano, M., Di Matteo, G., & Mannina, L. (2022). NMR-based approaches in the study of foods. *Molecules, 27*(22), 7906. https://doi.org/10.3390/molecules27227906

Spiteri, M., Jamin, E., Thomas, F., Rebours, A., Lees, M., Rogers, K. M., & Rutledge, D. N. (2015). Fast and global authenticity screening of honey using 1H-NMR profiling. *Food Chemistry, 189*, 60–66. https://doi.org/10.1016/j.foodchem.2014.11.099

Spraul, M., Schütz, B., Humpfer, E., Mörtter, M., Schäfer, H., Koswig, S., & Rinke, P. (2009). Mixture analysis by NMR as applied to fruit juice quality control. *Magnetic Resonance in Chemistry, 47*(S1). https://doi.org/10.1002/mrc.2528

Straadt, I. K., Aaslyng, M. D., & Bertram, H. C. (2011). Assessment of meat quality by NMR—an investigation of pork products originating from different breeds. *Magnetic Resonance in Chemistry, 49*(S1). https://doi.org/10.1002/mrc.2805

Tabago, M. K. A. G., Calingacion, M. N., & Garcia, J. (2021). Recent advances in NMR-based metabolomics of alcoholic beverages. *Food Chemistry: Molecular Sciences, 2*, 100009. https://doi.org/10.1016/j.fochms.2020.100009

Tarachiwin, L., Ute, K., Kobayashi, A., & Fukusaki, E. (2007). ^{1}H NMR based metabolic profiling in the evaluation of Japanese green tea quality. *Journal of Agricultural and Food Chemistry, 55*(23), 9330–9336. https://doi.org/10.1021/jf071956x

Tociu, M., Todasca, M.-C., Bratu, A., Mihalache, M., & Manolache, F. (2018). Fast approach for fatty acid profiling of dairy products fats using 1 H-NMR spectroscopy. *International Dairy Journal, 83*, 52–57. https://doi.org/10.1016/j.idairyj.2018.03.005

Tristán, A. I., Abreu, A. C., Aguilera-Sáez, L. M., Peña, A., Conesa-Bueno, A., & Fernández, I. (2022). Evaluation of ORAC, IR and NMR metabolomics for predicting ripening stage and variety in melon (Cucumis melo L.). *Food Chemistry, 372*, 131263. https://doi.org/10.1016/j.foodchem.2021.131263

Van, Q. N., Issaq, H. J., Jiang, Q., Li, Q., Muschik, G. M., Waybright, T. J., Lou, H., Dean, M., Uitto, J., & Veenstra, T. D. (2008). Comparison of 1D and 2D NMR spectroscopy for metabolic profiling. *Journal of Proteome Research, 7*(2), 630–639. https://doi.org/10.1021/pr700594s

Wang, Y., Zhang, T., Xu, J., & Du, W. (2011). Comparison of the binding affinity of chlorogenic acid with two serum albumins. *International Journal of Biological Macromolecules, 48*(1), 81–86. https://doi.org/10.1016/j.ijbiomac.2010.09.015

Wei, F., Furihata, K., Miyakawa, T., & Tanokura, M. (2014). A pilot study of NMR-based sensory prediction of roasted coffee bean extracts. *Food Chemistry, 152*, 363–369. https://doi.org/10.1016/j.foodchem.2013.11.161

Yap, I. K. S., Brown, I. J., Chan, Q., Wijeyesekera, A., Garcia-Perez, I., Bictash, M., Loo, R. L., Chadeau-Hyam, M., Ebbels, T., De Iorio, M., Maibaum, E., Zhao, L., Kesteloot, H., Daviglus, M. L., Stamler, J., Nicholson, J. K., Elliott, P., & Holmes, E. (2010). Metabolome-Wide Association study identifies multiple biomarkers that discriminate North and South Chinese populations at differing risks of cardiovascular disease: INTERMAP study. *Journal of Proteome Research, 9*(12), 6647–6654. https://doi.org/10.1021/pr100798r

Yuan, Y., Zhao, Y., Yang, J., Jiang, Y., Lu, F., Jia, Y., & Yang, B. (2017). Metabolomic analyses of banana during postharvest senescence by 1H-high resolution-NMR. *Food Chemistry, 218*, 406–412. https://doi.org/10.1016/j.foodchem.2016.09.080

Zhang, J., Chen, H., Fan, C., Gao, S., Zhang, Z., & Bo, L. (2020). Classification of the botanical and geographical origins of Chinese honey based on 1H NMR profile with chemometrics. *Food Research International, 137*, 109714. https://doi.org/10.1016/j.foodres.2020.109714

Chapter 4
Computer Vision Systems

Shubham Chandrakar, M. Suji, and R. Mahendran

4.1 Introduction

Food is a crucial source that must be harvested, processed, packed, and transported worldwide with enhanced shelf life without compromising its qualities. Due to the fast-moving pace of the world, the increase in population, and the demand for new food innovations, the requirement of a high level of processing rate of about 70% (FAO) will result, which in turn increases the wastage and creates harmful environmental impacts simultaneously (Redlingshöfer & Soyeux, 2012). According to FAO, food waste is reported to be 1.3 billion tons globally, whereas fruits and vegetables contribute about 0.5 billion tons (FAO, 2019). Given reducing wastage, food quality plays a major role by validating the specific parameters in foods from time to time. The other elements, such as supply chain traceability, consumer preferences, pre & postharvest handling, variety, food packaging, processing using non-thermal/thermal technologies, coatings/encapsulation, and storage conditions according to the nature of the product, as well as indicate the provision of quality (Aung & Chang, 2014; Feldmann & Hamm, 2015; Kyriacou & Rouphael, 2018; Silvestre et al., 2011).

Food quality involves the assessment of the internal and external quality of fruits and vegetables, such as size, shape, colour, texture, insect damage, decaying, and damages or punctures on tissues, skin, peels, etc., due to the impact after a fall, abrasion, compression, surface mechanical damage, cracks, bruises, crushing, mechanical vibration, and so on (Soltani Firouz & Sardari, 2022; Tscheuschner, 1987). In the bakery, milk, meat, fried food, aquatic, and spice processing sectors, the evaluation of the topping rate on pizza, characteristics of crumbs, melting

S. Chandrakar · M. Suji · R. Mahendran (✉)
National Institute of Food Technology, Entrepreneurship and Management, Thanjavur (NIFTEM-T), Thanjavur, Tamil Nadu, India
e-mail: mahendran@niftem-t.ac.in

C. K. Sunil et al. (eds.), *Emerging Non-destructive Technologies for Food Quality Analysis*, https://doi.org/10.1007/978-3-031-95048-3_4

properties of cheese, the colour of meat, marbling effect, torn skin, defects on chips, and drying rate, etc. to be included (Narendra & Hareesh, 2010; Singh, 2013). These qualities, in turn, are to be evaluated without affecting the actual parameters or conditions of produce or food products. The destructive technologies include sample preparation, solvent extraction, high expense, and slow and complex (Millicent, 2019).

To simplify the purpose and non-destructive detection (NDD) inspection of food quality attributes, ultrasound, optical, x-ray, acoustic, electromagnetic, chemical technologies, visible/near-infrared spectroscopy, multispectral imaging, sensing techniques, and computer vision are used. It is classified into transportable, experimental, and product categories. Usually, the specific wavelength of light gets absorbed into the food materials and results in excitation, rotation, and vibration of its associated molecules, which can emit signals to identify the quality or defects present in the food products. In 1930, the application of NDD started by categorising soybeans and peanuts in a mixture and its evolution in food analysis happened in the twentieth century (Abasi et al., 2018; Chen et al., 2021; Mohd Ali & Hashim, 2022).

The study of computer vision creates the theoretical and technical framework for automatically extracting and analysing meaningful data about a food item from a picture and image sequencing (Gunasekaran, 1996). It includes capturing, processing, and analysis of objects. The fundamentals in identifying foods' required categorisation and quantification mainly depend on algorithms and procedures (Narendra & Hareesh, 2010). The terminologies involved are digital images, a 2-dimensional image, indicated by I (x,y) where I represent amplitude and (x,y) are the spatial coordinates that could be documented after clicking, the pixel is the single point in an image arranged closely in rows and columns. The grey shades of colour exhibit the number of bits involved. As an example, the values of pixels can be stored in an 8-bit integer with a grey level of 0 to 255 (Gunasekaran, 1996). The parts of a computer vision system include hardware and software systems, including illumination devices, cameras, A/D, computers, colour detectors, computer languages, image acquisition, image processing, and assessment of required quality parameters (Mahendran, 2012). Nowadays, advancements in computer systems allow the development of cost-effective technology and the investment in new research ideas (Brosnan & Sun, 2004). The technique has a research background of over 40 years. It finds applications in robotics, food processing, medicine, etc (Mahendran, 2012). This chapter focuses on image acquisition, image processing, equipment for classification, application, and future scope. The following is a brief explanation of the components of CVS:

4.2 Image Acquisition

The initial step is to generate images to collect the details of the contents after selecting an appropriate image acquisition technique. The step involves the conversion of electronic to numerical form under the sensor. Sensors were developed with various configurations upon crucial research progress. To inspect the quality, characterisation, and classification of foods, charge-coupled devices (CCD), and for low energy & cost, fast clicking, complementary metal oxide silicon (CMOS) is the frequently used sensor that contains photoelectric effects. It is mainly useful in detecting photons and, upon merging/combining, provides an electrical signal to photons that drop on the pixel surface. Electrical signals are produced by the intensity of light from the surface. Other sources like computerised X-ray microtomography, magnetic resonance imaging, and scanning electron microscopy are used for inspecting the crumb's porous nature, the structural arrangement of grains, and high magnification, yet it holds drawbacks as it is cost insufficient with additional care during handling. The A/D converter transforms these signals to 8- or 16-bit data. Thus, a digital image is created and can be stored. The colour digital camera can be used by following the standards of colour and kept in a dark box view to protect against light and reflection. The maximum resolution of the image is in the pixel ranges that could be fixed and attachable to the device's USB portal. On the other hand, the factors responsible for the precision are lightning, electromagnetic energy, camera quality, etc. As an unavoidable point, the essential attributes for selecting a proper machine vision system involve opting for preferential dimensional parts, the accuracy of results, and the rapidity of processing area to avoid improper handling of materials and choices of light. The visible spectrum light works under image acquisition after striking a partially reflecting and partially absorptive surface, creating pictures. The dispersed photons are then collected in the camera lens and transformed into electrical impulses using a tube or CCD. CCD is an array where the result corresponds to less light-sensitive components or line-scan types in which high image resolution could be created quickly. The single device (camera and image capturing system) can connect wirelessly or with wires directly to the personal computer. On light detection, the sensor on the camera can be specifically designed to handle three basic colours: red, green, and blue (RGB). The camera settings, the time of photograph capturing, the source of light, and the image analysis to obtain necessary elements for decision-making are all controlled by the software (Abdullah et al., 2016; Amani et al., 2023; Judal et al., 2015; Mahajan et al., 2015; Mahendran, 2012; Patel et al., 2015; Pourdarbani et al., 2012; Swami et al., 2010). Bhargava and Bansal (2021), Mahendran (2012) explains the grading of oranges, mangoes, etc., using a 550D Canon camera under the resolution of 0.03 mm/pixel on 100 images. Upon 640 × 480 size, resolution at 0.28 mm/pixel, the oranges were examined for defects using CCD. A thermal camera was also used to click 250 images with the size 240 × 320 for maturity and size-based categorisation of mangoes. The length-width ratio of potatoes is identified using a Canon power shot camera upon creating 57 images. Software such as Canon remote capture might be useful in transferring the clicked image to the computer without any change in size.

4.2.1 Illumination

Chen (2018), Mahajan (2015), Lukinac (2019), Pourdarbani (2012), Saldaña (2013), Wu et al. (2013), portray that the type, execution, and grade of illumination with adjustment qualities like distinct or clear light is inevitable to view the object in different directions according to the requirement. Light positioning could be either of two types: front lighting for surface view and backlighting for viewing the proportions on borders or part of surface characterisation. A structured illumination device would be better, leading to the obtainment of correct results, less time for processing, reduced difficulties involved in succeeding steps, favourable outcomes, and cost-effective analysis. The geometry of the sample is important. The irregular lightning effect could be neglected with the assistance of computation and calibration work. For example, an angle of 45 degrees vertically might be suitable for an object to restrict reflections. The unnecessary glow and mirroring effects can be avoided by creating the polarisation techniques and their associated filters. The light sources are fluorescent, incandescent bulbs, x-ray tubes, light-emitting diodes (LED), and light-emitting capacitors (LEC). The wider application involves fluorescence to produce constant light dissipation at a specific wavelength with high or profound intensity. Incandescent bulbs can be used for darker environments as they provide bright light at night. LEDs are frequently used nowadays as they are economically convenient, release less heat, and are strong. Depending on the angle of incidence, lightning methods are classified into direct lightning, where the light source is directly passed over the object—vertical, ring & angular illumination; diffusive lightning, in which the object is round—flat, coaxial, and dome-shaped; lateral light; shallowly; and collimated lighting, a laser light. The modelling scenario for computation is approachable, resulting in a good precision value. Using computer-aided design (CAD), one study has been developed to place the light source and camera properly in place. The mathematical programs and simulation works have also been included under the provision of illumination. The lightning on apple slices and potato chips resulted in identifying the colour effects due to drying and frying techniques. Furthermore, UV-IF (ultraviolet-infrared) sources were used to find defects and contaminants in apples. The internal defects and colour of strawberries, peaches, and cucumbers, upon the spectral ranges of 400 to 1000 nm, gave precise results of 86%.

4.2.2 Imaging System

This comprises a camera, sensors, filters, data analysis, and other electrical components that process the signals. An imaging system must carefully select The procedure and inputs (Loce, 2017). The different imaging systems involved are listed below:

4.2.2.1 Monochrome Imaging

A monochrome camera is used for high resolution, speed, and spatial arrangements, which is beneficial over the colour camera. Though the colour camera is used for appropriate visualisation, algorithms could be given only to monochrome images (MCI) as they are well receivable. It could also distinct light features (Patel et al., 2021). It needs a single fragment chip. The resolution could be set in 480 or 1024 lines or above (Mahendran, 2012). The grey level is the intensity of MCI. The grey level interval is known as greyscale values numerically from 0—minimum level (black) to L—maximum level (white), and the in-between values represent grey (Gunasekaran, 1996). Using the monochrome cameras, the sorting of mangoes has been achieved with an accuracy of 98% and 89%. The clear vision of muscle texture was notable using MCI. The external defects of mangoes were identified in a short period using a high accuracy range of 98% (Du & Sun, 2004; Patel et al., 2021; Xiao et al., 2022).

4.2.2.2 Colour Imaging

Daley et al. (1993) explain that colour imaging (CLI) includes the entire 256 grey scales in a band with its associated parameters such as saturation, hue, intensity, etc., which can be identifiable using a single chip of CCD. The CMYK (cyan, magenta, yellow and black) model and the L*a*b* model were also used along with the RGB model. The colour changes on apples at various locations were identifiable, and the indication of healthy parts and bruises were easily achievable with the indication of different colours with near-infrared, green, hue and red. The visualisation of quality checking on poultry products was also developed (Throop et al., 1993; Vyawahare et al., 2013).

4.2.2.3 Hyperspectral Imaging

A hyperspectral image is made out of a collection of obtained monochromatic images. Usually, hyperspectral imaging (HSI) consists of a group of monochromatic images over a considerable range that corresponds to continuous wavelengths that might show a high degree of correlation or typically include duplicate information. The main parts needed for developing an HSI system are light sources of different wavelengths, technology of interest, and detectors. The repeatedly used devices for wavelength selection are imaging spectrographs, liquid crystal tunable filters (LCTF), and acousto-optic tunable filters (AOTF). An imaging spectrograph could define the spectra detail of a particular part in an image of an object. The AOTF is a gadget that filters the illumination according to sound waves propagating through a crystal and selects one particular colour from tones of colours. Throughout the tellurium dioxide crystal, light and sound waves pass through it. With selective filtering, one specific colour at a time could be allowed. LCTF is used to block the other colours and choose a particular colour. It assists in sorting fruits or identifying their defects (Lorente et al., 2012). Multispectral imaging (MSI) is an uncomplicated and quick execution of the HSI system with fewer bands in numbers expressible (Fracarolli et al., 2020). The defects in apples are easily indefinable using the MSI technique, categorising from highest to lowest (Kleynen et al., 2005). For classifying citrus fruits, lemons and mandarins with external surface damage, 93% and 94% accuracy have been obtained with MSI (Aleixos et al., 2002). MSI aims to obtain a specific spot identification on images of the object and spectral detailing, which can be used for instantaneous results in the food processing sectors. Line-scan and area-scan methods could collect images of less wavelength (Qin et al., 2013).

4.2.2.4 Thermal Sensing

Thermal sensing progresses through infrared (IR) radiation. The three wave regions are known as mid, long and short, where imaging under low-temperature sensitivity is captured under the extended region. In contrast, detection under high temperature is expressed under the mid-level region and is often suitable in food applications for output/result; complexity upon coinciding with reflections will be on the short region. The wavelength of 3–15 μm is the (mid to longest IR) most commonly used range. It is useful in identifying the seed quality, such as detecting the growth of fungi, insects, foreign materials, etc., without direct contact with the object; changes in potato upon freezing with defining about ice nucleation; yield of citrus fruits, detection of bruises on fruits, maturity of fruits, and so on (ElMasry et al., 2020; Vadivambal & Jayas, 2011).

4.2.3 Frame Grabbers

The other name of the frame grabber is the vision processor board. A frame grabber is used to capture the image from the camera. It has the functions of image acquisition, processing, and camera control. The capture of images could be done using analog or digital signals. The image will be divided into 2-dimensional grids. The camera control can produce precise signals and capture the effects of images. Frame grabbers manage the image's data, lighting, and variations. At present, the transferring of data to the computer's memory is at a faster rate with the help of software drivers. It consists of parts like a digitiser, converters, array of data collection, block of memory, A/D converters, look-up table—mathematical function, image buffer for modifying the image, image buffer control for creating a fixed pixel, and data transferring system; Peripheral Component Interconnect (PCI) bus. The system has been utilised to add expansion cards to interface the network and USB ports, with a capacity of 32 bits and bandwidth of 133 MB/s. It can communicate with the central processing unit (CPU). The faster action will be valuable in developing the real-time detection parameters on food applications (Abdullah, 2007; Harris & Harris, 2016; Judal et al., 2015; Ma et al., 2016).

4.3 Image Processing

Image processing and analysis are the main considerations for computer vision. Image processing operates a set of images with high-quality images to sort out defects such as alterations in shape, improper camera focus and directions, noises, and uneven lighting. Image analysis is used to differentiate the objects from the background and create content particulars. The above-said parameters consist of three levels, e.g. low, intermediate, and high-level processing. It includes image acquisition and pre-processing. Image pre-processing indicates processing at the initial stages without image distortions, neglecting noises with filters, thereby increasing the quality attributes of the image. Image processing mainly provides the appropriate image details for analysis, storage, transmittance, its associated attributes and existing formations (Brosnan & Sun, 2004). The digitalised form of value must be converted into a quantifiable range related to the object using a low-level image processing technique and calibration for precise values. Mechanical fruits' into real surface mechanical damage under hyperspectral imaging (to determine the internal surface of the fruits Xiao et al., 2022).

Walther et al. (2023) state that the mid-level processing of visual perception is used to create portrayals of dimensions of objects (Kaganami & Beiji, 2009; Karthick, 2014; Lukinac et al., 2018; Majumder, 1986; Swami et al., 2010) defines that mid-level processing uses the results of low-level processing for creating images with edge linking for yielding the image's pixel at its edges, segmentation for creating accurate data and can be classified into thresholding, edge-based segmentation, and region-based segmentation. Threshold is the process of drawing a line on the brightness scale. Pixels brighter than the line are a part, and darker pixels relative to the line are further other parts. The non-success rate of achieving the algorithm, such as size issues, disconnected edges, and incomplete separation. Edge-based segmentation detects the pixels on the borderline. With the surrounding pixel, edges are discovered. It can be developed as a binary categorisation at the pixel region by recognising each pixel as an on/off edge. Region-based segmentation (RBS) is a group of pixels with markings with the principle of Gray value differences. Examples of RBS are joining pixels, joining the pictures for a specific objective, separating peaks, detecting colourful areas, and segregating the pixels of similarities. The other segmentation method includes a gradient-based technique resembling edge detection methods grounded in an image's gradient. The process of computing the image gradient is preferred due to its ability to display the boundaries of local contrast in gradient images, making it easier to detect object edges.

In the food sector, classification-based segmentation is the second most often utilised technique. It is pixel-oriented, and its variables are often derived from the characteristics of the picture (colour, shape, and texture). Next, as the input for classification, a matrix containing each pixel as an observer is generated. Each observer is then classified using a learning model based on its variables. After identifying and separating objects from the background through image segmentation, the next step involves measuring the characteristics of these objects, known as object measurements, under high-level processing. These measurements are crucial in computer vision systems as they provide valuable information for understanding, interpreting images, and classifying objects. The other important parameter is artificial neural network (ANN). An ANN is a computer programme designed to mimic the functioning of the human brain. It comprises linked processing units that speak to one another via weighted connections. ANN is a potent food science technique that analyses and simulates food safety and quality. In order to categorise several types of pistachio nuts, examine the visual traits and categorise vegetables, cereal grains, chicken carcasses, etc. ANNs have been used. Fuzzy logic is a method of describing information (context-dependent) rather than exact or black-and-white. It is applied in computers to manage the properties of control systems and other systems. The fuzzy logic implementation process consists of three basic steps: fuzzification, fuzzy inference, and defuzzification. Fuzzification is a process that converts membership functions or clear, exact data into fuzzy data. Depending on the application's requirements, these functions might be bell-shaped, trapezoidal, or triangular. Fuzzy Inference is where the fuzzy outputs are obtained by combining membership functions with control rules.

Defuzzification is the fuzzy output that needs to be converted back to precise data for control purposes. Defuzzification achieves this using methods such as Mean of Maximum, Center of Gravity, or Height. The above steps help make fuzzy logic applicable in real-world scenarios. As fuzzy technologies evolve, new fuzzy control strategies are being developed to improve performance and efficiency. A study was conducted to develop an intelligent automated control system (ACS) that would govern the performance of rice whitening machines by utilising machine vision and fuzzy logic approaches. Important quality indicators, such as the percentage of broken kernels and the degree of milling, were regarded as input variables, and the discharge pressure of the whitening machine was the fuzzy inference unit's output variable. The evaluation's findings showed that the ACS was 89.2% accurate in identifying the machine's ideal operating parameters. In another research work, a fruit sorting and grading system was developed. At first, the pictures of the fruits were taken with a digital camera. Following image processing, these photos are sorted and rated using fuzzy logic approaches. Fruits are categorised according to size and form, supporting further visual characteristics like texture and colour that lessen the likelihood of similar-looking fruits being misclassified. Overall, the system's results are quite encouraging and indicate that it may effectively automate the grading and sorting of agricultural goods. Moreover, fuzzy logic concepts and neural networks could mimic the nervous system (Khan et al., 2016; Patel et al., 2015; Zareiforoush et al., 2016). Meanwhile, the entire objective of segmentation is to divide the image into areas of interest on the image for processing further.

Image processing classification includes image enhancement for differentiating and filtering the images, storage with rectifications, analysis for selecting and categorising the images, compression, and synthesis of reconstructed images. The advanced uses of image processing techniques include defect indication and granulation analysis. Range imaging and volumetric methods provide new avenues for exploring three-dimensional space, while image sequence analysis provides insights into temporal events by analysing dynamic processes. Image processing has spread into many other domains, emphasising its importance in developing science and resolving challenging issues (Valous & Sun, 2012). A boundary was set (threshold) to identify the bruises on apple skin. At the end of the result, 79.5% of the time, the detection was correct in detecting the bruised area. Other applications include automatic grading and categorisation of dates. The method retrieved these attributes from RGB pictures of the dates. Then, using a neural network classifier trained by professionals, the dates were divided into three quality ratings. The system exhibited 80% accuracy in sorting data samples when evaluated. In another study, the damaged peels of oranges have few characteristics such as hue, saturation, and intensity for categorising peels from diseased ones (AKodagali & Balaji, 2012). An essential part of computer vision systems is image analysis, which makes it possible to analyse and comprehend visual data. Computer vision systems can identify and analyse things inside pictures using techniques like object measurement and image segmentation. This allows for the extraction of useful data for subsequent processing and decision-making.

4.4 Equipment/Instrumentation in Computer Vision System

A general overview of equipment and instrumentation commonly used in computer vision systems includes cameras and lighting systems, optical components, image processing units, sensors and actuators, and data acquisition and storage systems (Fig. 4.1 and Table 4.1). Every component is essential to the system's ability to gather, handle, and evaluate visual data for various uses. The brief notes on each part are listed below.

4.4.1 Camera and Lighting Systems

In food processing, measurement is essential for quality assessment and inspection. It can be done via computer vision, conventional tools like colourimeters, or visual examination. Trained inspectors frequently do visual assessments, although they are often subjective, time-consuming, and variable because of lighting. Quantitative measurements can be obtained using conventional devices such as spectrophotometers and colourimeters, but they may not work effectively when working with tiny or irregular forms or non-homogeneous materials. The requirement for sample

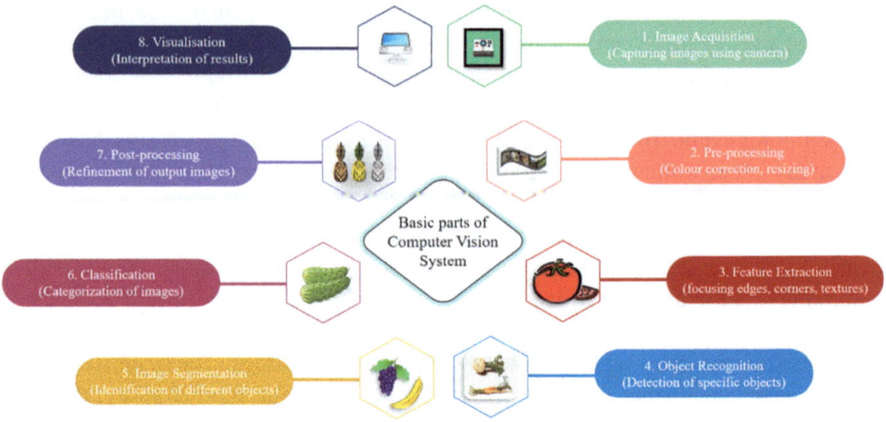

Fig. 4.1 Basic parts in a computer vision system

characterisation has led to the development of automatic computer vision-based pixel-based colour assessment systems, notwithstanding the drawbacks of these conventional approaches. These automated technologies overcome conventional techniques' limitations, which provide a quick and non-invasive assessment of colour dispersion inside foods (Wu et al., 2013). Depth cameras primarily function as 3D measuring devices, with early research focusing on tasks like 3D registration and reconstruction. Devices like Kinect use near-infrared laser projectors and cameras to triangulate points in 3D space, providing outputs such as IR, RGB, and inverse-depth images. The IR camera decodes projection patterns to capture 3D scenes, while the RGB camera textures the 3D points or recognises image content. Kinect's depth image, representing "inverse depth," is constructed through triangulation from the IR image and projector. However, it may exhibit limitations, such as vertical stripes of missing depth data, likely due to correlation windowing effects (Fossati et al., 2013).

A new method for using infrared (IR) systems has surfaced, which improves their effectiveness and makes optimisation possible by using a sequence of IR monochromatic pictures. This development makes it easier to identify and assess characteristics that are difficult to detect using traditional techniques. The capacity of these methodologies to offer vital insights into the intrinsic qualities of the items under analysis has led to a recent boom in their utilisation. Moreover, these systems can analyse goods in portions of the electromagnetic spectrum that are invisible to the human eye, such as the near-infrared (NIR), IR, and UV wavelengths. In sorting procedures, human error is a major risk that machine vision technology successfully mitigates (Guzmán et al., 2013). In a study conducted on apple sorting, lighting setup with eight LED lights arranged in two lines at a 40 to 50° angle from vertical. The two pairs of visible and NIR LED lights are positioned symmetrically above the samples. Halogen light with a 40 nm slit for specific apple bands from 540 to 1320 nm was also used in other studies. The 150 W halogen lamps light the apples uniformly for bruise detection. A 45° angle could be maintained for

Table 4.1 Equipment/instrumentation in computer vision system

S. No.	Food products	Application	References
1.	Peach	Utilisation of multiple wavelength camera, lamp: halogen	Cubero et al. (2011)
2.	Watermelon	To identify the three-dimensional space	Cubero et al. (2011)
3.	Mushroom	Visible and near-infrared (VIS/NIR) hyperspectral imaging (445–945 nm) leads to browning	Blasco et al. (2017)
4.	Orange	Different angle views at 0.4 m/s	Blasco et al. (2017)
5.	Fruits	RGB cameras are not able to detect the defective product	Blasco et al. (2009)
6.	Alcohol	Adulteration by near IR imaging	Cen and He (2007)
7.	Potato	Softmax regression model excels in detecting sample sizes with a 94.4% success rate, the application with appearance classification achieving only 14.5% accuracy for the lowest quality group under depth imaging	Su et al. (2020)
8.	Olive fruits	Identification of external damages from the wavelength of 760 to 1000 nm with an IR imaging system	Guzmán et al. (2013)
9.	Fuji apples	C-mount zoom lens identified early defects	Mohammadi Baneh et al. (2018)
10.	Apples	Viewing the three-dimensional image with optical filters at 840 nm	Mohammadi Baneh et al. (2018)
11.	Mangoes	RGB digital camera and LED flood lighting with GPU unit for detecting the quality	Koirala et al. (2019)
12.	Agricultural activities	For saving water, improvising the yield and automation in agricultural activities with FPGA	Mathurkar et al. (2014)
13.	Bread	Vibrational activities for identifying the mechanical damage	Abasi et al. (2018)
14.	Food odours, identification of volatiles	Electronic-chemical sensors, Piezoelectric crystal sensors	Di Rosa et al. (2017)
15.	Pear fruit	Actuators for identifying the resonance frequency around 700 Hz	Hiruta et al. (2021)

consistent lighting of halogen lamps. A quartz halogen lamp might be suggested for adjusting the power on imaging (Mohammadi Baneh et al., 2018). Online applications benefit much more from strobe lighting, which may freeze motion and eliminate fuzzy visuals from moving objects. Choosing a light repetition rate for strobe illumination in machine vision that corresponds with the object's motion is crucial. For machine vision applications, a strobe unit should have the capacity to operate continuously at a minimum of 30 repetitions per second. Strobe illumination is utilised to gather information from a moving situation in real-time. It is imperative to synchronise the strobe firing with the camera and image collection process to guarantee a clean image under strobe illumination. When analysing individual grain kernels, which travelled at a modest pace of 6.5 m/min, strobe illumination was used to prevent picture blur (Gunasekaran, 1996).

4.4.2 Optical Components

The camera consists of an optical lens, and its selection depends upon the viewing field and the working distance with a dimension of 12 mm as a focal length is used (Labudzki et al., 2014). The lens is affixed before the camera to focus on the object (Jayas & Singh, 2012). The filters are designed to fit the camera with less minimum spectra. On the other hand, two filters require two cameras or a replacement filter to achieve the spectral ranges. The filter system has to be developed to facilitate the fastest exchange without variations. The program ensures filters run within each spectral band's scheduled exposure period by coordinating their operation with picture capture. The programming language C was used for all software development. Some of the application's key features are the ability to capture hyperspectral images or perform live capture, support for one or two synchronised filters, acquisition of dark and white references for hyperspectral image correction, and calibration options for calculating exposure time using different light sources. The camera and filters are automatically initialised (Gómez-Sanchis et al., 2014). Under the illuminations of larger regions, mirrors, lenses, and prisms take place. Light guides better serve smaller areas. A convex reflection grating and two spherical mirrors make up a standard light guide on prism grating. Light is directed from an entry slit by the bottom mirror onto the grating, where it is scattered into different wavelengths. The top mirror reflects the scattered light onto a detector, creating a continuous spectrum. The grating disperses the light from the entry slit after it has been collimated by a front lens, allowing the middle wavelength to pass through symmetrically. A spread of wavelengths is produced when shorter and longer wavelengths are distributed above and below the centre axis, respectively. After that, a rear lens projects the scattered light onto a detector, producing a two-dimensional picture with one dimension representing spatial information and the other representing spectrum information (Qin, 2012; Shahin et al., 2012).

4.4.3 Image Processing Units

The Graphics Processing Unit (GPU) is an essential component of many computer vision systems because it handles parallel processing workloads effectively. Various image processing and computer vision methods, including image filtering, feature extraction, object identification, and classification, are accelerated by the GPU in the systems. The CUDA (Compute Unified Device Architecture) Toolkit and NVIDIA's DNN (Deep Neural Network) library are essential resources for speeding up deep learning calculations with hardware and software. CUDA is the go-to option for GPU programming, especially with NVIDIA GPUs, because of its extensive deep-learning libraries and robust NVIDIA support. However, large dataset compilation takes much time when using GPUs for training. Errors in data annotation and collection might also happen and affect the performance of the model (Chen, 2022). Another platform, Field Programmable Gate Array (FPGA), stands out as an optimal platform due to its numerous benefits, such as flexibility, parallel processing capability, energy efficiency, and the ability to be reconfigured as needed (Gupta & Kaur, 2022). Under a machine vision system, FPGA works on the Synchronous Dynamic Random Access Memory (SDRAM) data and creates signals to the devices by the results (Nie, 2021).

4.4.4 Sensors and Actuators

According to (Din et al., 2019), an accelerometer sensor measures variations in an object's or device's speed. The modifications are sent to a GPS receiver to change the modified navigation. A gyroscope is a device that monitors an axis's rotational speed and direction from many angles. Drive motors in a computer vision system or actuators are physically movable or modifiable lenses, cameras, and other system components. The system may control the viewpoint or perspective from which images are acquired and provide precision for creating the best imaging circumstances involving analysis or monitoring.

4.4.5 Data Acquisition and Storage Systems

Interconnected sensors that continuously report on the state of connected devices are a component of data acquisition systems (DAC), making deploying Internet of Things (IoT) technologies easier. The output can be created using data acquisition cards such as voltage, current, pressure, temperature, or sound. A solid-state drive (SSD) is used for a fast-paced level of data storage and easy accessibility while processing (Mashaly, 2021; Raval, 2021). Meanwhile, the amount of data, the required access speed, and the cost play a role in the storage technology selection.

4.5 Brief Introduction to Different Types of Imaging Techniques

Different types of imaging techniques such as Hyperspectral imaging, Magnetic Resonance Imaging, Biospeckle Imaging, Thermal, X-ray, Ultrasound and SEM are used for the food application (Fig. 4.2 and Table 4.2). These imaging techniques and their food applications are discussed in the below sections.

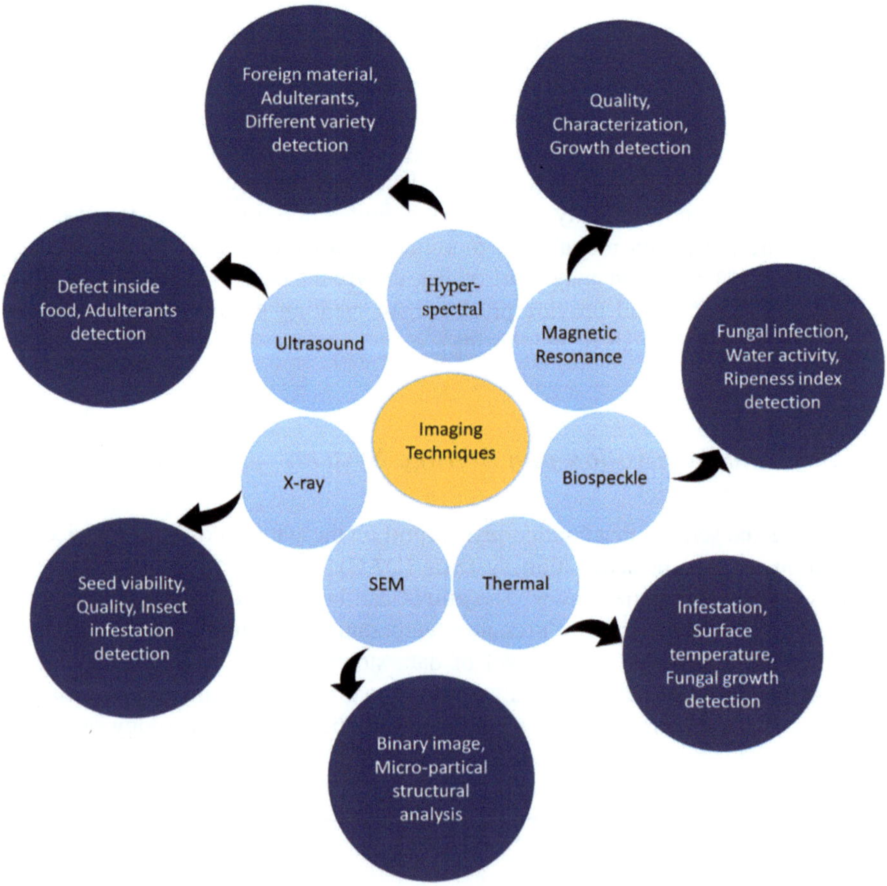

Fig. 4.2 Different types of imaging techniques and their applications

Table 4.2 Imaging techniques applications

Imaging techniques	Food/agricultural products	Applications	References
Hyper Spectral Imaging Techniques	Fresh-cut vegetables	Foreign material detection	Tunny et al. (2023)
	Minced chicken meat	Adulterated Starch detection	Yang et al. (2023)
	Sweet maize seed	Identification of a different variety	Wang and Song (2023)
	Beef	Differentiate fresh and frozen-thawed beef	Pu et al. (2023)
	Frozen pork	Lipid oxidation detection	Cheng et al. (2023)
	Lettuce cultivars	Nutrient content detection	Eshkabilov et al. (2021)
	Fish Fillet	Authenticate and detect substitution and mislabelling	Qin et al. (2020)
	Tomato	Bacterial and target spot disease detection	Abdulridha et al. (2020)
Magnetic resonance imaging	Meat and meat product	Quality parameter analysis	Caballero et al. (2023)
	Pumpkin	Variation of water content over drying time	Ripoli et al. (2021)
	Foodstuffs	Characterisation	Gruwel et al. (2010)
	Grapes wine	Ripening and growth of berries	Andaur et al. (2004)
	Soybean seed	Water distribution observation during the hydration process	Pietrzak et al. (2002)
Biospeckle imaging	Eggs	For quality evaluation	Kumar et al. (2024)
	Apple Slices	Monitoring the quality of slices during drying	Arefi et al. (2023)
	Soybean seed	To detect fungal-infected seeds	Kaler et al. (2023)
	Papaya leaf	To analyse water activity in the leaf during photosynthesis	Kumar and Nirala (2023)
	Tomato	To find the ripeness index	Pavarin et al. (2023)
Thermal imaging	Mango (Alphonso and Totapuri)	To detect infestation by fruit fly (*Bctrocera dorsalis*)	Shoba et al. (2024)
	Mango	To analyse the surface temperature	Pugazhendi et al. (2023)
	Egg	To detect egg freshness	Zhang et al. (2023)
	Pineapple	To characterise different cultivars in an altered storage situation	Mohd Ali et al. (2022)
	Apple	To detect fungal growth	

(continued)

Table 4.2 (continued)

Imaging techniques	Food/agricultural products	Applications	References
			Lipińska et al. (2022)
	Rice grain and flour	To ensure safety and quality.	Estrada-Pérez et al. (2021)
	Chili pepper	For harvesting and robotic inspection	Hespeler et al. (2021)
	Minced mutton	To find adulteration of minced mutton with pork	Zheng et al. (2021)
	Potato tubers	To identify fungal infection	Farokhzad et al. (2020)
	Olive oil	To find Adulteration state	Izquierdo et al. (2020)
	Tea	To identify diseases	Yang et al. (2019)
Ultrasound Imaging	Olive oil	Quality assessment	Zarezadeh et al. (2021)
	Milk	To detect adulterants	George et al. (2021)
	Edible oil	To detect adulterants	George et al. (2017)
	Fruit juices	To detect microbial contamination	Elvira et al. (2014)
	Cheese and poultry	To detect foreign matters and defects inside food materials	Cho and Irudayaraj (2003)
X-ray Imaging	Tomato	To check seed viability	Hong et al. (2023)
	Apple	Inspection of quality	Tempelaere et al. (2023)
	Mango	Internal and external quality evaluation	Vani Ashok et al. (2023)
	Sterculia seeds	Internal defect detection	Xue et al. (2022)
	U. brizantha seeds	For seed quality classification	Medeiros et al. (2020)
	Agricultural produce	Internal quality determination	Kotwaliwale et al. (2014)
	Wheat and pea	Detection of insect infestation	Neethirajan et al. (2008)
Scanning electron microscopy	Essential oil of lemon peel	The analysis of the bursting of oil glands	Rajan et al. (2023)
	Raspberry powder	Microparticle structural analysis	Przybył et al. (2021)
	rice	Representation of the binary image of the starchy structure	Zareiforoush et al. (2015)

4.5.1 Hyperspectral Imaging System

Hyperspectral imaging simultaneously captures images and detailed spectral data from food samples (Ravikanth et al., 2017). Hyperspectral technology, called chemical or spectroscopic imaging, combines regular imaging with spectroscopy to gather spatial and spectral details from an object using radiometry, resulting in images (Feng & Sun, 2012). In hyperspectral imaging, the term "hyper" implies "excessive" and indicates a greater number of measured wavelength bands. The data captured by a spectral imager forms a three-dimensional cube, incorporating two spatial and a third spectral dimensions. The values can be recorded and transformed into radiometric quantities allied to scene features (such as reflectance, radiance emissivity, etc.) by appropriate calibration processes (Wilcock et al., 2004). Hyperspectral imaging operates through three distinct methods to develop a spectral image of an object. Point scanning involves capturing the spectrum point by point, where the imaging system moves from one point to another along different axes, typically facilitated by a robotic phase. Area scanning by HI maintains a fixed field of view over an object and captures images at individual wavelengths one by one (Li et al., 2018). Line measurement involves moving the detector and the object, but the sample is moved line by line in this method. The resulting hypercube is saved in BIL (band interleaved by line) format. In the food industry, using line measurements with conveyor belts is convenient. A typical setup for hyperspectral imaging includes a desktop, lens, spectrograph, data acquisition system, and stage for movement. Each component's performance affects the system's accuracy. The illumination, camera and spectrograph cover the visible and near-infrared range. The system spreads incoming light lines onto spectral and spatial matrices, which are converted digitally using a CCD (charge-coupled device). The camera and spectrograph scan the product line-by-line, combining spatial matrices to create a three-dimensional (3D) dataset of spectral and spatial information.

4.5.2 Magnetic Resonance Imaging

This technique offers a notable advantage: it enables the examination of both 2-D and 3-D internal images without invasive or destructive methods (Luypaert & Van Cauteren, 1991). When analysed, these images provide a thorough understanding of the composition and physical makeup of food materials. The equipment detects signals that vary depending on the water content and magnetic properties of different foods. Due to its distinctive characteristics, MRI is a valuable real-time tool for internal process control systems in the food industry (Divya et al., 2012). MRI functions similarly to nuclear magnetic resonance (NMR) spectroscopy, utilising NMR principles to generate images of foods. When field gradients are introduced alongside a constant magnetic field, the spins' resonance frequency becomes linked to their position along an axis. This position-dependent frequency generates a spectrum, producing the image and its characteristics (Prakash., J et al., 2019). The basic principle of generating images using NMR theory includes applying linear field gradients to the sample in addition to the static magnetic field (Bo). For example, when a linear field gradient is applied in the x direction, the spins' resonance frequency depends on the position of the volume element of interest in relation to x. This resonance frequency variation results in a signal-intensity spectrum dependent on position. A typical MRI laboratory setup includes an imaging probe, magnet, radio frequency transmitter and receiver coil, shim and gradient power supplies, and a computer workstation (Spiller & Marciani, 2019).

4.5.3 Biospeckle Imaging

Biospeckle is a non-destructive optical technique that utilises coherent laser light to analyse biological materials. It detects static and variable speckle patterns, known as biospeckle activity (BA). The speckle pattern forms when coherent wavefronts reflected from an object's surface interfere with each other (Pandiselvam et al., 2020). When a laser illuminates a sample, each point of reflection serves as a secondary source, creating a random interference pattern known as "speckle." Moreover, if the sample exhibits biological or physiological activity, the resulting speckle patterns show temporal changes, known as dynamic or biospeckle patterns. The speckle pattern forms when coherent wavefronts reflected from an object's surface interfere. When a laser illuminates a sample, each point of reflection serves as a secondary source, creating a random interference pattern known as "speckle." Moreover, if the sample exhibits biological or physiological activity, the resulting speckle patterns show temporal changes, known as dynamic or biospeckle patterns (Thakur et al., 2024). This method has diverse applications in assessing food quality, identifying defects in fruits and vegetables, detecting bacterial growth, evaluating seed viability, and monitoring tissue perfusion variations. It offers advantages such as non-contact analysis, objective interpretation of results, and a relatively low-cost experimental setup (Pandiselvam et al., 2020).

4.5.4 Thermal Imaging

All substances emit and absorb thermal radiation by their surface temperature. The method of gauging this temperature and presenting it as a visual image is termed thermography. Thermal imaging (TI) is applicable for monitoring temperature fluctuations within food cores across different processing scenarios (Takacs et al., 2020). Thermal imaging visualises various foods based on the heat distributed or emitted across their faces, allowing for resolution measurement without physical contact (Rahkonen & Jokela, 2003). Thermal imaging offers data regarding the mechanical and physical properties of food and abnormal heat conditions and dimensions (Giorleo & Meola, 2002). Thermal imaging is additionally employed to assess the ripeness level of different fruits, aiding in pinpointing the ideal harvesting moment, which is crucial for maintaining their quality. It captures diverse temperatures across a defined area, transforming them into a thermal map or thermogram. This high-resolution thermal map forms the basis for visualising and evaluating objects with distinct thermal gradients. With an image detection rate of approximately 50–60 images per second, thermal imaging can detect alterations in thermal conditions within food substances (Rahkonen & Jokela, 2003). Thermal imaging works based on the concept that all matter emits infrared radiation across different parts of the electromagnetic spectrum, such as near, mid-wave, short-wave, long-wave, and extreme infrared. It can detect radiation with varying wavelengths, from short waves to long-wave infrared. Long-wave infrared radiation is more sensitive at room temperature. In contrast, mid-wave infrared radiation shows the highest sensitivity at maximum temperatures (around 400 °C), making them a preferred option for applications in the food industry (Takacs et al., 2020).

4.5.5 X-ray Imaging

This technique is gaining traction in agriculture and food quality assessment. It offers a non-invasive method for assessing food quality, employing real-time processing of images and proving to be suitable for monitoring food products online within the food industry (Herremans et al., 2014). In a real-time imaging configuration, X-rays pass through food products at the finish of the processing line. Detectors produce a sequential image of the food products for visual examination. These images are stored within a computer system and processed to ascertain their adequacy or inadequacy (Magwaza & Opara, 2014). The system automatically rejects any contaminated products based on the decision. The process begins with scanning, analysis, and decision-making. Refinement within a similar framework can aid in identifying different contaminants, thus enhancing customer safety (Pathmanaban et al., 2019).

4.5.6 Ultrasound Imaging

This technique employs mechanical sound waves resulting from molecules oscillating within a particular propagation medium. The wave's intensity can be classified as high or low, depending on how their energy is distributed. While this technique has traditionally been used primarily in the medical field, it is also employed in the food industry (Yaqoob et al., 2020). Ultrasound waves are transmitted through the vibration of particles and can penetrate opaque materials, providing detailed insights into the physical properties of the food. Low-power ultrasound is commonly used as a non-invasive analytical method to assess the structural and physico-chemical properties of foods and detect foreign substances without direct contact. This method serves as a robust tool for online quality assessment. During food testing, the velocity of ultrasound is responsive to molecular arrangements and intermolecular interactions, making it applicable for determining structural parameters, molecular processes and other physical states (Awad et al., 2012). Materials like bones, glass, and other metal fragments can be readily identified in poultry and cheeses because they interfere with ultrasound rays, altering the time of flight and velocity of the ultrasound waves (Cho & Irudayaraj, 2003).

4.5.7 Scanning Electron Microscopy

This technique is a valuable electron microscopy system that provides complete visual representations of high-resolution objects. This is primarily employed to study surface phenomena in materials. When exposed to the high-energy electron beam in SEM, a sample yields information about its morphology, topology, chemical composition, and crystallographic information. Thus, it is a useful method for material characterisation. It generates sample images by electron beam scatter on matter. Numerous signals are generated due to electron matter scattering (Akhtar et al., 2018).

4.6 Applications in Non-destructive Food Quality and Analysis

4.6.1 Application in 3D Food Printing

3D food printing, a rising technology for personalised food production, faces challenges due to the diverse rheological properties of food printing materials. Consistent extrusion, stemming from variations in elasticity, is a common issue. To address this, Ma et al. (2023) gave a computer vision-based method to measure extrusion rate and width in real-time, enabling control of nozzle motion for pneumatic 3D food printers. A texture analyser was employed to conduct compression tests for method development to replicate the real printing process. This method significantly enhances extrusion accuracy, achieving 97.9–100% rates and preventing under-extrusion of various materials such as processed cheese, cookie dough and chocolate spread. It also permits customisation of filament width with a deviation from the target of less than 8%. With its straightforward setup and user-friendly interface, this CV-based approach can streamline food printing applications, minimising trial and error when working with new materials. However, under-extrusion was significantly reduced with nozzle acceleration, leading to improved layer completion.

4.6.2 *Application in Grading Fruits*

4.6.2.1 Grading of Apple and Banana

The traditional method of visually grading fruits has posed a major challenge for the agricultural sector because it is labour-intensive and prone to inconsistencies in classification and inspection. However, the integration of automated defect detection using computer vision and machine learning presents a hopeful solution for transforming visual inspection procedures by employing a machine vision system that leverages sophisticated deep learning techniques and stacking ensemble methods; a non-invasive and economical approach is offered for automating the evaluation of fruit freshness and appearance. Various deep learning models, including MobileNetV2, ResNet, DenseNet, EfficientNet and NASNet, were explored to identify the most suitable model for grading fruit. Through training and evaluation on two datasets consisting of bananas and apples, the system achieved an average accuracy of 99.2% for apples and 98.6% for bananas using the EfficientNet model (Ismail & Malik, 2022).

4.6.2.2 Grading of Mango

The system utilises factors such as mango shape, maturity ripening stage, texture, defects and colour to differentiate mango varieties and assess their quality. Mango maturity ripening period is determined through CNN analysis, while image processing techniques and Computer Vision are employed to extract shape, defects and texture features. The extracted characteristics are then inputted into a Random Forest classifier to recognise different types of mangoes and assess their quality, dividing them into three categories: Unfit, Average, and Good. The system was validated on a dataset specifically made for this study, encompassing three different varieties: Banganapalli, Rumani and Neelam, the most prevalent in Tamil Nadu. Using features obtained from CNN, the system boosts efficiency notably, attaining a 93.23% accuracy rate in identifying varieties and a 95.11% accuracy rate in grading quality (Gururaj et al., 2023).

4.6.3 *Application on Detection of External Defects and Quality*

4.6.3.1 Detection of Defects in Tomato

The system utilises factors such as mango shape, maturity ripening stage, texture, defects and colour to differentiate mango varieties and assess their quality. Mango maturity ripening period is determined through CNN analysis, while image processing techniques and Computer Vision are employed to extract shape, defects and texture features. The extracted characteristics are then inputted into a Random Forest classifier to recognise different types of mangoes and assess their quality, dividing them into three categories: Unfit, Average, and Good. The system was validated on a dataset specifically made for this study, encompassing three different varieties: Banganapalli, Rumani and Neelam, the most prevalent in Tamil Nadu. Using features obtained from CNN, the system boosts efficiency notably, attaining a 93.23% accuracy rate in identifying varieties and a 95.11% accuracy rate in grading quality (Gururaj et al., 2023).

4.6.3.2 Detection of Defects in Apples

A deep-learning setup incorporating CNN and a cost-efficient CV module was utilised to spot faulty apples on a four-line fruit sorting machine running at 5 fruits per second. The CNN-driven classification model showcased remarkable performance, attaining an accuracy, specificity and recall of 96.5%, 92.9%, and 100% on the testing dataset. Conversely, a conventional image processing technique, dependent on counting candidate defective regions and a support vector machine classifier, produced subpar outcomes, with an accuracy, specificity and recall of 87.1%, 83.3%, and 90.9%, respectively. To validate its efficacy, the CNN-based model was integrated into simple software and assessed with an independent dataset of 200 apples, achieving an accuracy of 92% while processing six images of apple fruit in under 72 milliseconds (Fan et al., 2020).

4.6.3.3 Detection of Defects in Coffee Beans

There is a growing demand in the industry for well-organised and non-dangerous ways to pick out good quality coffee beans. Factors like colour, shape, morphology, and size play a crucial role in identifying the best beans, but traditional visual or mechanical inspection methods must meet these demands. So, Garcia et al. (2019) introduce a method that combines machine learning and image processing, incorporated with an Arduino Mega board, to assess these key features while choosing green coffee beans of high quality. The system encompasses logical processes, supervised learning algorithms and image processing, all coded in MATLAB and transferred to the Arduino board. Results indicate that this approach is highly effective in categorising individual green coffee beans based on their visual traits, and it can handle multiple beans within a particular image. Their statistical analysis demonstrates the method's ability to identify defects and assess quality accurately.

4.6.3.4 Detection of Defects in Carrots

Grading carrots manually takes a lot of time and effort. Xie et al. (2019) developed algorithms using machine vision to identify surface defects like green shoulders, bending, fibrous roots, surface cracks, and breakage to make it more efficient. They used techniques like binarisation and edge detection on different colour components to detect these defects. Then, they defined five measures to quantify these defects. They tested the algorithm on 720 randomly selected carrot images and found high accuracy rates for detecting each defect, with an overall recognition rate of 90.9%. This algorithm could greatly improve carrot grading and classification processes in the future.

4.6.4 Application on Detection of Seed Varieties

4.6.4.1 Detection of Corn Seed Variety

Automated sorting of seed varieties is crucial for producers to uphold variety purity and optimise crop yield. Traditional methods relying on computer vision and basic feature extraction often fail to achieve high accuracy in classification. Javanmardi et al. (2021) introduce a novel approach employing a deep convolutional neural network (CNN) as a versatile feature extractor. Models trained with features extracted by the CNN outperform those based solely on basic features when classifying corn seed varieties. Notably, the CNN-ANN (artificial neural network) classifier exhibits the highest performance, classifying 2250 test instances in 26.8 seconds with an accuracy of 98.1%, precision of 98.2%, recall of 98.1%, and F1-score of 98.1%.

4.6.4.2 Detection of Tea Variety

In the study conducted by Chen et al. (2008), computer vision was employed to classify various types of tea, including "Biluochun," "Tunchaoqing," "Maofeng," "Queshe," and "Maoshan Changqing," based on their external characteristics such as colour and shape. Due to significant variations in colour among different tea varieties, both HIS and RGB colour spaces were utilised to extract comprehensive colour features for tea leaf identification. 120 tea samples, comprising 20 samples per variety, were utilised in the experiment. Twelve variables were extracted for more testing, including six texture characteristic parameters based on the statistical moment and six based on spectrum characteristics. All sample data underwent standardisation to ensure comparability due to variations in orders of magnitude and dimension among the variables. Due to the varying dimensions and orders of magnitude among these variables, standardisation was performed on all sample data to ensure comparability. The presence of correlations between the texture characteristic variables and extracted colour could introduce redundancy in information. Principal Component Analysis (PCA) was employed as a statistical method to address this issue by transforming multiple indicators into a reduced set of comprehensive indicators. PCA facilitated the extraction of vectors by projecting from a multidimensional spectral information space to a lower-dimensional space, following the maximum direction of covariance. These vectors were then utilised as inputs for the pattern recognition model. The Linear Discriminant Analysis (LDA) method was utilised to establish the classification model, with 60 samples allocated as training sets and the remaining 60 as prediction sets. This approach facilitated the creation of a robust model for accurately classifying tea varieties based on their external characteristics.

4.6.5 Application on Detection of Plant and Fruit Disease/Pest

4.6.5.1 Detection of Citrus Fruit Pests

Capturing images of citrus fruits at three periods: (1) prior to pest infestation, (2) at the onset of infestation, and (3) after the eighth day of the second stage, under natural lighting (7000–11,000 lux). 1519 images were collected across all stages. For image classification, 70% of the images were allocated for network training, while 10% and 20% were reserved for validation and testing. Four pre-trained CNN models-ResNet-50, GoogleNet, VGG-16, and AlexNet-were employed, along with the RMSProp, Adam optimisation algorithms and SGDm, to detect and categorise well fruits versus those contaminated with the Mediterranean fly. During the pest outbreak stage, assessment of the models indicated that the VGG-16 model, paired with the SGDm algorithm, exhibited the highest efficiency, achieving F1 scores and detection accuracy of 98.36% and 98.33%, respectively. Furthermore, evaluation at the final stage revealed that the AlexNet model, combined with the SGDm algorithm, delivered the most favourable outcomes, achieving F1 scores and detection accuracy of 99.34% and 99.33%, respectively. Remarkably, using the SGDm optimisation algorithm, the AlexNet model boasted the shortest network training time, completing in 323 seconds (Hadipour-Rokni et al., 2023).

4.6.5.2 Detection of Apple Fruit Diseases

A cost-effective approach for diagnosing apple diseases involves utilising a neural network to classify fruits into four categories: scab, bitter rot, black rot, and healthy. This method relies on colour and texture features extracted from the images. A multi-layer perceptron neural network processes these features as input, with its output corresponding to the defined classes. Sixty per cent of the images were used to train the network, while the remaining images were kept for testing. The accuracy of the proposed method was evaluated using various structures of single-layer and two-layer neural networks. It was found that employing a two-layer structure with eight neurons in both layers achieved the optimal accuracy of 73.7% (Azgomi et al., 2023).

4.6.5.3 Detection of Rice and Potato Plant Diseases

Deep learning-based computer vision techniques, such as Convolutional Neural Network (CNN) and traditional machine learning-based image classification methods, are being applied to classify rice and potato plant leaf diseases. Rice leaf diseases, including bacterial blight, blast, brown spot, and tungro, are diagnosed, while potato leaf images are categorised into three classes: healthy leaves, early blight, and late blight diseases. Sharma et al. (2022) utilise a dataset comprising 5932 rice leaf images and 1500 potato leaf images. The proposed CNN model successfully learns hidden patterns from the raw images and achieves high accuracy rates of 99.58% for rice images and 97.66% for potato leaves. The findings demonstrate that the proposed CNN model outperforms other machine learning image classifiers, such as Support Vector Machine (SVM) and K-Nearest Neighbors (KNN).

4.6.5.4 Detection of Tomato Plant Diseases

The examination of tomato leaf samples with disorders is conducted to assist farmers in identifying diseases based on symptoms which can be shown earlier. First, the tomato leaf samples are adjusted to a standard size of 256 × 256 pixels, and Histogram Equalization is used to improve their clarity. Following this, K-means clustering is utilised to divide the data space into Voronoi cells. The edges of the leaf sample are outlined using contour tracing. Various methods extract important characteristics from the leaf samples, including Principal Component Analysis, Grey Level Co-occurrence Matrix and Discrete Wavelet Transform. These extracted characteristics are then analysed using machine learning techniques such as Convolutional Neural Network (CNN), K-Nearest Neighbor (K-NN), and Support Vector Machine (SVM). The effectiveness of the proposed model is evaluated using K-NN (97%), CNN (99.6%), and SVM (88%) on samples of diseased tomatoes (Harakannanavar et al., 2022).

4.6.6 Application in the Detection of Meat/Animal Product Freshness

4.6.6.1 Detection of Egg Quality

Detecting egg quality is crucial for food processing and consumption. This study aimed to use hyperspectral imaging (HSI), multivariate analysis, and image processing to identify fresh eggs, eggs with scattered yolks, and eggs with cracks in their shells. Yao et al. (2022) captured images of egg samples using HSI in the visible-near infrared range. They normalised the spectral data and optimised wavelength selection. They then built a model to quantify egg freshness using Extreme Gradient Boosting (XGBoost), achieving a prediction accuracy (R2p) of 0.91 and a root mean square error (RMSEP) of 4.64. They proposed an algorithm to enhance image contrast, remove noise, and segment yolks based on morphological features, achieving a 97.33% accuracy in identifying eggs with scattered yolks. Additionally, they developed a method to enhance cracks, segment them using double thresholds, and classify cracked eggs using XGBoost, achieving a 93.33% identification accuracy. These suggest that HSI can effectively detect egg quality without damaging the eggs.

4.6.6.2 Detection of the Freshness of Pork

Detecting the freshness of frozen meat in real-time without thawing is crucial. This study looked into using fluorescence hyperspectral imaging (HSI) to assess the features of frozen pork with no thawing. Zhuang et al. (2022) constructed models using partial least squares regression (PLSR) using fluorescence spectra to forecast indicators such as pH, total volatile basic nitrogen L*, a*, and b*. They then contrasted these models with those created using visible/near-infrared (Vis/NIR) hyperspectral imaging (HSI) of the identical samples. They employed competitive adaptive reweighted sampling to identify crucial fluorescence wavelengths associated with every indicator. The models built using fluorescence spectra showed high correlation coefficients of prediction (Rp) for pH, TVB-N, L*, a*, and b*. Except L*, the models based on fluorescence HSI outperformed those based on Vis-NIR HSI. Additional enhancements were observed by incorporating selected key wavelengths. These findings suggest that fluorescence HSI can reliably evaluate freshness indicators of frozen pork without requiring prior thawing.

4.6.6.3 Detection of Freshness of Meat

The key components in meat, such as myoglobin, oxymyoglobin, fat, water, and collagen, absorb light similarly in the visible to near-infrared (NIR) spectral range. This means that variations in meat structure and composition can cause differences in how light is absorbed. By using fibre-optic probes and integrating sphere setups, Peyvasteh et al. (2020) observed changes in the freshness of pork samples at room temperature by measuring the relative absorbance of these key components. Applying principal component analysis (PCA) to the absorbance spectra allowed for a more detailed examination of freshness stages, which conventional reflectance spectra analysis could not provide. The findings suggest that combining optical-NIR spectroscopy with PCA offers great potential for assessing meat quality and monitoring changes in the absorbance of key components like oxymyoglobin, myoglobin, fat, water, and collagen across different spectral ranges.

4.7 Conclusion and Future Scope

Computer vision systems are a powerful tool for assessing and analysing non-destructive food quality. By leveraging various imaging techniques and advanced image processing methods, these systems enable accurate and efficient evaluation of food products, ensuring compliance with quality standards and meeting consumer expectations. Furthermore, integrating computer vision systems into the food industry facilitates automation and streamlines processes, increasing productivity and cost-effectiveness. With the ability to detect defects, assess freshness, and identify contaminants, these systems enhance food safety and quality assurance measures. As technology advances, future research can focus on enhancing the speed, accuracy, and versatility of computer vision systems for a wider range of applications in the food industry. Additionally, integrating artificial intelligence and machine learning algorithms can further improve the capabilities of these systems for automated food inspection and quality control.

References

Abasi, S., Minaei, S., Jamshidi, B., & Fathi, D. (2018). Dedicated non-destructive devices for food quality measurement: A review. *Trends in Food Science & Technology, 78*, 197–205. https://doi.org/10.1016/j.tifs.2018.05.009

Abdullah, M. Z. (2007). Computer vision and infrared techniques for image acquisition in the food and beverage industries * *Adapted from Computer vision technology for food quality evaluation, ed. Da-Wen Sun, Chapter 1, Image acquisition systems, pp. 3–35. Copyright (2007), wit. In *Computer vision technology in the food and beverage industries* (pp. 3–26). Elsevier. https://doi.org/10.1533/9780857095770.1.3

Abdullah, M. Z., et al. (2016). Image acquisition systems. In *Computer vision technology for food quality evaluation* (pp. 3–43). Elsevier. https://doi.org/10.1016/B978-0-12-802232-0.00001-3

Abdulridha, J., Ampatzidis, Y., Kakarla, S. C., & Roberts, P. (2020). Detection of target spot and bacterial spot diseases in tomato using UAV-based and benchtop-based hyperspectral imaging techniques. *Precision Agriculture, 21*(5), 955–978. https://doi.org/10.1007/s11119-019-09703-4

Akhtar, K., Khan, S. A., Khan, S. B., & Asiri, A. M. (2018). Scanning electron microscopy: Principle and applications in nanomaterials characterization. In *Handbook of materials characterization* (pp. 113–145). Springer International Publishing. https://doi.org/10.1007/978-3-319-92955-2_4

Akodagali, J., & Balaji, S. (2012). Computer vision and image analysis based techniques for automatic characterization of fruits a review. *International Journal of Computer Applications, 50*(6), 6–12. https://doi.org/10.5120/7773-0856

Aleixos, N., Blasco, J., Navarrón, F., & Moltó, E. (2002). Multispectral inspection of citrus in real-time using machine vision and digital signal processors. *Computers and Electronics in Agriculture, 33*(2), 121–137. https://doi.org/10.1016/S0168-1699(02)00002-9

Amani, H., et al. (2023). *Rapid quality assessment of bakery products using machine vision.* PhD thesis. https://phd.mater.uni-mate.hu/552/

Andaur, J. E., Guesalaga, A. R., Agosin, E. E., Guarini, M. W., & Irarrázaval, P. (2004). Magnetic resonance imaging for nondestructive analysis of wine grapes. *Journal of Agricultural and Food Chemistry, 52*(2), 165–170. https://doi.org/10.1021/jf034886c

Arefi, A., Sturm, B., Raut, S., von Gersdorff, G., & Hensel, O. (2023). NIR laser-based imaging techniques to monitor quality attributes of apple slices during the drying process: Laser-light backscattering & biospeckle imaging techniques. *Food Control, 143*, 109289. https://doi.org/10.1016/j.foodcont.2022.109289

Ashok, V., Bharathi, R. K., & Sheela, N. (2023). An automatic non-destructive external and internal quality evaluation of mango fruits based on color and X-ray imaging with machine learning and deep learning based classification models. *Inteligencia Artificial, 26*(72), 223–243. https://doi.org/10.4114/intartif.vol26iss72pp223-243

Aung, M. M., & Chang, Y. S. (2014). Traceability in a food supply chain: Safety and quality perspectives. *Food Control, 39*, 172–184. https://doi.org/10.1016/j.foodcont.2013.11.007

Awad, T. S., Moharram, H. A., Shaltout, O. E., Asker, D., & Youssef, M. M. (2012). Applications of ultrasound in analysis, processing and quality control of food: A review. *Food Research International, 48*(2), 410–427. https://doi.org/10.1016/j.foodres.2012.05.004

Azgomi, H., Haredasht, F. R., & Safari Motlagh, M. R. (2023). Diagnosis of some apple fruit diseases by using image processing and artificial neural network. *Food Control, 145*, 109484. https://doi.org/10.1016/j.foodcont.2022.109484

Bhargava, A., & Bansal, A. (2021). Fruits and vegetables quality evaluation using computer vision: A review. *Journal of King Saud University - Computer and Information Sciences, 33*(3), 243–257. https://doi.org/10.1016/j.jksuci.2018.06.002

Blasco, J., Aleixos, N., Cubero, S., Juste, F., Gómez-Sanchis, J., Alegre, V., & Moltó, E. (2009). Computer vision developments for the automatic inspection of fresh and processed fruits. In *First International Workshop on Computer Image Analysis in Agriculture* (pp. 21–24). Leibniz Institute for Agricultural Engineering Potsdam Bornim (ATB).

Blasco, J., Munera, S., Aleixos, N., Cubero, S., & Molto, E. (2017). Machine vision-based measurement systems for fruit and vegetable quality control in postharvest. In *Measurement, modeling and automation in advanced food processing* (pp. 71–91). Springer. https://doi.org/10.1007/10_2016_51

Brosnan, T., & Sun, D.-W. (2004). Improving quality inspection of food products by computer vision—a review. *Journal of Food Engineering, 61*(1), 3–16. https://doi.org/10.1016/S0260-8774(03)00183-3

Caballero, D., Pérez-Palacios, T., Caro, A., & Antequera, T. (2023). Use of magnetic resonance imaging to analyse meat and meat products non-destructively. *Food Reviews International, 39*(1), 424–440. https://doi.org/10.1080/87559129.2021.1912085

Cen, H., & He, Y. (2007). Theory and application of near infrared reflectance spectroscopy in determination of food quality. *Trends in Food Science & Technology, 18*(2), 72–83. https://doi.org/10.1016/j.tifs.2006.09.003

Chen, T. C. (2022). Research on food safety sampling inspection system based on deep learning. *Food Science and Technology, 42*, e29121. https://doi.org/10.1590/fst.29121

Chen, Q., Zhao, J., Liu, M., & Cai, J. (2008). Nondestructive Identification of Tea (Camellia sinensis L.) varieties using FT-NIR spectroscopy and pattern recognition. *Czech Journal of Food Sciences, 26*(5), 360–367. https://doi.org/10.17221/1125-cjfs

Chen, Y., Wu, J., & Cui, M. (2018). Automatic classification and detection of oranges based on computer vision. In *2018 IEEE 4th International Conference on Computer and Communications (ICCC)* (pp. 1551–1556). IEEE. https://doi.org/10.1109/CompComm.2018.8780680

Chen, Q., Lin, H., & Zhao, J. (2021). *Advanced nondestructive detection technologies in food*. Springer Singapore. https://doi.org/10.1007/978-981-16-3360-7

Cheng, J., Sun, J., Xu, M., & Zhou, X. (2023). Nondestructive detection of lipid oxidation in frozen pork using hyperspectral imaging technology. *Journal of Food Composition and Analysis, 123*, 105497. https://doi.org/10.1016/j.jfca.2023.105497

Cho, B.-K., & Irudayaraj, J. M. K. (2003). Foreign object and internal disorder detection in food materials using noncontact ultrasound imaging. *Journal of Food Science, 68*(3), 967–974. https://doi.org/10.1111/j.1365-2621.2003.tb08272.x

Cubero, S., Aleixos, N., Moltó, E., Gómez-Sanchis, J., & Blasco, J. (2011). Advances in machine vision applications for automatic inspection and quality evaluation of fruits and vegetables. *Food and Bioprocess Technology, 4*(4), 487–504. https://doi.org/10.1007/s11947-010-0411-8

Daley, W. D., Carey, R., & Thompson, C. (1993). In F. Y. Wu & B. M. Dawson (Eds.), *Poultry grading/inspection using color imaging* (pp. 124–132). https://doi.org/10.1117/12.144816

de Medeiros, A. D., da Silva, L. J., Ribeiro, J. P. O., Ferreira, K. C., Rosas, J. T. F., Santos, A. A., & da Silva, C. B. (2020). Machine learning for seed quality classification: An advanced approach using merger data from FT-NIR spectroscopy and X-ray imaging. *Sensors, 20*(15), 4319. https://doi.org/10.3390/s20154319

Di Rosa, A. R., Leone, F., Cheli, F., & Chiofalo, V. (2017). Fusion of electronic nose, electronic tongue and computer vision for animal source food authentication and quality assessment—A review. *Journal of Food Engineering, 210*, 62–75. https://doi.org/10.1016/j.jfoodeng.2017.04.024

Din, I. U., Guizani, M., Rodrigues, J. J. P. C., Hassan, S., & Korotaev, V. V. (2019). Machine learning in the Internet of Things: Designed techniques for smart cities. *Future Generation Computer Systems, 100*, 826–843. https://doi.org/10.1016/j.future.2019.04.017

Divya, S., Thyagarajan, D., & Sujatha, G. (2012). Magnetic resonance imaging technology for process control and quality maintenance in food quality operation. *International Journal of Engineering and Technology, 4*(6), 441–449.

Du, C.-J., & Sun, D.-W. (2004). Recent developments in the applications of image processing techniques for food quality evaluation. *Trends in Food Science & Technology, 15*(5), 230–249. https://doi.org/10.1016/j.tifs.2003.10.006

ElMasry, G., ElGamal, R., Mandour, N., Gou, P., Al-Rejaie, S., Belin, E., & Rousseau, D. (2020). Emerging thermal imaging techniques for seed quality evaluation: Principles and applications. *Food Research International, 131*(November 2019), 109025. https://doi.org/10.1016/j.foodres.2020.109025

Elvira, L., Durán, C. M., Urréjola, J., & Montero de Espinosa, F. R. (2014). Detection of microbial contamination in fruit juices using non-invasive ultrasound. *Food Control, 40*, 145–150. https://doi.org/10.1016/j.foodcont.2013.11.033

Eshkabilov, S., Lee, A., Sun, X., Lee, C. W., & Simsek, H. (2021). Hyperspectral imaging techniques for rapid detection of nutrient content of hydroponically grown lettuce cultivars. *Computers and Electronics in Agriculture, 181*, 105968. https://doi.org/10.1016/j.compag.2020.105968

Estrada-Pérez, L. V., Pradana-López, S., Pérez-Calabuig, A. M., Mena, M. L., Cancilla, J. C., & Torrecilla, J. S. (2021). Thermal imaging of rice grains and flours to design convolutional systems to ensure quality and safety. *Food Control, 121*, 107572. https://doi.org/10.1016/j.foodcont.2020.107572

Fan, S., Li, J., Zhang, Y., Tian, X., Wang, Q., He, X., Zhang, C., & Huang, W. (2020). Online detection of defective apples using computer vision system combined with deep learning methods. *Journal of Food Engineering, 286*. https://doi.org/10.1016/j.jfoodeng.2020.110102

FAO. (2019). *Moving forward on food loss and waste reduction food and agriculture* (1st ed.). Food and Agriculture Organization of the United Nations (FAO).

Farokhzad, S., Modaress Motlagh, A., Ahmadi Moghadam, P., Jalali Honarmand, S., & Kheiralipour, K. (2020). Application of infrared thermal imaging technique and discriminant analysis methods for non-destructive identification of fungal infection of potato tubers. *Journal of Food Measurement and Characterization, 14*(1), 88–94. https://doi.org/10.1007/s11694-019-00270-w

Feldmann, C., & Hamm, U. (2015). Consumers' perceptions and preferences for local food: A review. *Food Quality and Preference, 40*, 152–164. https://doi.org/10.1016/j.foodqual.2014.09.014

Feng, Y.-Z., & Sun, D.-W. (2012). Application of hyperspectral imaging in food safety inspection and control: A review. *Critical Reviews in Food Science and Nutrition, 52*(11), 1039–1058. https://doi.org/10.1080/10408398.2011.651542

Fossati, A., Gall, J., Grabner, H., Ren, X., & Konolige, K. (Eds.). (2013). *Consumer depth cameras for computer vision*. Springer London. https://doi.org/10.1007/978-1-4471-4640-7

Fracarolli, J. A., Pavarin, F. F. A., Castro, W., & Blasco, J. (2020). Computer vision applied to food and agricultural products. *Revista Ciência Agronômica, 51*(5). https://doi.org/10.5935/1806-6690.20200087

García, M., Candelo-Becerra, J. E., & Hoyos, F. E. (2019). Quality and defect inspection of green coffee beans using a computer vision system. *Applied Sciences (Switzerland), 9*(19). https://doi.org/10.3390/app9194195

George, T., Rufus, E., & Alex, Z. C. (2017). Artificial neural network based ultrasonic sensor system for detection of adulteration in edible oil. *Journal of Engineering Science and Technology, 12*(6), 1568–1579.

George, T., Paul, J., & Babu Pukkunnen, E. (2021). Non destructive testing of milk adulteration using ultrasound. *Annals of the Romanian Society for Cell Biology, 25*(4), 8912–8919.

Giorleo, G., & Meola, C. (2002). Comparison between pulsed and modulated thermography in glass–epoxy laminates. *NDT & E International, 35*(5), 287–292. https://doi.org/10.1016/S0963-8695(01)00062-7

Gómez-Sanchis, J., Lorente, D., Soria-Olivas, E., Aleixos, N., Cubero, S., & Blasco, J. (2014). Development of a hyperspectral computer vision system based on two liquid crystal tuneable filters for fruit inspection. Application to detect citrus fruits decay. *Food and Bioprocess Technology, 7*(4), 1047–1056. https://doi.org/10.1007/s11947-013-1158-9

Gruwel, M. L. H., Latta, P., Matwiy, B., & Tomanek, B. (2010). Characterization of food stuffs using magnetic resonance elastography. *Food Research International, 43*(8), 2087–2092. https://doi.org/10.1016/j.foodres.2010.07.015

Gunasekaran, S. (1996). Computer vision technology for food quality assurance. *Trends in Food Science and Technology, 7*(8), 245–256. https://doi.org/10.1016/0924-2244(96)10028-5

Gupta, I., & Kaur, D. P. (2022). FPGA based feature extraction in real time computer vision-a comprehensive survey. In *2022 International Conference on Signal and Information Processing (IConSIP)* (pp. 1–7). IEEE. https://doi.org/10.1109/ICoNSIP49665.2022.10007508

Gururaj, N., Vinod, V., & Vijayakumar, K. (2023). Deep grading of mangoes using convolutional neural network and computer vision. *Multimedia Tools and Applications, 82*(25), 39525–39550. https://doi.org/10.1007/s11042-021-11616-2

Guzmán, E., Baeten, V., Pierna, J. A. F., & García-Mesa, J. A. (2013). Infrared machine vision system for the automatic detection of olive fruit quality. *Talanta, 116*, 894–898. https://doi.org/10.1016/j.talanta.2013.07.081

Hadipour-Rokni, R., Askari Asli-Ardeh, E., Jahanbakhshi, A., Esmaili Paeen-Afrakoti, I., & Sabzi, S. (2023). Intelligent detection of citrus fruit pests using machine vision system and convolutional neural network through transfer learning technique. *Computers in Biology and Medicine, 155*, 106611. https://doi.org/10.1016/j.compbiomed.2023.106611

Harakannanavar, S. S., Rudagi, J. M., Puranikmath, V. I., Siddiqua, A., & Pramodhini, R. (2022). Plant leaf disease detection using computer vision and machine learning algorithms. *Global Transitions Proceedings, 3*(1), 305–310. https://doi.org/10.1016/j.gltp.2022.03.016

Harris, S. L., & Harris, D. M. (2016). I/O systems. In *Digital design and computer architecture* (pp. 531.e1–531.e64). Elsevier. https://doi.org/10.1016/B978-0-12-800056-4.00015-7

Herremans, E., Verboven, P., Defraeye, T., Rogge, S., Ho, Q. T., Hertog, M. L. A. T. M., Verlinden, B. E., Bongaers, E., Wevers, M., & Nicolai, B. M. (2014). X-ray CT for quantitative food microstructure engineering: The apple case. *Nuclear Instruments and Methods in Physics Research Section B: Beam Interactions with Materials and Atoms, 324*, 88–94. https://doi.org/10.1016/j.nimb.2013.07.035

Hespeler, S. C., Nemati, H., & Dehghan-Niri, E. (2021). Non-destructive thermal imaging for object detection via advanced deep learning for robotic inspection and harvesting of chili peppers. *Artificial Intelligence in Agriculture, 5*, 102–117. https://doi.org/10.1016/j.aiia.2021.05.003

Hiruta, T., Sasaki, K., Hosoya, N., Maeda, S., & Kajiwara, I. (2021). Firmness evaluation of postharvest pear fruit during storage based on a vibration experiment technique using a dielectric elastomer actuator. *Postharvest Biology and Technology, 182*, 111697. https://doi.org/10.1016/j.postharvbio.2021.111697

Hong, S.-J., Park, S., Lee, C.-H., Kim, S., Roh, S.-W., Nurhisna, N. I., & Kim, G. (2023). Application of X-Ray imaging and convolutional neural networks in the prediction of tomato seed viability. *IEEE Access, 11*, 38061–38071. https://doi.org/10.1109/ACCESS.2023.3265998

Ismail, N., & Malik, O. A. (2022). Real-time visual inspection system for grading fruits using computer vision and deep learning techniques. *Information Processing in Agriculture, 9*(1), 24–37. https://doi.org/10.1016/j.inpa.2021.01.005

Izquierdo, M., Lastra-Mejías, M., González-Flores, E., Cancilla, J. C., Aroca-Santos, R., & Torrecilla, J. S. (2020). Deep thermal imaging to compute the adulteration state of extra virgin olive oil. *Computers and Electronics in Agriculture, 171*, 105290. https://doi.org/10.1016/j.compag.2020.105290

Javanmardi, S., Miraei Ashtiani, S. H., Verbeek, F. J., & Martynenko, A. (2021). Computer-vision classification of corn seed varieties using deep convolutional neural network. *Journal of Stored Products Research, 92*. https://doi.org/10.1016/j.jspr.2021.101800

Jayas, D. S., & Singh, C. B. (2012). Grain quality evaluation by computer vision. In *Computer vision technology in the food and beverage industries* (pp. 400–421). Elsevier. https://doi.org/10.1533/9780857095770.3.400

Judal, A., et al. (2015). Role of machine vision system in food quality and safety evaluation. *International Journal of Advance Research and Innovation, 3*(4), 18–22. https://doi.org/10.51976/ijari.341504

Kaganami, H. G., & Beiji, Z. (2009). Region-based segmentation versus edge detection. In *2009 Fifth International Conference on Intelligent Information Hiding and Multimedia Signal Processing* (pp. 1217–1221). IEEE. https://doi.org/10.1109/IIH-MSP.2009.13

Kaler, N., Bhatia, V., & Mishra, A. K. (2023). Deep learning-based robust analysis of laser bio-speckle data for detection of fungal-infected soybean seeds. *IEEE Access, 11*, 89331–89348. https://doi.org/10.1109/ACCESS.2023.3305273

Karthick, S. (2014). A survey based on region based segmentation. *International Journal of Engineering Trends and Technology, 7*(3), 143–147. https://doi.org/10.14445/22315381/ijett-v7p262

Khan, A., et al. (2016). *Vision based classification of fresh fruits using fuzzy logic*. IEEE.

Kleynen, O., Leemans, V., & Destain, M.-F. (2005). Development of a multi-spectral vision system for the detection of defects on apples. *Journal of Food Engineering, 69*(1), 41–49. https://doi.org/10.1016/j.jfoodeng.2004.07.008

Koirala, A., Walsh, K. B., Wang, Z., & McCarthy, C. (2019). Deep learning for real-time fruit detection and orchard fruit load estimation: Benchmarking of 'MangoYOLO'. *Precision Agriculture, 20*(6), 1107–1135. https://doi.org/10.1007/s11119-019-09642-0

Kotwaliwale, N., Singh, K., Kalne, A., Jha, S. N., Seth, N., & Kar, A. (2014). X-ray imaging methods for internal quality evaluation of agricultural produce. *Journal of Food Science and Technology, 51*(1), 1–15. https://doi.org/10.1007/s13197-011-0485-y

Kumar, N., & Nirala, A. K. (2023). A novel computational method for dynamic laser speckle and its application to analyze water activity during photosynthesis in papaya leaf. *Optik, 274*, 170518. https://doi.org/10.1016/j.ijleo.2023.170518

Kumar, N., Nirala, A. K., & Patel, H. B. (2024). Quality evaluation of eggs during shelf-life ambient storage by laser speckle technique. *Optik, 307*, 171820. https://doi.org/10.1016/j.ijleo.2024.171820

Kyriacou, M. C., & Rouphael, Y. (2018). Towards a new definition of quality for fresh fruits and vegetables. *Scientia Horticulturae, 234*, 463–469. https://doi.org/10.1016/j.scienta.2017.09.046

Labudzki, R., Legutko, S., & Raos, P. (2014). The essence and applications of machine vision. *Tehnicki Vjesnik, 21*(4), 903–909.

Li, X., Li, R., Wang, M., Liu, Y., Zhang, B., & Zhou, J. (2018). Hyperspectral imaging and their applications in the nondestructive quality assessment of fruits and vegetables. In *Hyperspectral imaging in agriculture, food and environment*. IntechOpen. https://doi.org/10.5772/intechopen.72250

Lipińska, E., Pobiega, K., Piwowarek, K., & Błażejak, S. (2022). Research on the use of thermal imaging as a method for detecting fungal growth in apples. *Horticulturae, 8*(10), 972. https://doi.org/10.3390/horticulturae8100972

Loce, R. P. (2017). In R. P. Loce, R. Bala, & M. Trivedi (Eds.), *Computer vision and imaging in intelligent transportation systems*. Wiley. https://doi.org/10.1002/9781118971666

Lorente, D., Aleixos, N., Gómez-Sanchis, J., Cubero, S., García-Navarrete, O. L., & Blasco, J. (2012). Recent advances and applications of hyperspectral imaging for fruit and vegetable quality assessment. *Food and Bioprocess Technology, 5*(4), 1121–1142. https://doi.org/10.1007/s11947-011-0725-1

Lukinac, J. L., Jukić, M. J., Mastanjević, K. M., & Lučan, M. L. (2018). Application of computer vision and image analysis method in cheese-quality evaluation: A review. *Ukrainian Food Journal, 7*(2), 192–214. https://doi.org/10.24263/2304-974x-2018-7-2-4

Lukinac, J., Mastanjević, K., Mastanjević, K., Nakov, G., & Jukić, M. (2019). Computer vision method in beer quality evaluation—A review. *Beverages, 5*(2), 38. https://doi.org/10.3390/beverages5020038

Luypaert, R., & Van Cauteren, M. (1991). Principles of magnetic resonance spectroscopy and imaging. In *Magnetic resonance imaging and spectroscopy in sports medicine* (pp. 3–22). Springer. https://doi.org/10.1007/978-3-642-75686-3_1

Ma, J., Sun, D.-W., Qu, J.-H., Liu, D., Pu, H., Gao, W.-H., & Zeng, X.-A. (2016). Applications of computer vision for assessing quality of agri-food products: A review of recent research advances. *Critical Reviews in Food Science and Nutrition, 56*(1), 113–127. https://doi.org/10.1080/10408398.2013.873885

Ma, Y., Potappel, J., Chauhan, A., Schutyser, M. A. I., Boom, R. M., & Zhang, L. (2023). Improving 3D food printing performance using computer vision and feedforward nozzle motion control. *Journal of Food Engineering, 339*. https://doi.org/10.1016/j.jfoodeng.2022.111277

Magwaza, L. S., & Opara, U. L. (2014). Investigating non-destructive quantification and characterization of pomegranate fruit internal structure using X-ray computed tomography. *Postharvest Biology and Technology, 95*, 1–6. https://doi.org/10.1016/j.postharvbio.2014.03.014

Mahajan, S., Das, A., & Sardana, H. K. (2015). Image acquisition techniques for assessment of legume quality. *Trends in Food Science & Technology, 42*(2), 116–133. https://doi.org/10.1016/j.tifs.2015.01.001

Mahendran, R. (2012). Application of computer vision technique on sorting and grading of fruits and vegetables. *Journal of Food Processing & Technology, 10*, 2175–7110. https://doi.org/10.4172/2157-7110.S1-001

Majumder, D. (1986). Pattern recognition, image processing and computer vision in fifth generation computer systems. *Sadhana, 9*(September), 139–156.

Mashaly, M. (2021). Connecting the twins: A review on digital twin technology & its networking requirements. *Procedia Computer Science, 184*, 299–305. https://doi.org/10.1016/j.procs.2021.03.039

Mathurkar, S. S., Patel, N. R., Lanjewar, R. B., & Somkuwar, R. S. (2014). Smart sensors based monitoring system for agriculture using field programmable gate array. In *2014 International Conference on Circuits, Power and Computing Technologies, ICCPCT 2014* (pp. 339–344). https://doi.org/10.1109/ICCPCT.2014.7054914

Millicent. (2019). Scientia horticulturae destructive and non-destructive techniques used for quality evaluation of nuts: A review. *Scientia Horticulturae, 247*(November 2018), 138–146. https://doi.org/10.1016/j.scienta.2018.12.008

Mohammadi Baneh, N., Navid, H., & Kafashan, J. (2018). Mechatronic components in apple sorting machines with computer vision. *Journal of Food Measurement and Characterization, 12*(2), 1135–1155. https://doi.org/10.1007/s11694-018-9728-1

Mohd Ali, M., & Hashim, N. (2022). Non-destructive methods for detection of food quality. In *Future foods* (pp. 645–667). Elsevier. https://doi.org/10.1016/B978-0-323-91001-9.00003-7

Mohd Ali, M., Hashim, N., Abd Aziz, S., & Lasekan, O. (2022). Characterisation of pineapple cultivars under different storage conditions using infrared thermal imaging coupled with machine learning algorithms. *Agriculture, 12*(7), 1013. https://doi.org/10.3390/agriculture12071013

Narendra, V. G., & Hareesh, K. S. (2010). Quality inspection and grading of agricultural and food products by computer vision-a review. *International Journal of Computer Applications, 2*(1), 43–65. https://doi.org/10.5120/612-863

Neethirajan, S., Jayas, D. S., White, N. D. G., & Zhang, H. (2008). Investigation of 3D geometry of bulk wheat and pea pores using X-ray computed tomography images. *Computers and Electronics in Agriculture, 63*(2), 104–111. https://doi.org/10.1016/j.compag.2008.01.019

Nie, X. (2021). Research on embedded machine vision inspection system based on FPGA. *Forest Chemicals Review*, 545–556.

Pandiselvam, R., Mayookha, V. P., Kothakota, A., Ramesh, S. V., Thirumdas, R., & Juvvi, P. (2020). Biospeckle laser technique—A novel non-destructive approach for food quality and safety detection. *Trends in Food Science & Technology, 97*, 1–13. https://doi.org/10.1016/j.tifs.2019.12.028

Patel, K. K., et al. (2015). Image processing tools and techniques used in computer vision for quality assessment of food products: A review. *International Journal of Food Quality and Safety | Year-2015 |, 1*(December), 1–16. https://www.jakraya.com/journal/ijfqs

Patel, K. K., Kar, A., & Khan, M. A. (2021). Monochrome computer vision for detecting common external defects of mango. *Journal of Food Science and Technology, 58*(12), 4550–4557. https://doi.org/10.1007/s13197-020-04939-9

Pathmanaban, P., Gnanavel, B. K., & Anandan, S. S. (2019). Recent application of imaging techniques for fruit quality assessment. *Trends in Food Science & Technology, 94*, 32–42. https://doi.org/10.1016/j.tifs.2019.10.004

Pavarin, F. F. A., Fracarolli, J. A., & Falcão, A. X. (2023). Biospeckle laser for assessing tomatoes ripeness indexes. *Revista Ciência Agronômica, 15*. https://doi.org/10.5935/1806-6690.20240022

Peyvasteh, M., Popov, A., Bykov, A., & Meglinski, I. (2020). Meat freshness revealed by visible to near-infrared spectroscopy and principal component analysis. *Journal of Physics Communications, 4*(9), 095011. https://doi.org/10.1088/2399-6528/abb322

Pietrzak, L. N., Frégeau-Reid, J., Chatson, B., & Blackwell, B. (2002). Observations on water distribution in soybean seed during hydration processes using nuclear magnetic resonance imaging. *Canadian Journal of Plant Science, 82*(3), 513–519. https://doi.org/10.4141/P01-150

Pourdarbani, R., Ghassemzadeh, H. R., Seyedarabi, H., Zaare Nahandi, F., & Moghaddam Vahed, M. (2012). The effect of camera and light source characteristics on image quality in machine vision application in food industry. *Journal of Agricultural Science, 4*(10). https://doi.org/10.5539/jas.v4n10p169

Prakash, J., Vignesh, K., Anusuya, T., Kalaivani, T., Ramachandran, C., Sudha, R. R., Rubab, M., Khan, I., Elahi, F., Oh, D.-H., & DevanandVenkatasubbu, G. (2019). Application of nanoparticles in food preservation and food processing. *Journal of Food Hygiene and Safety, 34*(4), 317–324. https://doi.org/10.13103/JFHS.2019.34.4.317

Przybył, K., Koszela, K., Adamski, F., Samborska, K., Walkowiak, K., & Polarczyk, M. (2021). Deep and machine learning using SEM, FTIR, and texture analysis to detect polysaccharide in raspberry powders. *Sensors, 21*(17), 5823. https://doi.org/10.3390/s21175823

Pu, H., Yu, J., Sun, D.-W., Wei, Q., Shen, X., & Wang, Z. (2023). Distinguishing fresh and frozen-thawed beef using hyperspectral imaging technology combined with convolutional neural networks. *Microchemical Journal, 189*, 108559. https://doi.org/10.1016/j.microc.2023.108559

Pugazhendi, P., Balakrishnan Kannaiyan, G., Anandan, S. S., & Somasundaram, C. (2023). Analysis of mango fruit surface temperature using thermal imaging and deep learning. *International Journal of Food Engineering, 19*(6), 257–269. https://doi.org/10.1515/ijfe-2022-0302

Qin, J. (2012). Hyperspectral and multispectral imaging in the food and beverage industries. In *Computer vision technology in the food and beverage industries* (pp. 27–63). Elsevier. https://doi.org/10.1533/9780857095770.1.27

Qin, J., Chao, K., Kim, M. S., Lu, R., & Burks, T. F. (2013). Hyperspectral and multispectral imaging for evaluating food safety and quality. *Journal of Food Engineering, 118*(2), 157–171. https://doi.org/10.1016/j.jfoodeng.2013.04.001

Qin, J., Vasefi, F., Hellberg, R. S., Akhbardeh, A., Isaacs, R. B., Yilmaz, A. G., Hwang, C., Baek, I., Schmidt, W. F., & Kim, M. S. (2020). Detection of fish fillet substitution and mislabeling using multimode hyperspectral imaging techniques. *Food Control, 114*, 107234. https://doi.org/10.1016/j.foodcont.2020.107234

Rahkonen, J., & Jokela, H. (2003). Infrared radiometry for measuring plant leaf temperature during thermal weed control treatment. *Biosystems Engineering, 86*(3), 257–266. https://doi.org/10.1016/S1537-5110(03)00138-7

Rajan, A., Boopathy, B., Radhakrishnan, M., Rao, L., Schlüter, O. K., & Tiwari, B. K. (2023). Plasma processing: A sustainable technology in agri-food processing. *Sustainable Food Technology, 1*(1), 9–49. https://doi.org/10.1039/D2FB00014H

Raval, M. S. (2021). Computer vision and machine learning in agriculture. In *Data science in agriculture and natural resource management.* Springer. https://doi.org/10.1007/978-981-16-5847-1_5

Ravikanth, L., Jayas, D. S., White, N. D. G., Fields, P. G., & Sun, D.-W. (2017). Extraction of spectral information from hyperspectral data and application of hyperspectral imaging for food and agricultural products. *Food and Bioprocess Technology, 10*(1), 1–33. https://doi.org/10.1007/s11947-016-1817-8

Redlingshöfer, B., & Soyeux, A. (2012). Food losses and wastage as a sustainability indicator of food and farming systems. In *Producing and reproducing farming systems. New modes of organisation for sustainable food systems of tomorrow.* 10th European IFSA Symposium, Aarhus, Denmark, 1–4 July 2012, unpaginated.

Ripoli, C., Adiletta, G., Russo, P., Funicello, N., Iannone, G., Di Matteo, M., & De Pasquale, S. (2021). Study of pumpkin drying through magnetic resonance imaging. *Frontiers in Sustainable Food Systems, 5*. https://doi.org/10.3389/fsufs.2021.644829

Saldaña, E., Siche, R., Luján, M., & Quevedo, R. (2013). Review: Computer vision applied to the inspection and quality control of fruits and vegetables. *Brazilian Journal of Food Technology, 16*(4), 254–272. https://doi.org/10.1590/S1981-67232013005000031

Shahin, M. A., Hatcher, D. W., & Symons, S. J. (2012). Development of multispectral imaging systems for quality evaluation of cereal grains and grain products. In *Computer vision technology in the food and beverage industries* (pp. 451–482). Elsevier. https://doi.org/10.1533/9780857095770.3.451

Sharma, R., Singh, A., Jhanjhi, N. Z., Masud, M., Jaha, E. S., & Verma, S. (2022). Plant disease diagnosis and image classification using deep learning. *Computers, Materials and Continua, 71*(2), 2125–2140. https://doi.org/10.32604/cmc.2022.020017

Shoba, S., Pandiyarajan, T., & Shashidhar, K. C. (2024). Nondestructive detection of fruit fly (Bactrocera dorsalis) infestation in Alphonso and Totapuri varieties of mango fruits by thermal imaging technique. *Journal of Food Process Engineering, 47*(4). https://doi.org/10.1111/jfpe.14610

Silvestre, C., Duraccio, D., & Cimmino, S. (2011). Food packaging based on polymer nanomaterials. *Progress in Polymer Science, 36*(12), 1766–1782. https://doi.org/10.1016/j.progpolymsci.2011.02.003

Singh. (2013). Advances in computer vision technology for foods of animal and aquatic origin-a review. *Journal of Meat Science and Technology, 1*(2), 40–49.

Soltani Firouz, M., & Sardari, H. (2022). Defect detection in fruit and vegetables by using machine vision systems and image processing. *Food Engineering Reviews, 14*(3), 353–379. https://doi.org/10.1007/s12393-022-09307-1

Spiller, R., & Marciani, L. (2019). Intraluminal impact of food: New insights from MRI. *Nutrients, 11*(5), 1147. https://doi.org/10.3390/nu11051147

Su, Q., Kondo, N., Al Riza, D. F., & Habaragamuwa, H. (2020). Potato quality grading based on depth imaging and convolutional neural network. *Journal of Food Quality, 2020*, 1–9. https://doi.org/10.1155/2020/8815896

Swami, et al. (2010). Quality inspection of food products by computer vision. *International Journal of Agricultural Engineering, 3*(1), 184–187.

Takacs, R., Jovicic, V., Zbogar-Rasic, A., Geier, D., Delgado, A., & Becker, T. (2020). Evaluation of baking performance by means of mid-infrared imaging. *Innovative Food Science & Emerging Technologies, 61*, 102327. https://doi.org/10.1016/j.ifset.2020.102327

Tempelaere, A., Van Doorselaer, L., He, J., Verboven, P., Tuytelaars, T., & Nicolai, B. (2023). Deep learning for apple fruit quality inspection using X-Ray imaging. In *Proceedings - 2023 IEEE/CVF International Conference on Computer Vision Workshops, ICCVW 2023* (pp. 552–560). https://doi.org/10.1109/ICCVW60793.2023.00062

Thakur, P. S., Krejcar, O., Bhatia, V., & Prakash, S. (2024). Deep learning based processing framework for spatio-temporal analysis and classification of laser biospeckle data. *Optics & Laser Technology, 169*, 110138. https://doi.org/10.1016/j.optlastec.2023.110138

Throop, J. A., Aneshansley, D. J., & Upchurch, B. L. (1993). In J. A. DeShazer & G. E. Meyer (Eds.), *Near-IR and color imaging for bruise detection on Golden Delicious apples* (pp. 33–44). https://doi.org/10.1117/12.144042

Tscheuschner, H.-D. (1987). N. N. Mohsenin: Physical properties of plant and animal materials. Structure, physical characteristics and mechanical properties. 2. Aufl. 891 Seiten, zahlr. Abb. und Tab. Gordon and Breach Science Publishers, New York u. a. 1986. Preis: 140.—£. *Food/Nahrung, 31*(7), 702. https://doi.org/10.1002/food.19870310724.

Tunny, S. S., Kurniawan, H., Amanah, H. Z., Baek, I., Kim, M. S., Chan, D., Faqeerzada, M. A., Wakholi, C., & Cho, B.-K. (2023). Hyperspectral imaging techniques for detection of foreign materials from fresh-cut vegetables. *Postharvest Biology and Technology, 201*, 112373. https://doi.org/10.1016/j.postharvbio.2023.112373

Vadivambal, R., & Jayas, D. S. (2011). Applications of thermal imaging in agriculture and food industry—A review. *Food and Bioprocess Technology, 4*(2), 186–199. https://doi.org/10.1007/s11947-010-0333-5

Valous, N. A., & Sun, D.-W. (2012). Image processing techniques for computer vision in the food and beverage industries. In *Computer vision technology in the food and beverage industries* (pp. 97–129). Elsevier. https://doi.org/10.1533/9780857095770.1.97

Vyawahare, A., Rao, K. J., & Pagote, C. N. (2013). Computer vision system for colour measurement - fundamentals and applications in food industry: A review. *Research and Reviews: Journal of Food and Dairy Technology, 1*(2), 22–31.

Walther, D. B., Farzanfar, D., Han, S., & Walther, D. B. (2023). The mid-level vision toolbox for computing structural properties of real-world images. *Frontiers in Computer Science, 5*, 1–10. https://doi.org/10.3389/fcomp.2023.1140723

Wang, Y., & Song, S. (2023). Variety identification of sweet maize seeds based on hyperspectral imaging combined with deep learning. *Infrared Physics & Technology, 130*, 104611. https://doi.org/10.1016/j.infrared.2023.104611

Wilcock, A., Pun, M., Khanona, J., & Aung, M. (2004). Consumer attitudes, knowledge and behaviour: A review of food safety issues. *Trends in Food Science & Technology, 15*(2), 56–66. https://doi.org/10.1016/j.tifs.2003.08.004

Wu, et al. (2013). Colour measurements by computer vision for food quality control—A review. *Trends in Food Science & Technology, 29*(1), 5–20. https://doi.org/10.1016/j.tifs.2012.08.004

Xiao, Z., Wang, J., Han, L., Guo, S., & Cui, Q. (2022). Application of machine vision system in food detection. *Frontiers in Nutrition, 9*. https://doi.org/10.3389/fnut.2022.888245

Xie, W., Wang, F., & Yang, D. (2019). Research on carrot surface defect detection methods based on machine vision. *IFAC-PapersOnLine, 52*(30), 24–29. https://doi.org/10.1016/j.ifacol.2019.12.484

Xue, Q., Miao, P., Miao, K., Yu, Y., & Li, Z. (2022). X-ray-based machine vision technique for detection of internal defects of sterculia seeds. *Journal of Food Science, 87*(8), 3386–3395. https://doi.org/10.1111/1750-3841.16237

Yang, N., Yuan, M., Wang, P., Zhang, R., Sun, J., & Mao, H. (2019). Tea diseases detection based on fast infrared thermal image processing technology. *Journal of the Science of Food and Agriculture, 99*(7), 3459–3466. https://doi.org/10.1002/jsfa.9564

Yang, F., Sun, J., Cheng, J., Fu, L., Wang, S., & Xu, M. (2023). Detection of starch in minced chicken meat based on hyperspectral imaging technique and transfer learning. *Journal of Food Process Engineering, 46*(4). https://doi.org/10.1111/jfpe.14304

Yao, K., Sun, J., Chen, C., Xu, M., Zhou, X., Cao, Y., & Tian, Y. (2022). Non-destructive detection of egg qualities based on hyperspectral imaging. *Journal of Food Engineering, 325*, 111024. https://doi.org/10.1016/j.jfoodeng.2022.111024

Yaqoob, M., Aggarwal, P., Aslam, R., & Rehal, J. (2020). Extraction of bioactives from citrus. In *Green sustainable process for chemical and environmental engineering and science* (pp. 357–377). Elsevier. https://doi.org/10.1016/B978-0-12-817388-6.00015-5

Zareiforoush, H., Minaei, S., Alizadeh, M. R., & Banakar, A. (2015). Potential applications of computer vision in quality inspection of rice: A review. *Food Engineering Reviews, 7*(3), 321–345. https://doi.org/10.1007/s12393-014-9101-z

Zareiforoush, H., Minaei, S., Alizadeh, M. R., Banakar, A., & Samani, B. H. (2016). Design, development and performance evaluation of an automatic control system for rice whitening machine based on computer vision and fuzzy logic. *Computers and Electronics in Agriculture, 124*, 14–22. https://doi.org/10.1016/j.compag.2016.01.024

Zarezadeh, M. R., Aboonajmi, M., & Ghasemi Varnamkhasti, M. (2021). Fraud detection and quality assessment of olive oil using ultrasound. *Food Science & Nutrition, 9*(1), 180–189. https://doi.org/10.1002/fsn3.1980

Zhang, J., Lu, W., Jian, X., Hu, Q., & Dai, D. (2023). Nondestructive detection of egg freshness based on infrared thermal imaging. *Sensors, 23*(12), 5530. https://doi.org/10.3390/s23125530

Zheng, M., Zhang, Y., Gu, J., Bai, Z., & Zhu, R. (2021). Classification and quantification of minced mutton adulteration with pork using thermal imaging and convolutional neural network. *Food Control, 126*, 108044. https://doi.org/10.1016/j.foodcont.2021.108044

Zhuang, Q., Peng, Y., Yang, D., Wang, Y., Zhao, R., Chao, K., & Guo, Q. (2022). Detection of frozen pork freshness by fluorescence hyperspectral image. *Journal of Food Engineering, 316*, 110840. https://doi.org/10.1016/j.jfoodeng.2021.110840

Chapter 5
X-Ray, CT, and MRI

Mathangi Raja Sekhar, Sruthileena Bommi, B. N. Vinutha, M. A. Yogesh, C. K. Sunil, Ashish Rawson, S. Shanmugasundaram, and D. V. Chidanand

5.1 Introduction

Ensuring food quality throughout the supply chain is paramount importance in the context of food safety and consumer health. Traditionally, destructive methods involving sample dissection and chemical analysis have been employed for food quality analysis. However, these techniques are time-consuming, labour-intensive, and limit the number of samples evaluated. While there are well-recognized ways of making such analyses as the gas chromatography-mass spectroscopy (GC-MS), high performance liquid chromatography (HPLC) and Fourier-transform infrared spectroscopy (FTIR) offer unparalleled accuracy in determining specific food quality attributes, their application often presents limitations in high-throughput food processing environments. These sophisticated methods typically necessitate sample preparation steps that are destructive or time-consuming, hindering real-time quality control and process integration. The developments have facilitated the widespread adoption of non-destructive quality evaluation methods. Non-destructive quality evaluation methods offer a paradigm shift by enabling the analysis of food samples without causing any physical or chemical alteration. This offers several benefits, firstly the ability to analyse a higher volume of samples, which is essential for maintaining robust quality control in high-production environments. Secondly, non-destructive quality evaluation methods are often rapid, allowing for real-time

M. R. Sekhar
ICAR-Central Arid Zone Research Institute, Regional Research Station, Leh, UT of Ladakh, India

S. Bommi · B. N. Vinutha · M. A. Yogesh · C. K. Sunil · A. Rawson · S. Shanmugasundaram · D. V. Chidanand (✉)
National Institute of Food Technology, Entrepreneurship and Management (NIFTEM-T), Thanjavur, Tamil Nadu, India
e-mail: chidanand@iifpt.edu.in

C. K. Sunil et al. (eds.), *Emerging Non-destructive Technologies for Food Quality Analysis*, https://doi.org/10.1007/978-3-031-95048-3_5

monitoring of food quality during processing, enabling adjustments to be made on the fly. Consequently, the need for efficient non-destructive quality assessment methods is growing. This has led to the emergence of a new wave of sophisticated analytical tools that offer rapid, in-line, and non-destructive evaluation of food quality attributes. These advancements hold immense potential for optimizing quality control processes across the food industry, minimizing product waste, and ensuring consumer safety and satisfaction. In the domine of non-destructive food quality evaluation, several imaging techniques offer valuable insights.

X-ray radiography provides a two-dimensional image based on the varying X-ray absorption of different food components. Denser areas appear darker, allowing for the detection of internal defects like cracks, voids, or foreign objects. However, this technique offers limited information about specific tissues or chemical composition.

Computed tomography (CT) builds upon X-ray radiography by capturing multiple images from different angles. These images are then computationally reconstructed to generate a three-dimensional representation of the food sample. This allows for more detailed visualization of internal structures, including the distribution of fat, air pockets, and even subtle textural variations.

Magnetic resonance imaging (MRI) utilizes a different approach altogether. It exposes the food sample to a strong magnetic field and radio waves, causing specific atomic nuclei within the food to resonate. MRI can reveal details about the water distribution and chemical composition of food by examining these resonances. This is very helpful for determining ripeness and freshness as well as internal quality changes that may not be apparent using other techniques.

Each technique offers unique advantages and limitations. X-ray radiography is simple, relatively inexpensive, and rapid, making it suitable for high-throughput screening. CT provides superior detail but is more expensive and time-consuming. MRI offers unparalleled chemical information but requires specialized equipment and expertise. By strategically selecting the appropriate technique or even combining them, researchers and food quality professionals gain a comprehensive understanding of the internal characteristics of food products without damaging the sample. The present chapter discusses theory and principle, and basic concepts of X-ray, CT and MRI, Equipment/instrumentation—Image acquisition-processing of data/image- Applications in non-destructive food quality and analysis; hazards.

5.2 X-Ray—Theory and Principle

X-ray production at the atomic level involves a two-step interaction between high-energy electrons and the target atom. They are Bremsstrahlung Radiation and Characteristic X-rays.

5.2.1 *Bremsstrahlung Radiation*

The process starts outside the target atom. A device called an X-ray tube generates a stream of high-energy electrons. These electrons are accelerated by a strong voltage difference, giving them a lot of kinetic energy. This stream of energetic electrons is then directed towards a metal target within the X-ray tube (Bombardment). These electrons collide with the atoms in the target material. During these collisions, kinetic energy of the incoming electrons is transferred to the electrons within the target atom. This can happen in two ways, in the first or direct collision, the incoming electron might directly collide with an inner-shell electron (closer to the nucleus) of the target atom, knocking it out of its orbit. This creates a vacancy in the inner electron shell. In the second, even if there is no direct collision, the strong electric field of the fast-moving electron can still interact with and disrupt the inner-shell electrons, causing them to jump to higher energy levels or even escape the atom altogether. Regardless of how the inner-shell electron is ejected, the energetic interaction itself can also cause the incoming electron to decelerate suddenly. This sudden deceleration of a charged particle (electron) generates electromagnetic radiation in the form of X-rays. This type of X-ray emission is called Bremsstrahlung radiation, which translates to "braking radiation" in German (Fig. 5.1). The energy of these X-rays varies depending on the initial energy of the incoming electrons.

5.2.2 *Characteristic X-Rays*

During electron emission, one or more inner shell electrons are knocked of the target atom, leaving it in an excited state with an empty place in its inner electron shell. Then electrons from outer shells (higher energy levels) of the target atom are "attracted" to the vacancy in the inner shell. These outer electrons can jump down to fill the empty spot in the lower energy level. Jumping down to a lower energy level is energetically favourable for the electron. However, to maintain the overall energy

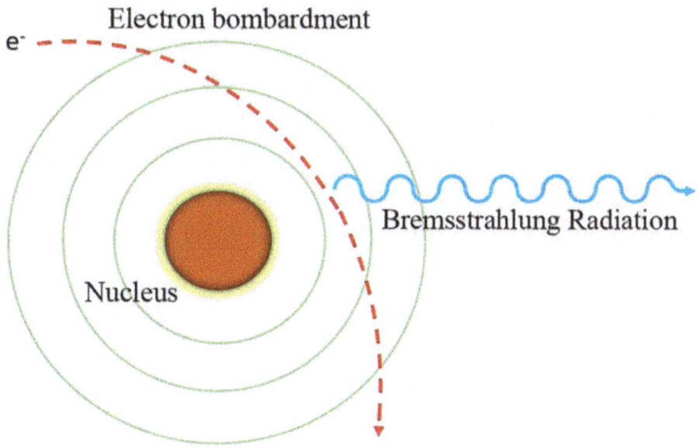

Fig. 5.1 X-rays produced through Bremsstrahlung

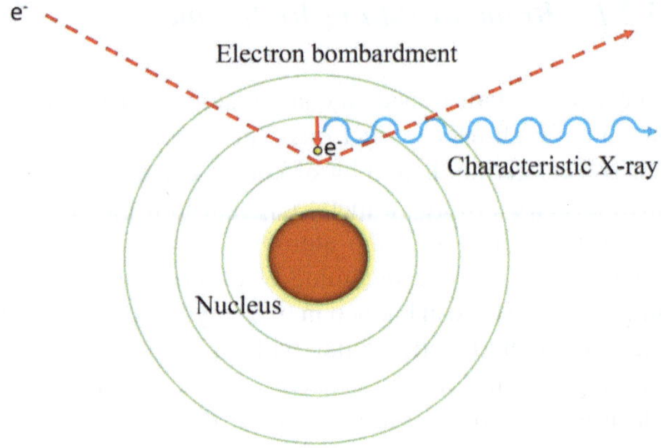

Fig. 5.2 Production of characteristic X-rays

balance of the atom, the excess energy carried by the electron (due to its previous higher energy level) needs to be released. This excess energy is carried away in a particular wavelength of electromagnetic radiation known as X-ray photon (Fig. 5.2). The energy of the emitted X-ray photon depends only on the target element and the transition that occurred. Unlike Bremsstrahlung radiation, characteristic X-rays have very specific and discrete energies.

5.2.3 *Properties of X-Rays*

X-rays, discovered by Wilhelm Conrad Röntgen in 1895, revolutionized our ability to peer into the unseen world. These enigmatic rays hold the key to visualizing our bones, detecting hidden fractures, and even fighting diseases. But what exactly are X-rays, and how do their properties make them so valuable?

5.2.3.1 Nature of X-Rays

X-rays belong to the electromagnetic spectrum, existing alongside familiar forms of radiation like visible light and radio waves. However, X-rays reside in a much higher energy realm compared to their counterparts. This difference lies in their wavelength (λ), the distance between two consecutive peaks in the wave. X-rays possess wavelengths ranging from 0.01 to 10 nm (Sprawls, 1994). According to the well-known equation relating energy (E) and wavelength (Saran & Fahim, 2020).

$$E = hc/\lambda \tag{5.1}$$

where h is Planck's constant (6.63×10^{-34} J·s) and c is the speed of light (3×10^{8} m/s), shorter wavelengths translate to higher energy for X-rays. This inherent high energy empowers X-rays to penetrate through materials that readily absorb lower energy radiation, like visible light.

5.2.3.2 Penetration Depth

The ability of X-rays to penetrate various materials is crucial for their applications. This penetration depth depends not only on the X-ray's energy (determined by wavelength) but also on the density (ρ) of the material it encounters. Denser materials, like bones, offer greater resistance to X-ray passage compared to less dense tissues like muscles. This relationship can be described by the Beer-Lambert Law:

$$I = I_0 e^{(-\mu\rho x)} \tag{5.2}$$

where I_0 represents the initial intensity of the X-ray beam, I is the intensity of the transmitted beam, μ is the material's attenuation coefficient (dependent on both material and X-ray energy), and x is the thickness of the material traversed. Essentially, this equation tells us that the intensity of the X-ray beam (I) exponentially weakens (e raised to a negative power) as it travels through a material (x) with a specific density (ρ). Materials with higher attenuation coefficients (μ) absorb X-rays more effectively, resulting in a lower transmitted intensity (I).

By manipulating the energy (via wavelength) of the X-ray beam, its penetration depth can be controlled. Lower energy X-rays, with longer wavelengths, are readily absorbed by soft tissues, making them unsuitable for visualizing bones. Conversely, higher energy X-rays, with shorter wavelengths, penetrate deeper, allowing us to see through soft tissues and image bones. This selective absorption is the foundation of X-ray imaging techniques like radiography, where bones appear white due to their high X-ray attenuation and soft tissues appear grey due to their lower attenuation.

5.2.4 Evolution of 2D X-Ray Systems & Features

X-ray technology got a captivating journey since its accidental discovery in 1895. The key milestones are given as following.

- 1895: Röntgen's discovery sparks the X-ray revolution. Early systems were bulky and captured images on photographic plates requiring long exposures (Röntgen, 1896).
- Early 1900s: Fluoroscopy enables real-time imaging, while specialized units emerge for dental X-rays (International Atomic Energy Agency, 2009).
- Mid-twentieth Century: Film becomes the standard, offering advantages but requiring processing chemicals and exposing personnel (Bushong,2012)

- Late twentieth Century: The digital revolution arrived. Digital Radiography (DR) offers faster processing, real-time viewing, and reduced radiation with electronic image capture (Bushong,2012).
- Present—Future: Flat-panel detectors push boundaries with superior resolution and speed, minimizing radiation dose. Dual-Energy X-ray Absorptiometry (DXA) utilizes dual beams for bone mineral assessment (Bushong,2012).

The developments being continued with advancements in processing, AI, and detector technology. X-ray systems remain a cornerstone of medical imaging and industrial food inspection, promising a brighter future for different applications.

5.2.5 X-Ray Equipment

X-rays, with their power to penetrate and reveal hidden structures, are a cornerstone of medical imaging and scientific exploration. But how exactly are these enigmatic waves generated? Let us delve into the heart of the X-ray machine, exploring its key components and their functionalities.

5.2.5.1 X-Ray Tube

The X-ray tube is the workhorse of the X-ray machine. It's a sealed, evacuated glass enclosure housing two crucial elements (Bhargava et al., 2013): Cathode is a negatively charged electrode, often a filament made of tungsten, is heated by an electric current, causing it to emit electrons (Fig. 5.3). Anode is a positively charged

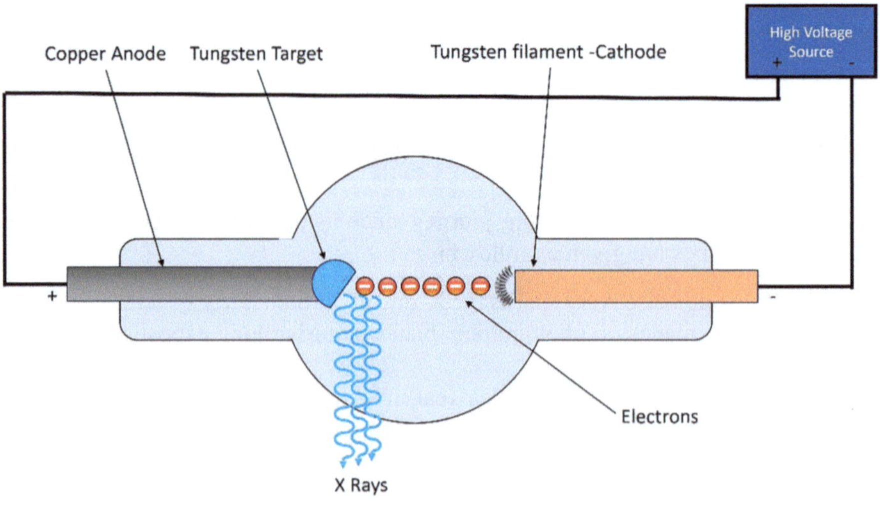

Fig. 5.3 X ray tube

target, typically made of a high-atomic-number metal like tungsten, is where the X-ray generation happens. Electrons from the cathode accelerate towards the anode at high speeds due to the potential difference created by a high voltage transformer.

5.2.5.1.1 Functionality

When the electrons strike into the anode's target, their kinetic energy is abruptly converted into X-ray photons and heat (William et al., 2015). The energy of the X-rays depends on the voltage applied to the X-ray tube (higher voltage = higher energy X-rays) (Bhargava et al., 2013). The remaining heat is dissipated through a cooling system within the anode to prevent damage (William et al., 2015).

5.2.5.2 High Voltage Transformer

The high voltage transformer plays a critical role in X-ray generation. It takes a relatively low input voltage and steps it up to a much higher voltage (typically tens of thousands of volts) required for efficient X-ray production (Bhargava et al., 2013). This high voltage creates the necessary acceleration force for the electrons, ultimately influencing the energy of the generated X-rays (William et al., 2015).

5.2.5.3 Control Panel

The control panel acts as the central command center for the X-ray machine. It allows the operator to (Bushberg 2011):

- Set the desired X-ray tube voltage and current, thereby controlling the energy and intensity of the X-ray beam.
- Adjust exposure time, determining the duration of X-ray emission.
- Select imaging parameters specific to the examination being conducted.
- Monitor system status and ensure safe operation.

5.2.5.4 Collimator

The collimator is a device that controls the shape and direction of the X-ray beam. It typically consists of lead or tungsten plates with strategically placed slits or apertures (Bhargava et al., 2013). By adjusting the collimator, the operator can focus the X-ray beam onto the specific area of interest, minimizing unnecessary radiation exposure to surrounding tissues (Nickoloff & Boone, 2012).

5.2.5.5 Beam Filtration System

The beam filtration system is another crucial component for ensuring image quality and patient safety. Filters are placed in the path of the X-ray beam to remove unwanted low-energy X-rays (William et al., 2015). These low-energy X-rays contribute little to image formation and can unnecessarily increase radiation dose to the patient (Bhargava et al., 2013).

5.2.6 X-Ray Image Acquisition and Processing

X-ray imaging is a cornerstone of medical diagnosis, allowing us to visualize internal structures. This process involves converting the invisible energy of X-rays into clear digital images. This conversion relies on a well-coordinated interplay between the physics of X-ray generation and the technology of image capture and processing.

5.2.6.1 X-Ray Acquisition

The process begins within the X-ray tube. As described in prior sections, electrons slam into the anode target, generating X-rays. These X-rays then travel outwards, interacting with the patient's body. Denser tissues, like bones, absorb more X-rays, while softer tissues like muscles allow them to pass through more readily (Bushberg 2011). This differential absorption forms the foundation of the image. On the opposite side of the patient lies the X-ray detector, a digital imaging panel or film cassette. The first X-ray detector employed by Roentgen in 1895 was a sheet of paper coated with barium platinocyanide. This chemical glowed when struck by X-rays, revealing the existence of these previously unknown rays (Haff & Toyofuku, 2008). In digital detectors, a scintillator layer converts the incoming X-rays into visible light photons (Berger et al., 2018). These light photons are then captured by a photodiode array, transforming them into an electrical signal. The intensity of the signal corresponds to the number of X-rays that passed through a particular region of the patient, reflecting the varying tissue densities (Berger et al., 2018). This electrical signal is then digitized and converted into a two-dimensional image by the X-ray system's computer.

5.2.6.2 Image Processing

Due to inherent limitations in detector response, the initial digital image may not directly translate to optimal diagnostic information. Image processing techniques are employed to refine and enhance the information content within the image, facilitating accurate diagnosis. Common processing steps include: *Noise Reduction*: X-ray images can be susceptible to electronic noise, appearing as grainy artifacts. Noise reduction algorithms help to remove this unwanted noise, improving image clarity (Ren et al., 2023). *Contrast Enhancement*: Subtle differences in tissue density can be difficult to discern in the raw image. Contrast enhancement techniques manipulate the pixel values to amplify these differences, making anatomical structures stand out more clearly (Choudhary & Gawade, 2016). *Spatial Filtering*: Sharpening filters can be applied to accentuate edges and boundaries within the image, further aiding in the visualization of anatomical details (Gonzalez & Woods, 2008). These processing steps, often automated by the X-ray system software, play a vital role in transforming the raw data into a high-quality image that radiologists can use for accurate diagnosis.

5.2.6.2.1 Image Quality Parameters

Three parameters define optimal X-ray image quality: spatial resolution, which determines the ability to distinguish fine details; signal-to-noise ratio (SNR), indicating the clarity of the image; and contrast, representing the difference in X-ray absorption between the target and its environment.

5.2.7 Applications of X-Ray in Non-destructive Food Quality and Analysis

In the domine of food quality control, ensuring consumer safety and product integrity is paramount. X-rays, familiar from medical imaging, are proving to be a valuable tool in this arena. Their ability to penetrate food products without causing damage makes them ideal for non-destructive quality assessment (Nicolaï et al., 2022). Scientific literature reveals a range of applications for X-ray technology in this domain. Figure 5.4 shows an X ray imaging system of food using a conveyor. The application of X ray in foods is shown in Table 5.1.

One key application lies in detecting foreign objects and contaminants within food. X-rays interact differently with materials of varying densities. Denser materials, like metal fragments or glass shards, will absorb more X-ray energy compared to the surrounding food. This difference in absorption is translated into an image, allowing us to pinpoint the location of these contaminants (Haff & Toyofuku, 2008).

Furthermore, X-ray imaging can assess internal defects within food products. Fruits, for instance, can be inspected for bruises or internal cracks, while nuts can be screened for voids or insect damage—all invisible to the naked eye (Kotwaliwale et al., 2014). X-ray technology can even be used to determine a food item's ripeness or measure its fat content based on subtle variations in density (Kotwaliwale et al., 2014).

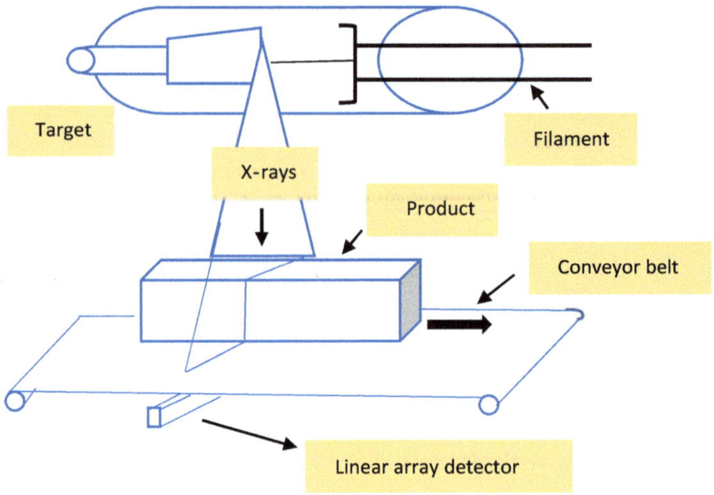

Fig 5.4 X-ray imaging of food on conveyor

Table 5.1 Application of X-rays in foods

X-ray medium/detector type/ operating conditions	Application/ commodity	Objective of imaging & results	References
Soft X-ray & Semiconductor detector	Pomegranate	Determination of internal defects in pomegranate	Payel and Sunil (2014)
Soft X-ray & Operating condition at 30 °C and 10 °C	Wheat	Detection of fungal infection in wheat	Narvankar et al. (2009)
X-ray & Operating condition at 20 °C and 55% RH	Crambe abyssinica seed	Determination of seed quality	de Medeiros et al. (2021)
X-ray & operated at ambient temperature	Fried chicken nuggets	Determination of fat distribution in fried food	Adedeji (2011)
X-ray micro-CT & Operated at room temperature	Meat, fish, chicken	Freezing visualization of foods	Mousavi et al. (2007)
Soft X-ray & digital X-ray images	Wheat	Inspection of wheat for infestation by the granary weevil.	Haff and Slaughter (2004)

Food producers can ensure consistent quality throughout their products, minimizing waste and safeguarding consumer health by implementing X-ray based techniques. However, it is important to acknowledge that X-rays are ionizing radiation, and proper safety protocols need to be followed during operation (European Food Information Council (EUFIC). https://dl.cdn-anritsu.com/anritsu-infivis/en-us/XRayPDFs/Anritsu_X-Ray_101_Presentation.pdf). Nevertheless, the research community is actively developing new X-ray applications for food quality assessment, making it a promising technology for the future of food safety.

5.3 Computer Tomography

CT or CAT scan (Computed Axial Tomography) is a non-invasive diagnostic technique employed to provide cross sectional images of the human body. The words tomography is derived from the 'tomos', which is Greek for 'slice', and 'graphos' for 'imaging'. It is based on the having of a sequence of X-ray pictures taken round the body and the application of computers to produce cross sectional images (slices) of the bones, blood vessels and the other soft tissues in the body.

The first useful Computed Tomography (CT) scanner was developed in 1971 by Sir Godfrey Hounsfield an electrical engineer, and by Dr. Allan Cormack, a South African medical physicist. Sir Godfrey Hounsfield initiated the design of the CT scanner in 1967 at EMI Laboratories, UK, and in 1969, he had a primitive model of the device working. The first clinical CT scan was performed on October 1, 1971, on a patient at Atkinson Morley's Hospital in London who had a suspected brain tumour (Thomas, 2012). This initial scan was of the human brain, marking a significant advancement in diagnosing brain tumours and other neurological conditions, providing much clearer images than conventional X-rays. In recognition of their groundbreaking work. Hounsfield and Cormack shared the Nobel Prize in Physiology or Medicine in the year 1979. Additionally, Hounsfield Units (HU), used in CT imaging to quantify radiodensity, were named in honour of Sir Godfrey Hounsfield. The invention of the CT scanner revolutionized medical imaging, allowing for much more detailed and accurate diagnosis of a wide range of medical conditions (Bushong, 1988).

Two principle goals were defined for the development of the CT scanning technique. First, during the examination, there is a radiographic tube that produces X-rays which, while rotating axially around the patient, transmit the rays and an array of detectors placed on the other side of the patient. Second, the digitized data from the detectors is fed into mathematical sets of instructions or computer algorithms that produce kinetic axial tomographic images of the body. This technique was originally referred to as computer-assisted tomographic scanning, or CAT scanning (Friedland & Thurber, 1996). CT scanning has allowed us to view structures that are invisible on traditional radiographs by revealing very subtle differences in soft-tissue density, even those less than 0.5%. However, its spatial resolution is not as high as that of conventional radiographs. This is notable because many experts initially thought that improving the spatial resolution of radiographs would address the issue of inaccessibility. Only a few researchers understood that the solution lay in a different approach: increasing the contrast in the radiographic image depending on the density of the soft tissues.

5.3.1 CT Theory

Traditional computed tomography (CT) involves the rotation of an X-ray tube around a sample, with the resulting attenuation recorded on a detector. The spatial resolution is limited by the X-ray tube's focal point, while the actual resolution depends on the sample's size and magnification. Several different exposures are obtained around the perimeter of the sample, often at 180-degree intervals. They usually incorporate charge-coupled device (CCD) camera to improve the radiographic images. From these projections, 2D slice images are generated, which can be superimposed to create a 3D volume (Cosmi & Bernasconi, 2013; Wang, 2004). The CT image formation process consists of three stages: image scanning, reconstruction, and display. In CT images, voxels represent the basic structural units, while pixels are the fundamental units displayed on the screen. The clarity of the image is determined by the resolution of both voxels and pixels.

5.3.2 Some Key Concepts of CT

- Attenuation: When X-rays pass through different tissues, they are absorbed or scattered to varying degrees. This is known as attenuation. Different tissues have different attenuation coefficients, which allows CT to differentiate between various structures.
- Hounsfield Units (HU): To quantify the attenuation of different tissues, Hounsfield Units (HU) are used. Water is assigned a value of 0 HU, while bone has a much higher value, typically around 1000 HU.
- Slice Thickness: CT scans are typically performed in a series of slices or cross-sections. The thickness of these slices can be adjusted to provide different levels of detail. Thinner slices offer higher resolution but may require longer scan times.
- Windowing: To optimize the visualization of specific tissues or organs, CT images can be windowed. This involves adjusting the display parameters to enhance the contrast of the desired structures.

5.3.3 CT Equipment

CT Equipment consists of following components: *X-ray Tube*: Generates high-energy X-ray radiation that passes through the patient. *Detectors*: Measure the intensity of X-rays that pass through the patient. *Patient Table*: Accompanies the patient through the scan and manipulates the scanner. *Gantry*: Houses the X-ray tube and detectors and rotates around the patient. *Computer System:* Controls the scanner's operation, processes data, reconstructs images, and stores and displays images.

The primary instrumentation features for producing high quality images are collimators filters, grid, and cooling system. Collimators to narrow the X-ray beam to improve image quality and reduce scatter radiation. Filters to remove low-energy X-rays to improve image contrast. Grid to absorbs scattered radiation to reduce image artifacts. Cooling System to cool the X-ray tube to prevent overheating.

5.3.4 Developments in CT Equipment

Early CT Scanners (1970s) are Single-slice scanners with limited image quality and long scan times. Multi-Slice CT Scanners (1990s) consists multiple detector rows for faster scan times and improved image quality. Helical CT Scanners (1990s) consists continuous table movement for reduced artifacts and improved image quality. Cone-Beam CT Scanners (2000s) consists wider X-ray beam for faster scan times and improved image quality. Dual-Source CT Scanners (2000s) consists two X-ray sources and two detector rings for even faster scan times and improved image quality. Flat-Panel CT Scanners (2010s) consists Solid-state detectors for high sensitivity and fast response times. AI-Powered CT Scanners (2020s) incorporate artificial intelligence for improved image quality, reduced artifacts, and assisted diagnosis.

5.3.5 Instrumentation & Image Acquisition of CT

The fundamental principle behind CT is the ability to reconstruct a 3D image from multiple 2D projections. This is achieved through the following steps:

5.3.5.1 X-Ray Source

A CT scanner consists of an X-ray source that emits a narrow beam of X-rays and a ring of detectors positioned opposite the source. The X-ray source in a CT scanner is typically a rotating anode tube. This type of tube consists of a cathode, which emits electrons, and a rotating anode, which is a target for the electrons. When the electrons strike the anode, X-rays are produced through a process called bremsstrahlung as explained in previous sections.

5.3.5.1.1 Characteristics of an X-Ray Source in CT

High Heat Capacity: The anode must be able to withstand the high temperatures generated during X-ray production, as the electrons strike it at high speeds. This is achieved through rapid rotation of the anode, which helps to spread the heat over a larger area. *High X-ray Output:* The X-ray source must produce a high intensity of X-rays to ensure adequate signal-to-noise ratio in the images. This is achieved by using high voltages and currents to accelerate the electrons. *Collimation:* The X-ray beam is collimated to produce a narrow, fan-shaped beam that passes through the patient. This helps to reduce scatter radiation and improve image quality.

5.3.5.2 Detectors

CT detectors are responsible for measuring the intensity of X-rays that pass through the patient. They are arranged in a ring around the patient and are synchronized with the rotation of the X-ray source. X-ray detection in CT systems involves the conversion of high-energy X-ray quanta into lower-energy forms, such as optical photons or electron-hole pairs. This down-conversion process generates numerous secondary quanta, typically thousands for each primary X-ray quantum. Detector materials like phosphors, scintillating ceramics, or pressurized xenon gas are employed for this conversion. These materials ultimately produce an electrical signal, either a current or voltage. Electronic amplifiers condition this signal, and an analog-to-digital converter (ADC) transforms it into a digital number. The dynamic range of signals produced in CT is substantial. It spans from the nearly complete transmission of X-rays through air to their significant attenuation by large patients with metal implants. This range can be as wide as a factor of almost 105. To ensure accurate image reconstruction, even at the lowest signal levels, the ADC must be capable of detecting subtle modulations of a few percent. Therefore, the overall dynamic range of CT detectors approaches a factor of one million, equivalent to a 20-bit ADC.

5.3.5.2.1 Types of CT Detectors

Ionization Chambers: These detectors are filled with a gas that is ionized by the incoming X-rays. The ionization current is then measured to determine the X-ray intensity. *Scintillation Detectors:* These detectors use a scintillator material that emits light when struck by X-rays. The light is then detected by a photomultiplier tube or a solid-state detector. *Semiconductor Detectors:* These detectors use a semiconductor material that generates an electrical signal when struck by X-rays.

5.3.5.2.2 Characteristics of CT Detectors

High Sensitivity: Detectors must be able to detect low levels of X-ray radiation to ensure accurate image reconstruction. *Fast Response Time:* Detectors must be able to respond quickly to changes in X-ray intensity to capture the rapidly rotating X-ray beam. *Low Noise:* Detectors must have low noise levels to minimize image artifacts and improve image quality.

5.3.5.3 Data Acquisition

As the X-ray source rotates around the patient, the detectors measure the intensity of the X-rays that pass through the body. Data acquisition is a critical step in the CT imaging process, involving the collection of X-ray attenuation data from multiple angles around the patient. This data is then used to reconstruct the 3D image. The efficiency and accuracy of data acquisition significantly impact the quality of the CT scan. Data Acquisition Process involves the following steps. *Beam Attenuation*: The X-ray beam passes through the patient's body. Different tissues attenuate the X-rays to varying degrees, depending on their density and composition. *Detector Measurement*: A ring of detectors positioned opposite the X-ray source measures the intensity of the X-rays that pass through the patient. *Data Collection*: The detector measurements are recorded for each rotation angle, creating a set of projections.

5.3.5.3.1 Factors Affecting Data Acquisition

Rotation Speed: The speed at which the X-ray source rotates determines the scan time. Faster rotation speeds can reduce scan time but may compromise image quality if the detectors cannot keep up with the data acquisition rate. *Detector Sensitivity:* The sensitivity of the detectors determines the accuracy of the attenuation measurements. More sensitive detectors can improve image quality but may also increase noise. *Collimation:* The collimation of the X-ray beam affects the spatial resolution of the images. Narrower collimation can improve spatial resolution but may also increase image noise. *Pitch:* The pitch of the CT scan is the ratio of table movement to X-ray tube rotation. A higher pitch reduces scan time but may also reduce image quality, especially for thin structures. *Reconstruction Algorithm:* The reconstruction algorithm used to convert the projection data into an image affects the quality of the CT scan. Advanced reconstruction algorithms can improve image quality but may require more computational resources.

5.3.5.3.2 Data Acquisition Techniques

Helical CT: In helical CT, the patient table moves continuously during the scan, allowing for faster acquisition times. *Spiral CT*: Similar to helical CT, spiral CT involves continuous table movement. However, the X-ray source rotates at a constant speed, resulting in a spiral trajectory. *Cone-Beam CT*: Cone-beam CT uses a wider X-ray beam, allowing for faster scan times and improved image quality for certain applications. *Dual-Source CT*: Dual-source CT scanners have two X-ray sources and two detector rings, enabling faster scan times and improved image quality for some applications.

5.3.5.4 Image Reconstruction/Image Processing

5.3.5.4.1 X-Ray Signals

X-ray imaging is a technique that uses X-rays as tools for making internal structures of the object being visualized to be visible. It involves production of beam of X-rays, transmission of the beam through the matter of the object and finally, the measurement of the energy of X-rays that comes out of the object. The penetration of X-rays actually through an item depends on the atomic contacts within an object. Every molecule in the object has a particular cross-section through which an X-ray will interact and hence reduce the flux of the X-ray beam passing through the object.

The Lambert-Beer law quantifies this attenuation, expressing the transmission probability as an exponentially decreasing function of the material's thickness and attenuation coefficient. The attenuation coefficient is a measure of how effectively a material absorbs X-rays.

A mathematical algorithm, typically based on the Radon transform, is used to reconstruct the 3D image from the collected data. This algorithm essentially reverses the process of X-ray attenuation to create a cross-sectional view of the body.

5.3.5.4.2 Image Reconstruction from Two-Dimensional Projection Data

CT image reconstruction involves reconstructing a two-dimensional image section from projection measurements. This process entails collecting line measurements from various angles around the patient. These measurements are stored in a sinogram, which represents the raw projection data. The sinogram is subsequently processed to derive the original image attenuation. This involves applying mathematical techniques, such as filtered back projection, to reconstruct the image from the projection data. Filtered back projection involves summing a weighted combination of the projection measurements to calculate the attenuation at each point within the field of view.

5.3.5.4.3 Image Reconstruction from Three-Dimensional Projection Data

For helical CT, which acquires three-dimensional volumes, specialized reconstruction algorithms are necessary. These algorithms, such as single-detector row spiral CT and multidetector row spiral CT, address the challenges associated with reconstructing images from data acquired across multiple planes.

Single-detector row spiral CT: This technique involves interpolating missing gantry measurements to create a complete set of projection data for each image plane. Two common interpolation methods are 360LI and 180LI.

Multidetector row spiral CT: This technique involves reconstructing images from data acquired by multiple detector rows. It employs methods such as z-interpolation or z-filtering to account for the cone-beam angle between detector rows.

5.3.5.5 Applications of CT in Non-destructive Food Quality and Analysis

Computer Tomography (CT) has emerged as a powerful tool for non-destructive food quality analysis due to its ability to provide detailed 3D images of internal structures without compromising the integrity of the product. CT applications in food are listed in Table 5.2.

Internal Defects: Void Detection: CT has been successfully used to detect internal voids in fruits and vegetables, such as those caused by diseases or pests. Crack Detection: CT can identify cracks or fissures within food products, which can affect their shelf life and quality. Foreign Object Detection: CT is highly effective in detecting foreign objects, including metal, glass, or stones, that may be present in food products.

Fruit Maturity: Internal Structure Analysis: CT can assess the internal structure of fruits, such as the presence and size of seeds, to determine maturity level. Density

Table 5.2 Computed tomography applications in food products

CT medium/detector type/operating conditions	Application/commodity	Object and objective of imaging & results	References
Vortex Automated Single Pass CT	Caryopses of sorghum grain	Phenotypic variability was observed significant for all the traits such as embryo volume, endosperm hardness, endosperm texture, endosperm volume, pericarp volume, and kernel volume.	Crozier et al. (2022)
CT using Algorithms	Pear fruit (Conference & Cepuna)	Detection of internal distortion	Van De Looverbosch et al. (2020)
X-ray Computed Tomography	Apples (Braeburn)	Detection of browning and assessing the storage conditions of the apples	Van Dael et al. (2019)
X-ray Computed Tomography	pomegranate	Investigated structural arrangements and the proportion of edible to inedible parts of pomegranate fruit.	Magwaza and Umezuruike (2014)
CT with RT 16 Elite, multi-detector	Asparagus	Classification of asparagus according to their undesirable fibres and overall quality analysis was studied	Donis-González et al. (2016)
X-ray microtomography (μCT)	Italian biscuits and breadstick varieties	Overall microstructure of the products was studied	Frisullo et al. (2010)

Measurement: CT can measure the density of fruits, which is often correlated with maturity and quality.

Moisture Content: Attenuation Coefficient Analysis: CT can estimate the moisture content of food products based on their attenuation coefficients. Density-Moisture Relationship: By combining density measurements with other parameters, CT can accurately determine moisture content

Density and Volume: Accurate Measurements: CT provides precise measurements of density and volume, which are essential for quality control and packaging applications.

Food Composition: Component Distribution: CT can visualize the distribution of different components within food products, such as fat, water, and solids.

Texture Analysis: CT can assess the texture of food products based on their internal structure and density.

Process Monitoring: Cooking and Drying: CT can monitor the cooking and drying processes to ensure uniform heating and moisture removal. Freezing: CT can assess the quality of frozen food products, including the presence of ice crystals and the extent of freezing.

5.4 Magnetic Resonance Imaging

MRI is an imaging technique used primarily with an aim of attaining high resolution inside the object in two or three planes. Due to the rising use of medical imaging systems during the 1980s, it has been accepted internationally as being one of the most effective tools for diagnosing and evaluating clinical conditions. The authors point that the magnetic resonance imaging is the window on the internal condition of bodies of the agricultural substance and foodstuff that enables one to conduct over them. But none of the image windows displays all the information. With both methods, the dependence of the image quality and visibility of structure is very wide substantial bearing in mind the characteristics of the imaging involved. Segments will be exposed to radiation and the time will be taken to complete imaging.

5.4.1 Principle

The MRI phenomenon and its use in studying the food materials, wherein the macroscale characteristics are usually, a function of the intrinsic response to the magnetic properties of the specific atomic nuclei in the field. For instance, a hydrogen nucleus is a proton that behaves like a charged particle in circular motion and one that has spin and rotational motion results in the formation of a polar field (Kauten & McCarthy, 2005). Food is divided into liquids to solutions, gels and other forms like emulsion, foam, suspensions, and solid depending on the content of water or from the chemical composition aspect. Food is made up of water, carbohydrate lipids and proteins.

Thus, MRI being an analytical technique, has its root in absorption and emission of energy in the radio frequency band, which is a segment of the electromagnetic spectrum. The generation of the radio frequency can manipulate the magnetic resonance images which is used to excite the nuclear spins for the purpose of acquiring the image. The use of these parameters for the pulse sequence results in better images with higher accentuation (McCarthy, 1994). However, as indicated by the researchers, although using RF field at the resonant frequency; that is the electric field or the magnetic field at the frequency that makes the spinning nuclei in their equilibrium magnetic moments and to transmit the signal which is a function of that line integrals of the magnetic resonance of the object. This radiation describes the frequencies distribution, and if applying Fourier transform to these signals, one can get the image of freedom of orientation of the magnetization (Rinck, 2001).

5.4.2 *Components of MRI*

A typical MRI system is shown in Fig. 5.5. The three main components of MRI system are:

1. Large Magnets

   Magnets are the heart of the system. The MRI magnet gives a completely uniform magnetic field output. By analysing the basic concept and the working principle of MRI, it is concluded that MRI is not only a powerful diagnostic tool but it is also a technology that gives a completely uniform magnetic field output. This can be achieved theoretically by creating a surface current with spherical symmetry and having sinusoidal spatial dependence (Jin, 1998). The primary coil is expected to generate a large magnetic field (B0) that should be uniform across the volume of interest. It is always advantageous to be able to have a strong magnetic field as it promises a better signal to noise ratio (SNR) in spatial domains. When magnets cannot deliver the needed results as mentioned above, magnets can be used to treat a variety of health conditions, although, in some instances, the magnets may not be effective in achieving the wanted outcomes.

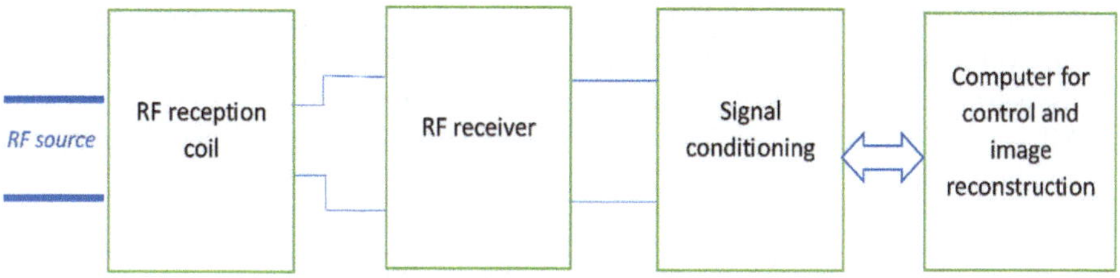

Fig. 5.5 Block diagram of a typical MRI system

Clinical imaging systems normally do not operate at a field strength above 3 Tesla. However, there are some experimental systems that are said to possess main magnets of 4T or 7T.

2. Coil (Gradient Coil)

   The gradient coils are used to encode spatially the positions of the nuclei spins in the body of MRI System by altering the value of the local magnetic field which produce precessional frequencies that vary with rank and position (Jin, 1989). A gradient coil is made of three coils that produce the x, y and z gradient. The field that the gradient coil is going to create should be ideally a linear field within the imaging volume. The gradient coils are switched on and off more rapidly for imaging applications require here more swiftly.

3. RF receive and RF transmit coils

   In MRI, the coils that are used to excite the magnetization are known as RF coils, but are also called transmit coils. For the signals originating from the relaxing nuclei spins there are two clear receiver coils. A single coil may be used for transmitting signals and for receiving signals also. To boost the SNR, there are different coils in most of the situations. Its use depends on properties which are suitable for each mode of operation in a particular environment. Transmit coils are also applied to generate a uniform B1 field over the region of interest, while receive coils require it.

MRI Techniques Some types of MRI techniques used in food analysis are:

- Standard MRI: This is the conventional MRI used in visualizing the internal structure of the food items to determine their texture, moisture content and fat distribution.
- Magnetic Resonance Imaging (MRI) T1 and T2 Relaxation Times: These techniques determine the time taken by protons of the various food components to regain their equilibrium once they have been knocked off balance by a magnetic field. It can be used to distinguish between various food types depending on its moisture and fat content.
- Diffusion-Weighted MRI: This method helps in determining the diffusion of water molecules in the products to determine its microstructure and qualities especially the fruits and vegetables.
- Chemical Shift Imaging: This technique also allows for differentiation between different fat types and water content and is essential for quality assessments of meat and dairy products.
- Functional MRI (fMRI): fMRI is now used mainly to investigate brain functioning, yet its fundamentals can be applied in analysing the metabolism of food and measuring its freshness and quality at different stages.
- MRI-based Spectroscopy: This method enables one to assess specific chemical compounds.

Table 5.3 MRI application in food

Product	Application	Reference
Apple	Apple transformation mechanisms in cider-making technologies may be investigated using time-domain NMR and quantitative MRI. Assess water core distribution inside apple	Rondeau-Mouro et al. (2015), Melado-Herreros et al. (2013)
Watermelon	Study the internal quality of watermelon/melons	Sun et al. (2010)
Carrot	Studied the effect of freeze-drying on micro-structure and rehydration traits of carrots	Voda et al. (2012)
Cucumber/ Courgette	To determine internal freeze damage	Kotwaliwale et al. (2014)
Pork	Determine intramuscular and subcutaneous fat in pork carcass	Monziols et al. (2006)
Chicken	Determine the size and volume of the chicken carcasses and chicken breasts	Davenel et al. (2000)
Beef	Studied the effect of the denaturation of the proteins throughout the cooking.	Ishiwatari et al. (2013)

5.4.3 Applications

Magnetic resonance imaging seems to provide a unique opportunity for investigation of various food products, which makes it a suitable technique, preferred over other comparable methods. The MRI food applications are shown in Table 5.3. MRI offers excellent resolution and contrast that allows elucidating all the aspects of the internal structure, composition, and properties of the analysed foods without their destructiveness (Zhang et al., 2015, 2022).

This is one of the most important areas due to increasing concern of quality and safety in food processing industries. Raman imaging as a technique analogous to MRI has been widely applied for evaluation of quality, authentication, and pesticide residue analysis in foods (Yaseen et al., 2017). MRI can give important data on specific property of food material such as water content, fat distribution and micro structural changes which are hall marks of fresh food quality.

As for food analysis, MRI has been most advantageous in the examination of porous media which is characteristic of most foods. MRI has been used to investigate the flow behaviour and transport phenomena in porous media at high resolution for characterizing the food products during processing and storage (Jiang et al., 2013).

5.5 Conclusion

X-ray imaging is effectively applied in food safety when it comes to the detection of foreign objects such as metals, glasses, or bones in addition to imaging the inside of the food. This technique is fast and non-destructive, making it suitable for applications where multiple samples can be processed at one time, such as in an industrial setting. It can show density variations in food products, which is useful in analysing for packing or identifying spoilt foods, but it cannot offer chemical or molecular information. They may also cause deterioration of the quality of foods that are rich in certain sensitive ingredients by exposing them to X-rays for long.

MRI is particularly good at differentiating between various types of soft tissue and can provide non-destructive measurement of moisture, fat, and protein content. It enhances high-resolution internal imaging of food structures, textural properties, and even water distribution, which is important for food preservation. MRI machines are highly costly and are not commonly installed in large factories. Compared to other techniques, scanning can be slower and it is not the best method for finding dense foreign materials.

CT has a higher resolution in 3D images than MRI and is more suitable for studying the intricate structures of food commodities as well as identifying internal flaws such as air voids, cracks, or voids. It is especially useful in the analysis of food porosity, which has an impact on texture and shelf-life, and in determining product homogeneity. As with X-rays, CT is significantly more concerned with the physical structure than with what the material is made of. It is also more expensive and necessitates the use of more advanced software for reconstructing the image.

Each imaging method—X-ray, MRI, and CT—offers unique strengths for food analysis. X-ray imaging is best suited for rapid detection of foreign objects and density variations, MRI excels in non-invasive analysis of moisture, fat, and texture in soft foods, and CT provides detailed 3D insights into food structures. The choice of method depends on the specific food analysis goal, balancing factors like resolution, speed, and cost.

References

Adedeji, A. A., Ngadi, M. O. (2011). Microstructural Properties of Deep-Fat Fried Chicken Nuggets Coated with Different Batter Formulation. *International Journal of Food Properties, 14*(1) , 68–83. https://doi.org/10.1080/10942910903131423

Berger, M., Yang, Q., & Maier, A. (2018). X-ray imaging. In *Medical imaging systems: An introductory guide* (pp. 119–145). SpringerOpen.

Bhargava, S., Singh, G., & Gupta, P. (2013). X-ray generation and its applications. *International Journal of Engineering Research and Technology, 2*(4), 319–324.

Bushong, S. C., Goerner, F. (2012). Radiologic science for technologists *Elsevier Health Sciences*

Cosmi, F., Bernasconi, A. (2013). Micro-CT investigation on fatigue damage evolution in short fibre reinforced polymers. *Composites Science and Technology, 79*, 70–76

Bushberg, J. T. (2011). The essential physics of medical imaging. In J. T. Bushberg, X-ray Production, X-ray Tubes, and X-ray Generators (p. 171). *Lippincott Williams & Wilkins.*

Bushong, S. C. (1988). *Radiologic science for technologists. Andrew Allen*

Choudhary, R., & Gawade, S. (2016). Survey on image contrast enhancement techniques. *International Journal of Innovative Studies in Sciences and Engineering Technology, 2*(3), 21–25.

Crozier, D., Riera-Lizarazu, O., & Rooney, W. L. (2022). Application of X-ray computed tomography to analyze the structure of sorghum grain. *Plant Methods, 18*(1), 3.

Davenel, A., Seigneurin, F., Collewet, G., & Remignon, H. (2000). Estimation of poultry breast meat yield: Magnetic resonance imaging as a tool to improve the positioning of ultrasonic scanners. *Meat Science, 56*(2), 153–158.

de Medeiros, A. D., Bernardes, R. C., da Silva, L. J., de Freitas, B. A. L., dos Santos Dias, D. C. F., & da Silva, C. B. (2021). Deep learning-based approach using X-ray images for classifying Crambe abyssinica seed quality. *Industrial Crops and Products, 164*, 113378. ISSN 0926-6690. https://doi.org/10.1016/j.indcrop.2021.113378

Donis-González, I. R., Guyer, D. E., & Pease, A. (2016). Postharvest noninvasive classification of tough-fibrous asparagus using computed tomography images. *Postharvest Biology and Technology, 121*, 27–35. https://doi.org/10.1016/J.POSTHARVBIO.2016.07.012

European Food Information Council (EUFIC). (n.d.). *X-Ray in food inspection*. Retrieved from https://dl.cdn-anritsu.com/anritsu-infivis/en-us/XRayPDFs/Anritsu_X-Ray_101_Presentation.pdf

Friedland, G. W., & Thurber, B. D. (1996). The birth of CT. *AJR. American Journal of Roentgenology, 167*(6), 1365–1370.

Frisullo, P., Conte, A., & Del Nobile, M. A. (2010). A novel approach to study biscuits and breadsticks using X-ray computed tomography. *Journal of Food Science, 75*(6), E353–E358.

Gonzalez, R. C., & Woods, R. E. (2008). *Digital image processing* (3rd ed.). Pearson Prentice Hall.

Haff, R. P., & Slaughter, D. (2004). Real-time X-ray inspection of wheat for infestation by the granary weevil, Sitophilus granarius (L). *Transactions of the American Society of Agricultural Engineers, 47*, 531–537. https://doi.org/10.13031/2013.16022s

Haff, R. P., & Toyofuku, N. (2008). X-ray detection of defects and contaminants in the food industry. *Sensing and Instrumentation for Food Quality and Safety, 2*, 262–273.

International Atomic Energy Agency. (2009). *Industrial radiography using x and gamma rays* (IAEA Nuclear Safety Standards Series No. NS-R-2). Author. Retrieved from https://www-pub.iaea.org/MTCD/Publications/PDF/Pub1466_web.pdf

Ishiwatari, N., Fukuoka, M., & Sakai, N. (2013). Effect of protein denaturation degree on texture and water state of cooked meat. *Journal of Food Engineering, 117*(3), 361–369.

Jiang, L., et al. (2013). Measurement of two-phase flow in porous medium using high-resolution magnetic resonance imaging. *Chinese Journal of Chemical Engineering, 21*(1), 85–93.

Jin, J. (1989). *Electromagnetic analysis and design in magnetic resonance imaging*. CRC Press LLC.

Jin, J. M. (1998). Electromagnetics in magnetic resonance imaging. *IEEE Antennas and Propagation Magazine, 40*(6), 7–22.

Kauten, R., & McCarthy, M. J. (2005). Applications of NMR imaging in processing of foods. In A. G. Gaonkar (Ed.), *Food processing: Recent developments* (p. 2). Elsevier Science BV.

Kotwaliwale, N., Singh, K., Kalne, A., Jha, S. N., Seth, N., & Kar, A. (2014). X-ray imaging methods for internal quality evaluation of agricultural produce. *Journal of Food Science and Technology, 51*, 1–15.

Magwaza, L. S., & Umezuruike, L. O. (2014). Investigating non-destructive quantification and characterization of pomegranate fruit internal structure using X-ray computed tomography. *Postharvest Biology and Technology, 95*, 1–6.

McCarthy, M. J. (1994). *Magnetic resonance imaging in foods*. Chapman & Hall.

Melado-Herreros, A., Muñoz-García, M. A., Blanco, A., Val, J., FernándezValle, M. E., & Barreiro, P. (2013). Assessment of watercore development in apples with MRI: Effect of fruit location in the canopy. *Postharvest Biology and Technology, 86*, 125–133.

Monziols, M., Collewet, G., Bonneau, M., Mariette, F., Davenel, A., & Kouba, M. (2006). Quantification of muscle, subcutaneous fat and intramuscular fat in pig carcasses and cuts by magnetic resonance imaging. *Meat Science, 72*(1), 146–154.

Mousavi, R., Miri, T., Cox, P., & Fryer, P. (2007). Imaging food freezing using X-ray microtomography. *International Journal of Food Science & Technology, 42*, 714–727. https://doi.org/10.1111/j.1365-2621.2007.01514.x

Narvankar, D. S., Singh, C. B., Jayas, D. S., & White, N. D. G. (2009). Assessment of soft X-ray imaging for detection of fungal infection in wheat. *Biosystems Engineering, 103*(1), 49–56. ISSN 1537-5110. https://doi.org/10.1016/j.biosystemseng.2009.01.016

Nickoloff, E. L., & Boone, J. M. (2012). *Radiation interactions in medical and biological imaging*. Medical Physics Publishing.

Nicolaï, B., De Ketelaere, B., Dizon, A., Wouters, N., Postelmans, A., Saeys, W., et al. (2022). Nondestructive evaluation: Detection of external and internal attributes frequently associated with quality and damage. In *Postharvest handling* (pp. 399–433). Academic.

Payel, G., & Sunil, C. K. (2014). Quality analysis of pomegranate by non-destructive soft X-ray method. *Journal of Food Processing & Technology, 5*(341), 2.

Ren, K., Gu, Y., Luo, M., Chen, H., & Wang, Z. (2023). Deep-learning-based denoising of X-ray differential phase and dark-field images. *European Journal of Radiology, 163*, 110835.

Rinck, P. A. (2001). *Magnetic resonance in medicine*. Blackwell.

Rondeau-Mouro, C., Deslis, S., Quellec, S., Bauduin, R. (2015). Assessment of TD-NMR and quantitative MRI methods to investigate the apple transformation processes used in thecider-making technology. *Magn. Reson. Food Sci*, 127–140

Röntgen, W. C. (1896). On a new kind of rays. *Science, 3*(59), 227–231.

Saran, R., & Fahim, I. (2020). X-ray generation and manipulation. *Nature Reviews Physics, 2*(2), 130–144.

Sprawls, P. (1994). Physical principles of x-ray imaging. *Radiologic Technology, 65*(2), 107–125.

Sun, T., Huang, K., Xu, H., & Ying, Y. (2010). Research advances in nondestructive determination of internal quality in watermelon/melon: a review. *Journal of Food Engineering, 100*(4), 569–577.

Thomas, A. M. (2012). The developmet of computed-assisted tomography. In G. Thompson (Ed.), *Nobel Prizes that changed medicine* (pp. 151–170). Imperial College.

Van Dael, M., et al. (2019). Combination of shape and X-ray inspection for apple internal quality control: In silico analysis of the methodology based on X-ray computed tomography. *Postharvest Biology and Technology, 148*, 218–227.

Van De Looverbosch, T., et al. (2020). Nondestructive internal quality inspection of pear fruit by X-ray CT using machine learning. *Food Control, 113*, 107170.

Voda, A., Homan, N., Witek, M., Duijster, A., van Dalen, G., van der Sman, R., et al. (2012). The impact of freeze-drying on mi- crostructure and rehydration properties of carrot. *Food Research International, 49*, 687–693.

Wang, M. P. (2004). CT detection technology *Shanghai, China: Fudan University Press* , 10–35.

William R. Hendee, Murray Eisenberg, David R. McCoubrey (2015) Radiation Physics for Medical Imaging *Fourth Edition. Medical Physics Publishing*

Yaseen, T., Sun, D.-W., & Cheng, J.-H. (2017). Raman imaging for food quality and safety evaluation: Fundamentals and applications. *Trends in Food Science & Technology, 62*, 177–189.

Zhang, Y., et al. (2015). Preclinical diagnosis of magnetic resonance (MR) brain images via discrete wavelet packet transform with Tsallis entropy and generalized eigenvalue proximal support vector machine (GEPSVM). *Entropy, 17*(4), 1795–1813.

Zhang, L., et al. (2022). Motion correction in MRI using deep learning and a novel hybrid loss function. https://doi.org/10.48550/arXiv.2210.14156.

Chapter 6
Hyperspectral Imaging

Princess Tiffany D. Mendoza, Paul R. Armstrong, and Kaliramesh Siliveru

Recent changes in consumer preferences and pressing health and environmental issues are among the major factors influencing the growing demand for high quality foods (Petrescu et al., 2019). With this demand, more technologies are being developed to produce food with high nutritional quality, maintain product quality and safety from production to distribution, and meet the required food standards in inspection and quality control (Qu et al., 2015; Wu & Sun, 2013). Consequently, non-destructive sensing technologies are also being explored for accurate, reliable, efficient, rapid, and cost-effective evaluation of food quality.

Traditional methods for measuring quality parameters involve wet chemistry techniques which are destructive, slow, time-consuming, and costly. Near-infrared (NIR) spectroscopy and computer vision are sensing techniques that allow fast and non-destructive quality measurement of various food products whether for on-/at-line or off-line applications. In NIR spectroscopy, the information about the sample chemical composition will be obtained based on the absorbed light of the sample. Meanwhile, computer vision uses physical attributes and image analysis to measure different physical characteristics of the food. Combining both spectroscopy (chemical composition) and imaging (spatial features) techniques, hyperspectral imaging (HSI) allows simultaneous assessment and spatial mapping of the different components within the product. Initially designed for remote sensing applications, hyperspectral imaging has evolved from a promising laboratory technique to an effective tool commercially applied for food quality analysis. It has caught enormous interest from academia and industry as it has been widely explored in the agricultural, food, and pharmaceutical industries for several years (Lind & Murhed, 2012).

P. T. D. Mendoza · K. Siliveru (✉)
Department of Grain Science and Industry, Kansas State University, Manhattan, KS, USA
e-mail: kaliramesh@ksu.edu

P. R. Armstrong
USDA-ARS, Center for Grain and Animal Health Research, Manhattan, KS, USA

C. K. Sunil et al. (eds.), *Emerging Non-destructive Technologies for Food Quality Analysis*, https://doi.org/10.1007/978-3-031-95048-3_6

Hyperspectral imaging, which is also referred to as spectral imaging, is now being used for plant phenotyping, for field applications while mounted on an unmanned aerial vehicle (UAV) and in combination with other sensors (i.e., LiDAR, RGB, and/or thermal), and for commercial food processing lines. Over the years, it has been explored and successfully applied to actual operations.

A major advantage of hyperspectral imaging is its non-destructive and non-invasive approach, allowing the further use of the samples. Breeders can use the seeds as plant materials and food processors can use the sample for processing. The samples can be utilized for the assessment of other parameters as the integrity of the sample composition is still intact. Second, it requires minimal to no sample preparation. The sample can be analyzed as-is and in real time. It also allows simultaneous assessment of different parameters, eliminating the use of multiple instruments and a large number of samples. Since it is fast and efficient, it is applicable for high-throughput analysis. Moreover, it is cost-effective as compared to the traditional wet chemistry methods, saving time, labor, and other resources. Its major advantage over single-point NIR spectroscopy is it allows visualization of the distribution of chemical composition in a sample which aids in spotting areas of high or low concentrations or sections with defects and contamination.

Hyperspectral imaging has offered numerous benefits for an extensive range of agricultural and food products and diverse applications from postharvest to processing to consumer distribution. Several instrument manufacturers have developed spectral imaging instruments specific to the needs of the customers and tailored to fit their specific products, in combination with advanced machine learning algorithms, as a portable and practical device or for automatic sorting and quality control in processing lines. However, this technology is still facing challenges and limitations. Compared to conventional spectroscopy, which is a standard analytical method, spectral imaging is not yet considered a standard in food quality applications. Among the major challenges are the instrument cost, complexity in data processing and analysis, non-homogenous food samples with difficulty in acquiring reference data and achieving representative sampling, and the lack of standard guideline methods for model development and validation.

6.1 Theory and Principle of Hyperspectral Imaging

6.1.1 Spectroscopy and Imaging

The interaction between light and the molecules in the sample serves as the core principle in vibrational spectroscopy, such as near-infrared spectroscopy. The sensitivity of NIR spectroscopy to the C–H, N–H, and O–H absorption bonds, which are related to the food components, produces a resultant spectrum that captures the molecular information associated with specific wavelengths (McClure, 2006). By measuring the average absorption spectrum of the whole sample, NIR instruments can relate this data to the sample's chemical composition with the aid of chemometrics and multivariate analysis. From its first application in food and agriculture in the 1960s, NIR spectroscopy is now widely used not only in the laboratory but also in the field, in grain elevators, and in processing plants for diverse applications in food and agriculture (Delwiche, 2004; Williams & Norris, 2001). NIR spectroscopy is currently used as a standard method to measure grain moisture, protein, and oil contents (AACC, 1999). Despite its wide range of applications in food and agriculture, NIR spectroscopy is still limited by its scope of measurement, which is challenging for samples that are not homogenous or when contaminants are present in small sections of the food (Lu & Park, 2015).

Utilizing the red, green, and blue (RGB) bands, a computer vision determines the characteristics of a sample, including color, shape, size, and texture. It has been useful in commercial applications to replace human visual inspection and allow automation, particularly to detect foreign materials and classify food according to its physical characteristics. With the recent updates on artificial intelligence in computer vision applications, this technology has been applied extensively in food quality applications (De Baerdemaeker, 2023). However, it faces challenges in classifying samples with the same colors, detecting internal defects that are not visible to the human eyes, and measuring chemical and biological sample characteristics (ElMasry & Sun, 2010).

The combination of spectroscopy, statistics, and imaging techniques has resulted to the development of spectral imaging which surpasses the limitations and challenges of those two methods and provides more applications in food and agriculture.

6.1.2 Fundamentals of Hyperspectral Imaging

A hyperspectral image, which is also referred to as hypercube, is a three-dimensional dataset consisting of two spatial dimensions and one spectral dimension (Lind & Murhed, 2012), such that each pixel in the image has a spectrum across the wavelength range. As compared to an RGB image that has three bands (red, green, and blue) and a multispectral image that has few discrete wavebands (typically less than 10), a hyperspectral image has multiple contiguous bands (ten to hundreds) that correspond to the wavelength range of the instrument. Each pixel has a full spectrum in a hyperspectral image. For example, a hyperspectral image ($106 \times 206 \times 125$) of rice seeds in 419–1007 nm has an image size of 106×206 pixels with 125 wavebands (which spans from 419 up to 1007 nm wavelength) (Fig. 6.1). Recently, the high dimensionality and redundancy of data in hyperspectral imaging and the cost of HSI instruments have resulted in many studies exploring the development of multispectral imaging systems using the selected wavebands obtained from HSI to create a portable and more practical instrument to be used in the field or the market.

Similar to other spectroscopic methods, hyperspectral imaging can be also performed in reflectance, transmission, fluorescence, and Raman measurement modes, wherein the major difference is in the arrangement of the major components

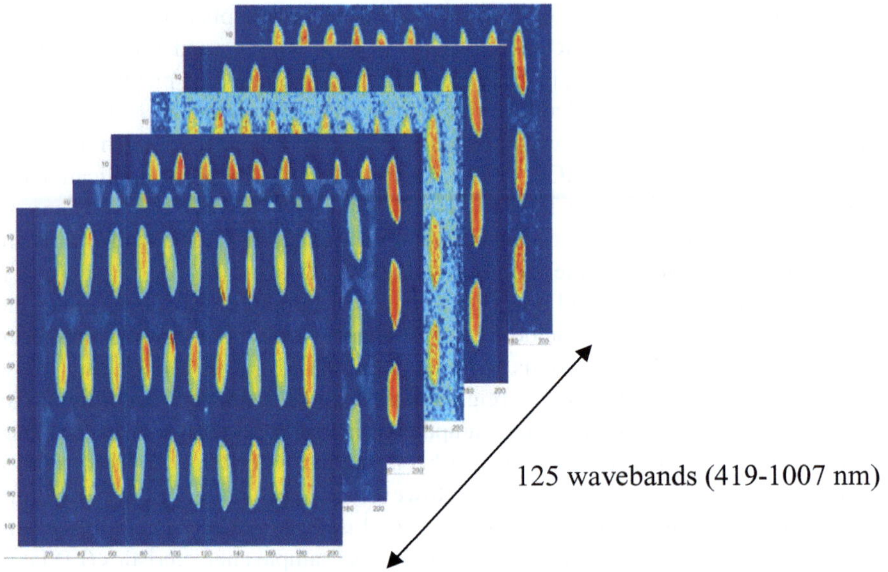

Fig. 6.1 A sample hyperspectral image (106 × 206 pixels and 125 wavebands) of rice seeds

(camera, light sources, and spectrograph) and the sample (ElMasry & Sun, 2010; Lu et al., 2020). Many applications in HSI are centered on diffuse reflectance and transmittance, however, fluorescence and Raman modes have been frequently used recently, such as in food quality and safety inspection (Lu et al., 2020).

When light strikes a sample, some of it is reflected, transmitted, or absorbed. In reflectance hyperspectral imaging, the amount of light reflected by the sample is measured across the wavelength range of the spectrometer. Reflectance can be diffuse or specular, although diffuse reflectance is more frequently used because of the limited sample information provided by specular reflectance. Reflectance HSI is commonly applied for nondestructive analysis of food in terms of chemical composition, ripeness, and freshness of fruits, vegetables, grains, and meat, among others. In contrast, transmittance hyperspectral imaging measures the amount of light passed through the sample across the wavelength range, thereby obtaining information on the internal characteristics of the sample. It is typically used to analyze the internal and external composition of liquid and ground food samples like oil, drinks, dairy, and baked products (Lohumi et al., 2015).

In fluorescence hyperspectral imaging, a light source (usually at longer wavelengths, such as ultraviolet light) excites the sample, which will then emit fluorescence signals. The analysis of the sample is based on its fluorescence emission properties when exposed to certain wavelengths, which may typically vary in food products with contamination and other biochemical changes (Kim et al., 2001; Shi et al., 2020). As such, fluorescence HSI is utilized in food quality applications to detect spoilage, defects, and other contaminants in grains, fruits, vegetables, and meat.

When a sample is illuminated with a high-energy monochromatic light beam (laser, for example), Raman signals emitted by the sample are detected which then provides the molecular structure information of the sample (Achata et al., 2018). With its high sensitivity and specificity, Raman hyperspectral imaging is used for different food quality applications, particularly to identify compounds, detect contaminants, and evaluate food authenticity and adulteration (Lohumi et al., 2015; Qin et al., 2017).

In terms of spectral range, which is the wavelength region covered by a spectrometer, HSI systems cover ultraviolet, visible, near-infrared, or infrared depending on the application requirement. The typical spectral ranges for food applications are visible (380–800 nm), visible and near-infrared (VNIR) (400–1000 nm), near-infrared (900–1700 nm), and shortwave infrared (SWIR) (900–2500 nm). Visible (VIS) HSI instruments are commonly used for the evaluation of color, surface properties, and other external parameters of fruits, vegetables, grains, and meat. Near-infrared (NIR) HSI systems are effective in the assessment of chemical composition (moisture, protein, and oil contents), internal contamination, and defects which make it widely used in food and agricultural applications. Some physico-chemical properties are better characterized when using both VIS and NIR and thus, other studies explore using the VNIR range. SWIR HSI system spans a wide range, so it can be used for different applications in food quality evaluation.

6.2 Equipment/Instrumentation

6.2.1 Acquisition Modes

Depending on the application, hyperspectral images can be acquired using point scan (whiskbroom), line scan (push broom), and area scan.

The point-scan method involves measuring the spectral information of each pixel or point in the image, repeating this process until all the points in the scene have been scanned, and storing the hyperspectral images in band interleaved by pixel format (Qin, 2010; ElMasry & Sun, 2010). Although it provides higher spectral resolution which is particularly useful for some applications requiring detailed analysis, this method takes time to acquire data (Amigo & Grassi, 2019).

In the line-scan method, which collects data more rapidly than the point-scan, works by measuring the spectral information of each line of the entire scene and recording the hyperspectral images in band interleaved by line format (Qin, 2010; ElMasry & Sun, 2010; Amigo & Grassi, 2019). With its ability to capture images of moving samples, the line-scan method is among the preferred technologies for online food inspection in industrial applications since food products typically have linear movement, such as in conveyor belt systems along a production line.

In the area-scan, this method collects the spectral data of the entire scene on a wavelength basis at one time, with full spatial information and with hyperspectral images being acquired in band sequential format. As compared to the point-scan and line-scan methods, the sample or detector does not move in the area-scan method (ElMasry & Sun, 2010; Qin, 2010).

Snapshot imaging, also referred to as snapshot hyperspectral imaging, is one of the recent developments in spectral imaging technologies. In this method, the entire hyperspectral image with the spatial and spectral data of the sample is acquired using a single shot, typically at a few wavelengths from 400 to 900 nm (Amigo & Grassi, 2019).

6.2.2 Components of Hyperspectral Imaging System

In food quality applications, line-scan or push broom hyperspectral imaging is commonly used. The basic components of a line-scan HSI are the light source, wavelength-dispersion devices, area detector, translation stage, and a data acquisition system.

In sensing technologies, particularly vision-based and spectroscopy systems, the light source is a vital data carrier as it provides illumination or excitation on the sample and affects the performance and reliability of the instrument (Qin et al., 2013). The choice of which light source to use depends on the intended application and spectral range, while the light source configuration will vary according to the acquisition mode. The most common light sources are halogen lamps, light-emitting diodes (LEDs), and lasers. Tungsten and other halogen lamps cover a broad wavelength range (ultraviolet, VIS, and NIR), are commercially available, and are relatively low-cost (Amigo & Grassi, 2019). However, they are relatively inefficient with their short life span and their large amount of high-temperature heat generated (Lawrence et al., 2007). On the other hand, LEDs are the cheaper substitute as a light source in the visible to near-infrared range HSI in food applications, with major benefits in energy efficiency, life span, maintenance cost, speed, and reliability, among others (Amigo & Grassi, 2019). While halogen lamps and LEDs are typically used for reflectance or transmission HSI, lasers are utilized to excite samples in fluorescence and Raman spectral imaging instruments because of their high power and directional energy along with other distinct characteristics (Qin et al., 2013; Amigo & Grassi, 2019).

In hyperspectral imaging, one of the vital components is the wavelength-dispersing element where the emitted broadband light from the light source or the reflected broadband light from the sample passes through before reaching the detector (Amigo & Grassi, 2019). Examples of these are imaging spectrographs, variable and tunable filters, Fourier transform imaging spectrometers, and single shot imagers. Each of these varies in its features, configurations, spectral range, and applications.

The quality of hyperspectral images in a spectral imaging system is directly affected by the performance of the detector, which works by converting the collected incident light into electrical signals to measure the light intensity (Huang et al., 2017). Solid-state detectors, mainly charge-coupled devices (CCDs) and complementary metal-oxide-semiconductor (CMOS) cameras, dominate the market. Although Silicon (Si) is preferred due to its cost and temperature range, indium gallium arsenide (InGaAs) and other light-sensitive materials are also used for the small photodiodes that make up the CCD sensors (Amigo & Grassi, 2019).

6.3 Imaging and Processing

The analysis and processing of hyperspectral data need spatial and spectral pre-processing, machine learning algorithms, chemometrics, and statistical tools to obtain the essential information considering its high data dimensionality and the redundant information across the wavebands. Recently, various software, programming platforms, and other tools are explored to analyze hyperspectral data and to facilitate the model development. Typically, a raw hyperspectral image will be calibrated with dark reference and white reference, checked for dead pixels and other erroneous data, and pre-processed using spatial (e.g., background subtraction, object segmentation, and selection of region of interest) and spectral (such as smoothing, normalization, and baseline correction) pretreatment techniques. Pre-processed hyperspectral data are often reduced in dimensionality using techniques such as feature selection methods and principal component analysis. Finally, data will be processed using various methods of image processing and data analysis, particularly involving classification, regression, identification, and visualization which are mostly used in food quality evaluation applications.

6.3.1 Spatial and Spectral Pre-processing

The output image of hyperspectral imaging, similar to RGB and other image types, depends on the light source, sample geometry, and camera optics which can cause spatial issues and variability (Amigo & Santos, 2019). Raw hyperspectral data goes through a series of pre-processing steps. The first step is calibration using dark subtraction (image obtained when the lens is covered) and white reference (using a material with known intensity). Some instrument manufacturers have built-in calibration into their software while others produce raw hyperspectral images. The resulting calibrated hyperspectral image has reflectance, transmittance, or absorbance values relative to the reference material used.

In food quality evaluation, the common spatial pre-processing steps are background removal, object segmentation, and selection of the region of interest (ROI). Background removal and ROI selection, which remove noise and other unwanted areas in the image, are usually done simultaneously. For instance, in a sample of grains, the area (the sampling plate or cup) surrounding the grains will be subtracted from the image, then each kernel will be segmented from one another, and the area around a kernel will be selected as ROI. In other food products, segmentation occurs

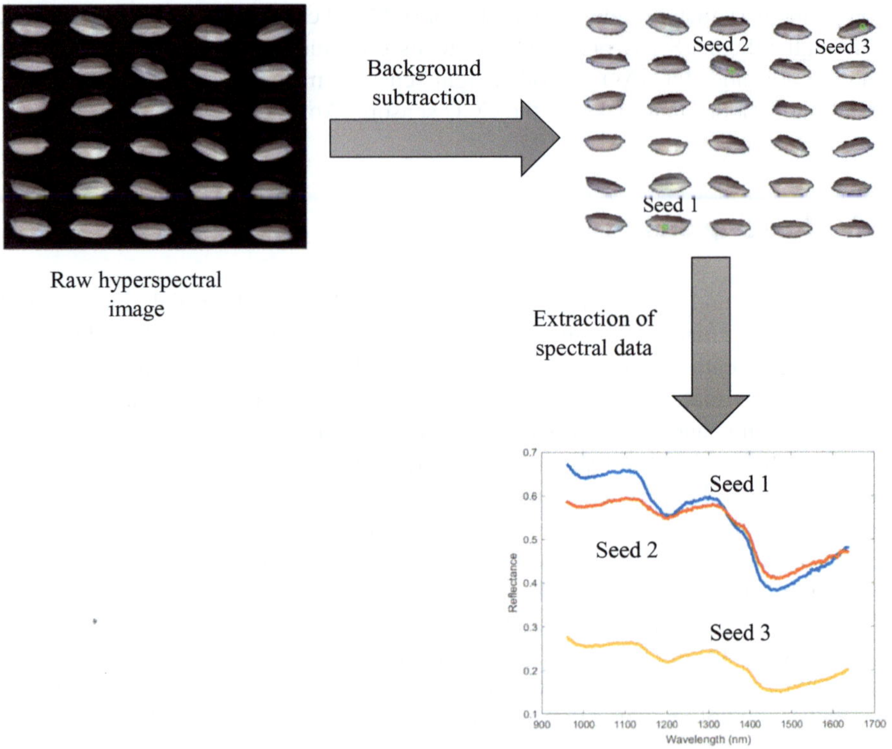

Fig. 6.2 A sample hyperspectral data pre-processing of rice seeds sample involving background subtraction and extraction of spectral data

within the object depending on the objective, such as fat, lean, and blood sections in meat, or germ and endosperm in a grain. These pre-preprocessing steps can be achieved in different ways, by manual selection of the ROI, using histograms, and other thresholding (e.g., K-means, PCA, or wavelength selection) (Amigo & Santos, 2019; Vidal & Amigo, 2012).

Checking and managing dead pixels, spiked points, and missing scan lines, which can be caused by errors due to detector or light source, are crucial before processing the hyperspectral images since they can affect the performance of the multivariate models developed (Vidal & Amigo, 2012). One of the ways to correct them is by replacing them with the mean or mean spectra of the neighboring points (Amigo & Santos, 2019).

From the entire sample area or a specific ROI, spectral data is extracted from each pixel in the image or as a mean spectrum of the entire region. As shown in Fig. 6.2, the entire seed is selected as a region of interest where the mean spectrum was extracted.

Spectral pretreatment methods, including baseline correction, detrending, multiplicative scatter correction, Savitzky-Golay derivatives, smoothing, and standard normal variate, which are typical in preprocessing spectral data, can be used in minimizing scattering effects and other unwanted spectral variation and enhancing the spectral signal-to-noise ratio (Amigo & Santos, 2019).

Due to redundant information, large data size, longer processing time, and complicated data analysis of hyperspectral images, different techniques of dimensionality reduction are usually applied while preserving the relevant information in the hyperspectral data. Some of the common ones in the assessment of food quality are principal component analysis (PCA), successive projection algorithm (SPA), competitive adaptive reweighting sampling (CARS), multidimensional scaling, stepwise regression, and other feature selection algorithms. The resulting wavebands selected are typically used to develop a multispectral imaging instrument which is more practical in terms of cost and portability.

6.3.2 Model Development and Data Analysis

After pre-processing is the development of model calibration using the pre-processed spectral data as input in training the instrument. Depending on the objective, reference data will be needed along with the spectral data to build the model which will be used to predict the specific properties of a sample or to classify food based on certain indices. Various supervised and unsupervised algorithms can be explored to build the calibration model. The performance of model is assessed with the use of the coefficient of determination (R^2), root mean square error (RMSE), standard error (SE), classification accuracy, or other statistical parameters. Analyzing spatial data, spectral data, or both can be done using chemometrics, descriptive statistics, visualization, and other data analytics.

Multivariate regression methods are mostly used to predict composition, including moisture, protein, oil, starch, fiber, and other physicochemical properties. Although partial least squares regression (PLSR) is the most common in food quality prediction, other methods such as artificial neural networks (ANN), principal component regression (PCR), and support vector regression (SVR) are also explored in various applications. In terms of classification, several studies utilize partial least squares discriminant analysis (PLS-DA). Still, other algorithms, namely linear discriminant analysis (LDA), convolutional neural networks (CNN), k-means clustering, and SVM, among others, are increasingly applied.

6.4 Applications in Non-destructive Food Quality Analysis

With its versatility, hyperspectral imaging has been widely explored and recently adopted in various food quality evaluation applications. In quality control, it has been used for the classification and evaluation of ripeness in fruits or freshness in meat and fish products. The majority of the studies conducted were focused on the prediction and analysis of composition, such as moisture, protein, oil, fiber, and starch, among others, with the added feature of visualizing the distribution of this predicted parameter within a sample. As a technique that is more powerful than conventional machine vision, HSI has been useful in detecting physical or biological contaminants in fresh produce and verifying authenticity and adulteration in grains and processed food products.

6.4.1 Meat and Fish Products

In the meat industry, the high demand for superior-quality meat products poses intense pressure on the producers since meats are highly perishable. In general, meat quality is composed of sensory characteristics, chemical composition, physicochemical properties, health-protecting properties, nutritional characteristics, and safety attributes (Damez & Clerjon, 2012).

Consumers predominantly depend on the appearance of meat when selecting and buying meat as it provides insights into the product's condition, freshness, taste, and tenderness—all of which are related to chemical composition and nutritional characteristics. In hyperspectral imaging, most applications concerning meat quality are centered on sensory characteristics and chemical composition (Wu et al., 2022). Color, fat content, moisture content and water-holding capacity, pH, tenderness, marbling, microbial activity, and adulteration are among the major quality parameters being measured or classified in meat and meat products.

Hyperspectral imaging has been recognized as an effective technique in evaluating meat freshness across different types, including beef (Park et al., 2023; Yang et al., 2017a), pork (Pu et al., 2015; Li et al., 2015), chicken breast fillets (Falkovskaya & Gowen, 2020), and yak (Dong et al., 2022). In most of these studies, HSI predicted the total volatile basic nitrogen (TVB-N), which is a key indicator for freshness in diverse meat types (Dong et al., 2022; Yang et al., 2017a; Khulal et al., 2017; Guo et al., 2018) whether in the conventional reflectance HSI or the fluorescence HSI (Lee et al., 2018). Typically operating within the 400–1000 nm range, hyperspectral imaging non-destructively classifies meat as fresh or frozen based on its spectral characteristics, providing valuable insights for traders, regulators, and consumers alike. Some of the recent applications of hyperspectral imaging in meat and meat products are presented in Table 6.1.

Moisture, fat, and protein contents are among the major parameters affecting meat quality. They are extensively explored for prediction using hyperspectral imaging in beef or ground beef, chicken breast or flesh, duck, lamb, pork, and venison (He et al., 2023; Kamruzzaman et al., 2016; Tang et al., 2023; Qin et al., 2021; Dixit et al., 2021). Other physicochemical and sensory parameters, such as color, tenderness, juiciness, and pH, are affected by moisture, fat, and protein contents and are also predicted using HSI.

In poultry meat, hyperspectral imaging has been used for quality classification as a tool for quality grading to discriminate inferior quality chicken before processing

Table 6.1 Applications of hyperspectral imaging for prediction and classification of different quality attributes in meat with the corresponding model and spectral range used

Product	Property	Model	Spectral Range (nm)	Reference
Beef	Color (L*, a*, b*)	PLSR, LS-SVM	400–1000	Kamruzzaman et al. (2016)
	Color (L*, a*)	SVM, MLR	400–1000	Liu et al. (2018)
	Fat	PLSR	880–1720	Zhao et al. (2017)
	Freshness	PLS-DA; SVM	400–1000	Park et al. (2023)
	IMF	PLSR, DCNN	548–1701	Dixit et al. (2021)
	pH	PLSR, DCNN	548–1701	Dixit et al. (2021)
	Moisture content	SVM, MLR	400–1000	Liu et al. (2018)
	Moisture content	PLSR	880–1720	Zhao et al. (2017)
	TVC	LS-SVM	400–1000	Yang et al. (2017a)
Chicken	Ash	PLSR	900–1700	He et al. (2023)
	Foreign materials	GMM	1000–1700	Campos et al. (2023)
	Moisture content	PLSR	900–1700	He et al. (2023)
	pH	PLSR	400–900	Jia et al. (2017)
	Protein	PLSR	900–1700	He et al. (2023)
	TVB-N	PCA-BPANN	430–960	Khulal et al. (2017)
	TVC	PLSR, SVM	400–1000	Jin et al. (2023)
Duck	IMF	PLSR	950–1650	Qin et al. (2021)
Fish	Fat	PLSR	550–1700 (line scan)	Dixit and Reis (2023)
	Fat	PLSR	470–630 (snapshot)	Dixit and Reis (2023)
	Fat	PLSR	670–950 (VNIR)	Dixit and Reis (2023)
	Fat	PLSR	899–1694	Zhu et al. (2014)
	Hardness	PLS, SVM, BP-NN	400–1100	Wang et al. (2021)
	Hemoglobin concentration	Least squares (spectral unmixing)	430–1000	Skjelvareid et al. (2017)
	Moisture	PLSR	899–1694	Zhu et al. (2014)
	Tenderness	PLSR	400–1000	Ma et al. (2017)
	TVB-N	PLSR, LS-SVM	400–1000	Cheng et al. (2014)
	TVB-N	LDNN	430–1010	Moosavi-Nasab et al. (2021)
	TVB-N	PLSR	400–1000	Yu et al. (2021)
	TVB-N	PLSR	900–1700	Yu et al. (2021)
Ham	Moisture	SIMPLS	900–1700	Garrido-Novell et al. (2015)
	NaCl content	SIMPLS	900–1700	Garrido-Novell et al. (2015)
Lamb	Color	PLSR, LS-SVM	400–1000	Kamruzzaman et al. (2016)

(continued)

Table 6.1 (continued)

Product	Property	Model	Spectral Range (nm)	Reference
	IMF, pH	PLSR, DCNN	548–1701	Dixit et al. (2021)
Pork	Color	PLSR; LS-SVM	400–1000	Kamruzzaman et al. (2016)
	Color (L*, a*, b*)	ANN	400–1000	Tang et al. (2023)
	Fat	ANN	400–1000	Tang et al. (2023)
	Fresh and frozen thawed	PNN	400–1000	Pu et al. (2015)
	Moisture content	ANN	400–1000	Tang et al. (2023)
	offal adulteration	PLSR	400–1000	Jiang et al. (2021)
	Protein	ANN	400–1000	Tang et al. (2023)
	TVB-N	PLSR, LS-SVM	400–1000	Yang et al. (2017b)
	TVB-N	LS-SVM	400–1000	Guo et al. (2018)
	TVB-N	PLSR, LS-SVM	400–1000	Lee et al. (2018)
	TVB-N	PCA-BPANN	430–960	Li et al. (2015)
Venison	IMF, pH	PLSR, DCNN	548–1701	Dixit et al. (2021)
Yak	TVB-N	PCR, SVR, PLSR	400–1000	Dong et al. (2022)

GMM Gaussian mixture model, *IMF* intramuscular fat, *LDNN* linear deep neural network, *PNN* probabilistic neural network

or packaging, and to serve as the basis for market pricing (Yang et al., 2021). Several studies were also made exploring the use of HSI as a quality control to detect contamination, microbial spoilage, and tumors in chicken carcasses during the production process in a slaughterhouse or processing facility (Campos et al., 2023; Jin et al., 2023). Tested for commercial use, a real-time hyperspectral imaging system was developed for dual-purpose (wholesomeness inspection and contaminant detection) quality control during the processing of poultry (Park & Yoon, 2015).

Similarly, the quality of fish and other seafood products is examined using hyperspectral imaging. Several studies determined moisture and fat contents, and some showed their distribution within the sample, such as in salmon (Zhu et al., 2014). Although the typical pushbroom HSI system was used in these applications, snapshot HSI has been proven successful in the evaluation of the fat content in salmon fillets (Dixit & Reis, 2023). Textural properties, including Warner-Bratzler shear force (WBSF), hardness, gumminess, and chewiness, were also assessed using hyperspectral imaging of grass carp fillets during vacuum freeze-drying (Ma et al., 2017) or during conventional refrigerated storage (Wang et al., 2021). Moreover, TVB-N has been analyzed in fish which involves grass carp (Cheng et al., 2014), rainbow trout, (Moosavi-Nasab et al., 2021), and tilapia (Yu et al., 2021). As one of the major quality parameters for fish grading, the amount and distribution of blood inside the fish fillets have been examined by HSI (Skjelvareid et al., 2017). Along with this successful application, HSI is now being applied commercially to measure the amount of residual blood inside the fish before filleting which serves as the basis for classification before fish processing (Jackson, 2023).

Other applications of HSI on meat (pork, beef, poultry, and fish) and meat products are more focused on the assessment of physicochemical properties, characterization of fresh and frozen-thawed products, and evaluation of tenderness as these are the most common quality indices used in the market. Most of these studies evaluate raw materials or finished products.

6.4.2 *Fruits and Vegetables*

Consumers generally inspect the visual appearance of fruits and vegetables to determine their quality. As such, commercial grading instruments typically assess color, size, shape, and external abnormalities in fruits and vegetables. However, the internal composition and hidden defects may not be reflected in the visual appearance and may be challenging to evaluate by machine vision systems. In this regard, the strengths of hyperspectral imaging have transformed the non-destructive evaluation of the quality of fruits and vegetables, gearing towards automated sorting and quality control. Several studies have explored using HSI directly in the field and monitoring crop quality even before harvesting. Ultimately, researchers aim to use this technology to develop a portable tool in the market for rapid and real-time quality inspection or an integration with a smartphone application for consumers to evaluate the produce they are buying. Presented in Table 6.2 are some of the recent applications of HSI in fruit and vegetable quality evaluation.

The applicability of near-infrared hyperspectral imaging in the evaluation and characterization of fruits and vegetables from production to postharvest operations, primarily to measure chemical composition, classify ripeness and maturity, and detect defects. The most common quality parameters in terms of composition are moisture content and dry matter (Dong & Guo, 2015; Vega Díaz et al., 2021; Wendel et al., 2018), pH (Dong & Guo, 2015; Ma et al., 2021), and soluble solids content (Dong & Guo, 2015; Gutiérrez et al., 2019; Ma et al., 2021; Rungpichayapichet et al., 2017) in various fruits and vegetables. Other quality parameters such as firmness, elasticity, and mechanical damage can be correlated to their chemical composition. In many of these studies, HSI was utilized to visualize the distribution of a specific quality parameter within a single crop or an orchard and to integrate with other non-destructive sensing instruments.

Ripeness and maturity are conventionally assessed by experts and experienced researchers using objective tests and visual inspection (Zhang et al., 2016), which may be prone to subjective judgment. As fruits and vegetables are highly perishable in nature, fast evaluation of ripening and maturity is critical in minimizing postharvest losses and facilitates efficient classification even from the field (Davur et al., 2023). In combination with machine learning algorithms, HSI was applied to non-destructively evaluate spectral data and classify ripeness or maturity according to certain composition and indices, such as in strawberries (Zhang et al., 2016; Gao

Table 6.2 Applications of hyperspectral imaging for prediction and classification of different quality attributes in fruits and vegetables with the corresponding model and spectral range used

Product	Property	Model	Spectral Range (nm)	Reference
Achacha	Ripeness	ANN, KNN, PLS-DA, PLSR, SVM, SVR	400–780	Nguyen and Liou (2022)
Apple	Firmness	BP-NN, LS-SVM, PLSR	900–1700	Dong and Guo (2015)
	Moisture content	BP-NN, LS-SVM, PLSR	900–1700	Dong and Guo (2015)
	pH	BP-NN, LS-SVM, PLSR	900–1700	Dong and Guo (2015)
	SSC	BP-NN, LS-SVM, PLSR	900–1700	Dong and Guo (2015)
Avocado	Dry matter	MLR, PCR, PLSR, SVMR	400–1000	Vega Díaz et al. (2021)
	Ripeness	CNN	388.9–1005.33	Davur et al. (2023)
Banana	Carotenoids	PLSR	400–1000	Permata et al. (2017)
Grapes	Anthocyanin	SVM	400–1000	Gutiérrez et al. (2019)
	SSC	PLS-DA, PLSR	400–1000	Benelli et al. (2021)
	TSS	SVM	400–1000	Gutiérrez et al. (2019)
Kiwi	pH	PLSR	1002–2300	Ma et al. (2021)
	SSC	PLSR	1002–2300	Ma et al. (2021)
Mango	Dry matter	CNN, PLS-DA, PLSR	411.3–867.0	Wendel et al. (2018)
	Firmness	MLR, PLSR	450–998	Rungpichayapichet et al. (2017)
	Moisture content	PLSR	880–1720	Pu and Sun et al. (2017)
	TA	MLR, PLSR	450–998	Rungpichayapichet et al. (2017)
	TSS	MLR, PLSR	450–998	Rungpichayapichet et al. (2017)
Persimmon	Ripeness	LDA, LS-SVM, SIMCA	400–1000	Wei et al. (2014)
Soybean	Color	PLSR	400–1000	Huang et al. (2014)
	Moisture content	PLSR	400–1000	Huang et al. (2014)
Strawberry	Ripeness	CNN	370–1015	Gao et al. (2020)
		SVM	380–1030, 874–1734	Zhang et al. (2016)
Tomato	Ripeness	PLS-DA	500–950	Zhu et al. (2015)

SSC soluble solids content, *BP-NN* backpropagation neural network, *MLR* multiple linear regression, *PCR* principal component regression, *PLSR* partial least squares regression, *SVMR* support vector machine regression, *TSS* total soluble solids, *TA* titratable acidity, *LDA* linear discriminant analysis, *SIMCA* soft independence modeling of class analogy, *LS-SVM* least squares support vector machines

et al., 2020), tomatoes (Zhu et al., 2015), bananas (Permata et al., 2017), achachas (Nguyen & Liou, 2022), persimmons (Wei et al., 2014), avocadoes (Davur et al., 2023), and grapes (Benelli et al., 2021).

Most applications of HSI on fruits and vegetables deal more with fresh, cold-stored, or minimally processed produce. However, HSI was also shown to be successful in the quality evaluation of processed fruits and vegetables. For example, in drying mango slices, moisture content distribution and color retention on the mango slices were measured using HSI to compare three drying methods (hot air, microwave-vacuum, and a combination of both) (Pu & Sun, 2017). Similar parameters were also investigated in another study which involved the drying of vegetable soybeans (Huang et al., 2014).

6.4.3 Grains, Cereals, and Oilseeds

The quality of cereals and grains is of paramount importance as staple foods for the majority of the global population, especially in addressing food and nutrition security, providing livelihood, and boosting a nation's economy. In food, among the first applications of NIR spectroscopy is the prediction of grain composition. NIR spectroscopy has been considered a standard method and is widely applied commercially in grain elevators and grain processing facilities. Likewise, hyperspectral imaging has been widely explored in grains and cereals, and its commercial applications are continuously expanding with the addition of advanced machine learning algorithms and in combination with other sensors (Table 6.3).

Moisture, protein, and oil contents are the major quality parameters being measured by hyperspectral imaging in grains, cereals, oilseeds, and grain products, particularly in wheat, corn, rice, soybean, sorghum, oats, barley, and wheat flour, as these are used as the basis for pricing and determining the end-use quality. Aside from these parameters, hyperspectral imaging is also used for the detection of biological and physical contamination, adulteration, and damage. Recent advances in machine learning algorithms have benefited the classification of grain according to variety, hardness, and color, among others. One of the major advantages of hyperspectral imaging over typical single-point spectroscopy in grain applications is the ability of HSI to visualize the distribution within a sample of grains, an individual kernel, or a sample of flour, which is vital in locating defects, areas with extremely high or low concentrations, and non-uniformity in ground samples.

In wheat, hyperspectral imaging has been applied to measure protein content (Caporaso et al., 2018; Zhang et al., 2023; Morales-Sillero et al., 2018), moisture content (Zhang et al., 2023), and starch content (Zhang et al., 2023). HSI was also explored to classify wheat according to hardness (Erkinbaev et al., 2019), variety or wheat types (Dreier et al., 2022; Vermeulen et al., 2018; Mahesh et al., 2011); detect fungal contamination (Barbedo et al., 2015; Senthilkumar et al., 2016), adulteration (Verdú et al., 2016), sprout damage (Barbedo et al., 2018), insect damage (Singh

Table 6.3 Applications of hyperspectral imaging for prediction and classification of different quality attributes in cereals, grains and grain products with the corresponding model and spectral range used

Product	Attributes	Model	Spectral Range (nm)	Reference
Wheat	Foreign material	SVM, NB, k-NN	1000–1600	Ravikanth et al. (2015)
	Hardness	ANN, PLSR	1000–2500	Erkinbaev et al. (2019)
	Insect damage	BPNN, QDA	700–1100	Singh et al. (2010)
	Moisture	PLSR, PCR, SVMR, MLR	969–2174	Zhang et al. (2023)
	Mycotoxin	PCA	528–1785	Barbedo et al. (2015)
	Protein	PLSR, LS-SVM	1000–1700	Caporaso et al. (2018)
		PLSR, PCR, SVMR, MLR	969–2174	Zhang et al. (2023)
		PLSR	1120–2424	Morales-Sillero et al. (2018)
	Starch	PLSR, PCR, SVMR, MLR	969–2174	Zhang et al. (2023)
	Variety	PLS-DA	1100–2400	Vermeulen et al. (2018)
Bakery Products	Hardness	PLSR	938–1630	Amigo et al. (2021)
	Protein	PLSR	400–1000, 1000–2500	Olakanmi et al. (2024)
	Shelf life	PLSR, PLS-DA	935–1720	Sricharoonratana et al. (2021)
	Staling	PLSR	938–1630	Amigo et al. (2021)
	Storage time	PLSR, PLS-DA	935–1720	Sricharoonratana et al. (2021)
Corn	AFB1	PLS-DA	1100–1700	Kandpal et al. (2015)
	Moisture	PLS, CNN, LSTM and CNN-LSTM	968–2575	Zhang et al. (2022)
	Variety	CNN	900–1700	Fu et al. (2023)
		SVM, PLS-DA	400–1000	Yang et al. (2015)
Oat	Contaminants	PCA, PLS-DA	900–1700	Erkinbaev et al. (2017)
	Gluten	PCA, PLS-DA	900–1700	Whitworth et al. (2010)
Rice	Moisture	LS-SVR	871–1766	Sun et al. (2017)
	Variety	k-NN, SVM CNN	380–1030, 874–1734	Qiu et al. (2018)
	Variety		874–1734	Kong et al. (2013)

(continued)

Table 6.3 (continued)

Product	Attributes	Model	Spectral Range (nm)	Reference
		PLS-DA, SIMCA, k-NN, SVM, RF		
Soybeans	Adulteration	PLS-DA	885–1729	Li et al. (2024)
	Pest infestation	QDA	960–1700	Chelladurai et al. (2014)
	Variety	CNN	400–1000	Zhu et al. (2020)

PLSR partial least square regression, *PCR* principal component regression, *SVMR* support vector machine regression, *MLR* multiple linear regression, *SVM* support vector machine, *NB* Naïve Bayes, *k-NN* k-nearest neighbors, *ANN* artificial neural network, *QDA* quadratic discriminant analysis, *PCA* principal component analysis, *PLS-DA* partial least squares discriminant analysis, *LS-SVR* least-squares support vector regression, *CNN* convolutional neural network, *RF* random forest, *SIMCA* soft independence modeling of class analogy, *LS-SVM* least squares support vector machines

et al., 2010), or foreign material contamination (Ravikanth et al., 2015); separate from other grains (Erkinbaev et al., 2017), in bulk seeds, single kernels, or flour.

One of the challenges in the quality evaluation of grain products is the detection of gluten. In wheat flour production, the addition of dried gluten happens in the absence of high or insufficient flour protein content. NIR hyperspectral imaging was used to detect the presence of added dried gluten in wheat flour (Whitworth et al., 2010). By visual inspection, gluten in flour was not recognizable but HSI was able to differentiate gluten from flour as evident in the difference in the mean spectra and as shown in the higher absorbance values from darker shades in the sample, signifying the presence of gluten. This application also opens opportunities to utilize hyperspectral imaging to discriminate impurities or measure the composition of products having ground samples or mixed powders.

Eating quality of bakery products is usually difficult to test using instruments. Conventional methods are destructive and time-consuming, while RGB imaging is limited to physical characteristics. As such, hyperspectral imaging was also explored for bread and other bakery product quality applications. HSI was used to evaluate the quality of a baked sponge and estimate its shelf-life by measuring the water absorption band and water content reduction (Marshall et al., 2015) and by classifying it according to microbial infection (Sricharoonratana et al., 2021). Moreover, it was also used for quality monitoring by classifying bread according to protein content (Olakanmi et al., 2024), and by measuring bread hardness and investigating the staling process (Amigo et al., 2021).

Such applications were also applied to corn (Fu et al., 2023; Zhang et al., 2022; Yang et al., 2015; Kandpal et al., 2015; Liu et al., 2016), rice (Sun et al., 2017; Qiu et al., 2018; Kong et al., 2013; Meng et al., 2023; Wu et al., 2020; Liu et al., 2016), soybeans (Zhu et al., 2020; Chelladurai et al., 2014; Liu et al., 2016; Zhu et al., 2019; Li et al., 2024), and other cereals and grains (Mendoza et al., 2023) during postharvest handling, including drying, storage, and processing, and during quality evaluation of raw and finished grain products.

6.4.4 Dairy and Other Products

In the early years, a few research on the use of HSI on dairy products were done due to the homogenous nature of milk (liquid and powdered) and the difficulty of analyzing dairy products, such as cheese because of heat build-up caused by the lamp and of the higher time to acquire images (Manley, 2014). Recent advancements in instrumentation, particularly faster acquisition methods, improved light sources and detectors, and lower cost, applications of HSI on milk and dairy products have increased in terms of the product composition (e.g., protein content, fat content, and contaminants) and on the quality of dairy products during or after processing. For milk samples, HSI was used in the discrimination of different milk powder samples, including the prediction of dispersibility and solubility and the detection of melamine (Munir et al., 2018; Lim et al., 2016). Successful application of hyperspectral imaging has also been reported in the characterization of commercially available milk and yogurt products with varying fat composition (Abildgaard et al., 2015) and in the detection of adulteration in milk (Kimbahune et al., 2016). In cheese, the determination of quality parameters such as meltability, cheese defects, and texture during ripening, and prediction of enzyme treatment and fat content were considered feasible using hyperspectral imaging (Darnay et al., 2017). During the processing of milk and dairy products, various quality parameters are monitored and evaluated. Possible uses of HSI in processing operations are monitoring the optical properties of milk during coagulation and curd formation; characterizing the moisture distribution in a sample and evaluating drying parameters during spray drying; and providing an estimation of particle size distribution during dairy processing.

Aside from dairy, the quality evaluation of eggs such as their freshness, bubble formation, or scattered yolk is challenging as these parameters are internal. With hyperspectral imaging, non-destructive and rapid assessment of these internal attributes was made possible (Zhang et al., 2015). This may overcome the limitations of RGB imaging computer vision which is widely applied commercially.

In oil, many of the applications are on the prediction of composition, detection of rancidity, and detection of authentication and adulteration, which will be beneficial for regulation and trading if applied commercially. In combination with feature selection algorithms, HSI has been proven successful in classifying different sesame oil brands (Deng et al., 2013) and varieties (Xie et al., 2014), and in predicting acidity, peroxide index, and humidity content of olive oil (Gila et al., 2012).

6.5 Future Trends

Hyperspectral imaging has continuously evolved over the years, from being a tool in remote sensing to a device for benchtop and laboratory analytical measurements and a commercially applied instrument for in-line and in-field applications. Along with the increasing demand for high quality foods and the changing consumer food preferences, advances in instrumentation and artificial intelligence are also keeping up, with spectral imaging addressing the challenges and gearing toward more opportunities for practical and effective use in the food industry. As such, here are some of the future considerations for the development of hyperspectral imaging.

With the high instrument cost and complexity of data processing and analysis in hyperspectral imaging, several studies have utilized the results from feature selection and model development to create a portable hyperspectral imaging device. This practical approach with the goal of lesser cost, faster data processing, and reliable data analysis can cater more applications. This will be convenient for quality testing in the field, in real-time automated processing lines for production and quality control, or inside the food storage facilities.

The performance of the models developed is usually limited to the range of samples used and may not depict actual conditions. With this, it is vital to build calibration models with samples having a wide range of variability and to test these models using independent set of samples, those that were not included in the calibration and validation set. With the emergence of more advanced machine learning algorithms for classification and prediction, many of the studies focus more on improving accuracy and less on the robustness of the model.

Another potential development is the integration with RGB camera and other non-destructive devices, which has been successfully applied in plant phenotyping and precision agriculture using unmanned aerial vehicles and robotic technology in the field. Few studies have already explored this multi-modal approach in combination with machine learning and feature selection techniques (Mishra & Xu, 2023; Qin et al., 2023; Garillos-Manliguez & Chiang, 2021) and highlighted its potential for non-destructive food inspection.

In NIR spectroscopy, standard guideline methods that cover instrument calibration and maintenance, sample preparation, data pre-processing, model development, evaluation of model performance, among others, are already established for different food products. As hyperspectral imaging becomes a potential analytical tool for food quality applications in commercial and research use, a standard method is needed that will highlight the guidelines and protocols. This can serve as a reference for data comparability and consistency and guide future instrument users and other stakeholders.

References

AACC. (1999). *Near-infrared methods: Guidelines for model development and maintenance–AACC method 39–00*. Approved Methods of the American Association of Cereal Chemists.

Abildgaard, O. H., Kamran, F., Dahl, A. B., Skytte, J. L., Nielsen, F. D., Thomsen, C. L., Andersen, P. E., Larsen, R., & Frisvad, J. R. (2015). Non-invasive assessment of dairy products using spatially resolved diffuse reflectance spectroscopy. *Applied Spectroscopy, 69*(9), 1096–1105. https://doi.org/10.1366/14-07529

Achata, E. M., Esquerre, C., Gowen, A. A., & O'Donnell, C. P. (2018). Feasibility of near infrared and Raman hyperspectral imaging combined with multivariate analysis to assess binary mixtures of food powders. *Powder Technology, 336*, 555–566. https://doi.org/10.1016/j.powtec.2018.06.025

Amigo, J. M., & Grassi, S. (2019). Configuration of hyperspectral and multispectral imaging systems. In J. M. Amigo (Ed.), *Data handling in science and technology* (Vol. 32, pp. 17–34). Elsevier. https://doi.org/10.1016/B978-0-444-63977-6.00002-X

Amigo, J. M., & Santos, C. (2019). Preprocessing of hyperspectral and multispectral images. In J. M. Amigo (Ed.), *Data handling in science and technology* (Vol. 32, pp. 37–53). Elsevier.

Amigo, J. M., Olmo, A., Engelsen, M. M., Lundkvist, H., & Engelsen, S. B. (2021). Staling of white wheat bread crumb and effect of maltogenic α-amylases. Part 3: Spatial evolution of bread staling with time by near-infrared hyperspectral imaging. *Food Chemistry, 353*, 129478. https://doi.org/10.1016/j.foodchem.2021.129478

Barbedo, J. G. A., Tibola, C. S., & Fernandes, J. M. C. (2015). Detecting Fusarium head blight in wheat kernels using hyperspectral imaging. *Biosystems Engineering, 131*, 65–76. https://doi.org/10.1016/j.biosystemseng.2015.01.003

Barbedo, J. G. A., Guarienti, E. M., & Tibola, C. S. (2018). Detection of sprout damage in wheat kernels using NIR hyperspectral imaging. *Biosystems Engineering, 175*, 124–132. https://doi.org/10.1016/j.biosystemseng.2018.09.012

Benelli, A., Cevoli, C., Ragni, L., & Fabbri, A. (2021). In-field and non-destructive monitoring of grapes maturity by hyperspectral imaging. *Biosystems Engineering, 207*, 59–67. https://doi.org/10.1016/j.biosystemseng.2021.04.006

Campos, R. L., Yoon, S.-C., Chung, S., & Bhandarkar, S. M. (2023). Semisupervised deep learning for the detection of foreign materials on poultry meat with near-infrared hyperspectral imaging. *Sensors, 23*, 7014. https://doi.org/10.3390/s23167014

Caporaso, N., Whitworth, M. B., & Fisk, I. D. (2018). Protein content prediction in single wheat kernels using hyperspectral imaging. *Food Chemistry, 240*, 32–42. https://doi.org/10.1016/j.foodchem.2017.07.048

Chelladurai, V., Karuppiah, K., Jayas, D. S., Fields, P. G., & White, N. D. G. (2014). Detection of Callosobruchus maculatus (F.) infestation in soybean using soft X-ray and NIR hyperspectral imaging techniques. *Journal of Stored Products Research, 57*, 43–48. https://doi.org/10.1016/j.jspr.2013.12.005

Cheng, J. H., Sun, D. W., Zeng, X. A., & Pu, H. B. (2014). Non-destructive and rapid determination of TVB-N content for freshness evaluation of grass carp (*Ctenopharyngodon idella*) by hyperspectral imaging. *Innovative Food Science and Emerging Technologies, 21*, 179–187. https://doi.org/10.1016/j.ifset.2013.10.013

Damez, J. L., & Clerjon, S. (2012). Recent advances in meat quality assessment. In Y. H. Hui (Ed.), *Handbook of meat and meat processing* (pp. 161–175). https://doi.org/10.1201/b11479-12

Darnay, L., Králik, F., Oros, G., Koncz, Á., & Firtha, F. (2017). Monitoring the effect of transglutaminase in semi-hard cheese during ripening by hyperspectral imaging. *Journal of Food Engineering, 196*, 123–129. https://doi.org/10.1016/j.jfoodeng.2016.10.020

Davur, Y. J., Kämper, W., Khoshelham, K., Trueman, S. J., & Bai, S. H. (2023). Estimating the ripeness of Hass avocado fruit using deep learning with hyperspectral imaging. *Horticulturae, 9*, 599. https://doi.org/10.3390/horticulturae9050599

De Baerdemaeker, J. (2023). *Artificial intelligence in the agri-food sector: Applications, risks and impacts*. European Parliamentary Research Service. https://doi.org/10.2861/516636

Delwiche, S. R. (2004). Analysis of small grain crops. In C. A. Roberts, J. Workman Jr., & J. B. Reeves III (Eds.), *Near-infrared spectroscopy in agriculture* (Vol. 44, pp. 269–320). https://doi.org/10.2134/agronmonogr44.c11

Deng, S., Xu, Y., Li, L., Li, X., & He, Y. (2013). A feature-selection algorithm based on support vector machine-multiclass for hyperspectral visible spectral analysis. *Journal of Food Engineering, 119*(1), 159–166. https://doi.org/10.1016/j.jfoodeng.2013.05.024

Dixit, Y., & Reis, M. M. (2023). Hyperspectral imaging for assessment of total fat in salmon fillets: A comparison between benchtop and snapshot systems. *Journal of Food Engineering, 336*, 111212. https://doi.org/10.1016/j.jfoodeng.2022.111212

Dixit, Y., Al-Sarayreh, M., Craigie, C. R., & Reis, M. M. (2021). A global calibration model for prediction of intramuscular fat and pH in red meat using hyperspectral imaging. *Meat Science, 181*, 108405. https://doi.org/10.1016/j.meatsci.2020.108405

Dong, J., & Guo, W. (2015). Nondestructive determination of apple internal qualities using near-infrared hyperspectral reflectance imaging. *Food Analytical Methods, 8*, 2635–2646. https://doi.org/10.1007/s12161-015-0169-8

Dong, K., Guan, Y., Wang, Q., Huang, Y., An, F., Zeng, Q., Luo, Z., & Huang, Q. (2022). Non-destructive prediction of yak meat freshness indicator by hyperspectral techniques in the oxidation process. *Food Chemistry: X, 17*, 100541. https://doi.org/10.1016/j.fochx.2022.100541

Dreier, E. S., Sorensen, K. M., Lund-Hansen, T., Jespersen, B. M., & Pedersen, K. S. (2022). Hyperspectral imaging for classification of bulk grain samples with deep convolutional neural networks. *Journal of Near Infrared Spectroscopy, 30*(3), 107–121. https://doi.org/10.1177/09670335221078356

ElMasry, G., & Sun, D.-W. (2010). Chapter 1 - Principles of hyperspectral imaging technology. In D.-W. Sun (Ed.), *Hyperspectral imaging for food quality analysis and control* (pp. 3–43). Academic Press. https://doi.org/10.1016/B978-0-12-374753-2.10001-2

Erkinbaev, C., Henderson, K., & Paliwal, J. (2017). Discrimination of gluten-free oats from contaminants using near-infrared hyperspectral imaging technique. *Food Control, 80*, 197–203. https://doi.org/10.1016/j.foodcont.2017.04.036

Erkinbaev, C., Derksen, K., & Paliwal, J. (2019). Single kernel wheat hardness estimation using near infrared hyperspectral imaging. *Infrared Physics & Technology, 98*, 250–255. https://doi.org/10.1016/j.infrared.2019.03.033

Falkovskaya, A., & Gowen, A. (2020). Literature review: Spectral imaging applied to poultry products. *Poultry Science, 99*(7), 3709–3722. https://doi.org/10.1016/j.psj.2020.04.013

Fu, Z., Fangyuan, Z., Shunqing, W., Lantao, L., Qiang, L., Sanling, F., Xinyue, W., Qingfeng, L., & Yakun, Z. (2023). Hyperspectral imaging combined with CNN for maize variety identification. *Frontiers in Plant Science, 14*. https://doi.org/10.3389/fpls.2023.1254548

Gao, Z., Shao, Y., Xuan, G., Wang, Y., Liu, Y., & Han, X. (2020). Real-time hyperspectral imaging for the in-field estimation of strawberry ripeness with deep learning. *Artificial Intelligence in Agriculture, 4*, 31–38. https://doi.org/10.1016/j.aiia.2020.04.003

Garillos-Manliguez, C. A., & Chiang, J. Y. (2021). Multimodal deep learning and visible-light and hyperspectral imaging for fruit maturity estimation. *Sensors, 21*, 1288. https://doi.org/10.3390/s21041288

Garrido-Novell, C., Garrido-Varo, A., Pérez-Marín, D., Guerrero-Ginel, J. E., & Kim, M. (2015). Quantification and spatial characterization of moisture and NaCl content of Iberian dry-cured ham slices using NIR hyperspectral imaging. *Journal of Food Engineering, 153*, 117–123. https://doi.org/10.1016/j.jfoodeng.2014.09.035

Gila, D. M., Marchal, P. C., García, J. G., & Ortega, J. G. (2012, September). Hyperspectral imaging for determination of some quality parameters for olive oil. In *18th International Conference on Automation and Computing (ICAC)* (pp. 1–6). IEEE.

Guo, T., Huang, M., Zhu, Q., Guo, Y., & Qin, J. (2018). Hyperspectral image-based multi-feature integration for TVB-N measurement in pork. *Journal of Food Engineering, 218*, 61–68. https://doi.org/10.1016/j.jfoodeng.2017.09.003

Gutiérrez, S., Tardaguila, J., Fernández-Novales, J., & Diago, M. P. (2019). On-the-go hyperspectral imaging for the in-field estimation of grape berry soluble solids and anthocyanin concentration. *Australian Journal of Grape and Wine Research, 25*, 127–133. https://doi.org/10.1111/ajgw.12376

He, H.-J., Wang, Y., Ou, X., Ma, H., Liu, H., & Yan, J. (2023). Rapid determination of chemical compositions in chicken flesh by mining hyperspectral data. *Journal of Food Composition and Analysis, 116*, 105069. https://doi.org/10.1016/j.jfca.2022.105069

Huang, M., Wang, Q., Zhang, M., & Zhu, Q. (2014). Prediction of color and moisture content for vegetable soybean during drying using hyperspectral imaging technology. *Journal of Food Engineering, 128*, 24–30. https://doi.org/10.1016/j.jfoodeng.2013.12.008

Huang, L., Zhou, Y., Meng, L., Wu, D., & He, Y. (2017). Comparison of different CCD detectors and chemometrics for predicting total anthocyanin content and antioxidant activity of mulberry fruit using visible and near infrared hyperspectral imaging technique. *Food Chemistry, 224*, 1–10.

Jackson, L. (2023). *Vision quest: Behind Norwegian researchers' high-tech approach to fish welfare assessments*. Accessed from https://www.globalseafood.org/advocate/vision-quest-behind-norwegian-researchers-high-tech-approach-to-fish-welfare-assessments/

Jia, B., Yoon, S., Zhuang, H., Wang, W., & Li, C. (2017). Prediction of pH of fresh chicken breast fillets by VNIR hyperspectral imaging. *Journal of Food Engineering, 208*, 57–65. https://doi.org/10.1016/j.jfoodeng.2017.03.023

Jiang, H., Yu, R., Chen, Q., Wang, J., & Li, X. (2021). Near-infrared hyperspectral imaging for detectiong and visualization of offal adulteration in ground pork. *Spectrochimica Acta Part A: Molecular and Biomolecular Spectroscopy, 249*, 119307. https://doi.org/10.1016/j.saa.2020.119307

Jin, S., Liu, X., Wang, J., Pan, L., Zhang, Y., Zhou, G., & Tang, C. (2023). Hyperspectral imaging combined with fluorescence for the prediction of microbial growth in chicken breasts under different packaging conditions. *LWT—Food Science and Technology, 181*, 114727. https://doi.org/10.1016/j.lwt.2023.114727

Kamruzzaman, M., Makino, Y., & Oshita, S. (2016). Online monitoring of red meat color using hyperspectral imaging. *Meat Science, 116*, 110–117. https://doi.org/10.1016/j.meatsci.2016.02.004

Kandpal, L. M., Lee, S., Kim, M. S., Bae, H., & Cho, B. K. (2015). Short wave infrared (SWIR) hyperspectral imaging technique for examination of aflatoxin B1 (AFB1) on corn kernels. *Food Control, 51*, 171–176. https://doi.org/10.1016/j.foodcont.2014.11.020

Khulal, U., Zhao, J., Hu, W., & Chen, Q. (2017). Intelligent evaluation of total volatile basic nitrogen (TVB-N) content in chicken meat by an improved multiple level data fusion model. *Sensors and Actuators B: Chemical, 238*, 337–345. https://doi.org/10.1016/j.snb.2016.07.074

Kim, M. S., Chen, Y. R., & Mehl, P. M. (2001). Hyperspectral reflectance and fluorescence imaging system for food quality and safety. *Transactions of the ASAE, 44*(3), 721–729. https://doi.org/10.13031/2013.6099

Kimbahune, S., Ghouse, S. M., Mithun, B. S., Shinde, S., & Jha, A. K. (2016, May). Hyperspectral sensing based analysis for determining milk adulteration. In *Hyperspectral imaging sensors: Innovative applications and sensor standards 2016* (Vol. 9860, pp. 44–51). SPIE.

Kong, W., Zhang, C., Liu, F., Nie, P., & He, Y. (2013). Rice seed cultivar identification using near-infrared hyperspectral imaging and multivariate data analysis. *Sensors, 13*(7), 8916–8927. https://doi.org/10.3390/s130708916

Lawrence, K. C., Park, B., Windham, W. R., Heitschmidt, G. W., & Thai, C. N. (2007). Evaluation of LED and tungsten-halogen lighting for fecal contaminant detection. *Applied Engineering in Agriculture, 23*(6), 811–818. https://doi.org/10.13031/2013.24048

Lee, H., Kim, M. S., Lee, W. H., & Cho, B. K. (2018). Determination of the total volatile basic nitrogen (TVB-N) content in pork meat using hyperspectral fluorescence imaging. *Sensors and Actuators B: Chemical, 259*, 532–539. https://doi.org/10.1016/j.snb.2017.12.102

Li, H., Chen, Q., Zhao, J., & Wu, M. (2015). Nondestructive detection of total volatile basic nitrogen (TVB-N) content in pork meat by integrating hyperspectral imaging and colorimetric sensor combined with nonlinear data fusion. *LWT—Food Science and Technology, 63*(1), 268–274. https://doi.org/10.1016/j.lwt.2015.03.052

Li, X., Wang, D., Yu, L., Ma, F., Wang, X., Pérez-Marín, D., Li, P., & Zhang, L. (2024). Origin traceability and adulteration detection of soybean using near infrared hyperspectral imaging. *Food Frontiers, 5*, 237–244. https://doi.org/10.1002/fft2.345

Lim, J., Kim, G., Mo, C., Kim, M. S., Chao, K., Qin, J., Fu, X., Baek, I., & Cho, B.-K. (2016). Detection of melamine in milk powders using near-infrared hyperspectral imaging combined with regression coefficient of partial least square regression model. *Talanta, 151*, 183–191. https://doi.org/10.1016/j.talanta.2016.01.035

Lind, R., & Murhed, A. (2012). Chapter 5 - Computer vision in food processing: An overview. In D. W. Sun (Ed.), *Computer vision technology in the food and beverage industries* (Woodhead Publishing Series in Food Science, Technology and Nutrition) (pp. 133–149). Woodhead Publishing. https://doi.org/10.1533/9780857095770.2.133

Liu, Y., Xie, H., Chen, Y., Tan, K., Wang, L., & Xie, W. (2016). Neighborhood mutual information and its application on hyperspectral band selection for classification. *Chemometrics and Intelligent Laboratory Systems, 157*, 140–151. https://doi.org/10.1016/j.chemolab.2016.07.009

Liu, Y., Sun, D., Cheng, J., & Han, Z. (2018). Hyperspectral imaging sensing of changes in moisture content and color of beef during microwave heating process. *Food Analytical Methods, 11*(9), 2472–2484. https://doi.org/10.1007/s12161-018-1234-x

Lohumi, S., Lee, S., Lee, H., & Cho, B.-K. (2015). A review of vibrational spectroscopic techniques for the detection of food authenticity and adulteration. *Trends in Food Science & Technology, 46*(1), 85–98.

Lu, R., & Park, B. (2015). Introduction. In B. Park & R. Lu (Eds.), *Hyperspectral imaging technology in food and agriculture* (Food Engineering Series). Springer. https://doi.org/10.1007/978-1-4939-2836-1_1

Lu, Y., Saeys, W., Kim, M., & Peng, Y. (2020). Hyperspectral imaging technology for quality and safety evaluation of horticultural products: A review and celebration of the past 20-year progress. *Postharvest Biology and Technology, 170*, 111318.

Ma, J., Sun, D., Qu, J., & Pu, H. (2017). Prediction of textural changes in grass carp fillets as affected by vacuum freeze drying using hyperspectral imaging based on integrated group wavelengths. *LWT - Food Science and Technology, 82*, 377–385. https://doi.org/10.1016/j.lwt.2017.04.040

Ma, T., Xia, Y., Inagaki, T., & Tsuchikawa, S. (2021). Non-destructive and fast method of mapping the distribution of the soluble solids content and pH in kiwifruit using object rotation near-infrared hyperspectral imaging approach. *Postharvest Biology and Technology, 174*, 111440. https://doi.org/10.1016/j.postharvbio.2020.111440

Mahesh, S., Jayas, D. S., Paliwal, J., & White, N. D. G. (2011). Identification of wheat classes at different moisture levels using near-infrared hyperspectral images of bulk samples. *Sensing and Instrumentation for Food Quality and Safety, 5*, 1–9. https://doi.org/10.1007/s11694-010-9104-2

Manley, M. (2014). Near-infrared spectroscopy and hyperspectral imaging: Non-destructive analysis of biological materials. *Chemical Society Reviews, 43*, 8200–8214. https://doi.org/10.1039/c4cs00062e

Marshall, S., Kelman, T., Qiao, T., Murray, P., & Zabalza, J. (2015). Hyperspectral imaging for food applications. In *23rd European Signal Processing Conference, (EUSIPCO 2015)* (pp. 1–5). IEEE. https://doi.org/10.1109/EUSIPCO.2015.7362906

McClure, W. F. (2006). Introduction. In Y. Ozaki, W. F. McClure, & A. A. Christy (Eds.), *Near-infrared spectroscopy in food science and technology*. Wiley. https://doi.org/10.1002/9780470047705.ch1

Mendoza, P. T. D., Armstrong, P. R., Peiris, K. H., Siliveru, K., Bean, S. R., & Pordesimo, L. O. (2023). Prediction of sorghum oil content using near-infrared hyperspectral imaging. *Cereal Chemistry, 100*(3), 775–783. https://doi.org/10.1002/cche.10656

Meng, Y., Yuan, W., Aktilek, E. U., Zhong, Z., Wang, Y., Gao, R., & Su, Z. (2023). Fine hyperspectral classification of rice varieties based on self-attention mechanism. *Ecological Informatics, 75*, 102035. https://doi.org/10.1016/j.ecoinf.2023.102035

Mishra, P., & Xu, J. (2023). Multimodal close range hyperspectral imaging combined with multiblock sequential predictive modelling for fresh produce analysis. *Journal of Near Infrared Spectroscopy, 31*(3), 141–149. https://doi.org/10.1177/09670335231173142

Moosavi-Nasab, M., Khoshnoudi-Nia, S., Azimifar, Z., et al. (2021). Evaluation of the total volatile basic nitrogen (TVB-N) content in fish fillets using hyperspectral imaging coupled with deep learning neural network and meta-analysis. *Scientific Reports, 11*, 5094. https://doi.org/10.1038/s41598-021-84659-y

Morales-Sillero, A., Pierna, J. A. F., Sinnaeve, G., Dardenne, P., & Baeten, V. (2018). Quantification of protein in wheat using near infrared hyperspectral imaging: Performance comparison with conventional near infrared spectroscopy. *Journal of Near Infrared Spectroscopy, 26*(3), 186–195. https://doi.org/10.1177/0967033518780506

Munir, M. T., Wilson, D. I., Yu, W., & Young, B. R. (2018). An evaluation of hyperspectral imaging for characterising milk powders. *Journal of Food Engineering, 221*, 1–10. https://doi.org/10.1016/j.jfoodeng.2017.10.001

Nguyen, N. M. T., & Liou, N.-S. (2022). Ripeness evaluation of Achacha fruit using hyperspectral image data. *Agriculture, 12*, 2145. https://doi.org/10.3390/agriculture12122145

Olakanmi, S. J., Jayas, D. S., Paliwal, J., Chaudhry, M. M. A., & Findlay, C. R. J. (2024). Quality characterization of fava bean-fortified bread using hyperspectral imaging. *Food, 13*, 231. https://doi.org/10.3390/foods13020231

Park, B., & Yoon, S. C. (2015). Real-Time hyperspectral imaging for food safety. In B. Park & R. Lu (Eds.), *Hyperspectral imaging technology in food and agriculture* (Food Engineering Series). Springer. https://doi.org/10.1007/978-1-4939-2836-1_13

Park, S., Hong, S.-J., Kim, S., Ryu, J., Roh, S., & Kim, G. (2023). Classification of fresh and frozen-thawed beef using a hyperspectral imaging sensor and machine learning. *Agriculture, 13*, 918. https://doi.org/10.3390/agriculture13040918

Permata, Y. C., Handayani, W., & Saputro, A. H. (2017). Carotenoid profile map system based on hyperspectral technique in banana (Musa sp.). In *2017 5th International Conference on Instrumentation, Control, and Automation (ICA)* (pp. 163–166). https://doi.org/10.1109/ICA.2017.8068433

Petrescu, D. C., Vermeir, I., & Petrescu-Mag, R. M. (2019). Consumer understanding of food quality, healthiness, and environmental impact: A cross-national perspective. *International Journal of Environmental Research and Public Health, 17*(1), 169. https://doi.org/10.3390/ijerph17010169

Pu, Y. Y., & Sun, D. W. (2017). Combined hot-air and microwave-vacuum drying for improving drying uniformity of mango slices based on hyperspectral imaging visualisation of moisture content distribution. *Biosystems Engineering, 156*, 108–119. https://doi.org/10.1016/j.biosystemseng.2017.01.006

Pu, H., Sun, D. W., Ma, J., & Cheng, J. H. (2015). Classification of fresh and frozen-thawed pork muscles using visible and near infrared hyperspectral imaging and textural analysis. *Meat Science, 99*, 81–88. https://doi.org/10.1016/j.meatsci.2014.09.001

Qin, J. (2010). Hyperspectral imaging instruments. In D. W. Sun (Ed.), *Hyperspectral imaging for food quality analysis and control* (pp. 129–172). Academic.

Qin, J., Chao, K., Kim, M. S., Lu, R., & Burks, T. F. (2013). Hyperspectral and multispectral imaging for evaluating food safety and quality. *Journal of Food Engineering, 118*(2), 157–171. https://doi.org/10.1016/j.jfoodeng.2013.04.001

Qin, J., Kim, M. S., Chao, K., Schmidt, W. F., Cho, B. K., & Delwiche, S. R. (2017). Line-scan Raman imaging and spectroscopy platform for surface and subsurface evaluation of food safety and quality. *Journal of Food Engineering, 198*, 17–27.

Qin, F. L., Wang, X. C., Ding, S. R., Li, G. S., & Hou, Z. C. (2021). Prediction of Peking duck intramuscle fat content by near-infrared spectroscopy. *Poultry Science, 100*(8), 101281. https://doi.org/10.1016/j.psj.2021.101281

Qin, J., Hong, J., Cho, H., Van Kessel, J. S., Baek, I., Chao, K., & Kim, M. S. (2023). A multimodal optical sensing system for automated and intelligent food safety inspection. *Journal of the ASABE, 66*(4), 839–849.

Qiu, Z., Chen, J., Zhao, Y., Zhu, S., He, Y., & Zhang, C. (2018). Variety identification of single rice seed using hyperspectral imaging combined with convolutional neural network. *Applied Sciences, 8*(2), 212. https://doi.org/10.3390/app8020212

Qu, J. H., et al. (2015). Applications of near-infrared spectroscopy in food safety evaluation and control: A review of recent research advances. *Critical Reviews in Food Science and Nutrition, 55*, 1939–1954. https://doi.org/10.1080/10408398.2013.871693

Ravikanth, L., Singh, C. B., Jayas, D. S., & White, N. D. G. (2015). Classification of contaminants from wheat using near-infrared hyperspectral imaging. *Biosystems Engineering, 135*, 73–86. https://doi.org/10.1016/j.biosystemseng.2015.04.007

Rungpichayapichet, P., Nagle, M., Yuwanbun, P., Khuwijitjaru, P., Mahayothee, B., & Müller, J. (2017). Prediction mapping of physicochemical properties in mango by hyperspectral imaging. *Biosystems Engineering, 159*, 109–120. https://doi.org/10.1016/j.biosystemseng.2017.04.006

Senthilkumar, T., Jayas, D. S., White, N. D. G., Fields, P. G., & Gräfenhan, T. (2016). Detection of fungal infection and Ochratoxin A contamination in stored wheat using near-infrared hyperspectral imaging. *Journal of Stored Products Research, 65*, 30–39. https://doi.org/10.1016/j.jspr.2015.11.004

Shi, W., Koo, D. E. S., Kitano, M., et al. (2020). Pre-processing visualization of hyperspectral fluorescent data with spectrally encoded enhanced representations. *Nature Communications, 11*, 726. https://doi.org/10.1038/s41467-020-14486-8

Singh, C. B., Jayas, D. S., Paliwal, J., & White, N. D. G. (2010). Identification of insect-damaged wheat kernels using short-wave near-infrared hyperspectral and digital colour imaging. *Computers and Electronics in Agriculture, 73*(2), 118–125. https://doi.org/10.1016/j.compag.2010.06.001

Skjelvareid, M. H., Heia, K., Olsen, S. H., & Stormo, S. K. (2017). Detection of blood in fish muscle by constrained spectral unmixing of hyperspectral images. *Journal of Food Engineering, 212*, 252–261. https://doi.org/10.1016/j.jfoodeng.2017.05.029

Sricharoonratana, M., Thompson, A. K., & Teerachaichayut, S. (2021). Use of near-infrared hyperspectral imaging as a nondestructive method of determining and classifying the shelf life of cakes. *LWT - Food Science and Technology, 136*, 110369. https://doi.org/10.1016/j.lwt.2020.110369

Sun, J., Lu, X., Mao, H., Wu, X., & Gao, H. (2017). Quantitative determination of rice moisture based on hyperspectral imaging technology and BCC-LS-SVR algorithm. *Journal of Food Process Engineering, 40*(3), e12446. https://doi.org/10.1111/jfpe.12446

Tang, X., Rao, L., Xie, L., Yan, M., Chen, Z., Liu, S., Chen, L., Xiao, S., Ding, N., Zhang, Z., & Huang, L. (2023). Quantification and visualization of meat quality traits in pork using hyperspectral imaging. *Meat Science, 196*, 109052. https://doi.org/10.1016/j.meatsci.2022.109052

Vega Díaz, J. J., Sandoval Aldana, A. P., & Reina Zuluaga, D. V. (2021). Prediction of dry matter content of recently harvested 'Hass' avocado fruits using hyperspectral imaging. *Journal of the Science of Food and Agriculture, 101*(2), 897–906. https://doi.org/10.1002/jsfa.10697

Verdú, S., Vásquez, F., Grau, R., Ivorra, E., Sánchez, A. J., & Barat, J. M. (2016). Detection of adulterations with different grains in wheat products based on the hyperspectral image technique: The specific cases of flour and bread. *Food Control, 62*, 373–380. https://doi.org/10.1016/j.foodcont.2015.11.002

Vermeulen, P., Suman, M., Fernández Pierna, J. A., & Baeten, V. (2018). Discrimination between durum and common wheat kernels using near infrared hyperspectral imaging. *Journal of Cereal Science, 84*, 74–82. https://doi.org/10.1016/j.jcs.2018.10.001

Vidal, M., & Amigo, J. M. (2012). Pre-processing of hyperspectral images: Essential steps before image analysis. *Chemometrics and Intelligent Laboratory Systems, 117*, 138–148. https://doi.org/10.1016/j.chemolab.2012.05.009

Wang, Q. X., Su, L. H., Zou, J., Chen, N. X., Wu, T., & Yang, L. (2021). Research on hardness detection method of crisped grass carp based on visible - near infrared hyperspectral technology.

Journal of Physics Conference Series, 1757(1), 012002. https://doi.org/10.1088/1742-6596/1757/1/012002
Wei, X., Liu, F., Qiu, Z., et al. (2014). Ripeness classification of astringent persimmon using hyperspectral imaging technique. *Food and Bioprocess Technology, 7*, 1371–1380. https://doi.org/10.1007/s11947-013-1164-y
Wendel, A., Underwood, J., & Walsh, K. (2018). Maturity estimation of mangoes using hyperspectral imaging from a ground-based mobile platform. *Computers and Electronics in Agriculture, 155*, 298–313. https://doi.org/10.1016/j.compag.2018.10.021
Whitworth, M. B., Millar, S. J., & Chau, A. (2010). Food quality assessment by NIR hyperspectral imaging. *Proc. SPIE 7676, Sensing for Agriculture and Food Quality and Safety II*. https://doi.org/10.1117/12.852170
Williams, P. S., & Norris, K. H. (2001). *Near-infrared technology in the agricultural and food industries* (2nd ed.). American Association of Cereal Chemists.
Wu, D., & Sun, D.-W. (2013). Advanced applications of hyperspectral imaging technology for food quality and safety analysis and assessment: A review — Part I: Fundamentals. *Innovative Food Science & Emerging Technologies, 19*, 1–14. https://doi.org/10.1016/j.ifset.2013.04.014
Wu, N., Jiang, H., Bao, Y., Zhang, C., Zhang, J., Song, W., Zhao, Y., Mi, C., He, Y., & Liu, F. (2020). Practicability investigation of using near-infrared hyperspectral imaging to detect rice kernels infected with rice false smut in different conditions. *Sensors and Actuators B: Chemical, 308*, 127696. https://doi.org/10.1016/j.snb.2020.127696
Wu, X., Liang, X., Wang, Y., Wu, B., & Sun, J. (2022). Non-destructive techniques for the analysis and evaluation of meat quality and safety: A review. *Foods (Basel, Switzerland), 11*(22), 3713. https://doi.org/10.3390/foods11223713
Xie, C., Wang, Q., & He, Y. (2014). Identification of different varieties of sesame oil using near-infrared hyperspectral imaging and chemometrics algorithms. *PLoS One, 9*(5), e98522. https://doi.org/10.1371/journal.pone.0098522
Yang, X., Hong, H., You, Z., & Cheng, F. (2015). Spectral and image integrated analysis of hyperspectral data for waxy corn seed variety classification. *Sensors, 15*(7), 15578–15594. https://doi.org/10.3390/s150715578
Yang, D., He, D., Lu, A., Ren, D., & Wang, J. (2017a). Detection of the freshness state of cooked beef during storage using hyperspectral imaging. *Applied Spectroscopy, 71*(10), 2286–2301. https://doi.org/10.1177/0003702817718807
Yang, Q., Sun, D., & Cheng, W. (2017b). Development of simplified models for nondestructive hyperspectral imaging monitoring of TVB-N contents in cured meat during drying process. *Journal of Food Engineering, 192*, 53–60. https://doi.org/10.1016/j.jfoodeng.2016.07.015
Yang, Y., Wang, W., Zhuang, H., Yoon, S. C., & Jiang, H. (2021). Prediction of quality traits and grades of intact chicken breast fillets by hyperspectral imaging. *British Poultry Science, 62*(1), 46–52. https://doi.org/10.1080/00071668.2020.1817326
Yu, H., Qing, L., Yan, D., Xia, G., Zhang, C., Yun, Y., & Zhang, W. (2021). Hyperspectral imaging in combination with data fusion for rapid evaluation of tilapia fillet freshness. *Food Chemistry, 348*, 129129. https://doi.org/10.1016/j.foodchem.2021.129129
Zhang, W., Pan, L., Tu, S., Zhan, G., & Tu, K. (2015). Non-destructive internal quality assessment of eggs using a synthesis of hyperspectral imaging and multivariate analysis. *Journal of Food Engineering, 157*, 41–48. https://doi.org/10.1016/j.jfoodeng.2015.02.013
Zhang, C., Guo, C., Liu, F., Kong, W., He, Y., & Lou, B. (2016). Hyperspectral imaging analysis for ripeness evaluation of strawberry with support vector machine. *Journal of Food Engineering, 179*, 11–18. https://doi.org/10.1016/j.jfoodeng.2016.01.002
Zhang, L., Zhang, Q., Wu, J., Liu, Y., Yu, L., & Chen, Y. (2022). Moisture detection of single corn seed based on hyperspectral imaging and deep learning. *Infrared Physics & Technology, 125*, 104279. https://doi.org/10.1016/j.infrared.2022.104279
Zhang, J., Guo, Z., Ren, Z., Wang, S., Yue, M., Zhang, S., Yin, X., Gong, K., & Ma, C. (2023). Rapid determination of protein, starch and moisture content in wheat flour by near-infrared

hyperspectral imaging. *Journal of Food Composition and Analysis, 117*, 105134. https://doi.org/10.1016/j.jfca.2023.105134

Zhao, M., Esquerre, C., Downey, G., & O'Donnell, C. P. (2017). Process analytical technologies for fat and moisture determination in ground beef - a comparison of guided microwave spectroscopy and near infrared hyperspectral imaging. *Food Control, 73*, 1082–1094. https://doi.org/10.1016/j.foodcont.2016.10.023

Zhu, F., Zhang, Y., Shao, Y., He, Y., & Ngadi, M. (2014). Mapping of fat and moisture distribution in Atlantic Salmon using near-infrared hyperspectral imaging. *Food and Bioprocess Technology, 7*, 1208–1214. https://doi.org/10.1007/s11947-013-1228-z

Zhu, Q., He, C., Lu, R., Mendoza, F., & Cen, H. (2015). Ripeness evaluation of 'Sun Bright' tomato using optical absorption and scattering properties. *Postharvest Biology and Technology, 103*, 27–34. https://doi.org/10.1016/j.postharvbio.2015.02.007

Zhu, S., Zhou, L., Zhang, C., Bao, Y., Wu, B., Chu, H., Yu, Y., He, Y., & Feng, L. (2019). Identification of soybean varieties using hyperspectral imaging coupled with convolutional neural network. *Sensors, 19*(19), 4065. https://doi.org/10.3390/s19194065

Zhu, S., Zhang, J., Chao, M., Xu, X., Song, P., Zhang, J., & Huang, Z. (2020). A rapid and highly efficient method for the identification of soybean seed varieties: Hyperspectral images combined with transfer learning. *Molecules, 25*, 152. https://doi.org/10.3390/molecules25010152

Chapter 7
Multispectral Imaging Technique: Principles and Recent Applications in Food Processing

Tej Bhan Singh, Yuvaraj Ghosh, Riya Kundu, and Kambhampati Vivek

7.1 Introduction

Multispectral imaging (MSI) is a sophisticated technique that involves capturing images at multiple wavelengths across the electromagnetic spectrum. Unlike traditional RGB imaging, which captures only three primary colors (red, green, and blue), multispectral imaging expands this range to include additional bands beyond the visible spectrum, such as ultraviolet (UV) and infrared (IR). Spectral imaging is inherently three-dimensional, relying on data categorized within the wavelength domain. It is segmented into two main types such as hyperspectral imaging (HSI) and multispectral imaging (MI), distinguished by their treatment of wavebands. HSI involves capturing images with numerous continuous wavebands, often ranging from tens to thousands, while MSI captures images with fewer discrete wavebands, typically less than 10 and upto 20. MSI, data is stored in distinct wavebands, resulting in a photograph for each pixel composed of independent data points (Qin et al., 2013). HSI is the evolved version of MSI, which offers more detailed information by capturing images across a wider range of wavelengths with higher resolutions but it has disadvantages of cost and complexity. MSI enables the acquisition of valuable information beyond what the human eye can perceive, allowing for enhanced analysis and interpretation of objects. The technology has wide applications like in agriculture, it aids in crop monitoring, disease detection, yield prediction, and optimization of agricultural practices. By analyzing crop health

T. B. Singh
Department of Allied Health Sciences, School of Health Sciences and Technology, UPES University, Dehradun, Uttarakhand, India

Y. Ghosh · R. Kundu · K. Vivek (✉)
Department of Food Process Engineering, National Institute of Technology Rourkela, Rourkela, Odisha, India
e-mail: kambhampativ@nitrkl.ac.in

C. K. Sunil et al. (eds.), *Emerging Non-destructive Technologies for Food Quality Analysis*, https://doi.org/10.1007/978-3-031-95048-3_7

and vitality based on spectral signatures, it assists farmers in making informed decisions regarding irrigation, fertilization, and pest management, ultimately enhancing crop yield and quality. Environmental studies utilize it for assessing vegetation health, monitoring water quality, and mapping ecosystems. In medicine, multispectral imaging aids in diagnosing diseases, assessing tissue viability, and guiding surgical procedures. Remote sensing applications involve multispectral imaging for land use classification, disaster assessment, and mineral exploration. Furthermore, it is employed in art conservation to analyze pigments, detect forgeries, and preserve cultural heritage (Liang, 2012; Vozel et al., 2019; Karlas et al., 2021). Multispectral imaging offers valuable insights across diverse domains, driving advancements in research, industry, and conservation efforts. Its ability to capture detailed information across multiple wavelengths enables more precise analysis and decision-making, ultimately leading to improved efficiency, productivity, and consumer satisfaction.

Multispectral imaging also finds applications in a food industry. In the food industry, multispectral imaging holds significant promise and has become an invaluable tool for various applications ranging from quality control to safety assessment. By examining food products across multiple wavelengths, this technology can provide detailed insights into various aspects such as freshness, ripeness, contamination, and structural integrity. One of the primary applications of multispectral imaging in the food industry is quality inspection. With the ability to detect minor differences in color, texture, and composition, MSI systems can rapidly assess the quality of food products, identifying any defects or abnormalities that may compromise their integrity. For example, it can detect bruising in fruits, mold growth on surfaces, or foreign objects in packaged goods, facilitating early intervention and minimizing waste. Moreover, multispectral imaging plays a crucial role in ensuring food safety. By detecting contaminants such as bacteria, pesticides, or chemical residues, it helps prevent the distribution of contaminated products, safeguarding public health. Additionally, it enables the monitoring of food processing and packaging operations, ensuring compliance with hygiene and safety standards. In the pursuit of improving human lifestyles, it is crucial to enhance consumer awareness about modern technologies, safety, and security, particularly in the realm of food. Given the vital role food plays in our well-being, it is essential to recognize the emergence of food frauds alongside technological advancements. To address these challenges, scientists have developed various technologies, such as platform tests for milk and comprehensive physical and chemical analyses of raw materials in industries. Today, Artificial Intelligence and Machine Learning (AI-ML) are playing pivotal roles in streamlining human life, extending their applications to the food industry to mitigate contamination risks and conduct thorough analyses, both physical, chemical, and microbiological of food products. Moreover, with the aid of AI-ML, imaging techniques are now employed to study foods, effectively addressing issues like food frauds and bacterial infections in food processing.

The fast, reliable, and accurate results provided by MSI have proven instrumental in preventing food spoilage. However, in recent years, numerous cases of food fraud have been reported, such as the deliberate adulteration of meat products, such as mixing beef with pork, which not only compromises quality but also impacts socio-economic beliefs (Su & Sun, 2018). MSI stands as an emerging technology, proving to be a boon for the food processing industry. It offers a non-destructive means to evaluate various aspects crucial to food quality, including assessing for adulteration, contamination, authenticity, and safety, thus enhancing the overall quality of food products (Wu & Sun, 2013). Its applications extend to quality control, sorting and grading, as well as measuring ripeness and maturity in fruits and vegetables processing units. MSI aids in detecting foreign materials, analyzing chemical composition, identifying pathogens, ensuring traceability, and authentication (Su & Sun, 2018). Utilizing MSI, spatial and spectral data are obtained by integrating computer vision with optical techniques within the visible and near-infrared (NIR) range, providing insights into surface metabolites optimization (Su & Sun, 2018; Spyrelli et al., 2020). Unlike conventional methods such as metal detectors and X-rays that can only detect metals, imaging technology offered by MSI can identify biological level contaminations and quantitatively assess the presence of bacterial contamination, making it a versatile and indispensable tool in food analysis. Therefore, this chapter covers the know-how of multispectral imaging applied in food processing for assessing food quality and safety, detecting defects and contaminants, and monitoring various processing operations including packaging inspections and drying and dehydration processes.

7.2 Theory and Principles of Multispectral Imaging

MSI surpasses the limitations of traditional imaging methods, which are restricted to capturing images solely within the visible light spectrum. MSI goes beyond this constraint by collecting data across a broader range of wavelengths, extending beyond what is visible to the human eye, rather than solely relying on visible light (Amigo, 2019). MSI gathers information from diverse wavelengths across the electromagnetic spectrum, enabling a more comprehensive analysis of the object or subject matter. MSI represents a cutting-edge approach employed in various fields, particularly in the food quality and safety detection in food industries. The technology leverages multiple discrete bands, and allows for the comprehensive acquisition of spectral information from objects under scrutiny. The term spectrum in the multispectral imaging refers to the process of separating visible light into its constituent wavelengths, resulting in the perception of various colors. These colors are discerned by the human eye's sensitivity to different wavelengths of light. The human eye is most sensitive to light with wavelengths between approximately 400–700 nanometers (nm), which corresponds to the range of colors we perceive as violet, blue, green, yellow, orange, and red. This range is often referred to as the visible spectrum. Light outside of this range, such as ultraviolet (UV) or infrared (IR) light, is typically not visible to the human eye without the aid of specialized equipment. As a result, both multispectral and hyperspectral imaging have evolved and have sophisticated high-resolution visible sensors capable of functioning across numerous spectral bands. These bands are optimized to detect specific wavelengths of light, including blue (450–515 nm) and near-infrared (NIR, 750–900 nm), green (515.520–590.600 nm), and red (600–630–680.690 nm) (Nicolis & Gonzalez, 2021). Mishra et al. (2024) proposed that hyperspectral imaging (HSI) faces challenges due to its slower image processing and complex data handling. To mitigate these bottlenecks, multispectral imaging (MSI) was introduced into the system. MSI offers greater sensitivity to specific wavelengths, enabling faster analysis of targeted subjects. MSI systems utilize specialized sensors to detect light at specific wavelengths, capturing data across multiple spectral bands. Each band corresponds to a distinct range of wavelengths, enabling targeted imaging of desired features. Scenes are illuminated with light sources emitting across these bands, and the MSI sensor captures images representing light intensity at each wavelength. These images are processed and combined to create a multispectral image, offering comprehensive scene analysis beyond RGB. Analysis techniques include comparing spectral signatures, identifying features, and quantitative measurements for properties like chemical composition. MSI operates by capturing data within a limited range of the electromagnetic spectrum, typically between 3 and 10 bands, resulting in RGB color imagery resembling human vision. Unlike HSI, which captures images across a broader spectrum ranging from 10 to 100 bands, MSI focuses on specific wavelengths for targeted analysis. This narrow focus allows MSI to offer higher spatial resolution due to its reduced dataset compared to HSI. The simplified spectra produced by MSI facilitate easier interpretation of data. Additionally, MSI sensors are often more cost-effective than HSI sensors due to their simpler working mechanisms. Examples of multispectral imaging (MSI) applications span various fields, including the utilization of Landsat satellites for environmental monitoring. In the food industry, MSI is employed for tasks such as assessing microbiological quality, identifying adulteration (e.g., mixing minced beef with chicken meat), detecting fecal contaminants in poultry processing lines, and identifying tumors on the surface of chicken breasts (Amigo, 2019; Mishra et al., 2024).

7.3 Instrumentation and Setup

MSI systems typically include components such as a light source to illuminate the scene as shown in Fig. 7.1, a wavelength dispersive device (such as filters or prisms) to separate light into different wavelengths, and a photodetector to capture the spectral data. The core imaging sensor unit, which includes the photodetector along with associated electronics for signal processing and control is considered as the heart of MSI system. The photodetector is capable of detecting light across various wavelengths and comprised of conventional image sensors viz., charge-coupled devices (CCDs) or complementary metal-oxide-semiconductor (CMOS) detectors, for acquiring multispectral data. Additionally, electronic tuning devices may be included to adjust the spectral bands being captured. Infrared cameras like InGaAs (Indium Gallium Arsenide), InSb (Indium Antimonide), HgCdTe (Mercury

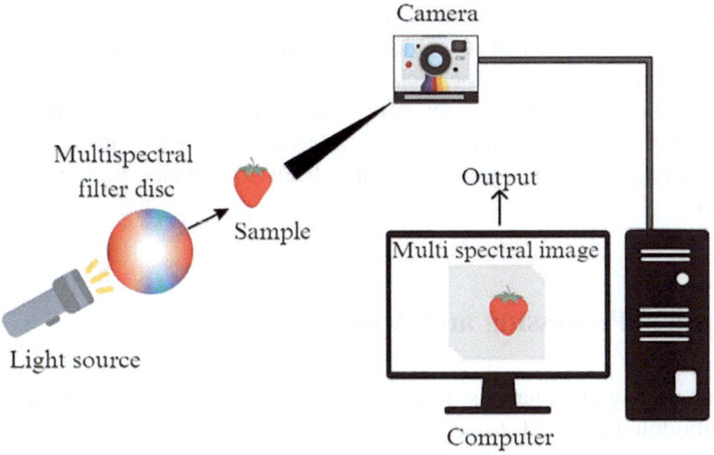

Fig. 7.1 Schematic diagram of multispectral imaging for fruit sample

Cadmium Telluride), and QWIP (Quantum Well Infrared Photodetector) focal plane arrays may also be used in MSI systems for capturing infrared spectra (Tran, 2005; Qin et al., 2013). MSI encompasses a range of non-invasive techniques utilized for analyzing various quality parameters to combat food poisoning, contamination, and fraud. This sophisticated equipment comprises an array of essential elements, including cameras, light sources, filters, optical components, spectral calibration tools, image processing software, data storage management systems, and integration and control electronics. The cameras employed in MSI systems encompass diverse types such as visible light cameras, near-infrared (NIR) cameras, infrared cameras, ultraviolet cameras, and multispectral frame cameras. These cameras play a pivotal role in capturing detailed images of food products, packaging, and scenes under examination. Light sources serve the purpose of illuminating the target area, ensuring optimal image capture. Typically, halogen lights are utilized for broadband illumination, while quartz tungsten halogen lamps are employed to generate a smooth spectrum spanning from visible to infrared regions. Filters are employed to selectively pass wavelengths through the cameras, enhancing the specificity of the imaging process. These filters are often integrated into the camera lens, enabling precise control over the wavelengths captured. Each component within the MSI setup is carefully selected based on the specific imaging requirements, including the desired spectral range and resolution. Once an image is captured, further analysis is conducted, including spectral calibration, image processing, and data storage, to extract meaningful insights and ensure the accuracy of the analysis results. This comprehensive approach enables MSI systems to effectively address challenges related to food safety and quality, providing valuable tools for food industry professionals in their efforts to safeguard consumer health and combat food-related risks. Morales et al. (2020) devised a design strategy focusing on modularity and cost efficiency, centering around a rotating filter wheel. Traditionally, sensors constitute a significant portion of the expenses in multispectral cameras. By employing a single sensor capable of capturing all wavelengths, rather than employing multiple sensors for each filter, a notable reduction in overall costs is achieved. Moreover, this approach offers added flexibility; users can readily swap out optic filters within the rotating wheel, interchange wheels with varying numbers of filter slots, or even integrate multiple wheels to enhance the device's spectral resolution.

7.4 Image Processing and Analysis

Image processing and analysis in multispectral imaging involve a set of techniques and methodologies used to extract meaningful information from multispectral images. Typically, a spectral image is structured as a cube, where the first two dimensions denote spatial information (such as rows and columns of pixels), while the third dimension encompasses a sequence of spectral images acquired at varying wavelengths. Multispectral imaging system used in various segments such as agriculture fields, food analysis, microbial analysis, and compositional analysis of food etc. (Ma et al., 2023). The initial steps crucial for achieving spectral accuracy and enabling quantitative analysis involve sensor corrections aimed at extracting geometrically consistent data from raw measurements. This essential aspect of the measurement process includes noise reduction, geometric correction, anomaly detection, lens distortion, and normalization to account for variations in illumination and sensor response across different spectral bands. Noise correction procedures are employed to mitigate systematic errors inherent in multispectral sensors, thereby ensuring accurate and reliable data for subsequent analysis (Jiang et al., 2019). Image registration is an important step followed by the preprocessing. It is a common method for registering multispectral images that involves selecting one channel as the reference, or target channel, and aligning all other image channels to match it. It is a process of aligning multiple multispectral images acquired from different sensors or at different times to a common coordinate system. Image registration ensures spatial alignment, allowing for accurate comparison and analysis of the multispectral data (Zitová & Flusser, 2003) reported three key methods for assessing image registration accuracy viz., localization error, matching error, and back-projection error. Image registration is followed by feature extraction, which involves identifying and extracting specific features or regions of interest within the multispectral images. Various techniques can be applied to analyze the resulting spectral data, including feature extraction, spectral unmixing, and classification. These methods allow for the extraction of valuable information about the imaged object, enabling tasks such as material identification and anomaly detection (Zahra et al., 2023). Machine learning (ML) necessitates the extraction of numerous features from images to optimize its parameters for generating high-quality results. MSI images offer ample information for this purpose. Consequently, the fusion of spectral imaging technology with ML has become prevalent for detecting defects in food products (Deng et al., 2023). In multispectral images, each pixel is characterized by its spectral response across different wavelengths, forming a spectral signature. These spectral signatures represent the unique characteristics of each pixel across multiple spectral bands. Using these spectral features, pixels or regions within multispectral images can be categorized into different classes or categories based on their spectral properties. This process, known as spectral classification or spectral clustering, which allows for the identification of distinct features or materials within the image. Furthermore, segmentation, the process of partitioning an image into meaningful regions or segments, can be conducted in a multidimensional feature space where each dimension corresponds to a spectral band or channel. Clustering algorithms, such as k-means or hierarchical clustering, can then be applied to this multichannel feature space to group pixels with similar spectral characteristics together. This approach allows for the identification of distinct objects or materials within the image based on their spectral properties, enabling tasks such as land cover classification, object detection, and environmental monitoring in multispectral imagery. Quantitative analysis involves measuring and quantifying the properties of objects or features within the multispectral images. This may include calculating vegetation indices, estimating concentrations of specific substances (e.g., chlorophyll content), or assessing spatial patterns and distributions within the scene (Levenson & Mansfield, 2006). Finally, visualization techniques are used to visually represent multispectral data in a meaningful and intuitive manner. False-color composites, color mapping, and spectral signatures are commonly employed for visual interpretation of multispectral images. Visualization aids in the identification of patterns, trends, and anomalies within the data, facilitating interpretation and decision-making (Offroy & Duponchel, 2016).

7.5 Applications in Food Processing

Multispectral imaging offers a wide range of applications in the food processing industry and transformed various aspects of the food processing industry, revolutionizing inspection, sorting, and quality control practices (Table 7.1). By capturing and analyzing images across multiple wavelengths, multispectral imaging systems offer valuable insights into the composition, quality, and safety of food products. A primary application of multispectral imaging lies in quality inspection and grading, where detailed images captured by multispectral cameras allow for the detection of defects, blemishes, and irregularities that could compromise product quality. Utilizing spectral signatures, these systems differentiate between various quality grades and efficiently sort food products accordingly. Moreover, multispectral imaging plays a crucial role in enhancing food safety standards by detecting contaminants, pathogens, and foreign objects in food items, thereby ensuring the integrity and safety of the products along the production line. Additionally, the technology facilitates accurate sorting and grading of food products based on various parameters such as quality, size, and ripeness, optimizing efficiency and consistency in food

Table 7.1 Applications of multispectral imaging in food processing

S. No.	Product	Spectral coverage (nm)	Significant observations	Reference
1.	Maize	375, 405, 435, 450, 470, 505, 525, 570, 590, 630, 645, 660, 700, 780, 850, 870, 890, 940, 970	Discrimination of haploid and diploid maize kernels via multispectral imaging	De La Fuente et al. (2017)
2.	Apple	780, 860, 950	Online detection of bruises in apples	Huang et al. (2015)
3.	Castor seed	375, 405, 435, 450, 470, 505, 525, 570, 590, 630, 645, 660, 700, 780, 850, 870, 890, 940, and 970	Germination ability	Olesen et al. (2015)
4.	Tomato	405, 435, 450, 470, 505, 525, 570, 590, 630, 645, 660, 700, 780, 850, 870, 890, 910, 940 and 970	Determination of lycopene and phenolic compounds content in intact tomatoes	Liu et al. (2015)
5.	Soyabean seed	405, 435, 450, 470, 505, 525, 570, 590, 630, 645, 660, 700, 780, 850, 870, 890, 910, 940 and 970	Non-destructive discrimination of conventional and glyphosate-resistant soybean seeds and their hybrid descendants using multispectral imaging and chemometric methods	Liu et al. (2016a)
6.	Rice seed	405, 435, 450, 470, 505, 525, 570, 590, 630, 645, 660, 700, 780, 850, 870, 890, 910, 940 and 970	Rapid and nondestructive determination of transgenic rice seeds from its non-transgenic counterparts	Liu et al. (2014a)
7.	Sunflower seed	435, 470, 525, 630, 645, 660, 700, 850, 910, and 970	Rancidity degree	Ma et al. (2015)
8.	Tomato	375, 405, 435, 450, 470, 505, 525, 570, 590, 630, 645, 660, 700, 780, 850, 870, 890, 940 and 970	Varietal identification of Tomato	Shrestha et al. (2015)
9.	Carrot slices	405, 435, 450, 470, 505, 525, 570, 590, 630, 645, 660, 700, 780, 850, 870, 890, 910, 940 and 970	Real-time and non-destructive determination of colour change and moisture distribution using multispectral imaging during the hot air dehydration of carrot slices	Liu et al. (2016b)
10.	Strawberry	405, 435, 450, 470, 505, 525, 570, 590, 630, 645, 660, 700, 780,	Firmness, Soluble solids content	Liu et al. (2014b)

(continued)

Table 7.1 (continued)

S. No.	Product	Spectral coverage (nm)	Significant observations	Reference
		850, 870, 890, 910, 940, and 970		
11.	Apple	680, 880, 905, 940	Firmness evaluation	Qin et al. (2013)
12.	Chicken	520, 560	Feces detection	Qin et al. (2013)
13.	Citrus	730, 830	Canker detection	Lleó et al. (2009), Qin et al. (2012)
14.	Peach	450, 675, 800	Maturity evaluation	Lleó et al. (2009)
15.	Tea	580, 680, 800	Category sorting	Wu et al. (2008)
16.	Tomato	530, 595, 630, 850	Maturity evaluation	Hahn (2002)
17.	Rice seed samples	405, 435, 450, 470, 505, 525, 570, 590, 630, 645	Varietal Identification	Liu et al. (2014b)
18.	Carrot samples	430–970	Varietal Identification	Gaikwad (2021)
19.	Citrus orchard and peach orchard	494–760	Disease Detection	Pourazar et al. (2019
20.	Tulsi, Peppermint, Bael, Lemon balm, Catnip, and Stevia	Five bands with range of 460 nm to 1560 nm known as Red (R), Blue (B), Green (G), (NIR) and Spectral Bands Shortwave Infrared (SWIR)	Classification of medicinal plant leaves utilizing multi-spectral and texture features through a machine learning (ML) methodology	Naeem et al. (2021)

processing operations. Multispectral imaging also enables the assessment of moisture content and freshness of perishable items, aiding in storage optimization and waste reduction. Furthermore, it enables authentication and traceability by analyzing unique spectral signatures, verifying the authenticity of food items, and tracking their origin throughout the supply chain. Packaging inspection in food processing facilities is another crucial application of multispectral imaging, where it detects defects, leaks, or seal integrity issues, ensuring proper packaging and product protection. Mishra et al. (2024) demonstrated the potential of multispectral imaging in rapidly and non-destructively analyzing food contamination, spoilage, and related factors, particularly in detecting aflatoxin contamination in almond kernels. Moreover, multispectral imaging allows real-time monitoring of food processing operations, facilitating timely adjustments to optimize productivity, quality, and resource utilization. Additionally, it aids in monitoring and controlling drying and dehydration processes, ensuring uniformity and consistency in product quality while minimizing energy consumption and waste. By harnessing the capabilities of multispectral imaging systems, food manufacturers can improve product quality, safety compliance, and consumer satisfaction in an increasing competitive market landscape.

7.5.1 Food Quality Assessment

Quality refers to the characteristics of a food product that satisfy customer expectations and comply with regulatory standards. Food industry utilizes various technologies to assess and maintain the quality of food products. These technologies encompass a wide range of approaches, from traditional methods to advanced analytical techniques. Spectroscopy, chromatography, imaging techniques, E-Nose and E-Tongue, Rapid microbial techniques, smart and intelligent packaging, and block chain technology and IoT. are widely used for quality evaluation in food industry (Ghasemi-Varnamkhasti et al., 2018),(Gomes et al., 2022). Among these, imaging techniques are highly valuable for evaluating the quality of food products due to non-destructive evaluation, real time monitoring, versatility and adaptability, comprehensive analysis, and standardized evaluation (Vinothkanna et al., 2024). Imaging techniques used in food industry includes, MSI, HIS, digital imaging, X-ray imaging, and ultrasonic imaging. Each of these imaging techniques has its advantages and applications in the food industry, depending on the specific requirements of quality evaluation, product type, and processing stage. Multispectral imaging technique offers several advantages compared to other imaging techniques used in the food industry due to the fast, accurate and low cost with enhanced productivity thus making it a valuable tool for quality assessment and inspection in food industry.

MSI offers valuable insights into the ripeness, freshness, and overall condition of fruits and vegetables. Ripeness assessment is one of the key areas where multispectral imaging outsmarts other imaging techniques. It provides a non-destructive method to discern the maturity of fruits and vegetables by analyzing spectral signatures associated with different ripening stages. This enables precise classification, empowering producers to optimize harvesting schedules and minimize post-harvest losses effectively. In addition, multispectral imaging addresses the crucial aspect of evaluating freshness. It detects subtle changes in color, texture, and surface features as fruits and vegetables age, facilitating the timely removal of items that may compromise produce quality. This not only improves product quality but also enhances processing efficiency, leading to reduced waste and optimized resource utilization across the food supply chain. MSI system models developed for pomegranate fruit quality had a great ability to predict pH and TSS accurately using a MSI system equipped with visible/NIR spectroscopy in 400–1100 nm range. While for assessing celeriac and carrots quality 430–970 nm range was used (Clemmensen et al., 2012).

In the assessment of grain quality, MSI proves to be indispensable in evaluating critical parameters such as colour and defects. Because, laboratory based chemical detection methods are not suitable for rapid inspection and quality assurance protocols. The presence of green immature kernels downgrades grain shipments due to their adverse effects on the appearance and processing quality of the grain. The capability of MSI is to capture detailed data across multiple wavelengths enables a comprehensive analysis, ensuring that only high-quality products. Gray to black discoloration of the wheat kernel caused by mildew, adversely affects the color of the flour, resulting in a less vibrant and duller product of inferior value and quality. Sprout damage poses a severe threat to the quality of end-products, particularly in wheat, rye, canola and barley (Su & Sun, 2018). The starchy regions affected by sprout damage play a crucial role in reducing semolina yield during milling, results in the production of undesirable durum flour with poor color quality. This is unsuitable for production of high-quality pasta thus leads to considerable economic losses (Shahin et al., 2012). Therefore, these four wavelengths (494, 578, 639, 678 nm) were chosen for the development of a multispectral system for Fusarium-damaged kernels of wheat. A multispectral system could be used to identify chlorophyll in barley kernels at 580, 671, 730 nm.

In meat industry MSI can assess the quality of meat products by analyzing factors such as marbling, color, fat content, and texture. Different wavelengths can be used to detect variations in these attributes, allowing for objective and standardized grading according to industry standards. Determine the freshness of meat by analyzing its color, surface moisture, and other visual attributes. Changes in color and appearance over time can indicate spoilage or degradation. Identify contaminants such as foreign objects, bone fragments, or other defects in meat products. By analyzing spectral signatures and texture patterns, multispectral systems can distinguish between normal meat tissue and unwanted materials, ensuring product safety and compliance with regulatory requirements. Detect pathogens such as bacteria or parasites on the surface of meat products. By using specific wavelengths that interact with microbial cells, multispectral systems can differentiate between contaminated and uncontaminated areas, enabling targeted sanitation measures and reducing the risk of foodborne illness. MSI integrated into automated sorting and grading systems could classify meat products based on quality attributes and specifications. By analyzing multiple spectral bands simultaneously, multispectral systems can sort meat products according to desired criteria such as fat content, muscle maturity, or visual appearance, optimizing product consistency and market value. The spectral intensities at 494 nm from multispectral imaging of pork sausages with different cooking temperatures exhibited significant variation. This variation occurred within the range of 510–650 nm, which is associated with heme pigments such as hemoglobin, myoglobin, and their oxides (Feng et al., 2018).

7.5.2 Food Safety Inspection

Food safety includes practices and regulations aimed at ensuring that food is safe for human consumption. It involves preventing contamination throughout the food supply chain. While food safety inspection involves the systematic examination and evaluation of food production, processing, handling, and distribution facilities, as well as food products themselves, to ensure compliance with food safety regulations and standards. These inspections are conducted by trained inspectors employed by regulatory agencies or third-party organizations and may occur at various stages of the food supply chain, from farm to fork (Bendeković et al., 2015). Food safety inspections include identifying and mitigating potential hazards, verifying adherence to Good Manufacturing Practices (GMPs) and Hazard Analysis and Critical Control Points (HACCP) principles, assessing hygiene and sanitation practices, and ensuring accurate labeling and traceability of food products. Inspection findings may lead to corrective actions, enforcement measures, or recommendations to improve food safety practices and prevent foodborne illnesses (Todd, 2020). Food safety inspections can effectively be performed using image techniques, including multispectral imaging. This technique captures images across various wavelengths, allowing for the detection of adulteration and contaminants, assessment of food quality, inspection of packaging, evaluation of freshness, and authentication of food products. This approach offers a non-invasive and potentially automated method to identify hazards, monitor quality parameters, and ensure compliance with food safety regulations. Using this technology food inspector can enhance the effectiveness and efficiency of food safety inspections, which leads to safeguarding public health and ensuring the safety of the food supply (Feng & Sun, 2012). Inspection methods, including multispectral imaging and other advanced technologies, are employed to detect adulterants and contaminants in food products. Adulteration involves the addition of inferior, harmful, or unauthorized substances to food products, which can pose risks to consumer health and safety. These substances may include contaminants, toxins, allergens, or unapproved additives, and their presence in food products can lead to adverse health effects such as foodborne illnesses, allergic reactions, or toxicity. Several studies demonstrate the effectiveness of MSI in various food safety applications. (Liu et al. 2022) achieved a 93.33% accuracy in detecting the harmful mycotoxin Zearalenone (ZEN) in maize. (Bandara et al. 2020) successfully detected tartazine-colored rice flour adulteration in turmeric powder using MSI. Lee et al. (2016) and Zhu et al. (2017) detected virus infections in watermelon seeds and tobacco samples with classification accuracies exceeding 83% using MSI. Ma et al. (2014) estimated aerobic plate count (APC) in cooked pork sausages, achieving an R^2 of 0.89 by combining MSI with partial least square regression (PLS-R) analysis. Similarly, Wang et al. (2015), Jiang et al. (2016), and Chu et al. (2017) demonstrated the practicality of MSI in detecting aflatoxin B1 levels in maize and peanut kernels using principal component analysis (PCA) and support vector machine (SVM). These findings underscore the versatility and efficacy of MSI in detecting various food adulterations and contaminants, including almond, cumin, and wheat flour adulterations (Florian-Huaman et al., 2022).

7.5.3 Sorting and Grading

Sorting and grading are important processes in food processing aimed at categorizing and organizing food products based on various attributes such as size, shape, color, quality, and ripeness. These processes help ensure uniformity, consistency, and quality control throughout the food supply chain. Sorting involves separating food products based on predetermined criteria which ensures that only products meeting certain standards or specifications move further along the processing line, while defective or substandard items are removed. While grading involves assigning a quality grade to food products based on specific standards or criteria. This can include factors such as size, color, texture, flavor, and overall appearance. Grading helps establish the value of the product and assists consumers in making informed purchasing decisions. Both sorting and grading can be effectively done using imaging techniques, which have become increasingly popular and efficient in modern food processing operations. Imaging techniques utilize cameras and sensors to capture detailed images of food products, which are then analyzed using software algorithms to make sorting and grading decisions based on predetermined criteria. Methods like colour sorting, size sorting, and defects detection are popular in the food industries. Manual sorting by visual inspection is labor-intensive, sluggish, unreliable, and prone to errors. Therefore, to achieve higher accuracy and consistency in sorting and grading operations, leading to improved product quality, reduced waste, and increased efficiency in the production process MSI is employed. The technology has revolutionized sorting and grading processes in the food industry, offering unparalleled precision and efficiency across various applications. A primary use of MSI lies in sorting and grading fruits and vegetables, where it leverages its capability to capture data across various wavelengths to assess color, size, and external characteristics. The defective cucumbers consistently allow more light to pass through them within a specific range of wavelengths compared to normal cucumbers in the region of 675–1000 nm. This observation suggests that the difference in transmittance could potentially be used as a method to distinguish defective cucumbers from normal ones (Lu et al., 2017). Laser-based multispectral scattering imaging had shown promising results in the online sorting and grading of apples based on their firmness. Online sorting and grading with multispectral scattering profiles and its function parameters correlated well for blueberries and apples for determining firmness of fruits (Mendoza et al., 2011).

In the grain industry, size and shape play crucial roles in determining the grade and market value of grains. Multispectral imaging offers precise measurements, allowing producers to accurately adhere to quality standards. Automated sorting is indispensable for ensuring the efficiency and consistency of food processing operations. Significant advancements in sorting procedures have been achieved with the integration of MSI technology. The application of a multispectral imaging system for differentiating between haploid and diploid maize seeds was reported by De La Fuente (2017). Fully automated LED-pulsed multispectral system utilized a combination of three visible and three near-infrared LEDs, each with peak emission wavelengths set at 470–1070 nm, tailored specifically for sorting wheat kernels (ElMasry et al., 2019). By precisely classifying food items based on multiple criteria such as size, shape, ripeness, and color, MSI enhances overall quality control measures. The technology enables the accurate identification of minute differences in attributes, ensuring that only products of the highest caliber reach consumers. Further, the potential of automated sorting is maximized through the combination of MSI with robotics and machine learning techniques. Machine learning algorithms continuously improve the accuracy and efficiency of sorting activities over time, while robotics expedite the sorting process, reducing the need for human intervention and minimizing the risk of errors. Consequently, food processing lines benefit from enhanced productivity and cost savings due to increased efficiency and reliability. Colour sorting is another vital application of MSI in the food sector, enabling the recognition and classification of items based on their color attributes. MSI systems excel at capturing fine details about color at different wavelengths, allowing them to identify even the smallest color variations that may indicate contamination or quality issues. By photographing food products at multiple light wavelengths and comprehensively examining their color composition, MSI systems can promptly detect color changes that may signify flaws or anomalies in the products. This sophisticated color perception is essential for ensuring food safety and quality, as it enables companies to promptly detect and separate tainted products before they reach the market.

7.5.4 Composition Analysis

Composition analysis in the food industry involves determining the chemical composition and nutritional content of food products. It includes identifying and quantifying macronutrients, micronutrients, moisture, ash content, fibers, additives, and contaminants. This analysis is crucial for nutritional labeling, quality control, product development, regulatory compliance, allergen detection, and contaminant monitoring (Ross et al., 2018; Md Noh et al., 2020). Composition analysis can be conducted using imaging techniques, although it may be more common to use spectroscopic techniques for this purpose (Singh et al., 2023). Imaging techniques, such as multispectral imaging or hyperspectral imaging, can provide valuable information about the spatial distribution of components within a food sample, allowing for non-destructive analysis and visualization of its composition. Food products consist of various chemical components, one of which is water. These components serve as the fundamental elements that contribute to the structure and composition of the food. Water plays a crucial role in facilitating various biochemical reactions within the food, in addition to serving as a key component in providing the food's texture, moisture, and overall structure. Multispectral imaging is instrumental in determining complex structures within samples. This technique captures images of food samples in a three-dimensional spectral datacube (x, y, λ), representing both spatial and spectral data for every pixel. Each pixel in the image provides information not only about its spatial location but also its spectral characteristics across different wavelengths. By analyzing these spectral signatures, multispectral imaging identifies and characterize various components or structures within the sample, offering valuable insights into its composition (Singh et al., 2023). Spectral data collected through techniques such as multispectral imaging can reveal intricate structural characteristics of samples. These characteristics are associated with strong absorptions of pigments and overtones, or combination modes of molecular bonds such as C–H (aliphatic and aromatic), C–O (carboxyl), N–H (amide and amine), and O–H (hydroxyl) functional groups. Multispectral imaging is particularly valuable for determining chemical constituents like soluble solids content, monosaccharides, disaccharides, and antioxidants, providing comprehensive insights into the composition of the food samples. Based on multispectral imaging, the soluble solid content of pears was determined by integrating Partial Least Squares Regression (PLSR) with variable selection methods in the range of 400–1000 nm. The SPA-PLSR model showcased a relatively modest predictive capability while CARS-SPA-PLSR model achieved better performance for determining soluble solid content in pears (Fan et al., 2015). The MSI technique was also examined for its potential in predicting the levels of glucose, fructose, or sucrose in potato, kiwi, and sugar beet samples (Rady et al., 2015; Pan et al., 2016). Multispectral imaging has been extensively utilized for measuring moisture content (MC) in a variety of food products, including various fruits and vegetables like peanut, mango, apple, sugar beet, potato, sweet potato, and carrot. This technique involves the use of eight specific feature wavelengths i.e., 977, 1068, 1138, 1202, 1319, 1366, 1399, and 1470 nm for accurate and reliable determination of moisture content in these food items (Su & Sun, 2018). MSI system coupled with chemometric analysis, has proven effective in determining anthocyanin contents in mulberry and sweet potato. This approach also shows promising results for noninvasively measuring lycopene content and total polyphenols content (TPC) (Su et al., 2017). Furthermore, multispectral imaging has successfully determined oil content in palm oils, protein content in wheat, and pH values in litchi and apple fruits (Budiman & Seminar, 2022). In specific cases, such as the determination of insoluble and soluble dietary fiber in celeries, genetic synergy interval partial least square regression (GA-Si-PLSR) models achieved the highest accuracies. Additionally, the techniques used for visualizing caffeine content in coffee beans, tannins in persimmons, and starch content in potato and sweet potato tubers, PLSR models in combination with wavelength selection methods like RF, SPA, and FMCIA demonstrated superior performance (Pu et al., 2015; Zhang et al., 2017). (Lie et al. 2014) reported on the efficacy of multispectral imaging for assessing firmness, total soluble solids (TSS) content, and ripeness stage in strawberry fruit. The study utilized 19 wavelengths spanning from 405 to 970 nm as a nondestructive technology for this purpose. To create spatial distribution maps of chemical compositions over the surface of pineapple slices, filter-based or light-based multispectral imaging systems can be employed. These systems utilize various filters or light sources to capture images of the pineapple slices at different wavelengths across the electromagnetic spectrum. By analyzing the spectral information obtained from these images, the spatial distribution of chemical compositions such as moisture content, sugar content, acidity, and ripeness can be mapped out (Mollazade et al., 2023).

7.5.5 Process Modelling

Process modeling in the food industry involves creating mathematical or computational representations of food processing operations. These models simulate various aspects of food production, including unit operations, process flows, temperature profiles, ingredient interactions, and product quality parameters. Process modeling aims to understand, optimize, and control food manufacturing processes to improve efficiency, quality, and safety while minimizing costs and waste. Imaging techniques like multispectral imaging and hyperspectral imaging can be used for process modeling in the food industry, although these techniques are typically used more for monitoring and analysis rather than direct process modeling. Generally, these imaging techniques including computer vision can provide valuable data for understanding and optimizing food processing operations. An automated multispectral image processing method relies on an unsupervised machine learning approach, specifically Gaussian Mixture Models, and introduces a novel unsupervised scheme for optimizing the segmentation process through spectral band selection. This methodology is designed to achieve high-throughput analysis and automated processing of food data samples, particularly focusing on meat and meat products (Tsakanikas et al., 2015). The study conducted by Li et al. (2021) developed a classification method by integrating MSI with machine learning-based classifiers, showcasing the potential of machine learning and feature fusion techniques for meat classification using multispectral imaging. Multispectral imaging facilitates analysis through statistical methods and machine learning algorithms to discern trends and patterns in processing parameters over time. It enables real-time monitoring of all processing parameters, enabling swift detection and rectification of deviations from the ideal state. By calibrating multispectral imaging systems with ideal conditions before production, any deviations can be promptly identified and corrected, ensuring consistent quality and efficiency throughout the production process. MSI enables the collection of spatially resolved spectral data, capturing information across multiple wavelengths of light. By analyzing these spectral signatures, process engineers can monitor key parameters such as moisture content, chemical composition, and physical characteristics of food products in real-time. Multispectral imaging facilitates the development of accurate mathematical models and simulations that represent the complex dynamics of food processing systems. These models can be used to optimize process parameters, predict outcomes, and identify potential areas for improvement. Additionally, multispectral imaging allows for the detection of deviations from ideal process conditions, enabling timely adjustments to maintain product quality and consistency. (Bandara et al. 2019) developed a multispectral imaging system tailored for assessing meat quality. The system comprises six LEDs with nominal wavelengths ranging from 405 nm to 740 nm. Light emitted from these LEDs is directed onto the meat sample, positioned within a dark chamber, through an integrating hemisphere. This setup allows for the comprehensive evaluation of meat quality using multispectral imaging techniques. Yang et al. (2017) also developed a model for non-destructive monitoring TVB-N contents in cured meat during drying process. Achieving effective multispectral inspection involves analyzing a selected region of interest within the chicken breast area using images captured between 580 nm and 620 nm. These wavelengths exhibit the most significant difference between the average spectra of wholesome and unwholesome chicken samples. By focusing on this specific spectral range, multispectral imaging can efficiently distinguish between desirable and undesirable characteristics in chicken products. Ensuring accurate quality assessment and adherence to food safety standards is crucial, as it guarantees compliance with regulatory standards and industry guidelines, thereby ensuring safe consumption (Wu & Sun, 2013).

7.5.6 Authentication and Traceability

Ensuring authenticity and traceability of food products in the food industry, is important for safeguarding consumer health. Various imaging techniques involves capturing and analyzing visual data to verify the identity, origin, and movement of food products throughout the supply chain. MSI can capture detailed images of food products, that serve as digital records which help customers verify the authenticity of products by comparing them to reference images or databases of authentic products. Complex supply chains, limited transparency, potential for fraud, data management issues, and the need for standardized systems are some of the challenges requires robust technologies. MSI has emerged as a powerful technology for addressing these challenges. Authentication is the process of verifying the genuineness of a food product, is critical in combating food fraud and ensuring compliance with regulatory standards. MSI enables precise authentication by analyzing specific characteristics of food items, such as composition, origin, and production methods. By capturing images at multiple wavelengths across the electromagnetic spectrum, MSI can detect subtle differences in food properties that may indicate adulteration or misrepresentation. Spectral imaging has demonstrated significant potential for applications in screening food authenticity in recent years (Pu et al., 2014). Liu (2021) used multispectral imaging coupled with chemometric methods to detect the authenticity of Thai jasmine rice. While traceability, on the other hand, involves tracking the movement of food products throughout the supply chain. MSI facilitates traceability by providing detailed information about the origin, processing history, and distribution paths of individual food items. Through techniques like barcode scanning or RFID tagging, MSI enables the creation of robust traceability systems that enhance food safety and enable rapid recalls in the event of contamination incidents. This also maintain the quality and improve the safety of fresh produce (Gupta et al., 2023). Imaging techniques, are capable of distinguishing between agricultural products originating from different regions. MSI plays a crucial role in generating integrated spectroscopic fingerprints for food geographical origin traceability by capturing comprehensive spectral data. However, the effectiveness of these methods may be compromised when agricultural products are adulterated with the same type of product but from different geographic origins. The combination of near-infrared hyperspectral imaging with multivariate analysis has demonstrated excellent performance in detecting the geographical origin traceability of agricultural products (Li et al., 2024). Utilizing multispectral imaging technology enables automated detection and traceability of food products to be achieved. Tea quality assessment relied on subjective judgments by professional tea tasters, which could be inconsistent and inaccurate due to their reliance on sensory perceptions. To address this, imaging techniques offer a more objective means of evaluation. Quantitative analysis and geographical traceability of black tea can be enhanced through the application of imaging techniques (Ren et al., 2013).

7.5.7 Packaging Inspection

Packaging inspections in the food industry refer to the systematic examination and evaluation of packaging materials used to store, protect, and transport food products. These inspections are conducted to ensure that the packaging meets regulatory requirements, industry standards, and specific quality criteria to maintain the safety, integrity, and freshness of the food (Paiva, 2013). In addition to the general regulations that apply to conventional foods, specialty foods often have additional statutory and regulatory requirements to meet. Products like infant formula are subject to specific standards set by the Food and Drug Administration (FDA), encompassing compositional and labeling requirements. Further, stringent Good Manufacturing Practices (GMPs) are imposed, including mandatory testing protocols before the formula is approved for distribution (Pascall et al., 2022). Visual inspection involves manual examination to detect defects like tears or misprints. Dimensional analysis ensures packaging meets size requirements and functions properly. Leak testing assesses seals and closures for any leaks using methods like pressure or vacuum decay. Seal integrity assessment methods such as polarized light and laser scatter are extensively employed in the packaging industry to effectively detect faults in heat seals of food packaging. These techniques leverage the principles of light polarization and laser scattering to scrutinize seal quality and integrity. By analyzing the polarization patterns or scattered light signatures, these methods can identify defects or inconsistencies in the seals, ensuring that they meet stringent quality standards (Huang et al., 2014). X-ray inspection provides non-destructive imaging to detect foreign objects or defects within packaging. Optical inspection uses cameras and algorithms to detect defects, contaminants, or inconsistencies. A line scan camera, comprising 410 lines by 320 pixels, with 168 channels covering wavelengths ranging from 890 to 1728 nm, was employed to conduct scans of the food packaging (Benouis et al., 2021). Gas analysis monitors gas composition in modified atmosphere packaging for freshness. Sensory evaluation assesses packaging odor or taste for any signs of contamination or deterioration. Barcode and Radio Frequency Identification (RFID) scanning verifies accurate tracking and authentication throughout the food supply chain. These methods collectively safeguard packaged food products, ensuring they meet regulatory standards, maintain integrity, and protect consumer health. A hyperspectral-multispectral line-scan imaging system is employed for the purpose of distinguishing between healthy and diseased chickens (Xiong et al., 2015). Similarly, MSI systems was used to detect the defects in packed potato by inspecting seven distinct types of defects using a multispectral camera operating within a spectral range of 676–952 nm. Employing preprocessing techniques including simple thresholding and morphological filtering, can narrowed down the analysis to three specific wavelengths (690, 757, and 927 nm) to compose a new image. Through this approach, they achieved a model accuracy rate of 94.36% in identifying defective potatoes (Zhang et al., 2019). Implementing fault detection algorithms based on multispectral bands necessitates significant computational resources. However, in the context of food packaging, where processing occurs rapidly with multiple trays handled per second, it becomes necessary to conduct detection algorithms in real-time. These algorithms must be capable of swiftly assessing tray quality as trays circulate along the production line. This demands efficient processing techniques and optimized algorithms that can quickly analyze multispectral data streams, identify potential faults or defects in packaging, and provide timely feedback for corrective actions. Real-time fault detection in food packaging minimizes production downtime and waste (Gültekin et al., 2022). Therefore, MSI play a critical role in upholding consumer confidence and ensuring the safety and quality of food products from production to consumption and prevent human poisoning.

7.5.8 Storage Conditions

Storage of foods is an important operation in the food industry for preserving the quality and safety. It facilitates the maintenance of food quality by preventing spoilage, degradation, and microbial growth, thereby preserving flavor, texture, appearance, and nutritional value. Proper storage conditions, including temperature and humidity control, as well as appropriate packaging, minimize the risk of spoilage and foodborne illnesses. Mango, a significant tropical fruit, is particularly vulnerable to chilling injury when subjected to low temperatures. Optimal storage conditions for mangoes typically fall within the range of 12 °C to 13 °C (Wazed et al., 2023). Compliance with food safety regulations necessitates adherence to specific storage practices, such as maintaining designated temperatures and labeling requirements. Effective storage practices also contribute to optimized inventory management, ensuring adequate stock levels while minimizing waste and associated costs. By extending the shelf life of food products and reducing losses due to spoilage or deterioration, proper storage ultimately supports efficient operations and cost savings within the food industry (Allai et al., 20232023). During storage, perishable food commodities undergo changes in texture, color, and internal chemical composition, ultimately resulting in spoilage. MSI was used to detect defects on the matured red tomatoes and classify bruised and non-bruised apples with an accuracy more than 90% (Yang et al., 2014; Huang et al., 2015). Similarly, Li et al. (2016)

achieved a high accuracy of 96.6% in detecting bi-color skin defects in peaches using multispectral imaging. MSI system can be used to predict the chilling injury in mangoes during storage. Zhu et al. (2019) explored the suitability of imaging techniques for assessing the freshness of spinach leaves during storage at various temperatures. They utilized visible-SWNIR (380–1030 nm) and NIR reflectance (874–1734 nm) hyperspectral imaging systems. Additionally, Babellahi et al. (2020) assessed chilling injuries in green bell peppers using HSI covering the spectral range of 400–2500 nm. They also developed a Partial Least Squares Regression (PLSR) model to predict storage duration at 4 °C, aiding in preemptive measures to prevent further damage. Moreover, Yan et al. (2017) employed an HSI device (380–1000 nm) to quantify soluble and insoluble dietary fiber in fresh-cut celery, a widely consumed vegetable. The spatial distribution of fiber content was visualized till 28 days of storage. The apple industry encounters difficulties in monitoring and mitigating postharvest losses caused by invasive pests during extended storage periods. Therefore, Khaled et al. (2023) developed a near-infrared hyperspectral imaging system and utilized to develop machine learning models for predicting the quality attributes of apples during a 20-week storage period. Partial least squares regression (PLSR) and support vector regression (SVR) methods were employed for this purpose.

7.5.9 Drying and Dehydration

High water content in food products creates favorable conditions for microbial growth, increasing the risk of contamination and spoilage. Excessive moisture content not only poses challenges during transportation and storage but also leads to increased transportation costs and microbial contamination risks. Therefore, drying and dehydration are essential processes in food preservation, primarily aimed at reducing the moisture content of food products to enhance their shelf life and stability. By removing moisture, these processes inhibit the growth of microorganisms, including bacteria, molds, and yeasts, which thrive in moisture-rich environments. Additionally, reducing water activity impedes enzymatic activity, slowing down biochemical reactions that lead to spoilage, such as oxidation and hydrolysis. Further, drying and dehydration help prevent physical deterioration of food products, such as texture loss, clumping, and caking, which can occur when water interacts with food components like starches and proteins. These processes also reduce the weight and volume of food, making them more convenient and cost-effective to transport and store. Additionally, drying and dehydration enhance the flavor concentration and intensity of food products by removing water, resulting in more concentrated taste profiles. Multispectral imaging expedites the determination of moisture distribution in various food products. It discerns regions of high and low moisture within food matrices, aiding in the identification of areas drying too quickly or slowly. By analyzing reflected light intensity at specific wavelengths associated with water absorption, multispectral imaging generates moisture maps of the food surface. These maps inform the adjustment of drying parameters for uniform drying across food matrices, providing real-time data. For instance, multispectral imaging effectively determined moisture distribution in carrot slices (Liu et al., 2016b), tea leaves (Liu et al., 2021), and beef samples (Liu et al., 2016a). Moreover, multispectral imaging finds utility in determining temperature distribution within food products and around the drying chamber. Drying parameters such as temperature, relative humidity, airflow can be optimized and after analyzing the relation between drying parameters and the moisture distribution, optimal conditions can be developed for uniform drying condition. Multispectral imaging detects potential drying defects such as surface cracks, blistering and localized overheating which can hinder uniform drying. Multispectral imaging takes images of food products at different wavelengths including near infrared and visible region of electromagnetic spectrum. Yonus et al. (2020) introduced a nondestructive approach for measuring the water status in hot-air dried mushrooms using multispectral imaging (MSI) technology in conjunction with chemometric methods. The average reflected spectrum intensity of mushrooms was measured using the MSI system at various drying periods. A consistent increase in reflectance intensity was observed across wavelengths, although each drying period (120–420 minutes) to exhibited lower intensity compared to the previous stage. Similarly, Ren et al. (2023) investigated the nondestructive detection of water status and distribution in corn kernels during hot air drying utilizing multispectral imaging techniques. In this study the chemometric methods employed to construct prediction models included backpropagation neural network, least-squares support vector machine, and partial least squares. Among these, the backpropagation neural network demonstrated the most effective prediction performance for both total and free water contents.

7.6 Recent Advances and Emerging Trends

In the era of artificial intelligence and machine learning (AI-ML), the advanced technology has become widespread and can easily penetrate in various fields. Incorporating AI-ML into multispectral imaging (MSI) systems has proven to be transformative, enhancing the capability to analyze extensive spectral data in real-time (van Ineveld et al., 2022). Initially designed for measuring physical parameters of plants and seeds, recent innovations in imaging technology have expanded the utility of MSI within the food processing chain. For instance, MSI is now employed for assessing the oil content of seeds, determining moisture levels, monitoring water absorption, and profiling nutrient composition in food components (Karlas et al., 2021). In the food processing industry, MSI has emerged as a versatile tool with applications ranging from quality control to product authentication. It plays a vital role in ensuring food safety and integrity by detecting contaminants, identifying adulterants, and assessing product quality. Additionally, MSI aids in optimizing processing parameters, improving product consistency, and enhancing overall efficiency. Recent advancements in sensor technology have facilitated the development of portable MSI devices, including cameras and infrared cameras, tailored specifically for food processing applications. Moreover, the integration of system-on-chip technology has streamlined data acquisition, processing, and visualization across various electromagnetic spectra (Valdes-Garcia et al., 2020). By leveraging MSI systems alongside hyperspectral imaging (HSI), food processors can achieve more accurate and comprehensive analysis of food products. This integration enables the fusion of high-resolution spatial images with detailed spectral information, resulting in enhanced product characterization and real-time decision-making (Sara et al., 2021). Ultimately, the utilization of MSI in food processing holds tremendous potential for improving product quality, safety, and efficiency across the entire food supply chain.

7.7 Future Perspectives and Challenges

The future prospects of MSI in the food processing industry hold immense promise but are accompanied by several challenges. As awareness regarding food safety and hygiene continues to grow, regulatory agencies emphasize strict monitoring of food quality. MSI, hailed as a boon for agriculture and food sectors, presents an opportunity to enhance food safety measures and optimize resource utilization. The integration of MSI with artificial intelligence and machine learning (AI-ML) technologies offers unprecedented efficiency in screening and analyzing food samples within short timeframes. However, to realize its full potential, further scientific research and collaboration are essential to refine MSI methodologies and achieve higher-quality results (Ma et al., 2023). While MSI has demonstrated benefits in various aspects of food processing, challenges persist. Visualizing comprehensive sample views is hindered by limitations in the ultraviolet (UV) and near-infrared (NIR) regions, where light penetration is minimal. Overcoming this obstacle requires innovative approaches and interdisciplinary collaboration, involving experts in food science, optics, electronics, computer programming, automation control, and mechanical engineering. Cost considerations, including hardware and software expenses, pose another challenge, along with the need for precise timing, accuracy, and result validation. To address these challenges, investments in automation, research, and advancements in high-resolution, real-time imaging systems are crucial (Su & Sun, 2018). Moreover, MSI generates vast amounts of data due to its high-resolution images with multiple spectral bands, necessitating efficient data processing and management strategies. Handling this data efficiently requires significant storage space and processing time, underscoring the importance of developing streamlined data processing pipelines (Pinheiro Claro Gomes et al., 2022). Despite these challenges, the potential of MSI to revolutionize food processing is undeniable. With continued innovation, collaboration, and technological advancements, MSI holds the promise of improving food safety, quality assurance, and resource optimization in both laboratory and industrial settings.

7.8 Conclusion

Multispectral imaging emerges as a versatile tool with significant potential to revolutionize various aspects of food processing. This chapter covers the fundamental principles of multispectral imaging and its intricate components, showcasing its capability to provide detailed insights into the composition, quality, and safety of food products. MSI has wide applications ranging from quality assessment to defect detection and process monitoring, the technique demonstrates its efficacy in enhancing food processing operations including packaging inspections, drying and dehydration. Further, recent advancements such as integration with machine learning and the development of portable devices signal a promising future for this technology in the food industry. Despite challenges such as calibration and standardization, the benefits of multispectral imaging in ensuring food quality assurance and safety are undeniable. As the field continues to evolve, it is essential for researchers, manufacturers, and regulators to collaborate in utilizing the full potential of multispectral imaging for the betterment of food processing operations and products in food industries, ultimately contributing to improved consumer health and satisfaction.

References

Allai, F. M., Azad, Z. R. A. A., Mir, N. A., & Gul, K. (2023). Recent advances in non-thermal processing technologies for enhancing shelf life and improving food safety. *Applied Food Research, 3*(1),100258. https://doi.org/10.1016/j.afres.2022.100258

Amigo, J. M. (2019). Hyperspectral and multispectral imaging: Setting the scene. In *Data handling in science and technology* (Vol. 32, pp. 3–16). Elsevier.

Babellahi, F., Paliwal, J., Erkinbaev, C., Amodio, M. L., Chaudhry, M. M. A., & Colelli, G. (2020). Early detection of chilling injury in green bell peppers by hyperspectral imaging and chemometrics. *Postharvest Biology and Technology, 162*, 111100.

Bandara, W. G. C., Prabhath, G. W. K., Dissanayake, D. W. S. C. B., Herath, H. M. V. R., Godaliyadda, G. M. R. I., Ekanayake, M. P. B., ... & Madhujith, T. (2019). A multispectral imaging system to assess meat quality. In 2018 IEEE *Region 10 Humanitarian Technology Conference (R10-HTC)* (pp. 1–6). IEEE

Bandara, W. G. C., Prabhath, G. W. K., Dissanayake, D. W. S. C. B., Herath, V. R., Godaliyadda, G. M. R. I., Ekanayake, M. P. B., & Madhujith, T. (2020). Validation of multispectral imaging for the detection of selected adulterants in turmeric samples. Journal of *Food Engineering, 266*, 109700.

Bendeković, J., Naletina, D., & Nola, I. (2015). Food safety and food quality in the supply chain. *Trade Perspectives, 151*, 1–13.

Benouis, M., Medus, L. D., Saban, M., Ghemougui, A., & Rosado-Muñoz, A. (2021). Food tray sealing fault detection in multi-spectral images using data fusion and deep learning techniques. *Journal of Imaging, 7*(9), 186.

Budiman, R., & Seminar, K. B. (2022). The estimation of nutrient content using multispectral image analysis in palm oil (Elaeis guineensis Jacq). *IOP Conference Series: Earth and Environmental Science, 974*(1), 12062.

Chu, X., Wang, W., Yoon, S.-C., Ni, X., & Heitschmidt, G. W. (2017). Detection of aflatoxin B 1 (AFB 1) in individual maize kernels using short-wave infrared (SWIR) hyperspectral imaging. *Biosystems Engineering, 157*, 13–23.

Clemmensen, L. H., Dissing, B. S., Hyldig, G., & Løje, H. (2012). Multispectral imaging of wok-fried vegetables. *Journal of Imaging Science and Technology, 56*, 20404-1.

De La Fuente, G. N., Carstensen, J. M., Edberg, M. A., & Lübberstedt, T. (2017). Discrimination of haploid and diploid maize kernels via multispectral imaging. *Plant Breeding, 136*(1), 50–60.

Deng, D., Liu, Z., Lv, P., Sheng, M., Zhang, H., Yang, R., & Shi, T. (2023). Defect detection in food using multispectral and high-definition imaging combined with a newly developed deep learning model. *Processes, 11*(12), 3295.

ElMasry, G., Mandour, N., Al-Rejaie, S., Belin, E., & Rousseau, D. (2019). Recent applications of multispectral imaging in seed phenotyping and quality monitoring—An overview. *Sensors, 19*(5), 1090.

Fan, S., Huang, W., Guo, Z., Zhang, B., & Zhao, C. (2015). Prediction of soluble solids content and firmness of pears using hyperspectral reflectance imaging. *Food Analytical Methods, 8*(8), 1936–1946.

Feng, Y.-Z., & Sun, D.-W. (2012). Application of Hyperspectral Imaging in Food Safety Inspection and Control: A Review. *Critical Reviews in Food Science and Nutrition, 52*(11), 1039–1058. https://doi.org/10.1080/10408398.2011.651542

Feng, C. H., Makino, Y., Oshita, S., & Martín, J. F. G. (2018). Hyperspectral imaging and multispectral imaging as the novel techniques for detecting defects in raw and processed meat products: Current state-of-the-art research advances. *Food Control, 84*, 165–176.

Florian-Huaman, J., Cruz-Tirado, J. P., Barbin, D. F., & Siche, R. (2022). Detection of nutshells in cumin powder using NIR hyperspectral imaging and chemometrics tools. *Journal of Food Composition and Analysis, 108*, 104407.

Gaikwad, S. (2021). *Literature review on multi-spectral imaging for fruits and vegetable*. Available at SSRN 3905180.

Ghasemi-Varnamkhasti, M., Apetrei, C., Lozano, J., & Anyogu, A. (2018). Potential use of electronic noses, electronic tongues and biosensors as multisensor systems for spoilage examination in foods. *Trends in Food Sci and Technol, 80*, 71–92.

Gomes, W. P. C., Gonçalves, L., da Silva, C. B., & Melchert, W. R. (2022). Application of multispectral imaging combined with machine learning models to discriminate special and traditional green coffee. *Computers and Electronics in Agriculture, 198*, 107097.

Gültekin, Ö., Cinar, E., Özkan, K., & Yazıcı, A. (2022). Real-time fault detection and condition monitoring for industrial autonomous transfer vehicles utilizing edge artificial intelligence. *Sensors, 22*(9), 3208.

Gupta, N., Soni, G., Mittal, S., Mukherjee, I., Ramtiyal, B., & Kumar, D. (2023). Evaluating traceability technology adoption in food supply chain: A game theoretic approach. *Sustainability, 15*(2), 898.

Hahn, F. (2002). AE—Automation and Emerging Technologies: Multi-spectral prediction of unripe tomatoes. *Biosystems Engineering, 81*(2), 147–155.

Huang, H., Liu, L., & Ngadi, M. O. (2014). Recent developments in hyperspectral imaging for assessment of food quality and safety. *Sensors, 14*(4), 7248–7276.

Huang, W., Li, J., Wang, Q., & Chen, L. (2015). Development of a multispectral imaging system for online detection of bruises on apples. *Journal of Food Engineering, 146*, 62–71.

Jiang, J., Qiao, X., & He, R. (2016). Use of near-infrared hyperspectral images to identify moldy peanuts. *Journal of Food Engineering, 169*, 284–290.

Jiang, J., Zheng, H., Ji, X., Cheng, T., Tian, Y., Zhu, Y., et al. (2019). Analysis and evaluation of the image preprocessing process of a six-band multispectral camera mounted on an unmanned aerial vehicle for winter wheat monitoring. *Sensors, 19*(3), 747.

Karlas, A., Masthoff, M., Kallmayer, M., Helfen, A., Bariotakis, M., Fasoula, N. A., Schäfers, M., Seidensticker, M., Eckstein, H.-H., Ntziachristos, V., & Wildgruber, M. (2021). Multispectral optoacoustic tomography of peripheral arterial disease based on muscle hemoglobin gradients-a pilot clinical study. *Annals of Translational Medicine, 9*(1), 36. https://doi.org/10.21037/atm-20-3321

Khaled, A. Y., Ekramirad, N., Donohue, K. D., Villanueva, R. T., & Adedeji, A. A. (2023). Non-destructive hyperspectral imaging and machine learning-based predictive models for physicochemical quality attributes of apples during storage as affected by codling moth infestation. *Agriculture, 13*(5), 1086.

Lee, H., Kim, M. S., Lim, H.-S., Park, E., Lee, W.-H., & Cho, B.-K. (2016). Detection of cucumber green mottle mosaic virus-infected watermelon seeds using a near-infrared (NIR) hyperspectral imaging system: Application to seeds of the "Sambok Honey" cultivar. *Biosystems Engineering, 148*, 138–147.

Levenson, R. M., & Mansfield, J. R. (2006). Multispectral imaging in biology and medicine: Slices of life. *Cytometry Part A: The Journal of the International Society for Analytical Cytology, 69*(8), 748–758.

Li, J., Chen, L., Huang, W., Wang, Q., Zhang, B., Tian, X., et al. (2016). Multispectral detection of skin defects of bi-colored peaches based on vis–NIR hyperspectral imaging. *Postharvest Biology and Technology, 112*, 121–133.

Li, A., Li, C., Gao, M., Yang, S., Liu, R., Chen, W., & Xu, K. (2021). Beef cut classification using multispectral imaging and machine learning method. *Frontiers in Nutrition, 8*, 755007.

Li, X., Wang, D., Yu, L., Ma, F., Wang, X., Pérez-Marín, D., et al. (2024). Origin traceability and adulteration detection of soybean using near infrared hyperspectral imaging. *Food Frontiers, 5*, 237–244.

Liang, H. (2012). Advances in multispectral and hyperspectral imaging for archaeology and art conservation. *Applied Physics A,106*(2), 309–323. https://doi.org/10.1007/s00339-011-6689-1

Liu, C., Liu, W., Lu, X., Chen, W., Yang, J., & Zheng, L. (2014a). Nondestructive determination of transgenic Bacillus thuringiensis rice seeds (Oryza sativa L.) using multispectral imaging and chemometric methods. *Food Chemistry, 153*, 87–93.

Liu, C., Liu, W., Lu, X., Ma, F., Chen, W., Yang, J., & Zheng, L. (2014b). Application of multispectral imaging to determine quality attributes and ripeness stage in strawberry fruit. *PLoS One, 9*(2), e87818. https://doi.org/10.1371/journal.pone.0087818

Liu, C., Liu, W., Chen, W., Yang, J., & Zheng, L. (2015). Feasibility in multispectral imaging for predicting the content of bioactive compounds in intact tomato fruit. *Food Chemistry, 173*, 482–488.

Liu, C., Liu, W., Lu, X., Chen, W., Chen, F., Yang, J., & Zheng, L. (2016a). Non-destructive discrimination of conventional and glyphosate-resistant soybean seeds and their hybrid descendants using multispectral imaging and chemometric methods. *The Journal of Agricultural Science, 154*(1), 1–12.

Liu, C., Liu, W., Lu, X., Chen, W., Yang, J., & Zheng, L. (2016b). Potential of multispectral imaging for real-time determination of colour change and moisture distribution in carrot slices during hot air dehydration. *Food Chemistry, 195*, 110–116.

Liu, W., Xu, X., Liu, C., & Zheng, L. (2021). Nondestructive detection of authenticity of Thai jasmine rice using multispectral imaging. *Journal of Food Quality, 2021*, 1–8.

Liu, W., Deng, H., Shi, Y., Liu, C., & Zheng, L. (2022). Application of multispectral imaging combined with machine learning methods for rapid and non-destructive detection of zearalenone (ZEN) in maize. *Measurement, 203*, 111944

Lleó, L., Barreiro, P., Ruiz-Altisent, M., & Herrero, A. (2009). Multispectral images of peach related to firmness and maturity at harvest. *Journal of Food Engineering, 93*(2), 229–235.

Lu, Y., Huang, Y., & Lu, R. (2017). Innovative hyperspectral imaging-based techniques for quality evaluation of fruits and vegetables: A review. *Applied Sciences, 7*(2), 189.

Ma, F., Yao, J., Xie, T., Liu, C., Chen, W., Chen, C., & Zheng, L. (2014). Multispectral imaging for rapid and non-destructive determination of aerobic plate count (APC) in cooked pork sausages. *Food Research International, 62*, 906–908. https://doi.org/10.1016/j.foodres.2014.05.010

Ma, F., Wang, J., Liu, C., Lu, X., Chen, W., Chen, C., Yang, J., & Zheng, L. (2015). Discrimination of Kernel Quality Characteristics for Sunflower Seeds Based on Multispectral Imaging Approach. *Food Analytical Methods, 8*(7), 1629–1636. https://doi.org/10.1007/s12161-014-0038-x

Ma, F., Yuan, M., & Kozak, I. (2023). Multispectral imaging: Review of current applications. *Survey of Ophthalmology, 68*(5), 889–904. https://doi.org/10.1016/j.survophthal.2023.06.004

Md Noh, M. F., Gunasegavan, R. D.-N., Mustafa Khalid, N., Balasubramaniam, V., Mustar, S., & Abd Rashed, A. (2020). Recent techniques in nutrient analysis for food composition database. *Molecules, 25*(19), 4567.

Mendoza, F., Lu, R., Ariana, D., Cen, H., & Bailey, B. (2011). Integrated spectral and image analysis of hyperspectral scattering data for prediction of apple fruit firmness and soluble solids content. *Postharvest Biology and Technology, 62*(2), 149–160.

Mishra, G., Panda, B. K., Ramirez, W. A., Jung, H., Singh, C. B., Lee, S.-H., & Lee, I. (2024). Detection of aflatoxin contamination in single kernel almonds using multispectral imaging system. *Journal of Food Composition and Analysis, 125*, 105701. https://doi.org/10.1016/j.jfca.2023.105701

Mollazade, K., Hashim, N., & Zude-Sasse, M. (2023). Towards a multispectral imaging system for spatial mapping of chemical composition in fresh-cut pineapple (Ananas comosus). *Food, 12*(17), 3243.

Morales, A., Guerra, R., Horstrand, P., Diaz, M., Jimenez, A., Melian, J., et al. (2020). A multispectral camera development: From the prototype assembly until its use in a UAV system. *Sensors, 20*(21), 6129.

Naeem, S., Ali, A., Chesneau, C., Tahir, M. H., Jamal, F., Sherwani, R. A., & Ul Hassan, M. (2021). The classification of medicinal plant leaves based on multispectral and texture feature using machine learning approach. *Agronomy, 11*(2), 263. https://doi.org/10.3390/agronomy11020263

Nicolis, O., & Gonzalez, C. (2021). 19 - Wavelet-based fractal and multifractal analysis for detecting mineral deposits using multispectral images taken by drones. In S. Gaci, O. Hachay, & O. Nicolis (Eds.), *Methods and applications in petroleum and mineral exploration and engineering geology* (pp. 295–307). Elsevier. https://doi.org/10.1016/B978-0-323-85617-1.00017-5

Offroy, M., & Duponchel, L. (2016). Chapter 15 - Super-resolution in vibrational spectroscopy: From multiple low-resolution images to high-resolution images. In C. Ruckebusch (Ed.), *Resolving spectral mixtures* (Vol. 30, pp. 477–517). Elsevier. https://doi.org/10.1016/B978-0-444-63638-6.00015-2

Olesen, M. H., Nikneshan, P., Shrestha, S., Tadayyon, A., Deleuran, L. C., Boelt, B., & Gislum, R. (2015). Viability prediction of Ricinus cummunis L. seeds using multispectral imaging. *Sensors, 15*(2), 4592–4604. https://doi.org/10.3390/s150204592

Paiva, C. L. (2013). Quality management: Important aspects for the food industry. *Food Industry, 9*, 191–192.

Pan, L., Lu, R., Zhu, Q., Tu, K., & Cen, H. (2016). Predict compositions and mechanical properties of sugar beet using hyperspectral scattering. *Food and Bioprocess Technology, 9*(7), 1177–1186.

Pascall, M. A., DeAngelo, K., Richards, J., & Arensberg, M. B. (2022). Role and importance of functional food packaging in specialized products for vulnerable populations: Implications for innovation and policy development for sustainability. *Food, 11*(19), 3043.

Pinheiro Claro Gomes, W., Gonçalves, L., Barboza da Silva, C., & Melchert, W. R. (2022). Application of multispectral imaging combined with machine learning models to discriminate special and traditional green coffee. *Computers and Electronics in Agriculture, 198*, 107097. https://doi.org/10.1016/j.compag.2022.107097

Pourazar, H., Samadzadegan, F., & Dadrass Javan, F. (2019). Aerial multispectral imagery for plant disease detection: Radiometric calibration necessity assessment. *European Journal of Remote Sensing, 52*(sup3), 17–31. https://doi.org/10.1080/22797254.2019.1642143

Pu, H., Sun, D. W., Ma, J., Liu, D., Cheng, J. H. (2014). Using wavelet textural features of visible and near-infrared hyperspectral image to differentiate between fresh and frozen–thawed pork. *Food Bioproc Technol 7*:3088–99.

Pu, H., Kamruzzaman, M., & Sun, D. W. (2015). Selection of feature wavelengths for developing multispectral imaging systems for quality, safety and authenticity of muscle foods-a review. *Trends in Food Science & Technology, 45*(1), 86–104.

Qin, J., Burks, T. F., Zhao, X., Niphadkar, N., & Ritenour, M. A. (2012). Development of a two-band spectral imaging system for real-time citrus canker detection. *Journal of Food Engineering, 108*(1), 87–93.

Qin, J., Chao, K., Kim, M. S., Lu, R., & Burks, T. F. (2013). Hyperspectral and multispectral imaging for evaluating food safety and quality. *Journal of Food Engineering, 118*(2), 157–171. https://doi.org/10.1016/j.jfoodeng.2013.04.001

Rady, A., Guyer, D., & Lu, R. (2015). Evaluation of Sugar Content of Potatoes using Hyperspectral Imaging. *Food and Bioprocess Technology, 8*(5), 995–1010. https://doi.org/10.1007/s11947-014-1461-0

Ren, G., Wang, S., Ning, J., Xu, R., Wang, Y., Xing, Z., et al. (2013). Quantitative analysis and geographical traceability of black tea using Fourier transform near-infrared spectroscopy (FT-NIRS). *Food Research International, 53*(2), 822–826.

Ren, L., Liu, W., Liu, C., & Zheng, L. (2023). Nondestructive detection of water status and distribution in corn kernels during hot air drying using multispectral imaging. *Journal of the Science of Food and Agriculture, 103*(6), 3139–3145.

Ross, G. M. S., Bremer, M. G. E. G., & Nielen, M. W. F. (2018). Consumer-friendly food allergen detection: Moving towards smartphone-based immunoassays. *Analytical and Bioanalytical Chemistry, 410*, 5353–5371.

Sara, D., Mandava, A. K., Kumar, A., Duela, S., & Jude, A. (2021). Hyperspectral and multispectral image fusion techniques for high resolution applications: A review. *Earth Science Informatics, 14*(4), 1685–1705. https://doi.org/10.1007/s12145-021-00621-6

Shahin, M. A., Hatcher, D. W., & Symons, S. J. (2012). Development of multispectral imaging systems for quality evaluation of cereal grains and grain products. In *Computer vision technology in the food and beverage industries* (pp. 451–482). Woodhead Publishing.

Shrestha, S., Deleuran, L. C., Olesen, M. H., & Gislum, R. (2015). Use of multispectral imaging in varietal identification of tomato. *Sensors, 15*(2), 4496–4512.

Singh, S. S., Lim, L. T., & Manickavasagan, A. (2023). Imaging and spectroscopic techniques for microstructural and compositional analysis of lignocellulosic materials: a review. *Biomass Conversion and Biorefinery, 13* (1), 499–517. https://doi.org/10.1007/s13399-020-01075-4

Spyrelli, E. D., Doulgeraki, A. I., Argyri, A. A., Tassou, C. C., Panagou, E. Z., & Nychas, G.-J. E. (2020). Implementation of Multispectral Imaging (MSI) for microbiological quality assessment of poultry products. *Microorganisms, 8*(4), 552. https://doi.org/10.3390/microorganisms8040552

Su, W., & Sun, D. (2018). Multispectral imaging for plant food quality analysis and visualization. *Comprehensive Reviews in Food Science and Food Safety, 17*(1), 220–239. https://doi.org/10.1111/1541-4337.12317

Su, W.-H., He, H.-J., & Sun, D.-W. (2017). Non-destructive and rapid evaluation of staple foods quality by using spectroscopic techniques: A review. *Critical Reviews in Food Science and Nutrition, 57*, 1039–1051.

Todd, E. (2020). Food-borne disease prevention and risk assessment. *In Int J Environ Res Public Health, 17*(14), p 5129. MDPI.

Tran, C. D. (2005). Principles, instrumentation, and applications of infrared multispectral imaging, an overview. *Analytical Letters, 38*(5), 735–752. https://doi.org/10.1081/AL-200047754

Tsakanikas, P., Pavlidis, D., & Nychas, G. J. (2015). High throughput multispectral image processing with applications in food science. *PLoS One, 10*(10), e0140122.

Valdes-Garcia, A., Pepeljugoski, P., Duran, I., Plouchart, J.-O., Yeck, M., & Liu, H. (2020). Portable multi-spectral imaging: Devices, vertical integration, and applications. In *2020 IEEE International Electron Devices Meeting (IEDM)* (pp. 17.1.1–17.1.4). IEEE. https://doi.org/10.1109/IEDM13553.2020.9372014

van Ineveld, R. L., Collot, R., Román, M. B., Pagliaro, A., Bessler, N., Ariese, H. C. R., Kleinnijenhuis, M., Kool, M., Alieva, M., de Sousa, C., Lopes, S. M., Wehrens, E. J., & Rios, A. C. (2022). Multispectral confocal 3D imaging of intact healthy and tumor tissue using mLSR-3D. *Nature Protocols, 17*(12), 3028–3055. https://doi.org/10.1038/s41596-022-00739-x

Vinothkanna, A., Dar, O. I., Liu, Z., & Jia, A.-Q. (2024). Advanced detection tools in food fraud: A systematic review for holistic and rational detection method based on research and patents. *Food Chemistry, 446*, 138893. https://doi.org/10.1016/j.foodchem.2024.138893

Vozel, B., Lukin, V., & Bazi, Y. (2019). Editorial to Special Issue "Multispectral Image Acquisition, Processing, and Analysis." *In Remote Sensing (Vol. 11, Issue 19)*. https://doi.org/10.3390/rs11192310

Wang, J., Wu, X., Wang, C., Shao, N., Dong, P., Xiao, R., & Wang, S. (2015). Magnetically assisted surface-enhanced Raman spectroscopy for the detection of Staphylococcus aureus based on aptamer recognition. *ACS Applied Materials & Interfaces, 7*, 20919–20929.

Wazed, M. A., Sheikh, M. A. M., Akhtaruzzaman, M., Awal, M. S., & Mozumder, N. H. M. R. (2023). Quality evaluation and shelf life analysis of fruit juice cocktail containing Mango (Mangifera indica) and Pineapple (Ananas comosus). *Journal of Agriculture and Food Research, 14*, 100773. https://doi.org/10.1016/j.jafr.2023.100773

Wu, D., & Sun, D. W. (2013). Advanced applications of hyperspectral imaging technology for food quality and safety analysis and assessment: A review—Part II: Applications. *Innovative Food Science & Emerging Technologies, 19*, 15–28.

Wu, D., Yang, H., Chen, X., He, Y., & Li, X. (2008). Application of image texture for the sorting of tea categories using multi-spectral imaging technique and support vector machine. *Journal of Food Engineering, 88*(4), 474–483.

Xiong, Z., Sun, D. W., Pu, H., Zhu, Z., & Luo, M. (2015). Combination of spectra and texture data of hyperspectral imaging for differentiating between free-range and broiler chicken meats. *LWT-Food Science and Technology, 60*(2), 649–655.

Yan, L., Xiong, C., Qu, H., Liu, C., Chen, W., & Zheng, L. (2017). Non-destructive determination and visualisation of insoluble and soluble dietary fibre contents in fresh-cut celeries during storage periods using hyperspectral imaging technique. *Food Chemistry, 228*, 249–256.

Yang, C. C., Kim, M. S., Millner, P., Chao, K., Cho, B. K., Mo, C., et al. (2014). Development of multispectral imaging algorithm for detection of frass on mature red tomatoes. *Postharvest Biology and Technology, 93*, 1–8.

Yang, Q., Sun, D. W., & Cheng, W. (2017). Development of simplified models for nondestructive hyperspectral imaging monitoring of TVB-N contents in cured meat during drying process. *Journal of Food Engineering, 192*, 53–60.

Younas, S., Liu, C., Qu, H., Mao, Y., Liu, W., Wei, L., et al. (2020). Multispectral imaging for predicting the water status in mushroom during hot-air dehydration. *Journal of Food Science, 85*(4), 903–909.

Zahra, A., Qureshi, R., Sajjad, M., Sadak, F., Nawaz, M., Khan, H. A., & Uzair, M. (2023). Current advances in imaging spectroscopy and its state-of-the-art applications. *Expert Systems with Applications, 238*, 122172.

Zhang, C., Jiang, H., Liu, F., & He, Y. (2017). Application of near-infrared hyperspectral imaging with variable selection methods to determine and visualize caffeine content of coffee beans. *Food and Bioprocess Technology, 10*(1), 213–221.

Zhang, W., Zhu, Q., Huang, M., Guo, Y., & Qin, J. (2019). Detection and classification of potato defects using multispectral imaging system based on single shot method. *Food Analytical Methods, 12*, 2920–2929.

Zhu, H., Chu, B., Zhang, C., Liu, F., Jiang, L., & He, Y. (2017). Hyperspectral imaging for presymptomatic detection of tobacco disease with successive projections algorithm and machine-learning classifiers. *Scientific Reports, 7*, 4125.

Zhu, S., Feng, L., Zhang, C., Bao, Y., & He, Y. (2019). Identifying freshness of spinach leaves stored at different temperatures using hyperspectral imaging.*Food, 8*(9), 356.

Zitova, B., & Flusser, J. (2003). Image registration methods: A survey.*Image and Vision Computing, 21*(11), 977–1000.

Chapter 8
Backscattering Imaging

Abhishek Dasore and Norhashila Hashim

8.1 Introduction

Light backscattering imaging (LBI) is a cost-effective and efficient tool for online monitoring of agricultural and food products, offering low instrumentation costs, quick processing times, and high accuracy without the need to physically contact samples. It integrates seamlessly into postharvest and food production processes, providing deep penetrations into samples. The scientific foundation of LBI traces back to the late nineteenth century with the discovery of Rayleigh scattering. However, the significant expansion of LBI applications for assessing agro-food materials occurred in early twenty-first century due to recent advances in computing technology. LBI plays a fundamental role in understanding the optical characteristics of agricultural and food products, which can range from transparent to semi-transparent or opaque. These materials interact with light through processes such as absorption, transmittance, and reflectance, as illustrated in Fig. 8.1. Reflectance can occur in different forms, including regular (specular) reflectance, external diffuse

A. Dasore
Department of Biological and Agricultural Engineering, Faculty of Engineering, Universiti Putra Malaysia, Serdang, Selangor, Malaysia

Department of Mechanical Engineering, Koneru Lakshmaiah Education Foundation, Guntur, Andhra Pradesh, India

N. Hashim (✉)
Department of Biological and Agricultural Engineering, Faculty of Engineering, Universiti Putra Malaysia, Serdang, Selangor, Malaysia

SMART Farming Technology Research Centre (SFTRC), Faculty of Engineering, Universiti Putra Malaysia, Serdang, Selangor, Malaysia

Institute of Aquaculture and Aquatic Sciences (I-AQUAS), Universiti Putra Malaysia, Batu 7, Jalan Kemang 5, Negeri Sembilan, Port Dickson, Malaysia
e-mail: norhashila@upm.edu.my

C. K. Sunil et al. (eds.), *Emerging Non-destructive Technologies for Food Quality Analysis*, https://doi.org/10.1007/978-3-031-95048-3_8

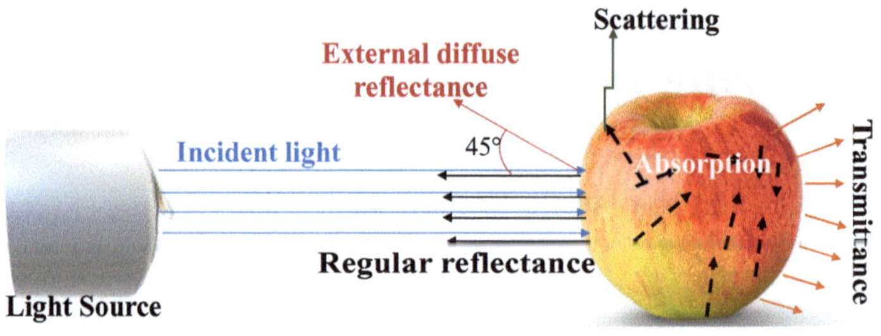

Fig. 8.1 Spatial incident light distribution within fruit

reflectance, and scattering. Regular reflectance follows the law of reflection, where the angle of incidence equals the angle of reflection, while external diffuse reflectance occurs at a fixed 45° to the incident beam, providing information about surface color and texture (Alhamdan & Atia, 2017). Only a small percentage (4–5%) of incident light is reflected by regular and external diffuse reflectance, while the majority penetrates the tissue and is scattered by internal components of biological food materials (Lu et al., 2020).

Phenomena such as scattering occur when photons change direction upon encountering scattering particles, whereas absorption takes place if the photon's energy matches that of the particle according to quantum theory. The direction change of scattered photons depends on the optical properties of the scattering particle, elucidated by the scattering phase function ($\rho(\theta)$), unique to each particle based on size, shape, and orientation. The anisotropy factor (g) governs the fraction of photons scattered in specific directions, ranging from -1 to 1. This factor is computed using the function, indicated in Eq. (8.1) (Mollazade et al., 2012):

$$\rho(\theta) = \frac{1 - g^2}{4\pi\left[(1 + g^2 - 2g cos\theta)^{3/2}\right]} \tag{8.1}$$

where, θ represents the angle between the photon direction before and after a scattering event, within the range of [0, π]. Here, g value specifies the type of scattering: $g = -1$ signifies total backward scattering, $g = 0$ represents isotropic scattering (equal scattering of photons in all directions), and $g = 1$ denotes total forward scattering. For most food and plant materials, g values typically fall between 0.7 and 0.9, indicating a prevalence of forward scattering (Lu et al., 2020).

Absorption is described by the absorption coefficient (μ_a), while the scattering is characterized by the scattering coefficient (μ_s) and the anisotropy factor (g). By understanding μ_a, μ_s, and g, it becomes feasible to describe photon transfer within biological materials. In scenarios dominated by scattering (i.e., $\mu_s \gg \ > \mu_a$), the anisotropy factor can be integrated into the scattering coefficient, yielding a new optical parameter termed the reduced scattering coefficient. This approach streamlines the characterization of light interaction with plant materials, such as fruits and vegetables, utilizing μ_a and μ_s' as pivotal parameters, as indicated in Eq. (8.2).(Qin & Lu, 2007)

$$\mu_s' \equiv (1-g)\mu_s \tag{8.2}$$

The complex structure of biological tissue, featuring strong scattering and intricate optical parameters, presents challenges for investigating light propagation. Theoretical solutions for these complex geometries are often elusive, prompting the development of various methods like diffusion approximation theory and numerical simulations, widely employed in the literature. These models rely on essential optical properties including refractive index, absorption coefficient, and reduced scattering coefficient.

8.1.1 *Diffusion Approximation Theory*

Diffusion approximation theory plays a crucial role in understanding light propagation within biological tissues, especially in the wavelength range of 500–1300 nm, where scattering is a dominant mechanism influencing turbid materials. To model the transport of photons under these conditions, the diffusion approximation simplifies the complex radiative transfer equation, which is derived from the principle of energy conservation (Lu et al., 2016). The radiative transfer equation, characterized by an integrodifferential form with six variables, is generally challenging to solve analytically except under specific conditions.

In practical scenarios, particularly with biological materials such as fruits and vegetables where scattering is prevalent, the radiative transfer equation simplifies to a diffusion equation, represented by Eq. (8.3).

$$\frac{\partial \varphi\left(\vec{r},t\right)}{\partial t} = c\left[D\nabla^2\varphi\left(\vec{r},t\right) - \mu_a\varphi\left(\vec{r},t\right) + S\left(\vec{r},t\right)\right] \tag{8.3}$$

where, ∇^2 represents Laplacian operator; $\varphi\left(\vec{r},t\right)$ denotes photon flux; $S\left(\vec{r},t\right)$ indicates the number of photons emitted at location $\vec{r}$ and time t per unit volume per unit time; and D signifies diffusion coefficient, calculated as $[3(\mu_a + (1 - g)\mu_s)]^{-1}$, which can be expressed using absorption and reduced scattering coefficients as indicated in Eq. (8.4).

$$D = \frac{1}{3\left(\mu_a + \mu_s'\right)} \tag{8.4}$$

This diffusion equation has been effectively solved analytically for significant lighting scenarios, illustrated in Fig. 8.2. These scenarios include continuous-wave

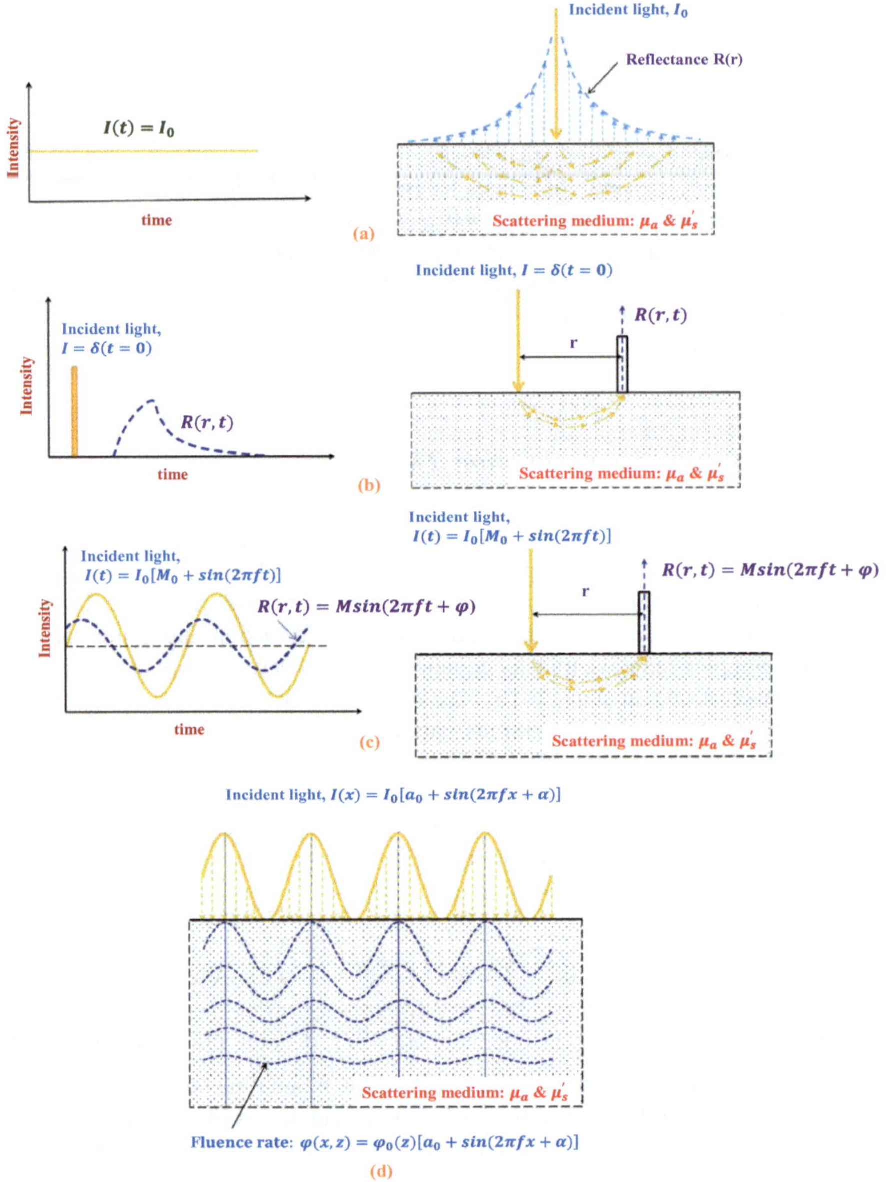

Fig. 8.2 Four scenarios of light illumination on a semi-infinite turbid medium, each with analytically solvable solutions to the diffusion approximation equation: (**a**) continuous-wave point lighting (time-independent), (**b**) short-pulsed lighting, (**c**) frequency-modulated point lighting, and (**d**) spatially-modulated area lighting (time-independent). Here, *I* denote incident light intensity and *f* represents frequency in either the time (t) or spatial domain. (Reprinted from "Measurement of optical properties of fruits and vegetables: A review," by Lu et al. (2020). Copyright (2020), with permission from Elsevier)

or steady-state point lighting (Fig. 8.2a), pulsed point lighting (Fig. 8.2b), frequency-modulated point lighting (Fig. 8.2c), and spatially-modulated area lighting (Fig. 8.2d). The resulting analytical solutions provide the theoretical basis for various advanced techniques used in optical imaging and analysis, including spatially resolved (SR), time resolved (TR), Frequency-domain (FD), and spatial-frequency domain (SFD) (Tuchin, 2007).

8.1.2 Numerical Simulations

In the estimation of optical properties for biological materials, the diffusion equation serves as a foundational tool used alongside techniques like SR, TR, and SFD methods, each offering unique perspectives on how light interacts with biological tissues to reveal their optical properties. The SR method, originally applied in medical contexts, uses parallel optical fibers to measure diffuse reflection, enabling computation of absorption and scattering coefficients. This technique is instrumental for tissue characterization and predictive modeling, offering detailed optical property maps and supporting non-invasive evaluation of biological tissues (Wang et al., 2017; Yu et al., 2024). Similarly, the TR method involves injecting short light pulses into a medium, analyzing the temporal distribution of re-emitted photons to extract absorption and scattering coefficients with high resolution (Torricelli et al., 2008; Rizzolo et al., 2010). On the other hand, SFD imaging utilizes spatially modulated lighting to capture diffuse reflectance and generate detailed 2D or 3D maps of optical properties for comprehensive tissue characterization (Hu et al., 2020a).

Despite their effectiveness in nondestructive evaluation, optical techniques face challenges when applied to biological tissues due to the significant ratio between scattering (μ_s) and absorption (μ_a), particularly when assessing radiance from a considerable distance from the light source, typically beyond one mean free path. In the visible/near-infrared (Vis/NIR) spectrum, biological tissues exhibit substantial absorption by chromophores like chlorophyll, carotenoids, anthocyanins, and water, which can hinder accurate optical assessment. To address these challenges, researchers utilize numerical methods such as adding-doubling (AD) and Monte Carlo (MC) simulations. These techniques enable the simulation of light propagation within biological tissues, providing a more accurate representation of their optical behavior. By incorporating the effects of scattering and absorption, AD and MC simulations overcome limitations posed by high absorption levels in specific spectral ranges, thereby enhancing understanding of tissue optical properties (Lu et al., 2020).

The AD method, initially developed in atmospheric sciences, was later adapted for tissue optics to solve the radiative transfer theory (RTT) in slab geometry. Prahl et al. (1993) extended this method to create the inverse adding-doubling (IAD) method, which efficiently computes absorption (μ_a) and reduced scattering $\left(\mu_s'\right)$ properties with high accuracy comparable to MC simulations. The IAD method combines iterative parameter estimation with integrating sphere (IS) measurements, making it a valuable technique for precise determination of optical properties in various biological materials. Despite its sensitivity to light loss through sample holders and boundaries, the IAD method has been extensively applied to calculate absorption and scattering properties in fruits (Zhang et al., 2019; He et al., 2016), vegetables (Wang et al., 2014; Wieme et al., 2022), milk (Aernouts et al., 2015), and meat (Zamora-Rojas et al., 2014). Its high accuracy has established IAD as a reference method for developing other optical techniques like MC and SR methods used in absorption and scattering property estimation (Qin & Lu, 2009; Cen & Lu, 2010).

On the other hand, MC simulations offer precise and versatile methods for quantifying light propagation properties, particularly in scenarios where experimental data collection is challenging due to the diverse range of possible conditions. MC methods involve probabilistic tracing of photons through absorbing and scattering media, providing valuable estimates of photon distributions based on tissue optical properties (Hu et al., 2020b). Classic MC algorithms, such as those used for multi-layered media, simulate light propagation in semi-infinite slabs. Meanwhile, structured finite mesh approaches allow for more detailed descriptions of biological structures. Within MC simulations, the movement of individual photons entails several steps, including determining parameters like step size, deflection angle, and absorbed fraction based on the absorption and scattering characteristics of the medium (Pratx & Xing, 2011). Despite their accuracy, MC simulations can be computationally demanding, necessitating the simulation of a large number of photons (typically 10^4 to 10^6), for reliable results (Tuchin, 2007). To address this challenge and achieve computational efficiency, various techniques have been developed, including hybrid MC methods, perturbation MC, scaling MC, variance reduction, and parallel computing (Xie & Guo, 2020). MC simulations have played a pivotal role in estimating absorption and scattering properties across diverse biological materials, such as kiwifruit, apple, and meat (Baranyai & Zude, 2009; Huang et al., 2018). They have also been instrumental in optimizing measurement systems and providing detailed insights into light distribution within fruits (Zamora-Rojas et al., 2014). The accuracy and flexibility of MC simulations have firmly established them as the 'golden standard' in optical simulations, facilitating the study of light-matter interactions across a wide range of applications (Carles et al., 2019).

8.2 Imaging Systems

The effectiveness of backscattering imaging heavily relies on selecting the appropriate system components. Each LBI system consists of a light source and an imaging unit, as depicted in Fig. 8.3, with laser diodes commonly chosen for their ability to deliver high-intensity light at specific wavelengths, such as 658 nm, facilitating deeper penetration into fruit flesh for detailed scattering data (Mozaffari et al., 2022). Laser characteristics, including wavelength, incident angle, power

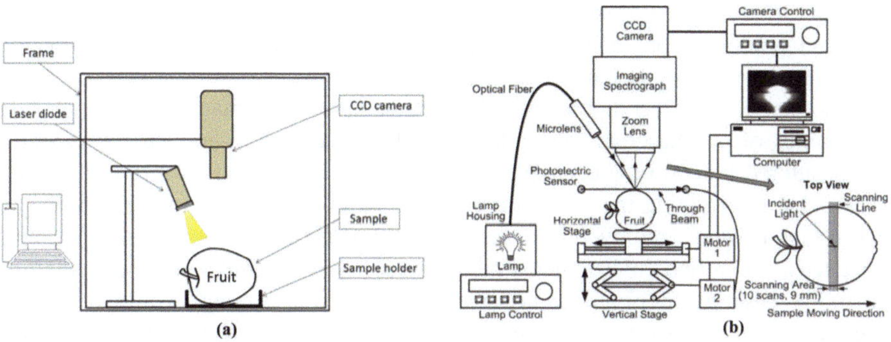

Fig. 8.3 Visual illustrations of backscattering imaging systems for acquiring SR scattering image from a fruit sample: (**a**) Laser LBI. (Reprinted with permission from Ali et al., 2020b); (**b**) Hyperspectral LBI. (Reprinted with permission from Qin & Lu, 2008)

output, and beam diameter, must be carefully considered for optimal sample illumination and signal generation. Identifying sensitive wavelengths, crucial for revealing internal quality indices, can be achieved through methods like a complete search across a spectral range or leveraging specific wavelengths suggested in literature reviews (Qing et al., 2007b, 2008). However, these methods can be time-consuming and costly. Alternatively, NIR spectroscopy findings can guide wavelength selection based on strong correlations with qualitative parameters, employing spectrum preprocessing and selection techniques like standard normal variate, Savitzky–Golay, correlation coefficient, and genetic algorithms (GA) for optimal selection. Data mining techniques like support vector machines, simulated annealing, and sensitivity analysis, further enhance the process, refining methods tailored to specific fruit types based on internal quality indices and measurement conditions (Witten & Frank, 2005). Validation of NIR results against current conditions is crucial, as appropriate wavelengths may vary for different fruit quality parameters. Additionally, economic factors may influence selection, with more common wavelengths like 650 nm often preferred over less common ones like 600 nm (Picotronic GmbH, 2023).

Regarding the incident angle, it is generally preferable to keep them as small as possible for symmetrical scattering patterns, facilitating analysis (Hashim et al., 2013). Excessively small angles, however, may cause unwanted reflectance effects that interfere with camera signal detection. In cases where symmetry is less critical, larger incident angles can be considered (Udomkun et al., 2014), influencing laser light penetration depth within agricultural products. Power output is another crucial consideration. Higher laser power typically results in stronger backscattered signals, leading to improved imaging quality and reliability, particularly over extended distances or when dealing with surfaces that have low reflectivity. However, higher power levels necessitate strict adherence to safety protocols to mitigate potential hazards associated with increased energy levels (DIN EN 60825-1, 2022). Additionally, the laser beam diameter significantly impacts spatial resolution and energy density on the target surface. A smaller beam diameter enhances spatial resolution and concentrates more energy onto a confined area, advantageous for specific imaging scenarios, optimizing system effectiveness.

Another essential component of LBI systems is the imaging unit, typically consisting of a charge-coupled device (CCD) camera. The camera captures backscattered light from the sample surface, with specifications like resolution and sensitivity determining the image quality. To optimize imaging, in LBI systems, a zoom lens is often employed to adjust the field of view and focus. These systems operate in controlled environments like darkrooms to minimize noise and interference. Precision positioning techniques, such as radial averaging, profile averaging, spectral averaging, and pixel binning, ensure accurate alignment of components for consistent and reproducible results. Radial averaging divides backscattering profiles into circular rings, averaging pixel values within each ring to determine the image center based on maximum light intensity (Peng & Lu, 2005). Profile averaging leverages the symmetry of scattering profiles around the incident point (Qin & Lu, 2005), while spectral averaging improves spectral resolution and signal-to-noise ratio by averaging three consecutive wavelengths (Qin & Lu, 2006). Pixel binning reduces image resolution by combining adjacent pixel information, increasing pixel intensity per area (Lu et al., 2006). These techniques collectively enhance image quality, reduce noise, and maintain consistent imaging conditions in light backscattering systems used for agricultural and food product assessment.

LBI systems are controlled by a dedicated computer, which operates the camera and stores captured images and data. Specialized image processing software like Image-Pro Insight or MATLAB is used to analyze and interpret backscattering images, enhancing image quality, removing noise, and extracting quantitative data (Romano et al., 2011). Optical components such as collimating lenses ensure uniform light distribution on the sample surface, while objective lenses optimize image clarity and resolution by focusing backscattered light onto the CCD camera sensor. Additional components like grabber boards facilitate image data transfer to the computer for further processing. Reference standards and calibration tools are used to validate and calibrate the system for accurate measurement and analysis. A limitation of this technique is that the acquired images convey information solely within the selected wavelength, restricting the data to the specific wavelength used (Mozaffari et al., 2022).

8.3 Backscattering Image Processing

8.3.1 Image Analysis

After acquiring images using LBI, the subsequent processing involves a systematic approach of image preprocessing, segmentation, feature extraction, and predictive modeling tailored to extract and quantify food quality. LBI images often contain noise due to environmental factors or equipment, which is addressed by applying preprocessing techniques such as median filtering or Gaussian smoothing to improve image quality. Additionally, enhancements like contrast adjustment and histogram equalization are employed to enhance visual quality, making features more distinguishable (Siyum et al., 2023). Segmentation techniques are then used to identify and extract specific regions of interest within LBI images relevant to the food quality attributes under study. This involves using threshold-based methods or clustering algorithms to separate food components from the background. Further analysis includes extracting textural features using techniques like Grey-Level Co-Occurrence Matrix (GLCM) or wavelet transforms to gain insights into the internal structure and composition of food samples (Zheng et al., 2006). Statistical features such as mean, standard deviation, skewness, and kurtosis are computed from LBI images to quantify image characteristics related to food quality.

Also, machine learning algorithms, including artificial neural networks (ANNs) or regression models, are trained using these extracted features along with corresponding reference data (e.g., moisture content, firmness) to develop calibration models. These calibrated models are utilized to predict or classify food quality attributes based on new LBI images, providing quantitative assessments of properties like moisture content, soluble solids, or ripeness (Romano et al., 2011). Model performance is validated using cross-validation techniques to assess accuracy and generalizability, with fine-tuning of parameters and feature selection methods to optimize analysis accuracy. The results obtained from LBI image analysis are interpreted to draw conclusions about food quality attributes and their variations. Comprehensive reports summarizing findings from LBI image processing are generated, offering actionable insights for food producers, researchers, or quality control personnel. This systematic approach enables efficient and accurate assessment of food quality characteristics critical for agricultural and food industry applications using LBI images.

For an instance, in a specific research study by Ali et al. (2017), both seeded and seedless watermelon samples underwent analysis using LBI to monitor changes in quality throughout postharvest storage. The LBI setup utilized included a CCD camera paired with a laser diode emitting light at a wavelength of 658 nm. During the image acquisition process, precautions were taken to minimize interference from ambient light by enclosing the setup within a closed frame covered with a black cloth. For each fruit sample, a series of six backscattering images were captured to derive the mean pixel values associated with specific backscattering parameters. To extract the mean pixel values of these parameters, image segmentation techniques were applied using MATLAB software. The primary objective of image segmentation was to delineate the region of interest (ROI) and eliminate any saturated pixels present in the backscattering images. Following image smoothing with a Gaussian filter to reduce noise, the ROI was carefully selected based on a determined threshold value derived from histogram analysis as depicted in Fig. 8.4.

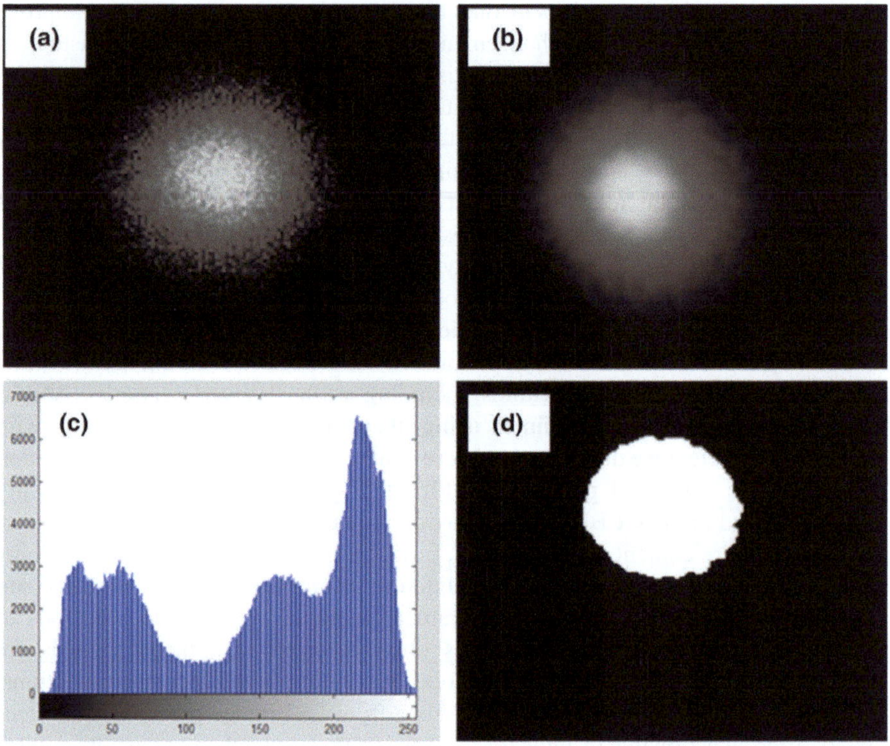

Fig. 8.4 Analysis of LBI employing image segmentation techniques: (**a**) original backscattering image, (**b**) image smoothing with Gaussian filter, (**c**) distribution of image pixel histogram profile, (**d**) region of interest (ROI) of the segmented image. (Reproduced from "Determination of the difference on color changes of watermelons by laser light backscattering imaging," by Mohd Ali et al. (2017). Copyright (2020), with permission from Springer Nature)

8.3.2 Processing of Spectral Scattering Imaging

During image preprocessing, the main focus is to enhance the signal-to-noise ratio (SNR) of the acquired images (Qing et al., 2007a). Techniques such as radial averaging divide each backscattering profile is into concentric rings around the incident center, with pixel intensities averaged within each ring to reduce noise. Other methods include profile averaging, spectral averaging, and pixel binning, all aimed at improving image quality and reducing unwanted noise artifacts (Peng & Lu, 2006). Additionally, corrections are made to compensate for light scattering distortion caused by fruit surface curvature, often using the Lambertian cosine law to adjust for light intensity distortion, ensuring more accurate backscattered light measurements (Qing et al., 2007a).

Segmentation of the image is crucial to isolate the region of interest (ROI) containing the relevant reflected photons. Techniques like thresholding and cluster

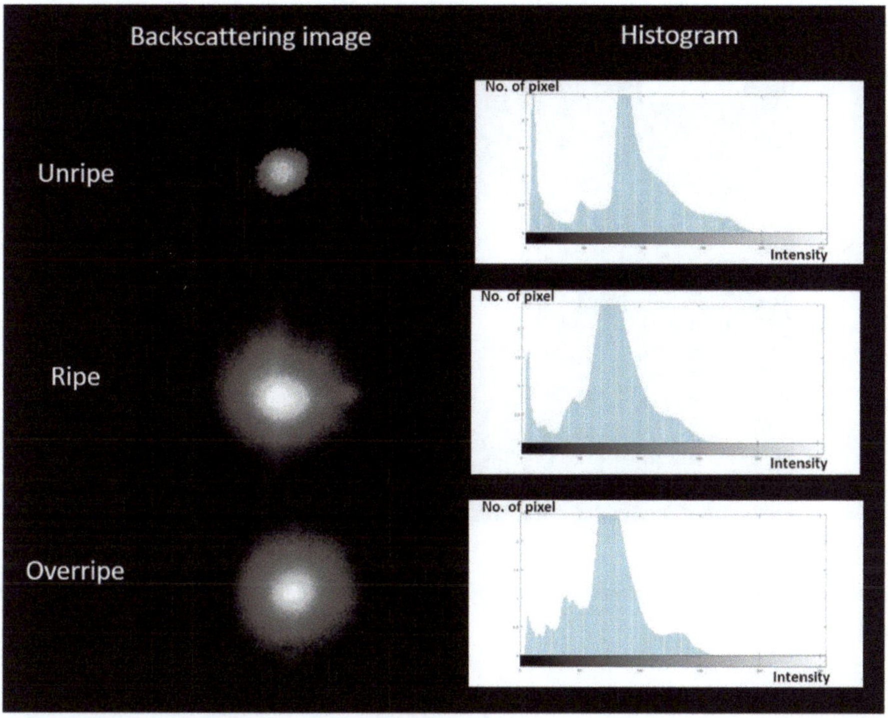

Fig. 8.5 Raw backscattering images and histograms at 658 nm depicting pixel intensities distribution of oil palm fresh fruit bunches at unripe, ripe, and overripe stages of maturity. (Reprinted from "Combination of laser-light backscattering imaging and computer vision for rapid determination of oil palm fresh fruit bunches maturity," by Mohd Ali et al. (2020a). Copyright (2020), with permission from Elsevier)

analysis differentiate the ROI from the background and areas where the CCD may be saturated (Tu et al., 2001). Feature extraction involves deriving a concise feature vector from backscattering images, utilizing methods like radial averaging to convert profiles to 1D, extracting statistical characteristics, and fitting mathematical functions to describe scattering behavior (Baranyai & Zude, 2009). Figure 8.5 displays backscattering images and histograms at 658 nm, illustrating pixel intensity distributions of oil palm fresh fruit bunches (FFB) at different maturity stages (Ali et al., 2020a). The brightness variations in the images correspond to FFB maturity levels, indicating changes in pixel intensity. Histograms depict pixel intensity distributions in grayscale images, while backscattering images show patterns of light scattering. Higher maturity levels exhibit fewer pixels at higher intensity values, resulting in larger illuminated areas.

To enhance feature representation in quality prediction models, preprocessing steps like normalization and calibration, refine profiles. Statistical features such as mean and standard deviation are extracted from scattering profiles using radial averaging (Hornberg, 2006). Gaussian-Lorentzian curves are fitted to the scattering

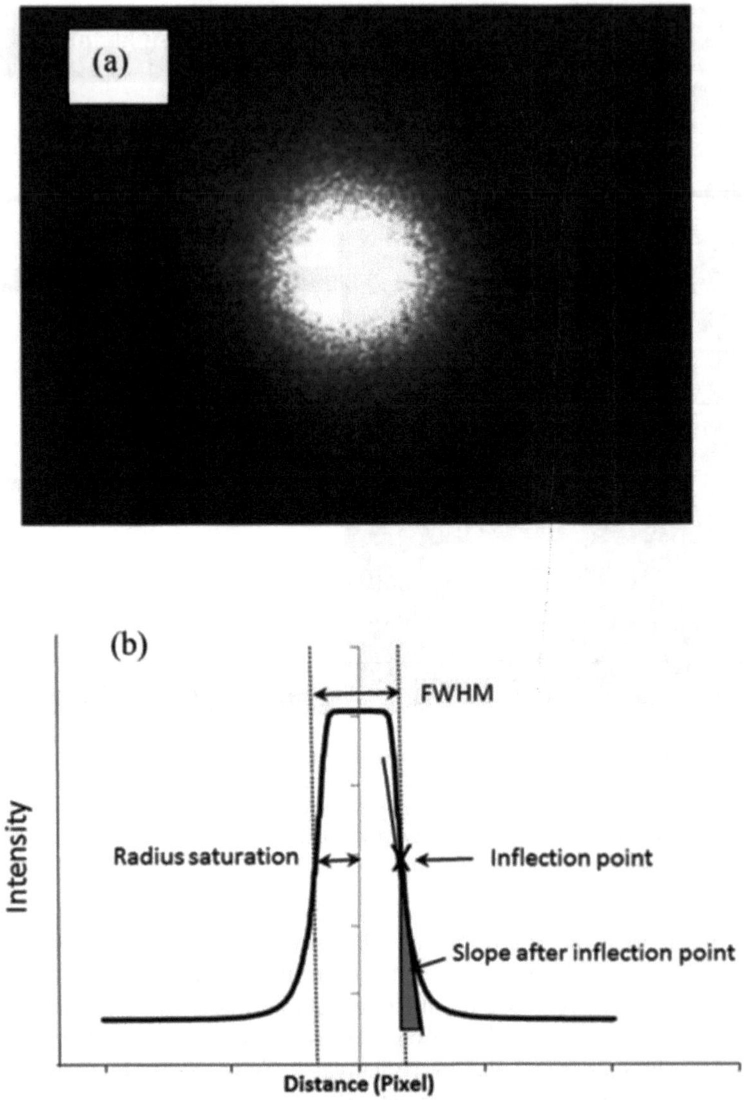

Fig. 8.6 Gaussian-Lorentzian cross product distribution model for backscattering profiles. (**a**) A Raw Backscattering Image Acquired Using a 785 nm Laser and (**b**) The Backscattering Profile and Parameters (Hashim et al., 2014)

profiles to extract meaningful model parameters including, inflection point (IP), the slope after inflection point (SA), the full-width-half-maximum (FWHM) and the saturation radius (Rsat) that describe the scattering behavior as illustrated in Fig. 8.6.

Feature selection techniques such as principal component analysis (PCA) and Fisher's linear discriminant analyses (FLDA) are applied to optimize the feature set and avoid overfitting, reducing the dimensionality of extracted features (Omid et al., 2010). Finally, calibration models are developed using statistical methods such as partial least squares regression (PLSR) or artificial neural networks (ANN) (Karray & Silva, 2004) to establish a relationship between the selected features and the desired quality parameters, enabling accurate prediction and assessment of fruit attributes such as firmness or sweetness (Omid et al., 2009).

8.4 Applications of Backscattering Imaging Techniques

LBI techniques have found widespread application in assessing agricultural and food products quality, including defect detection, quality assessment, postharvest process monitoring and evaluation of mechanical properties. Table 8.1 provides an overview of LBI technique applications in agriculture and food. LBI has proven instrumental in assessing soluble solids content (SSC) and firmness in fruits and vegetables, linking these attributes with maturity and shelf life. Researchers Qing et al. (2008) and Tu et al. (2006), utilized LBI to evaluate SSC and firmness in apples across different growth stages and locations, using spectral images captured with laser diodes at specific wavelengths (680, 780, 880, 940, and 980 nm). They applied statistical methods like PLSR, SMLR, and PCR to accurately predict SSC and firmness based on fruit absorption and scattering properties. Moreover, LBI has been utilized to monitor changes in drying characteristics of fruits (Romano et al., 2008). For instance, in the work of Onwude et al. (2018), who examined pumpkin slices drying at 70 °C, at increasing drying time and decreasing moisture content, as illustrated in Fig. 8.7 led to a decrease in backscattered image size and an expansion of the illuminated area. Additionally, decreasing moisture content over time was accompanied by a reduction in diffuse reflectance and backscattered area. This behavior is attributed to water's transparent nature between solids, leading to reduced distances between solids as moisture diminishes.

In separate investigations, Baranyai and Zude (2008, 2009) used MC simulation and LBI techniques to examine light propagation in apple and kiwifruit tissue. Their findings highlighted the considerable influence of the anisotropy factor on the morphology of backscattering profiles compared to the scattering coefficient. Significant differences ($p < 0.01$) in the anisotropy factor were observed between overripe and premium quality kiwifruits in terms of textural attributes. Furthermore, significant differences ($p < 0.05$) were found in the backscattering characteristics of apples that had been bruised during storage. The importance of mathematical, statistical, and modeling approaches for LBI was underscored, emphasizing LBI's versatility in assessing fruit and vegetable quality across pre- and post-harvest applications (ElMasry & Nakauchi, 2016).

Hashim et al. (2013) applied LBI to monitor chilling injuries in bananas stored at 6 °C for 2 days, achieving a strong coefficient of determination ($R^2 > 0.90$) between extracted backscattering properties and reference quality attributes (visual evaluation, chlorophyll index, and water content). Similarly, Zulkifli et al. (2019), extended

Table 8.1 An overview of the applications of LBI for agricultural and food products

Wavelength (nm)	Material	Calibration model	Application	Findings	Reference
635	Apple (Aksu Fuji)	CNN (Convolutional Neural Network)	Apple defect detection	The use of an 11-layer CNN combined with LBI achieves a 92.5% recognition rate for apple defects, surpassing traditional algorithms. This non-destructive method is efficient and ideal for industrial grading.	Ang et al. (2020)
650	Cavendish bananas	Statistical calculations	Predict the ripeness of Cavendish bananas	The LBI parameters exhibit strong correlations (>62%) with banana maturity levels, leading to a highly accurate predictive model (99.44%). Moreover, a significant portion (99.42%) of LBI parameters directly influence banana ripeness, with real-world testing achieving an 87.63% accuracy in predicting banana ripeness.	Kuala et al. (2024)
658	Sweet Potatoes	PLSR, PCA, LDA	Monitoring and Classifying Sweet Potatoes quality changes under different storage conditions	LBI effectively monitors sweet potato quality under varied storage conditions, showing significant impacts on backscattering properties and quality parameters. SSC performs well across all storage conditions ($R = 0.56–0.66$; $RMSE = 0.76–1.10$), with 15 °C storage yielding favorable results ($R > 0.50$) in all parameters. PCA and LDA achieve high accuracies (>90% and >80%, respectively) in classifying samples based on backscattering variations.	Sanchez et al. (2020)
658, 705	Theobroma cacao (cocoa pods)	Linear regression	Evaluation of cocoa pod color and maturity stages	LBI parameters at 658 nm and 705 nm correlate with cocoa pod maturity stages, with strong linear regression observed for chroma (C) ($R^2 = 0.755$ for 658 nm, 0.800 for 705 nm). Classification achieved 90% accuracy (658 nm) and 95% accuracy (705 nm) in	Lockman et al. (2019)

				identifying unripe samples. LBI at 705 nm outperformed 658 nm for evaluation and classification.	
658	Watermelon	PLS regression	Color evaluation and grading of seedless watermelons	Backscattering imaging with 658 nm laser diode effectively classifies storage days with >94% accuracy compared to computer vision system.	Ali et al. (2020b)
850	Potatoes	SVM, LDA, MDD	Automatic separation of potatoes from foreign materials	Effective feature selection using laser backscattering imaging achieved >96% accuracy in separating potatoes from non-potatoes of different origins and varieties.	Geng et al. (2021)
650	apricot	PLSR, ANN, PCA-ANN	Predicting firmness and total soluble solids (TSS) during apricot ripening	LBI successfully predicts apricot quality during ripening. ANN model outperformed PLSR, achieving $R^2CV = 0.974$, RMSECV = 3.482 for firmness and $R^2CV = 0.963$, RMSECV = 1.146 for TSS.	(Mozaffari et al. 2022)
633, 850	Apple	PLS, SVM, and ANN	Estimating mealiness in apple fruit during storage	LBI effectively estimates apple fruit mealiness. An ensemble model achieved the highest performance (R = 0.682, RMSE = 7.281), with an ANN model combining both wavelengths also performing well (R = 0.634, RMSE = 7.621), surpassing single wavelength models.	Iida et al. (2023)
450, 520, 635	Mango	MLR	Monitoring moisture content during drying	The study revealed notable impacts of drying temperatures and ripeness stages on moisture content, showing a decrease in moisture ratio with higher drying temperatures and a shorter drying time with increased ripeness. The model at 635 nm demonstrated high accuracy in predicting moisture content ($R^2 = 0.9247$, RMSE = 0.0771), indicating variations in laser transmission properties based on moisture levels and the potential for	Bai et al. (2021)

(continued)

Table 8.1 (continued)

Wavelength (nm)	Material	Calibration model	Application	Findings	Reference
				predicting moisture in unknown maturity states.	
520	SunGold™ and Hayward Kiwifruit	Flexible discriminant analysis (FDA)	Detecting chilling injury damage	LBI calibration achieved high true-positive (TP) classification rates (91.5% for μ_s', 65.6% for μ_a). 'SunGold™' kiwifruit with severe injury features, showed 92% TP using LBI parameters directly and 75% with predicted μ_a and μ_s' after pre-classification. 'Hayward' kiwifruit had lower CI classification accuracy (58% with LBI parameters, 35% with predicted μ_a and μ_s') due to surface trichome interference.	Yang et al. (2021)
658	Banana	LDA	Maturity detection	LBI parameters (grey level intensity, backscattering area size) strongly correlate with banana color and soluble solids ($R^2 > 0.70$). LDA achieves 94.2% ripening stage classification accuracy. LBI is promising for automated fruit quality control.	Zulkifli et al. (2019)
658	Oil palm fresh fruit	PLSR, PCA, LDA, QDA, PCA	Maturity detection	The integration of computer vision and backscattering imaging achieved R^2 values exceeding 0.80 for oil content and color values (L*, a*, b*), with maturity level classification accuracies surpassing 85%. This combination of optical techniques demonstrates high potential as a promising non-destructive tool for evaluating oil palm fresh fruit bunch (FFB) quality based on maturity levels.	Ali et al. (2020a)

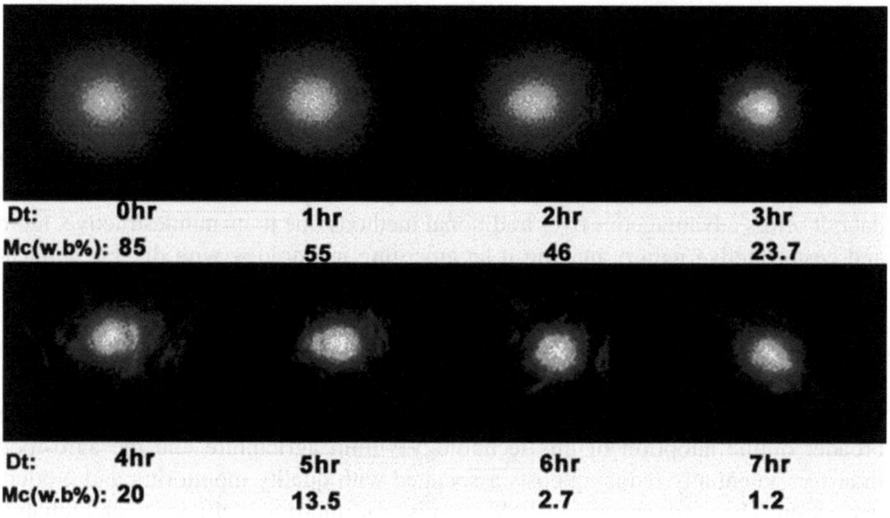

Fig. 8.7 Backscattering profiles of pumpkin at different drying time and moisture content during drying temperature at 70 °C (Onwude et al., 2018)

LBI analysis to predict and classify ripening stages of bananas during storage, demonstrating high accuracy in sorting fruit into ripe and unripe categories ($R^2 > 0.90$) using both LDA and PCA. These two studies collectively highlight LBI's effectiveness in delivering accurate and valuable insights across different multivariate data analysis methods, particularly in fruit quality assessment and monitoring.

Adebayo et al. (2016) studied the optical properties of bananas to predict and classify fruit quality based on ripeness stages using backscattered images at different wavelengths. The study found strong correlations between optical properties and ripening stages, achieving high prediction accuracies in chlorophyll and elasticity using visible wavelengths (532, 660, 785 nm). Recently, Ang et al. (2020) demonstrated a robust classification model using LBI to identify apple defects with a recognition accuracy exceeding 90% through Convolutional Neural Network (CNN) analysis. LBI has also been explored extensively in monitoring quality changes in various fruits postharvest, including papaya drying (Udomkun et al., 2014); plum and pear storage (Mollazade et al., 2013; Kalaj et al., 2016; Zude-Sasse et al., 2019); watermelons storage (Ali et al., 2017); citrus fruits decay detection (Lorente et al., 2013); cacao maturity estimation (Lockman et al., 2019); and oil palm fresh fruit bunches analysis (Ali et al., 2020a). These studies collectively showcase LBI's effectiveness and versatility in enhancing fruit and vegetable quality assessment and monitoring across diverse agricultural products.

8.5 Conclusions

This chapter presents a comprehensive overview of LBI techniques, focusing on their applications in assessing internal quality attributes of fruits and vegetables. LBI utilizes a monochromatic laser diode for enhanced penetration. LBI stands out among other imaging systems by providing spatial, spectral, and multi-constituent data. It offers advantageous over traditional methods due to its nondestructive, rapid and cost-effective nature, making it an emerging technology with diverse applications in agriculture and food quality and safety.

Looking ahead, beyond applications in agriculture and food quality and monitoring, there is a need to advance the development of low-cost LBI systems tailored for specific applications by identifying optimal wavelengths or wavebands. Enhancements in preprocessing efficiency and model robustness, could facilitate broader online adoption of this technology within agriculture and the agro-food industry, potentially reducing costs associated with quality monitoring and product inspection. These advancements hold promise for improving efficiency and reliability in LBI technology across various domains.

Acknowledgment The authors are grateful to Universiti Putra Malaysia for the financial support provided through the Grant Putra Berimpak research funding (GPB/2023/9783900) awarded for this project.

References

Adebayo, S. E., Hashim, N., Abdan, K., Hanafi, M., & Mollazade, K. (2016). Prediction of quality attributes and ripeness classification of bananas using optical properties. *Scientia Horticulturae, 212*, 171–182. https://doi.org/10.1016/j.scienta.2016.09.045

Aernouts, B., Beers, R. V., Watté, R., Huybrechts, T., Jordens, J., Vermeulen, D., Gerven, T. V., Lammertyn, J., & Saeys, W. (2015). Effect of ultrasonic homogenization on the Vis/NIR bulk optical properties of milk. *Colloids and Surfaces B: Biointerfaces, 126*, 510–519. https://doi.org/10.1016/j.colsurfb.2015.01.004

Alhamdan, A. M., & Atia, A. (2017). Non-destructive method to predict Barhi dates quality at different stages of maturity utilising near-infrared (NIR) spectroscopy. *International Journal of Food Properties, 20*(S3), S2950–S2959. https://doi.org/10.1080/10942912.2017.1387794

Ali, M. M., Hashim, N., Bejo, S. K., & Shamsudin, R. (2017). Quality evaluation of watermelon using laser-induced backscattering imaging during storage. *Postharvest Biology and Technology, 123*, 51–59. https://doi.org/10.1016/j.postharvbio.2016.08.010

Ali, M. M., Hashim, N., & Hamid, A. S. (2020a). Combination of laser-light backscattering imaging and computer vision for rapid determination of oil palm fresh fruit bunches maturity. *Computers and Electronics in Agriculture, 169*, 105235. https://doi.org/10.1016/j.compag.2020.105235

Ali, M. M., Hashim, N., Bejo, S. K., & Shamsudin, R. (2020b). Comparison of laser backscattering imaging and computer vision system for grading of seedless watermelons. *Journal of Food Measurement and Characterization, 14*, 69–77. https://doi.org/10.1007/s11694-019-00268-4

Ang, W., Juanhua, Z., & Taiyong, R. (2020). Detection of apple defect using laser-induced light backscattering imaging and convolutional neural network. *Computers & Electrical Engineering, 81*, 106454. https://doi.org/10.1016/j.compeleceng.2019.106454

Bai, J-W., Zhang, L., Cai, J-R., Wang, Y-C., & Tian, X-Y. (2021). Laser light backscattering image to predict moisture content of mango slices with different ripeness during drying process *Journal of Food Process Engineering 44*(12) https://doi.org/10.1111/jfpe.v44.1210.1111/jfpe.13900

Baranyai, L., & Zude, M. (2008). Analysis of laser light migration in apple tissue by Monte Carlo simulation. *Progress in Agricultural Engineering Sciences, 4*, 45–59. https://doi.org/10.1556/progress.4.2008.3

Baranyai, L., & Zude, M. (2009). Analysis of laser light propagation in kiwifruit using backscattering imaging and Monte Carlo simulation. *Computers and Electronics in Agriculture, 69*(1), 33–39. https://doi.org/10.1016/j.compag.2009.06.011

Carles, G., Zammit, P., & Harvey, A. R. (2019). Holistic Monte-Carlo optical modelling of biological imaging. *Scientific Reports, 9*, 15832. https://doi.org/10.1038/s41598-019-51850-1

Cen, H., & Lu, R. (2010). Optimization of the hyperspectral imaging-based spatially-resolved system for measuring the optical properties of biological materials. *Optics Express, 18*(16), 17412–17432. https://doi.org/10.1364/OE.18.017412

DIN EN 60825-1. (2022). *Safety of laser products - Part 1: Equipment classification and requirements* (IEC 60825–1:2014).

ElMasry, G. M., & Nakauchi, S. (2016). Image analysis operations applied to hyperspectral images for non-invasive sensing of food quality—A comprehensive review. *Biosystems Engineering, 142*, 53–82. https://doi.org/10.1016/j.biosystemseng.2015.11.009

Geng, J., Min, H., & Rao, X. (2021). Separation of clods and stones from harvested potatoes using laser backscattering imaging technique. *Journal of Food Measurement and Characterization, 15*, 3262–3273. https://doi.org/10.1007/s11694-021-00896-9

Hashim, N., Pflanz, M., Regen, C., Janius, R. B., Rahman, R. A., Osman, A., Shitan, M., & Zude, M. (2013). An approach for monitoring the chilling injury appearance in bananas by means of backscattering imaging. *Journal of Food Engineering, 116*(1), 28–36. https://doi.org/10.1016/j.jfoodeng.2012.11.018

Hashim, N., Janius, R. B., Rahman, R. A., Osman, A., Shitan, M., & Zude, M. (2014). Changes of backscattering parameters during chilling injury in bananas. *Journal of Engineering Science and Technology, 9*(3), 314–325.

He, X., Fu, X., Rao, X., & Fang, Z. (2016). Assessing firmness and SSC of pears based on absorption and scattering properties using an automatic integrating sphere system from 400 to 1150 nm. *Postharvest Biology and Technology, 121*, 62–70. https://doi.org/10.1016/j.postharvbio.2016.07.013

Hornberg, A. (2006). *Handbook of machine vision*. Wiley-VCH Verlag GmbH & Co. https://doi.org/10.1002/9783527610136

Hu, D., Lu, R., & Ying, Y. (2020a). Spatial-frequency domain imaging coupled with frequency optimization for estimating optical properties of two-layered food and agricultural products. *Journal of Food Engineering, 277*, 109909. https://doi.org/10.1016/j.jfoodeng.2020.109909

Hu, D., Sun, T., Yao, L., Yang, Z., Wang, A., & Ying, Y. (2020b). Monte Carlo: A flexible and accurate technique for modeling light transport in food and agricultural products. *Trends in Food Science & Technology, 102*, 280–290. https://doi.org/10.1016/j.tifs.2020.05.006

Huang, Y., Lu, R., Hu, D., & Chen, K. (2018). Quality assessment of tomato fruit by optical absorption and scattering properties. *Postharvest Biology and Technology, 143*, 78–85. https://doi.org/10.1016/j.postharvbio.2018.04.016

Iida, D., Kokawa, M., & Kitamura, Y. (2023). Estimation of apple mealiness by means of laser scattering measurement. *Food and Bioprocess Technology, 16*, 2483–2496. https://doi.org/10.1007/s11947-023-03068-3

Kalaj, Y. R., Mollazade, K., Herppich, W., Regen, C., & Geyer, M. (2016). Changes of backscattering imaging parameter during plum fruit development on the tree and during storage. *Scientia Horticulturae, 202*, 63–69. https://doi.org/10.1016/j.scienta.2016.02.029

Karray, F., & Silva, C. D. (2004). *Soft computing and tools of intelligent systems design: Theory and applications*. Pearson Addison Wesley.

Kuala, S. I., Juliastuti, E., & Dwivany, F. M. (2024). Manufacture and performance test of banana ripe detection tool using laser light backscattering imaging. *AIP Conference Proceedings, 2957*, 050005. https://doi.org/10.1063/5.0184012

Lockman, N. A., Hashim, N., & Onwude, D. I. (2019). Laser-based imaging for cocoa pods maturity detection. *Food and Bioprocess Technology, 12*, 1928–1937. https://doi.org/10.1007/s11947-019-02350-7

Lorente, D., Zude, M., Regen, C., Palou, L., Gómez-Sanchis, J., & Blasco, J. (2013). Early decay detection in citrus fruit using laser-light backscattering imaging. *Postharvest Biology and Technology, 86*, 424–430. https://doi.org/10.1016/j.postharvbio.2013.07.021

Lu, R., Qin, J., & Peng, Y. (2006). *Measurement of the optical properties of apples by hyperspectral imaging for assessing fruit quality* (p. 066179). American Society of Agricultural and Biological Engineers. https://doi.org/10.13031/2013.21203

Lu, Y., Li, R., & Lu, R. (2016). Gram–Schmidt orthonormalization for retrieval of amplitude images under sinusoidal patterns of illumination. *Applied Optics, 55*(25), 6866. https://doi.org/10.1364/AO.55.006866

Lu, R., Beers, R. V., Saeys, W., Li, C., & Cen, H. (2020). Measurement of optical properties of fruits and vegetables: A review. *Postharvest Biology and Technology, 159*, 111003. https://doi.org/10.1016/j.postharvbio.2019.111003

Mohd Ali, M., Norhashila, H., Bejo, S. K., & Roshnah, S. (2017). Determination of the difference on color changes of watermelons by laser light backscattering imaging. *Journal of Food Science and Technology, 54*, 3650–3657. https://doi.org/10.1007/s13197-017-2826-y

Mollazade, K., Omid, M., Tab, F., & Mohtasebi, S. (2012). Principles and applications of light backscattering imaging in quality evaluation of agro-food products: A review. *Food and Bioprocess Technology, 5*, 1465–1485. https://doi.org/10.1007/s11947-012-0821-x

Mollazade, K., Omid, M., Tab, F. A., Kalaj, Y. R., Mohtasebi, S. S., & Zude, M. (2013). Analysis of texture-based features for predicting mechanical properties of horticultural products by laser light backscattering imaging. *Computers and Electronics in Agriculture, 98*, 34–45. https://doi.org/10.1016/j.compag.2013.07.011

Mozaffari, M., Sadeghi, S., & Asefi, N. (2022). Prediction of the quality properties and maturity of apricot by laser light backscattering imaging. *Postharvest Biology and Technology*. Volume 186, 111842. https://doi.org/10.1016/j.postharvbio.2022.111842

Omid, M., Mahmoudi, A., & Omid, M. H. (2009). An intelligent system for sorting pistachio nut varieties. *Expert Systems with Applications, 36*(9), 11528–11535. https://doi.org/10.1016/j.eswa.2009.03.040

Omid, M., Mahmoudi, A., & Omid, M. H. (2010). Development of pistachio sorting system using principal component analysis (PCA) assisted artificial neural network (ANN) of impact acoustics. *Expert Systems with Applications, 37*(10), 7205–7212. https://doi.org/10.1016/j.eswa.2010.04.008

Onwude, D. I., Hashim, N., & Dana, A. A. (2018). Backscattering imaging as a monitoring system for moisture content and colour changes in Pumpkin during drying process. *ASM Science Journal, 13*(S4), 27–35.

Peng, Y., & Lu, R. (2005). Modeling multispectral scattering profiles for prediction of apple fruit firmness. *Transactions of the ASAE, 48*(1), 235–242. https://doi.org/10.13031/2013.17923

Peng, Y., & Lu, R. (2006). Improving apple fruit firmness predictions by effective correction of multispectral scattering images. *Postharvest Biology and Technology, 41*(3), 266–274. https://doi.org/10.1016/j.postharvbio.2006.04.005

Picotronic GmbH. (2023). *Laser searching from Picotronic GmbH*. https://www.picotronic.de/picopage/en/product/search. Accessed 24 April 2024.

Prahl, S. A., Gemert, M. J., & Welch, A. J. (1993). Determining the optical properties of turbid media by using the adding–doubling method. *Applied Optics, 32*(4), 559–568. https://doi.org/10.1364/AO.32.000559

Pratx, G., & Xing, L. (2011). Monte Carlo simulation of photon migration in a cloud computing environment with MapReduce. *Journal of Biomedical Optics, 16*(12), 125003. https://doi.org/10.1117/1.3656964

Qin, J., & Lu, R. (2005). *Determination of the optical properties of turbid materials by hyperspectral diffuse reflectance.* ASAE paper no 053068. ASAE.

Qin, J., & Lu, R. (2006). Hyperspectral diffuse reflectance imaging for rapid, noncontact measurement of the optical properties of turbid materials. *Applied Optics, 45*(32), 8366–8373. https://doi.org/10.1364/AO.45.008366

Qin, J., & Lu, R. (2007). Measurement of the absorption and scattering properties of turbid liquid foods using hyperspectral imaging. *Applied Spectroscopy, 61*(4), 388–396. https://doi.org/10.1366/000370207780466190

Qin, J., & Lu, R. (2008). Measurement of the optical properties of fruits and vegetables using spatially resolved hyperspectral diffuse reflectance imaging technique. *Postharvest Biology and Technology, 49*(3), 355–365. https://doi.org/10.1016/j.postharvbio.2008.03.010

Qin, J., & Lu, R. (2009). Monte Carlo simulation for quantification of light transport features in apples. *Computers and Electronics in Agriculture, 68*(1), 44–51. https://doi.org/10.1016/j.compag.2009.04.002

Qing, Z., Ji, B., & Zude, M. (2007a). Predicting soluble solid content and firmness in apple fruit by means of laser light backscattering image analysis. *Journal of Food Engineering, 82*(1), 58–67. https://doi.org/10.1016/j.jfoodeng.2007.01.016

Qing, Z., Ji, B., & Zude, M. (2007b). Wavelength selection for predicting physicochemical properties of apple fruit based on near-infrared spectroscopy. *Journal of Food Quality, 30*(4), 511–526. https://doi.org/10.1111/j.1745-4557.2007.00139.x

Qing, Z., Ji, B., & Zude, M. (2008). Non-destructive analyses of apple quality parameters by means of laser-induced light backscattering imaging. *Postharvest Biology and Technology, 48*(2), 215–222. https://doi.org/10.1016/j.postharvbio.2007.10.004

Rizzolo, A., Vanoli, M., Spinelli, L., & Torricelli, A. (2010). Sensory characteristics, quality and optical properties measured by time-resolved reflectance spectroscopy in stored apples. *Postharvest Biology and Technology, 58*(1), 1–12. https://doi.org/10.1016/j.postharvbio.2010.05.003

Romano, G., Baranyai, L., Gottschalk, K., & Zude, M. (2008). An approach for monitoring the moisture content changes of drying banana slices with laser light backscattering imaging. *Food and Bioprocess Technology, 1*, 410–414. https://doi.org/10.1007/s11947-008-0113-7

Romano, G., Nagle, M., Argyropoulos, D., & Müller, J. (2011). Laser light backscattering to monitor moisture content, soluble solid content and hardness of apple tissue during drying. *Journal of Food Engineering, 104*(4), 657–662. https://doi.org/10.1016/j.jfoodeng.2011.01.026

Sanchez, P. D., Hashim, N., Shamsudin, R., & Nor, M. Z. (2020). Laser-light backscattering imaging approach in monitoring and classifying the quality changes of sweet potatoes under different storage conditions. *Postharvest Biology and Technology, 164*, 111163. https://doi.org/10.1016/j.postharvbio.2020.111163

Siyum, Z. H., Pham, T. T., Vozáry, E., Kaszab, T., Nguyen, L. L., & Baranyai, L. (2023). Monitoring of banana's optical properties by laser light backscattering imaging technique during drying. *Journal of Food Measurement and Characterization, 17*, 5268–5287. https://doi.org/10.1007/s11694-023-02019-y

Torricelli, A., Spinelli, L., Contini, D., Vanoli, M., Rizzolo, A., & Zerbini, P. E. (2008). Time-resolved reflectance spectroscopy for non-destructive assessment of food quality. *Sensing and Instrumentation for Food Quality and Safety, 2*, 82–89. https://doi.org/10.1007/s11694-008-9036-2

Tu, K., Jancsók, P., Nicolaï, B., & Baerdemaeker, J. D. (2001). Use of laser-scattering imaging to study tomato-fruit quality in relation to acoustic and compression measurements. *International*

Journal of Food Science & Technology, 35(5), 503–510. https://doi.org/10.1046/j.1365-2621.2000.00407.x

Tu, K., Chen, Y., Ren, K., Shao, X., Dong, Q., & Pan, L. (2006). Modeling apple quality changes based on laser scattering image analysis under simulated shelf life conditions. *Acta Horticulturae, 712*, 371–380. https://doi.org/10.17660/ActaHortic.2006.712.42

Tuchin, V. (2007). *Tissue optics: Light scatteringmethods and instruments for medical diagnosis* (Vol. PM166, 2nd ed.). SPIE Press Monograph.

Udomkun, P., Nagle, M., Mahayothee, B., & Müller, J. (2014). Laser-based imaging system for non-invasive monitoring of quality changes of papaya during drying. *Food Control, 42*, 225–233. https://doi.org/10.1016/j.foodcont.2014.02.010

Wang, W., Li, C., & Gitaitis, R. D. (2014). Optical properties of healthy and diseased onion tissues in the visible and near-infrared spectral region. *Transactions of the ASABE, 57*(6), 1771–1782. https://doi.org/10.13031/trans.57.10815

Wang, A., Lu, R., & Xie, L. (2017). Improved algorithm for estimating the optical properties of food products using spatially-resolved diffuse reflectance. *Journal of Food Engineering, 212*, 1–11. https://doi.org/10.1016/j.jfoodeng.2017.05.005

Wieme, J., Mollazade, K., Malounas, I., Zude-Sasse, M., Zhao, M., Gowen, A., Argyropoulos, D., Fountas, S., & Beek, J. V. (2022). Application of hyperspectral imaging systems and artificial intelligence for quality assessment of fruit, vegetables and mushrooms: A review. *Biosystems Engineering, 222*, 156–176. https://doi.org/10.1016/j.biosystemseng.2022.07.013

Witten, I. H., & Frank, E. (2005). *Data mining: Practical machine learning tools and techniques*. Morgan Kaufmann.

Xie, D., & Guo, W. (2020). Measurement and calculationmethods on absorption and scattering properties of turbid food in Vis/NIR range. *Food and Bioprocess Technology, 13*, 229–244. https://doi.org/10.1007/s11947-020-02402-3

Yang, Z., Li, M., East, A. R., & Zude-Sasse, M. (2021). Application of absorption and scattering properties obtained through image pre-classification method using a laser backscattering imaging system to detect kiwifruit chilling injury. *Foods, 10*(7), 1446. https://doi.org/10.3390/foods10071446

Yu, X., Wenxi, L., Jingwu, M., Jinghao, H., Wenbo, L., Jie, K., Bin, L., Han, Z., & Wei, T. (2024). Principles, developments, and applications of spatially resolved spectroscopy in agriculture: A review. *Frontiers in Plant Science, 14*, 1–23. https://doi.org/10.3389/fpls.2023.1324881

Zamora-Rojas, E., Garrido-Varo, A., Aernouts, B., Pérez-Marín, D., Saeys, W., Yamada, Y., & Guerrero-Ginel, J. E. (2014). Understanding near infrared radiation propagation in pig skin reflectance measurements. *Innovative Food Science & Emerging Technologies, 22*, 137–146. https://doi.org/10.1016/j.ifset.2014.01.006

Zhang, M., Li, C., & Yang, F. (2019). Optical properties of blueberry flesh and skin and Monte Carlo multi-layered simulation of light interaction with fruit tissues. *Postharvest Biology and Technology, 150*, 28–41. https://doi.org/10.1016/j.postharvbio.2018.12.006

Zheng, C., Sun, D.-W., & Zheng, L. (2006). Recent applications of image texture for evaluation of food qualities—a review. *Trends in Food Science & Technology, 17*(3), 113–128. https://doi.org/10.1016/j.tifs.2005.11.006

Zude-Sasse, M., Hashim, N., Hass, R., Polley, N., & Regen, C. (2019). Validation study for measuring absorption and reduced scattering coefficients by means of laser-induced backscattering imaging. *Postharvest Biology and Technology, 153*, 161–168. https://doi.org/10.1016/j.postharvbio.2019.04.002

Zulkifli, N., Hashim, N., Abdan, K., & Hanafi, M. (2019). Application of laser-induced backscattering imaging for predicting and classifying ripening stages of "Berangan" bananas. *Computers and Electronics in Agriculture, 160*, 100–107. https://doi.org/10.1016/j.compag.2019.02.031

Chapter 9
Biospeckle Imaging

T. Seerangurayar, T. Senthilkumar, and V. Chelladurai

9.1 Introduction

Agricultural products, such as fruits, vegetables, grains, and meat, are vital sources of nutrients essential to human health. However, these products are highly perishable, with a short shelf life, and are prone to rapid deterioration during handling, transportation, and processing. Fresh produce quality is often compromised due to spoilage caused by physiological, biological, biochemical, and microbiological factors (Pandiselvam et al., 2020). The market value of fresh products and consumers' purchasing decisions are strongly influenced by their quality. In recent years, there has been an increasing emphasis on the internal quality of fruits over their external appearance. Consumers typically seek high-quality fruits with desirable textures and flavors, free from contaminants, and rich in nutritional and health-promoting properties (Rizzolo et al., 2010). Therefore, ensuring the safety and quality of fresh products from farm to table is critical, necessitating rapid and non-destructive quality evaluation methods.

Advances in information technology have enabled the development of imaging techniques that serve as highly effective systems for assessing the quality and safety of fresh products (Bhargava & Bansal, 2021). Various mechanical, electrochemical, and electromagnetic non-destructive evaluation techniques are widely employed to detect compositional, textural, and visual characteristics due to their efficiency and ease of use (Pandiselvam et al., 2020). Traditional methods for analyzing the physical and chemical attributes of fresh produce are often destructive. However, recent scientific and technological advancements have spurred the adoption of

T. Seeranguryar · T. Senthilkumar
Faculty of Sustainable Design Engineering, University of Prince Edward Island, Charlottetown, PEI, Canada

V. Chelladurai (✉)
Bannari Amman Institute of Technology, Erode, Tamil Nadu, India

C. K. Sunil et al. (eds.), *Emerging Non-destructive Technologies for Food Quality Analysis*, https://doi.org/10.1007/978-3-031-95048-3_9

non-destructive methods. Techniques such as visible/near-infrared spectroscopy, hyperspectral imaging, ultrasonics, acoustics, and electronic nose technology have gained prominence in the field (Samuel et al., 2017). Despite these advancements, the complex mechanical, physical, and chemical properties of fresh produce pose significant challenges in achieving rapid and accurate internal quality assessments. Consequently, the development of reliable, accurate, and non-destructive methods for evaluating product quality during postharvest and packing stages is essential to maintain product excellence.

Biospeckle laser imaging technology has emerged as a promising solution to this challenge. This rapid, non-invasive, portable, and cost-effective technique evaluates biological tissues, ensuring the quality and safety of fresh produce (Retheesh et al., 2016). Unlike conventional imaging methods, biospeckle technology utilizes the dynamic patterns of light scattered from biological tissues, which reflect physiological activities within the produce. This technique has shown great potential in assessing quality parameters such as firmness, ripeness, and internal defects without damaging the sample. Its adaptability to different types of agricultural products further highlights its versatility and reliability. Biospeckle technology provides real-time data, enabling better decision-making during storage, transportation, and retail operations. By integrating this technology into the supply chain, postharvest losses can be reduced, and consumer satisfaction enhanced through the delivery of high-quality products. As the demand for sustainable and efficient quality assessment methods continues to grow, biospeckle imaging offers an innovative solution that aligns with the needs of modern agriculture and food safety protocols.

9.2 Theory and Principles of Biospeckle

When a biological material is exposed to a coherent light, like laser light, the biospeckle phenomenon occurs (Zdunek et al., 2014). The numerous interactions between coherent light and intricate biological material result in variations in the speckle pattern (bright and dark patches), and these changes are captured in the time domain. A camera records these changes, which are then sent to a processing system. Biospeckle activity can be measured using two different setups: forward-scattering speckles and back-scattering speckles. The back-scattering system captures the reflected light after the laser light reaches the sample's surface. In contrast, the laser light in the forward-scattering system travels through the sample and arrives at the camera (Rahmanian et al., 2020).

Biological samples typically exhibit two types of speckle patterns: a static pattern that is caused by the tissue's stationary particles and a variable pattern that is caused by the tissues' moving components. Of these, the variable speckle pattern of the biological material often known as "biospeckle" is unique, whereas the inorganic components are linked to static patterns (Pandiselvam et al., 2020). The term "biospeckle activity" describes the degree of movement of a speckle pattern. Due to the rough surfaces of most agricultural products, biospeckle equipment requires laser light with a wavelength greater than 600 nm for illumination (Ansari et al., 2018; Ansari & Nirala, 2012, 2013). The biospeckle activity is dependent on the inner and outer tissues (Mulone et al., 2014). Biospeckle can offer information about the biological and physical features of the tissue, which makes it useful for a variety of applications in the field of food quality management and safety assurance.

Numerous studies have examined the use of laser biospeckles in agricultural products, including monitoring plant maturity (Kumari & Nirala, 2019; Pieczywek et al., 2018b), detecting fungal infection (Braga Jr et al., 2005; Koley & Nirala, 2020; Pieczywek et al., 2018a), evaluating apple quality characteristics (Kumari & Nirala, 2016; Retheesh et al., 2018b; Zdunek & Cybulska, 2011; Zdunek & Herppich, 2012), detecting chilling and freezing damages in orange (Rahmanian et al., 2020), evaluating seed viability (Braga et al., 2003; Singh et al., 2020), detecting the progress of germination and simulated sprouting damages in wheat kernels (Sutton & Punja, 2017), identifying apple mealiness (Arefi et al., 2016), analyzing the quality of meat (Amaral et al., 2013), and monitoring food emulsions (da Silva Jr et al., 2010).

9.3 Components of Biospeckle Laser Imaging

The basic components required for biospeckle laser imaging are laser light, a detector like a Charge Coupled Devices (CCDs) camera (usually with a lens that is greatly defocused), and a computer with a framing device that can take sequence images at a consistent interval. Several mathematical techniques are used to estimate the biospeckle activity from the time series photos in order to facilitate rapid and nondestructive sampling applications. The experimental setup for measuring biospeckle activity is shown in Fig. 9.1.

In the biospeckle computing process, factors such as "the laser's wavelength, the distance and angle amongst the targets and the detector, polarization, light intensity, and the detector objective aperture" must be considered. When a diffuse target is illuminated by coherent laser light, the backscattered light will appear on the detector as a speckle pattern, which is a bright and dark interference pattern. Depending on the movement of the object, the speckle phenomenon will change over time and aid in the visualization of the tissues (Sujayasree et al., 2022).

9.4 Methods to Analyze Biospeckle Activity

The speckle activity is commonly measured in three approaches such as "spatial and temporal contrast, time history of the speckle pattern (THSP) and spatial-temporal speckle correlation technique" (Zdunek et al., 2014). Several techniques can be used to analyze the biospeckle activity, however, most of them use the "correlation of irradiance as a function of time". The Fuji's method, Weighted Generalized

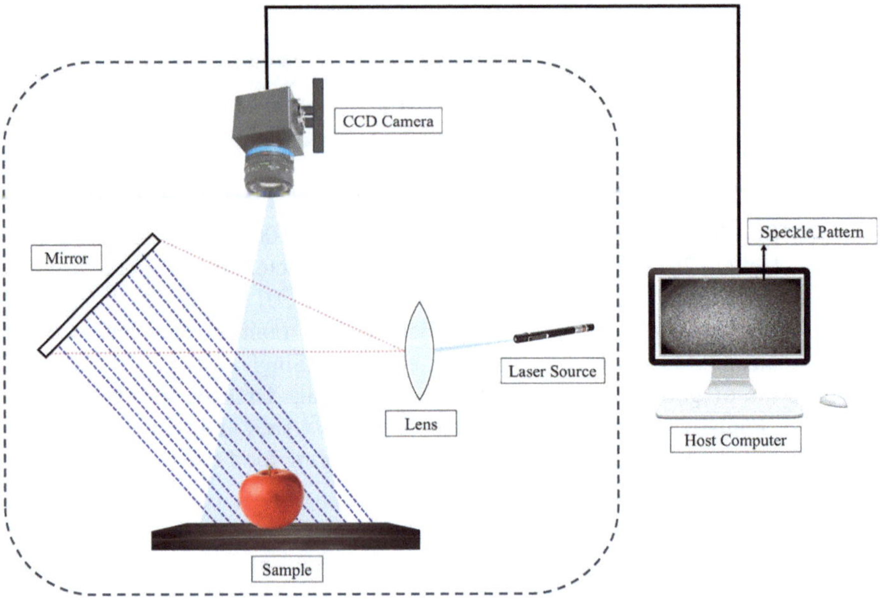

Fig. 9.1 Schematic diagram of biospeckle laser imaging method of analysis

Difference (WGD), and Generalized Difference (GD) techniques show patterns of biological action, however, Inertia Moment (IM) and Absolute Value Difference (AVD) are the numerical methods of application (Pandiselvam et al., 2020). These analytical techniques can be further divided into two categories: qualitative and quantitative techniques. In the quantitative analysis, the AVD, cross-correlation, and IM approaches are frequently used. For qualitative analysis, Fujii, temporal difference (TD), laser speckle temporal contrast analysis (LASTCA), laser speckle contrast analysis (LASCA), and generalized differences are commonly used (Pandiselvam et al., 2020; Retheesh et al., 2016). The biospeckle laser's (BSL) output numerical data is further processed to produce graphical and visual information.

9.5 Spatial Analysis of Biospeckle Activity

9.5.1 Fujii Method

This method employs a scanning process to analyze the image action. Despite being created for blood flow monitoring, this method has been widely used in biospeckle-based research in recent years. This method develops a strategy based on the total absolute disparities among the intensities of subsequent image frames, even though the mean of each frame estimate is computed (Ribeiro et al., 2014). The weighting method suggested by Fujii, however, shows a nonlinear correlation, which leads to progressively fewer steady regions of reduced activity where noise is strengthened and can be misinterpreted as a motion signal (Retheesh et al., 2018a).

9.5.2 Generalized Difference (GD)

The GD method can be defined as the cumulative sum of absolute values of the pixel strength variations between each frame. This method was developed as an option to Fuji's approach. The weighting process has been eliminated from this computation process (Minz & Nirala, 2016).

9.5.3 Time History Speckle Pattern (THSP)

THSP is a matrix composed of different progressive images of dynamic biospeckles. In this, a column indicates the biospeckle's behaviour over time, while rows show the geographical profile. To create THSP, a similar segment is taken out of each image and placed next to it. The segments are arranged in a sequential order, one after the other. The biospeckle action time scale is shown by the width of the THSP, which is equal to the number of images used. As a result, THSP is composed of dark-scale pixels arranged in horizontal lines that show the evolution of biospeckle intensity variations over time (Pandiselvam et al., 2020).

9.5.4 Inertia Moment (IM)

A popular numerical technique called the Inertia Moment (IM) is based on the origin of a definite matrix called the THSP, which is followed by the co-occurrence matrix (COM), an intermediary image (Ansari & Nirala, 2016). The IM of the co-event association is appropriate in evaluating the distribution of the IM values over the major diagonal. It is calculated by multiplying the total matrix numbers by the squared distance from the main diagonal. It is expected that stronger biospeckle action is recognized by a higher inertia moment (Sujayasree et al., 2022).

9.5.5 Absolute Value Difference (AVD)

As an alternative to the Inertia Moment technique, this approach is the "first statistic moment order." The fundamental idea of AVD is that the difference summation is the most significant information analyzed, and the square utilized in the IM approach can amplify the deviations over time in a biased manner (Braga et al., 2011). In certain cases, the AVD approach shows significantly better biospeckle activity analysis, especially when there are no high-frequency data in the THSP matrix.

9.5.6 Laser Speckle Temporal Contrast Analysis (LASTCA)

This method relies on estimating temporal or spatial contrast. In the temporal arrangement, the ratio of each frame pixel's mean value to standard deviation is calculated. It is possible to reveal the darkened subsurface objects by using the sequential properties of dynamic speckles. Furthermore, it was noted that the optical speckle phenomena still constrained the image resolutions that could be achieved using this method. In any case, this method is easy to apply and can be utilized as a rapid image screening tool for material flaw detection and safety evaluations (Nothdurft & Yao, 2005).

9.5.7 Laser Speckle Contrast Analysis (LASCA)

An online processing method called Laser Speckle Contrast Analysis (LASCA) uses a single image to analyze sample activity by looking at the degree of grain blurring. It is one of the rapid and affordable online techniques for mapping biospeckle activity and identifying changes. To depict the activity, the index contrast (C) is utilized. When image analysis is done in segments, the resolution of the produced image is decreased. The LASCA technique has certain limitations, including the requirement for camera adjustment and its slow speeds to investigate variations in the speckle boiling effect. Extra caution is needed when interpreting LASCA maps. The LASCA maps have a reverse color scale in contrast to LASTCA or Fujii maps because the areas with lower speeds have more contrast and vice versa. The constraints of LASCA can be solved by using multiple images to make online perception without requiring continuous modification and resolution loss (Pandiselvam et al., 2020).

9.5.8 *Motion History Image (MHI)*

This "real-time imaging" method creates a sequence of steps based on the most recent movement of an image progression. The MHI technique is suggested as a substitute for the standard virtual methods of evaluating biospeckle images. In this case, the "pixel intensity" is determined by the movement captured at a particular location, while the brighter values correspond to more recent motion. The primary treatment is applied to the photos that are placed inside the buffer (Godinho et al., 2012). The processing method creates a "threshold of the silhouette" by calculating the variations between two successive photos. The final MHI image is generated by applying this method of weighing the threshold images placed inside the buffer in relation to each image's "lifetime". When evaluating biological and nonbiological targets, MHI produced better results than LASCA. Furthermore, the MHI produced similar findings to the popular offline techniques, such as GD and Fujii's (Sujayasree et al., 2022).

9.6 Biospeckle Applications in Agri-Food Products

9.6.1 *Apples*

9.6.1.1 Mealiness

The mealiness in apples is an undesirable parameter which affects the marketability of apples. Mealiness is a combined attribute of crispiness, hardness, juiciness and starchiness and is usually measured manually or through time-consuming and destructive testing methods (Barreiro et al., 1998; Harker et al., 2002). Arefi et al. (2016) artificially induced mealiness and semi-mealiness in the apples by storing them at 20 °C and 95% RH for 26 days and 3 °C and 95% RH, for 16 days. The control samples were stored at 0 ± 1 °C and 85 ± 5% RH for 0, 30, 60, 120, and 150 days. A total of 540 control, 100 semi-mealy and 120 mealy apples were used in this study. The apples were subjected to imaging using a closed chamber biospeckle imaging system comprised of two 3 mW low-power diode lasers with wavelengths of 680 and 780 nm, and a CCD camera operated at 20 frames per second. The apples subjected to biospeckle imaging were instrument tested to obtain stiffness, juiciness, soluble solids content and moisture content. The biospeckle activity is captured in video format with a total of 500 frames per sample. The commonly used Time-history speckle pattern (THSP) was used to study the biospeckle activity and the resolution of the THSP images was 400 × 500 pixels. The first and second-order statistical features were extracted from the indices of the THSP images. The statistical features were subjected to the STEPDISC feature selection function using SPSS software, and selected features were used as inputs for Artificial Neural Networks (ANN) classification models. The classification models were able to classify the control apples with a classification accuracy of 81.7% and mealy and semi-mealy apples with a classification accuracy of 77.3% and 70.9%, respectively.

9.6.1.2 Chlorophyll Content

A biospeckle imaging system was used to detect the chlorophyll content in apples. The biospeckle imaging system used 1 diode laser (3 mW) at 670 nm and a CCD camera. The study used 32 apples from Pinova and Topaz varieties to acquire biospeckle activity in video format at 15 frames per second for a total 210 frames for every sample. The biospeckle activity of the apples was evaluated by correlation coefficient. There was a negative correlation between maximum fluorescence and biospeckle activity was observed. It is found that biospeckle activity increases with a decrease in chlorophyll content and vice versa. The laser diode of 670 nm was a red laser light and absorbed the chlorophyll content in apples; for further monitoring of chlorophyll content in apples, the diode in the wavelength range between 800 and 880 nm can be the best choice (Zdunek & Herppich, 2012).

9.6.1.3 Fungal Infection

Fungal infection in apples needs to be identified in an early stage to make corrective actions and to separate the healthy ones from the infected ones to avoid further damage. The apples infected by bull's eye rot can be seen by visual eyes only after 4–5 days of infection. A feasibility study utilized the biospeckle imaging system to detect fungal infection in apples. Apples artificially infected with *Pezicula malicorticis* (H. Jacks.) and control apples were subjected to biospeckle imaging. The biospeckle imaging system used 3 laser diodes at 473 nm, 532 nm, and 830 nm, and beam expander to acquire wider angle images using a CCD camera. The videos were captured at 60 frames per second. The acquired biospeckle activity was evaluated using correlation coefficient, fujii index, moment of inertia and frequency. The apple fungal infection as early as day 1 was detected by biospeckle imaging using all the 4 evaluation methods, except moment of inertia. The laser beam at 532 nm images provided better results compared to 473 and 830 nm wavelengths laser beam (Pieczywek et al., 2018a).

9.6.1.4 Apple Slices Quality Prediction

Arefi et al. (2023) deployed biospeckle imaging to detect the quality of apple slices during the drying process. The apple slices' quality, such as moisture content, vitamin C, and soluble solids content (SSC), were measured at 980 nm and 1450 nm in the near-infrared region. The biospeckle images of apple slices were acquired during the drying process at 60 °C and 9 different drying periods starting from 0 min to 195 min. The initial moisture content, vitamin C, and SSC of apple slices were 87.56 ± 1.08%, 9.37 ± 0.83 mg/100 g fresh weight (FW), and 10.95 ± 0.95%, respectively. After 195 min of drying at 60 °C, these parameters changed significantly: the final moisture content, vitamin C, and SSC were 14.87 ± 1.40%, 3.47 ± 0.77 mg/100 g FW, and 65.69 ± 3.2%, respectively. Two cameras were used in this study: one CMOS camera to capture images in the wavelength range of 400–1050 nm and an InGaAs camera to capture images in the wavelength range of 400–1750 nm. The biospeckle activity was more initially and gradually reduced during the drying process for both 980 and 1450 nm. The Gaussian Process Regression (GDR) and Partial Least Square Regression (PLSR) models were developed to predict moisture content, vitamin C and SSC in apples slices with an R^2 value between 0.8 and 0.96. The wavelength 1450 nm performed better than 980 nm.

9.6.1.5 Firmness

Apple firmness is usually measured using destructive methods. Jamshidi and Arefi (2016) used non-destructive, fast and accurate biospeckle imaging method to detect firmness as an alternative to traditional methods. The laser beams at 680 and 780 nm were used along with a CCD camera to observe biospeckle activity in apples. The features moment of inertia (IM), absolute value of differences (AVD), wavelets, and texture were extracted from biospeckle images and used as inputs for Pearson's correlation coefficients. The texture features provided better pearson's correlation coefficients to predict firmness in apples.

9.6.2 Orange

9.6.2.1 Chilling and Freezing Disorders

The oranges were stored at no chilling, chilled at 1 °C, placed in a freezer at −7 °C and −20 °C to stimulate sound, chilled, moderately frozen and completely frozen oranges. The chilling and freezing assessment of oranges was done by experts and used as a reference to compare biospeckle imaging data. The four differently treated oranges were subjected to biospeckle imaging using diode lasers at 638 and 655 nm and a CMOS camera operated at 30fps. The biospeckle activity was recorded using both forward and back-scattering methods. The biospeckle activity increased from sound to chilled oranges and decreased from chilled to moderately frozen and moderately frozen to completely frozen oranges. This reduced biospeckle activity in the oranges can be attributed to less water motility in the frozen oranges. Twelve features: two moments of inertia (IM), four absolute values of differences (AVD), four regular values of differences (RVD), and two modified AVD (NUMAD) were extracted and used as inputs for soft independent modelling of class analogy (SIMCA), linear discriminant analysis (LDA), quadratic discriminant analysis (QDA), artificial neural networks (ANN), and support vector machines (SVM) classifiers. The images obtained from the forward-scattering mode provided better classification accuracy than the back-scattering mode. The ANN and SVM classifiers using features from forward scattering-mode images provided 100% and 98.1% classification accuracies, respectively for all four differently treated oranges (Rahmanian et al., 2020).

9.6.2.2 Orange Storage Period

Orange storage period was predicted using a non-destructive biospeckle imaging technique. A 10-mW helium-neon laser beam with a CCD camera acquired biospeckle activity in oranges. The images' statistical cumulant and moment of inertia were used for further analysis using the statistical Tukey's test. The moment of inertia and statistical cumulant decreased with an increase in the storage period. Based on the obtained result, biospeckle imaging can be an effective tool for detecting the orange storage period (Rabelo et al., 2005).

9.6.2.3 Green-Orange Scar Region

The scar region of the green oranges was detected using the biospeckle imaging method. A biospeckle imaging system with a 10 mW helium-neon laser beam at 632.8 nm and a CCD was used to acquire biospeckle activity. This study used motion history images (MHI) to identify biospeckle activity at scar regions. The MHI method is compared with traditional Fujii, GD, TD and tLASCA methods. The results obtained by MHI are better than other methods and can be used as a tool for scar region detection in green oranges (Retheesh et al., 2018a).

9.6.3 Potato

9.6.3.1 Anti-browning Activities in Fresh-cut Potatoes

The biospeckle imaging of the chemically treated (citric acid (CA-0.5% and 1.0%) and citric acid + sodium chloride (CS-0.5% and 1.0%)) fresh-cut potato tubers were acquired for eight consecutive days and stored at cold storage. The biospeckle imaging system used in this study comprised of a 10 mW helium-neon laser beam at 632.8 nm and a CMOS camera. The effects of chemical treatment on fresh-cut potato tubers on physiological activities such as the respiration rate and weight loss at cold storage have been studied for eight consecutive days and used as a reference to compare the biospeckle activity. Moment of inertia (IM) with and without frequency filtering from the captured images were obtained and used for further analysis. The respiration rate of CS-treated samples was higher than that of

CA-treated samples. The results of the biospeckle imaging followed the same pattern and correlated with IM curves (Minz et al., 2015).

9.6.3.2 Mechanical Properties

The biospeckle activity was used to detect the effect of storage on the textural properties of potatoes. The mechanical, rheological properties, uniaxial compression and creep tests were carried out on potatoes subjected to imaging and used as a reference to compare the biospeckle activity. A 5-mW low-power laser diodes at 650 nm, 780 nm, and 980 nm coupled with a CMOS camera was used to acquire biospeckle activity of potatoes stored for 6 months. The moment of inertia, absolute value of differences, autocorrelation function, and energy of wavelet transform features were extracted from the images. The secant modulus of elasticity, maximum force, and elastic modulus were predicted accurately at 650 nm compared to the other two wavelengths. The creep parameters were predicted accurately at 780 nm compared to 980 nm (Omidi-Arjenaki et al., 2024).

9.6.3.2.1 Blackspot Bruise

The blackspot bruise is an undesirable quality causing huge economic losses for the farmers and processors; the blackspot bruise is a subsurface defect occurring because of impact during handling. The biospeckle imaging was deployed to detect the difference in biospeckle activity between healthy and bruised potatoes. 214 fresh potatoes procured from the market were divided into 46 control samples, and the remaining were used for creating blackspot bruises artificially using an impact method (Fig. 9.2). The potatoes subjected to impact were divided into normal potatoes and blackspot potatoes because some potatoes didn't develop bruises even after the applied impact.

The biospeckle imaging system used in this study consists of a 635 nm, 50 mW low-power laser beam, beam expander of 20 × 20 mm, and a CCD camera. The biospeckle activity for the control samples was less than the impacted samples. However, there is no difference in biospeckle activity between normal and blackspot bruised samples (Gao & Rao, 2019).

9.6.3.3 Other Studies

The biospeckle imaging system was also used limitedly to detect defects and other quality parameters associated with beans, strawberries, blueberries, wheat, mango, Indian gooseberries, beef and ice cream, and the details are provided in Table 9.1.

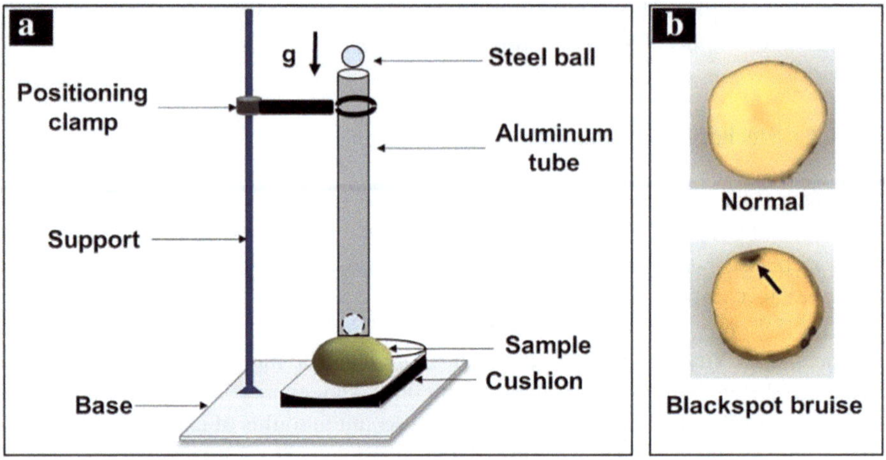

Fig. 9.2 Artificial blackspot bruise creation using steel ball impact (**a**) and the difference between normal and blackspot bruise potatoes (**b**). (Reproduced from Gao & Rao, 2019)

9.7 Summary

Biospeckle laser imaging has emerged as a promising tool for quality assessment and safety assurance in the food and agricultural industries. This chapter delves into this innovative technique's theoretical underpinnings, technological advancements, and practical applications. By analyzing dynamic biological activities, biospeckle imaging enables the detection of critical parameters like firmness, ripeness, texture, microbial growth, and physiological changes in various food products. Its versatility, adaptability to different wavelengths and analytical methods, and non-destructive nature make it suitable for diverse agricultural commodities and conditions.

The application of the biospeckle method in food and agricultural product quality analysis was studied elaborately by global researchers. To illustrate the breadth of applications and effectiveness of biospeckle techniques, the following Table 9.1 summarizes key findings from recent studies. It provides a comprehensive overview of the food materials investigated, the specific biospeckle methods employed, the wavelengths used, and the notable observations or results achieved. This tabular representation is a valuable reference, connecting the theoretical and experimental details discussed earlier with tangible outcomes, thereby showcasing the practical impact of biospeckle laser imaging in the agri-food sector.

Table 9.1 Application of biospeckle laser technique in food products

Application	Food materials	Wavelength	Biospeckle features extraction methods	Observation/Result	Reference
Bruise detection	Apple	632.8 nm	Fujii method, Generalized Difference, and Laser speckle contrast analysis	Low biospeckle activity was observed in damaged areas in all three methods. Among them, a significantly low biospeckle value was detected in the GD method (82.82 ± 12.86)	Yan et al. (2017)
	Apple	632.8 nm	Co-occurrence matrix, Inertia moment, Absolute value difference, Generalized difference, Fujii method, Biospeckle activity (BA) value, Granulometric size distribution (GSD), Grey-level co-occurrence matrix (GLCM), Parameterized generalized difference, Alternative generalized difference (AGD) and Parameterized global average Fujii method	The inertia moment method offered maximum biospeckle activity difference (521.99) for bruised and fresh apples	Kumari and Nirala (2016)
	Apples, guavas, mangoes, and cucumbers	632.8 nm	Inertia moment (IM) and Cross-correlation (CC)	Mechanical damages and pesticides in vegetables were effectively detected. The percentage of CC difference between normal and damaged was higher in apples (25–30%).	Samuel et al. (2017)
	Potato	635 nm	Time history of the speckle pattern-based Inertia moment method (THSP-based IM)	Early skinning injury can be recognized by biospeckle imaging with an accuracy of 88.10%.	Gao et al. (2016)
	Potato	635 nm	Time history of speckle patterns-Wavelet entropy (THSP-WE), Correlation of contiguous speckle patterns-Wavelet entropy, Correlation of contiguous speckle patterns-Modified	CCSP-WE method had better robustness and discrimination ability to predict blackspot bruise.	Gao and Rao (2019)

(continued)

Table 9.1 (continued)

Application	Food materials	Wavelength	Biospeckle features extraction methods	Observation/Result	Reference
			wavelet entropy (CCSP-MWE) methods		
Scar detection	Green orange	632.8 nm	Fujii method, Generalized differences (GD), Temporal difference (TD) and Laser speckle contrast analysis (LASCA)	The motion history image (MHI) algorithm accurately detects the scar area. The quality of MHI results depends on the proper adjustments of buffer size and threshold level.	Retheesh et al. (2018a)
Mealiness identification	Apple	680 and 780 nm	Inertia moment, Absolute value of differences, and Autocorrelation	Samples were classified as fresh, semi-mealy, and mealy based on their stiffness and juiciness. The best classification accuracy was demonstrated at 780 nm wavelength for fresh (81.7%) and semi-mealy (70.9%), whereas for mealy apples, 680 nm wavelength offered the best result (77.3%)	Arefi et al. (2016)
Fungal infection detection	Apple	632.8 and 635 nm	Correlation coefficient	The biospeckle activity increased in the fungal-affected samples. The biospeckle activity decreased during shelf life, irrespective of storage conditions.	Adamiak et al. (2012)
	Apple	473, 532 and 830 nm	Correlation coefficient, Fujii index, Inertia moment (IM), and New frequency analysis method	All four biospeckle activity indices demonstrated a high level of performance in detecting fungus. The best result was obtained in the 532 nm biospeckle images.	Retheesh et al. (2018b)
	Citrus fruits	650 nm	Correlation coefficient (CC), temporal contrast (TC), Fujii index, Numerical analysis of the average differences (NUMAD), and Frequency index (FI)	The biospeckle images detected fungal infection earlier than RGB images. The ANN-based one-day-in-advance, two-days-in-advance and three-days-	Yang et al. (2024)

				in-advance prediction models' discriminant accuracy was 93.9%, 89.3%, and 86.4%, respectively. In comparison, the SVM-based model's accuracy levels were 90.2%, 83.9%, and 81.4%, respectively.	
	Strawberries	650 nm	Generalized differences (GD) and Fujii methods	Fungal infection can be detected early on the surface of the fruit and distinguish a healthy areas from damaged/bruised ones.	Mulone et al. (2014)
	Blueberry	632.8 nm	Fujii and Generalized differences (GD) methods	GD method worked well in identifying infested areas	Janaszek-Mańkowska et al. (2022)
	Soybean	632:8 nm	Random time history of speckle pattern (RTHSP) and Co-occurrence matrix (COM)	Biospeckle activity had a strong correlation (r = 0.93 to 0.97) with the lesion length of the infected plant area	Singh et al. (2021)
	Bean	632.8 nm	Inertia moment (IM) and Frequency analysis	Fungal infection detected. *Aspergillus flavus* fungal infected seeds showed a significantly higher IM value than those infected by *Fusarium oxysporum.*	Rabelo et al. (2011)
	Beans	632 nm	Generalized differences, Fujii's algorithm and Inertia moment	Among the three fungi specimens, *Aspergillus* and *Colletotrium* fungal infection in beans was detected, while *Sclerotinia* was not detected.	Braga Jr et al. (2005)
Detection of Blue mold disease	Indian gooseberry	632.8 nm	Co-occurrence Matrix, Inertia Moment, Absolute Value of Differences, Fujii, Inverse Fujii, Alternative Fujii, Parameterized Fujii, Generalized Difference (GD), Parameterized GD, Alternative GD (AGD), Cubic	CAD provides the highest value of bioactivity difference between non-infected & bright yellow (3463.81), non-infected & blue (4250.29) and bright yellow & blue (786.48) regions	Koley and Nirala (2020)

(continued)

Table 9.1 (continued)

Application	Food materials	Wavelength	Biospeckle features extraction methods	Observation/Result	Reference
			Absolute Difference (CAD), qFujii, Parameterized Geometrical Mean of Temporal Difference (PGMTD)		
Detecting maturity and ripe stages	Mango, guava and banana	632.8 nm	Histograms, RGB and HSB colour space, Inertia moment (IM), Absolute value difference, Autocovariance, Granulometric size distribution, gray level co-occurrence matrix, Parameterized geometric mean of generalized difference (GD), Image sequencing mean of parameterized GD, and Squared temporal difference (STD)	IM offered the best differentiation between the maturity and ripe stages of the three fruits. The highest BA difference (1562) was observed in bananas among the three fruits.	Kumari and Nirala (2019)
Freshness identification based on chlorophyll content	Apples	670 nm	Correlation coefficient	Biospeckle activity linearly decreased for increasing the chlorophyll content of the apple. The low R^2 (~ 0.3) result is associated with biospeckle activity being hidden by chlorophyll absorption at 670 nm, which suggests the right laser wavelength selection is essential.	Zdunek and Herppich (2012)
Detection of quality changes during storage	Apples	632 nm	Correlation coefficient	The biospeckle activity of apples decreased for starch degradation and increased for chlorophyll degradation. The biospeckle activity and starch content had a significant correlation ($R = 0.849$, $p < 0.01$), whereas firmness, titratable acid and soluble solids content did not have any correlation.	Zdunek and Cybulska (2011)
	Apples		Cross-correlation coefficient		

		635, 690, 830 and 1060 nm		Biospeckle at 635 nm, 690 nm and 830 nm had high potential to predict biochemical changes	Kurenda et al. (2014)
Chilling and freezing disorder identification	Orange	638 and 655 nm	Graphical (Fujii algorithm) and numerical (Time history of speckle pattern (THSP)) methods	Both back-scattering and forward-scattering systems were used. The forward-scattering method with the ANN classifier provided the best prediction accuracy (100%) for chilled, moderate freezing, and extreme freezing.	Rahmanian et al. (2020)
Storage quality	Orange	10 mW helium-neon laser	Statistical cumulant and the moment of inertia	Moment of inertia values and statistical cumulant decreased with an increase in the storage period	Rabelo et al. (2005)
Effect of minimal processing	Cut apple	632.8 nm	Speckle grain size and Degree of polarization (DOP)	Speckle grain size and DOP had a correlation to total soluble solids and titratable acidity of the cut apple.	Minz and Nirala (2016)
Determination of mechanical Properties (texture)	Potato	650, 780 and 980 nm	Inertia moment (IM), Absolute value of differences (AVD), Autocorrelation function (AF), and Energy of wavelet transform	An adaptive neuro-fuzzy inference system with the genetic algorithm was used to develop the mechanical properties of potato models. The best correlation coefficients of 0.812, 0.773, and 0.750 were achieved for retarded compliance, maximum force and retarded viscosity, respectively.	Omidi-Arjenaki et al. (2024)
Measuring browning effect	Fresh cut potato	632.8 nm	Inertia moment (IM) and Wavelet transform (time-frequency analysis)	IM method with filtration can be able to separate different physiological changes (respiration rate and weight loss).	Minz et al. (2015)
Predicting Seed Viability	Peas	632.8 nm	Inertia moment (IM) and Generalized differences (GD)	Distinguish viable and non-viable seeds	Men et al. (2015)

(continued)

Table 9.1 (continued)

Application	Food materials	Wavelength	Biospeckle features extraction methods	Observation/Result	Reference
Detecting Sprouting	Wheat seeds	532 nm	Mask centroid (longitudinal centre of the biospeckle activity in the masked region)	A novel whole seed grading protocol was developed based on the biospeckle measurement. This simulated seed grading protocol can differentiate sprouting damaged from healthy seeds	Sutton and Punja (2017)
Determination of Seed priming	Chickpea	632.8 nm	Saturation, Contrast and Homogeneity	Biospeckle activity had a strong correlation with seed priming treatments (germination percentage ($R = 0.88$) and mean germination time ($R = 0.92$))	Singh et al. (2020)
Determination of Beef aging	Beef	632 nm	Inertia Moment and its variation, and Absolute values of differences (AVD)	The tenderness (Warner–Bratzler Shear Force (WBSF)) and color values are studied to analyze beef quality. The aging process had a correlation of 0.6146 with WBSF and 0.7973 with aging time. The colour parameters such as hue angle, redness intensity and percentage of metmyoglobin had the R-value of 0.7953, 0.8120, and 0.9119, respectively.	Amaral et al. (2013)
Predicting textural properties during the transient process	Ice creams	633 nm	Time history of the speckle pattern	Texture testing was coherent with the speckle behaviour and viscoelastic properties (ice melting and air loss) of the systems. A higher Pearson coefficient ($R^2 > 0.8$) suggests that experimental values and adjusted functions are strongly correlated.	da Silva Jr et al. (2010)

9.8 Conclusion

The application of biospeckle laser imaging has opened new avenues for the non-destructive analysis of food and agricultural products, ensuring the quality, safety, and marketability of these perishable goods. By leveraging coherent laser light to analyze biological activities, this method provides a rapid, accurate, and non-invasive solution for evaluating both the internal and external properties of produce. This chapter extensively discusses the theoretical principles, experimental setups, and various analytical methods associated with biospeckle imaging. The ability of this technology to detect a broad spectrum of attributes—ranging from textural properties and firmness to physiological changes such as chlorophyll content and fungal infections—has been highlighted through numerous studies. One of the primary advantages of biospeckle imaging is its adaptability across diverse food products, including apples, oranges, potatoes, and seeds. This versatility is further enhanced by its integration with advanced data processing techniques such as Artificial Neural Networks (ANN), Support Vector Machines (SVM), and Gaussian Process Regression (GPR), which improve its predictive accuracy and reliability. Additionally, the technology's ability to provide real-time, non-destructive assessments makes it ideal for use in storage, transportation, and retail environments.

Despite its numerous benefits, the biospeckle method is not without challenges. Variability in biological tissue responses, the need for precise calibration of imaging systems, and the influence of environmental conditions can affect the consistency of results. Addressing these limitations requires ongoing research and development, particularly in standardizing experimental protocols and optimizing the mathematical models used for data interpretation.

The adoption of biospeckle imaging in the agri-food sector is expected to grow as the demand for sustainable and efficient quality control methods increases. Its potential extends beyond current applications, with opportunities for integration into automated quality monitoring systems, robotics, and IoT-enabled supply chains. Furthermore, advancements in laser technology, imaging sensors, and computational algorithms are likely to enhance the resolution and accuracy of biospeckle analyses.

This chapter aims to provide a comprehensive understanding of the biospeckle method, offering valuable insights for students, researchers, and industry professionals. Bridging the gap between fundamental principles and practical applications, it underscores the transformative potential of this emerging technology. As policymakers and stakeholders prioritize food quality and safety, biospeckle imaging stands out as a pivotal tool in achieving these goals. Its alignment with global efforts to reduce food waste, ensure nutritional value, and promote consumer satisfaction further solidifies its relevance in modern agricultural and food industries.

References

Adamiak, A., Zdunek, A., Kurenda, A., & Rutkowski, K. (2012). Application of the biospeckle method for monitoring bull's eye rot development and quality changes of apples subjected to various storage methods—Preliminary studies. *Sensors, 12*(3), 3215–3227.

Amaral, I. C., Braga, R. A., Jr., Ramos, E. M., Ramos, A. L. S., & Roxael, E. A. R. (2013). Application of biospeckle laser technique for determining biological phenomena related to beef aging. *Journal of Food Engineering, 119*(1), 135–139.

Ansari, M. Z., & Nirala, A. (2012). Biospeckle techniques in quality evaluation of Indian fruits. *World Academy of Science, Engineering and Technology, 6*, 11–20.

Ansari, M. Z., & Nirala, A. K. (2013). Biospeckle activity measurement of Indian fruits using the methods of cross-correlation and inertia moments. *Optik, 124*(15), 2180–2186.

Ansari, M. Z., & Nirala, A. (2016). Assessment of biospeckle activity of lemon fruit. *Agricultural Engineering International: CIGR Journal, 18*(2), 190–200.

Ansari, M., Mujeeb, A., & Nirala, A. (2018). Assessment of biological leaf tissue using biospeckle laser imaging technique. *Laser Physics, 28*(6), 065608.

Arefi, A., Moghaddam, P. A., Hassanpour, A., Mollazade, K., & Motlagh, A. M. (2016). Non-destructive identification of mealy apples using biospeckle imaging. *Postharvest Biology and Technology, 112*, 266–276.

Arefi, A., Sturm, B., Raut, S., von Gersdorff, G., & Hensel, O. (2023). NIR laser-based imaging techniques to monitor quality attributes of apple slices during the drying process: Laser-light backscattering & biospeckle imaging techniques. *Food Control, 143*, 109289.

Barreiro, P., Ortiz, C., Ruiz-Altisent, M., De Smedt, V., Schotte, S., Andani, Z., et al. (1998). Comparison between sensory and instrumental measurements for mealiness assessment in apples. A collaborative test. *Journal of Texture Studies, 29*(5), 509–525.

Bhargava, A., & Bansal, A. (2021). Fruits and vegetables quality evaluation using computer vision: A review. *Journal of King Saud University-Computer and Information Sciences, 33*(3), 243–257.

Braga, R. A., Jr., Rabelo, G. F., Granato, L. R., Santos, E. F., Machado, J. C., Arizaga, R., et al. (2005). Detection of fungi in beans by the laser biospeckle technique. *Biosystems Engineering, 91*(4), 465–469.

Braga, R. A., Dal Fabbro, I. M., Borem, F. M., Rabelo, G., Arizaga, R., Rabal, H. J., & Trivi, M. (2003). Assessment of seed viability by laser speckle techniques. *Biosystems Engineering, 86*(3), 287–294.

Braga, R., Nobre, C., Costa, A., Sáfadi, T., & Da Costa, F. (2011). Evaluation of activity through dynamic laser speckle using the absolute value of the differences. *Optics Communications, 284*(2), 646–650.

da Silva Jr, E., Da Silva, E. R. T., Muramatsu, M., & da Silva Lannes, S. C. (2010). Transient process in ice creams evaluated by laser speckles. *Food Research International, 43*(5), 1470–1475.

Gao, Y., & Rao, X. (2019). Blackspot bruise in potatoes: Susceptibility and biospeckle activity response analysis. *Journal of Food Measurement and Characterization, 13*, 444–453.

Gao, Y., Geng, J., Rao, X., & Ying, Y. (2016). CCD-Based skinning injury recognition on potato tubers (Solanum tuberosum L.): A comparison between visible and biospeckle imaging. *Sensors, 16*(10), 1734.

Godinho, R., Silva, M., Nozela, J., & Braga, R. (2012). Online biospeckle assessment without loss of definition and resolution by motion history image. *Optics and Lasers in Engineering, 50*(3), 366–372.

Harker, F., Maindonald, J., Murray, S., Gunson, F., Hallett, I., & Walker, S. (2002). Sensory interpretation of instrumental measurements 1: Texture of apple fruit. *Postharvest Biology and Technology, 24*(3), 225–239.

Jamshidi, B., & Arefi, A. (2016). Comparison of different feature extraction techniques in biospeckle images for nondestructive assessment of apple firmness. In *Paper presented at the proceedings of the 3rd Iranian International NDT Conference*. Olympic Hotel.

Janaszek-Mańkowska, M., Ratajski, A., & Słoma, J. (2022). Biospeckle activity of highbush blueberry fruits infested by spotted wing drosophila (Drosophila suzukii Matsumura). *Applied Sciences, 12*(2), 763.

Koley, C., & Nirala, A. K. (2020). Detection, differentiation and mapping of different states of blue mold disease of Indian gooseberry (Emblica officinalis G.) using a biospeckle technique. *European Journal of Plant Pathology, 158*(4), 925–937.

Kumari, S., & Nirala, A. (2016). Biospeckle technique for the non-destructive differentiation of bruised and fresh regions of an Indian apple using intensity-based algorithms. *Laser Physics, 26*(11), 115601.

Kumari, S., & Nirala, A. (2019). Biospeckle image processing algorithms for non-destructive differentiation between maturity and ripe stages of Indian climacteric fruits and evaluation of their ripening period. *Laser Physics, 29*(7), 075601.

Kurenda, A., Zdunek, A., Schlüter, O., & Herppich, W. B. (2014). VIS/NIR spectroscopy, chlorophyll fluorescence, biospeckle and backscattering to evaluate changes in apples subjected to hydrostatic pressures. *Postharvest Biology and Technology, 96*, 88–98.

Men, S., Yan, L., & Qian, H. (2015). Research article application of bio-speckle activity to assess seed viability. *Advance Journal of Food Science and Technology, 8*(3), 214–218.

Minz, P. D., & Nirala, A. (2016). Laser speckle technique to study the effect of chemical pre-treatment on the quality of minimally processed apples. *Laser Physics, 26*(4), 045602.

Minz, P. D., Ansari, M. Z., & Nirala, A. K. (2015). Effect of antibrowning agents on fresh-cut potato tubers using frequency filtering of biospeckle images. *Laser Physics, 25*(5), 055601. https://doi.org/10.1088/1054-660X/25/5/055601

Mulone, C., Budini, N., Vincitorio, F. M., Freyre, C. E., López, A. J., & Ramil, A. (2014). Biospeckle activity evolution of strawberries. *SOP Transactions on Applied Physics, 1*(2), 65–73.

Nothdurft, R., & Yao, G. (2005). Imaging obscured subsurface inhomogeneity using laser speckle. *Optics Express, 13*(25), 10034–10039.

Omidi-Arjenaki, O., Ghanbarian, D., Mollazade, K., & Naderi-Boldaji, M. (2024). Assessing mechanical properties of stored potato tubers using biospeckle imaging: Reference tests and biospeckle activity analysis. *Potato Research*. https://doi.org/10.1007/s11540-024-09741-0

Pandiselvam, R., Mayookha, V., Kothakota, A., Ramesh, S., Thirumdas, R., & Juvvi, P. (2020). Biospeckle laser technique–A novel non-destructive approach for food quality and safety detection. *Trends in Food Science & Technology, 97*, 1–13.

Pieczywek, P., Cybulska, J., Szymańska-Chargot, M., Siedliska, A., Zdunek, A., Nosalewicz, A., et al. (2018a). Early detection of fungal infection of stored apple fruit with optical sensors–Comparison of biospeckle, hyperspectral imaging and chlorophyll fluorescence. *Food Control, 85*, 327–338.

Pieczywek, P. M., Nowacka, M., Dadan, M., Wiktor, A., Rybak, K., Witrowa-Rajchert, D., & Zdunek, A. (2018b). Postharvest monitoring of tomato ripening using the dynamic laser speckle. *Sensors, 18*(4), 1093.

Rabelo, G. F., Braga Júnior, R. A., & Fabbro, I. (2005). Laser speckle techniques in quality evaluation of orange fruits. *Revista Brasileira de Engenharia Agrícola e Ambiental, 9*, 570–575.

Rabelo, G. F., Enes, A. M., Junior, R. A. B., & Dal Fabbro, I. M. (2011). Frequency response of biospeckle laser images of bean seeds contaminated by fungi. *Biosystems Engineering, 110*(3), 297–301.

Rahmanian, A., Mireei, S. A., Sadri, S., Gholami, M., & Nazeri, M. (2020). Application of biospeckle laser imaging for early detection of chilling and freezing disorders in orange. *Postharvest Biology and Technology, 162*, 111118.

Retheesh, R., Samuel, B., Radhakrishnan, P., Nampoori, V., & Mujeeb, A. (2016). Use of laser biospeckle for the evaluation of fruit ripening. *Journal of Pure Applied and Industrial Physics, 6*(5), 65–70.

Retheesh, R., Ansari, M. Z., Radhakrishnan, P., & Mujeeb, A. (2018a). Application of qualitative biospeckle methods for the identification of scar region in a green orange. *Modern Physics Letters B, 32*(09), 1850113.

Retheesh, R., Thomas, D., Ansari, M. Z., Varghese, B., Radhakrishnan, P., & Mujeeb, A. (2018b). Application of laser biospeckle technique for the analysis of artificially introduced local dynamics in apple fruit. *Laser Physics, 28*(11), 115601.

Ribeiro, K., Barreto, B., Pasqual, M., White, P., Braga, R., & Dupuy, L. (2014). Continuous, high-resolution biospeckle imaging reveals a discrete zone of activity at the root apex that responds to contact with obstacles. *Annals of Botany, 113*(3), 555–563.

Rizzolo, A., Vanoli, M., Spinelli, L., & Torricelli, A. (2010). Sensory characteristics, quality and optical properties measured by time-resolved reflectance spectroscopy in stored apples. *Postharvest Biology and Technology, 58*(1), 1–12.

Samuel, B., Retheesh, R., Ansari, M. Z., Nampoori, V., Radhakrishnan, P., & Mujeeb, A. (2017). Cross-correlation and time history analysis of laser dynamic specklegram imaging for quality evaluation and assessment of certain seasonal fruits and vegetables. *Laser Physics, 27*(10), 105601.

Singh, P., Chatterjee, A., Bhatia, V., & Prakash, S. (2020). Application of laser biospeckle analysis for assessment of seed priming treatments. *Computers and Electronics in Agriculture, 169*, 105212.

Singh, P., Chatterjee, A., Rajput, L. S., Kumar, S., Nataraj, V., Bhatia, V., & Prakash, S. (2021). Biospeckle-based sensor for characterization of charcoal rot (Macrophomina Phaseolina (Tassi) Goid) disease in soybean (Glycine Max (L.) Merr.) crop. *IEEE Access, 9*, 31562–31574.

Sujayasree, O., Pandiselvam, R., Chaitanya, A., & Kothakota, A. (2022). Application of biospeckle laser technique for assessment of fruit quality. In *Nondestructive quality assessment techniques for fresh fruits and vegetables* (pp. 61–83). Springer.

Sutton, D. B., & Punja, Z. K. (2017). Investigating biospeckle laser analysis as a diagnostic method to assess sprouting damage in wheat seeds. *Computers and Electronics in Agriculture, 141*, 238–247.

Yan, L., Liu, J., & Men, S. (2017). The biospeckle method for early damage detection of fruits. *Modern Physics Letters B, 31*(19–21), 1740034.

Yang, S., Li, C., Li, X., Jiang, J., Zhao, Y., Wang, X., et al. (2024). Early detection of fungal infection in citrus using biospeckle imaging. *Computers and Electronics in Agriculture, 225*, 109293.

Zdunek, A., & Cybulska, J. (2011). Relation of biospeckle activity with quality attributes of apples. *Sensors, 11*(6), 6317–6327.

Zdunek, A., & Herppich, W. B. (2012). Relation of biospeckle activity with chlorophyll content in apples. *Postharvest Biology and Technology, 64*(1), 58–63.

Zdunek, A., Adamiak, A., Pieczywek, P. M., & Kurenda, A. (2014). The biospeckle method for the investigation of agricultural crops: A review. *Optics and Lasers in Engineering, 52*, 276–285.

Chapter 10
Thermal Imaging Applications in Food Processing

D. Ethaya, H. Cubarani, R. Himavanth, and S. Anandakumar

10.1 Introduction

Thermal imaging technology has become an essential tool in food processing and safety, offering a non-contact, fast, and non-destructive approach. The ability of thermal cameras to detect temperature variations across surfaces provides crucial information for monitoring and controlling critical points in food production, ensuring compliance with safety regulations and maintaining product quality (Gowen et al., 2010). Key applications include detecting contaminants, assessing cooking uniformity, monitoring storage conditions, and identifying spoilage. The integration of thermal imaging into food safety protocols enhances efficiency, reliability, and traceability, ensuring the overall quality and safety of food products.

The foundation of thermal imaging technology lies in the discovery of infrared radiation by Sir William Herschel in 1800. Through his experiments with sunlight, Herschel measured the temperature of different colors and found that the highest temperature was beyond the visible red spectrum. He referred to this invisible radiation as "calorific rays," now known as infrared radiation (Shawal Jadin et al., 2011). Infrared radiation shares many properties with visible light, such as reflection, refraction, and transmission. All objects with a temperature above absolute zero (−273 °C) emit infrared radiation, and its behavior—emissivity, absorptivity, transmissivity, and reflectivity—varies across different materials. These principles have paved the way for the application of thermal imaging in various fields, including

D. Ethaya · H. Cubarani · R. Himavanth · S. Anandakumar (✉)
Department of Food Packaging and Storage Technology, National Institute of Food Technology, Entrepreneurship and Management—Thanjavur (NIFTEM-T), Ministry of Food Processing Industries, GOI, Thanjavur, Tamil Nadu, India
e-mail: anand@iifpt.edu.in

C. K. Sunil et al. (eds.), *Emerging Non-destructive Technologies for Food Quality Analysis*, https://doi.org/10.1007/978-3-031-95048-3_10

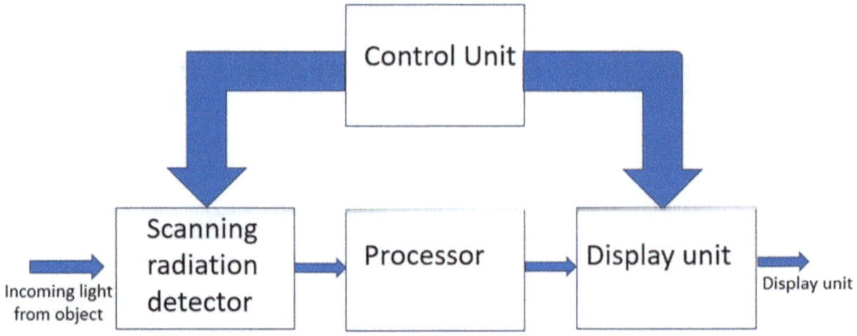

Fig. 10.1 Working principle of thermal imaging camera

food safety, where precise temperature monitoring is crucial. This book chapter explores the application of thermal imaging in various stages of food processing, from raw material inspection to final product quality assessment.

10.2 Principles

A thermal imaging system works by capturing infrared radiation emitted by objects and converting it into a visual display as represented in Fig. 10.1. The process begins with a lens that focuses the infrared light from the objects in view onto an infrared detector. This detector scans the focused light and generates electrical signals based on the intensity of the detected radiation. These signals are then processed by a signal-processing unit, which translates the data into a thermal image. The final display presents different colors representing varying levels of infrared emission, making it possible to identify temperature variations across the observed surface (ElMasry et al., 2020).

Thermal image or thermogram is a 2-dimensional interpreted temperature data acquired by the infrared detector by recording the infrared radiation emitted by objects. It operates by capturing, processing, and interpreting data primarily within the thermal infrared range of the electromagnetic spectrum. The radiation emitted by an object depends on its temperature and its emissivity (ε).

10.3 Theory

The precision of the resulting thermograms or thermal maps, which use invisible infrared radiation from an object to create a pseudo image of the temperature gradients on the body's surface, primarily depends on the object's ability to emit infrared radiation (Ishimwe et al., 2014). All thermal imaging systems operate by perceiving temperature differences between two items, such as an object and a

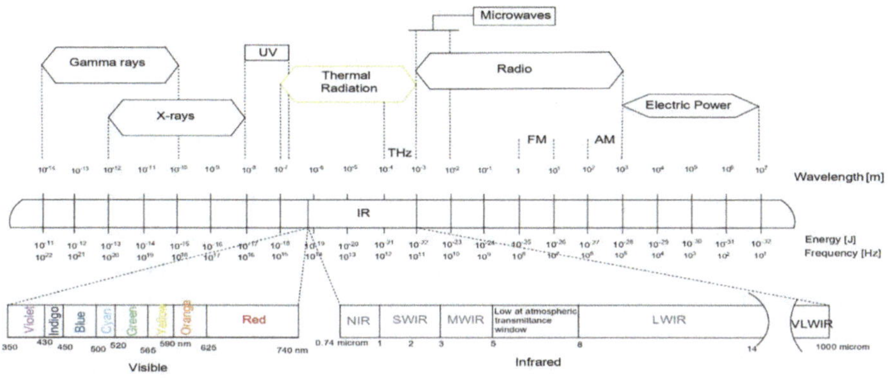

Fig. 10.2 Electromagnetic spectrum with wavelength and frequencies

Table 10.1 Infrared regions and their wavelength

Sub division in infrared region	Wavelength (μm)
Near infrared (NIR)	0.75–1
Shortwave infrared (SWIR)	1.3–3
Mid-wave infrared (MWIR)	3–8
Far-infrared or long-wave infrared (FIR or LWIR)	8–14
Extreme infrared	14–1000

background, and converting that difference to a visual image. Here the IR distribution is shown when all bodies of the picture are not equally heated The majority of the properties of invisible infrared radiation are comparable to those of visible light, which has the ability to be transmitted, reflected, focused, and refracted. Consequently, items that are hotter than absolute zero (−273.15 °C or −459 °F) release this kind of infrared radiation, which is a portion of the electromagnetic spectrum that is invisible and falls between 0.74 and 1000μm in wavelength (Massaro et al., 2020).

As illustrated in Fig. 10.2, the electromagnetic spectrum is divided into various sub-regions, such as gamma rays, X-rays, ultraviolet (UV) radiation, visible light (VIS), infrared (IR) radiation, microwaves, and radio waves (FM and AM). Ultraviolet (UV) light, with wavelengths ranging from approximately 10–350 nm, cannot be detected by the human eye. Visible light, with wavelengths extending from violet (around 350 nm) to red (up to 800 nm), is utilized in various applications such as illumination, communication, and analytical processes. Infrared (IR) light, positioned beyond the visible spectrum and spanning wavelengths from 800 nm to 100μm, is categorized into near-infrared (NIR), mid-infrared (MIR), and far-infrared (FIR) regions (Alinaghi et al., 2023).

The sequence presented in Table 10.1 follows an arrangement of increasing wavelength and decreasing frequency (Gowen et al., 2010). Mid-wave infrared imaging systems exhibit peak sensitivity at elevated temperatures, such as 400 °C, making them potentially suitable for monitoring changes during food processing in various food industry applications. Whereas near infrared that usually measures the diffuse reflected radiation than emitted.

On the other hand, where mid wave and long wave are sensitive only at room temperature which are not suitable for such applications. But they are called the first and second thermal imaging. IR cameras are based on two laws, the total radiation law and the Stefan—Boltzmann's law (Akula et al., 2011). The Total Radiation Law states that any object with a temperature above absolute zero emits infrared radiation, with the intensity increasing as the temperature rises. Stefan-Boltzmann's Law expressed in the Eq. (10.1) further explains the total energy emitted by an object is proportional to the fourth power of its absolute temperature. In thermal imaging, these principles are essential for measuring temperature differences, as infrared cameras detect and interpret the emitted radiation to create temperature-based images for analysis.

$$W = \varepsilon \sigma T^4 \tag{10.1}$$

Where,

W—Total radiant energy
ε—Emissivity
T—Temperature
σ—Stefan Boltzmann's constant

10.4 Types of Thermography

There are many types of thermography such as active (that depends on heat conductivity principle) and passive (that depends on emissivity principle), which as applied in food industry. Lock in thermography (Modulated thermography), Pulse thermograph, Vibro thermography is used for types of active thermography and used for different applications (Shawal Jadin et al., 2011).

Active thermography	**Passive thermography**
Active thermography applies external heat or energy to food products to detect defects, contamination, or inconsistencies. It helps assess cooking uniformity, identify foreign objects, and inspect packaging integrity, ensuring food safety, quality control, and compliance with industry standards (Cuibus, 2013).	Passive thermography monitors naturally emitted infrared radiation to assess food temperature without contact. It is used for detecting spoilage, monitoring cold storage, ensuring proper cooking, and preventing bacterial growth. This method enhances food safety by providing real-time, non-destructive temperature assessments in processing and storage (Ferreira, 2020).

10.5 Methods of Thermography and Their Characteristics Features

Pulse Thermography (PT) Pulse thermography in food processing employs a transient thermal stimulus to evaluate heat diffusion properties, enabling precise detection of structural inconsistencies, contamination, and processing anomalies. This non-destructive technique enhances quality control by assessing thermal conductivity variations, ensuring uniform cooking, packaging integrity, and safety compliance (Baranowski et al., 2009).

Lock-in Thermography Utilizes periodic thermal excitation to analyze phase and amplitude variations in heat propagation. This highly sensitive, non-destructive technique detects subsurface defects, moisture distribution, and structural inconsistencies, ensuring product uniformity, packaging integrity, and adherence to food safety and quality standards (Bicanic et al., 2015).

Vibro Thermography Employs mechanical vibrations to generate localized heat at defect sites, enabling precise detection of cracks, voids, and structural inconsistencies. This non-destructive technique enhances quality control by identifying processing anomalies, packaging defects, and textural inconsistencies, ensuring food safety and product integrity (Gowen et al., 2010).

10.6 Processing—Measurement and Analysis

In order to highlight regions of interest in the image, image processing is done to increase the contrast in the TI. To show the compositional contrast between pixels in a picture, greyscale or color mapping with intensity scaling is frequently utilized. Thermal image processing is typically an iterative process (Akula et al., 2011). Some common steps are shown in Fig. 10.3.

Pre-processing is typically used to remove superfluous information from thermal images and prepare the data for subsequent processing. TI's pre-processing stages include dead pixel rectification, noise reduction, picture filtering, and deblurring. The pre-processed TI histogram shows first-order statistical properties such mean pixel value, standard deviation, skew, kurtosis, and entropy (Gowen et al., 2010). This information can be used to identify regions of interest in an image (for example, histogram-based thresholding). The goal of classifying thermal imaging data is to use the information in the TI to identify areas or objects that have certain characteristics (such as bulk material or flaws) Wiecek (2005). Thermal image processing can benefit from a variety of supervised and unsupervised data mining approaches that have been used in color, multispectral, and hyperspectral image processing.

Fig. 10.3 Flow chart working of thermographic camera

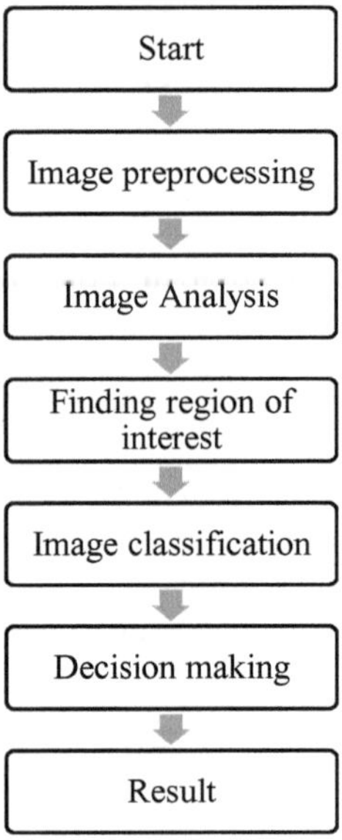

10.6.1 Thermal Imaging System

Thermal imaging systems usually include the following components: a camera, an optical system (e.g., focusing lens, collimating lenses, filters), a detector array (e.g., microbolometers), signal processing, and an image processing system (Rogalski & Chrzanowski, 2014). Thermal imaging, unlike multi- and hyperspectral imaging, does not require illumination. TI detectors work on the idea that when heated by incoming infrared radiation, their temperature rises and is detected by any temperature-dependent mechanism, such as thermoelectric voltage, resistance, or pyroelectric voltage (Rogalski, 2003). The TI camera system converts radiation emitted from a sample in a specific waveband (e.g., 8–12 m) into an electrical signal, which is subsequently processed into an image. The resulting image displays the sample's temperature profile since it is a matrix of many color levels, each of which denotes a distinct temperature (Gowen et al., 2010). Computer software may be used in conjunction with the thermal image to improve interpretation. Two wavelength ranges have been the primary focus of common thermal imaging systems: 8–14 m long wavelength infrared (LWIR) and 3–5 m middle wavelength infrared (MWIR). LWIR is commonly used for food applications since the wavelength range is determined by the temperature range (Rogalski, 2003). Numerous image processing methods can be used to further analyze the resulting thermograph.

10.6.2 Thermal Camera

In 1965, the first thermal imaging camera was made available for inspecting high-voltage power lines. Ever since, the use of thermal imaging cameras for industrial purposes has grown to be a crucial part of the market for FLIR (Forward-looking infrared) systems, which are essentially high voltage power line (Abdulshahed et al., 2015). Since then, there has been a significant advancement in thermal imaging technology, leading to the development of compact thermal imaging cameras that resemble digital photo cameras. The different thermal imaging cameras and their specifications are given in Table 10.2. They are currently a very significant tool for industrial applications since they are simple to use and generate sharp, high-resolution images in real time. They can identify irregularities that are typically undetectable to the unaided eye, enabling remedial action to be taken before expensive systems completely fail.

10.6.3 Thermal Sensors

The development of uncooled thermal detectors primarily uses two types of detectors: ferroelectric and microbolometer. The ferroelectric phase transition in specific dielectric materials is used by ferroelectric detectors. Small temperature changes at this phase transition result in significant changes in electrical polarization (Rai et al., 2017). Ferroelectric detectors use Barium Strontium Titanate (BST) as the material of choice. The Table 10.3 gives the purpose, wavelength, waveband and spatial resolution of various sensors used for detecting the IR radiation emitted from object.

10.6.4 Lens

Lenses function through refraction, where light changes direction as it moves through materials of different optical densities. Based on their shape, lenses can either converge (convex lenses) or diverge (concave lenses) light rays (Chernonozhkin et al., 2017). In thermal imaging, traditional glass cannot be used because it blocks infrared wavelengths, so materials like germanium and chalcogenide glass are preferred to allow accurate infrared transmission and imaging (Muthukumar et al., 2018).

Germanium is commonly used in thermal imaging instead of glass due to its low thermal radiation transmittance. It is a gray-white metalloid that reflects visible light but is nearly transparent to infrared radiation. However, its high cost makes lens size a crucial factor. The f-number of a lens, which is the ratio of its focal length to the diameter of its entrance pupil, affects both cost and performance. A higher f-number reduces lens cost but allows less radiation to reach the detector. For uncooled

Table 10.2 Various Thermal Imaging devices and their specifications

IR camera	Specifications			Uses
	Resolution	Temperature range	Features	
Bosch GTC 400 C	160 × 120 pixels	−10 °C to +400 °C	Integrated digital camera, Wi-Fi connectivity, easy-to-use interface, overlay of thermal and visible images, compatible with Bosch Measuring Master App for detailed analysis.	Enhances food processing by ensuring cooking uniformity, monitoring storage conditions, detecting leaks, and verifying packaging integrity. Its thermal imaging aids in food safety, HACCP compliance, and contamination detection, with real-time analysis via Wi-Fi and the Bosch Measuring Master App.
FLIR systems	320 × 240 pixels	−20 °C to +250 °C	MSX image enhancement, Wi-Fi connectivity, built-in digital camera, thermal fusion, and picture-in-picture.	Food processing by ensuring precise temperature monitoring, detecting cold chain inefficiencies, verifying cooking uniformity, and inspecting packaging integrity. Its MSX image enhancement, Wi-Fi connectivity, and thermal fusion provide high-resolution, real-time analysis for improved food safety and quality control.
Fluke Ti480 PRO	640 × 480 pixels	−20 °C to +1000 °C	Laser Sharp Auto Focus, advanced manual focus, Fluke Connect compatibility, rugged design, and multiple image modes (IR-Fusion, Picture-in-Picture, Auto Blend).	Ensures precise temperature monitoring in food processing with Laser Sharp Auto Focus, IR-Fusion, and Fluke Connect compatibility. It detects cooking inconsistencies, cold chain issues, and packaging defects, enhancing food safety and quality control.

(continued)

Table 10.2 (continued)

IR camera	Specifications			Uses
	Resolution	Temperature range	Features	
Seek Thermal CompactPRO	320 × 240 pixels	−40 °C to +330 °C	Smartphone attachment, adjustable focus, high and low gain settings, various color palettes, lightweight and portable design.	Portable, smartphone-compatible thermal camera ideal for food processing. With adjustable focus and high/low gain settings, it detects temperature variations, ensures cooking uniformity, monitors cold storage, and verifies packaging integrity, enhancing food safety and quality control.
Testo 890	640 × 480 pixels	−20 °C to +1200 °C	Panorama image assistant, HDR image function, high temperature measurement, replaceable lenses, fully radiometric video recording.	High-precision thermal camera for food processing, featuring HDR imaging, high-temperature measurement, and radiometric video recording. It ensures cooking uniformity, monitors cold storage, detects contamination, and inspects packaging integrity, improving food safety and quality control
Hikvision DS-2TD2617B	160 × 120 pixels (thermal), 4 MP (optical)	−20 °C to +150 °C	Bi-spectrum camera (thermal and optical), smart analytics (fire detection, temperature anomaly detection), IP67 rating, PoE support, H.265+ compression.	It ensures temperature anomaly detection, fire prevention, and cold storage monitoring. With IP67 protection and smart analytics, it enhances food safety, equipment monitoring, and quality control in production facilities.

(continued)

Table 10.2 (continued)

IR camera	Specifications			Uses
	Resolution	Temperature range	Features	
Optris PI 640	640 × 480 pixels	−20 °C to +1500 °C	High-speed thermographic camera, USB and GigE interfaces, software for detailed analysis, customizable optics, suitable for research and industrial monitoring.	Customizable optics and advanced software, it ensures precise temperature monitoring, cooking uniformity, contamination detection, and packaging inspection, enhancing food quality control and safety in industrial food production.
Bullard T4X	320 × 240 pixels	−40 °C to +1200 °C	Super Red-Hot colorization, digital zoom, Electronic Thermal Throttle, lightweight and durable design, specifically designed for firefighting.	Detecting overheating equipment, fire hazards, and temperature inconsistencies. Its Super Red-Hot colorization and digital zoom help monitor ovens, grills, and storage areas, ensuring safety and quality control.

cameras, a low f-number is essential to maintain sensitivity, while cooled cameras can tolerate a higher f-number by increasing exposure time to maintain radiation throughput.

In thermal imaging, infrared lenses focus emitted heat onto a sensor. Materials like germanium, zinc selenide, and chalcogenide glass allow infrared transmission. The sensor converts the focused radiation into electrical signals, processed into a thermal image. This enables applications like food safety monitoring, cooking uniformity, and contamination detection (Gowen et al., 2010).

Thermal imaging in food processing requires specialized lenses made from materials that efficiently transmit infrared radiation, as standard glass blocks IR wavelengths. Common materials include germanium, zinc selenide (ZnSe), chalcogenide glass, and sapphire. Germanium is widely used due to its excellent infrared transmission and durability. Zinc selenide provides a broader wavelength range, making it useful for various thermal applications. Chalcogenide glass offers a cost-effective alternative, particularly for large-scale production. Sapphire lenses, known for their high durability and scratch resistance, are ideal for harsh environments. These materials ensure accurate thermal imaging, supporting food safety, quality control, cooking consistency, contamination detection, and packaging inspection in food processing (Rai et al., 2017).

Table 10.3 Sensors used in thermographic camera for absorbance of IR Radiation

Name of sensor	Wavelength (μm)	Wave band	Spatial resolution (m)	Primary purpose	Uses
AATSR/ ENVISAT	7.34	Band 10	2000	Low-level water vapor, winds, and sulfur dioxide (SO_2).	Moisture detection, baking quality, storage monitoring, and gas emission control
ABI/ GOES-R	12.3	Band 15	2000	Total water, ash, SST	Monitors moisture, ash contamination, and surface temperature
Aster	8.92–9.27 10.25–10.95	Band12 Band13	90 90	Surface temperature Surface temperature	Monitors surface temperature in food processing, ensuring optimal drying, cooking consistency, and storage conditions
AVHRR	3.0–5.4	Band 9	5	Identification of oil spills, fire mapping, and geological area detection.	Monitor heating efficiency, detect contamination, and ensure proper cooking and drying
Daedalus (AMS) AA3607DS	8.5–12.5	Band 10		Monitoring water algae, conducting forest inventories, assessing crop health, detecting fires, and mapping oil spills.	Detect contamination, and ensure proper thermal processing, enhancing food safety and quality control
Landsat 8	10.30–11.30 11.50–12.50	Band 10 Band 11	100 100	Thermal mapping and estimated soil moisture Improved thermal mapping and estimated soil moisture	Thermal mapping, monitoring soil moisture for agriculture, and optimizing drying and storage conditions

10.7 Factors Affecting the Quality of Thermal Images

The factors influencing the thermal image quality of the imaging system are categorized into three groups: atmospheric, software, and object/acquisition parameters (ElMasry et al., 2020). The factors that impact the creation of precise thermal images are ambient temperature, air temperature, humidity level, wind velocity, solar radiation, and shadows from surrounding objects. The precision of thermal images can be affected by software elements, such as the selected temperature range and the color palette used for rendering. Additionally, a third group of factors influencing thermal images and their analysis involves the configuration of objects and the hardware setup.

Atmospheric factors	Software factors	Object/acquisition factors
• Ambient temperature • Air temperature • Humidity level • Wind velocity • Solar radiation	• Temperature range • Color palette	• Camera wavelength range • Angle of observation • Distance from object • Object emissivity

Factors impacting camera performance include camera wavelength range, angle of observation, distance from object, object emissivity, material properties between camera and object, object moisture and thermal properties, external optics temperature and transmission, object emissivity, and filter types (Akula et al., 2011). From a technical standpoint, the functionality of a completely calibrated thermal imaging setup can be characterized by various attributes including image quality, detector sensitivity, electronic noise, thermal response, geometric resolution, accuracy, spectral range, frame rate, and integration time (ElMasry et al., 2020).

10.8 Applications in Non-destructive Food Quality and Analysis

Thermal imaging as a non-destructive and non-contact method for recording temperature by measuring the infrared radiation emitted from a body's surface (Bejo-Khairunniza et al., 2016). Thermal imaging serves as a two-dimensional, non-contact diagnostic technique for capturing surface temperatures of materials. This technique is advantageous for non-destructive quality evaluation, providing insights into product properties such as discontinuities, structure, dimensions, physical and mechanical properties, composition, chemical analysis, stress response, and abnormal heat sources. Due to their portability and real-time imaging capabilities, TI systems are highly suitable for a wide range of application (Gowen et al., 2010) as given in Table 10.4.

Table 10.4 Thermal Imaging applications

S. No	Food product	Application	References
1	Chicken	IR imaging combined with a neural network approach is used to determine the internal temperature of meat immediately after cooking.	Tao (2000)
2	Food simulant	Controlled heating and cooling procedures applied to the surface of food products.	Foster et al. (2006)
3	Wheat grain	Identification of infestations by six developmental stages of *Cryptolestes ferrugineus* (four larval instars, pupae, and adults) inside wheat kernels.	Manickavasagan et al. (2008)
4	Apple	Identification of bruises in apples during forced convective heat transfer and assessment of their surface quality.	Varith et al. (2003)
5	Tomato	Bruise damage (soft areas).	Linden et al. (2003)
6	Potato, cauliflower	Ice nucleation & freezing of plants.	Fuller and Wisniewski (1998)
7	Wheat	Identification of Wheat class	Manickavasagan et al. (2010a)
8	Meat	Quality defects of Pig meat Assessment of pork and raw ham quality defects in Slaughter line	Gariepy et al. (1989)
9	Walnut	The temperature-time profile and surface temperature distribution of walnut kernels during radio frequency processing.	Wang et al. (2006)
10	Citrus drying	Determining the duration of drying for citrus fruits.	Fito et al. (2004)

10.8.1 Grain Processing Industries

Thermal imaging has a wide range of applications in agriculture, covering both pre-harvest and post-harvest activities, as outlined in Table 10.5. It offers a promising alternative by detecting the heat generated from insect respiration, which creates temperature differences compared to the grain itself. This technology allows for the detection of eggs, pupae, larvae, and adult insects inside grain kernels without the need for destructive sampling or complex procedures as shown in Fig. 10.4. Thermal imaging to detect *Cryptolestes ferruginous* infestation in wheat kernels, achieving high classification accuracy for infested and uninfested kernels using advanced image processing techniques (Gowda & Alagusundaram, 2020).

Hot spots are high-temperature regions in grain storage bins caused by respiration of fungi, insects, or the grain itself, leading to spoilage. The fungal activity, particularly from low-temperature *Penicillium species*, can initiate heating in grains during winter, reaching significant temperatures. Thermal imaging offers a superior alternative, providing non-contact temperature distribution analysis across the grain bulk (Manickavasagan et al., 2006).

Table 10.5 Detection of defects in stored grains using thermal imaging

Application	Description	References
Detection of sprout-damaged kernel	Thermal imaging identifies sprouted kernels by detecting temperature variations at the germ end, which indicate sprouting or sprout breakage.	Vadivambal et al. (2011)
Detection of foreign bodies in grains	Identifies foreign materials like dirt, stones, and insects using thermal imaging which reveals temperature differences compared to the grains combined with image processing techniques.	Gowda and Alagusundaram (2020)
Detection of insect infestation	Detecting insects inside grain kernels by identifying heat from insect respiration. Thermal imaging is non-destructive and effective for all life stages of insect.	Gowda and Alagusundaram (2020)
Detection of fungal infection	Thermal imaging detects temperature variations caused by fungal activity, offering non-invasive quality assessment.	Jayas and Singh (2012)
Classification of grain	Thermal imaging offers rapid online classification, overcoming the subjectivity of visual inspection. Thermal imaging aids in the classification and grading of grains by analyzing temperature patterns to differentiate between grain varieties and quality.	Manickavasagan et al. (2010a)

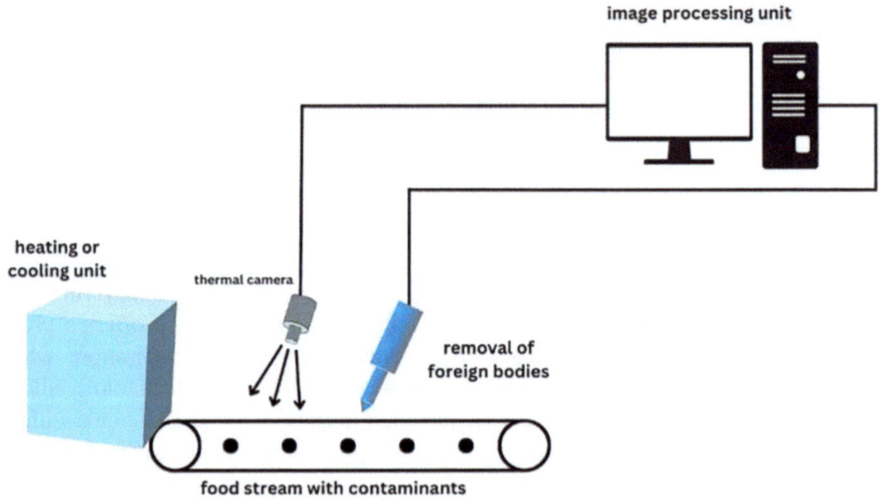

Fig. 10.4 Thermal imaging techniques used for evaluation of contaminants

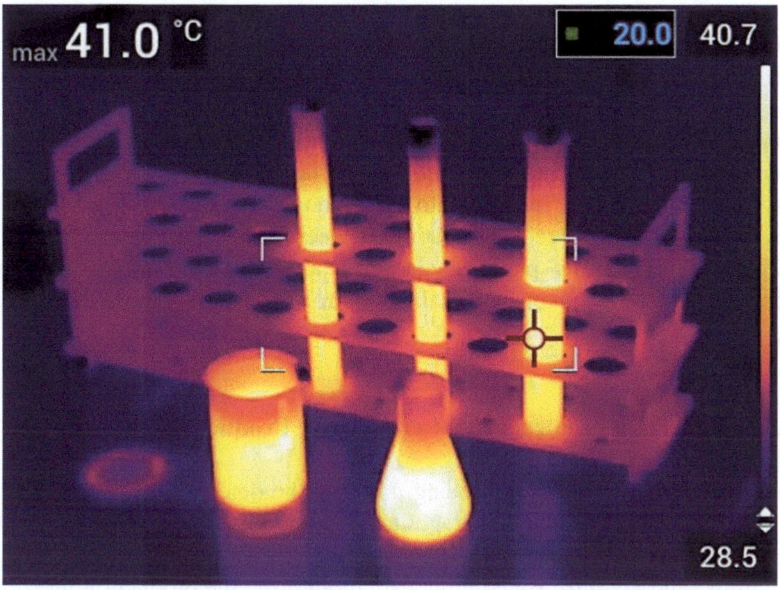

Fig. 10.5 Thermovision scanning of Milk

10.8.2 Dairy Industries

Continuous monitoring of storage environments using thermal imaging detects temperature fluctuations that could compromise product freshness and shelf life. By identifying thermal anomalies, dairy processors can implement corrective actions to uphold cold chain integrity and minimize product losses during storage and distribution as shown in Fig. 10.5. This ensures that dairy products remain fresh and safe for consumption, reducing waste and maintaining quality throughout the production process (Lei, 2018).

10.8.2.1 Quality Control in Cheese Aging

Thermal imaging role in monitoring cheese ageing processes assess the temperature gradients within ageing chambers and thermal imaging facilitates non-invasive quality control. The higher temperature leads to faster ageing and thereby produces a quick breakdown of proteins and fats, resulting in intense flavour profile. The lower temperature slow down the ageing process and creates a milder cheese with more delicate texture. This capability enables cheese manufacturers to optimize ageing conditions, ensuring consistent texture and flavor development while meeting consumer preferences (Rosca & Rosca, 2011).

Monitoring cheese ripening is a complex task, as it involves evaluating various biochemical changes throughout the process. Traditionally, cheese maturity has been assessed using conventional sensory analysis methods, which are often destructive, time-consuming, and costly. To overcome these limitations, NIR-HS imaging, combined with advanced data analytical techniques, offers a non-destructive and efficient approach for real-time monitoring of cheese ripening. This imaging technology provides valuable insights into the homogeneity of maturation at different stages, enabling a more precise evaluation of cheese quality. Additionally, analyzing heterogeneity in ripening patterns can serve as a powerful tool for improving decision-making and enhancing quality control in cheese production (Alinaghi et al., 2023).

10.8.2.2 Detection of Foreign Matter in Dairy Products

Thermal imaging is increasingly being used for the detection of foreign matter in dairy products. This technique leverages temperature differences to identify contaminants such as plastic, metal, or organic matter that might be present in products like cheese, milk, and yoghurt. By continuously monitoring the production line, thermal imaging can quickly spot anomalies, ensuring that contaminants are detected before the products reach consumers (Yaqoob et al., 2021).

10.8.2.3 Early Detection of Health Issues

The application of Infrared Thermography (IRT) in detecting early signs of health issues in dairy cattle, focusing particularly on stress indicators such as changes in eye temperature shown in Fig. 10.6. Moreover, previous studies have underscored IRT's efficacy in detecting infectious diseases like bovine viral diarrhea virus and respiratory ailments through facial scans, suggesting its role in enhancing early intervention strategies.

In the dairy industry, ensuring cows' udder health is crucial for producing safe and high-quality milk. Mastitis, a common udder infection, poses significant economic challenges due to reduced milk yield and treatment costs. Thermal Imaging is a valuable tool for detecting mastitis early by assessing temperature changes in the udder shown in Fig. 10.6, farmers can quickly pinpoint potential health issues and take corrective actions before the condition worsen (Aytekin & Coşkun, 2021). The various thermographic cameras used for detecting health issues in animals are given in Table 10.6.

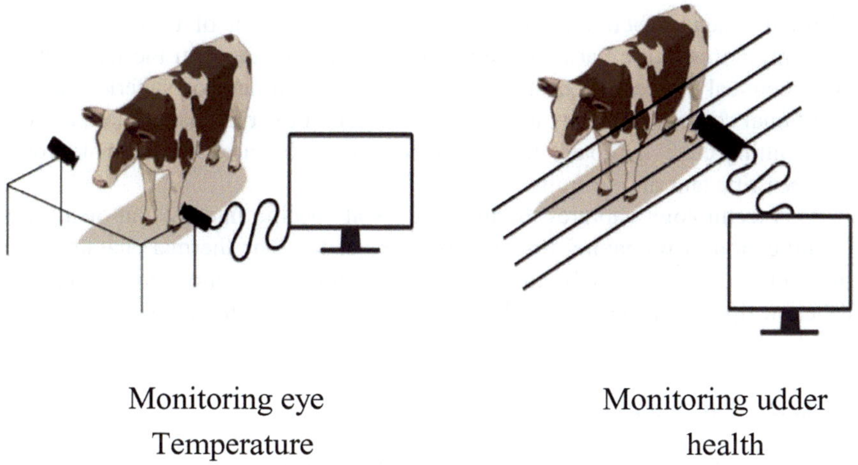

Fig. 10.6 IRT detection of early stage health issues in farm animals

Table 10.6 Application of IRT cameras in detection of Cow health

Name of the IRT camera	Application	References
FLIR Inframetrics 760	Employed to assess daily fluctuations in Uncooled Sensor Technology (UST) readings of dairy cows in connection with mastitis.	Berry et al. (2003)
ThermaCam545	Utilized to monitor variations in teat temperature during machine milking.	Vegricht et al. (2007)
ThermaCam P25	Employed for the early identification of mastitis in dairy cows.	Colak et al. (2008)
IRI 4010	Intramammary infusion of E. coli endotoxin during late lactation and its impact on UST changes.	Castro-Costa et al. (2014)
FLIR 760 IR scanner	The effects of E. coli infusion on UST variations.	Metzner et al. (2014)

10.8.3 Application of Thermal Imaging in Fruits and Vegetables

Thermal imaging plays a key role in measuring the surface temperature of apples, assisting in identifying defects and inadequate ripeness. Proper packaging and storage conditions, monitored through thermal imaging, help reduce spoilage and maintain the freshness of fruits and vegetables before juice extraction. Thermal imaging allows for the detection of surface defects, such as bruises, cracks, or cuts, which might not be visible to the naked eye. These defects can impact the overall quality and safety of the juice (Badia-Melis et al., 2016).

10.8.3.1 Ripeness and Maturity Detection

Thermal imaging facilitates the non-invasive assessment of these parameters by detecting temperature variations that correspond to changes in the fruit's physical and chemical properties, ensuring optimal harvest times for superior flavor and nutritional content. Maturity and ripeness detection of fruits using thermal imaging is a cutting-edge approach that leverages the differences in thermal properties between ripe and unripe fruits.

This technology captures thermal images at specific intervals, allowing for the identification of immature, ripe, and overripe states. The thermal characteristics of the fruit's surface are analyzed, revealing temperature variations that correlate with different ripeness levels (Sumriddetchkajorn & Intaravanne, 2013).

10.8.3.2 Packaging Integrity of Fruit Juice

This thermal image depicts an automated bottle filling process, allowing for the removal of bottles that are either over-filled or under-filled. When dealing with bottles or jars made from dark-colored glass or plastic, thermal imaging cameras prove to be significantly more effective than visible light cameras. Thermal imaging detects defects such as leaks or improper seals by visualizing temperature changes indicative of air infiltration or product leakage, ensuring packaged juices remain fresh and safe throughout distribution and storage (Pathmanaban et al., 2023).

10.8.4 Application of Thermal Imaging in Meat Industries

In the meat processing industries, thermal imaging has become an invaluable tool for ensuring product quality, safety, and efficiency (Ghosh et al., 2016a). In this industries, maintaining product quality and safety is paramount. Variations in temperature can affect texture, taste, and safety. Now a day many non-invasive technique has emerged and shown great potential in quality and safety assessment for fish and other meat products (Xiong et al., 2017). Thermal imaging provides a means to monitor and control Quality, Safety Compliance and Process Optimization and also provide a Reduction in processing cost by Minimizes waste by ensuring products are processed to the desired specifications.

10.8.4.1 Monitoring and Control of Cooking and Cooling Processes

One of the primary applications of thermal imaging in meat processing is monitoring and controlling cooking and cooling processes. Traditional temperature measurement methods, such as thermocouples, provide point measurements which may not represent the entire product. In contrast, thermal imaging offers a comprehensive view of temperature distribution, allowing for precise control of these processes (Kor & Icier, 2016). Thermal cameras can be installed in cooking and cooling chambers to continuously monitor the temperature of meat products. By using this camera, we can check whether the chicken is cooked or dried out properly (Barbin et al., 2012).

This real-time data helps in adjusting process parameters to ensure even cooking and cooling by determining the temperature distribution of the products (Döner et al., 2020) and for instance, ensuring that meat products reach the necessary internal temperatures to kill pathogens while preventing overcooking, which can affect texture and taste.

10.8.4.2 Quality Assurance and Consistency

Thermal imaging plays a crucial role in quality assurance by ensuring consistency in meat products. Consistency in texture, color, and moisture content is vital for customer satisfaction and brand reputation. It can detect variations in these parameters by analyzing the thermal profile of meat products. The study of (Kor & Icier, 2016) showed that the thermal imaging non-contact temperature measurement method is faster and can greatly reduce the errors of point measurement. For example, during the grilling or roasting of meat, thermal imaging can ensure uniform temperature distribution, leading to consistent browning and doneness.

10.8.4.3 Detection of Contaminants and Defects

Thermal imaging has now evolved as a solution for the detection of microbial growth, which is a serious issue in food industry (Lipinska et al., 2019). By analyzing thermal images, processors can identify and address these issues before the products reach consumers. Additionally, thermal imaging can help in identifying potential contamination source. For example, areas of the processing equipment that consistently show higher temperatures may indicate bacterial growth or other contaminants that need to be cleaned or removed (Jayas & Singh, 2012). This is more advantageous than any other traditional method as it overcomes its limitation such as expensive and longer processing time.

Employing a thermal imaging camera equipped with an uncooled microbolometric detector, a method was developed for assessing microbiological quality of mortadella (Lipińska et al., 2022). In this technology the samples contaminated and non-contaminated where distinguished based on the heat emission by the microorganism. Lower speeds of heat emission was found in the sample contaminated in comparison to the other samples. Thermal imaging is a highly potential technology to detect the microbial quality and can also be used in common in the near future, before which its precision issue with the results must be corrected.

10.8.4.4 Verification of Packaging Integrity

Thermal imaging can detect temperature differences on the surface of packaging, allowing for the identification of even minor seal defects. This improves the overall quality of your packaging, (Vadivambal et al., 2011). In the meat processing industry, the integrity of packaging is crucial for ensuring product safety and extending shelf life. Thermal imaging can be used to verify the integrity of vacuum-sealed packages and other types of packaging. Any leaks or defects in the packaging material can be detected as temperature anomalies, allowing for timely corrective actions.

Future Trends and Developments: The future of thermal imaging in the meat processing industries looks promising, with ongoing advancements in technology and applications. Emerging trends include the integration of thermal imaging with artificial intelligence (AI) and machine learning (ML) (He et al., 2021). This AI and ML algorithms can analyze thermal images in real-time, providing actionable insights and predictive maintenance recommendations. This can further enhance quality control, energy efficiency, and overall productivity in processing facilities. Additionally, as the cost of thermal imaging technology continues to decrease, it is expected to become more accessible to small and medium-sized processing facilities (Hou et al., 2022). This democratization of technology will enable more businesses to leverage its benefits and improve their processes.

Thermal imaging has demonstrated its effectiveness as a useful tool in the meat and fish processing industries given in Table 10.7 and offers numerous applications that enhance quality control, safety, and process optimization. By providing detailed real-time, non-contact temperature data, thermal imaging helps processors ensure

Table 10.7 Significance of IRT in processed meat

S. No	Product	Application of thermal imaging	Description	References
1	Pork sausages	Quality control	Detection of temperature variations to ensure uniformity and consistency.	Dissing et al. (2013)
2	Cooked chicken	Process monitoring	Tracking of change in temperature during processing (cooking, chilling) to optimize operations	Barbin et al. (2012)
3	Salmon fillets	Freshness assessment	Detection of spoilage—related temperature patterns to assess the freshness of product.	Cheng et al. (2015)
4	Beef steaks	Contamination detection	Detection of abnormal temperature patterns on the surface to identify contamination.	Liu et al. (2017)
5	Minced lamb meat	Safety and compliance	Monitoring critical control points to verify that products meet safety standards.	Kamruzzaman et al. (2013)

consistent product quality, detect defects, and optimize energy use. As technology continues to advance, the meat processing industries are likely to see even more innovative applications of thermal imaging, leading to further improvements in product quality and operational efficiency.

10.8.5 Application of Thermal Imaging in Baked Products

Applying thermal imaging technology in the context of baked products offers significant advantages in quality control, process optimization, and product development. This technology can be employed at various stages of production, from raw material inspection to final product analysis. The control of temperature is the prime important for baking which will directly affects the quality attributes of baked products and the efficiency of energy utilization (Therdthai et al., 2002). This comprehensive application of thermal imaging ensures consistency in product quality, and enhances efficiency in production processes.

In the baking industry, maintaining precise temperature control is crucial for achieving desired product characteristics such as texture, color, and taste. Traditional methods of temperature measurement, such as thermocouples or infrared guns, have limitations in capturing temperature distributions across large surfaces or inside products. Thermal imaging overcomes these limitations by determining temperature profiles, enabling better decision-making during baking processes (Comba et al., 2021).

10.8.5.1 Benefits of Thermal Imaging in Process Control

One of the important parameters that affects the quality of baking product is its temperature. The study by (Sritham et al., 2024) on temperature measurement of baking product have shown that the surface temperature can be found using the emissivity value with the use of thermal imaging camera and infrared sensor as shown in Fig. 10.7. For example Crust and crumb temperatures of bread during

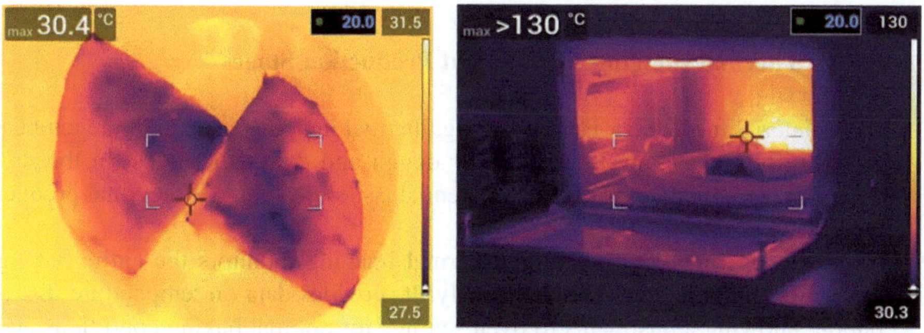

Fig. 10.7 Thermal imaging of baked product

Table 10.8 Significance of IRT in manufacturing of baked goods

Specific product	Application	Description	References
Biscuits	Detection of foreign bodies	Detect the foreign bodies like wood, metal etc. and ensuring the product quality	Senni et al. (2014)
Bread	Monitoring and controlling bread making process	Measurement of surface temperature by using the emissivity value of the bread while baking	Sritham et al. (2024)
Bread production process	Prevent heat loss	This will scan the heat emission by the energy system in processing of bread	Antić et al. (2015)
Wheat (raw material)	Classification of wheat varieties which will be used for different baked products	Temperature rise and drop will be varied for different varieties of wheat	Manickavasagan et al. (2010a)

baking is determined by these techniques (Antić et al., 2015). Thermal imaging serves as the potential technique for real-time, non-contact temperature measurement to monitor and control the bread-baking process.

Detection of Uneven Heating: Thermal imaging can identify areas within ovens or on baking trays where heat distribution is uneven (Vadivambal & Jayas, 2010). This information helps in optimizing oven settings or tray placements to ensure uniform baking of products and by continuously monitoring the temperature of products as they bake, thermal imaging helps maintain consistency across batches. It alerts operators to deviations from set parameters, preventing undercooked or overcooked products.

Early Detection of Defects: Thermal imaging can detect anomalies such as air pockets or inconsistent baking within products (Takacs et al., 2020). This early detection allows for immediate corrective actions, reducing waste and maintaining product quality Standards as shown in Table 10.8.

Verification of Product Specifications: Thermal imaging can verify whether products meet temperature requirements specified in baking protocols (Manickavasagan et al., 2010a). This ensures compliance with regulatory standards and customer expectations.

10.8.5.2 Thermal Imaging Throughout Production Stages

1. Raw Material Inspection: Before baking, thermal imaging can assess the temperature uniformity of raw ingredients or dough and detect the defect in the raw materials. This helps in identifying potential issues that could affect final product quality.
2. Oven Monitoring: During baking, thermal imaging monitors the entire baking surface or multiple trays simultaneously. It provides data on temperature distribution, allowing adjustments to oven settings for optimal baking conditions and also prevent the oven from heat loss.

3. Cooling and Packaging: After baking, thermal imaging can ensure products cool uniformly before packaging. It helps in determining when products have reached suitable temperatures for safe handling and packaging (Ghosh et al., 2016a).

10.8.5.3 Energy Efficiency with Thermal Imaging

The bakery production line is considered to be a large energy consumer and by optimizing oven settings based on thermal imaging data which will analyze the overheat emission, bakeries can reduce energy consumption while maintaining product quality (Antić et al., 2015). This contributes to cost savings and sustainability efforts. By monitoring the thermal profiles of ovens, thermal imaging ensures uniform heating, leading to consistent baking quality while minimizing overbaking or underbaking, which directly reduces energy consumption.

Additionally, it identifies steam leaks in pipes and valves, a common source of energy inefficiency in bakeries, thereby enabling timely repairs and cutting energy waste.

10.9 Conclusion

In food processing, thermal imaging has become a flexible and non-invasive technique that offers several benefits for tracking and enhancing the safety and quality of food products. Better process control in a variety of applications, including baking, grain, dairy and meat processing. This will provides precise temperature distribution data in real-time. Its capacity to identify irregularities, like uneven cooking, or possible risks, such contamination, improves productivity and lowers waste. In a similar vein, it helps to maintain ideal processing conditions during frying or drying, reducing energy use and the possibility of over processing. Additionally, the technology improves customer satisfaction by identifying thermal or physical flaws in final items, which is crucial for quality assurance. In addition to quality control, thermal imaging can be used in predictive maintenance of food processing machinery to detect overheated parts or possible malfunctions, which lowers operating expenses and downtime. Its widespread adoption is still hampered by issues including the expensive cost of the equipment and the requirement for qualified operators. It is anticipated that further developments in thermal imaging technology and equipment cost reductions will increase its availability throughout the food sector. Overall, thermal imaging represents a forward-looking approach to enhancing food quality, safety, and efficiency, making it an indispensable tool for the evolving demands of modern food processing.

References

Abdulshahed, A. M., Longstaff, A. P., Fletcher, S., & Myers, A. (2015). Thermal error modelling of machine tools based on ANFIS with fuzzy c-means clustering using a thermal imaging camera. *Applied Mathematical Modelling, 39*(7), 1837–1852.

Akula, A., Ghosh, R., & Sardana, H. K. (2011). Thermal imaging and its application in defence systems. In *AIP conference proceedings* (Vol. 1391, No. 1, pp. 333–335). American Institute of Physics.

Alinaghi, M., Nilsson, D., Singh, N., Höjer, A., Saedén, K. H., & Trygg, J. (2023). Near-infrared hyperspectral image analysis for monitoring the cheese-ripening process. *Journal of Dairy Science, 106*(11), 7407–7418.

Antić, D., Krstić, I., & Bijelić, B. (2015). Thermo-visual diagnostics for increasing energy efficiency in bakery production lines. *eNergetics,* 64–69.

Aytekin, İ., & Coşkun, G. (2021). Early detection of mastitis by using infrared thermography in Holstein-Friesian dairy cows via classification and regression tree (CART) analysis. *Selcuk Journal of Agricultural and Food Sciences, 35*(2), 115–124.

Badia-Melis, R., Qian, J. P., Fan, B. L., Hoyos-Echevarria, P., Ruiz-García, L., & Yang, X. T. (2016). Artificial neural networks and thermal image for temperature prediction in apples. *Food and Bioprocess Technology, 9*(7), 1089–1099.

Baranowski, P., Mazurek, W., Witkowska-Walczak, B., & Sławiński, C. (2009). Detection of early apple bruises using pulsed-phase thermography. *Postharvest Biology and Technology, 53*(3), 91–100.

Barbin, D., Elmasry, G., Sun, D. W., & Allen, P. (2012). Near-infrared hyperspectral imaging for grading and classification of pork. *Meat Science, 90*(1), 259–268.

Bejo-Khairunniza, S., Azman, N., & Jamil, N. (2016). Paddy grading using thermal imaging technology. *International Food Research Journal, 23*(December), S245–S248.

Berry, R. J., Kennedy, A. D., Scott, S. L., Kyle, B. L., & Schaefer, A. L. (2003). Daily variation in the udder surface temperature of dairy cows measured by Infrared Thermography: Potential for mastitis detection. *Canadian Journal of Animal Science, 83*(4), 687–693.

Bicanic, D., Streza, M., Dóka, O., Valinger, D., Luterotti, S., Ajtony, Z., et al. (2015). Non-destructive measurement of total carotenoid content in processed tomato products: Infrared lock-in thermography, near-infrared spectroscopy/chemometrics, and condensed phase laser-based photoacoustics—Pilot study. *International Journal of Thermophysics, 36*, 2380–2388.

Castro-Costa, A., Caja, G., Salama, A. A. K., Rovai, M., Flores, C., & Aguiló, J. (2014). Thermographic variation of the udder of dairy ewes in early lactation and following an Escherichia coli endotoxin intramammary challenge in late lactation. *Journal of Dairy Science, 97*(3), 1377–1387.

Cheng, J. H., Sun, D. W., Zeng, X. A., & Liu, D. (2015). Recent advances in methods and techniques for freshness quality determination and evaluation of fish and fish fillets: A review. *Critical Reviews in Food Science and Nutrition, 55*(7), 1012–1225.

Chernonozhkin, S. M., Weyrauch, M., Goderis, S., Oeser, M., McKibbin, S. J., Horn, I., et al. (2017). Thermal equilibration of iron meteorite and pallasite parent bodies recorded at the mineral scale by Fe and Ni isotope systematics. *Geochimica et Cosmochimica Acta, 217*, 95–111.

Colak, A., Polat, B., Okumus, Z., Kaya, M., Yanmaz, L. E., & Hayirli, A. (2008). Short communication: Early detection of mastitis using infrared thermography in dairy cows. *Journal of Dairy Science, 91*(11), 4244–4248.

Comba, L., Biglia, A., Aimonino, D. R., Barge, P., Tortia, C., & Gay, P. (2021). Automatic inspection of baked goods based on cost-effective RGB-D camera. In *2021 IEEE international workshop on metrology for agriculture and forestry (MetroAgriFor)* (pp. 108–113). IEEE.

Cuibus, L. (2013). *Applications of infrared thermography in the food industry.*

Dissing, B. S., Papadopoulou, O. S., Tassou, C., Ersbøll, B. K., Carstensen, J. M., Panagou, E. Z., & Nychas, G. J. (2013). Using multispectral imaging for spoilage detection of pork meat. *Food and Bioprocess Technology, 6*, 2268–2279.

Döner, D., Çokgezme, Ö. F., Cevik, M., Engin, M., & Icier, F. (2020). Thermal image processing technique for determination of temperature distributions of minced beef thawed by ohmic and conventional methods. *Food and Bioprocess Technology, 13*(11), 1878–1892.

ElMasry, G., ElGamal, R., Mandour, N., Gou, P., Al-Rejaie, S., Belin, E., & Rousseau, D. (2020). Emerging thermal imaging techniques for seed quality evaluation: Principles and applications. *Food Research International, 131*, 109025.

Ferreira, D. S. (2020). Thermal imaging as a tool in food analysis. *Journal of Spectral Imaging, 9*, 1–4. https://doi.org/10.1255/jsi.2020.a7

Fito, P. J., Ortolá, M. D., De los Reyes, R., & De los Reyes, E. (2004). Control of citrus surface drying by image analysis of Infrared Thermography. *Journal of Food Engineering, 61*(3), 287–290.

Foster, A. M., Ketteringham, L. P., Purnell, G. L., Kondjoyan, A., Havet, M., & Evans, J. A. (2006). New apparatus to provide repeatable surface temperature-time treatments on inoculated food samples. *Journal of Food Engineering, 76*(1), 19–26.

Fuller, M. P., & Wisniewski, M. (1998). The use of infrared thermal imaging in the study of ice nucleation and freezing of plants. *Journal of Thermal Biology, 23*(2), 81–89.

Gariepy, C., Amiot, J., & Nadai, S. (1989). Ante-mortem detection of PSE and DFD by infrared thermography of pigs before stunning. *Meat Science,* 25(1):37–41. https://doi.org/10.1016/0309-1740(89)90064-8

Ghosh, P. A. Y. E. L., Rana, S. S., Nayak, A. N. K. I. T. A., & Pradhan, R. C. (2016a). Quality evaluation of food by thermal imaging. *International Journal of Processing and Post Harvest Technology, 7*, 126–133.

Gowda, N., & Alagusundaram, K. (2020). Use of thermal imaging to improve the food grains quality during storage. *International Journal of Current Agricultural Research, 7*(1), 34–41.

Gowen, A. A., Tiwari, B. K., Cullen, P. J., McDonnell, K., & O'Donnell, C. P. (2010). Applications of thermal imaging in food quality and safety assessment. *Trends in Food Science & Technology, 21*(4), 190–200.

He, Y., Deng, B., Wang, H., Cheng, L., Zhou, K., Cai, S., & Ciampa, F. (2021). Infrared machine vision and infrared thermography with deep learning: A review. *Infrared Physics & Technology, 116*, 103754.

Hou, F., Zhang, Y., Zhou, Y., Zhang, M., Lv, B., & Wu, J. (2022). Review on infrared imaging technology. *Sustainability, 14*(18), 11161.

Ishimwe, R., Abutaleb, K., & Ahmed, F. (2014). Applications of thermal imaging in agriculture—A review. *Advances in Remote Sensing, 3*(03), 128.

Jayas, D. S., & Singh, C. B. (2012). Grain quality evaluation by computer vision. In *Computer vision technology in the food and beverage industries*. Woodhead Publishing Limited.

Kamruzzaman, M., Sun, D. W., ElMasry, G., & Allen, P. (2013). Fast detection and visualization of minced lamb meat adulteration using NIR hyperspectral imaging and multivariate image analysis. *Talanta, 103*, 130–136.

Kor, G., & Icier, F. (2016). Thermal imaging during infrared final cooking of semi-processed cylindrical meat product. *Infrared Physics & Technology, 79*, 242–251.

Lei, L. (2018). *Cold food chain: Infrared Thermography applied to the evaluation of insulation anomalies in refrigerated vehicles for the transport of food & Exploration of cold approach in Infrared Thermography for Non-Destructive Testing*.

Lipinska, E., Pobiega, K., Szymankiewicz, K., Rudziak-Talalaj, A., & Klimkiewicz, M. (2019). Application of the thermographic method for the monitoring growth of selected microorganisms. *Zeszyty Problemowe Postępów Nauk Rolniczych, 599*.

Lipińska, E., Pobiega, K., Piwowarek, K., & Hać-Szymańczuk, E. (2022). Research on thermal imaging usage as a method for detecting Bacillus subtilis bacteria in mortadella. *Applied Sciences, 12*(21), 11252.

Liu, Y., Pu, H., & Sun, D. W. (2017). Hyperspectral imaging technique for evaluating food quality and safety during various processes: A review of recent applications. *Trends in Food Science & Technology, 69*, 25–35.

Manickavasagan, A., Jayas, D. S., White, N. D. G., & Jian, F. (2006). T i s g s d h s, *22*(6), 891–897.

Manickavasagan, A., Jayas, D. S., & White, N. D. G. (2008). Thermal imaging to detect infestation by Cryptolestes ferrugineus inside wheat kernels. *Journal of Stored Products Research, 44*(2), 186–192.

Manickavasagan, A., Jayas, D. S., White, N. D. G., et al. (2010a). Wheat class identification using thermal imaging. *Food and Bioprocess Technology, 3*, 450–460.

Massaro, A., Panarese, A., & Galiano, A. (2020). Infrared Thermography applied on fresh food monitoring in automated alerting systems. In *2020 IEEE international workshop on metrology for industry 4.0 & IoT* (pp. 554–558). IEEE.

Metzner, M., Sauter-Louis, C., Seemueller, A., Petzl, W., & Klee, W. (2014). Infrared Thermography of the udder surface of dairy cattle: Characteristics, methods, and correlation with rectal temperature. *Veterinary Journal, 199*(1), 57–62.

Muthukumar, P., Kumar, A., Raju, N. N., Malleswararao, K., & Rahman, M. M. (2018). A critical review on design aspects and developmental status of metal hydride based thermal machines. *International Journal of Hydrogen Energy, 43*(37), 17753–17779.

Pathmanaban, P., Gnanavel, B. K., Anandan, S. S., & Sathiyamurthy, S. (2023). Advancing post-harvest fruit handling through AI-based thermal imaging: Applications, challenges, and future trends. *Discover Food, 3*(1), 27.

Rai, M., Maity, T., & Yadav, R. K. (2017). Thermal imaging system and its real time applications: A survey. *Journal of Engineering Technology, 6*(2), 290–303.

Rogalski, A. (2003). Infrared detectors: Status and trends. Progress in quantum 685. *Electronics, 27*, 59–210.

Rogalski, A., & Chrzanowski, K. (2014). Infrared devices and techniques (revision). *Metrology and Measurement Systems, 21*(4), 565–618.

Rosca, A., & Rosca, D. (2011). Thermovision applications in food industry. *Progress of Cryogenics and Isotopes Separation, 14*(2), 195–201.

Senni, L., Ricci, M., Palazzi, A., Burrascano, P., Pennisi, P., & Ghirelli, F. (2014). On-line automatic detection of foreign bodies in biscuits by Infrared Thermography and image processing. *Journal of Food Engineering, 128*, 146–156.

Shawal Jadin, M. S., Taib, S., Kabir, S., & Yusof, M. A. B. (2011). Image processing methods for evaluating infrared thermographic image of electrical equipments. In *Proceedings of the progress in electromagnetics research symposium*. PIERS Proceedings, Marrakesh, MOROCCO, March 20–23.

Sritham, E., Nunak, N., Veng, T., Chaishome, J., Tuppadung, Y., Nunak, T., & Suesut, T. (2024). Non-contact temperature measurement based on infrared radiation for bread baking process. *Journal of Hygienic Engineering and Design, 47*, 3–8.

Sumriddetchkajorn, S., & Intaravanne, Y. (2013). Two-dimensional fruit ripeness estimation using thermal imaging. In *ICPS 2013: International conference on photonics solutions, 8883 (June 2013)* (p. 88831C).

Takacs, R., Jovicic, V., Zbogar-Rasic, A., Geier, D., Delgado, A., & Becker, T. (2020). Evaluation of baking performance by means of mid-infrared imaging. *Innovative Food Science & Emerging Technologies, 61*, 102327.

Tao, Y. (2000). Combined IR imaging-neural network method for the estimation of internal temperature in cooked chicken meat. *Optical Engineering, 39*(11), 3032.

Therdthai, N., Zhou, W., & Adamczak, T. (2002). Optimization of the temperature profile in bread baking. *Journal of Food Engineering, 55*(1), 41–48.

Vadivambal, R., & Jayas, D. S. (2010). Non-uniform temperature distribution during microwave heating of food materials—A review. *Food and Bioprocess Technology, 3*, 161–171.

Vadivambal, R., Chelladurai, V., Jayas, D. S., & White, N. D. G. (2011). Determination of sprout-damaged barley using thermal imaging. *Agricultural Engineering International: CIGR Journal, 13*(2), 1–6.

Van Linden, V., Vereycken, R., Bravo, C., Ramon, H., & De Baerdemaeker, J. (2003). Detection technique for tomato bruise damage by thermal imaging. *Acta Hortic, 599*, 389–394. https://doi.org/10.17660/ActaHortic.2003.599.49

Varith, J., Hyde, G. M., Baritelle, A. L., Fellman, J. K., & Sattabongkot, T. (2003). Non-contact bruise detection in apples by thermal imaging. *Innovative Food Science and Emerging Technologies, 4*(2), 211–218.

Vegricht, J., Machálek, A., Ambroź, Brehme, U., & Rose, S. (2007). Milking-related changes of teat temperature caused by various milking machines. *Research in Agricultural Engineering, 53*(4), 121–125.

Wang, S., Tang, J., Sun, T., Mitcham, E. J., Koral, T., & Birla, S. L. (2006). Considerations in design of commercial radio frequency treatments for postharvest pest control in in-shell walnuts. *Journal of Food Engineering, 77*(2), 304–312.

Wiecek, B. (2005). Review on thermal image processing for passive and active thermography. In *Proceedings of the 2005 IEEE engineering in medicine and biology 27th annual conference, Shanghai, China*, September 1–4, 2005.

Xiong, Z., Sun, D. W., Pu, H., Gao, W., & Dai, Q. (2017). Applications of emerging imaging techniques for meat quality and safety detection and evaluation: A review. *Critical Reviews in Food Science and Nutrition, 57*(4), 755–768.

Yaqoob, M., Sharma, S., & Aggarwal, P. (2021). Imaging techniques in Agro-industry and their applications, a review. *Journal of Food Measurement and Characterization, 15*(3), 2329–2343.

Chapter 11
Terahertz Spectroscopy

N. Sharath Kumar, C. K. Sunil, Ashish Rawson, D. V. Chidanand, and S. Shanmugasundaram

11.1 Introduction

With the growing proportion of consumer segments being diverted to nutrition-sensitive and health-conscious, it becomes necessary to design and install food safety monitoring systems that are real-time, non-destructive, and provide complete information on quality, safety, and participation in the product purchase decision. Conventional methods for food quality and safety assurance come with the lacuna of being laborious, tedious in sample preparation, destructive, and time-consuming, eventually making the entire testing and analysis process tiring, cumbersome and may often come with human errors (Hassoun & Karoui, 2017; El-Mesery et al., 2019). The requirement of online assessment with rapid results for evaluating food products is highly demanded in industrial applications. Emerging tools and techniques of non-destructive techniques (NDT) of food analysis, which comes with the advantage of being rapid, selective, specific, and non-destructive in nature, utilize the qualitative and quantitative signals emitted by the food matrices either from external or internal composition and provide the changes associated without compromising on the product integrity status (Ali & Hashim, 2022). The analyzed properties should provide insights into the internal quality composition of the foods, which can thereby assist in predicting the shelf life and consumer acceptability of food samples.

Regarding assessment and instrumentation, these non-destructive techniques (NDTs) help with the simultaneous measurement and analysis of food quality and

N. Sharath Kumar
ICAR-Central Institute of Temperate Horticulture, Srinagar, India

C. K. Sunil (✉) · A. Rawson · D. V. Chidanand · S. Shanmugasundaram
National Institute of Food Technology, Entrepreneurship and Management (NIFTEM-T), Thanjavur, India
e-mail: sunil.ck@iifpt.edu.in

C. K. Sunil et al. (eds.), *Emerging Non-destructive Technologies for Food Quality Analysis*, https://doi.org/10.1007/978-3-031-95048-3_11

safety. On the other hand, they effectively minimize food loss and waste while detecting a broader range of food quality assessments, evaluations, and inspections that consider the material properties of the foods. Few of the established NDTs for food quality assessment include the principle of spectroscopy (Su et al., 2017), acoustic (Aboonajmi et al., 2015), and artificial sensors (Gopalakrishnan et al., 2021), all of which are coupled with advanced statistical data analysis, adaptive machine learning and chemometrics approach for decoding the received signal and provide established models for accurate prediction in real time scenario (Modupalli et al., 2021). The spectroscopic method of food quality evaluation is based on the interaction of electromagnetic radiation with the material systems or sample of interest with the principle of collisions (Sharma, 2017). The electromagnetic radiation spectrum allows one to visualize the interaction effects of wavelength spectra, which are classified based on wavelength, frequency, and energy domains. In the increasing wavelength path, it hovers from γ-rays, X-rays, ultraviolet rays, visible rays, infrared, microwaves, and radio waves (Su et al., 2017). Spectroscopic methods evolved from the measurements includes interaction of materials with wavelengths of UV–visible (Włodarska et al., 2019), fluorescence (Hassoun et al., 2019), near-infrared (NIR) (Manley, 2014), mid-infrared (MIR) (Alexandrakis et al., 2012), Raman spectroscopy (Wang et al., 2021), and nuclear magnetic resonance spectroscopy (NMR) (Fan & Zhang, 2019).

11.2 Terahertz Spectroscopy

The molecular spectroscopy techniques such as Terahertz spectroscopy utilizes the spectral frequency range of 0.1–20 terahertz (THz) corresponding to wavelength of 30–3000 μm and wavenumber of 3.3–33.3 cm^{-1} is believed to interact with physical, chemical and biological components as a non-ionizing source of radiation and elucidate the material properties both in structure and composition (Fu et al., 2022; Koch et al., 2023). Although few of the studies were conducted in the 1920s due to the lack of scientific understanding of sources, generation, and detection patterns, it left a terahertz gap with limited understanding and scope. However, recent advancements in synthesizing novel generators and detection tools have increased its space in the spectroscopic arena. It is a non-destructive, non-invasive, and safer technology for large-scale practical applications. It is also called submillimeter radiation, terahertz waves, terahertz light, T-rays, T-waves, T-light, T-lux, or THz. The terahertz spectrum falls between the spectral gap of microwaves and infrared radiation, which acts as an interface and combines the advantage of both radiations to understand the molecules' vibrational, torsional, and rotational patterns. This is also called the fingerprint region for identifying biological molecules (Pawar et al., 2013). One terahertz corresponds to a photon energy of 4 meV (33.3 cm^{-1}) with 300 μm wavelength and a wavenumber of 33.3 cm^{-1} (Qin et al., 2013). The application of terahertz spectroscopy spans across astronomical studies (Kawano, 2013; Risacher et al., 2015), medical imaging (Banerjee et al., 2020; Yu et al., 2019), pharmaceutical (Zeitler et al., 2017; Bawuah & Zeitler, 2021) food inspection and quality control (Afsah-Hejri et al., 2019; Ok et al., 2019), security checks (Tzydynzhapov et al., 2020; Vaseashta, 2014), environmental monitoring (Ren et al., 2019; Zhan et al., 2016), communication strategies (Chen et al., 2019; Song & Nagatsuma, 2011) biosensing technologies (Tian et al., 2023; Geng et al., 2017).

Terahertz spectroscopic techniques can sense the picosecond range with high temporal resolution. They are recognized for their ability to operate at lower energy transitions in the range of 1–100 meV range that includes the alterations inbound electrical charges, free charge plasmas, strongly confined charge plasma, excitons, transient molecular dipoles, phonons, and magnons in crystalline solids, weakly bonded molecular crystals, relaxational dynamics in aqueous liquids, and hydrated biological matter (Jepsen et al., 2011). Due to the ionizing effect, the relatively lower photonic energy of waves generated in the terahertz domain has minimal effect on damaging the biomolecules. The generated wavelength can penetrate diverse materials, including dielectric, conductive, and non-polar materials. The unique ability of terahertz lies in quantifying both amplitude and phase in coherently generated electromagnetic radiation, which in turn is dependent on the absorption coefficient, refraction index, and permissivity of the sample (Hua & Zhang, 2010). The time-resolved detection of the emitted electric field has been instrumental in providing the categories of terahertz applications such as terahertz time-domain spectroscopy (THz-TDS), terahertz time-resolved spectroscopy (THz-TRS) and terahertz emission spectroscopy (THz-TES). Terahertz Time domain spectroscopy is highly used for structural elucidation and quality assessments of biomolecules and food products. Terahertz spectroscopic application in combination with chemometrics can provide reliable estimates on the safety assessment of food commodities. This is mainly due to the vibrational match of the biomolecules in food with the terahertz domain, enabling it to identify the motion pattern and its relative change intensity quickly (Chen et al., 2015). These innumerous benefits of terahertz spectroscopy have made its innovation potential in the quality estimation of agricultural commodities by defect checking of foreign matter (Hu et al., 2023), adulterants (Liu et al., 2020), toxin generation in the foods (Chen et al., 2023), determination of antibiotic (Qin et al., 2016) and pesticide residues (Qu et al., 2018), detection of genetically modified organisms (Liu et al., 2018) and quality assessment of high-value commodities (Du et al., 2021).

11.2.1 Theory and Principles

A source (emitters or generators) emits terahertz radiation at a specific frequency, which is then modulated and scanned using filters and a lens system or collimator to precisely select a bandwidth or splits into pulsed and probed beams and passes through the sample of interest and is detected by detectors for spectral analysis that can be quantified. In short, radiation generation, modulation, and detection are highly investigated segments that aim to increase the signal intensity and spectral information gain. Recently, chemometrics has been used for enhanced quantification and making the obtained signal with higher prediction accuracy, whereas multivariate analysis is used for qualitative data to generate and validate models. The waves can be generated by optical rectification or heterodyne mixing, through excited charge carriers (photoconductive emission), or by employing microwave technology using Gunn diodes, Impatt diodes, and resonant tunneling diodes. The waves generated can have more extended bandwidth in continuous wave patterns, while others can generate short bursts of higher bandwidth waves (pulsed waves). The pulsed form is highly used due to its ability to preserve the time-gated phase information and can be used for the study of nonlinear optics due to its high intensity and short-pulsed form (Shen, 2011; Chen & Guo, 2017). The output spectrum assists in the determination of molecular vibrations, spin moments, and lattice change. Therefore, an understanding of the molecular events is precisely carried out. The time domain capture of the signal emitted from the samples is a spectral fingerprint for identifying and recognizing molecular patterns and vibrations. Though gas-induced lasers are the originally used methods for terahertz generation, improvements in optics, semiconductor materials, and electronics have broadened the scope of new materials that can be used as sources for terahertz generation. Broadly, the methods available for terahertz generation include the charge accelerator-based and laser-based. Laser-based terahertz radiation is acknowledged for its compactness and increased conversion efficiency compared to accelerator-based emission (Wen et al., 2013). The increase in applications of spectroscopy is possible due to the increase in several emission sources and mechanisms, namely charge acceleration by photoconductive antennas, optical rectification from femtosecond laser pulse in nonlinear crystals, laser-induced plasma, and charge transfer process which selectively uses electric and optical energy for the generation.

11.3 Instrumentation

The basic instrumentation and working principle of terahertz spectroscopy for the determination of various molecular events lies in the selection of appropriate sources and accessories to focus, filter, and alter its collinear properties, followed by its successful detection by the detectors and analysis of the reflected, transmitted or refracted waves by data analysis and imaging systems. Further, the generated source of information can be used for the modeling and data analysis to create various data sets. As previously mentioned, the emission of continuous and pulsed terahertz waveforms gets focused and screened by the filters and mirrors, which eventually pass through the matrix of interest and get through the detector system. The terahertz spectrum is delayed concerning the time domain, and the signal corresponding to each delayed waveform is measured and quantified. This way, the transmitted waves' phase and amplitude are recognized and converted to the frequency domain using Fourier transform infrared spectroscopy (FT-IR) or suitable software modules. The measurement techniques of the terahertz spectroscopy fall into three major areas—(a) Terahertz-Time domain spectroscopy (THz-TDS), (b) Time-resolved Terahertz spectroscopy (TRTS), (c) Terahertz emission spectroscopy (TES).

11.3.1 Terahertz-Time Domain Spectroscopy (THz-TDS)

It is one of the most important and widely used spectroscopic determination techniques in food analysis, material chemistry, quality control, and bio-molecular studies due to its coherency and broader measurement capability with reduced background noise detection. Initially reported by Grischkowsky and his associates in 1989 while examining water vapor, this spectroscopy measures the static properties of the materials and their complex interactions (Koch et al., 2023). The direct time-domain sampling of TDS assists in better frequency capture, and the electric field component of the terahertz wave can detect variation in the sample. The generated frequency capture selective bands after passing through the sample of interest allows an understanding of the materials' chemical profile through a pump beam or probe beam. The spectroscopy consists of a source, beam splitters, optical delay lines, collimating optics, a sample of interest, an optically gated detector, and a lock-in amplifier (Peiponen et al., 2012). Though at first, electrical field-based sources were used, the need for better resolution and enhanced material properties detection has given rise to the use of femtosecond-based optical pulses excited by lasers, which are compact and cost-effective. The laser pulse generated in the near-infrared region strikes the semiconducting material between the antennas and creates charged carriers, further accelerating the electric field. The flow of carriers creates current, which is a coherent band of electromagnetic radiation. On average, this THz pulse travels through the sample and toward the detecting system over several hundred femtoseconds. Mirrors, lenses, and polarizers assist in better focusing on linear radiation and achieve enhanced resolution. The presence of an optical delay line is essential to increase the optical path of the radiation before reaching the detector, similar to the Michelson interferometer present in FT-IR. The current generated after the induction of photo carriers by optical pulses gradually decays with time constant. The measured signal's time delay corresponds to the rate at which the THz wave traversed through the sample and reached the detector. These optically excited transient waves indicate changes in the dielectric properties due to a change in conductivity. Often, signal amplifiers are used to increase the signal-to-noise ratio (SNR) and amplify the transmitted waves that reach detectors, thus increasing the intensity of the waves that reach the detectors. The spectral density of the generated waves ranges from 100 GHz to 5 THz with the resolution of time that gets synchronized with the optical pulses wherein these waves are found to provide information on the crystalline vibrations in solids, intermolecular interactions in liquids, and rotational transitions in gases (Smith & Arnold, 2011). The detectors for the TDS spectroscopy are categorized as photoconductive and electro-optic methods, which directly influence the source of terahertz propagation. Detection by the electro-optic method permits wider bandwidth and offers reduced

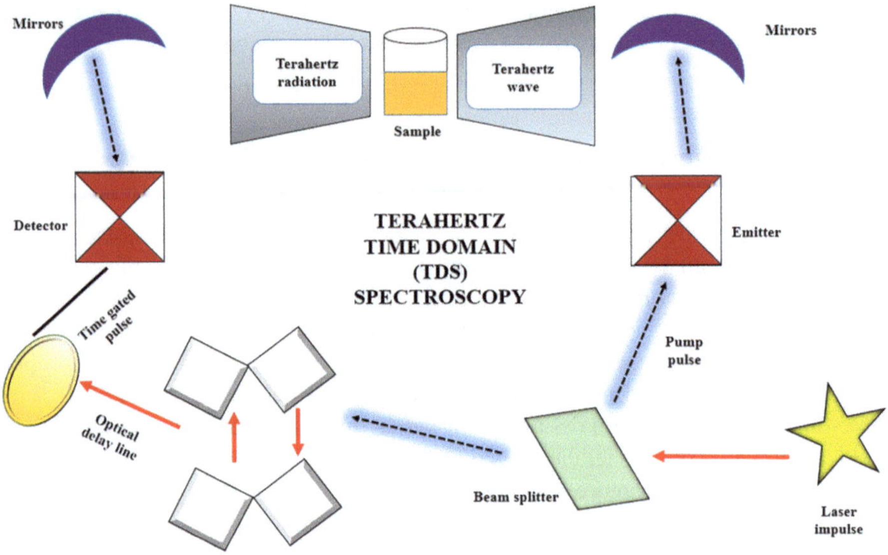

Fig. 11.1 Schematic representation of Terahertz Time domain spectroscopy

sensitivity to laser fluctuations. Figure 11.1 gives the schematic of Terahertz Time domain spectroscopy.

11.3.2 Time-Resolved Terahertz Spectroscopy (TRTS)

While the time domain terahertz spectroscopy measures the material properties as a function of time, time-resolved terahertz spectroscopy utilizes the spectrum measurements as a function of variable time measurements by quantification of absorption or emission. The TRTS spectroscopy is extremely useful for studying materials (transient state and carrier relaxation dynamics) at ultrafast time levels with subpicosecond temporal resolutions. An ultrafast laser pump pulse, having energies ranging from UV to mid-IR, is used for photoexcitation of the sample, leading to the emission of time-dependent terahertz radiation with delay time ranging from less than 100 femtoseconds to several nanoseconds. The subpicosecond time resolution is highly valued in the semiconductor and solar energy dynamics to quantify photoconductivity, measured as the proportion of reflected or transmitted waves. The frequency-resolved transfer function measures the photoconductivity of the sample. The time domain transmission is calculated as the time delay function between the probe beam and pump beams. The pump beam is used for timescale detection, and the probe beam is used for terahertz generation. Mechanisms such as scattering, trapping, recombination, and polariton development affect the performance of solar materials that can be evaluated using time resolved spectroscopy

(Pattengale et al., 2021). The limitation of TRTS is its sensitivity to background noise, which may hamper the time domain signal.

11.3.3 Terahertz Emission Spectroscopy (TES)

The precise quantification of shape, amplitude, polarity, and polarization of the transient electric field occurring in a picosecond time scale derived from the pulsed laser source is used for analyzing the mechanical, optical, and electric properties of the materials, which is used to elucidate the structural and electrical properties of the materials. The terahertz emission spectroscopy finds its advantage in characterizing graphene, TMDs, topological insulators, hybrid perovskites, and mixed-dimensional heterostructures, which is linear to time-domain spectroscopy with the advantage of utilizing the sample of interest as terahertz emitting material instead of specific semiconducting emitters. TES is more relevant in studying materials' nonlinear and photoelectric properties. The well-recognized nonlinear properties such as photon-drag effect, interfacial processes, high-order THz harmonics generation, and spin-to-charge conversion are being measured by the emitted spectrum, which is again dependent on crystalline symmetry, transient photocurrents, and adsorption/desorption. The non-equilibrium properties of materials can be measured by analyzing the emitted waves for events such as electrons, spin moment, and lattice detected from the generated spectrum. The topological properties, superconductivity, and quantum mechanical properties are some of the unique material characterizations carried out with terahertz emission spectroscopy. The instrumentation for the terahertz emission involves using a femtosecond laser for the source and electro-optic sampling for the detection. The generated femtosecond laser is divided into the pump (focused on carrying created radiation toward the sample) and a probe beam (focused on the detector) by the beam splitters.

11.4 Source of Generation

Terahertz wave generation has been found to increase its space in several applications due to innumerable and significant improvements in newer methods of source creation. The emission source has increased the physics, material chemistry, and radiation mechanism to understand the molecular events in biological molecules successfully.

11.4.1 Photoconductive Emission

The emission method has been used since 1990 for the terahertz generation. The method is valued for its simplicity, relatively lower cost, and reliability. Since the strength of the produced THz electric field may be raised significantly with increasingly high bias voltage amplitudes and relatively higher pump optical fluences, photoconductive THz emitters are particularly interesting. The accelerated charge carrier principle is followed for wave generation. The instrumentation consists of semiconductor material with photoconductive gaps, which generates free electrons in the photoconductive space upon illuminating with a light source that has higher photon energy than the bandwidth of the semiconductor. The generated free electrons lead to an instantaneous reduction in the medium's resistance. The generation of transient current by the acceleration and subsequent recapture of the carriers creates a quasi-single-cycle electromagnetic pulse at THz frequencies. The radiation generated in the terahertz domain is proportional to the generated free carrier and photocurrent.

The performance of the emitter is related to the output electric field, bandwidth, and signal-to-noise ratio (SNR). The selection of the semiconducting material is important as it should have lower resistance and greater breakdown voltage with high carrier capacity (Bacon et al., 2021). The manipulation of active and substrate materials and selecting suitable antenna designs are important considerations for higher photoconductive emissions. Few of the semiconducting materials include semi-insulating Gallium arsenide (SI-GaAs), Low Temperature-Gallium arsenide (LT-GaAs), doped series including -gold doped geranium (Au doped Ge), iron-doped geranium (Fe-doped Ge), Inon Gao-47As (Fe doped InGaAs). The major limitation of photoconductive emission lies in its use for lower field strength due to the saturation effect and the effect of larger created charge carriers resulting in joule heating and ohmic loss (Collier et al., 2016).

11.4.2 Optical Rectifications

A second-order nonlinear process for terahertz emission is the optical rectification process. An efficient terahertz emission depends on increasing the domain activity of the substrate and active region while reducing the thickness of the active site to achieve a higher response to the non-trivial electric field in a nonlinear spectrum coupled with higher conversion efficiency and greater field strength capable enough to quantify the nonlinear response. It is sometimes called electro-optical rectification due to its ability to change the polarization potential. Optical rectification of femtosecond laser pulses has emerged as one of the important sources of terahertz radiation generation at narrowband width that can result in higher resolution and imaging (Avestisyan et al., 2012). The end radiation with broader wavelengths (100 GHz–3 THz) and larger amplitudes of the optical rectification method finds its applications in estimating electron acceleration, coherent X-ray generation, and nonlinear spectroscopy. The frequency variation in the emitted optical pulse with respect to crystal morphology is critical for the efficient release profile. In addition, modification of the temperature profile of the crystals is shown to increase the conversion efficiency of the optical process of terahertz generation, as shown in the studies of Huang et al. (2013) in which cooling down the lithium niobate ($LiNbO_3$) crystals to its cryogenic temperature was able to have increased conversion efficiency of 3.8%. The requirement of larger second-order nonlinear susceptibilities, electro-optic crystals, and well-suited phase-matching properties are preliminary requirements of well-established optical rectification emission. Well-established electro-optic crystals such as ZnTe, GaP, ZnSe, GaP, and GaSe act as rectifying units for the frequency modulation of the terahertz radiation.

11.4.3 Laser-Based Plasma Generation

The plasma-based terahertz generation is conventionally categorized into femtosecond-based and sub-picosecond-based pulsed generation. The former leads to the generation of short and intense pulse generation of terahertz by femtosecond laser, which is believed to utilize the electrode gap distance between the selected photoconductors that leads to ionization and the creation of electrons that get accelerated by the provided strength of the electric field. Interaction of the femtosecond laser electric field with matter (solid, liquid, or gas) may lead to ionization and plasma generation, provided the optical strength is sufficient to separate the electrons and initiate the ionization in the medium. The generated plasma from the optical region's frequency conversion will radiate the terahertz waves. The latter process utilizes the sub-picosecond laser pulse to illuminate nonlinear crystals such as ZnTe. The nonlinear crystal domain of the material allows for photo mixing and generation of time-varying polarization that is used to generate the terahertz waves. The radiation emitted from laser-induced generation is influenced by the nature of the filament used for the ionization process. At the same time, the length of plasma movement is proportional to the terahertz radiation, and the medium of plasma generation can be solid, liquid, or gas. The gas-based laser generation can be single-color or dual-color based on the wavelengths selected for the optical field generation. The waves generated at the terahertz region can be coupled to an RF antenna to undergo the frequency coupling of the waves, leading to terahertz generation.

11.4.4 Free Electron Lasers

The need for intense, ultra-short pulsed, and coherent high energy terahertz radiation coupled with the ability to tune the wavelength (1 mm to 1 angstrom) and frequency

(1–10 THz) continuously have given rise to the need for free-electron laser for the terahertz generation (Tan et al., 2012). It can be visualized as relativistic electrons moving in a wiggling path created by a magnetic field or undulator. An electron moving in the electric field is known to emit radiation. Hence, the free electron laser works based on stimulated undulator radiation. The primary mechanism of terahertz wave generation lies in the ability to generate free electrons when sufficiently higher energy charged particles such as electrons positrons get inclined or wiggled under the influence of a magnetic field, then leads to synchrotron radiation, which gets generated as non-coherent emission, this can be converted to coherent by lock sync process wherein the free electrons generation gets increased linearly with the electric field (Vinokurov, 2011). Relatively higher energy output is possible with the positive interactive effect of the magnetic field, the generated free electrons, and the radiation matrix. The primary limitations include the larger size and system complexity, which makes it not a feasible and viable option for larger adoptions.

11.4.5 Quantum Cascade Lasers

The quantum-restricted confinement of sub-bands in the conduction band series for the electron transition and relaxation leads to the generation of terahertz wave and is highly acknowledged for its high power continuous spectra in a constricted space requirement. Both the free electron laser and quantum cascade lasers are known to emit terahertz in long band continuous wave spectrum; this can be viewed as a unipolar device, unlike the diode-based laser, which does not require any built-in gas setup. Kazarinov and Suris proposed the idea of inter-sub-band transitions in a super lattice-based structure in 1971 (Vitiello et al., 2015). The electrons passing through the built-in quantum sub-bands turn to emit photons in a cascading set of events. The size quantization and nature of semiconductor heterostructures and their active material allow for changes in energy transitions among the generated carriers or electrons, which are known to affect the electron's effective mass and barriers (Benveniste et al., 2008). The major advantage of quantum cascade laser generation is its ability to fine-tune the production of specific wavelengths using the combination strategy of semiconductor heterostructures and quantum confinement, a refinement and alternative to the conventionally used diode lasers. The cascading structural configuration may allow for electron recycling so that the electrons generated after reaching a certain threshold may generate an equivalent number of photons. Quantum cascade lasers have been able to detect frequency levels up to 1.2–5.4 THz (Walther et al., 2007). Detecting aluminum and oxygen atoms, hydroxyl radicals, nitrogen ions, and oxygen atoms with greater resolution is practically feasible in quantum cascade-based terahertz spectroscopy. However, the major limitation lies in its narrowed frequency range and the requirement of cryogenic conditions for its working.

11.5 Detectors

Devising an appropriate detector for detecting terahertz waves from pulse and probe mode is extremely important to understanding the various molecular geometries and reactions associated. It is rapidly gaining importance and significant scope in material characterization, imaging tools, cancer detection, biomedical fields, communication, and forecasting. An ideal detector in terahertz spectroscopy should receive and combine the emitted radiation from the sample of interest and project it into an understandable signal format. The data signal is converted into an electrical signal and can be used for imaging studies. The selected detector should be able to distinguish between different frequency ranges and provide a higher signal-to-noise ratio (SNR) with a lower response time to varying signals (Sizov & Rogalski, 2010). It should resist the polarization factor of photons and provide an account for both coherent and non-coherent radiation. Different sources of detectors differ in the reception of the physical properties of the receiving photons.

11.5.1 Thermal Detection

Thermal detectors are known to capture the energy stored in the accelerating photons and take up the heat energy. The materials that are known to behave as thermal detectors include Golay cells by thermal expansion upon heating, pyroelectric detectors by a change in electric potential upon heating, bolometers indicated by a change in electric resistance upon heating, glow discharge detectors indicated by ionization of the gas (Lewis, 2019).

11.5.2 Electron Transition Detection

The rectification of incoming terahertz radiation provides sufficient energy for the electrons to overcome the semiconducting barrier and transition from their native bound state. Semiconducting diodes such as Schottky diodes (Siles & Grajal, 2010) and resonant tunneling diodes (Asada & Suzuki, 2021) are used in electron transition detection, which varies in their sensitivity towards receiving signals.

11.5.3 Mixing Type Detection

The incoming terahertz radiation is combined with the reference radiation created by the local oscillator with specified frequency and power, which, upon nonlinear mixing, emits frequency corresponding to the sum and differences of individual frequency components. It can be employed as heterodyne mixing for non-coherent radiation (heterodyne) and homodyne mixing for coherent radiation. Semiconductor detectors hot electron bolometers are some of the commonly used mixers (Lin & Jarrahi, 2020).

11.5.4 Ultrashort Pulses Detection

The incoming pulsed terahertz waveform is received by employing detectors, which are viewed to produce harmonic flow (air-based detection), current flow generation (photoconductive-based), and polarization principle (electro-optic detection). The induction of current flow due to electron-hole pairs created by the incoming optical pulse is a direct measure of electric wavelength in the selected semiconducting material and is used to detect coherent terahertz radiation. Popular photoconductive materials such as GaAs, InGaAs, InGaAs/InAlAs:Be, are generally used (Löffler et al., 2005).

11.5.4.1 Electro-optic Detectors

Change in the polarization of reference signal upon receipt of terahertz radiation from the sample (pulsed and continuous mode) to the crystal produces characteristic detection of generated terahertz radiation leading to the electro-optic detectors. The signal capture at the detector is a measure of optical time delays (Benea-Chelmus et al., 2018). This is a classic reference for electro-optic-based detection wherein the electric field energy is converted and coded into optic-based detection in the optic phase. The rotation of the probe signal is proportional to the incoming electric field, and the quantification is done by employing a quarter-wave plate, Wollaston prism, photodetector diodes pair, and differential amplifiers. The selected detector material's phase matching and bandwidth response are important for detecting terahertz radiation accurately.

11.5.4.2 Air-Based Coherent Detection

In air-based coherent detection of terahertz radiation, incoming terahertz radiation is mixed with the probe pulse generated from the laser pump to detect the frequency component (Ho et al., 2010).

11.6 Imaging and Processing

The spectroscopic analysis of the sample generates ample information related to the material's physical, chemical, and structural character, which needs to be adjuvanted by combining necessary imaging techniques with an enhanced resolution to obtain the spatial complexity understanding of the materials. This has more relevance in understanding material physics and interaction kinctics and has greater scope in agriculture, biotechnology, medicine, and pharmaceuticals. The imaging studies need to convert the obtained wavelengths to pixel format. In addition, the system selected for the imaging needs to be in line with the phase, amplitude, and coherent structure and should effectively combine all the received signals without any loss. The wavelengths are read as individual objects for constructing an image with the required pixel strength and intensity. The amplitude and phase information must be converted to an image by image acquisition that can be processed successfully. The imaging-based terahertz spectroscopic detection can be detected by following three approaches- (a) Transmission imaging- the time data obtained by the sample scanning is converted into a 2D image, which then gets Fourier transformed to obtain the spectral information (Bowman et al., 2016). (b) Time delay mapping- optical path variations can be traced by mapping the delay in the time domain for the main peak in comparison with reference (Valušis et al., 2021) (c) Normalization of maximum peak function in the time domain with a reference function of absorption, reflection, and transmission. The signal peaks consist of absorption, reflection, refraction, transmission, or even the attenuated form, converted into multispectral and hyperspectral imaging that proactively provides complete information on the molecular configuration in the sample of interest. It is directly influenced by the source, processing, and detector selected, along with the nature and type of investigation attempted. The selected technique should provide complete voluminous information on the analyzed sample with good clarity and image clearance. The relatively higher wavelengths of terahertz radiation coupled with Abbe's diffraction limit sets the limitation for precise imaging and characterization. Therefore, real-time imaging techniques, blending of several techniques, and utilization of particular methods in synergy with improved tools such as machine learning, sensors, chemometrics, search algorithms, and artificial intelligence are in progress. An effective detector should combine imaging techniques with the highest sensitivity and provide maximum information on the internal characteristics of the sample. The imaging techniques are generally classified as active and passive-based imaging. Active imaging requires a source and detectors for the simultaneous scanning, detection, and identity imaging of the sample of interest, while passive imaging does not require any source or detectors; it works on the object's natural source of terahertz emission. Three types of imaging systems are used in terahertz spectroscopy-.

11.6.1 Conventional Imaging

The conventionally used imaging system runs on the raster scanning technique in which complete time domain waves are captured in the hyperspectral data cube. The essence lies in capturing the amplitude and time domain in the form of phase characterization of the sample. Image beam generated by illuminating the sample with femto optic laser pulse has elicited the transient birefringence in the system that the use of femtosecond probe pulse or polarization resolving optics can detect. Expanding the optical probe beam on the nonlinear crystal overlaps the terahertz image. It gets encoded onto the spatially varying polarization state of the femtosecond pulse's wavefront, and this spatial distribution of the femtosecond probe pulse gets detected using a charged coupled device camera (CCD). The relative complexity and cost associated with requiring a larger area for optical beams and the requirement of amplification with greater acquisition times for multidimensional scanning of the sample have made it necessary the need for real-time, on-site, and low-cost alternatives (Kanda et al., 2017).

11.6.2 Near Field Imaging

The problems of diffraction limits that limit the identification of sub-wavelengths have given rise to near-field imaging. The ray falling outside the aperture of the Rayleigh focal spot is the near field, and the evanescent field is found to exist in this near field, which is known to be used for the subwavelength imaging of the terahertz. The exponential decay of this particular evanescent waves is being prevented by imaging with techniques such as apertureless scattering-type scanning near-field optical microscopy (s-SNOM) and near-field photoconductive antennas (NFPA) (Tuniz & Kuhlmey, 2023). The basic requirement is the propagation and amplification of wavelengths falling in this evanescent field. The emergence of aperture-dependent imaging, aperture-non-independent imaging, tapered photoconductive probes, laser terahertz emission microscopy, and electro-optic sampling have increased the scope of understanding the spatial resolution in the range of millimeters to several micrometers range.

11.6.3 Tomography-Based Imaging

Tomography is a method of imaging that reconstructs an object's volume using a collection of projections made from the outside of the sample under investigation. The process consists of acquisition and inverse modulation from the acquired data series. Most of the used imaging techniques are classified as 2-D, the tomographic-based techniques are used to derive 3-D images of the sample spectrum, and the tomography technique is used (Guillet et al., 2014). As the nomenclature indicates, tomography refers to the cross-sectional examination of the sample, which reflects internal composition and arrangements. The possible delineation of the refractive index of the radiated waves helps in understanding diffraction tomography, which is based on Maxwell's hypothesis of the position function. The tomographic diffraction technique reconstructs the generation of inverted refractive index patterns based on the wave scattering principle. The external noise affects the image reconstruction process, and efforts are focused on increasing the signal-to-noise ratio and the noise factor.

11.6.4 Tip or Probe-Based Imaging

The diffraction limit of conventional imaging techniques has limited terahertz imaging for micro and nanoscale detection and imaging strategies. Coupling the incoming terahertz radiation with the sharp metal tips increases the spatial distribution of the received signal, which is influenced by the ratio of the receiving wave signal to the tip or probe diameter and the distance of the probe tip to the radiation. The emergence of scanning probe microscopy is a variant of probe-based imaging techniques in which detailed investigations on superconductivity, charge density, separation kinetics, and phase transitions can be studied with special reference to nonlinear spectroscopy. The requirement for higher-resolution imaging has increased the scope of tip-based imaging techniques. The classic example is the scanning tunneling microscope in which the applied DC voltage at the tip of the electrode induces the movement of electric current in a tunneling fashion, which is used to match the electronic state transition of the electrons. The optically induced femtosecond laser pulse gets probed with the time-resolved scanning tunneling microscope at a controlled delay time measurement (Mittleman, 2018). This can be used for nanoscale study of the material properties emitted from the terahertz radiation.

11.6.5 Real-Time Imaging Techniques

Though aperture-based imaging (SNOM) has made reasonable advancements by reducing the diffraction limits to a certain extent, resolution enhancement has a lot of research scope that needs to be improved. On the other end, scanning tunneling microscopy is limited by creating a high-power vacuum environment for its operation and the need for clean and conducting samples. This has made the technique a lesser approach for standard imaging methods. A susceptible single-pixel camera is used to scan the sample with reduced acquisition time and can be used for real-time imaging (Cong et al., 2023). The evolution of techniques such as fast optical delay lines, photoconductive antenna arrays, and electro-optical camera sampling are known to increase the real-time imaging resolution. The advantages of real-time imaging techniques are their cost advantage, reduced complexity, and adaptability to cater to real-time imaging in a non-destructive manner. Stantchev et al. (2020) showed that the highest modulation intensity was obtained when the emitted radiation was spatially controlled on the silicon and the complete inversion of total internal reflection was done. Real-time imaging is significant in diagnosing several ailments and accurately changes the body composition. Bajwa et al. (2017) performed a comparison study on the visible and reflective terahertz imaging on the water content for fully islanded skin flaps in reconstructive surgery to assess the flap viability. They found that terahertz imaging exhibited significant visual differences in the developed image sets.

11.7 Applications in Non-destructive Food Quality and Analysis

Food quality and safety are important health concerns debated and discussed in many prefects. They have evolved from conventional destructive-based analysis to non-destructive, online, and real-time monitoring. The increasing global population and adjacent need to provide nutritional security with limited resources have increased the problems of adulteration, quality substandardization, food fraudulent issues, and unsafe food handling and presentation to the end consumers that may have a deteriorating health impact (Accorsi et al., 2019). This has increased the requirement for safe, quality, and reliable food throughout the entire supply chain until it reaches the consumer, encompassing the concepts of traceability, risk detection, and hazard management. Had the quality standards been subsidized and given less importance, it would have had a devastating impact on the socio-economic and living standards of the public, ultimately having an impact on the economy and global outfit. Adhering to strict quality standards by mandatory specifying the quality assurance systems with the legally laid out regulations is the only viable alternative for maintaining stability and sustainability in the supply chain.

Quality is an important cumulative attribute of the food commodity that gets analyzed physically, physiologically, microbiologically, and organoleptically to determine the optimum characteristic and desirable consumption needs (Picó, 2015). It otherwise measures the degree of excellence of the product that can go for consumption or further processing. Measurement and quantification of quality parameters call for thorough inspection and quality testing procedures. A few of the quality attributes the consumers perceive include color, absence of foreign material, shape, size, moisture (in the case of dried foods such as chips), flavor, texture, and taste. These said quality parameters are found to be influenced by raw material characteristics and processing characteristics. While the raw material properties depend on the origin of food commodity (plant-based or animal-based), variety or breed, place and type of cultivation or method of breeding and maintenance, method of harvesting/slaughtering, and post-harvest practices, the post-harvest properties are found to be dependent on the type of processing, methods of packaging, storing and consuming. The conventionally followed quality analysis methods include their destructive and time-consuming nature, with the loss of significant quantity and the requirement of trained workforce for analysis and interpretation. Non-destructive based sampling and analysis of foods and food commodities are significantly seeing an increased insight and recommendations in the recent research space. The advantage is the real-time, on-site, and non-destructive way of sample receipt, handling, and analysis. The ability of terahertz waves to penetrate the packaging materials also assists in non-destructiveness (Wang et al., 2017). This is an added benefit for food manufacturers and safety professionals as it utilizes less time and space to provide the required results with higher accuracy potential.

Terahertz spectroscopy, as it falls in the spectrum gap of microwave and infrared, this unique terahertz spectrum can utilize the spectrum spatial benefits of both the radiation and being non-ionizing in nature (Xie et al., 2017). The ability of many of materials, including opaque nature, can exhibit a unique fingerprint spectrum either by absorption, transmission, or reflection in the terahertz domain that helps in understanding the surface and internal quality aspect of the foods (Jepsen et al., 2011; Blanchard et al., 2013). The terahertz-based spectroscopy can aid in measuring the food samples' physical, physiological, and structural properties in a non-destructive and non-invasive manner. The successful detection and quality analysis of foods by terahertz spectroscopy depends on the nature of the adulterant, contaminant, or foreign material being detected. It's material properties (size, density, surface characteristics), type of terahertz generating source, nature of waves (continuous or pulsed, pulsed or probed beam), accessories (beam splitters, lens, collimators, filters), detection methods (signal processing with its sensitivity, selectivity), statistical analysis and correlation modeling designs (support vector model (SVM), partial least square regression (PLSR), stepwise multiple linear regression (SMLR), stochastic neighbor embedding (*t*-SNE) and imaging strategies employed. In addition, fine-tuning properties such as increasing the signal-to-noise ratio, lowering the background noise, spatial resolution, and frequency change over by Fourier transform can aid in increasing the accuracy of the technique. Table 11.1 summarizes the application of terahertz technology in food safety and quality analysis.

11.7.1 Detection of Foreign Materials

Identifying foreign materials in the end product presents a challenging task for food manufacturers, from the grader to the processor. It is also a haunting area of identification for enforcement agencies. The major reason is the different nature of foreign materials and their unique response to different types of radiation. The transparency nature of metallic and foreign materials in the food matrix towards terahertz radiation is responsible for its detection property. The properties such as

Table 11.1 Applications of Terahertz spectroscopy for the food safety and quality analysis

Product	Terahertz principle	Detection of material used for	Specifications	Key findings	References
Walnut powder	Terahertz Time domain spectroscopy	Endogenous contaminants	Source—low-temperature GaAs photoconductive antenna Detector—electro-optic ZnTe crystal	Detection of endogenous contaminants in walnut kernel which can be differentiated with shell based contaminants	Wang et al. (2020)
Vegetable oils—Unrefined (sunflower oil, extra virgin olive oils, coconut oil, hemp oil, flax oil, camelina oil), refined (sunflower oil, rapeseed oil)	Terahertz Time domain spectroscopy	Deteriorated oils due to degradative reactions	Source—femtosecond laser wavelength—800 nm; pulse duration <90 fs; and 80 MHz pulse repetition rate	The selected technique was able to distinguish between healthy and deteriorated oils based on absorption coefficient and refractive index	Karaliūnas et al. (2018)
Pecan nuts slices of 1 mm, 2 mm, and 3 mm	Terahertz Time domain spectroscopy	Insect detection and moisture estimation through models	Source—photoconductive emission by Ti:sapphire femtosecond laser Bandwidth – 0.2-2.0THz	Only slices had the terahertz penetration as against the whole nuts Insect detection and moisture prediction was done in nutmeat, shell and inner separator based on absorption coefficients	Li et al. (2010)
Wheat flour, talc and sugar	Terahertz Time domain spectroscopy	Glass, ceramics, egg shells, metal, rubber and plastics	Source—Ti:sapphire femtosecond laser Frequency—0.5 THz	Lateral resolution was limited for the powder dimension of <2 mm glass, ceramics, egg shells, metal, rubber and plastics are detected	Herrmann et al. (2002)
Instant noodles	Terahertz Time domain	Aluminum and granite pieces; maggots and crickets	Source—solid-state oscillator (gunn diodes) Frequency—0.2 THz,	High density aluminum and granite pieces were detected in the sample	Lee et al. (2012)

	spectroscopy—continuous wave		Power—10 mW Detector—Schottky diode	Low density insects—maggots and crickets were detected	
Peanut oil	Terahertz Time domain spectroscopy	Aflatoxin B1	Source—LT-GaAs Ti:sapphire laser photoconductive antenna Detector—ZnTe electro-optical crystal Frequency—0.1 to 3.5 THz	RMSC was 18.2 ppm for 1.11 and 0.72 THz frequencies for aflatoxin B1 detection for step wise multiple linear regression (SMLR) model	Chen and Lijuan (2014)
Tea leaves	Terahertz Time domain spectroscopy	Insect fragments	Source—femtosecond laser device Frequency—0.05 to 4 THz	Detection of insect bodies was distinguishable with tea leaves *k*-nearest neighbor (KNN) and partial least squares discrimination analysis (PLS-DA) model was having prediction of 98 and 100% respectively	Sun et al. (2021)
Powdered milk	Terahertz Time domain spectroscopy	Insects and plastic fragments	Source—conical horn antenna Frequency—210 GHz Detector—Schottky diode Imaging—high-resistivity (HR) Si beam splitter (transmission and reflective mode)	Insects and plastic fragments were detected by continuous wave sub terahertz waves	Ok et al. (2014)
Chocolate bars, red ginseng roots, walnuts, dried lavers	Terahertz Time domain spectroscopy	Metal washes, rubber band, pepper seed and polystyrene pieces	Source—IMPact ionization Avalanche Transit-Time diode source Frequency—140 GHz Detector—subwavelength	Cracks and presence of foreign materials such as metal washes, rubber band, pepper seed and polystyrene pieces were detected	Ok et al. (2018)

(continued)

Table 11.1 (continued)

Product	Terahertz principle	Detection of material used for	Specifications	Key findings	References
			beam and a transmitted beam based.		
Polyethylene and milk powder	Terahertz Time domain spectroscopy – pulsed mode	Melamine (1,3,5-triazine-2,4,6-triamin)	Source—Laser based Frequency—82 MHz Wavelength—780 nm Detector—high-resistivity Si wafer.	Melamine fingerprint was established at 2.11 and 2.39 THz frequency spectra from absorption coefficients	Cui et al. (2009)
Flour, milk powder, and chocolate powder	Terahertz Time domain spectroscopy	Melamine (1,3,5-triazine-2,4,6-triamin)	Frequency—0.1–3 THz	Absorption peaks observed at 2, 2.26, and 2.6 THz Calibrated curve fitting enabled the linear regression coefficient (R^2) of >0.913 and limit of detection <13%.	Baek et al. (2014)
Flour	Terahertz Time domain spectroscopy	Talc powder	Source—Ti:sapphire laser Frequency—0.2–1.6 THz Detector—photodiode detector	Absorption peaks for talc powder was 0.95 1.36, and 1.56 THz based on absorption coefficient and refractive index	Zhao Xiao-li and Li Jiu-Sheng (2011)
Honey	Terahertz Time domain spectroscopy	Antibiotics and insecticides	Source—Ti:sapphire laser	Antibiotics and insecticides in honey 3.25, 3.98, 4.53, and 5.16 THz absorption peaks for tetracycline 1.064 THz absorption and fingerprint for sulfapyridine	Massaouti et al. (2013)
Infant milk powder	Terahertz Time domain spectroscopy	Tetracycline hydrochloride, chlortetracycline	Source—Femto Fiber pro NIR laser Frequency—0.3–1.8 THz	Characteristic peaks for different classes of tetracycline Prediction model was	Qin et al. (2014)

		hydrochloride, oxytetracycline hydrochloride, doxycycline		considered to be excellent score based on residual predictive deviation (RPD) value of >3	
Chicken, rice and soils	Terahertz Time domain spectroscopy	Chlortetracycline hydrochloride and tetracycline hydrochloride	Source—Two distributed feedback (DFB) lasers Frequency—0.4–1.1 THz	Characteristic fingerprint peaks for chlortetracycline hydrochloride at 0.49, 0.60, 0.76, 0.90, 1.03 THz and 0.52, 0.67, 0.79, 0.92, 1.05 THz for tetracycline hydrochloride respectively	Wang et al. (2018)
Wheat grains	Terahertz Time domain spectroscopy	Moisture content	Source—Titanium—Sapphire laser Frequency—0.1–2.0 THz	Characteristic absorption peaks were absorbed for control and wetted wheat grains at 12%, 14% and 18% moisture levels	Chua (2005)
Sugar	Terahertz Time domain spectroscopy (Attenuated type)	Crystallization monitoring	–	Characteristic phonon bands were observed and crystallization initiated at 50 hour and attained peak crystallization at 58.5 hours	May and Taday (2013)
Pork meat	Terahertz Time domain spectroscopy	Quality inspection	Frequency—0.2–1.0 THz	Deteriorated meat not stored in refrigerated condition showed characteristic absorption peak at 0.42–0.46 THz	Huang et al. (2015)
Preservatives (sorbic acid, potassium sorbate, benzoic acid, sodium benzoate, sodium diacetate, and sodium dehydroacetate)	Terahertz Time domain spectroscopy	Preservatives detection	Source—Ti:sapphire laser Frequency—0.2–2 THz	Absorption spectra obtained for the preservatives—(Sorbic acid—0.93, 1.05, 1.2, 1.35, 1.49 THz), (potassium sorbate –0.7, 0.97,	Ge et al. (2016)

(continued)

Table 11.1 (continued)

Product	Terahertz principle	Detection of material used for	Specifications	Key findings	References
				1.44 THz), (benzoic acid-0.64, 0.88, 1.08, 1.85 THz), (sodium benzoate –0.56, 0.83, 1.61 THz), (sodium diacetate-0.85, 1.08, 1.29, 1.58 THz), (sodium dehydroacetate—1.23, 1.64 THz)	

absorption, transmission, and reflection are used to study the defects and characterize them as per the defined principles. Cereal grains and pulses immediately after post-harvest operations such as shelling and cleaning get into packaging materials and are stored for several months in godowns. It becomes essential to check the defects in terms of holed grains, any foreign infestation, or microbial infestation so that the produce is preserved in due period of storage. Harnessing the property of penetration power of terahertz, it is possible to detect the presence of these materials without dislodging and dismantling the packaging material, and it is possible to assess the spoilage pattern of the commodity by checking the temperature of the grains, as the microbial infestation generates a certain degree of heat component during the storage. The heat thickness can be detected with resolution using terahertz imaging. Foreign materials, including small stones, glass fragments, wood chips, and metal screws, were identified using terahertz radiation with the reflective mode in wheat grain and flour, and the imaging intensity increased with the increasing frequency. The detection of foreign materials further also depends on the wave source, the wavelength of the terahertz radiation, the distance in reaching the product matrix, and the angle of exposure of the waves to the commodity (Qin et al., 2013). Ok et al. (2014) used Gaussian beam focusing for identifying the insects, metals, and polymeric material and found that the transmission mode was able to provide a better recognition domain as against the reflective mode, which may have an overlapping effect.

Further, the ejected frequency waves tend to get attenuated due to the particle size and orientation of the foreign materials being aimed at. The used terahertz at 0.2 THz was able to distinguish between denser (aluminum and granite materials) and lighter (insects—maggots and crickets) in the instant noodles (Lee et al., 2012). The geometric distribution of dense materials was detected with edge lengths ranging from 1 to 5 mm, while lighter materials can be visualized up to the length of 8–22 mm.

Organic particulates such as hair are detected by terahertz spectroscopy, in which the presence of hair follicles reduces the transmission spectrum. Metallic and non-metallic components are found to be contaminated during chocolate production. Terahertz radiation, owing to its ability to penetrate low-moisture and fat-rich regions, is able to detect contaminants (Schuster et al., 2011).

11.7.2 Ensuring Quality of Perishables

By employing its sandwiched radiation spectrum, Terahertz could reduce the deteriorating quality of various fresh, perishable, and commonly consumed staples. Tomato fruits, owing to their softer texture, are prone to damage when stored inappropriately, leading to moisture loss and shrinkage effects. Ogawa et al. (2006) observed a decreasing pattern for the reflective radiation when the damaged portion was kept for scanning. Terahertz has a water absorption coefficient of 200 cm^{-1} (Qin et al., 2013), which can be used for the determination of moisture gain or loss in many of the products where moisture transition is important (e.g., chips, fries, biscuits, wafers, grains, pulses, etc.). Wheat, rice, and its derived products are prone to carry the residues of methomyl insecticide commonly used in fields as broad-spectrum insecticides for the control of foliar pest attacks. Methomyl is found to be a toxic insecticide belonging to the carbamate group and has been banned in various countries owing to its toxicity and ability to induce carcinogenicity. Baek et al. (2015) identified the fingerprint frequency for the methomyl at 1 THz with a regression coefficient ($R^2 > 0.974$) and limit of detection ($<3.74\%$). An extension of detecting this residual methomyl in apples was carried out by Lee et al. (2016), in which they successfully detected the methomyl to a lowered concentration of 1000 ppm with more excellent selectivity and sensitivity. They used a metamaterial approach using a nanoslot antenna array and reflection imaging tools. This use of nano metamaterials enhanced the absorptivity range for the terahertz spectrum. The stored grains and cereals are prone to microbial attack and moisture-induced sprouting. As a result, grains get spoiled, leading to the development of off-flavor and off-odor, which may present a significant health risk if undetected. Employing the PCA-SVM model for detecting stored wheat grains was able to predict an accuracy of 95% for four types of classification (normal, worm-eaten, moldy, and sprouting wheat grains) (Ge et al., 2014). Classifying grains and occasionally segregating them during the storage period can assist in saving huge quantities of materials.

Transgenic crops are still not successfully adopted technology in many crops and suffer significant hesitation for its use in many countries; therefore, identifying the transgenic food crop from the admixtures is greatly adhered to by the consumer segment and serves to increase consumer confidence of the consuming materials. Transgenic beetroots were detected and identified by (Liu 2017) using terahertz transmission spectrum combined with hyper sausage neuron pattern recognition technique (HSN), which provided a recognition percentage of 95% and rejection of non-similar beetroots among the same category of beetroots by 95.75% respectively. The uniform particulate blending of powder particles is essential for processing operations, and any smaller deviation in the particle size of the flour or powder leads to scattering with a corresponding change in the shape, pattern, and orientation in the outgoing waves that eventually get detected and quantified.

11.7.3 Effect on Various Food Biomolecules and States

The various food biomolecules, such as carbohydrates, starch, fats, and proteins, have a specific absorption coefficient for the terahertz waves. Terahertz, on account of its wavelength and frequency with higher electric field strength, can induce phonon, rotational, and vibrational movements in the molecular spectrum, which can be useful in detecting the conformational structures of many of the important biomolecules such as starch, proteins, fats, etc. (Chen et al., 2015). Moisture contributed by free and bound water regimes has a higher degree of absorption coefficient for the terahertz radiation. In the analyzed wafer sample by Parasoglou et al. (2009), the water component had a higher absorption coefficient for the incoming terahertz below 0.6 THz, but those values reduced on increased terahertz radiation above 0.6 THz, which indicates a stronger water interaction with terahertz wave above the critical moisture content of wafer sample. Soft organic objects in flour and powders were detectable to the range of 500 μm with enhanced spatial resolution. The functional food segments are increasing their space for wider nutritional content and higher bioactive components. Therefore, it becomes essential to quantify the respective nutritional components for quality and quantity to increase consumer acceptance and confidence.

One such work was carried out by Lu et al. (2016) in which the team quantified the concentration of glutamic acid and glutamine amino acids, which are key contributors to enhanced amino acid functionality and protein contents. The corresponding amino acids had terahertz absorption spectra at 0.3 and 1.9 THz, which were reduced compared with the binary solutions blending pure amino acids. The reduced spectral intensity is attributed to hindrance substances such as cellulose, proteins, and cell saccharides in the complex foxtail millet flours. Further, the precise quantification of biomolecules such as starch and proteins in complex food systems are found to be linked with the available moisture; hence, the hydration, water absorption, and water-holding capacity of the molecules are important. One such aspect was noticed in the muscle determination in fish. The striated muscle with lower fat and high water content could absorb higher water content than the adipose tissue with high-fat content (Bowman et al., 2016)—furthermore, the reconstructed electric properties identified both tissues' time and frequency domains.

Similar results were derived in rats and pigs' skin, fat, and lean tissues due to the differential composition of moisture, fats, and lean components in the absorption of terahertz waves (He et al., 2006). The vegetable and nut-derived oils for consumption undergo quality deterioration owing to oxidation, leading to the increase in free fatty acids and a corresponding increase in the values peroxide values (PV), thus producing rancidity problems. The time delay for the occurrence of fatty acids was identified by terahertz spectroscopy in transmission by sunflower seed oil, peanut oil, soybean oil, and rapeseed oil were reported respectively (Jiusheng, 2010).

Chocolate confectioneries are valued for their unique melting profile at varied temperatures, fat crystallization, and sugar solubilization, all being affected by the physical states of the ingredients present, temperature, water activity, and the polymorphic stages of certain ingredients such as fats and sugars. The terahertz could predict the state of the sample with the help of linear and nonlinear classifiers. With this, the state of samples, such as creaminess, can be predicted. Yet another study to monitor the crystallization of sugar was done by using THz pulsed imaging and ATR THz-TDS configuration wherein the radiation was found to get reduced from 0th hour to 58.5 hours owing to the buildup of crystallization of sugar particles (May & Taday, 2013). In this way, the phase transition behavior of complex molecules can be elucidated. This assumes that the terahertz waves can be used to estimate the storage of different phase particles such as emulsions and gel systems. The food quality of Chinese green tea was accurately predicted in conjugation with the use of Back propagation artificial neural network (BPANN), naive Bayes, and least squares support vector machines (LS-SVM) (Xi-Ai et al., 2011).

11.7.4 Detection of Additives and Adulterants

The terahertz spectroscopic technique can detect various antibiotics, preservatives, and synthetic additives in pure forms and their isomers. The additives such as thiabendazole, o-phenylphenol, and EDTA have been detected by terahertz spectroscopy. Absorption coefficients and refractive index were calculated for the different TBHQ concentrations in the range of 0.2–2.2 THz frequency, which helps in the assessment of TBHQ additive in the oil, which is widely used as synthetic antioxidants (Zhang et al., 2011). The vibration and rotational oscillations with the structural confirmations were responsible for quantifying six selected preservatives (sorbic acid, potassium sorbate, benzoic acid, sodium benzoate, sodium diacetate, sodium dehydrogenate) by the use of terahertz spectroscopy in the frequency range of 0.2–2 THz (Ge et al., 2016). Melamine commonly used in milk and milk powders, flour, and chocolate powders for protein enrichment has been detected using terahertz spectroscopy (Baek et al., 2014). It is essential to quantify and identify these molecules as they are prone to lead to carcinogenicity, kidney stones, and other health complications. In close association with dairy products' fat, rich products are prone to get adulterated with low-grade materials as fats are highly priced entities. Terahertz spectroscopy could distinguish between high-fat and low-fat products with 88.62% sensitivity detection (Liu, 2016). Honey is a highly-priced commodity that is prone to adulteration due to economic benefits. Terahertz is successfully used to detect the commonly used adulterants in honey. Liu et al. (2016) identified the presence of high fructose corn syrup in acacia honey samples using time-domain spectroscopy with the chemometrics approach.

Similarly, the geographic origin of extra virgin olive oils was analyzed using THz spectroscopy with a chemometrics approach. Peaks at 6.3 THz were observed when adulterated turmeric powder was tested with terahertz spectroscopy, confirming the yellow chalk powder's presence (Nallappan et al., 2013). Antibiotic residues, which are an important concern in meat and meat products, have been detected in real-time scenarios (Redo-Sanchez et al., 2011). Often, the mixing of specific metabolites or analytes of adulterated interest cannot be easily identifiable in the product matrix due to the interaction effects. This was evident in the identification of antibiotic residues such as tetracycline hydrochloride and oxytetracycline hydrochloride seen mixed with the milk powder, where there was a deviation in the absorption peaks as compared with the pure peaks (Qin et al., 2014). The used terahertz radiation with chemometrics approach of partial least square regression (PLSR) was able to identify the peaks in the powder sample. However, it did not produce a significant effect in the liquid milk sample owing to the relative absorption capacity of the terahertz waves by water component. Therefore a modification by introducing attenuated total reflectance (ATR) was introduced which was successful enough to predict the tetracycline hydrochloride by model built in using simple linear regression (SLR) with greater values of R^2 in the range of 0.95–0.98 and lower values of RMSE 0.61–0.99 mg mL^{-1}, with the limits of detection (LOD) ranging from 0.45 to 1.29 mg mL^{-1} (Qin et al., 2017).

11.7.5 Microbial Detection and Bioprocessing Control

The unique ability of terahertz waves to detect the presence or growth of microorganisms inside the packed systems has been beneficial in detecting microbial contamination as the conventionally used plating techniques generally consume more time for its growth and assay; this procedure is handy. It becomes accessible for the detection of microbes during the shelf life period. Greater sensitivity toward the detection of microorganisms was done using a metamaterial-based terahertz approach in aqueous and ambient environments that can be used as microbial sensors for real-time monitoring (Park et al., 2014). The changes in the density and the dielectric constants of microbes are sensed by the designed metamaterials (Cr and Au deposited on Si substrate), leading to a change in the inductive-capacitive resonant frequency. This can be extended to develop lab-on-chip devices that are popularly booming for onsite quality and safety assessment strategies. Nut crops suffer higher microbial damage during storage due to host insects and microbial communities inside the sliced fragments, which can propagate and cause a decay in their quality. This aspect has been worked on by Li et al. (2010), in which the inner slices of pecans were sorted with healthy and infected nuts by identifying live and dead insect populations at terahertz frequency greater than 1.2 THz. The insects were separated by an inner separator, shell, and nutmeat component (Li et al., 2015). The absorption coefficients of microbes and their particulates (cell wall, proteins, genetic material) towards the incoming terahertz radiation have been instrumental in detecting the wave change formats and contribute to their peculiar signatures. The characteristic signatures are unique for living and dead cells and vary with bacterial species' shape and cell wall recognition domains (Yang et al., 2016). Further, the recognition pattern for the microbes by selected terahertz frequencies mainly depends on the intracellular hydration levels (Wang et al., 2023). The ability to preserve meat can be successfully detected by using terahertz radiation, which reduces water content and decomposition of tissue matrix due to microbial contamination (Huang et al., 2015).

Brown et al. (2004) observed signature fingerprints for *Bacillus subtilis* at a terahertz frequency of 1.2 THz. The signature spectrum for calcium dipiconilate (a component of spore coat) in spore-forming bacteria *Bacillus subtilis* was identified at a frequency of 1.54 THz (Yu et al., 2005). Since there is a mismatch between the wavelength of terahertz (300 μm at 1 THz) and a typical bacterial cell (1–3 μm), the problem of lower scattering often leads to poorer resolution. This can be rectified by the use of metamaterials, which are artificially devised subwavelength structures known to enhance, modify, or alter the electromagnetic interaction of terahertz radiation with the sample and offer considerable flexibility in tuning terahertz radiation (Yang et al., 2014; Xu et al., 2022). This has advanced the use of terahertz spectroscopy to develop biosensors (Wessel et al., 2013).

Aflatoxins, which are considered to be carcinogenic secondary metabolites, are known to contaminate food products they can route their infection starting in the field or in the animal in the case of milk, and continue to contaminate, causing substantial health concerns and major export barrier, in many food crops. Ge et al. (2016) detected the aflatoxin B1 using terahertz spectroscopy with a limit of detections ranging from 1 to 50 μg mL^{-1} by subjecting to linear regression models. Microbial cultures differ with the type of sugar, organic, and inorganic nutrients provided for successful bioconversion into valuable byproducts. It is possible to characterize the growth profile of microbes in response to variations in the quantity and quality of input materials. Traditionally followed colony counting and cell suspension detection, though it provides reasonable knowledge on substrate depletion and growth kinetics, it is time-consuming and requires a trained workforce for execution. Employing terahertz radiation onsite can aid in faster detection of growth characteristics. Fawole et al. (2015) used terahertz frequency to monitor the growth condition of yeast suspensions to selected sucrose variants (regular sugar and synthetic sweetener sucralose). They observed an increase in the volume of CO_2 bubbles in the sucralose batch compared to the sugar-constituted batch. The growth and lysis patterns of yeast suspensions in bioreactors have been investigated with the terahertz radiation in which the dielectric constant property of yeast cells helped modify the permissivity of the terahertz waves (Fawole et al., 2015).

11.8 Future Trends

Terahertz spectroscopy has been used for food safety and quality assessment in numerous commodities for the detection, identification, quantification, and assessment of the quality status of the produce. Though appreciable works have been highlighted here with respect to terahertz physics, instrumentation, working procedure, and potential applications, the technology still has to adapt itself to certain portfolios by which potential stakeholders can increase commercial acceptance. The technology has significantly impacted the detection of foreign materials and adulterants to a very low sensitivity and detection limit. Yet, the process is limited to research and laboratory needs for various reasons, the cost factor being the most important as there is a huge money factor spent on the source and detector manufacturing; devising cost-friendly yet equivalent materials can enhance its uptake. The extreme sensitivity response of terahertz to an aqueous system by ionization of polar molecules presents a significant challenge in terms of scattering loss. However, maintaining vacuum conditions with inert nitrogen or the use of a sub-terahertz spectrum is a temporary solution; the design, fabrication, and modulation of terahertz wavelengths need to be understood in depth, and more research has to be drawn into reducing the water and hydration-induced scattering loss. Some reported study claims that the absorption capacity of water is 105 times higher in the terahertz region compared to the visible spectrum. Further, sub-terahertz technology limits the spatial resolution obtained compared to the terahertz counterpart. The source or emitter design and principles have largely undergone improvements and transitions from charge accelerator with solid state electronic based (gunn oscillators, backward wave oscillators, and Schottky diodes) to electro-optic based methods (photoconductive effect, different-frequency generation, optical parametric oscillation) but still; these methods suffer with higher power consumption for pump operation making it not portable and easily adaptable. Lower resolution at <0.3 mm presents a limitation for the use of the terahertz spectrum; however, the use of the millimeter wavelength approach presently resolves it. Though higher frequency bandwidth is obtained using the optical rectification method, this is always at the expense of power intake. Therefore, detection and imaging need to be prioritized when optimizing the power consumption and usage of the effective terahertz source.

Future research should reduce the size and make the system more compatible with the existing space requirements of the food industry, as most of the presently investigated terahertz spectroscopy is bulky and established in laboratory conditions. Waves, passing through the sample, undergo an imaging technique, which at present relies on the raster scanning technique to collect individual pixel images, which further undergo processing, which is a matter of time concern. More research to be focused on increasing the bandwidth of data acquisition, increasing the speed of data acquisition. Synergy in the scanning area and spatial resolution acceptance needs to be established. Some works have proposed blending terahertz spectroscopy with the scanning tunneling microscope to increase the speed and spatiotemporal resolution response. Few works have been done in developing a micro-lens array coupled PCA array for faster data acquisition (Pradarutti et al., 2008). The spatial resolution of terahertz imaging is still in its infancy due to the diffraction limit that sets in at the sub-wavelength range. Improvement in the investigations of micro and nanofabrication and plasmonic nano-structured arrays needs to be enhanced to provide significant scope and space for the increased photon absorption and detection limit. The signal-to-noise (SNR) loss factor in imaging is a matter of concern that needs to be addressed by increasing the power output and devising specific materials that serve to enhance the conversion efficiency. Advanced data processing and machine learning algorithms must be focused on providing greater imaging solutions that can enhance the information derived from the material under investigation. Present means of establishing sensitivity factor turns around the use of several chemometrics tools and multivariate analysis; however, to predict a greater accuracy for any of the inspections requires huge data sets that need to be revalidated and built into a uniform and sturdy model that can be globally used. However, by far, the present methods used are limited to laboratory and research needs with smaller sets of data, which may lead to biases and can represent only a smaller amount of variability; therefore, further steps are needed to develop a spectral library for each of the commodity-specific identification, which will enhance the trust and validation of the constructed model.

Concerning food applications, the terahertz has increased its footprints in several aspects but still needs refinement to increase the trust and gain validation of the process. The detection limit for the microflora enumeration is comparatively lower than the conventional culturing system (Liu et al., 2018). Though attenuated to,tal reflectance (ATR) mode has been reported in many foods owing to transmission and water absorption loss, the system faces huge technical limitations factor is increased abruptly leading to operational losses, and inconsistent results. Scattering of terahertz waves needs to be further researched, especially with the powder and flour-based particles, as the particle dimensions (size, shape, length, presence of sharp edges) are known to account for scattering loss. Though pulsed mode is used to capture the internal composition of the food matrix, it is often faced with the challenge of reduced imaging resolution. In addition, the resolution is very limited in far-field imaging compared with near-field imaging. Captivating these drawbacks as a future source of research shall increase the scope and potential application of the research in the food industry.

11.9 Conclusion

Food safety is getting recognition and researchable importance due to its ability to influence public health on a larger scale. Food, being a ubiquitous requirement and an essential part of human health and wellbeing, is prone to contamination, adulteration, and quality deterioration due to a number of reasons, starting with non-scientific post-harvest handling, storage, and processing conditions that may provide a welcome step for the growth and proliferation of pathogenic microbes resulting in inducing toxicity and making the food unsafe for consumption and further use. Though conventionally followed subjective evaluation techniques for food quality assessment provide considerable results, those methods possess limitations due to their inherent characteristic of not being able to cater to the requirement of a faster, more accurate, and non-destructive approach for food quality and safety assessment. Another reason for contemplating the importance of a non-destructive approach is the ability to reduce or lower the influence of human bias and train incompetency to carry out routine analysis and provide interpretation. Terahertz spectroscopy in this domain has increased research competencies in various commodities aimed at different purposes. Its longer wavelength is resistant to scattering and reflection issues, coupled with its non-ionizing radiation spectrum capable enough to penetrate longer depths of the sample without being dismantled. Due to its inherent material and wave characteristics, the radiation spectra can detect biomolecular events not restricted to rotation, vibration, spin, and phonon. It is being adjuvanted by its capacity to penetrate opaque, low absorbable, and non-conductible materials such as clothes, paper, cardboard, wood, masonry, plastic, and ceramics (Wallace et al., 2008). The classical terahertz generation and propagation source has evolved from gas laser to electric-based and advancement to optic, which can generate desirable frequency and radiate with beam splitters, lens, and mirror components. Optical-based source emission requires the use of an optical delay line and lock amplifiers, the major objective being to increase the signal-to-noise ratio (SNR) and provide time-gated information that can be Fourier transformed to frequency component. Metamaterials and combinations can increase the sensitivity and detection limit for the incoming terahertz radiation propagating through the material. The wave can be propagated in pulsed or continuous form, with the former providing a more accurate determination of transient waves with excellent resolution and sensitivity. The spectral information is coupled with chemometrics, multivariate analysis, and advanced statistical techniques for the accurate determination and prediction of sensitivity and specificity with the generation of models (Wang et al., 2017).

Further, the application of terahertz has explicitly been addressed in relation to the detection aim and the principle employed. The material that can be analyzed for terahertz spectroscopy includes adulterant detection in agri-food commodities (grains, pulses, nuts, oils, fruits, vegetables, meat, marine products, milk), processed food products, toxin, contaminants, and residues detection. Future considerations and present limitations for the large-scale adoption are presented, which can be future research activity and consideration that should be taken up, and that requires a potential collaboration between agri-food and supply chain operations with the advanced physical and imaging techniques with the aid of advanced analysis tools to predict and establish robust models.

References

Aboonajmi, M., Jahangiri, M., & Hassan-Beygi, S. R. (2015). A review on application of acoustic analysis in quality evaluation of agro-food products. *Journal of Food Processing and Preservation, 39*(6), 3175–3188.

Accorsi, R., Tufano, A., Gallo, A., Galizia, F. G., Cocchi, G., Ronzoni, M., et al. (2019). An application of collaborative robots in a food production facility. *Procedia Manufacturing, 38*, 341–348.

Afsah-Hejri, L., Hajeb, P., Ara, P., & Ehsani, R. J. (2019). A comprehensive review on food applications of terahertz spectroscopy and imaging. *Comprehensive Reviews in Food Science and Food Safety, 18*(5), 1563–1621.

Alexandrakis, D., Downey, G., & Scannell, A. G. (2012). Rapid non-destructive detection of spoilage of intact chicken breast muscle using near-infrared and Fourier transform mid-infrared spectroscopy and multivariate statistics. *Food and Bioprocess Technology, 5*, 338–347.

Ali, M. M., & Hashim, N. (2022). Non-destructive methods for detection of food quality. In *Future foods* (pp. 645–667). Academic.

Asada, M., & Suzuki, S. (2021). Terahertz emitter using resonant-tunneling diode and applications. *Sensors, 21*(4), 1384.

Avestisyan, Y., Zhang, C., Kawayama, I., Murakami, H., Somekawa, T., Chosrowjan, H., et al. (2012). Terahertz generation by optical rectification in lithium niobate crystal using a shadow mask. *Optics Express, 20*(23), 25752–25757.

Bacon, D. R., Madéo, J., & Dani, K. M. (2021). Photoconductive emitters for pulsed terahertz generation. *Journal of Optics, 23*(6), 064001.

Baek, S. H., Lim, H. B., & Chun, H. S. (2014). Detection of melamine in foods using terahertz time-domain spectroscopy. *Journal of Agricultural and Food Chemistry, 62*(24), 5403–5407.

Baek, S. H., Kang, J. H., Hwang, Y. H., Ok, K. M., Kwak, K., & Chun, H. S. (2015). Detection of methomyl, a carbamate insecticide, in food matrices using terahertz time-domain spectroscopy. *Journal of Infrared, Millimeter and Terahertz Waves, 37*(5), 486–4974.

Bajwa, N., Au, J., Jarrahy, R., Sung, S., Fishbein, M. C., Riopelle, D., & Taylor, Z. D. (2017). Non-invasive terahertz imaging of tissue water content for flap viability assessment. *Biomedical Optics Express, 8*(1), 460–474.

Banerjee, A., Chakraborty, C., & Rathi, M., Sr. (2020). Medical imaging, artificial intelligence, internet of things, wearable devices in terahertz healthcare technologies. In *Terahertz biomedical and healthcare technologies* (pp. 145–165). Elsevier.

Bawuah, P., & Zeitler, J. A. (2021). Advances in terahertz time-domain spectroscopy of pharmaceutical solids: A review. *TrAC Trends in Analytical Chemistry, 139*, 116272.

Benea-Chelmus, I. C., Zhu, T., Settembrini, F. F., Bonzon, C., Mavrona, E., Elder, D. L., et al. (2018). Three-dimensional phase modulator at telecom wavelength acting as a terahertz detector with an electro-optic bandwidth of 1.25 terahertz. *ACS Photonics, 5*(4), 1398–1403.

Benveniste, E., Vasanelli, A., Delteil, A., Devenson, J., Teissier, R., Baranov, A., et al. (2008). Influence of the material parameters on quantum cascade devices. *Applied Physics Letters, 93*(13), 131108.

Blanchard, F., Doi, A., Tanaka, T., & Tanaka, K. (2013). Real-time, subwavelength terahertz imaging. *Annual Review of Materials Research, 43*(1), 237–259.

Bowman, T., El-Shenawee, M., & Campbell, L. K. (2016). Terahertz transmission vs reflection imaging and model-based characterization for excised breast carcinomas. *Biomedical Optics Express, 7*(9), 3756–3783.

Brown, E. R., Bjarnason, J. E., Chan, T. L. J., Lee, A. W. M., & Celis, M. A. (2004). Optical attenuation signatures of Bacillus subtillis in the THz region. *Applied Physics Letters, 84*(18), 3438–3440.

Chen, S. L., & Guo, L. J. (2017). Terahertz pulse detection techniques and imaging applications. In *Terahertz spectroscopy-A cutting edge technology*. IntechOpen.

Chen, M., & Lijuan, X. (2014). A preliminary study of aflatoxin B1 detection in peanut oil by terahertz time-domain spectroscopy. *Transactions of the ASABE, 57*(6), 1793–1799.

Chen, Z., Yuan, L., Hefferman, G., & Wei, T. (2015). Terahertz fiber Bragg grating for distributed sensing. *IEEE Photonics Technology Letters, 27*(10), 1084–1087.

Chen, Z., Ma, X., Zhang, B., Zhang, Y., Niu, Z., Kuang, N., et al. (2019). A survey on terahertz communications. *China Communications, 16*(2), 1–35.

Chen, G., Liu, B., Lu, P., & Peng, Y. (2023). Rapid determination of ochratoxin a in black tea using terahertz ultrasensitive biosensor. In *Photonics* (Vol. 11, No. 1, p. 9). MDPI.

Collier, C. M., Stirling, T. J., Hristovski, I. R., Krupa, J. D., & Holzman, J. F. (2016). Photoconductive terahertz generation from textured semiconductor materials. *Scientific Reports, 6*(1), 23185.

Chua, H.S. (2005). "Terahertz time-domain spectroscopy of crushed wheat grain," in IEEE MTT-S Int. *Microw. Symp. Dig., Jun., pp.* 2103–2106. https://doi.org/10.1109/mwsym.2005.1517162

Cong, M., Li, W., Liu, Y., Bi, J., Wang, X., Yang, X., et al. (2023). Biomedical application of terahertz imaging technology: A narrative review. *Quantitative Imaging in Medicine and Surgery, 13*(12), 8768.

Cui, Y., Mu, K., Wang, X., Zhang, Y., & Zhang, C. (2009, August). Measurement of mixtures of melamine using THz ray. In *International symposium on photoelectronic detection and imaging 2009: Terahertz and high energy radiation detection technologies and applications* (Vol. 7385, p. 73851E). International Society for Optics and Photonic.

Du, H., Chen, W., Lei, Y., Li, F., Li, H., Deng, W., & Jiang, G. (2021). Discrimination of authenticity of Fritillariae Cirrhosae Bulbus based on terahertz spectroscopy and chemometric analysis. *Microchemical Journal, 168*, 106440.

El-Mesery, H. S., Mao, H., & Abomohra, A. E. F. (2019). Applications of non-destructive technologies for agricultural and food products quality inspection. *Sensors, 19*(4), 846.

Fan, K., & Zhang, M. (2019). Recent developments in the food quality detected by non-invasive nuclear magnetic resonance technology. *Critical Reviews in Food Science and Nutrition, 59*(14), 2202–2213.

Fawole, O., Sinha, K., & Tabib-Azar, M. (2015). Monitoring yeast activation with sugar and zero-calorie sweetener using terahertz waves. In *Sensors, 2015 IEEE* (pp. 1–4). IEEE.

Fu, X., Liu, Y., Chen, Q., Fu, Y., & Cui, T. J. (2022). Applications of terahertz spectroscopy in the detection and recognition of substances. *Frontiers in Physics, 10*, 869537.

Ge, H., Jiang, Y., Xu, Z., Lian, F., Zhang, Y., & Xia, S. (2014). Identification of wheat quality using THz spectrum. *Optics Express, 22*(10), 12533–12544.

Ge, H., Jiang, Y., Lian, F., & Zhang, Y. (2016a). Terahertz spectroscopy investigation of preservative molecules. *Optik, 127*(12), 4954–4958.

Geng, Z., Zhang, X., Fan, Z., Lv, X., & Chen, H. (2017). A route to terahertz metamaterial biosensor integrated with microfluidics for liver cancer biomarker testing in early stage. *Scientific Reports, 7*(1), 16378.

Gopalakrishnan, K., Sharma, A., Emanuel, N., Prabhakar, P. K., & Kumar, R. (2021). Sensors for non-destructive quality evaluation of food. In *Food chemistry: The role of additives, preservatives and adulteration* (pp. 397–449). Scrivener Publishing LLC, Wiley.

Guillet, J. P., Recur, B., Frederique, L., Bousquet, B., Canioni, L., Manek-Hönninger, I., et al. (2014). Review of terahertz tomography techniques. *Journal of Infrared, Millimeter, and Terahertz Waves, 35*, 382–411.

Hassoun, A., & Karoui, R. (2017). Quality evaluation of fish and other seafood by traditional and nondestructive instrumental methods: Advantages and limitations. *Critical Reviews in Food Science and Nutrition, 57*(9), 1976–1998.

Hassoun, A., Sahar, A., Lakhal, L., & Aït-Kaddour, A. (2019). Fluorescence spectroscopy as a rapid and non-destructive method for monitoring quality and authenticity of fish and meat products: Impact of different preservation conditions. *LWT, 103*, 279–292.

He, M., Azad, A. K., Ye, S., & Zhang, W. (2006). Far-infrared signature of animal tissues characterized by terahertz time-domain spectroscopy. *Optics Communications, 259*(1), 389–392.

Herrmann, M., Tani, M., Watanabe, M., & Sakai, K. (2002). Terahertz imaging of objects in powders. *IEE Proceedings-Optoelectronics, 149*(3), 116–120.

Ho, I. C., Guo, X., & Zhang, X. C. (2010). Design and performance of reflective terahertz air-biased-coherent-detection for time-domain spectroscopy. *Optics Express, 18*(3), 2872–2883.

Hu, J., Zhan, C., Shi, H., Qiao, P., He, Y., & Liu, Y. (2023). Rapid non-destructive detection of foreign bodies in fish based on terahertz imaging and spectroscopy. *Infrared Physics & Technology, 131*, 104448.

Hua, Y. F., & Zhang, H. J. (2010). Qualitative and quantitative detection of pesticides with terahertz time-domain spectroscopy. *IEEE Transactions on Microwave Theory and Techniques, 58*(7), 2064–2070.

Huang, S.-W., Granados, E., Huang, W. R., Hong, K.-H., Zapata, L. E., & Kärtner, F. X. (2013). High conversion efficiency, high energy terahertz pulses by optical rectification in cryogenically cooled lithium niobate. *Optics Letters, 38*(5), 796–798.

Huang, X. B., Hou, D. B., Huang, P. J., Ma, Y. H., Li, X., & Zhang, G. X. (2015, November). The meat freshness detection based on terahertz wave. In *Selected papers of the photoelectronic technology committee conferences held June–July 2015* (Vol. 9795, pp. 748–753). SPIE.

Jepsen, P. U., Cooke, D. G., & Koch, M. (2011). Terahertz spectroscopy and imaging–Modern techniques and applications. *Laser & Photonics Reviews, 5*(1), 124–166.

Jiusheng, L. (2010). Optical parameters of vegetable oil studied by terahertz time-domain spectroscopy. *Applied Spectroscopy, 64*(2), 231–234.

Kanda, N., Konishi, K., Nemoto, N., Midorikawa, K., & Kuwata-Gonokami, M. (2017). Real-time broadband terahertz spectroscopic imaging by using a high-sensitivity terahertz camera. *Scientific Reports, 7*(1), 42540.

Karaliūnas, M., Nasser, K. E., Urbanowicz, A., Kašalynas, I., Bražinskienė, D., Asadauskas, S., & Valušis, G. (2018). Non-destructive inspection of food and technical oils by terahertz spectroscopy. *Scientific Reports, 8*(1), 18025.

Kawano, Y. (2013). Terahertz waves: A tool for condensed matter, the life sciences and astronomy. *Contemporary Physics, 54*(3), 143–165.

Koch, M., Mittleman, D. M., Ornik, J., & Castro-Camus, E. (2023). Terahertz time-domain spectroscopy. *Nature Reviews Methods Primers, 3*(1), 48.

Lee, Y. K., Choi, S. W., Han, S. T., Woo, D. H., & Chun, H. S. (2012). Detection of foreign bodies in foods using continuous wave terahertz imaging. *Journal of Food Protection, 75*(1), 179–183.

Lee, D. K., Kim, G., Son, J. H., & Seo, M. (2016). Highly sensitive terahertz spectroscopy of residual pesticide using nano-antenna. In *Terahertz, RF, millimeter, and submillimeter-wave technology and applications IX* (Vol. 9747, pp. 118–123). SPIE.

Lewis, R. A. (2019). A review of terahertz detectors. *Journal of Physics D: Applied Physics, 52*(43), 433001.

Li, B., Cao, W., Mathanker, S., Zhang, W., & Wang, N. (2010). Preliminary study on quality evaluation of pecans with terahertz time-domain spectroscopy. Paper presented at SPIE Infrared, Millimeter Wave, and Terahertz Technologies (p. 78543V1–78543V6)

Li, C., Li, M., & Jiang, L. (2015). Research on Lascorbic acid and thiamine based on wideband terahertz spectroscopy technique. *Guang Pu Xue Yu Guang Pu Fen Xi, 35*(3), 595–598.

Lin, Y. J., & Jarrahi, M. (2020). Heterodyne terahertz detection through electronic and optoelectronic mixers. *Reports on Progress in Physics, 83*(6), 066101.

Liu, J. (2016). Terahertz spectroscopy and chemometric tools for rapid identification of adulterated dairy product. *Optical and Quantum Electronics, 49*(1), 1.

Liu, J., (2017). Determination of transgenic organisms from non-transgenic using terahertz spectroscopy and chemometrics. *Optik 131,* 885–891. https://doi.org/10.1016/j.ijleo.2016.1

Liu, W., Zhang, Y., & Han, D. (2016). Feasibility study of determination of high-fructose syrup content of acacia honey by terahertz technique. In *SPIE, infrared, millimeter-wave, and terahertz technologie* (pp. 100300J1–100300J6).

Liu, J., Xie, H., Zha, B., Ding, W., Luo, J., & Hu, C. (2018). Detection of genetically modified sugarcane by using terahertz spectroscopy and chemometrics. *Journal of Applied Spectroscopy, 85*, 119–125.

Liu, W., Zhang, Y., Li, M., Han, D., & Liu, W. (2020). Determination of invert syrup adulterated in acacia honey by terahertz spectroscopy with different spectral features. *Journal of the Science of Food and Agriculture, 100*(5), 1913–1921.

Löffler, T., Hahn, T., Thomson, M., Jacob, F., & Roskos, H. G. (2005). Large-area electro-optic ZnTe terahertz emitters. *Optics Express, 13*(14), 5353–5362.

Lu, S., Zhang, X., Zhang, Z., Yang, Y., & Xiang, Y. (2016). Quantitative measurements of binary amino acids mixtures in yellow foxtail millet by terahertz time domain spectroscopy. *Food Chemistry, 211*, 494–501.

Manley, M. (2014). Near-infrared spectroscopy and hyperspectral imaging: Non-destructive analysis of biological materials. *Chemical Society Reviews, 43*(24), 8200–8214.

Massaouti, M., Daskalaki, C., Gorodetsky, A., Koulouklidis, A. D., & Tzortzakis, S. (2013). Detection of harmful residues in honey using terahertz time-domain spectroscopy. *Applied Spectroscopy, 67*(11), 1264–1269.

May, R., & Taday, P. F. (2013). Crystallization of sucrose monitored by terahertz pulsed spectroscopy. In *Paper presented at 2013 38th international conference on infrared, Millimeter, and terahertz waves (IRMMW-THz)* (pp. 1–1). IEEE.

Mittleman, D. M. (2018). Twenty years of terahertz imaging. *Optics Express, 26*(8), 9417–9431.

Modupalli, N., Naik, M., Sunil, C. K., & Natarajan, V. (2021). Emerging non-destructive methods for quality and safety monitoring of spices. *Trends in Food Science & Technology, 108*, 133–147.

Nallappan, K., Dash, J., Ray, S., & Pesala, B. (2013). Identification of adulterants in turmeric powder using terahertz spectroscopy. Paper presented at international conference on infrared, millimeter, and terahertz waves (pp. 1–2).

Ogawa, Y., Hayashi, S., Kondo, N., Ninomiya, K., Otani, C., & Kawase, K. (2006). Feasibility on the quality evaluation of agricultural. products with terahertz electromagnetic wave. American Society of Agricultural and Biological Engineering. In *Annual international meeting Oregon Convention Center*. ASABE.

Ok, G., Kim, H. J., Chun, H. S., & Choi, S.-W. (2014). Foreign-body detection in dry food using continuous sub-terahertz wave imaging. *Food Control, 42*, 284–289. https://doi.org/10.1016/j.foodcont.2014.02.021

Ok, G., Park, K., Lim, M. C., Jang, H. J., & Choi, S. W. (2018). 140-GHz subwavelength transmission imaging for foreign body inspection in food products. *Journal of Food Engineering, 221*, 124–131.

Ok, G., Shin, H. J., Lim, M. C., & Choi, S. W. (2019). Large-scan-area sub-terahertz imaging system for nondestructive food quality inspection. *Food Control, 96*, 383–389.

Parasoglou, P., Parrott, E. P. J., Zeitler, J. A., Rasburn, J., Powell, H., Gladden, L. F., & Johns, M. L. (2009, September). Quantitative moisture content detection in food wafers. In *2009 34th international conference on infrared, millimeter, and terahertz waves* (pp. 1–2). IEEE.

Park, S. J., Hong, J. T., Choi, S. J., Kim, H. S., Park, W. K., Han, S. T., et al. (2014). Detection of microorganisms using terahertz metamaterials. *Scientific Reports, 4*, 4988.

Pattengale, B., Ostresh, S., Schmuttenmaer, C. A., & Neu, J. (2021). Interrogating light-initiated dynamics in metal–organic frameworks with time-resolved spectroscopy. *Chemical Reviews, 122*(1), 132–166.

Pawar, A. Y., Sonawane, D. D., Erande, K. B., & Derle, D. V. (2013). Terahertz technology and its applications. *Drug Invention Today, 5*(2), 157–163.

Peiponen, K. E., Zeitler, A., & Kuwata-Gonokami, M. (Eds.). (2012). *Terahertz spectroscopy and imaging* (Vol. 171). Springer.

Picó, Y. (2015). Mass spectrometry in food quality and safety: An overview of the current status. *Comprehensive Analytical Chemistry, 68*, 3–76.

Pradarutti, B., Matthäus, G., Riehemann, S., Notni, G., Nolte, S., & Tünnermann, A. (2008). Highly efficient terahertz electro-optic sampling by material optimization at 1060 nm. *Optics Communications, 281*(19), 5031–5035.

Qin, J., Ying, Y., & Xie, L. (2013). The detection of agricultural products and food using terahertz spectroscopy: A review. *Applied Spectroscopy Reviews, 48*(6), 439–457.

Qin, J., Xie, L., & Ying, Y. (2014). Feasibility of terahertz time-domain spectroscopy to detect tetracyclines hydrochloride in infant milk powder. *Analytical Chemistry, 86*(23), 11750–11757.

Qin, J., Xie, L., & Ying, Y. (2016). A high-sensitivity terahertz spectroscopy technology for tetracycline hydrochloride detection using metamaterials. *Food Chemistry, 211*, 300–305.

Qu, F., Lin, L., He, Y., Nie, P., Cai, C., Dong, T., et al. (2018). Terahertz multivariate spectral analysis and molecular dynamics simulations of three pyrethroid pesticides. *Journal of Infrared, Millimeter, and Terahertz Waves, 39*, 1148–1161.

Redo-Sanchez, A., Salvatella, G., Galceran, R., Rold, E., Castellari, M., & Tejada, J. (2011). Assessment of terahertz spectroscopy to detect antibiotic residues in food and feed matrices. *Analyst, 136*(8), 1733–1738.

Ren, A., Zahid, A., Fan, D., Yang, X., Imran, M. A., Alomainy, A., & Abbasi, Q. H. (2019). State-of-the-art in terahertz sensing for food and water security–A comprehensive review. *Trends in Food Science & Technology, 85*, 241–251.

Risacher, C., Güsten, R., Stutzki, J., Hübers, H. W., Büchel, D., Graf, U. U., et al. (2015). First supra-THz heterodyne array receivers for astronomy with the SOFIA observatory. *IEEE Transactions on Terahertz Science and Technology, 6*(2), 199–211.

Schuster, F., Coquillat, D., Videlier, H., Sakowicz, M., Teppe, F., Dussopt, L., Giffard, B., Skotnicki, T., & Knap, W. (2011). Broadband terahertz imaging with highly sensitive silicon CMOS detectors. *Optics Express, 19*(8), 7827–7832.

Sharma, R. K. (2017). Chapter: 5. In B. R. Gurjar Pravindra Kumar (Ed.), *Environmental pollution monitoring, modeling and control*. Studium Press LLC. Executive Editor Govil, J. N.

Shen, Y. C. (2011). Terahertz pulsed spectroscopy and imaging for pharmaceutical applications: A review. *International Journal of Pharmaceutics, 417*(1–2), 48–60.

Siles, J. V., & Grajal, J. (2010). Physics-based design and optimization of Schottky diode frequency multipliers for terahertz applications. *IEEE Transactions on Microwave Theory and Techniques, 58*(7), 1933–1942.

Sizov, F., & Rogalski, A. (2010). THz detectors. *Progress in Quantum Electronics, 34*(5), 278–347.

Smith, R. M., & Arnold, M. A. (2011). Terahertz time-domain spectroscopy of solid samples: Principles, applications, and challenges. *Applied Spectroscopy Reviews, 46*(8), 636–679.

Song, H. J., & Nagatsuma, T. (2011). Present and future of terahertz communications. *IEEE Transactions on Terahertz Science and Technology, 1*(1), 256–263.

Stantchev, R. I., Yu, X., Blu, T., & Pickwell-MacPherson, E. (2020). Real-time terahertz imaging with a single-pixel detector. *Nature Communications, 11*(1), 2535.

Su, W. H., He, H. J., & Sun, D. W. (2017). Non-destructive and rapid evaluation of staple foods quality by using spectroscopic techniques: A review. *Critical Reviews in Food Science and Nutrition, 57*(5), 1039–1051.

Sun, X, J. Li, Y. Shen, W. Li, (2021). Non-destructive Detection of Insect Foreign Bodies in Finishing Tea Product Based on Terahertz Spectrum and Image, Front. *Nutr. 8*, 1–9. https://doi.org/10.3389/fnut.2021.757491

Tan, P., Huang, J., Liu, K., Xiong, Y., & Fan, M. (2012). Terahertz radiation sources based on free electron lasers and their applications. *SCIENCE CHINA Information Sciences, 55*, 1–15.

Tian, H., Huang, G., Xie, F., Fu, W., & Yang, X. (2023). THz biosensing applications for clinical laboratories: Bottlenecks and strategies. *TrAC Trends in Analytical Chemistry, 163*, 117057.

Tuniz, A., & Kuhlmey, B. T. (2023). Subwavelength terahertz imaging via virtual superlensing in the radiating near field. *Nature Communications, 14*(1), 6393.

Tzydynzhapov, G., Gusikhin, P., Muravev, V., Dremin, A., Nefyodov, Y., & Kukushkin, I. (2020). New real-time sub-terahertz security body scanner. *Journal of Infrared, Millimeter, and Terahertz Waves, 41*, 632–641.

Valušis, G., Lisauskas, A., Yuan, H., Knap, W., & Roskos, H. G. (2021). Roadmap of terahertz imaging 2021. *Sensors, 21*(12), 4092.

Vaseashta, A. (2014). New THz technologies and applications in applications in support of safety and security. In *THz and security applications: Detectors, sources and associated electronics for THz applications* (pp. 277–292). Springer Netherlands.

Vinokurov, N. (2011). Free electron lasers as a high-power terahertz sources. *Journal of Infrared, Millimeter, and Terahertz Waves, 32*, 1123–1143.

Vitiello, M. S., Scalari, G., Williams, B., & De Natale, P. (2015). Quantum cascade lasers: 20 years of challenges. *Optics Express, 23*(4), 5167–5182.

Wallace, V. P., MacPherson, E., Zeitler, J. A., & Reid, C. (2008). Three-dimensional imaging of optically opaque materials using nonionizing terahertz radiation. *JOSA A, 25*(12), 3120–3133.

Walther, C., Fischer, M., Scalari, G., Terazzi, R., Hoyler, N., & Faist, J. (2007). Quantum cascade lasers operating from 1.2 to 1.6 THz. *Applied Physics Letters, 91*(13).

Wang, K., Sun, D. W., & Pu, H. (2017). Emerging non-destructive terahertz spectroscopic imaging technique: Principle and applications in the agri-food industry. *Trends in Food Science & Technology, 67*, 93–105.

Wang, Y, Q. Wang, Z. Zhao, A. Liu, Y. Tian, J. Qin, (2018). Rapid qualitative and quantitative analysis of chlortetracycline hydrochloride and tetracycline hydrochloride in environmental samples based on terahertz frequency-domain spectroscopy, *Talanta 190*, 284–291.

Wang, Q., Hameed, S., Xie, L., & Ying, Y. (2020). Non-destructive quality control detection of endogenous contaminations in walnuts using terahertz spectroscopic imaging. *Journal of Food Measurement and Characterization, 14*, 2453–2460.

Wang, K., Li, Z., Li, J., & Lin, H. (2021). Raman spectroscopic techniques for nondestructive analysis of agri-foods: A state-of-the-art review. *Trends in Food Science & Technology, 118*, 490–504.

Wang, J., Peng, L., Han, D., Zheng, T., Chang, T., & Cui, H. L. (2023). Label-free detection and identification of single bacteria via terahertz near-field imaging. *Frontiers in Microbiology, 14*, 1195448.

Wen, H., Kim, K.-J., Zholents, A., Byrd, J., & Cavalleri, A. (2013). Preface to special topic: Intense terahertz sources for 718 time-resolved studies of matter. *Review of Scientific Instruments, 84*(2), 022501.

Wessel, J., Schmalz, K., Gastrock, G., Cahill, B. P., & Meliani, C. (2013, April). Contactless investigations of yeast cell cultivation in the 7 GHz and 240 GHz ranges. In *Journal of physics: Conference series* (Vol. 434, No. 1, p. 012033). IOP Publishing.

Włodarska, K., Szulc, J., Khmelinskii, I., & Sikorska, E. (2019). Non-destructive determination of strawberry fruit and juice quality parameters using ultraviolet, visible, and near-infrared spectroscopy. *Journal of the Science of Food and Agriculture, 99*(13), 5953–5961.

Xi-Ai, C., Guang-xin, Z., Ping-jie, H., Di-bo, H., Xu-sheng, K., & Ze-kui, Z. (2011). Classification of the green tea varieties based on support vector machines using terahertz spectroscopy. In *Paper presented at 2011 IEEE international instrumentation and measurement technology conference* (pp. 1–5). IEE.

Xie, J., Zhu, X., Zang, X., Cheng, Q., Chen, L., & Zhu, Y. (2017). Metamaterial-enhanced terahertz vibrational spectroscopy for thin film detection. *Optical Materials Express, 8*(1), 128–135.

Xu, R., Xu, X., & Lin, Y. S. (2022). Electrothermally tunable terahertz cross-shaped metamaterial for opto-logic operation characteristics. *iScience, 25*(4), 104072.

Yang, K., Liu, S., Arezoomandan, S., Nahata, A., & Sensale-Rodriguez, B. (2014). Graphene-based tunable metamaterial terahertz filters. *Applied Physics Letters, 105*(9).

Yang, X., Wei, D., Yan, S., Liu, Y., Yu, S., Zhang, M., et al. (2016). Rapid and label-free detection and assessment of bacteria by terahertz time-domain spectroscopy. *Journal of Biophotonics, 9*(10), 1050–1058.

Yu, B., Alimova, A., Katz, A., & Alfano, R. R. (2005). THz absorption spectrum of Bacillus subtilis spores. In *Terahertz and gigahertz electronics and photonics IV* (Vol. 5727, pp. 20–23). SPIE.

Yu, L., Hao, L., Meiqiong, T., Jiaoqi, H., Wei, L., Jinying, D., et al. (2019). The medical application of terahertz technology in non-invasive detection of cells and tissues: Opportunities and challenges. *RSC Advances, 9*(17), 9354–9363.

Zeitler, J. A., Rades, T., & Taday, P. F. (2017). Pharmaceutical and security applications of terahertz spectroscopy. In *Terahertz spectroscopy* (pp. 323–348). CRC Press.

Zhan, H., Zhao, K., Bao, R., & Xiao, L. (2016). Monitoring PM2.5 in the atmosphere by using terahertz time-domain spectroscopy. *Journal of Infrared, Millimeter, and Terahertz Waves, 37*, 929–938.

Zhang, M., Cai, H., & Shen, J. L. (2011). Terahertz spectroscopic testing of food additive tert-butylhydroquinone. *Guang Pu Xue Yu Guang Pu Fen Xi, 31*(7), 1809–1813.

Zhao Xiao-li and Li Jiu-sheng (2011). Diagnostic techniques of talc powder in flour based on the THz spectroscopy. *J. Phys.: Conf. Ser.* 276 012234.

Chapter 12
Ultrasonics

S. Guberavishnu, S. Surender, Pranav Nanda Kishore, Ashish Rawson, C. K. Sunil, D. V. Chidanand, and S. Shanmugasundaram

12.1 Introduction

Ultrasonics refers to the study of sound waves or the application of sound waves, where the frequency of the sound waves range above the human audible limits usually 20 kHz. includes the generation of high-frequency acoustic and elastic waves and the effects of the transmission, the interaction between the waves and the matter, the application to use the energy, used to make measurements, and at higher powers to process or to change the material.

These high-frequency sound waves have numerous applications in various fields which include food processing, chemistry, meteorology, material science, biology, seismology, and more. The study of ultrasonics has evolved since the beginning stage improved by advancements in piezoelectric and electronic technologies. The history of ultrasonics started when Pierre and Jacques Curie discovered the piezo-electric effect in 1880 which emerged as the foundation for the generation and detection of ultrasonic waves Paul Langevin developed the first practical ultrasonic transducer to detect the submarines during the First World War, post-war ultrasonics Expanded to Medical Diagnostics, Non-destructive technologies and other fields (Ensminger & Bond, 2024).

S. Guberavishnu · S. Surender · P. N. Kishore · A. Rawson (✉) · C. K. Sunil · D. V. Chidanand · S. Shanmugasundaram
National Institute of Food Technology Entrepreneurship and Management, Thanjavur, TN, India
e-mail: ashish.rawson@iifpt.edu.in

C. K. Sunil et al. (eds.), *Emerging Non-destructive Technologies for Food Quality Analysis*, https://doi.org/10.1007/978-3-031-95048-3_12

12.2 Theory of Ultrasonics

It is based on the basic concepts of acoustics, wave mechanics, and the behavior of sound waves at high frequencies about 20 kHz. It consists of wave propagation, frequency and wavelength, attenuation and absorption.

12.2.1 Wave Propagation

Ultrasonic wave propagation is defined as the propagation of high-frequency sound waves through different mediums, including liquids, solids, and gases. These waves are mechanical, meaning they need a medium to travel, which is not similar to electromagnetic waves. The transmission occurs when the ultrasonic waves make particles in the media fluctuate back and front around their counterpoise positions. This particle movement is transferred from one particle to another, it creates a wave that travels through the media.

The velocity at which ultrasonic waves transmit depends on two key qualities of the media such as density and elasticity. The density of a media is a determination of how much mass is contained in a given volume. A packed medium slows wave propagation because more mass should be moved. The elasticity, or the ability of the media to get back to its original condition after distortion, also plays a critical role. A more elastic medium allows faster transmission as it more effectively propagates the mechanical fluctuations of the wave.

12.2.2 Frequency and Wavelength

The sound waves with frequencies above the human audible limit, starting at 20 kHz and reaching the gigahertz range are known as ultrasonic sound waves. The speed of sound in a given medium is constant and increasing the frequency of the wave results in an equivalent decrease in its wavelength. The inverse relationship between the frequencies and wavelength is crucial in several applications. For example, in medical ultrasound imaging, higher frequencies (with low wavelengths) are used to create images with better resolution, it allows for more detailed imaging of small structures within the body. These higher frequencies also have low penetrative depth in the media after propagation, meaning the absorbance speed is faster by the tissues. In non-destructive testing (NDT), high-frequency ultrasonics are used to detect small faults or defects in materials, as the short wavelengths can deal with and produce small features, providing accurate detection capacities (Kinsler et al., 2000).

12.2.3 Attenuation and Absorption

Attenuation in ultrasonic waves is essentially an energy loss while negotiating a medium, usually by absorption and scattering. These attenuation mechanisms are particularly important to the understanding of ultrasound in diverse environments, from medical diagnostics to NDT. It depends upon frequency in ultrasonic waves and properties such as density, elasticity, and inhomogeneities of the medium. The main attenuation mechanism is absorption, wherein the transformation of sound energy into heat occurs as a result of the internal friction of particles. Particles start vibrating due to the ultrasonic wave, and this mechanical energy is transformed into thermal energy. Viscosity, density, and molecular structure affect absorption. The absorption in biological tissues is due to the content of water and complex organic molecules which produce frictional losses. The increase in frequency gives rise to faster oscillations which increases the energy conversion to heat. Attenuation is more rapid with higher frequency in ultrasonic waves compared to lower ones, on account of their shorter wavelengths.

The absorption coefficient is a measure of energy loss resulting from the absorption of energy in a medium, due to its interaction with the same medium. Scattering contributes to ultrasonic attenuation through the deflection of waves by particles or inhomogeneities, by which energy is redistributed, while absorption dissipates energy into heat. Scattering depends on the size of scatterers relative to the ultrasound wavelength. If scatterers are much smaller, it is called Rayleigh scattering and it is frequency dependent. The intensity of Rayleigh scattering is proportional to the fourth power of the frequency, which means higher frequencies scatter easier. In medical ultrasound imaging, strongly scattering tissue structures are common. The strong scattering increases image contrast and details but is responsible for rapid attenuation. Mie-scattering occurs for larger scatterers and is even less dependent on frequency than on geometric properties. Total ultrasonic wave attenuation depends on absorption and scattering, which are both highly dependent on frequency. While frequency is increased, absorption and scattering increase attenuation. In soft biological tissues, the attenuation coefficient increases linearly with frequency while in more complex media it may follow a power-law. High-frequency ultrasound gives a better resolution but shallow depth of penetration. Frequency that is in the range of 2–5 MHz, is generally regarded as low for imaging purposes and this range gives deeper scans such as abdominal imaging where the tissue can be better penetrated. Higher frequencies (7–15 MHz) are superficial imaging frequencies with emphasis on detail rather than depth. Attenuation is crucial in ultrasound applications. The choice of frequency represents a balance between resolution and penetration. Tissue attenuation controls this balance. High-frequency waves give the best resolution but are limited to superficial imaging because they are rapidly attenuated.

Lower frequencies penetrate deeper but have lower resolution. In NDT, attenuation impacts flaw detection; thick materials benefit from lower frequencies for better penetration, though this reduces sensitivity to small defects. High frequencies are used to detect small surface flaws where attenuation is less. In summary, attenuation from absorption and scattering influences ultrasonic wave propagation. Its frequency dependence involves careful designing in ultrasonic systems to balance resolution against penetration. Understanding these principles forms the very basis of applying ultrasonics in medical imaging, non-destructive testing, and material science (Szabo, 2013).

12.3 Principles of Ultrasonics

The principles of ultrasonics are the theoretical knowledge about ultrasonic waves put into practice in medical imaging, industrial testing, cleaning, food and chemical processing (Mason, 2016; Wang et al., 2021; Choudhary & Rawson, 2021; Thirunavookarasu et al., 2022). Ultrasonic waves can be described as sound waves at a frequency greater than that above which humans can usually above 20 kHz. A number of basic principles control the generation, propagation, and interaction of these waves with materials. Such principles have value in the design and optimization of an ultrasonic device for specific applications.

12.3.1 Piezoelectric Effect

Usually, ultrasonic devices operate based on the principle of the piezoelectric effect. In materials of a piezoelectric nature, mechanical tension creates an electric charge and vice versa, which causes deformation to take place by the applied electric field. It has been found that the generation of an ultrasonic wave due to ultrasonic vibration will take place for piezoelectric materials such as quartz or lead zirconate titanate when an alternating electrical current is applied. These ultrasonic waves can easily interact with the piezoelectric materials and therefore generate electrical signals. Medical ultrasonography converts ultrasonic waves from a piezoelectric device into energy that strikes the body. These same crystals detect the echoes reflected from tissues and organs and convert these back into electrical signals; the latter is processed to form an image. Among others, these properties of piezoelectric ceramics include piezoelectric constant, dielectric properties, and mechanical strength, which contribute a great deal to ensuring better performance in ultrasound applications.

12.3.2 Acoustic Impedance Matching

Acoustic impedance is the opposition of a medium to sound waves and its capability of transferring waves into another medium. The denser the material, the more it's going to conduct sound at any one given speed.

The acoustic impedances differ across the boundary of the medium, and hence, this reflects ultrasonic waves and reduces transmission efficiency. In doing so, the ultrasonic energy in question reflects the importance that there needs to be increased transmission of the same to the desired target medium by matching the acoustic impedance of the two media as closely as possible. In fact, it is one of the fundamental principles in designing ultrasonic transducers and the choice of coupling agents. An example of medical imaging can be mentioned, where a transducer uses coupling gel to bridge an impedance gap between a ceramic ultrasonic wave generator and human tissue. The impedance of this gel to sound is close to that of skin and, therefore, it's way much better for the transmission of ultrasonic waves into the body than air. Thus, a better-quality image is obtained. Besides the medical applications, cavitation is applied in a bunch of industrial applications. For example, ultrasonic welding requires that the acoustic impedance of the materials to be welded should be matched to give effective transfer of energy and quality welds.

12.3.3 Cavitation

Cavitation is the formation and propagation of ultrasound waves in a liquid medium having rapid pressure changes. Such pressure changes cause the appearance, growth, and implosive collapse of small bubbles in the liquid. The bubble collapse leads to an increase in the magnitude of local forces such that extremely high temperatures and pressures, and consequently mechanical and chemical effects, occur. Precisely this fact constitutes the basis for the techniques of ultrasonic cleaning and sonochemistry. Cavitation in general acts to dislodge particles from surfaces in ultrasonic cleaning. In ultrasonic cleaning, ultrasonic energy removes dirt grease, and other contaminants from the surface of articles submerged in a cleaning solution by the implosion of the bubbles so created during the interaction between ultrasonic waves and the liquid. This process is widely used in the cleaning of delicate parts and tools without causing any kind of damage in the fields of medical, automotive, and electronics. Sonochemistry: Sonochemistry explains the branch of chemistry whereby chemical reactions are caused by ultrasonic waves. During a process, such cavitations may form ultra-reactive radicals which, in many instances, cannot even exist at normal conditions. Further, it has applied uses to material synthesis, water process treatment, and the pharmaceutical industry in enhancing reaction rates and yields (Suslick, 1990).

12.3.4 NDT-Non-Destructive Testing

One of the major applications of ultrasonics in material science and engineering is ultrasonic NDT, which makes use of ultrasonic waves in order to study subsurface characteristics of a material for the presence of any internal flaws, such as cracks, voids, and inclusions, without causing damage. Wave reflection at material boundaries is the principle in ultrasonic NDT. It means that ultrasonic pulses penetrate through the body until they reach some boundary, such as a crack or material interface, upon which at least part of the waves reflect back to the transducer. The flaws can be detected and sized according to the amplitude, frequency, and return time of the waves. Most of these factors come into play in optimizations of frequency and angle of incidence for the wave's actions at boundaries and defects. High-frequency waves have far better resolution in spotting minute flaws, but their penetration depth is the bare minimum. The lower frequencies give less information but deeply penetrate. Ultrasonic NDT therefore has wide applications in major industries such as aerospace, automotive, and construction, among others, for critical components and structures where integrity and safety are most vital. It also applies well to composite materials and metallic components in welding and pipeline inspection because of its great sensitivity, reliability, and non-destructive nature (Krautkrämer & Krautkrämer, 2013).

12.3.5 Therapeutic Ultrasound

It is an approach based on the deposition of energy. The diagnostic uses low-intensity waves to produce an image while therapeutic ultrasonics involves intense waves focusing on a particular target located inside the body to result in either localized heating or mechanical effects. This focused energy can be used for ablation of tissues, fragmentation of stones, and enhancement of drug delivery, and therefore it becomes an extremely powerful tool for non-invasive medical treatment. The uses of High-Intensity Focused Ultrasound (HIFU) are in the treatment of solid tumours, uterine fibroids, and kidney stones. The major principle of Focused Ultrasonics includes focused energy on the affected tissue to minimize the damage to the surrounding healthy tissues. At the focus, the ultrasonic waves converge to cause rapid elevation in temperature, leading to thermal ablation in the target tissue. Alternatively, mechanical effects such as cavitation can be produced that disrupt cellular structure or render cell membranes permeable to facilitate target drug delivery. Clinical effectiveness with HIFU depends on the target tissue's acoustic characteristics, the frequency of the ultrasound waves applied, and the intensity at which they are applied, coupled with the time duration. Therefore, most of the HIFU treatments require real-time treatment with advanced imaging modalities such as MRI, or ultrasound-controlled HIFU, for effectiveness and safety (Kennedy, 2005).

12.4 Techniques Used for Non-destructive Analysis Using Ultrasonic

Measurement of ultrasonic properties of materials can be achieved through a variety or diversity of measurement systems. The selection of a specific measurement system is contingent upon several factors, including the type of substance being tested, the condition of the structure, equipment availability, initial equipment purchase cost, maintenance expenses, desired speed and precision, and whether measurements will be conducted online or in specialized labs. Pulse-echo techniques are the most commonly used method for measuring the properties of the ultrasonics in the materials due to their ease of use, non-invasive nature, and ability to be automated. The use of continuous waves is common in research laboratories that require precise measurements. The different signals used in ultrasound analysis are shown in Fig. 12.1. Pulse techniques may be less precise than continuous waves, but they are still accurate enough for most applications related to foods.

12.4.1 Pulse Technique

If there is a method that is even simpler than that described above, it is the pulse-echo method, which is by far the most common technique for ultrasound measurements. Efforts have been made to develop other advanced pulse techniques to enhance the precision of the measurements. Nevertheless, the conditions under which these are performed are still the same as the pulse-echo method, and this method only shall be described with some very important modifications (Thurston & Pierce, 2012; McSkimin, 1964).

Ordinarily, a testing setup includes a measuring cell with a sample, a signal source, a device for generating ultrasound and monitoring devices like an oscilloscope (Fig. 12.2). The signal generator generates an electrical signal of a certain frequency, amplitude, and time duration. It is transformed and radiated from the transducer in the form of an ultrasonic pulse, which passes through the entire sample until it reaches the end wall of the measuring cell, where it is reflected to the transducer (Fig. 12.3). The transducer behaves as a receiver at this point and receives

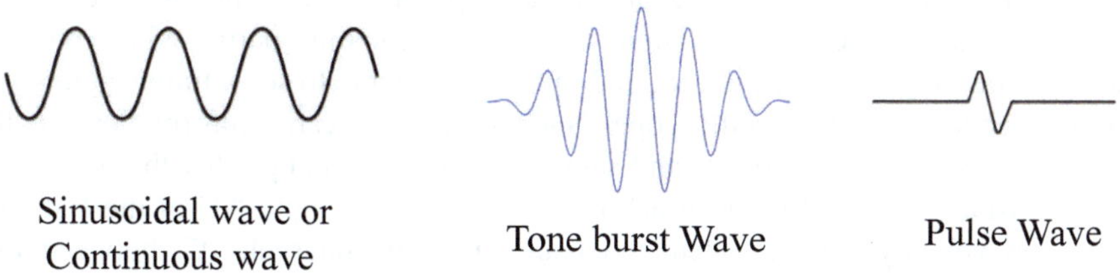

Fig. 12.1 Signals used in ultrasound analysis

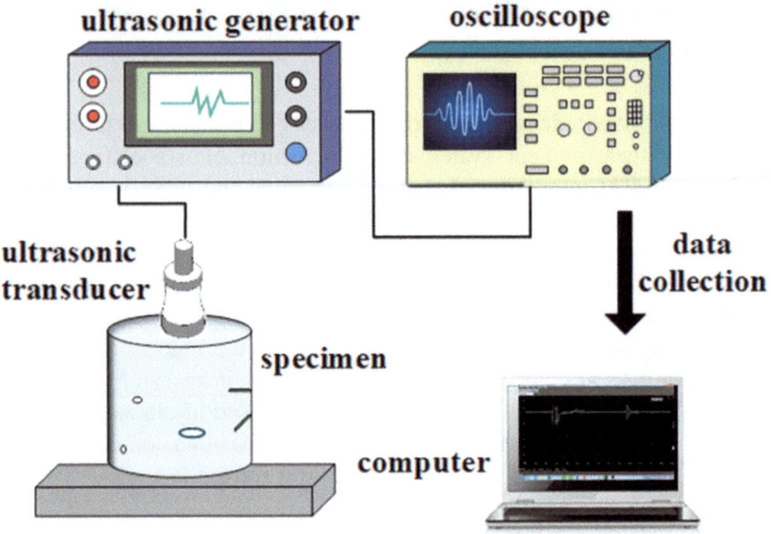

Fig. 12.2 Ultrasonic pulse-echo experiment setup (Tian et al., 2019). (Source: https://www.mdpi.com/1996-1073/12/24/4804#)

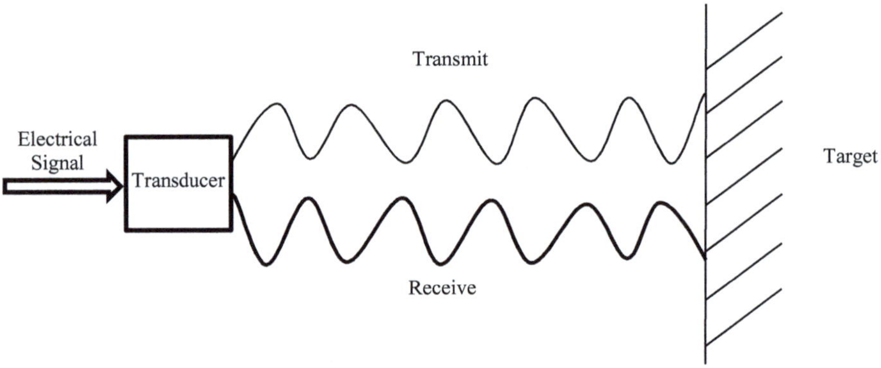

Fig. 12.3 Transducer converting the electrical pulse to an ultrasonic pulse

the converted ultrasonic pulse into an electrical pulse, which is output on the oscilloscope. As a result of the above description of how the oscillations are produced and observed, it is apparent that recording is only partial with respect to the structure of the wall of the cell, and thus it is possible to see a train or ghosts of multiple echoes on the oscilloscope. Such reflections carry information on the attenuation coefficient, velocity, and sample resistance, among other things.

The distance covered by each echo is twice the cell length, "d" of the last ball. Thus, the velocity can be calculated using the time interval, "t", between two successive echoes: $c = 2d/t$ where, c is the velocity of ultrasound, d is the cell length and t is the time. The cell length is typically determined by measuring the time taken for a medium with a known ultrasonic velocity, such as distilled water. A direct approach to determining the attenuation coefficient involves measuring the amplitude of the echo and the delay following the target signal, expressed as $A = A_0e^{-\alpha d}$ where A_0 is the initial amplitude of the signal, α is the attenuation coefficient, representing how quickly the signal diminishes and d is the distance the signal travels.

To determine the specific acoustic impedance of a sample, the amount of ultrasound energy reflected at the interface between the measuring cell and the sample is measured. Another common testing method is the transmission technique, which uses two ultrasonic transducers: one as a generator and the other as a receiver. The sample is placed between the transducers, and key details such as time delay and pulse energy spread through the sample are recorded. These parameters are then used to calculate the speed and propagation coefficient. The transmission method is particularly suited for soft samples that respond well to ultrasound, even when the sample pulse is partially masked by the larger transducer excitation pulse.

In many cases, especially for ultrasonic testing of food, it is essential to measure the frequency dependence of the sample's properties, as this provides more detailed information about the object. Pulsing techniques are used to scan frequencies in two ways. In the first method, a tone pulse from the generator is used to tune the transmitter to a specific frequency. The speed and ultrasonic coefficient are then determined by measuring the transit time and echo intensity at various frequencies.

The distance each echo has traversed equals two times the cell length 'd' of the immediate last ball. Hence velocity can be computed by considering the time 't' separating two successive echoes:

$$\mathbf{c = 2d/t}$$

Where,

- "c" is the Velocity of the ultrasound wave,
- "d" is one-way distance between the transducer and the reflector and
- "t" is the time it takes for the ultrasound to travel to the reflector and back.

The cell length is determined straightforwardly by measuring the time taken for a medium of known ultrasonic velocity, for example,

$$\mathbf{2d = C_w t_w}$$

Where,

- "t_w" is the time taken for the ultrasound to travel through the water and
- "C_w" is the known velocity of ultrasound in water.

A direct method of determining the attenuation coefficient is based on the measurement of the echo's amplitude alongside the delay following the target sign

$$\mathbf{A} = \mathbf{A_0 e}^{-\alpha d}$$

Where,

- "A_0" is the initial amplitude of the signal,
- α (alpha) is the attenuation coefficient, representing how quickly the signal diminishes and
- "d" is the distance the signal travels.

The determination of the specific acoustic impedance of a sample is done by measuring the amount of ultrasound energy reflected from the area between the measuring cell and the sample (McClements & Fairley, 1991).

Another method of conducting the test is the transmission method. This method usually involves two ultrasonic transducers, where one acts as a generator and the other as the receiver. The sample is sandwiched between the transducers, and the details like the time delay and spread of pulse energy over the sample are recorded. From these parameters, speed and the propagation coefficient are ascertained. The transmission method is applicable for soft samples, which are susceptible to ultrasound even though the sample pulse buried in the large transducer excitation pulse is portioned out. Many measurement requirements that involve food, especially for ultrasonic testing of the said food, would require measuring its properties' frequency dependence, as more information regarding the objects' properties is likely to be obtained. A tone pulse from the generator was used to tune the transmitter to a given frequency, after which the speed and the ultrasonic coefficient were determined by measuring the time of transit, while the intensity of echoes at different frequencies provided additional data on wave propagation characteristics (Javanaud & Robins, 1993).

12.4.2 Continuous-Wave Techniques

Many measuring ultrasound devices apply continuous wave ultrasound; however, an interferometer is the one the most used (McSkimin, 1964). The same basic components are used for continuous wave testing as for pulse testing: a transducer, a signal generator, an oscilloscope, and a measuring cell. However, here also the sequence is different. The sample is placed within a measuring cell between an ultrasound transducer and a mirror plate, and which plate is moved up and vertically through the sample. The signal generator produces an electromagnetic wave having a desired frequency and amplitude for transmission purposes. In essence, the transducer converts this ultrasonic wave into an ultrasound that travels through the sample and reflects back and forth between the reflector plate and the front plate of the transducer. Finally, standing waves are set up in the sample, and the signal strength picked up by the transmitter goes through cycles of maximums and minimums whenever the reflection plane is propelled up and down, alternating interference.

The factors of speed and deceleration are established from the analysis of the signal generated.

The distance between two adjacent peaks in a successive pattern is equal to X/2, where x is the wavelength of sound of the medium in which it is being calculated and the velocity can be determined. The amplitude reduces as the distance from the reflector to the transmitter increases owing to sample attenuation, boundary reflection, and diffraction effects. The coefficient of variation per cent is evaluated by plotting the graph of maximum vs. minimum amplitude distances with a sample and calibration material. It is possible to enhance the accuracy of measurements by taking the distance and the number of times the consecutive maxima are sampled.

The measurement frequency is defined by the resonance frequency of the piezoelectric crystal in the converter. While the transducer is typically operated at its fundamental resonant frequency, it can also function at its odd harmonic frequencies, such as 3f, 5f, and 7f. Not only does a converter work at its noise frequency, it can operate on other equivalents of noise frequency, but the power of generated signals decreases as frequency increases. So, a frequency sweep may be performed with the use of the continuous wave (cw) method with measurement of the resonant frequency, using the appropriate frequency input signals to drive the inverter. This procedure is more time- and energy-intensive than the Fourier transform method since the measurements have to be performed individually at every frequency.

12.4.3 Ultrasonic Imaging

Ultrasound imaging plays a pivotal role by offering images that show what's inside both living and non-living things in one, two, or even three dimensions (Fig. 12.4). Given its versatility, it naturally extends its benefits to the realm of food systems, offering insights that prove to be of great value. In the realm of imaging methods, short bursts of ultrasound are employed to distinguish between the echoes bouncing off the substance within the sample. Upon its journey, an ultrasound pulse comes

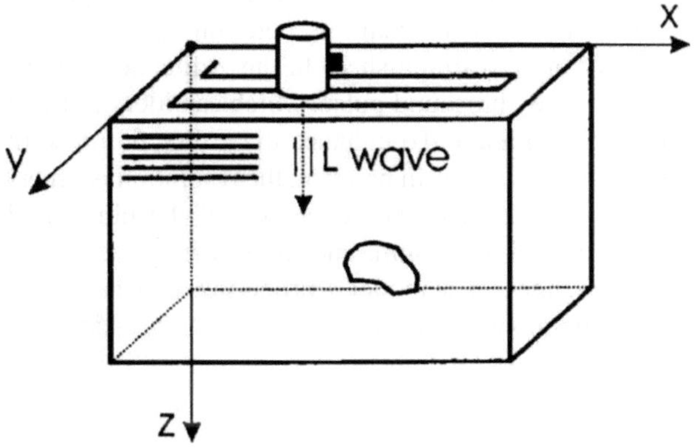

Fig. 12.4 Ultrasound can be used to create two- and three-dimensional images of food materials by physically moving the transducer (Spies & Jager, 2003). (Source:https://www.sciencedirect.com/science/article/abs/pii/S0041624X02004079)

across a boundary where two materials meet, each differing in how loudly they echo back. How long it takes for an echo to bounce back after being sent out hinges on how far it has to go before it hits something and comes back. Knowing how fast ultrasound waves move through a material lets one figure out the distance they've travelled based on how long they took to get there; by multiplying the speed (c) by the time (t), you get the distance ($d = ct$). Hence, it serves as a tool for pinpointing where things are kept within a folder. By looking at how loud the echoes are, one can learn about the resistance and pushback of different parts (Wells, 1981; Greguss, 1980).

There are several ways to create and show ultrasound images, with three common types being A-scan, B-scan, and C-scan. An A-scan measures the signal's intensity over time, which appears on an oscilloscope when an ultrasound pulse hits a sample and bounces back from an internal structure. The time it takes for the signal to return relates to the distance travelled, giving a one-dimensional image along the z-axis that helps detect objects. To gather more details about the internal structure, the transducer is moved in the x-y plane, taking an A-scan at each spot. A transducer is a mechanical device that can change its position using electric motors. The operator decides the area to scan and the distance between measurement points. This creates a series of A-scan data for each point in the x-y plane. The challenge is to present this data in a way that can be shown on a computer screen. B-scan and C-scan are two methods to do this. In a B-scan, a specific time-of-flight along the z-axis is chosen, and the signal's intensity is displayed on the screen in a grey scale. This is done for each A-scan in the x-y plane, resulting in a 2D image of the sample across the z-axis. In a C-scan, the focus is on a particular feature in the A-scan, measuring changes in flight time or amplitude as the transducer moves in the x-y direction. A greyscale is also used to show the flight time or range (Spies & Jager, 2003). The time it takes to capture all images varies based on how complex the imaging equipment is and how many measurements the operator chooses to take for each sample. Generally, obtaining a 100 × 100 A-scan image can take anywhere from a few seconds to a few minutes. Many ultrasound imaging devices have been created to quickly analyze systems that change rapidly. These devices utilize a mix of converters instead of a switch to produce the image. In the food industry, there are dynamic processes that can benefit from this type of imaging technology.

The clarity of an image is determined by the smallest distance at which two objects can be distinguished. In the z direction, the resolution is influenced by the length of the ultrasound pulse, which can identify two objects in water that are more than 1.5 mm apart. To enhance the z-direction resolution, a pulse of 10 ns can be generated. However, in practice, the resolution is often lower than this small distance because it also depends on the size of the ultrasound transducer, which typically ranges from 5 to 25 mm, and the shape of the beam it produces. This means that while changes in a model's properties can be detected quickly, the finer details within the model may not always be clearly visible.

12.5 Equipment Used for Non-destructive Analysis Using Ultrasonics

The main key components of ultrasonic measurement systems are the transducer, signal generator, digitizer, display, and measurement cell. An understanding of the operation of each of these components is a critical component of the interpretation of ultrasonic measurements and the design of the measurement systems.

12.5.1 On-Line Sensors

Online sensors are recognized for their ability to provide real-time information about food product conditions during processing. One of the most exciting future uses of ultrasound technology is internet-connected sensors. A web-based sensor should have several key features: it must measure quickly and accurately, be reliable and durable, remain affordable, and maintain hygiene while managing pressure and temperature during use. Additionally, it should have a data interface for on-site manufacturing and cleaning, allowing direct use of information by process control units. Ultrasound meets all these criteria, making it ideal for the food industry (Rizvi et al., 1993; Haley & Mulvaney, 1995; McClements, 1994).

A basic online ultrasonic sensor includes a transducer placed in a pipe through which food flows (Fig. 12.5). The transducer emits an ultrasonic pulse that travels through the food and reflects off the pipe's back wall, returning to the transducer. By measuring the time it takes for the echo to return and its amplitude, the speed of ultrasound and the feed rate can be calculated. Relevant equations are then applied to connect the ultrasonic measurements to the physical properties that food processors care about. A network of ultrasonic sensors can be set up to monitor food properties at various critical points in the process, giving manufacturers better control over the final product's characteristics. There are already several commercial ultrasonic sensors available that can analyze food material properties, primarily using ultrasonic velocity measurements at a single frequency to assess the composition of binary liquids and solutions. For more intricate tasks, like determining particle sizes or concentrations in emulsions, it is often necessary to analyze how ultrasonic properties change with frequency.

12.5.2 Transducer

Choosing the right transmitter is crucial when setting up an ultrasound exam. An ultrasonic transducer transforms electrical energy into mechanical energy, and in material testing, it typically features a piezoelectric crystal (Fig. 12.6). These piezoelectric materials generate an electric charge when they are bent, and they also

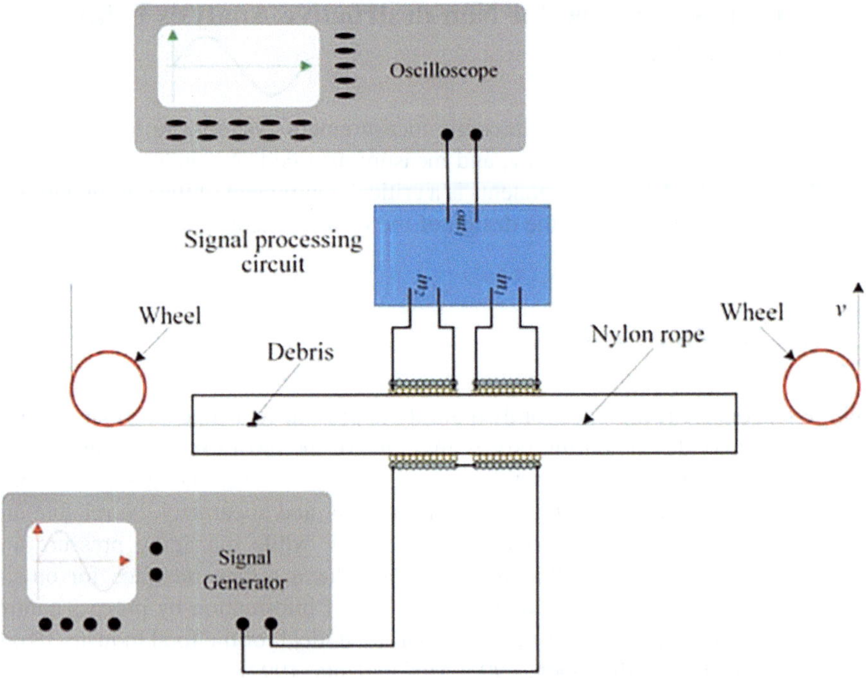

Fig. 12.5 Ultrasonic sensor fixed into the pipe (Wu et al., 2021). (Source: https://www.mdpi.com/1424-8220/21/22/7556)

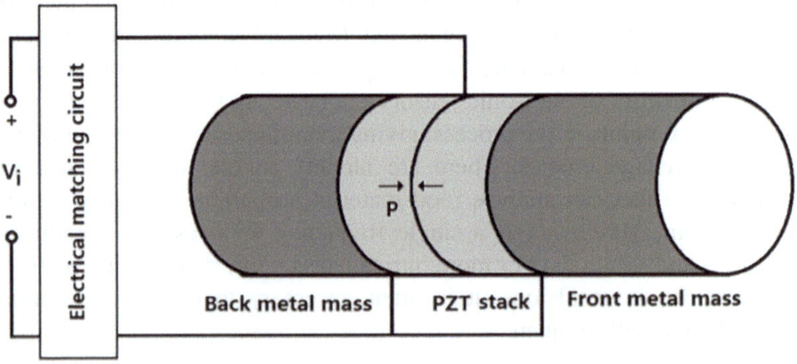

Fig. 12.6 Piezoelectric transducer (Lin & Zu, 2017). (Source: https://www.mdpi.com/1424-8220/17/2/329)

change shape when an electric charge is applied, functioning as both ultrasound generators and receivers. In industrial applications, these transducers are housed in metal casings, often with a glass faceplate for protection, and may include backing materials to absorb shocks. The design of the faceplate, including its acoustic impedance and thickness, is carefully selected to enhance the transducer's power output. When selecting a transducer, it is essential to consider its frequency range, diameter, and compatibility with audio signals.

The ultrasound frequency chosen for testing is based on the specific information needed from the material, and various testing conditions that influence measurement accuracy are also linked to the frequency used, which typically ranges from 0.1 to 100 MHz The frequencies that a transducer can generate are influenced by two main factors: the resonant frequency and the attenuation level of the piezoelectric crystal. The resonance frequency is affected by the thickness of the crystal; a thicker crystal results in a lower resonance frequency. The attenuation level also plays a role in determining the frequency range produced. An untuned transducer tends to have a prolonged "ring" when activated, focusing most of its energy in a narrow frequency band around the noise frequency, making it suitable for continuous-wave applications. This type of converter, known as a wide converter, is made from very thin glass and produces extremely short pulses, leading to a frequency range that is close to the harmonic frequency. Such converters are ideal for applications requiring narrow pulses or for conducting frequency scanning experiments using the Fourier transform method. To illustrate the difference between a modifier and a null modifier, consider a person striking a bell with a hammer; the bell rings for a long time, but if the edge is touched, the sound fades quickly due to the loss of energy (Greguss, 1980).

The size of the transducer affects the area of the material being tested in ultrasonic examinations. Typically, an ultrasound wave is perceived as moving through an object with a diameter matching that of the transducer. However, because the transducer is relatively small compared to the ultrasound wave, diffraction occurs. The ultrasound field can be categorized into two regions: the near field and the far field. The near field extends a certain distance from the transducer's front and is influenced by both the transducer's diameter and the ultrasound wavelength. Within this area, the ultrasound beam's diameter is roughly equal to that of the transducer. In contrast, the far field is where the ultrasound beam begins to spread out (Khimunin, 1972, 1975).

The resolution of an imaging device is determined by the minimum distance between two objects that can be distinctly identified. In the z-direction, resolution is influenced by the duration of the ultrasonic pulse; for instance, a 1-microsecond pulse can differentiate objects in water that are more than 1.5 mm apart. To improve resolution, pulses as short as 10 nanoseconds can be generated. In the x-y plane, resolution depends on the step size chosen by the operator between successive A-scans. However, in practice, the actual resolution is often lower than this theoretical minimum due to factors such as the diameter of the ultrasonic transducer (typically between 5 and 25 mm) and beam spreading. As a result, while variations in material properties over small distances can be detected, fine details of objects within the sample may not be clearly distinguished.

12.5.3 Signal Generator

The ultrasonic transducer receives its electrical input from a signal generator. The power input type varies based on the application and can be either continuous or pulsed. Various electronic components utilized in ultrasound tests are displayed. When setting up tests, parameters such as frequency, amplitude, and signal duration must be selected. The electrical input's frequency must align with the frequency range that the transducer can generate. For testing, electromagnetic waves are employed for Fourier transform tests, while short pulses and sound pulse tests use a beam with several chosen cycles. The pulse amplitude needs to be sufficiently high to produce a visible signal without damaging the sample.

12.5.4 Digitization Signal, Display, and Processing

Generating and transmitting an ultrasound wave through the sample, the resulting signal must be captured and analyzed. The signal that returns from the ultrasonic transducer is an analogue electrical signal, which can be shown directly on an analogue oscilloscope. However, it is preferable to convert this signal into a digital format before displaying it. This conversion is achieved using an analog-to-digital converter, which sends the data to a digital storage oscilloscope, a PC card, or another measuring device. When selecting an analog-to-digital converter, two key factors to consider are the speed of digitization and the resolution of the range. The sampling rate indicates how many points on the signal are sampled each second. To accurately represent the waveform, it should be sampled at a rate exceeding twice the highest frequency component of the analyzed signal. Typically, the maximum frequency in ultrasonic tests is about 100 MHz, so a sampling frequency of at least 200 MHz is recommended.

The digital rate indicates how quickly measurements can be taken between two points; a higher rate means a shorter measurement time. The digitizer's amplitude affects how well it can replicate the amplitude of the original analogue signal. Modern digital storage oscilloscopes typically have an 8-bit resolution. After the transmitter's signal is digitized, it can be shown on an oscilloscope or computer screen for analysis using various signal processing techniques, or it can be analyzed directly. To enhance the signal-to-noise ratio, averaging multiple signals can be beneficial. Digital smoothing can help eliminate excessive noise by replacing each signal value with the average of itself and its neighboring values. Filtering can be applied to remove unwanted frequency components by setting specific cutoff points. Cross-correlation techniques can help find the time interval between two signals. In ultrasound examinations, signals are usually represented in the time domain, showing amplitude over time. Understanding the frequency content of these signals is often important, and the Fourier transform is a mathematical method used to convert time-domain signals into the frequency domain (McSkimin, 1964; Challis et al., 1991).

This technique is based on the idea that a spatial signal can be expressed as a combination of waveforms with varying amplitudes and phases. By measuring the ultrasonic speed and material variation at different frequencies, one can effectively analyze the frequency content of broadband ultrasound pulses. This approach allows for the assessment of a material's ultrasonic properties across a broad frequency range more quickly than traditional ultrasonic pulse methods.

12.5.5 Measurement Cell

The design of the cell used for sample insertion is crucial for the success of an ultrasound application. It is important that the measuring cell has smooth, parallel walls that are thick enough and properly cooled. These parallel walls help avoid wave cancellation on the converter's surface, which can lead to either constructive or harmful interference in the signals. The walls need to be sufficiently thick to ensure that any ultrasound reverberations do not disrupt the signal being measured. Additionally, the cell should be made from chemically inert materials that can handle changes in temperature and pressure. Since the ultrasonic properties of materials can vary with temperature, it is essential to keep the measuring cell cool or to measure both the temperature and ultrasonic properties simultaneously for accurate adjustments (McClements, 1991).

12.6 Applications of Ultrasonics in Non-destructive Food Quality and Analysis

The utilization of ultrasonic analysis techniques as a non-destructive technique for the analysis of food quality has shown immense promise in recent years. This is due to the non-destructive nature of this technique and the relatively low cost of this technique compared to other analytical techniques. Various properties of the ultrasonic wave are measured and used as a basis to build predictive models for determining the properties of the target product. The properties measured are ultrasonic velocity, attenuation, time of flight etc. These properties can vary heavily depending on the material properties of the product. This variance can be used as a basis for online and non-destructive measurements of the target product.

Ultrasonic analysis has found many applications in the quality analysis of various products from very diverse product classes in the food industry. Some examples of applications in recent times, and some feasible methods which could take off in the future are as follows:

12.6.1 *Applications in the Meat Industry*

12.6.1.1 Ultrasonic Imaging of Live Cattle

Ultrasound scans, similar to those used for humans, can be used to check the quality of beef carcasses by checking the internal organs, fat and musculature of the live cow (Table 12.1). Specifically, the estimates of the back fat, fat near the 12th rib, and the fat near the longissimus and gluteus medius muscle. These scans can help to predict the cuttability and other quality parameters of the eventual carcass prior to slaughter. This can allow for reduced wastage and better livestock management (Paisley et al., 2007).

Similarly, the Ultrasound echo imaging techniques were utilized on live cattle to determine the beef marbling standard (Table 12.1). This standard is used to determine the grade of the beef carcass with respect to the amount and distribution of the intra-muscular fat in the meat. Since the value of the meat increases greatly with better marbling, a system to properly monitor the marbling of meat while the cattle are still alive is greatly beneficial to the farmers. Independent component analysis (ICA) and neural networks were utilized here to create an image recognition and analysis system. This system has shown similar and even greater accuracy in the correct estimation of the BMS of the cattle than the skilled and experienced inspectors (Fukuda et al., 2013).

12.6.1.2 Ultrasonic Monitoring of Chilling and Freezing Characteristics in Meat

Ultrasonic velocity has also been used for the characterization of the crystallization of fat in the back fat of pork samples during its cold storage (Table 12.1). This is an important parameter to monitor as this could affect the final texture and quality of the

Table 12.1 Application of ultrasound in non-destructive analysis/monitoring of meat

S. No.	Application	Type of ultrasound	References
1	Ultrasound imaging of living cattle	Ultrasound echo imaging	Paisley et al. (2007), Fukuda et al. (2013)
2	Monitoring of chilling of meat	Ultrasound velocity	Corona et al. (2014)
3	Monitoring of freezing of meat	Ultrasound pulse velocity	Mironovs et al. (2024), Grassi et al. (2024)
4	Testing of egg quality parameters	Ultrasonic phase velocities	Aboonajmi et al. (2014)
5	Detection of physical hazards in meat	Ultrasound echo imaging, pulse-echo technique	Schlager et al. (1991), Correia et al. (2008)
6	Monitoring of the dry salting	Ultrasonic velocity analysis	De Prados et al. (2016)
7	Determination of fat content & distribution in cured meat	Scanning acoustic microscopy	Corona et al. (2013)

pork meat. The results can be used to optimize the process parameters of the cold storage for a better final product. The system can also be used for continuous monitoring of the products (Corona et al., 2014).

Shock Freezing is a novel and popular method of freezing meat products (Table 12.1). This is due to the micro-crystallization of the water molecules leading to less cell ruptures and drip loss. However, the process parameters have to be maintained very strictly to ensure the proper crystal growth and prevent the growth of overly large ice crystals that would lead to cell ruptures and subsequently, tissue damage. Thus, Ultrasound propagation characteristics, specifically the ultrasound pulse velocity (UPV), waveform shapes, and intensity of ultrasonic signals, were measured and used to continuously monitor the meat characteristics while undergoing shock freezing (Mironovs et al., 2024).

Ultrasonic inspection using ultrasonic pulse velocity (UPV) in tandem with Near Infrared (NIR) analysis was proven to be a viable quality control measure during the freezing of meat products (Table 12.1). This method was used to monitor the formation of ice crystals in meat during the super-chilling process. This technique shows promise for better-controlled super-chilling for extending product shelf and for reducing drip losses. The simultaneous use of both UPV and NIR analysis leads to more precise and accurate monitoring to ensure better process control and thus higher quality final frozen meat products (Grassi et al., 2024).

12.6.1.3 Ultrasonic Analysis of Egg Quality

Ultrasound analysis has also been used for the non-destructive testing of poultry egg quality parameters (Table 12.1). Differences in Ultrasonic Phase Velocities (UPV) and peak values of ultrasound transmission times and frequency were used for the development of a model that can predict various parameters like albumin thickness, air cell height, and Haugh unit. These are very important parameters to determine the freshness and quality gradings of eggs. These are usually rather time-consuming and labour-intensive analyses and usually have to be conducted separately from each other. This method shows promise as it allows for rapid and non-destructive analysis of the eggs simultaneously predicting all the given parameters with reasonably high accuracy (Aboonajmi et al., 2014).

12.6.1.4 Ultrasonic Detection of Physical Hazards in Meat

Physical hazards pose a very real and dangerous threat to the food industry. These are physical objects that when consumed could cause direct damage to the body. The use of a small portable ultrasound device can be used to detect various foreign objects and contaminants in the meat sample that may be missed by visual analysis (Table 12.1). This allows for easy and effective quality control of the meat samples in the smaller-scale industries (Schlager et al., 1991).

The pulse-echo technique has been used to perform ultrasound scans of mechanically deboned chicken breasts to check for bone fragments in the meat (Table 12.1). This technique uses a Fast Fourier Transform Program to reduce noise in the signal. These bone fragments are a physical hazard that if left undetected could lead to harm to the consumers and subsequently a loss in reputation for the company, along with expensive legal troubles. Thus ensuring the proper monitoring of this is of utmost importance. This method shows promise as it can rapidly conduct quality control on the products and ensure that the hazardous products are properly detected and removed (Correia et al., 2008).

12.6.1.5 Ultrasonic Monitoring and Characterization of Salted and Cured Meats

Ultrasonic velocity analysis was also used for the online monitoring of the dry salting of pork muscle (Table 12.1). Low-intensity ultrasound was used for this method. The ultrasound system is able to monitor the saturation of the salt's penetration into the muscle fibers and the chemical changes that occur due to the same. This can ensure better quality salted products, and this same concept could possibly be used for the monitoring of the production of other salted or cured meat products (De Prados et al., 2016).

Scanning Acoustic Microscopy (SAM) is a non-destructive Ultrasonic analysis technique that was used to determine the Fat content and distribution in commercially sold dry-cured meat products like dry-cured ham and chorizo (Table 12.1). Air-coupled techniques, i.e. Without requiring the system to be in direct contact with the sample were proven to be effective at determining the characteristics of the sample (Corona et al., 2013).

12.6.2 Applications in the Edible Oil Industry

The authenticity of several edible vegetable oils like olive, sunflower, and corn can be tested by using ultrasound (Table 12.2). Since some of these oils are very expensive to produce and also highly sought after. Thus, there are a lot of bad actors who willingly adulterate the oils to increase their profits. These adulterations are usually hard to recognize, very time-consuming, labor-intensive and are destructive tests. Thus, the use of ultrasound analysis shows promises due to its rapid and non-destructive nature. The oils were distinguished by the measurement of the ultrasound velocity, attenuation and other frequency parameters. This data was then used to check for the purity of the oils and identify the presence of adulterants (Jiménez et al., 2023).

The authenticity of extra virgin olive oil was also tested by ultrasonic methods (Table 12.2). EVOO is a very expensive and sought-after product with very strict regulations and requirements. Thus the detection of adulterants in this case is very

Table 12.2 Application of ultrasound in non-destructive analysis/monitoring in other industry

S. No.	Industry	Application	Types of ultrasound	References
1	Edible oil industry	Testing the authenticity of olive oil	Ultrasound velocity and attenuation	Jiménez et al. (2023)
			Percentage of amplitude reduction, time of flight (TOF)	Zarezadeh et al. (2021)
2	Dairy	Analysis of fat content and dilution	Pulse-echo ultrasonic	Yang et al. (2021)
		Analysis of heat-induced coagulation	High-resolution ultrasonic resonator	Smyth et al. (1999)
		Analysis of rennet-induced pre-gelation and gelation	Ultrasonic velocity and attenuation	Dwyer et al. (2005)
3	Bakery	Rheology of bread dough	Ultrasonic velocity and attenuation	Lee et al. (2004)
		Rheology of biscuit dough	Low-frequency ultrasound	Kidmose et al. (2001)
		Analysis of breadcrumb	Ultrasonic machine vision system	Srivastava et al. (2015)
		Determine the physical properties of cake batters and cakes	Low-intensity ultrasonic systems	Gómez et al. (2008)

crucial and the benefits of ultrasonic analysis make it a promising candidate. In this case, the parameters measured were the percentage of amplitude reduction, time of flight (TOF), the difference between the first and second maximum amplitudes of the domain (in the time–amplitude diagram), and the ratio of the first and second maximum amplitude (Zarezadeh et al., 2021).

12.6.3 Applications in the Dairy Industry

The fat content and dilution of milk and cream products can be rapidly and non-destructively analyzed by the use of ultrasonic sensing systems (Table 12.2). A pulse-echo ultrasonic system is used to measure the ultrasonic velocity, and this was used to prepare a model to predict the fat content and dilution of the product. Since milk is consistently in high demand all the time, bad actors tend to adulterate and dilute the milk and cream to increase their profits. Ultrasound can be used to check for the adulterations of milk and cream products by quality control and regulatory bodies (Yang et al., 2021)

High-resolution ultrasonic resonator technique can be used to analyse the heat-induced coagulation of casein micelles in milk (Table 12.2). These coagulated micelles form a plaque on the pipelines and machineries over time which can lead to disruption of product flow and even contamination. So the control of this build-up is of paramount importance. The ultrasonic velocity and attenuation are measured in this system to predict the formation and accumulation of the coagulated micelles. This information can be used to optimize process parameters to reduce fouling of the piping and machinery by the coagulated milk residues. These systems also can be used to monitor the build-up of residue in the pipelines (Smyth et al., 1999).

Rennet-induced pre-gelation and gelation can also be analysed by ultrasonic techniques (Table 12.2). This gelation is of great importance in the process of cheese making. The characteristics of the final cheese are related to the level of gelation caused during the rennet activity stage of cheese production. Ultrasonic velocity and attenuation were measured and analysed along with NIR transmission and dynamic rheology measurements to predict the gelation levels in the sample. This method allows for the fast and non-destructive, online monitoring of the enzyme-aided coagulation step of the cheese production line (Dwyer et al., 2005).

12.6.4 Applications in the Bakery Industry

Ultrasonic techniques have shown promise as a method to test the rheological properties of fermented bread dough (Table 12.2). The ultrasonic velocity and attenuation of a longitudinal wave (P-wave) propagated through the dough samples were measured and analyzed to obtain the viscoelastic moduli of the dough. This data can be used for the optimization of the process criteria for the production of dough ideal consistency. This will be used to improve the final product qualities (Lee et al., 2004).

Low-frequency ultrasonic methods can also be used for the control of the rheological properties of biscuit wheat flour dough (Table 12.2). Here the ultrasonic velocity was measured and analysed to determine the viscoelastic properties of the dough. This can be used for the production of dough with perfect hydration, mixing ratio and other characteristics. This will lead to a better final biscuit product (Kidmose et al., 2001).

Quality analysis of bread samples was also tested using the ultrasonic method (Table 12.2). Here an ultrasonic machine vision system (UVS) was used to determine the breadcrumb features and freshness. This can be used for fast and accurate quality control for a better overall consumer experience (Srivastava et al., 2015).

Low-intensity ultrasonic systems can also be used to determine the physical properties of cake batters and cakes (Table 12.2). The ultrasound measurements were correlated with the density, viscosity, and rheology of the cake batters and the volume, symmetry, volume index, height and density of the cakes (Gómez et al., 2008).

12.7 Conclusion

The property of ultrasonics waves including the phenomenon of attenuation & absorption has made ultrasonics useful in different fields including medicine, non-destructive testing and analysis of food. Ultrasonication is finding increasing applications in various areas of food analysis and quality control including dairy industry, bakery industry, oil industry, poultry and meat industry etc. With improved efficiency and effectiveness, newer prospects have emerged for ultrasonication in the food industry that does not damage the product and evaluates its quality in the process. As the research continues, the potential of ultrasonics in meeting new demands as well as upgrading existing technologies is warranted.

References

Aboonajmi, M., Setarehdan, S. K., Akram, A., Nishizu, T., & Kondo, N. (2014). Prediction of poultry egg freshness using ultrasound. *International Journal of Food Properties, 17*(9), 1889–1899.

Challis, R. E., Harrison, J. A., Holmes, A. K., & Cocker, R. P. (1991). A wide bandwidth spectrometer for rapid ultrasonic absorption measurements in liquids. *Journal of the Acoustical Society of America, 90*(2), 730.

Choudhary, P., & Rawson, A. (2021). Impact of power ultrasound on the quality attributes of curd and its fermentation/gelation kinetics. *Journal of Food Process Engineering, 44*, e13698.

Corona, E., Garcia-Perez, J. V., Alvarez-Arenas, T. E. G., Watson, N., Povey, M. J., & Benedito, J. (2013). Advances in the ultrasound characterization of dry-cured meat products. *Journal of Food Engineering, 119*(3), 464–470.

Corona, E., García-Pérez, J. V., Santacatalina, J. V., Ventanas, S., & Benedito, J. (2014). Ultrasonic characterization of pork fat crystallization during cold storage. *Journal of Food Science, 79*(5), E828–E838.

Correia, L. R., Mittal, G. S., & Basir, O. A. (2008). Ultrasonic detection of bone fragment in mechanically deboned chicken breasts. *Innovative Food Science & Emerging Technologies, 9*(1), 109–115.

De Prados, M., Garcia-Perez, J. V., & Benedito, J. (2016). Ultrasonic characterization and online monitoring of pork meat dry salting process. *Food Control, 60*, 646–655.

Dwyer, C., Donnelly, L., & Buckin, V. (2005). Ultrasonic analysis of rennet-induced pre-gelation and gelation processes in milk. *Journal of Dairy Research, 72*(3), 303–310.

Ensminger, D., & Bond, L. J. (2024). *Ultrasonics: Fundamentals, technologies, and applications*. CRC Press. https://doi.org/10.1201/9780429286964

Fukuda, O., Nabeoka, N., & Miyajima, T. (2013, October). Estimation of marbling score in live cattle based on ICA and a neural network. In *2013 IEEE international conference on systems, man, and cybernetics* (pp. 1622–1627). IEEE.

Gómez, M., Oliete, B., García-Álvarez, J., Ronda, F., & Salazar, J. (2008). Characterization of cake batters by ultrasound measurements. *Journal of Food Engineering, 89*(4), 408–413.

Grassi, S., Jiménez, A., Ruiz, J., & González-Mohino, A. (2024). Use of ultrasound and NIRs as tools for monitoring ice formation in superchilled meat. *Journal of Food Engineering, 369*, 111957.

Greguss, P. (1980). *Ultrasonic imaging*. Focal Press.

Haley, T. A., & Mulvaney, S. J. (1995). Advanced process control techniques for the food industry. *Trends in Food Science & Technology, 6*, 103.

Javanaud, C., & Robins, M. M. (1993). Ultrasonic methods. In A. C. Pinder & G. Godfrey (Eds.), *Food process monitoring systems* (p. 129). Blackie Academic.

Jiménez, A., Rufo, M., Paniagua, J. M., González-Mohino, A., & Olegario, L. S. (2023). Authentication of pure and adulterated edible oils using non-destructive ultrasound. *Food Chemistry, 429*, 136820.

Kennedy, J. E. (2005). High-intensity focused ultrasound in the treatment of solid tumours. *Nature Reviews Cancer, 5*(4), 321–327.

Khimunin, A. S. (1972). Numerical calculation of the diffraction corrections for the precise measurement of ultrasound absorption. *Acustica, 27*(3), 173.

Khimunin, A. S. (1975). Numerical calculation of the diffraction corrections for the precise measurement of ultrasound phase velocity. *Acustica, 32*(3), 192.

Kidmose, U., Pedersen, L., & Nielsen, M. (2001). Ultrasonics in evaluating rheological properties of dough from different wheat varieties and during ageing. *Journal of Texture Studies, 32*, 321–334.

Kinsler, L. E., Frey, A. R., Coppens, A. B., & Sanders, J. V. (2000). *Fundamentals of acoustics*. Wiley.

Krautkrämer, J., & Krautkrämer, H. (2013). *Ultrasonic testing of materials*. Springer Science & Business Media.

Lee, S., Pyrak-Nolte, L. J., & Campanella, O. (2004). Determination of ultrasonic-based rheological properties of dough during fermentation. *Journal of Texture Studies, 35*(1), 33–52.

Mason, T. J. (2016). Ultrasonic cleaning: An historical perspective. *Ultrasonics Sonochemistry, 29*, 519–523.

McClements, D. J. (1991). Ultrasonic characterization of emulsions and suspensions. *Advances in Colloid and Interface Science, 37*, 33–52.

McClements, D. J. (1994). Ultrasonic NDT of foods and drinks. In W. J. McGonnagle (Ed.), *International advances in nondestructive testing* (Vol. 17, p. 63). Gordon and Breach.

McClements, D. J., & Fairley, P. (1991). Ultrasonic pulse echo reflectometer. *Ultrasonics, 29*(1), 58–62.

McSkimin, H. J. (1964). In W. P. Mason (Ed.), *Physical acoustics* (Vol. IA, p. 271). Academic.

Mironovs, V., Tatarinov, A., Abayev, A., & Zemchenkovs, V. (2024). Experimental system for investigating processes of shock freezing of meat. *Agronomy Research, 22*, 495–502.

Paisley, S., Loehr, C., & Niemala, F. (2007). *Ultrasound-based selection: Pitfalls and rewards*. Range beef cow symposium, paper 18.

Rizvi, S. S. H., Singh, R. K., Hotchkiss, J. H., Heldman, D. R., & Leung, H. K. (1993). Research needs in food engineering, processing, and packaging. *Food Technology, 47*, 26S.

Schlager, D., Sanders, A. B., Wiggins, D., & Boren, W. (1991). Ultrasound for the detection of foreign bodies. *Annals of Emergency Medicine, 20*(2), 189–191.

Smyth, C., Dawson, K., & Buckin, V. A. (1999). Ultrasonic analysis of heat-induced coagulation in calcium fortified milk. In *Trends in colloid and interface science XIII* (pp. 221–226). Springer.

Spies, M., & Jager, W. (2003). Synthetic aperture focusing for defect reconstruction in anisotropic media. *Ultrasonics, 41*(2), 125–131.

Srivastava, S., Vaddadi, S., & Sadistap, S. (2015). Quality assessment of commercial bread samples based on breadcrumb features and freshness analysis using an ultrasonic machine vision (UVS) system. *Journal of Food Measurement and Characterization, 9*(4), 525–540.

Suslick, K. S. (1990). Sonochemistry. *Science, 247*(4949), 1439–1445.

Szabo, T. L. (2013). *Diagnostic ultrasound imaging: Inside out*. Academic.

Thirunavookarasu, N., Kumar, S., & Rawson, A. (2022). Effect of ultrasonication on the protein–polysaccharide complexes: A review. *Journal of Food Measurement and Characterization, 16*, 4860–4879.

Thurston, R. N., & Pierce, A. D. (2012). *Ultrasonic measurement methods*. Elsevier.

Wang, Y., Ge, X., Ma, H., Qi, S., Zhang, G., & Yao, Y. (2021). Deep learning in medical ultrasound image analysis: A review. *IEEE Xplore, 9*, 54310–54324.

Wells, P. N. T. (1981). Pulse-echo diagnostic methods. In P. Papali (Ed.), *Medical physics* (p. 398). Elsevier.

Yang, Y., Wright, W. M., Hettinga, K. A., & van Ruth, S. M. (2021). Exploration of an ultrasonic pulse echo system for comparison of milks, creams, and their dilutions. *LWT, 136*, 110616.

Zarezadeh, M. R., Aboonajmi, M., & Ghasemi Varnamkhasti, M. (2021). Fraud detection and quality assessment of olive oil using ultrasound. *Food Science & Nutrition, 9*(1), 180–189.

Chapter 13
Electronic Nose and Electronic Tongue

Veerapandian Chandrasekar, Vijayaram Eyarkai Nambi, and Saravanan Shanmuga Sundaram

13.1 Introduction

Sensory analysis is a procedure to scientifically prompt, appraise, estimate and infer the human reactions and responses to a food product as they recognized through the five senses of colour, smell, texture, taste and sound. Sensory evaluation is done by methods of two conceptual such as 1. subjective and objective evaluation. Subjective sensory evaluation depends on views or human senses, whereas objective sensory evaluation uses instruments or defined processes. The objective methods are considered more reliable than subjective testing, since it does not allow for human bias or interpretation. Traditionally, sensory analysis is carried out by a group of people called sensory panelists in sensory booths. But it has time constraints and bottlenecks such as human error, bias and fatigue, and experience. Also, human panels cannot be used to sense toxic substances. It can be resolved with the use of technologies like electronic tongue and nose devices. Besides these, non-destructive sampling is adequate for E-nose and E-tongue analysis as they penetrate and interact with all kinds of food matrices such as solids, liquids and gases. These can be used in all applications like assess the sensory properties, quality analysis and control, optimize processes and product development and improve the inline and online

V. Chandrasekar (✉)
School of Sensory Sciences, National Institute of Food Technology, Entrepreneurship and Management, Thanjavur, India
e-mail: vchandrasekar@iifpt.edu.in

V. Eyarkai Nambi
Food Engineering Laboratory, National Institute of Food Technology, Entrepreneurship and Management, Thanjavur, India

S. Shanmuga Sundaram
Biosensor Laboratory, National Institute of Food Technology, Entrepreneurship and Management, Thanjavur, India

C. K. Sunil et al. (eds.), *Emerging Non-destructive Technologies for Food Quality Analysis*, https://doi.org/10.1007/978-3-031-95048-3_13

efficiency in manufacturing facilities, and automatically monitor the food regulations systems event at environmental and operational circumstances are subject to change due to easy hardware and software reconfiguration (Sha, 2023; Camacho-Olarte & Burgos, 2020; Wang et al., 2018). Electronic tongues and noses use sensors that work based on principles of piezoelectricity, electrochemistry, and optics. The chemical reaction between analyte and sensing electrode results in changes in voltage or current, which are detected by electrochemical sensors. The difference between reflectance, fluorescence, or absorption of the light by the analyte and the incidence light is measured by optical sensors. Similarly, the piezoelectric sensors employ crystal that vibrates in response to the analyte, producing a measurable electrical signal (Shaukat et al., 2023).

E-tongue, e-nose, and quantitative descriptive analysis (QDA) were used to predict sensory attributes and their relationship to apple juice quality (Bleibaum et al., 2002). With the use of an electronic panel (e-nose, e-tongue, and e-eye), the level of bitterness in virgin olive oil samples from various olive varietals was evaluated (Apetrei et al., 2010). The discrimination of the combined system performed better than that of the three systems separately, according to PCA. Strong correlations between the polyphenolic content and the e-tongue signals and the bitterness index produced by sensory panelists were shown by PLS regression models.

13.2 Electronic Nose

Any type of sensor that applies electronics principle to detect flavors or odors and is called electronic nose. The primary function of the E-nose is to identify a volatile compound's concentration as determined by olfactometry with sensor arrays and pattern recognition algorithms. A sensor array in e-nose generates a recognizable "fingerprint" of odor. E-nose consists of three systems such as sample administration, recognition, and computing the assessment (Behera et al., 2019; Calvini & Pigani, 2022; Rajamäki et al., 2006). In the 1990s, engineers at Warwick's School of Engineering invented, developed, and manufactured the first electronic nose (also known as e-nose) instruments for commercial use. Sweet potato flavors are determined using HERACLES GC Flash electronic nose (AlphaMos, France) to help the researchers to develop new variety with rich tastes and odors that are palatable to consumers (Ravi et al.,2022). There was a positive association between the E-nose and GC-MS data that offers to lay a foundation for the differentiation of various beef jerky formulations developed using native starter cultures for fermentation (Wen et al., 2021). An integrated soft E-nose methodology was created to increase efficiency and decrease complexity in efficient categorization of beef meat (Pulluri & Kumar, 2022). E-nose is used to identify, characterize, and grade fruits and vegetables of different cultivars and types by quickly and consistently evaluating simple and complicated volatile mixes. E-nose can be used to evaluate fruit, determine the freshness of fruits and vegetables, estimate product shelf life, discover faults and diseases, and monitor processes (Wilson, 2013). E-noses are widely used to monitor many smell-related processes, such as the fermentation of cocoa beans and tea (Yang et al., 2018), online fermentation monitoring of wine (Littarru et al., 2024). Assessment of chemical compounds like caffeine content in coffee samples (Aghdamifar et al., 2023), monitoring and predicting degree of coffee roasting process (Romani et al., 2012), To assess the physicochemical properties and flavor profiles of kiwi fruit (Wang et al., 2023), used for classification of fruits (Tyagi et al., 2023).

13.2.1 Principle of E-Nose

When the volatile molecules interact with the sensors, the conductivity and other electrically related characteristics are varied. The variation in the electric characteristics is identified using pattern recognition algorithms for classification or discrimination (Tan & Xu, 2020). E-nose is a type of fast GC equipment that distributes analyte molecules across the mobile and stationary phases based on their chemical structure (Fig. 13.1). The velocity of sample components in the column is determined by their equilibrium distribution between mobile (gaseous) and stationary (liquid) phases. Components with a higher affinity for the stationary phase move at a slower rate than those with a higher affinity for mobile phase. Hence, the separation of components is due to the varied migration rates caused by different partition coefficient values. The partition coefficient (K_s) is described using the Nernst equation (Eq. 13.1). The molecules with higher affinity to the stationary phase have the higher value of (K_s) and retention time (t_R).

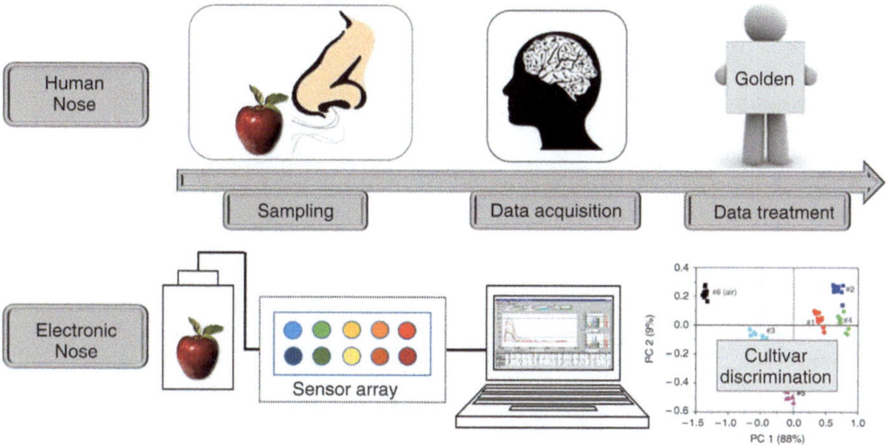

Fig. 13.1 Graphical illustration of E-nose and Olfactory system (Beghi et al., 2017)

$$K_s = \frac{C_L}{C_G} \tag{13.1}$$

Where,

C_L is the concentration of the substances in the stationary and
C_G is the concentration of the substances in mobile phases, respectively.

13.2.2 Working Mechanism of E-Nose

E-nose follows headspace analysis method. In the sample delivery system, a small quantity of samples is filled in a vial and sealed with a septum and kept for generation of vapour layer/ head space at a suitable high temperature. The low boiling chemicals start to move from the sample and the headspace layer, eventually approaching equilibrium. The time and temperature to generate head space is completely dependent on the sample and it is experimentally determined. Then the vapour layer or head space of the sample is taken by a syringe and inserted into the detector of the sensing system. The head space is introduced into the E-nose by an autosampler, and a carrier gas transports it along the stationary phase/column. In the detection system, when the head space contacts with the sensors, an electron or two electrons are removed from the head space to produce charged ions and electrical signals are generated. In the computing system, the electrical signals produced by various sensors are converted from analog to digital format using an A/D converter. The signals are processed by applying either signal amplification or noise reduction to produce a chromatograph/Spectrograph (Fig. 13.2).

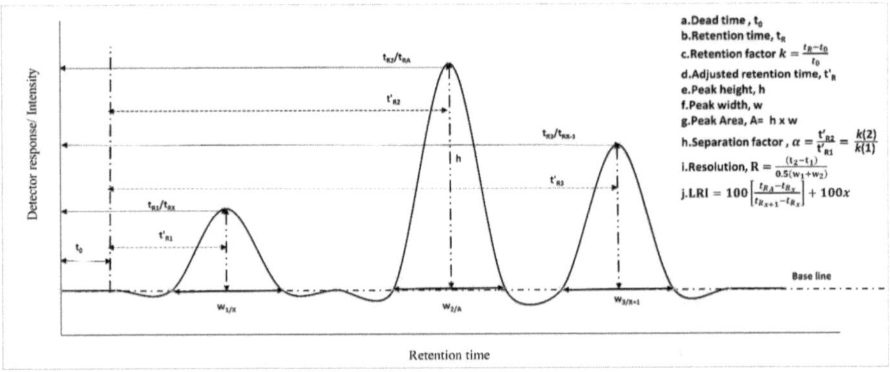

Fig. 13.2 Model chromatograph or spectrograph of E-nose and E-tongue

13.2.3 *Types of Sensors*

Electronic aroma detection sensors such as metal oxide semiconductors, piezoelectric sensors/ quartz crystal microbalance, optical sensors, and conducting polymers are used in electronic nose. Besides these other sensors like bulk acoustic wave, micro-electromechanical systems, surface acoustic wave, complementary metal oxide semiconductor, and carbon black composite were also used sensors (Wilson, 2013).

13.2.3.1 Metal Oxide Sensors

The working principle is identifying alterations in conductivity caused during adsorption of gas molecules in the column. Chemically adsorbed oxygen species are present in the oxide materials of sensors. These species interact with gassy particles on the surface of metal oxides modifies the oxide's conductivity. The amount of volatile organic compounds that have been adsorbed is represented by the change in conductivity, which is measured with an ammeter (Gardner & Bartlett, 1999). The oxides conductivity varies on the nature of VOC that reacts with the adsorbed oxygen on the semiconductor, the particle size of the metal oxide, and the temperature at which the sensing occurs (Maekawa et al., 2001).

13.2.3.2 Metal Oxide Semiconductor

The working principle of the Metal oxide semiconductor (MOSFET) is that molecules that have been separated from the column enter the sensor area by diffusion through a porous gas-sensitive material (gate material) made up of catalytic metal and get charged either positively or negatively (Eisele et al., 2001). These charged molecules alter the electric field inside the MOSFET and produce electric signal. Conducting polymers, such poly (ethylene-co-vinyl acetate), poly (styrene-co-butadiene), and poly (9-vinyl carbazole), are frequently utilized as gate materials. Catalytic metals, such as platinum (Pt), palladium (Pd), and iridium (Ir), can also be used (Chen et al., 2013).

13.2.3.3 Piezoelectric Sensors/Quartz Crystal Microbalance

Gas adsorption onto the polymer surface causes a change in mass on the surface of the sensor. The alteration in mass on the surface of the sensor is measured as resonance frequency by crystal impedance meters or transmission systems (Oh et al., 2011). The mass of the crystal is increased, and the oscillation frequency associated with the analyte concentration is changed by enabling the adsorption of gaseous components on the crystal through selective coating. There are several types of piezoelectric sensors, including quartz crystal microbalance (QCM), shear horizontal acoustic plate mode (SHAPM), flexural plate wave (FPW), surface acoustic wave (SAW), and bulk acoustic wave (BAW).

13.2.3.4 Conducting Polymers

The working principle of conductive polymer gas sensors is altering electrical resistance during adsorption of gases on the surface of the sensor. The physiochemical features of vapors cause different reactions in different polymers due to the properties of CPs such as doping concentration, water content, protonation level, and dopant ion size (Chen et al., 2013). CP sensors can react to volatile organic compounds (VOCs) more quickly at ambient temperature than MOS, which functions at high temperatures (Bai & Shi, 2007).

13.2.3.5 Optical Sensors

The optical sensors work based on stimulating a light source by the volatile analyte to produce a light signal. The lights signals are travelled from sensors to photo detectors using appropriate optics. The photodetector detects the light signals coming from the sensor by the principles of absorbance, fluorescence, polarization, refractive index, interference, scattering, and reflectance (Oh et al., 2011). Optical sensors can be equipped with a broad range of light sources, such as narrow-band solid-state light-emitting diodes (LEDs), broad spectrum band incandescent lamps, and highly coherent gas and semiconductor diode lasers (Chen et al., 2013). Other than these, amperometric, chemoresistor, chemocapacitor, gravimetric sensors are used in E-nose.

13.3 E-Tongue

Taste is a key factor in judging foods quality. Taste is the sense connected to the existence of compounds that arouse the tongue's taste buds. According to taste perception, basic tastes are divided into four categories: salinity, which shows an electrolyte balance in food; acidity, which signals decomposition; bitterness, which inhibits the consumption of toxic substances; and umami, which indicates the existence of amino acids. and sweetness might serve as a sign of nutrient supplies (Titova & Nachev, 2020).

The Electronic Tongue (E-Tongue) technique is made up of several sensors that react to chemical solutions, and it is made up of a measuring module, an electrochemical cell (sensor array), and a pattern identification arrangement that can identify both simple and complicated taste of molecules (Titova & Nachev, 2020). E tongue creates the taste profile by identifying all the flavor-producing molecules in liquid samples. The e-tongue evaluates the basic flavors of sweet, bitter, sour, and salt, and umami, metallic, sharp, and astringent tastes as well. The electronic tongue dissects a sample into its constituent parts to retrieve a signal signature. In a brief amount of time, the sensor array of an E-tongue produces multidimensional information and identifies the various patterns of chemical classes that give a flavor. The E-tongue measurement produces complicated data and, multivariate statistical analysis is used to process it (Di Rosa et al., 2020b). Examination the flavor and taste of Italian wine (Legin et al., 2003), aging in drinks (flavored milk, alcoholic or non-alcoholic drinks, and instant fruit juice), detection alcohol in beverages (Lvova et al., 2006), Sorting of mineral water varieties according to the water's hardness (Iiyama, 1995), and were performed by E-tongue.

13.3.1 Principle of E-Tongue

Taste characteristics of the samples are detected by sensors, which then convert them into analytical signals (electrochemical, optical, electrophysiological, etc.) and generate data. Data processing techniques like neural networks and chemometric methodologies are used to provide final sample information. The method of measuring taste molecules involves immersing the reference electrode and detective in an electrolytic solution and sample solution respectively to determine the electrode potentials (E). Then the reduction potential is calculated using the Nernst equation (Eq. 13.2) (Tan & Xu, 2020).

$$E = E^0 - \frac{RT}{zF} \ln Q \tag{13.2}$$

Where,

E is the reduction potential
E^0 is the standard potential
R is the universal gas constant
T is the temperature in kelvin
Z is the ion charge (moles of electrons)
F is the Faraday constant
Q is the reaction quotient

13.3.2 Working Mechanism of E-Tongue

E-tongue consists of three principal components such as sensory array, the signal emitters and receivers and data analysis (Fig. 13.3). Sample preparation is required for E-tongue as test substances must have adequate solubility. The solubility of test

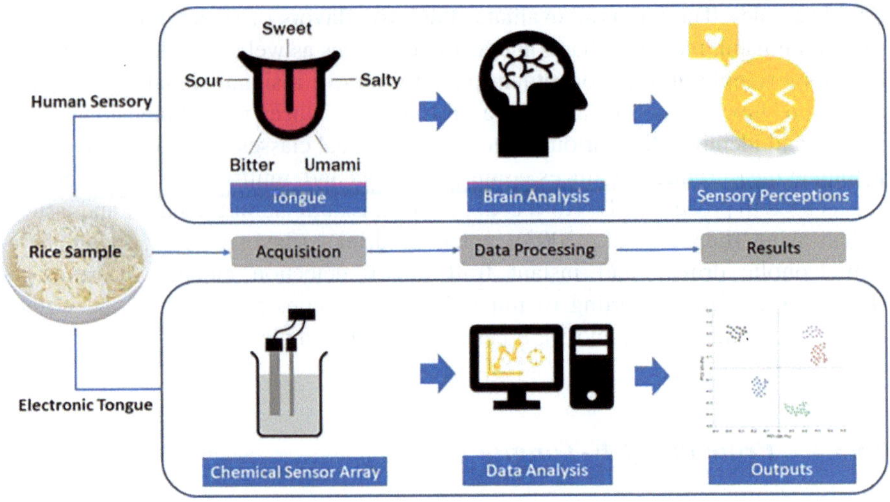

Fig. 13.3 Schematic diagram of E-tongue and human taste (Aznan et al., 2023)

substances can be increased by using co-solvents such as ethanol to release the molecules of the taste. The molecule of the distinct tastes interacts with the taste on the surface of the sensor and causes potential alterations. The change of potential difference between reference and measuring electrodes is measured to decide the concentration of a taste molecule in the solution. The change of potential is converted to electric signals using a transducer. The signals are processed by applying either signal amplification or noise reduction to produce a chromatograph/Spectrograph (Fig. 13.2).

13.3.3 Types of Sensors

The electronic tongue systems use a variety of sensors, ranging from gravimetric to optical (absorbance, luminescence, reflectance, etc.) to electrochemical (potentiometric, amperometric, voltammetric, impedimetric, and conductometric) and enzymatic biosensors. Potentiometric sensors such as ion selective electrodes (ISE) are used in the E-Tongue (Del Valle, 2012). The system consisting both selective and cross-sensitive sensors are ideal for detection. A cross-sensitivity of a sensor is a competence to respond to multiple analytes in the solution. This feature is seen in biological identification systems (Vlasov et al., 2005). Intricate data configurations given by gustative and olfactive receptors are contained of various, discrete cross-sensitive responses (del Valle, 2010).

13.3.3.1 Gravimetric Sensor

The gravimetric sensor operates by measuring the resonance frequency formed during the excitation of piezoelectric material, when molecules, vapors, or other species are deposited on the working surface of the resonator/piezoelectric material. To detect analytes selectively, the resonator surface is frequently precoated with a functional material that binds to the target molecules. Gravimetric sensors frequently employ a Quartz Crystal Microbalance (QCM).

13.3.3.2 Optical Sensor

The main concept of optical sensors is the transmission and reception of light. The light reflected or disturbed by the target object is converted to an output voltage.

13.3.3.3 Electrochemical Sensor

An electrochemical sensor works by causing taste molecules to react with the electrode's surface at the point of interaction. During the process, the electrode oxidizes or decreases the molecules of interest, producing an electrical signal via a transducer. The concentration of the interest molecules impacts the amount of electric signal generated. An electrochemical sensor is made up of two electrodes (a working electrode and a counter electrode), which allow charged molecules to pass through a thin layer of electrolytes.

13.3.3.4 Bio Sensor

A biosensor consists of a biological sensing element and a transducer. When the molecules interact with the biological substance, such as a specific enzyme or biological material result in change in pH of the biological materials. The change in pH can be detected as electrical response using a transducer.

13.4 Construction of Sensor

The sensors consist of variable resistors, diodes, coils are used amplifiers, integrated circuit. These components are used to construct the E-nose. A circuit diagram is designed first to show places of the components (Fig. 13.4). In the sensor circuit, gas sensors, load resistor and working voltage are connected in series. There are two input voltages such as Heater Voltage (V_H) and Circuit Voltage (V_C) and one output voltage (V_{out}) are given to the sensor (Table 13.1). The heater voltage supplies the

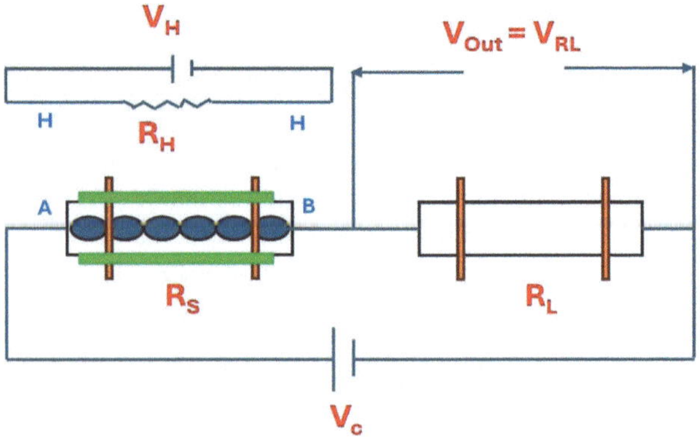

Fig. 13.4 Circuit diagram of E-Nose

working temperature to the sensor while the circuit voltage detects the voltage on the load resistance. The resistive load voltage (V_{RL}) is the output voltage (Vout) and it is measured between the load resistors (R_L). The gas sensing element consist of built in resistor between A and B and the heating element consist of a built-in resistor between H and H. When the concentration of the gas increases the resistance of the sensor decrease while the concentration decreases the resistance got decrease. So, the gas concentration is calculated from changes in the resistance. R_0 is the reference resistance of the sensor at a known concentration in fresh air or without the presence of other gases. R_S is the gas resistance of the sensor that varies based on the concentration of gas. From the circuit diagram, the current supplied to the circuit is estimated using the Eq. 13.3 and the output voltage at the load resistor using the value of current supplied to the circuit is estimated (Eq. 13.5). R_s is calculated using the Eq. 13.7 based on Ohm's Law.

$$I = \frac{V}{R} \tag{13.3}$$

$$I = \frac{V_C}{R_S + R_L} \tag{13.4}$$

$$V_{Out} = I \times R_L \tag{13.5}$$

Therefore,

$$V_{RL} = \frac{V_C}{R_S + R_L} \times R_L \tag{13.6}$$

$$R_S = \frac{V_C \times R_L}{V_{RL}} - R_L \tag{13.7}$$

The sensor resistance is varied when a target gas is introduced to the sensor that leads to the change of the load resistor. The response of the sensor to the gas is measured sensor response (Eq. 13.2). The sensor response (S) is defined as the ratio of the difference between the sensor resistance when exposed to gas and air to the sensor

Table 13.1 Sensors used in E-nose, E-tongue and volatile compounds detected

Product	Sensors used in E-Nose	Sensors used in E-Tongue	Measured volatile Compound	References
Taste composition of fermented shrimp paste			Nitrogen oxides hydride, sulfides Alcohol	Deng et al. (2022)
Different durian cultivars based	W1C W5S W3C W6S W5C; W1S W1W W2S; W2W W3S.	CA0, AAE and Gl1 were used to detect sourness, umami and sweet taste	Benzene. Oxides. Aroma and ammonia; hydrides; methyl; sulfides	Xiao et al. (2022)
Zhenyuan Daocai is a salted pickle				Liu et al. (2023)
Chinese soybean paste	W1C, W5S, W3C, W6S, W5C, W1S, W1W, W2S, W2W, W3S	Seven electrodes (SRS, GPS, STS, UMS, SPS, SWS, BRS), an Ag/AgCl reference electrode		Yu et al. (2022)
	TGS 2620		Alcohol, solvent vapors	Anonymous (2021)
	TGS 2612		Methane, propane, iso-butan	
	TGS 832		Chlorofluorocarbons	
	TGS 822		Organic solvent vapors	
	TGS 2603		Air contaminants (trimethylamine, methyl mercaptan, etc.)	
	TGS 2600		Air contaminants (hydrogen, ethanol, etc.)	
	TGS 813		Combustible gases	
	TGS 826		Ammonia	

resistance when exposed to air. The sensitivity (S') of the senor is defined as the rate of change in sensor response per unit change in gas concentration and is calculated using (Eq. 13.8). The time taken to reach 90% of its equilibrium value of sensor resistance after purging the gas to the sensor is called response time. The time taken to return to 90% of initial value of sensor resistance after venting the gas to the sensor is called recovery time.

$$S, \% = \frac{R_g - R_a}{R_a} \times 100 \tag{13.8}$$

13.4.1 Data Acquisition and Spectral Analysis

Measured electric signals are plotted as a chromatogram. Chromatogram is a graphical representation of time and signal strength. The horizontal axis displays the amount of time before the component reaches the detector, while the vertical axis displays the signal intensity. A line parallel to the horizontal axis appears in a condition where no chemicals are eluted from the column is known as **baseline**. The detector finds the concentration of the target compound in the elution band and creates a graph that looks like a bell-shaped curve is known as a **peak. The peak height** (h) is the distance between a peak's summit and the baseline. **The peak area** is the area under the peak and baseline. **The retention time (t_R)** is defined as the time between the sample injection point and the peak's apex. Also, it is the duration of analyte moves from a chromatographic column with a mobile phase and reaches the detector. **The dead time (t_0)** is the time it takes for non-retained compounds (compounds that do not interact with the stationary phase) to go from the injector to the detector. The difference between **retention time (t_R)** and **dead time (t_0)** is called the reduced or **adjusted retention time**. The flow rate of a mobile phase influences the retention time due to column length. As a result, a new parameter, the retention factor **(k)** is introduced (Eq. 13.3). **The retention factor (k)** relates the retention period of an analyte to the time required by an unretained component. It measures how long the analyte remains in the column. K values should be between 1 and 10 to get acceptable resolution within a tolerable time frame for the study. **The separation factor (α)** is the ratio of the retention factors of two molecules, kB and kA (Eq. 13.10). The effectiveness of separating two chemicals (A and B) is determined by the separation factor **(α)**. Two compounds get separated, when α is greater than 1. However, when the value of α increases, the separation of compound also increases. **Resolution (R)** is an additional measure that is also employed to characterize the separation (Eq. 13.11). **The peak widths** of two immediately adjacent peaks corresponding to compounds A and B are **wA** and **wB** respectively. Most intermolecular interaction metrics require appropriate context, and retention time data must be corrected for changes in column dimensions such as length, inner diameter, and stationary phase film thickness. The corrective procedure is to calculate the **Linera Retention Index** (Eq. 13.12) or **Kovats Retention Index (I)** for each analyte using Eq. 13.13. The "*x*" and "*x + 1*" denote the number of carbon atoms in the typical alkane eluting just before and after the analyte, respectively.

$$k = \frac{t_R - t_0}{t_0} \tag{13.9}$$

$$\alpha = \frac{k_1}{k_2} \tag{13.10}$$

$$R = \frac{(t_2 - t_1)}{0.5(w_1 + w_2)} \tag{13.11}$$

$$LRI = 100\left[\frac{t_{RA} - t_{Rx}}{t_{Rx+1} - t_{Rx}}\right] + 100x \tag{13.12}$$

$$I = 100\left[\frac{\log\left(t'_R\right)_A - \log\left(t'_R\right)_x}{\log\left(t'_R\right)_{x+1} - \log\left(t'_R\right)_x}\right] + 100x \tag{13.13}$$

13.4.2 Processing of Data

The data obtained from the electronic tongue and electronic nose is processed to summarize, assess, and interpret using a step-by-step data analysis approach (Fig. 13.5). Outliers are located and eliminated by preprocessing of the gathered data. Data points which deviate notably and exist outside from all other data known

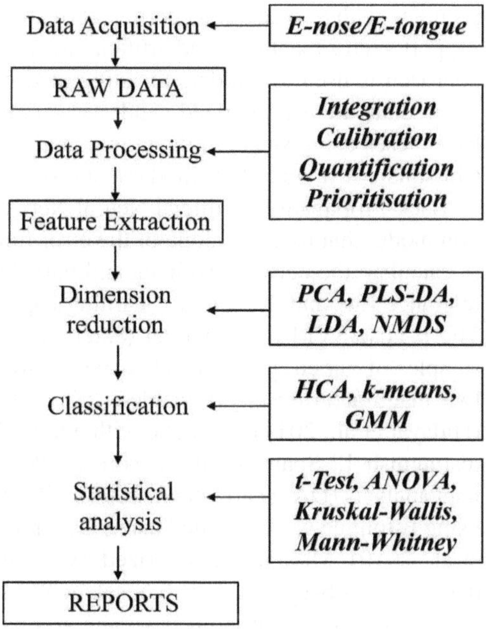

Fig. 13.5 Flow chart of data analysis

as outliers. Outliers are removed from the analysis without changing the result because they do not match most of the observations that were gathered.

Many mathematical techniques are used for data preprocessing, including multiplicative signal correction, orthogonal signal correction, Savitsky-Golay derivatization, mean centering, normalization, baseline correction, derivative and orthogonal signal correction. Through network analysis, multivariate data analysis, and graphical analysis, the data is reduced to visualization.

Chemometric techniques (differential analysis, hierarchical clustering, dimensionality reduction) are employed to comprehend the connections, patterns, and associations among individual samples and/or individual attributes. Multivariate data analysis generates a set of trained and untrained data analysis approaches. Untrained techniques are used when a database of samples is not practically obtained. The principal component analysis (PCA) is widely used untrained multidimensional analysis technique. MDA is a very useful tool when sensors have partial coverage sensitivity to certain compounds in a sample mixer. The neural network is the most well-known and widely used analytic method in statistical software packages for commercially available electronic nose. Overall, 18 accessible datasets were collected from the soft E-nose and a four-phase methodology was adopted to process the data. Initially, the data were gathered, then features were extracted using principal component analysis and moving average filter signal processing followed by pattern identification using effective machine learning and deep learning techniques such as support vector machine, k-nearest neighbors, artificial neural network, deep neural network, and extreme learning machine for classification and regression for the prediction and classification of the beef microbial population. Using -nearest neighbors and artificial neural networks yields 98% accuracy with a correlation coefficient of above 0.98 compared to the combination of support vector machines and artificial neural networks. The tenfold cross-validation approach is used to validate the suggested methodology (Pulluri & Kumar, 2022). Then data are summarized, and its variability is assessed to test fundamental hypotheses using standard deviation, analysis of variance, Wilcoxon rank sum test, Chi-square test, and other fundamental statistical calculations.

Back-Propagation Artificial Neural Network (BP-ANN) is a supervised recognition model that uses the slope of the error function as it changes with the goal value to calculate the network weights and bias. PCA is unable to achieve the high fault tolerance rate and continuous learning capabilities as BP-ANN (Wu et al., 2020). The E-nose-SVM-linear model (external validation dataset of 70 independent tea samples of varied quality levels from the seven brands, gathered over the course of two manufacturing days) enabled 100% of accurate full predictive classifications (Hidayat et al., 2019). e-tongue with a hybrid sensor array was used to identify and distinguish 12 Spanish red wines using PCA, SIMCA analysis, and PLS discriminant analysis (DA) (Parra et al., 2006). Fifty Belgian and Dutch beers were analyzed using E-tongue was classified using Canonical Correlation Analysis (CCA) (Polshin et al., 2010). Milk was categorized by producer based on potentiometric e-tongue data using PLS-DA analysis (Ciosek & Wróblewski, 2008). The authors used LDA was used to divide the four beverage groups based on the quantity of fruit juice in each category based on E-tongue data (Dias et al., 2011). PCA, CA, and ANN was used to categorise Honey samples from different floral and geographic sources using Alpha MOS e-tongue data (Wei et al., 2009).

13.5 E-Nose and E-Tongue Method Development and Validation

The process of selecting and optimizing the parameters that are required to measure a specific volatile or taste compounds in a product or total profiles of a product is called method development. The process of method development and validation involves series of steps (Fig. 13.6). The method is developed using the data extracted from the volatile profile and is being curate as needed. The processes sample or analyte preparation is adopted from existing literature, or modified the existing

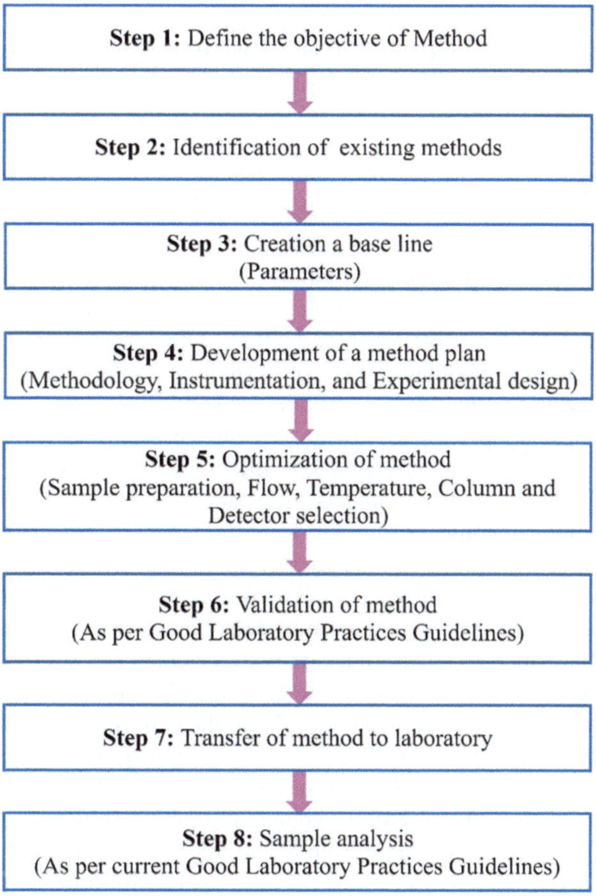

Fig. 13.6 Step by step approach of method development and validation

literature and used or newly developed procedure. The sample preparation wholly depends on the type of sample and nature of analyte. Similarly, the processes of procedure development are also adopted from literature and varied or improved by One Factor At a Time approach. One factor at a time is the process of one parameter is changed at a time and the set of terms is changed throughout parameter optimization, instead of using a trial-and-error approach.

The procedure for method is based on the instrument and physicochemical nature of the analyte. In the absence of a method for the analyte in the literature, the method for substances with the same chemical properties and structural similarity as the analyte is used.

13.5.1 Selection of Initial Conditions

Selection of initial conditions for keeping all analytes is the first step in developing a technique. This involves ensuring that no analyte has a capacity factor of less than 0.5 or more than 10–15. If the capacity factor is greater than 10–15, it may result in an analysis time that is too long and broad, hard-to-detect peaks; if it is less than 0.5, it may cause peak overlapping.

13.5.2 E-Nose Method Development

The electron ionization voltage, carrier gas, inlet temperature or injector port temperature, the injection volume, spilt ratio, carrier gas flow rate, column specification, column oven temperature, run time, detector time, detection temperature, acquisition rate and modulation period are required to optimised in hardware part. Also, the sample preparation is essential for E-nose analysis. Sample preparation is a process of making samples to convenient for analyses by grinding, dilution, filtration, centrifugation, precipitation, leaching, extraction, digestion, size exclusion, and homogenization. Samples preparation enables compatibility to further analysis, to simplify complex samples, to remove interferences from the matrix, to increase the speed of analysis, to acquire desired sensitivity and selectivity. Sample injection.

Ther are two types of sample injections (1) Split and (2) Splitless. The samples contain high concentration of detection compound then split the samples into ratio and injected to the injector port, whereas the splitless method is used when the samples contain low concentration of detection compound. The split ratio controls the amount of sample injected into the E-nose. The ratio of the amount of carrier gas flowing into the split vent to the column is known as the split ratio. It enhances chromatographic separation and retention time repeatability while enabling sufficient detection for all the chemicals of interest by preventing column overload. Split ratios typically range from 5:1 to 500:1. The reference solution, or artificial saliva, was made by dissolving 30 mM potassium chloride and 0.3 mM tartaric acid in distilled water for sensor conditioning and cleaning. The positively charged sensors (AE1, C00) were cleaned with 100 mM potassium chloride and 10 mM potassium hydroxide, while the negatively charged sensors (BT0, AN0) were cleaned with 100 mM hydrochloric acid diluted in 30% ethanol (Zhang et al., 2021). Standard chemicals quinine hydrochloride, sodium chloride, and monosodium glutamate reflect the four basic tastes of bitterness, sour, umami, and salty, with thresholds of 1E−3 mM, 0.01 mM, 0.05 mM, and 0.5 mM, respectively. The E-tongue‘s taste sensors have thresholds that are identical to human taste receptors.

13.5.3 E-Tongue Method Development

The choosing of sensory array is primary requirement for method development. The parameters like, sample preparation, number of times washing of E-tongue, number of s ample measurements required to produce a stable response signal, measurement time, and order of washing and sample measurements are optimized during method development (Li et al., 2016).

The output data from developed method is grouped into two sets. One set of data is used for training the classification models using various methods like fine tree, quadratic discriminant, quadratic support vector machine, cubic support vector machine, and k nearest neighbor and another set of data is used for testing the classification model. The models employed to categorize specific fruits into ripeness categories had a high f1-score and an average accuracy of 95%. After training, the best performing classification method is chosen using confusion matrix, the receiver operating characteristic (ROC) curve and precision recall curve (PRC) (Huang et al., 2021).

A confusion matrix or an error matrix (Fig. 13.7) is a method used to visualize classifier algorithm output. More specifically, it is a table that contrasts the number of actual ground truth instances of a certain class with the number of expected

Predicted

Actual or expected		Positive	Negative
	Positive	**True Positive**	**False Negative** *Type II Error*
	Negative	**False Positive** *Type I Error*	**True Negative**

Fig. 13.7 Confusion matrix for validation of methods

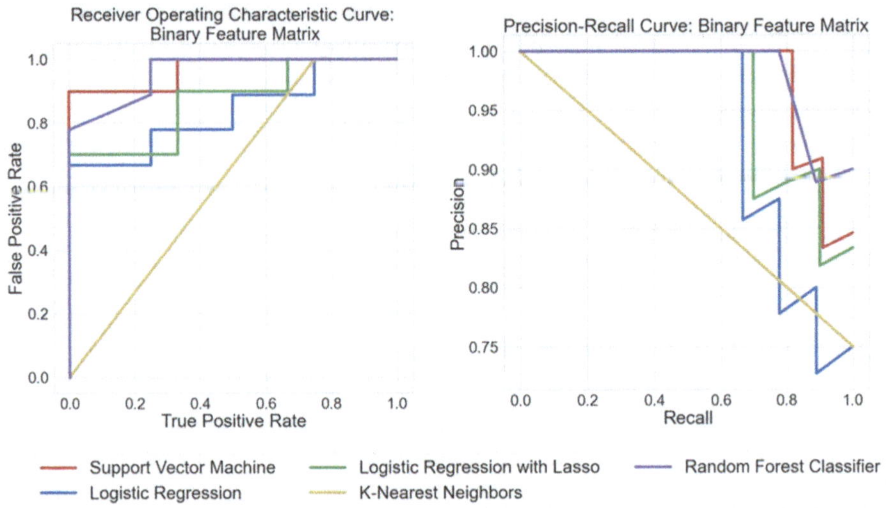

Fig. 13.8 ROC and PRC curves for classifiers trained on the binary feature matrix to Identify Discriminative Signatures of Volatile Organic Compounds (VOCs) from Bacteria and Fungi Using SPME-DART-MS (Arora et al., 2022)

instances for the class. One of the many assessment metrics used to determine the effectiveness of a classification model is confusion matrices. They can be used to calculate several additional model performance indicators, such as recall and precision. Because it offers a more comprehensive picture of model performance than merely accuracy, the confusion matrix is crucial. It draws attention to specific flaws in the model's operation, such false negatives, or defective batteries that are wrongly read as operational. These could have disastrous effects in industrial environments.

ROC curve (Fig. 13.8) is a visual depiction of the model's performance over all thresholds. Plotting the true positive rate (TPR) against the false positive rate (FPR) at each threshold setting creates the ROC curve. The true positive rate determines true-positive outcomes from all positive samples, whereas the false positive rate identifies false-positive outcomes from all negative samples. In ROC analysis, the area under the curve (AUC) represents test accuracy, which is proportional to the AUC. The diagonal line from [0, 0] to (1, 1) represents random classification. To accept a generalised classification model, the ROC curve must be higher than the diagonal line (Karami et al., 2020a, b). The PRC (Fig. 13.8) depicts the relationship between accuracy and recall, with the Y and X axes representing precision and recall at different thresholds, respectively. The larger the area under the curve, the better the precision and recall; high recall suggests a low false-negative rate, while high accuracy indicates a low false-positive rate (Semwal et al., 2021).

13.5.4 Validation

The process of confirming, proving and authorising the development method is applicable to measure a specific volatile or taste component in a product or total profiles of a product for which it is intended to develop and produces reliable findings over an extended period is called as method validation. A 100% recovery rate is not necessary if the results can be repeated. The only way to validate the validity of an analytical method is through laboratory research. Therefore, to ascertain whether a method is appropriate for the intended application, documenting of such successful research is a fundamental prerequisite. The testing data is then processed using the chosen classification method, and the resulting values are utilized to calculate true positive (TP), true negative (TN), false positive (FP), and false negative (FN). Method development and validation components include accuracy, precision, specificity, linearity, range, limit of detection (LOD), limit of quantification (LOQ), ruggedness, and robustness.

Specificity or selectivity is the ability to conclusively determine the analyte's value in the presence of potentially present components and it is calculated using (Eq. 13.14). The lack of specificity in any analytical method may be compensated for by additional supporting analytical procedure(s). Peak homogeneity or peak purity tests can be used to evaluate the selectivity of chromatographic techniques. A peak purity test reveals that no component of the sample has been co-eluted. Peak purity is assessed in this process using PDA or MS detectors. Resolution, plate count, tailing factor, and representative chromatograms with identified peaks should all be included in the validation report.

$$\text{Specificity/Selectivity} = \frac{\text{TN}}{\text{TN} + \text{FP}} \tag{13.14}$$

Linearity is ability of the data results said to be linear within a specified range and are precisely proportionate to the sample concentration. This illustrates how the detector response (peak area and height) and sample concentration are correlated linearly. Linearity is visually assessed by plotting signals against analyte concentration. The degree of linearity can be calculated quantitatively using data from the regression line and it expressed as correlation coefficients (R^2) and variance around the regression line's slope. In some cases, analytical solutions should be designated by the appropriate function.

The range of the analyte concentration in the sample (including these values) for which the analytical procedure's linearity, precision, and accuracy have been demonstrated to be sufficient is known as the analytical method's range. The range is dependent on how the process is intended to be used and is frequently established by linearity tests. The minimum stated limits of the assay method usually cover 80–120% of the test concentration. For content homogeneity, a minimum of 70–130% of the test concentration should be covered, depending on the kind of dosage form. ±20% beyond the specified range for dissolution testing; ±120% of the specification, beginning at the impurity reporting level, for impurity determination.

The accuracy of an analytical technique is determined by the degree of agreement between the value found and the value acknowledged as either a conventional true value or an acceptable reference value. The accuracy is calculated using Eq. 13.15. Sometimes, this is referred to as trueness.

$$\text{Accuracy} = \frac{\text{TP} + \text{TN}}{\text{TP} + \text{TN} + \text{FP} + \text{FN}} \tag{13.15}$$

Recall is another term for the true positive rate (TPR), which is the proportion of all real positives that were correctly identified as positive, and it is calculated using Eq. 13.16. False negatives are actual positives that were mistakenly classified as negatives; therefore, they show in the denominator. A hypothetical perfect model would have zero false negatives, resulting in a recall (TPR) of 1.0, or a 100% detection rate. In an imbalanced dataset with only 1–2 genuine positives, recall becomes less relevant and useful as a statistic.

$$\text{Recall or True Postive Rate or Sensitivity} = \frac{\text{TP}}{\text{TP} + \text{FN}} \tag{13.16}$$

The false positive rate (FPR) is the percentage of all genuine negatives that were mistakenly categorized as positives, also known as the probability of false alarm (Eq. 13.17). An ideal model would have zero false positives, resulting in an FPR of 0.0, or a 0% false alarm rate. In an imbalanced dataset with only 1–2 true negative occurrences, FPR is less meaningful and useful as a statistic.

$$\text{False Postive Rate} = \frac{FN}{\text{TP} + \text{FP}} \tag{13.17}$$

In analytical procedures, precision (Eq. 13.18) is defined as the degree of agreement (or scatter) between a set of measurements acquired by repeatedly sampling the same homogenous sample under established conditions. There are three distinct categories of accuracy: repeatability, moderate precision, and reproducibility. As a matter of fact, homogeneous samples are ideal for precision research. If a homogenous sample cannot be acquired, synthetic samples or a sample solution may be utilized for research. It is customary to convey the accuracy of an analytical procedure by using the variance, standard deviation, or coefficient of variation of a collection of observations. F-Measure (Eq. 13.19) generates a single score that addresses both precision and recall concerns in one number.

$$\text{Precision or Postive predictive value} = \frac{\text{TP}}{\text{TP} + \text{FP}} \tag{13.18}$$

$$\text{F} - \text{Measure} = 2 \times \left(\frac{\text{Precision} \times \text{Recall}}{\text{Precision} + \text{Recal}}\right) \tag{13.19}$$

Repeatability is the ability to convey precision over a brief period under the same operational conditions. Another name for repeatability is intra-assay precision.

The lowest concentration of analyte in a sample that can be identified, even if it cannot always be precisely measured, is known as the limit of detection (Eq. 13.20). There are multiple methods for determining the detection limit. The most straight forward method is evaluating the signal to noise ratio. This ratio is calculated by comparing the observed signals from samples with known low analyte concentrations to those from blank samples. An appropriate detection limit is typically indicated by a signal to noise ratio of 3:1 or 2:1. Another approach is considering the slope and standard deviation of the response.

$$\text{Limit of Detection, LOD} = \frac{3.3 \times \sigma}{S} \tag{13.20}$$

The Limit of Quantification of a specific analytical process is the smallest concentration of analyte in the sample that can be quantitatively determined with sufficient precision and accuracy. The most contributing factors are detector sensitivity and sample preparation precision. The quantification limit can be determined using the same method as the detection limit. S and σ values were calculated based on the detection limit. In addition to the LOD, the LOQ (Eq. 13.21) can be determined through visual inspection. The analyte concentration should be quantifiable at the LOQ level with a sufficient degree of accuracy and precision. A mean recovery at this level of 50–150% and a percentage RSD of ≤25% are usually required for LOQ acceptance.

$$\text{Limit of Quantification, LOQ} = \frac{10 \times \sigma}{S} \tag{13.21}$$

Where, σ is standard deviation of the response and S is the slope of the calibration curve.

Robustness is a metric that expresses an analytical procedure's resistance to small, deliberate changes in method parameters as well as its dependability under normal operating conditions. It is evaluated in part as the strategy is being developed. The robust study's objective is to identify the critical operational parameters required for the method's successful application. There should be enough control over these options and a warning in the method documentation. Studies on the robustness of HPLC systems involve adjusting process parameters, such as pH, flow rate, column temperature, and mobile phase composition, within an appropriate range. The system suitability parameters obtained for each circumstance are analyzed to identify the parameter that has the biggest impact on the technique.

System Suitability Testing key system components using the theory that system suitability testing, or SST, encompasses all procedures, equipment, and sample types. System appropriateness checks provide an additional layer of assurance that the process is yielding precise and accurate results on a given occasion. A system suitability test is carried out before and during analysis for each method that is used.

The primary SST factors considered are resolution (Rs), repeatability (% RSD of peak response and retention length), column efficiency (N), and tailing factor (Tf). The two SST factors are the retention factor (k) and the separation factor (α).

13.6 Factors Affecting E-Nose and E-Tongue Performance

The sample volume and analyte solubility are crucial for E-nose and E-tongue performance which determines the other conditions for K and β. K and β values should be chosen with the goal of minimizing that total to maximize detector response. The k value is influenced by temperature. It is recommended to maintain an oven temperature that is no more than 20 °C below the solvent boiling point. The analysis's consistency and repeatability can be enhanced by headspace factors such sample loop pressure, volume, and equilibration time as well as vial shaking. The main issues with utilizing e-noses and e-tongue in practical settings stem from the sensors' inconsistent responses due to the widely recognized impacts of drift, restricted repeatability, and sensitivity to different environmental factors.

13.7 Applications of E-Nose and E-Tongue

The electronic-nose and electronic-tongue are designed for use in different industrial production processes based on the specific types and categories of products, such as food, automotive, packaging, cosmetics, pharmaceuticals, and biomedical industries. They can also be applied for tasks like process design, monitoring freshness and ripeness, investigating shelf-life, assessing the authenticity of premium products, classifying smells and perfumes, detecting microbiological pathogens, and evaluating the environment. E-Nose and E-Tongue are precise analytical instruments for evaluating food's quality (Tan & Xu, 2020). They have shown a great deal of promise in the field of food authenticity, and there are several ways to boost their analytical effectiveness. Di Rosa et al. (2020a) discovered that a potent strategy for ensuring food authenticity was the combination of electronic sensors with chemometrics and pattern recognition techniques. The possibilities in sensor technology are constantly expanding.

Alcoholic beverages (beer and sparkling wines) are difficult to analyse using e-nose technology because they include ethanol and carbon dioxide. Following the dehydration and decoholization of the samples, the PCA and DFA analyses were employed in combination with a French e-nose maker known as AlphaMOS e-nose (FOX 4000) was used to differentiate between wines and beer tainted with off--flavors like octenol, 1-hexanol, ethyl acetate, 4-ethylphenol, and 2,3,4-trichloroanisole (TCA) and differentiating Ontario, Canada-made fruit and grape wines based on their aromatic signatures (McKellar et al., 2005; Ragazzo-Sanchez et al., 2009). Though it was unable to differentiate between wines produced by various Saccharomyces cerevisiae strains, the e-nose could identify between fermented and unfermented musts (García-Martínez et al., 2011). A specially designed e-tongue with a hybrid sensor array—voltammetric electrodes that were chemically altered with different electro-active substances was used to identify and distinguish 12 Spanish red wines based on characteristics such as vintage, grape variety, denomination, and origin because of the electrodes' cross-selectivity (Parra et al., 2006). E-tongue can predict 20 distinct sensory attributes of beer, including burnt, artificial, fruity, sour, sweet, and bitter (Rudnitskaya et al., 2010). E-nose was used to determine their physiochemical properties, including true extract, actual fermentation degree, alcohol content, pH, bitterness, color, polyphenol content, and carbon dioxide content of 50 Belgian and Dutch beers (Polshin et al., 2010). Using a potentiometric e-tongue, milk was categorized by producer and place of origin using an array of microelectrodes (Ag/AgCl) composed of epoxy-glass laminate and polyvinyl chloride (PVC) membranes with various additions as the chemosensitive layers (Ciosek & Wróblewski, 2008). Potentiometric e-tongues with 36 cross-sensibility lipid/polymeric membranes were used to assess commercial fruit juices (orange, pineapple, mango, and peach) from different brands and used divide the four beverage groups based on the quantity of fruit juice in each category (Dias et al., 2011). A microelectrode tongue could be used to differentiate oils based on their quality and provenance (Oliveri et al., 2009). Using PCA and a classification algorithm (K nearest neighbors, or KNN), the system was able to identify between different oils (olive and maize) or geographical origins (olive oils from distinct regions) (Oliveri et al., 2009). Honey samples from different floral and geographic sources were categorized using Alpha MOS e-tongue with seven potentiometric chemical sensors (Wei et al., 2009).

13.8 Conclusion

E-nose and E-tongue are emerging techniques in food sensory science. It is used for estimating and evaluating the taste of molecules in food. Generally, human panels are engaged in analysis of food development and production, quality. There are chances of exposing carcinogenic substances and risk associated with employing human panels for product testing. Hence, E-nose and E-tongue techniques are used as alternative techniques in sensory analysis. The creation of sensitive sensors has accounted for most advancements in electronic nose technology. This method misses the main problem with machine olfaction, which is that it simulates smell perception rather than detection. The interpretation of the signals obtained from the olfactory bulbs is what allows us to perceive scents. They mostly depend on the human brain. Any electronic nose that mimics human smell must also replicate how an aroma is perceived, necessitating a simulation of the full human olfactory system. Sensor technology has been significantly enhanced by e-noses and e-tongues. These techniques are widely employed; however, they do have some disadvantages. Their potential for limited sensitivity and selectivity in distinguishing between comparable compounds or flavor qualities is a major downside that could result in erroneous analysis, especially when dealing with complex samples. Furthermore, the sensors may react simultaneously with several different chemicals or flavor characteristics due to cross-sensitivity.

References

Aghdamifar, E., Sharabiani, V. R., Taghinezhad, E., Szymanek, M., & Dziwulska-Hunek, A. (2023). E-nose as a non-destructive and fast method for identification and classification of coffee beans based on soft computing models. *Sensors and Actuators B: Chemical, 393*, 134229. https://doi.org/10.1016/j.snb.2023.134229

Anonymous. (2021). *Figaro Engineering Inc. Product Information*. Available online: https://www.figarosensor.com/product/docs/TGS%202611C00(1013).pdf. https://www.figarosensor.com/?

Apetrei, C., Apetrei, I. M., Villanueva, S., De Saja, J. A., Gutiérrez-Rosales, F., & Rodriguez-Mendez, M. L. (2010). Combination of an e-nose, an e-tongue and an e-eye for the characterisation of olive oils with different degree of bitterness. *Analytica Chimica Acta, 663*(1), 91–97.

Arora, M., Zambrzycki, S. C., Levy, J. M., Esper, A., Frediani, J. K., Quave, C. L., Fernández, F. M., & Kamaleswaran, R. (2022). Machine learning approaches to identify discriminative signatures of volatile organic compounds (VOCs) from bacteria and fungi using SPME-DART-MS. *Metabolites, 12*(3), 232.

Aznan, A., Gonzalez Viejo, C., Pang, A., & Fuentes, S. (2023). Review of technology advances to assess rice quality traits and consumer perception. *Food Research International, 172*, 113105. https://doi.org/10.1016/j.foodres.2023.113105

Bai, H., & Shi, G. (2007). Gas sensors based on conducting polymers. *Sensors, 7*(3), 267–307.

Beghi, R., Buratti, S., Giovenzana, V., Benedetti, S., & Guidetti, R. (2017). Electronic nose and visible-near infrared spectroscopy in fruit and vegetable monitoring. *Reviews in Analytical Chemistry, 36*(4). https://doi.org/10.1515/revac-2016-0016

Behera, B., Joshi, R., Vishnu, G. K. A., Bhalerao, S., & Pandya, H. J. (2019). Electronic nose: A non-invasive technology for breath analysis of diabetes and lung cancer patients. *Journal of Breath Research, 13*(2), 024001.

Bleibaum, R. N., Stone, H., Tan, T., Labreche, S., Saint-Martin, E., & Isz, S. (2002). Comparison of sensory and consumer results with electronic nose and tongue sensors for apple juices. *Food Quality and Preference, 13*(6), 409–422.

Calvini, R., & Pigani, L. (2022). Toward the development of combined artificial sensing systems for food quality evaluation: A review on the application of data fusion of electronic noses, electronic tongues and electronic eyes. *Sensors, 22*(2), 577.

Camacho-Olarte, J., & Burgos, D. A. T. (2020). Real application for a data acquisition system from sensors based on embedded systems. *Engineering Proceedings, 2*(1), 25.

Chen, S., Wang, Y., & Choi, S. (2013). Applications and technology of electronic nose for clinical diagnosis. *Open Journal of Applied Biosensor 02*(02):39–50. https://doi.org/10.4236/ojab.2013.22005.

Ciosek, P., & Wróblewski, W. (2008). Miniaturized electronic tongue with an integrated reference microelectrode for the recognition of milk samples. *Talanta, 76*(3), 548–556.

del Valle, M. (2010). Electronic tongues employing electrochemical sensors. *Electroanalysis, 22*(14), 1539–1555.

Del Valle, M. (2012). Sensor arrays and electronic tongue systems. *International Journal of Electrochemistry, 2012*(1), 986025.

Deng, Y., Wang, R., Zhang, Y., Li, X., Gooneratne, R., & Li, J. (2022). Comparative analysis of flavor, taste, and volatile organic compounds in opossum shrimp paste during long-term natural fermentation using e-nose, e-tongue, and HS-SPME-GC-MS. *Food, 11*(13), 1938.

Di Rosa, A. R., Leone, F., & Chiofalo, V. (2020a). Chapter 7: Electronic noses and tongues. In Y. Pico (Ed.), *Chemical analysis of food* (Second ed., pp. 353–389). Academic. https://doi.org/10.1016/B978-0-12-813266-1.00007-3

Di Rosa, A. R., Leone, F., & Chiofalo, V. (2020b). Electronic noses and tongues. In *Chemical analysis of food* (pp. 353–389). Elsevier.

Dias, L. G., Peres, A. M., Barcelos, T. P., Morais, J. S., & Machado, A. A. S. C. (2011). Semi-quantitative and quantitative analysis of soft drinks using an electronic tongue. *Sensors and Actuators B: Chemical, 154*(2), 111–118.

Eisele, I., Doll, T., & Burgmair, M. (2001). Low power gas detection with FET sensors. *Sensors and Actuators B: Chemical, 78*(1–3), 19–25.

García-Martínez, T., Bellincontro, A., Peinado, R. A., Mauricio, J. C., Mencarelli, F., & Moreno, J. J. (2011). Discrimination of sweet wines partially fermented by two osmo-ethanol-tolerant yeasts by gas chromatographic analysis and electronic nose. *Food Chemistry, 127*(3), 1391–1396.

Gardner, J. W., & Bartlett, P. N. (1999). *Electronic noses: Principles and applications*. Oxford university press.

Hidayat, S. N., Triyana, K., Fauzan, I., Julian, T., Lelono, D., Yusuf, Y., Ngadiman, N., Veloso, A. C. A., & Peres, A. M. (2019). The electronic nose coupled with chemometric tools for discriminating the quality of black tea samples in situ. *Chem, 7*(3), 29.

Huang, Y., Doh, I.-J., & Bae, E. (2021). Design and validation of a portable machine learning-based electronic nose. *Sensors, 21*(11), 3923.

Karami, H., Rasekh, M., & Mirzaee-Ghaleh, E. (2020a). Application of the E-nose machine system to detect adulterations in mixed edible oils using chemometrics methods. *Journal of Food Processing and Preservation, 44*(9), e14696.

Karami, H., Rasekh, M., & Mirzaee-Ghaleh, E. (2020b). Qualitative analysis of edible oil oxidation using an olfactory machine. *Journal of Food Measurement and Characterization, 14*(5), 2600–2610. https://doi.org/10.1007/s11694-020-00506-0

Li, X., Gao, X., Liu, R., Wang, J., Wu, Z., Zhang, L., Li, H., Gui, X., Kang, B., & Shi, J. (2016). Optimization and validation of the protocol used to analyze the taste of traditional Chinese medicines using an electronic tongue. *Experimental and Therapeutic Medicine, 12*(5), 2949–2957.

Lvova, L., Paolesse, R., Di Natale, C., & D'Amico, A. (2006). Detection of alcohols in beverages: an application of porphyrin-based electronic tongue. *Sensors and Actuators B: Chemical 118*(1–2):439–47.

Littarru, E., Modesti, M., Alfieri, G., Pettinelli, S., Floridia, G., Bellincontro, A., Sanmartin, C., & Brizzolara, S. (2024). Optimizing the winemaking process: NIR spectroscopy and e-nose analysis for the online monitoring of fermentation. *Journal of the Science of Food and Agriculture, n/a*(n/a). https://doi.org/10.1002/jsfa.13336

Legin, A., Rudnitskaya, A., Lvova, L., Vlasov, Y., Di Natale, C., & D'amico, A. (2003). Evaluation of Italian wine by the electronic tongue: recognition, quantitative analysis and correlation with human sensory perception. *Analytica Chimica Acta 484*(1):33–44.

Liu, X., Wang, X., Cheng, Y., Wu, Y., Yan, Y., & Li, Z. (2023). Variations in volatile organic compounds in Zhenyuan Daocai samples at different storage durations evaluated using E-nose, E-tongue, gas chromatography, and spectrometry. *LWT, 173*, 114186.

Iiyama, S. (1995). Quantitative sensing of mineral water with multichannel taste sensor. *Sens. Materials*7:191–201.

Maekawa, T., Suzuki, K., Takada, T., Kobayashi, T., & Egashira, M. (2001). Odor identification using a SnO_2-based sensor array. *Sensors and Actuators B: Chemical, 80*(1), 51–58.

McKellar, R. C., Vasantha Rupasinghe, H. P., Lu, X., & Knight, K. P. (2005). The electronic nose as a tool for the classification of fruit and grape wines from different Ontario wineries. *Journal of the Science of Food and Agriculture, 85*(14), 2391–2396.

Oh, E. H., Song, H. S., & Park, T. H. (2011). Recent advances in electronic and bioelectronic noses and their biomedical applications. *Enzyme and Microbial Technology, 48*(6–7), 427–437.

Oliveri, P., Baldo, M. A., Daniele, S., & Forina, M. (2009). Development of a voltammetric electronic tongue for discrimination of edible oils. *Analytical and Bioanalytical Chemistry, 395*, 1135–1143.

Parra, V., Arrieta, Á. A., Fernández-Escudero, J. A., García, H., Apetrei, C., Rodríguez-Méndez, M. L., & de Saja, J. A. (2006). E-tongue based on a hybrid array of voltammetric sensors based on phthalocyanines, perylene derivatives and conducting polymers: Discrimination capability towards red wines elaborated with different varieties of grapes. *Sensors and Actuators B: Chemical, 115*(1), 54–61.

Polshin, E., Rudnitskaya, A., Kirsanov, D., Legin, A., Saison, D., Delvaux, F., Delvaux, F. R., Nicolaï, B. M., & Lammertyn, J. (2010). Electronic tongue as a screening tool for rapid analysis of beer. *Talanta, 81*(1–2), 88–94.

Pulluri, K. K., & Kumar, V. N. (2022). Development of an integrated soft E-nose for food quality assessment. *IEEE Sensors Journal, 22*(15), 15111–15122. https://doi.org/10.1109/JSEN.2022.3182480

Ragazzo-Sanchez, J. A., Chalier, P., Chevalier-Lucia, D., Calderon-Santoyo, M., & Ghommidh, C. (2009). Off-flavours detection in alcoholic beverages by electronic nose coupled to GC. *Sensors and Actuators B: Chemical, 140*(1), 29–34.

Rajamäki, T., Alakomi, H.-L., Ritvanen, T., Skyttä, E., Smolander, M., & Ahvenainen, R. (2006). Application of an electronic nose for quality assessment of modified atmosphere packaged poultry meat. *Food Control, 17*(1), 5–13.

Romani, S., Cevoli, C., Fabbri, A., Alessandrini, L., & Dalla Rosa, M. (2012). Evaluation of coffee roasting degree by using electronic nose and artificial neural network for off-line quality control. *Journal of Food Science, 77*, C960–C965. https://doi.org/10.1111/j.1750-3841.2012.02851.x

Ravi, R., Nandwani, D., & Nwosisi, S. (2022). Volatile profiles of cooked organic sweetpotato by electronic nose. *Organic Agriculture 12*(1):17–32. https://doi.org/10.1007/s13165-021-00376-w.

Rudnitskaya, A., Rocha, S. M., Legin, A., Pereira, V., & Marques, J. C. (2010). Evaluation of the feasibility of the electronic tongue as a rapid analytical tool for wine age prediction and quantification of the organic acids and phenolic compounds. The case-study of Madeira wine. *Analytica Chimica Acta, 662*(1), 82–89.

Semwal, R., Aier, I., Tyagi, P., & Varadwaj, P. K. (2021). DeEPn: A deep neural network based tool for enzyme functional annotation. *Journal of Biomolecular Structure and Dynamics, 39*(8), 2733–2743.

Sha, R. (2023). Graphene-based electrochemical sensors for environmental monitoring applications. *Encyclopedia of Materials: Electronics* 1–3:V3-199-V3-211. https://doi.org/10.1016/B978-0-12-819728-8.00130-3.

Shaukat, H., Ali, A., Bibi, S., Mehmood, S., Altabey, W. A., Noori, M., & Kouritem, S. A. (2023). Piezoelectric materials: Advanced applications in electro-chemical processes. *Energy Reports, 9*, 4306–4324.

Tan, J., & Xu, J. (2020). Applications of electronic nose (e-nose) and electronic tongue (e-tongue) in food quality-related properties determination: A review. *Artificial Intelligence in Agriculture, 4*, 104–115.

Titova, T., & Nachev, V. (2020). "Electronic tongue" in the food industry. *Food Science And Applied Biotechnology, 3*(1), 71–76.

Tyagi, P., Semwal, R., Sharma, A., Tiwary, U. S., & Varadwaj, P. (2023). E-nose: A low-cost fruit ripeness monitoring system. *Journal of Agricultural Engineering, 54*(1).

Vlasov, Y., Legin, A., Rudnitskaya, A., Di Natale, C., & D'amico, A. (2005). Nonspecific sensor arrays ("electronic tongue") for chemical analysis of liquids (IUPAC Technical Report). *Pure and Applied Chemistry, 77*(11), 1965–1983.

Wang, Y., Liu, M., Bao, Z., & Zhang, S. (2018). Short-term load forecasting with multi-source data using gated recurrent unit neural networks. *Energies, 11*(5), 1138.

Wang, Y., Wang, D., Lv, Z., Zeng, Q., Fu, X., Chen, Q., Luo, Z., Luo, C., Wang, D., & Zhang, W. (2023). Analysis of the volatile profiles of kiwifruits experiencing soft rot using E-nose and HS-SPME/GC–MS. *LWT, 173*, 114405. https://doi.org/10.1016/j.lwt.2022.114405

Wei, Z., Wang, J., & Liao, W. (2009). Technique potential for classification of honey by electronic tongue. *Journal of Food Engineering, 94*(3–4), 260–266.

Wen, R., Sun, F., Wang, Y., Chen, Q., & Kong, B. (2021). Evaluation the potential of lactic acid bacteria isolates from traditional beef jerky as starter cultures and their effects on flavor formation during fermentation. *LWT* 142:110982. https://doi.org/10.1016/j.lwt.2021.110982.

Wilson, A. D. (2013). Diverse applications of electronic-nose technologies in agriculture and forestry. *Sensors, 13*, 2295–2348.

Wu, Z., Wang, H., Wang, X., Zheng, H., Chen, Z., & Meng, C. (2020). Development of electronic nose for qualitative and quantitative monitoring of volatile flammable liquids. *Sensors, 20*(7), 1817.

Xiao, Z., Niu, M., & Niu, Y. (2022). Comparative study on volatile compounds and taste components of different durian cultivars based on GC-MS, UHPLC, HPAEC-PAD, E-tongue and E-nose. *Molecules, 27*(4), 1264.

Yang, X., Liu, Y., Mu, L., Wang, W., Zhan, Q., Luo, M., Tian, H., Lv, C., & Li, J. (2018). Discriminant research for identifying aromas of non-fermented Pu-erh tea from different storage years using an electronic nose. *Journal of Food Processing and Preservation, 42*(10), e13721.

Yu, S., Huang, X., Wang, L., Ren, Y., Zhang, X., & Wang, Y. (2022). Characterization of selected Chinese soybean paste based on flavor profiles using HS-SPME-GC/MS, E-nose and E-tongue combined with chemometrics. *Food Chemistry, 375*, 131840.

Zhang, X., Wu, H., Lin, L., Du, X., Tang, S., Liu, H., & Yang, H. (2021). The qualitative and quantitative assessment of xiaochaihu granules based on e-eye, e-nose, e-tongue and chemometrics. *Journal of Pharmaceutical and Biomedical Analysis, 205*, 114298.

Chapter 14
Biosensors

Anantika Sharma, B. O. Madhu, Ashok Suraj, C. Indumathi, C. K. Sunil, Ashish Rawson, D. V. Chidanand, and S. Shanmugasundaram

14.1 Introduction

In the field of food quality analysis, where each bite tells a story of safety, integrity, and nutritional value, biosensors stand as silent guardians, precisely decoding the tales hidden within our food. Imagine a world where the very essence of food quality could be unveiled with precision, where safety is assured, and integrity is preserved at every step. This is the world that biosensors have brought about, a technological and biological masterpiece of invention. Biosensors have become essential instruments in analytical research, transforming how different problems are viewed and addressed, especially in areas such as food quality analysis. Biosensors, which combine the precision of transducers with the elegance of biological recognition elements, are the perfect example of the union of biology and technology (Gavrilaş et al., 2022; Turner, 2013). Biosensors provide an effective method to identify and measure analytes with unmatched sensitivity and specificity.

Fundamentally, a biosensor is an analytical tool that combines a transducer and biological sensing element to find the presence or concentration of a certain analyte. By interacting with the target analyte, this biological sensing element which can be an enzyme, antibody, nucleic acid, or whole cell starts a biochemical response (Chambers et al., 2002). This biological signal is subsequently transformed by the transducer into a quantifiable and interpretable output, usually an electrical, optical, or thermal signal (Bhatia et al., 2024). The idea of biosensors is based on the high specificity of biological recognition components. These elements allow for the

A. Sharma · B. O. Madhu · A. Suraj · C. Indumathi · C. K. Sunil · A. Rawson · D. V. Chidanand · S. Shanmugasundaram (✉)
National Institute of Food Technology, Entrepreneurship and Management, Thanjavur, India
e-mail: ashoksuraj2021@iifpt.edu.in; sunil.ck@iifpt.edu.in; ashish.rawson@iifpt.edu.in; chidanand@iifpt.edu.in; sasu@iifpt.edu.in

C. K. Sunil et al. (eds.), *Emerging Non-destructive Technologies for Food Quality Analysis*, https://doi.org/10.1007/978-3-031-95048-3_14

targeted detection of analytes in complicated matrices, which makes them useful instruments for a variety of applications, such as the investigation of food quality.

The beauty of biosensors lies in their ability to navigate the complicated landscapes of food matrices, offering targeted detection among the complexity of contaminants, pathogens, and quality indicators. From ensuring freshness to preventing spoilage. Biosensors are essential for assuring food items' safety, integrity, and nutritional value when it comes to food quality analysis. Biosensors have several advantages over conventional laboratory-based techniques, including the potential to enable quick, on-site detection of pollutants, infections, and quality indicators. Biosensors give food producers, processors, and regulators the ability to monitor and evaluate food quality characteristics in real time, including freshness, spoilage, and nutritional content. This allows them to make well-informed judgments and punctually take corrective action. Moreover, biosensors contribute to enhancing consumer confidence by ensuring compliance with regulatory standards and safeguarding public health.

The first biosensor was developed in the 1950s by American biochemist L.L. Clark, whose groundbreaking work is the source of biosensor technology. The original biosensor, sometimes referred to as the oxygen electrode or Clark electrode, was developed to gauge blood oxygen levels (Heineman & Jensen, 2006). Later developments resulted in the combination of transducer technologies with biological recognition elements like enzymes and antibodies, creating a wide range of biosensors tailored for particular uses. Because of developments in materials science, nanotechnology, and biotechnology, biosensor technology has evolved dramatically over the years. These days, biosensors are an advanced instrument in the toolbox of analytical methods, having a wide range of uses in agriculture, healthcare, environmental monitoring, and, most importantly, food quality analysis.

As we dig deeper into the field of food quality analysis, biosensors emerge as guiding light of hope, illuminating the path towards a future where every bite is a indication to safety, integrity, and excellence, exploring the infinite possibilities of biosensor technology helps to shape the future of food by knowing its secrets. In the context of food safety, biosensors play a pivotal role in detecting foodborne pathogens, allergens, toxins, and chemical contaminants. It is critical to identify pathogens like Salmonella, Escherichia coli, and Listeria monocytogenes as soon as possible in order to stop foodborne infections and outbreaks. Because biosensors may provide data in real time or within hours, they provide advantages over traditional culture-based approaches. This means that prompt interventions can be made to limit hazards and stop the spread of contamination. Additionally, biosensors play a crucial role in detecting allergenic substances in food products, like gluten, soy, and peanuts, guaranteeing that labelling laws are followed and protecting customers with dietary allergies. To address this crucial component of food safety, biosensor technology's sensitivity and specificity are highlighted by its capacity to identify minuscule amounts of allergens, even in complex food matrices. Apart from microbiological infections and allergies, biosensors are utilized to identify chemical pollutants and adulterants present in food items. When quantities of pesticides, mycotoxins, heavy metals, and veterinary medication residues exceed permitted limits, there are serious health concerns to humans. Biosensors facilitate regulatory compliance and guarantee the quality of food items from farm to fork by providing quick screening methods for these pollutants. Biosensors provide quick evaluation of essential nutrients, vitamins, and micronutrients in food items in the field of nutritional analysis. Biosensors provide low-cost, high-throughput ways to verify nutritional quality and regulatory compliance, from measuring vitamin content in fortified meals to monitoring sugar levels in beverages.

With continued progress in materials science, data analytics, and nanotechnology, biosensors in food safety and quality testing have a bright future. The accessibility of food testing technology will be facilitated by the miniaturization of biosensor equipment, wireless communication advancements, and smartphone integration, especially in resource-constrained environments and developing nations. The efficiency and throughput of food testing laboratories will also be improved by the introduction of multiplexed biosensors, which can detect numerous analytes concurrently, cutting down on processing times and operating expenses. Furthermore, real-time data analysis and predictive modelling will be made possible by the integration of artificial intelligence and machine learning algorithms, giving food producers and regulatory bodies useful information for risk assessment and decision-making. Despite the numerous advantages offered by biosensors in food safety and quality testing, several challenges remain in their widespread adoption and implementation. Regulatory frameworks, defined protocols, and validation processes are some of the difficulties in ensuring the dependability, precision, and repeatability of biosensor-based assays. Furthermore, multidisciplinary cooperation between scientists, engineers, food manufacturers, and regulatory bodies is necessary for the integration of biosensors into the current food production and monitoring systems.

14.1.1 Evolution of Biosensors

Biosensors have gone through several evolutionary stages, divided into generations, which correspond to advances in technology and methodology, especially in the area of combining biological recognition elements with transducer technology.

14.1.1.1 First-Generation Biosensors

The first generation of biosensors, which mostly used biocomponents that were membrane-bound or membrane-entrapped, set the stage for future developments. Notably, innovators like Clark and Lyons developed the idea of the enzyme electrode, which streamlined experimental configurations by combining necessary parts into a single unit (Clark & Lyons, 1962). This generation mostly used direct detection approaches, like tracking changes in the amount of substrates or products produced by enzymes. In the conventional first-generation biosensor design, for example, glucose oxidase (GOx) functioned as a biological recognition element, allowing glucose monitoring through electrochemical detection of either an increase in H_2O_2 concentration or a drop in O_2 concentration (Schachinger et al., 2021). However, this method has drawbacks, such as the need to maintain consistency in O_2 concentrations between samples and being susceptible to influence by biological sample constituents such dopamine or ascorbic acid (Armstrong, 1991).

14.1.1.2 Second-Generation Biosensors

A significant advancement was marked by the second-generation biosensors, which placed an emphasis on removing inactive membrane layers and improving sensitivity. Reagentless measurement regimes were accomplished by directly fixing biocomponents at the transducer surface, which was a major advancement in the development of biosensors (Updike & Hicks, 1967). Increased sensitivity and specificity were also facilitated by advancements like the immobilization of cofactors or electron relays close to the active sites of enzymes.

According to Clemens et al. (1977), the integration of an electrochemical glucose biosensor into a bedside artificial pancreas serves as an example of how the second generation of biosensors increased analytical efficiency and yielded practical applications by introducing auxiliary enzymes and co-reactants into the biological layer.

14.1.1.3 Third-Generation Biosensors

Third-generation biosensors, which combine signal processing and chemical recognition at the biocomponent level, represent a paradigm change. By directly immobilizing biocomponents onto electronic components, such as a field-effect transistor (FET) gate, this transformational technique enables direct monitoring and amplification of surface property changes (Janata, 1990). This integration removes noise and enables statistical analysis using several gate sensors, resulting in increased precision and dependability. A significant improvement was made in the third generation when bioreceptor molecules were completely incorporated into the base sensing element, enabling direct interaction between the electrodes and enzymes without the need for intermediary stages (Naresh & Lee, 2021). When (Liedberg et al., 1983) employed surface plasmon resonance (SPR) technology in 1983 to detect dependent responses in real-time, this ground-breaking discovery became a symbol in the history of biosensors.

Biosensors have advanced with each new generation to overcome obstacles and increase their capabilities, driving innovation and transforming applications in a variety of industries. With their unmatched precision and versatility in non-destructive technologies for food quality analysis and beyond, biosensors continue to push limits from the simplicity of first-generation designs to the complexity of third-generation structures.

14.2 Principle

Biosensors are innovative devices that work on the basic principle of signal transduction and are applied in many different fields (Ziegler & Göpel, 1998). A biological sensing element and a transducer are the two fundamental parts of a biosensor. Biological materials such as nucleic acids, enzymes, antibodies, and whole cells are all included in the biological sensing element, often known as the bioelement. These bioelements interact with the material being studied, the analyte, to cause a particular biological reaction (Koyun et al., n.d.). The transducer, which skilfully transforms this biological reaction into an electrical signal to enable measurement and analysis, is essential to the operation of a biosensor. Deactivating the desired biological material such as a particular enzyme and placing it close to the transducer are common steps in this conversion process. The biological component causes an electrical reaction when it comes into contact with the analyte, producing a quantifiable result. In addition, the architecture of a biosensor commonly incorporates an electronic circuitry, comprising a signal conditioning unit, a processor or microcontroller, and a display unit. The signal conditioning unit is essential for improving and refining the electrical signal because it frequently houses an amplifier and filter circuitry (Perumal & Hashim, 2014). Figure 14.1 depicts the working principle of the biosensor.

14.3 Bio-recognition Elements

Biosensors use a recognition element and signal transduction to convert the interaction between the target and the recognition element into a signal (Morales & Halpern, 2018). Target analyte specificity is the main purpose of a biorecognition element on a biosensor. The target analyte is captured by the biomolecule while it rests on a sensor surface. The process of adding a biorecognition component to a modified sensing surface is known as immobilization. Figure 14.2 illustrates typical electrochemical biosensors incorporating various biorecognition elements. This approach is critical for developing a biosensor with high selectivity and durable reusability. The most important aspect is maintaining a perfect link between the

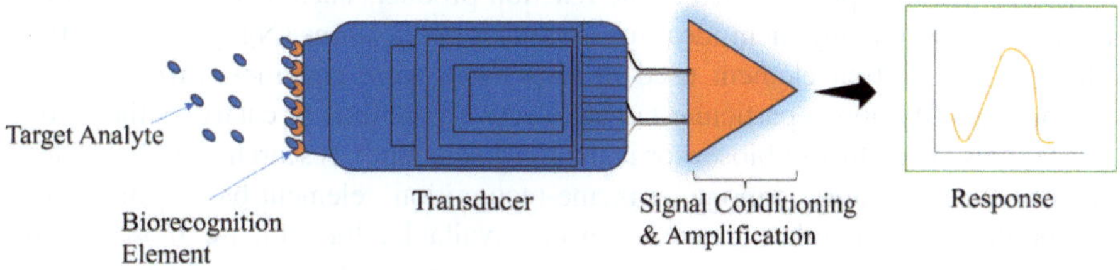

Fig. 14.1 Working principle of biosensor

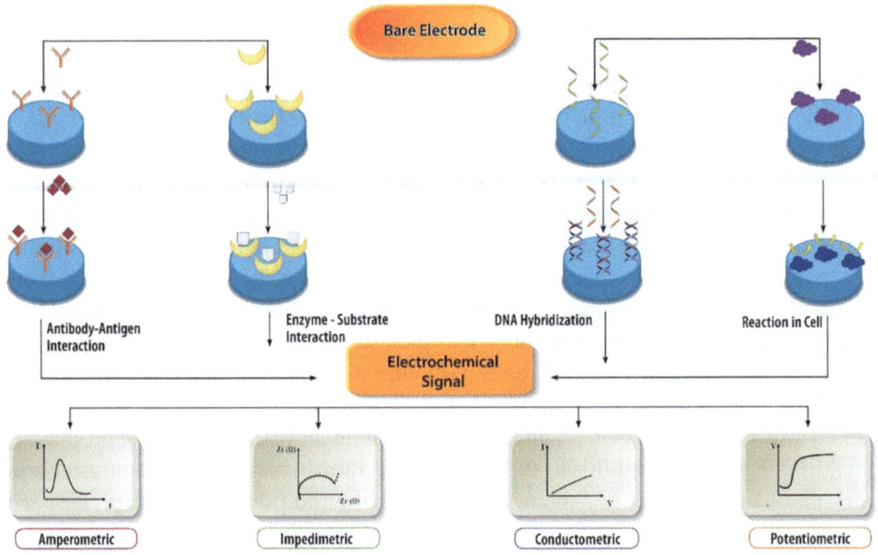

Fig. 14.2 Typical electrochemical biosensors with different biorecognition elements

target and the biorecognition element. In addition, the immobilization procedure needs to be quick and incorporate a lot of biorecognition components on the sensor surface. Reversible and irreversible immobilization are the two different forms of immobilization. The biorecognition element is immobilized irreversibly to prevent it from separating from the sensor surface. Procedures like covalent bonding, crosslinking, and trapping are used in this kind of immobilization. Although these methods are quite stable, they have some serious disadvantages, including toxicity linkers, loss of enzyme activity, and an expensive method that requires very pure enzyme. On the other hand, reversible immobilization enables the biorecognition component to be separated, making it suitable for regeneration.

14.3.1 Enzyme

Catalytic enzymes produce detectable reaction products such as protons, electrons, light, and heat, making them ideal for biosensor applications (Nguyen et al., 2019). The first recognition element used in biosensors were enzymes. Enzymatic biosensors quantify how a particular target selectively inhibits or catalyzes the activity of enzymes. The glucose biosensor is the most frequently researched and acclaimed sensor success story among enzyme-recognition element-based biosensors. Although there are other glucose sensors available, the glucose biosensor has remained relatively same in principle over time. Glucose interacts with an immobilized enzyme, and transduction is done amperometrically via an electrode.

Currently, most glucose biosensors use glucose oxidase as their identification element, which catalyzes the oxidation of glucose into gluconolactone (Chambers et al., 2002).

14.3.1.1 Immobilization of the Recognition Element

Enzyme immobilization is a technique used to attach enzymes to a solid support or matrix, thereby confining them in a certain space while allowing substrates to interact with them. Enzymes are simply adsorbed onto the surface of a solid support, such as activated carbon, silica gel, or ion-exchange resins. This method is relatively simple but may result in enzyme leaching (Hanefeld et al., 2009). Enzymes are covalently bonded to the surface of a support material through chemical reactions. This method provides strong attachment but may affect enzyme activity due to chemical modification. Enzymes are physically entrapped within a porous matrix, such as calcium alginate beads or polymeric membranes. This method protects the enzyme from harsh conditions but may reduce substrate accessibility (Subramanian et al., 1999). Enzymes are cross-linked to form a stable three-dimensional network within a support material, such as gelatin or polyacrylamide. This method maintains enzyme activity while providing stability and mechanical strength. Enzymes are immobilized in a water-immiscible organic solvent, often mixed with a hydrophilic polymer or surfactant. This method allows enzymatic reactions to occur in non-aqueous environments, expanding the range of potential applications. Enzymes are immobilized via specific affinity interactions between a ligand attached to the support and a complementary binding site on the enzyme. This method offers high specificity but requires the availability of suitable ligands (Guisan et al., 2020). The choice of immobilization method rests on factors such as the nature of the enzyme, the intended application, and the desired properties of the immobilized enzyme (e.g., stability, activity, reusability).

14.3.2 Antibodies

Antibodies (Abs) are glycoprotein molecules that are Y-shaped and consist of two heavy and two light polypeptide chains. B-cells create them to identify and neutralize antigens, which include a range of diseases. The bulk of antigens have a complex structure and comprise many epitopes that various lymphocytes can identify. The polyclonal antibody (PAbs) response resulted from each cell producing an antibody against a different epitope. This means that the produced PAbs are a collection of heterogeneous antibodies capable of recognizing and interacting with a wide range of epitopes within a single antigen (Crivianu-Gaita & Thompson, 2016). Kohler and Milstein first discovered a method for producing monoclonal antibodies (MAb) in 1975. Since a single B-lymphocyte clone is used to make MAbs, they have a monovalent affinity and are only able to recognize one particular, consistently identical epitope (a tiny antigen fragment) from an antigen (Foudeh et al., 2012). Yashini et al. (2022) created an electrochemical immunosensor for *Bacillus cereus* detection that is based on chitosan/gelatin. Cross-linking was employed to immobilize anti-Bacillus cereus polyclonal antibodies on an electrode modified with chitosan and gelatin. The invented immunosensor exhibited a LOD of 101 CFU/mL, LOQ of 10^3 CFU/mL, detection time.

14.3.2.1 Immobilization of the Recognition Element

Immobilizing antibodies onto a solid support allows for the capture and detection of specific antigens with high specificity and sensitivity. Antibodies are modified to introduce biotin molecules, which can then bind to streptavidin-coated surfaces. This method offers strong and specific immobilization but requires additional steps for biotinylating and may suffer from steric hindrance effects (Gao et al., 2022). Antibodies can be immobilized via their Fc region using protein A or protein G, which specifically bind to the Fc region of immunoglobulins (Choe et al., 2016). This method provides oriented and stable immobilization of antibodies while preserving antigen-binding activity. Antibodies are physically entrapped within a porous matrix, such as hydrogels or nanoparticles. This method protects antibodies from denaturation and may provide sustained release in drug delivery applications. Antibodies are covalently attached to the surface of a support material through chemical reactions. Common coupling chemistries include amine coupling, thiol coupling, and carbodiimide-mediated coupling. This method provides stable attachment and maintains antibody activity but requires careful optimization to prevent steric hindrance and maintain antigen-binding specificity.

14.3.3 Nucleic Acid

The primary foundation of traditional nucleic acid biosensors is single-stranded DNA's (ssDNA) inherent affinity for its complementary strand. Because of ssDNA's natural affinity, target-specific genes, such as those peculiar to bacteria, can be discovered. To accomplish this, these classic nucleic acid biosensors evaluate the hybridization of the single-stranded DNA (ssDNA) sample to a homologous strand fixed on the surface of the sensor chip (Gao, 2006). Moreover, when intercalating agents are integrated into the helical structure of a double-stranded oligonucleotide, they naturally bind to nucleic acids. By tracking the intercalation of various targets, a nucleic acid biosensor can quantify them by using dsDNA that is fixed onto the surface of the sensor chip.

14.3.3.1 Immobilization of the Recognition Element

High reactivity and appropriate orientation of the immobilized DNA probe would be encouraged by a decent DNA probe immobilization technique, enabling it to hybridize through its complementary DNA target (Teles & Fonseca, 2008). DNA probes are adhered to the working electrode surface by electrostatic adsorption of a negatively charged phosphate group of DNA on positively charged electrodes modified with film. According to earlier studies, the goal of using cationic polymeric films and chitosan for DNA probe adsorption is to produce a high density of charged positives and good biocompatibility (Rashid & Yusof, 2017). In order to bind covalently to the metal surface, the synthetic DNA probe is frequently attached to a particular functional group that is introduced to the electrode surface during the covalent bonding process, or to a group of thiols (Single bond H) or amines (NH2) at the end of 3′ or 5′. By using this method, the electrode surface of the DNA probe was attached in a very particular manner that can inhibit non-specific binding (Lu et al., 2007). Chemisorption and covalent attachment are widely utilized to immobilize DNA probes. Another technique for immobilizing DNA probes non-covalently on electrode surfaces is to combine avidin and biotin to create a compound. The avidin/streptavidin binding site is highly affinitized ($Ka = 1 \times 10^{-15}$ M) by the tiny biotin molecule (molecular weight = 2.44.31 g/mol), leading to nearly covalent attachment (Pividori et al., 2000).

14.3.4 Aptamer

Aptamer technology made it possible for nucleic acid biosensors to be extended to almost any kind of analyte because of the distinct three-dimensional structure of single-stranded nucleic acids (Wang et al., 2019). It is believed that these aptamers identify their target mostly based on their structure rather than their sequence. Since aptamer technology is still in its infancy, several aptamers have been chosen to target various targets, such as supramolecular structures, low molecular weight compounds, and entire organisms. Nevertheless, not all of these aptamers have been turned into biosensors yet. The word "aptamer" comes from the Latin "aptus," which meaning "to fit." Single-stranded DNA or RNA, or nucleic acid ligands, are extracted from oligonucleotide libraries using an in vitro selection method known as SELEX (Systematic Evolution of Ligands by EXponential enrichment) (Dwivedi et al., 2010).

14.3.4.1 Immobilization of the Recognition Element

Simple and straightforward immobilization methods, such as the utilization of short, thiolated linkers like cysteamine, 3-mercaptopropionic acid, 3,30-dithiodipropionic acid, Lomant's Reagent, and aromatic thiols, as well as the direct immobilization of thiolated aptamers (Oberhaus et al., 2020). Aptamers are often immobilized in a mixed self-assembled monolayer through co-immobilization or backfilling with short alkanethiols, such as mercaptohexanol. This technique tries to prevent unspecific interactions between aptamers and the gold surface by maintaining their upright orientation and tight packing, as well as covering vacant gold surfaces to prevent fouling (Zhou et al., 2010).

14.3.5 Whole Cell

whole cells. Biosensors, which are typically microbiological (algae, bacteria, fungus, yeast, and tissue culture cells), operate on the connection of electrodes to immobilized living cells and the detection of organic substances assimilated/produced by the cells (Ye et al., 2019). The capacity to generate vast quantities of microbial cells, a wide variety of enzymes, a long lifespan, high pH, and thermal stability are some of the benefits of these kinds of sensors. They also do not require extraction or purification. However, these sensors are slow and have little selectivity. Compared to enzymes and antibodies, using bacteria as a biorecognition factor is less expensive, permits more flexibility to specific environments, and can mobilize new compounds. Compared to plant or animal cells, microbial cells are highly stable and easier to handle. Electrochemical measures of CO_2 production and oxygen consumption, as well as measurements of pH and redox potential changes, are used to assess the condition of cells (Giese, 2002). Typically, this kind of biosensor has been employed in environmental monitoring and food analysis, such as the detection of pesticides, heavy metals, and organic contaminants. Diverse pathogens (whole cells) were used as a biomolecule to identify heavy metals (Bhavadharini et al., 2022).

14.3.5.1 Immobilization of the Recognition Element

Whole-cell immobilization involves confining entire living cells onto a solid support or within a matrix while maintaining their viability and metabolic activity. Cells are entrapped within a porous hydrogel matrix, such as alginate, agarose, or polyacrylamide. This method provides physical protection to the cells while allowing the exchange of nutrients and metabolites. Cells are encapsulated within microcapsules made of materials like alginate, chitosan, or liposomes. This method offers protection from harsh environments and allows for controlled release of metabolites. Cells are immobilized onto surfaces functionalized with specific ligands or antibodies that interact with cell surface receptors. This method offers site-specific immobilization and is often used in cell-based biosensors and tissue engineering. Cells are immobilized within a sol-gel matrix formed by the hydrolysis and polymerization of alkoxysilanes. This method offers a biocompatible environment and allows for the encapsulation of various types of cells, including bacteria, yeast, and mammalian cells.

14.3.6 *Phage*

Viruses known as bacteriophages infect bacteria and utilize the host bacterium as a factory to replicate themselves. Phage display is the term for the ability of bacteriophages to show proteins or peptides on their surface. Phage display makes it feasible to find peptides or proteins that have an affinity for many targets, including proteins, cells, and tiny molecules (Santos et al., 2020). Consequently, phage libraries which comprise a large variety of phages, each of which has a unique peptide or protein on its surface—are employed. There are various kinds of phage libraries that show various kinds of proteins or peptides, such as peptides, cellular proteins (from cDNA libraries), or antibody fragments like antigen binding fragments (Fab) and single chains variable fragments (scFv). An affinity selection process can be used to separate the phages with more affinity and specificity for a target from the enormous amount of phages in these phage libraries (Van Dorst, De Coen, et al., 2010a). These highly selective and affinity phages can be employed as a biosensor's target-specific recognition element. Apart from the target-specific phages, immediately usable peptides or proteins that are identified by phage display as efficient binders can also be employed as a recognition factor. Following that, peptides or proteins are produced chemically or by means of bacterial recombinant expression. The phages identify their host by utilizing specific receptor molecules on the exterior of the bacterial cell. Once the phages have located their specific receptors, they adhere to the bacterial cells and start the infection process. Only specific strains of a single species can become infected due to the highly precise interaction that phages can have with bacterial cells (Ahmad et al., 2019).

14.3.6.1 Immobilization of the Recognition Element

It was not necessary to immobilize the phage probe on a sensor surface because most research uses fluorescence or bioluminescence as the major techniques of detection. In investigations that used SPR or QCM to evaluate analyte binding, phage immobilization was achieved by interactions between phage primary amine functions and amine reactive groups (NHS ester, isocyanate, epoxy, etc.) on the sensor surface or thiol-gold chemisorption. Unlike other recognition elements, immobilizing phages does not require fluorescence or bioluminescence measurements, and can be achieved through simple dip coating (Torrance et al., 2006; Van Dorst, Mehta, et al., 2010b).

14.3.7 *MIP*

MIPs are materials that have been intentionally cross-linked and engineered to have recognition sites. MIPs have a narrow affinity for molecules that are closely similar to one another, but they can bind a target molecule selectively. They imitate the biological function of natural receptors like antibodies in this way (Pan et al., 2018). The process of molecular imprinting entails creating a "mould" around a template molecule made of polymerizable functional monomers. The template molecule and the monomer are allowed to form a bond. Next, utilizing cross-linkers, the resulting complexes proceed through copolymerization in the presence of an initiator. Following polymerization, a series of painstaking washing procedures are used to extract the template molecule, leaving behind particular recognition sites in the polymer that are identical to the template molecule in terms of size, shape, and functional group location. These recognition sites act functionally similar to the antibodies (Van Dorst, Mehta, et al., 2010b).

14.3.7.1 Immobilization of the Recognition Element

The specific transduction technique may require immobilizing MIPs on a sensor surface. The detecting MIP layer on the chip or sensor surface must be immobilized in order to use electrochemical and mass-based MIP sensors (also known as QCM-based sensors). By polymerizing MIPs on the sensor surface, this immobilization can be achieved instantly. As such, the pre-polymerization mix is applied to the sensor surface using spin coating, spray coating, or the simple drop or spread method (Van Dorst, Mehta, et al., 2010b). After then, UV, heat, or electro polymerization can be used to achieve the final MIP's polymerization. PVC coating is a common method for immobilizing MIPs on sensors. MIPs are mixed with PVC powder and dissolved in THF. When THF evaporates at room temperature, MIP is applied to the sensor surface. Finally, the MIPs are arranged in a flow-through cell since optical sensors which rely on fluorescence detection do not need to be directly immobilized. The displacement of MIP by the flow stream is inhibited by a small piece of glass wool or nylon net positioned at the bottom of the cell (Van Dorst, De Coen, et al., 2010a).

14.4 Types of Biosensors

14.4.1 Electrochemical Biosensors

Electrochemical biosensors are one of the widely researched and used biosensors, based on the electrochemical properties of the analyte and transducer. Electrochemical biosensors have high degrees of sensitivity, selectivity, and detection capabilities. Through an electrochemical reaction between the bioreceptor and analyte on the transducer surface, this biosensor generates detectable electrochemical signals in the form of voltage, current, impedance, and capacitance (Grieshaber et al., 2008). Figure 14.3 provides a schematic representation of various types of biosensors.

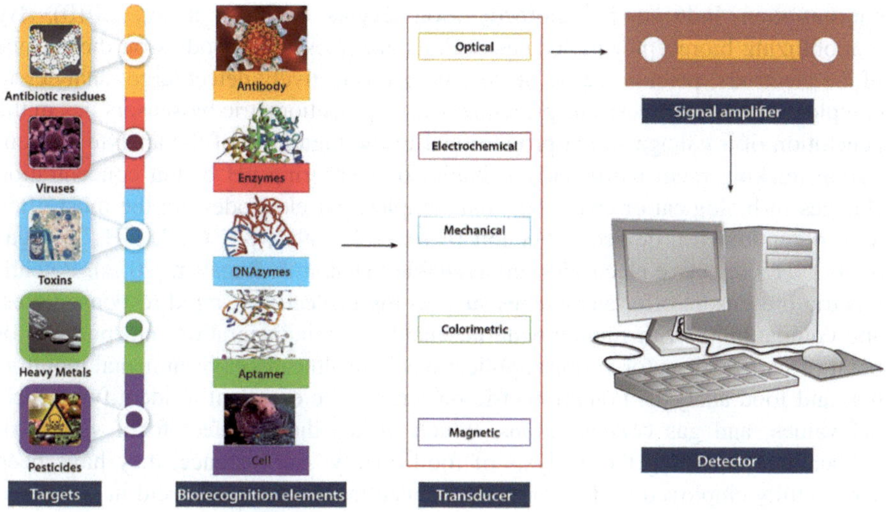

Fig. 14.3 Schematic representation of various types of biosensors

14.4.1.1 Amperometric Biosensors

Amperometric biosensors are an essential type of analytical instruments used in non-destructive food analysis methods. These biosensors work by detecting the current that is produced when analytes and biorecognition elements usually enzymes that have been immobilized on the sensor's surface electrochemically react. Amperometric biosensors rely on measuring the current resulting from a redox reaction (Bollella & Gorton, 2018; Davis, 1985). High sensitivity, quick reaction times, and the capacity to identify a wide range of analytes, such as sugars, organic acids, and pesticides (Alaejos & Montelongo, 2004; Artigues et al., 2017; Ciucu et al., 2002), are among the benefits that amperometric biosensors provide in the food analysis field. For instance, they are widely used in the food processing sectors to ensure quality control by measuring the levels of glucose in food items. They are also useful for identifying heavy metals and pesticides in food samples, which helps with the evaluation of food safety (Nomngongo et al., 2011; Wu et al., 2009). Amperometric biosensors are an effective and dependable way to evaluate food safety and quality without tampering with or harming the samples when integrated into non-destructive methods.

14.4.1.2 Potentiometric Biosensors

Potentiometric biosensors are analytical devices that measure changes in electrical potential at an electrode surface resulting from biochemical reactions. These biosensors typically utilize ion-selective electrodes (ISEs) to detect changes in ion concentration induced by biorecognition events (Bratov et al., 2010). By immobilizing bioreceptor molecules such as enzymes or antibodies on the surface of the electrode, potentiometric biosensors can selectively detect target analytes in complex sample matrices. The principle behind potentiometric biosensors lies in the generation of a voltage signal proportional to the logarithm of the analyte concentration, making them particularly suitable for measuring pH or ion concentration changes in biological or environmental samples. pH electrodes are the most often used potentiometric device; other ion (F-, I-, CN-, Na^+, K^+, Ca^{2+}, NH_4^+) or gas (CO_2, NH^3) selective electrodes are available (Thakur & Ragavan, 2013). Potentiometric biosensors offer advantages such as high selectivity, rapid response times, and compatibility with miniaturization, enabling their integration into portable or point-of-care devices for various applications in healthcare, environmental monitoring, and food analysis. Potentiometric biosensors are excellent in identifying ions, pH values, and gas concentrations, which makes them perfect for a variety of applications including the analysis of food quality. For instance, they have been successfully employed to determine the concentration of fumaric acid in powdered food products and urea concentrations in milk samples (Santini et al., 2012; Trivedi et al., 2009). A method to identify *Salmonella. typhimurium* at as low as 119 colony-forming units (CFUs) was developed by (Dill et al., 1999). Despite some difficulties with sample debris, they were able to obtain a promising recovery rate of 90% in a simulated real sample analysis employing chicken carcass washings spiked with Salmonella. Gehring et al. (1998) utilized potentiometric biosensors for live *Escherichia. Coli* O157:H7 test, exhibiting a limit of detection (LOD) of 2.5×10^4 cells/ml. With focus on identifying pathogens in food samples, this method demonstrates the sensitivity and adaptability of potentiometric biosensors, providing quick and dependable monitoring capabilities to guarantee food safety.

14.4.1.3 Impedimetric Biosensors

Impedimetric biosensors are advanced analytical tools that measure impedance changes at the electrode-electrolyte interface to identify and quantify biological interactions. The basic idea behind these biosensors is that when biomolecules or target analytes are present on the sensor surface, the electrical characteristics of the interface are changed, which results in variations in impedance (Chai & Oh, 2020). High sensitivity and specificity target analytes can be selectively captured and detected by impedimetric biosensors by immobilizing biorecognition elements on the sensor surface, such as enzymes, antibodies, or nucleic acids. Typically, methods like electrochemical impedance spectroscopy (EIS) are utilized to monitor the sensor's impedance spectrum as part of the detecting mechanism (Pejcic & De Marco, 2006). Label-free detection, real-time monitoring capabilities, and compatibility with miniaturization for portable or point-of-care applications are some benefits of impedimetric biosensors.

The technique of Impedimetric biosensor can be used to detect various mycotoxins, pesticides, herbicides Foodborne pathogenic bacteria and Allergens etc.

(Malvano et al., 2020), other than the solid sample (Calabrese et al., 2023) developed a non-Faradaic impedimetric biosensor for liquid and gas to sense three different volatile organic compound samples (1-octen-3-ol, trans-2-hexen-1-ol, and hexanal) involved in food spoilage. Impedimetric biosensors can achieve detection limits that are comparable to those of enzyme-linked immunosorbent assay (ELISA) and surface plasmon resonance (SPR) methods. By incorporating nanoparticles or employing microfluidic techniques, these biosensors are able to exceed the detection limitations of conventional immunoassays. To put it simply, they are incredibly sensitive and can identify compounds at very low quantities. Because of its adaptability, impedimetric biosensors are beneficial in a diversity of fields, such as environmental monitoring, food quality analysis, and healthcare, where precise substance detection is essential for accurate analysis and decision-making.

14.4.1.4 Voltammetric Biosensors

Voltammetric biosensors are a type of analytical instrument that uses current or potential changes as a function of applied voltage to identify and quantify biological interactions. Voltammetry, which measures the current response from an electrode when a voltage is applied, is the basis for these biosensors (Hung Tzang et al., 2001). Voltammetric biosensors have the ability to selectively capture and detect target analytes with high sensitivity and specificity by immobilizing biorecognition components such as enzymes, antibodies, or nucleic acids on the electrode surface. Following the oxidation or reduction of electroactive species produced during the biorecognition event, changes in current or potential are usually monitored as part of the detection process (Singh, Kumar, et al., 2020b). Voltammetric biosensors provide a number of benefits, including quick response times and suitability for miniaturization for moveable or point-of-care applications. These biosensors have a wide range of uses in pharmaceutical analysis, environmental monitoring, food safety evaluation, and biomedical diagnostics, demonstrating their adaptability and potential influence in numerous domains. Voltametric biosensors are crucial in the area of food quality analysis because they can identify a variety of substances, including acrylamides, nitrates, and tyramine (Dinçkaya et al., 2010; Kochana et al., 2016; Stobiecka et al., 2007). By quickly identifying potentially dangerous contaminants, these biosensors are effective tools for guaranteeing the safety and integrity of food products. Interestingly, by identifying adulterants like Sudan I and carmoisine (Elyasi et al., 2013; Shahidi & Branch, 2018), voltametric biosensors also help fight the widespread problem of food adulteration. This capacity is essential for maintaining regulatory standards in the food business and protecting consumer health. Voltametric biosensors make a substantial contribution to the continuous efforts to improve food quality and safety throughout the world through their sensitive and selective detection capabilities.

14.4.2 *Physical Biosensors*

14.4.2.1 Piezoelectric Biosensors

Piezoelectricity or piezoelectric effect is a physical phenomenon which relates to an ability of a material to generate voltage when mechanically stressed. It also has the same effect in the opposite scenario. When a piezoelectric material is exposed to alternating voltage, it results in mechanical stress or oscillation (Pohanka, 2018). Piezoelectric biosensors are novel analytical tools that observe and measure biological interactions by taking use of the piezoelectric effect. These biosensors make use of piezoelectric materials, which react to mechanical stress by producing an electric charge, such as quartz crystals or ceramics (Pramanik et al., 2013). Piezoelectric biosensors can selectively capture and detect target analytes with high sensitivity and specificity by immobilizing biorecognition components, such as enzymes, antibodies, or nucleic acids, onto the surface of the piezoelectric material. When the target analyte binds to the biorecognition element, it induces a variation in mass or viscoelastic properties at the sensor surface, leading to a corresponding alteration in the resonant frequency of the piezoelectric material (Pohanka, 2017; Skládal, 2016). The target analyte's presence and concentration are then determined by detecting and quantifying this shift in frequency. Larger molecular weight analytes induce a significant decrease in oscillation frequency, which makes piezoelectric immunosensors excellent at detecting them. However, the complexities of the procedure can make direct recognition employing antibodies on a piezoelectric platform difficult when dealing with low molecular weight analytes. In spite of this drawback, the technique is always quite good at identifying microorganisms, demonstrating its adaptability to a wide range of biosensing technology applications (Maraldo & Mutharasan, 2007a, b). Piezoelectric biosensors have various advantages, including label-free detection, real-time monitoring, and compatible with downsizing for portable or point-of-care use. These biosensors are versatile and have potential applications in a wide variety of sectors, including pharmaceutical analysis, environmental monitoring, food safety evaluation, and biological diagnostics (Huang et al., 2023; Kaur et al., 2023). Piezoelectric biosensors utilize the piezoelectric effect to detect changes in mass on the sensor surface. They are often employed in applications such as immunosensing and DNA detection.

14.4.2.2 Thermometric Biosensors

Thermometric biosensors are innovative analytical devices that take advantage of the principle of heat production or absorption resulting from biochemical reactions to detect and quantify biological interactions (Singh et al., 2020a). Usually, these biosensors work by immobilizing biorecognition components—like enzymes or antibodies—on the surface of a thermistor or other temperature-sensitive substance. The temperature change at the sensor surface is caused by the binding of the target analyte to the biorecognition element, and this change is then measured and quantified. Thermometric biosensors are simple, have quick reaction times, and can be miniaturized for use in point-of-care or portable applications (Yakovleva et al., 2013). These biosensors are versatile and have potential applications in a wide variety of sectors, including pharmaceutical analysis, environmental monitoring, food safety evaluation, and biological diagnostics (Ramanathan & Danielsson, 2001). Thermometric biosensors are essential instruments for assuring the safety and integrity of food products and are used extensively in food quality analysis. In particular, these biosensors are used in a variety of applications including milk quality inspection and food processing facility pathogen screening. Thermometric biosensors offer a quick and precise way to track temperature differences in milk, which might be an indication of contamination or spoiling (Kumar et al., 2012).

14.4.2.3 Surface Acoustic Wave (SAW) Biosensors

Surface acoustic wave (SAW) biosensors have emerged as indispensable tools in the realm of food quality analysis, offering rapid, sensitive, and non-destructive detection of various analytes. These biosensors harness the principles of acoustic wave propagation along the surface of a piezoelectric substrate, where interactions with immobilized biomolecules lead to measurable changes in wave characteristics. In the context of food quality analysis, SAW biosensors have demonstrated remarkable potential across multiple applications. For instance, they have been extensively utilized for the detection of foodborne pathogens, enabling timely intervention to prevent outbreaks of foodborne illnesses, and ensuring food safety standards are met. Hu et al. (2021) stated the successful detection of *Salmonella. Typhimurium* using a SAW biosensor with a self-assembled monolayer of anti-Salmonella antibody on a 36° YX $LiTaO_3$ substrate. This study highlights the efficacy of SAW biosensors in providing sensitive and specific detection of bacterial contaminants in food samples. Additionally, SAW biosensors have been instrumental in monitoring food spoilage indicators such as biogenic amines in real-time, facilitating early identification of deteriorating food products and minimizing food waste. Vasconcelos et al. (2021) developed a SAW sensor for the detection of biogenic amines based on sol–gel-derived titania thin films, showcasing the utility of SAW biosensors in ensuring food freshness and quality. With their high sensitivity, rapid response times, and potential for miniaturization, SAW biosensors hold great promise for revolutionizing food quality analysis by providing efficient, on-site detection solutions that enhance food safety and consumer confidence.

14.4.3 Optical Biosensors

An optical biosensor is a small analytical instrument that combines a biorecognition sensing element and an optical transducer system. The main objective of an optical biosensor is to produce a signal corresponding to the concentration of the chemical (analyte) being tested. A variety of biological materials, including complete cells, antigens, tissues, antibodies, receptors, nucleic acids, and enzymes, can be used by the optical biosensor as biorecognition components (Damborský et al., 2016).

Compared to conventional analytical methods, optical biosensors have a number of profits, including direct, real-time, label-free detection of a wide range of biological and chemical compounds. High specificity, sensitivity, small size, and affordability are some of its advantages. The type of biosensors that are most frequently reported on are optical biosensors. In optical biosensing, detection depends on how a biorecognition element interacts with the light field (Akgönüllü & Denizli, 2022). Label-free and label-based are the two primary types of optical biosensing. When using label-free sensing, the interaction between the transducer and the material being studied produces the observed signal. On the other hand, label-based sensing makes use of a label and produces the optical signal via luminescence, fluorescence, or colorimetric techniques (Balbinot et al., 2021; Chen & Wang, 2020). Label-assisted sensing enables the enzymatic oxidation of simple compounds, such as glucose, for detection. The most successful commercial use of biosensors to far has been in glucose analysis for blood glucose meters worn by diabetics. But when labeling is used in specific situations, such antibody-antigen interactions, it can change the binding characteristics and possibly induce systematic mistakes into biosensor analysis (Noy-Porat et al., 2021).

14.4.3.1 SPR

Since its discovery in 1902, surface plasmon resonance, or SPR, has grown to provide a thorough understanding of surface plasmon physics. In 1983, the first successful application of SPR for constructing biosensors to detect biomolecular interactions emerged. Pharmacia Biosensor AB, later renamed Biacore, introduced the first commercial SPR-based biosensor instrument (Damborský et al., 2016). Since several manufacturers now make SPR equipment, SPR has become the most widely used optical biosensing technique. When polarized light is shone at a particular angle on the surface of a metal (or other conducting material), surface photoluminescence (SPR) occurs at the interface of two media, usually glass and liquid. This results in the generation of surface plasmons, which at a particular angle known as the resonance angle reduce the intensity of reflected light. The mass on the surface determines the magnitude of this effect. By tracking changes in wavelengths, angle, or reflectivity over time, a sensorgram can be produced (Schasfoort, 2017).

Due to the SPR phenomenon, changes in the refractive index at the sensor surface that are proportionate to the concentration of biomolecules can be directly, label-free, and real-time monitored. Immobilizing one interacting molecule on the sensor surface allows for the measurement of ligand-analyte interactions. Through the use of surface plasmon-polaritons, a special kind of electromagnetic waves, SPR biosensors use optical spectroscopy to identify changes in refractive index brought on by binding events involving a target analyte. An SPR transducer plus a biological recognition element (such as an antibody, receptor protein, biomimetic material, or DNA) that may interact with a particular analyte make up SPR biosensors, a subset of affinity biosensors. Normally, one of these molecules often the biomolecular recognition element is present in a solution, while the other is immobilized on the sensor surface. The recognition element and analyte bind when the solution and sensor surface come into contact, changing the refractive index at the sensor surface (Khansili et al., 2018). By calculating changes in the properties of the interacting light wave, such as resonant wavelength, resonant angle, intensity, phase, and polarization, one can detect changes in the propagation constant of an SPP excited at the sensor surface (Mehrotra et al., 2019).

14.5 Applications in Non-destructive Food Quality and Analysis

At every point of the food supply chain, careful food handling and screening are required due to the rise in foodborne outbursts, growing consumer demand for better products, and an increase in food allergy instances. The current conventional methods require costly analytical equipment, skilled personnel, and extended sample preparation periods in order to identify food poisons, contaminants, allergies, and illnesses. Biosensors, which provide precise, selective, rapid, qualitative, and quantitative analyte detection, can help resolve these problems. On-site analyte detection is made possible by their non-destructive measuring techniques, portability, affordability, and simplicity of use. Biosensors' quick, precise, and affordable detection capabilities make them very promising for transforming food safety and quality evaluations. Biosensors may detect toxins, pollutants, allergens, and infections among other things in the food sector by utilizing a variety of transducer mechanisms. Figure 14.4 presents a list of some applications of biosensors in the food industry.

14.5.1 Detection of Biomolecules (Protein, Carbohydrates, Vitamins, Antioxidants etc)

Amperometric biosensors provide a flexible method for determining a food's antioxidant content. In one study, antioxidants in different fruits were identified using a biosensor based on the enzyme xanthine oxidase. The enzyme facilitated the production of H_2O_2 or O_2 radicals from the interaction between hypoxanthine and uric acid. The samples' antioxidants reduced these radicals, which decreased the observed current. The biosensor successfully detected gallic acid in many fruits, achieving a limit of detection (LOD) of 2.17μmol/L (Becker et al., 2019). In a different study, a polyphenol oxidase enzyme derived from bananas was used to

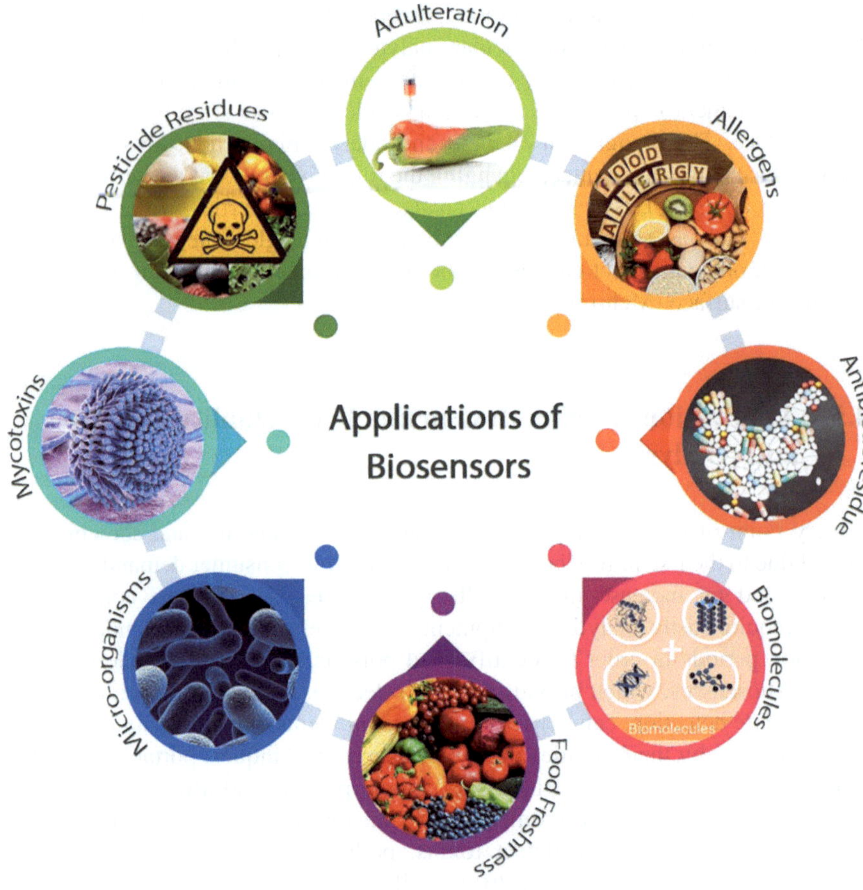

Fig. 14.4 Some applications of biosensors in the food industry

create a nylon membrane-based biosensor with a limit of detection of 0.2μM for the detection of polyphenolic chemicals in water samples, alcoholic beverages, and tea (Narang et al., 2011). Furthermore, an amperometric biosensor with a limit of detection of 0.83μM was developed to identify polyphenols in propolis samples. This biosensor was developed using the oxidation capability of a laccase enzyme that was immobilized on the electrodes. These results demonstrate how well amperometric biosensors work to calculate the antioxidant content of different food items (Mohtar et al., 2019).

A switch-on fluorescent biosensor was created by (Xu et al., 2015) to detect the antioxidant glutathione in food samples. The fluorescent probe used in this study was graphitic carbon nitride quantum dots (QDs). The presence of Hg^{2+} ions caused the QDs' negatively charged surface to initially exhibit quenched fluorescence. Glutathione molecules absorbed Hg^{2+} ions in the presence of glutathione, reviving the

QDs' fluorescence. With a LOD of 37 nM, the fluorescence intensity was directly correlated with the glutathione concentration. Tests were conducted on actual food samples, such as tomatoes, grapes, cucumbers, and spinach; all samples had recoveries greater than 94%. This biosensor demonstrates promising capabilities for detecting glutathione in food matrices.

14.5.2 Detection of Drugs or Antibiotic Residues

Antibiotics are widely used in cattle and aquaculture, but overuse can have long-term impacts on human health and the environment. Antibiotics can become a 'persistent' organic pollutant in water and soil due to repeated usage and continuous emissions. Antibiotics used in animals can leave residues in foods such poultry, beef, milk, honey, eggs, and fish. Residues of these antibiotic can cause allergies, immunopathological effects, nephropathy, hepatotoxicity, mutagenicity, carcinogenicity, bone marrow toxicity, reproductive problems, and anaphylactic shock. Antibiotic residues in food often lead to antimicrobial resistance, a growing global issue (Majdinasab et al., 2020). Antibiotics have posed a threat to human health in food items like grains, vegetables, milk, and surface water. Tetracycline consumption can cause bacterial drug resistance as well as harmful health effects like renal toxicity and intolerance. Therefore, it is crucial to find any antibiotic residue in meat, dairy, and other food products. Antibiotics are now frequently detected using colorimetric, fluorescent, and electrochemical techniques, HPLC, and surface-enhanced Raman scattering. However, the lengthy sample pre-treatment steps and specialised equipment needed for these methods make them time-consuming (Chu et al., 2020). To address the limits of traditional approaches, quick, affordable, sensitive, selective, and portable biosensors and bioassays have been developed to detect antibiotic residues in diverse matrices.

A biosensor using a bacterial mono-oxygenase enzyme immobilized on an activated glassy carbon electrode was used for the preliminary screening of Tetracycline in an animal food product (Besharati et al., 2019). According to their research, milk samples with little sample pretreatment could have Tetracycline detected at levels lower than the standards set by food control organisations. A multilayer electrode modified with a self-assembled thiourea monolayer (SATUM) followed by gold nanoparticles (AuNPs), mercapto-succinic acid (MSA) and antibody was investigated for the detection of ultra trace amount of a small molecule (chloramphenicol) in an impedimetric system for shrimp samples which caused the impedance to increase on formation of the antibody–antigen complex at the electrode surface (Chullasat et al., 2011).

A sensitive and effective ratiometric electrochemical aptasensor was created in order to identify tetracycline (TET) in milk. Aptasensor 1 and 2 were the two aptasensor that were combined to make the ratiometric electrochemical aptasensor. Gold nanoparticles (AuNPs) and ferrocene (Fc) combine to form the nanocomposite known as aptasensor 1. In the meantime, an AuNP nanocomposite and carbon nanofibers (CNFs) were used to create the aptasensor 2. By forming an Au-S link between AuNPs and the aptamer's thiol at the 5′ end, TET-aptamer was effectively immobilized onto the surface of screen-printed carbon electrodes (SPCEs), resulting in the creation of aptasensor 1 and 2. Additionally, a ratio was used to determine the detection findings. Under ideal conditions, differential pulse voltammetry (DPV) was used to measure the TET. With a detection limit of $3.3 \times 10^{-7} gL^{-1}$, the ratiometric aptasensor can quantitatively detect TET in the 10^{-8}–$10^{-3} gL^{-1}$ range using ratio calculation (Xu et al., 2017).

14.5.3 Detection of Adulteration

A three-electrode amperometric sensor in conjunction with micellar electrokinetic capillary chromatography allowed for the successful detection of melamine, a common food adulterant, in milk products. A carbon-disc working electrode was used to detect the oxidation current generated by melamine, and capillary electrophoresis was employed to aid the separation of components. A low detection limit of 2.1×10^{-6} g/mL was attained with this approach. Experiments conducted on samples of tampered milk and baby formula produced recovery rates higher than 83%, demonstrating the product's dependability in practical settings (Wang et al., 2010).

For the purpose of monitoring the food preservative benzoic acid, (Shan et al., 2007) developed a novel sensitive amperometric biosensor based on a composite matrix of polyaniline and polyacrylonitrile. For real-time monitoring of benzoic acid in milk, yoghurt, sprite, and cola samples, this biosensor was built on the inhibition of benzoic acid on the biocatalytic activity of the polyphenol oxidase enzyme to its substrate (catechol) with the detection of a limit of 2×10^{7} M. Benzoic acid was detected in beverage samples with a detection limit of 0.03 M using an amperometric biosensor based on the suppression of tyrosinase and polyphenol oxidase activity by benzoic acid.

14.5.4 Detection of Microorganisms

Quality control expenses amount to about 1.5–2% of the total sales of the food industry (Luong et al., 1997). It has been established that microbial contamination of food and water results in serious food-borne illnesses that have a substantial negative impact on human health. The most common bacteria known to cause foodborne illnesses are *Salmonella, Staphylococcus, E. Coli (O157:H7), Clostridium, Campylobacter, Shigella,* and *Bacillus cereus and L. monocytogenes.* Most food microorganisms are handily sent through the untreated water supply, half-cooked meat, milk, natural products, vegetables, and food. Utilization of normal facility areas may give higher probabilities of diseases (Arora et al., 2011). Several biosensors, including piezoelectric-based immunosensors, optical (SPR), and electrochemical biosensors, are frequently employed to identify microbes in food.

Yashini et al. (2022) created a chitosan/gelatin based electrochemical immunosensor to detect *Bacillus cereus*. Cross-linking was used to immobilize anti-Bacillus cereus polyclonal antibodies on an electrode modified with chitosan and gelatin. The invented immunosensor showed a LOD of 101 CFU/mL, LOQ of 103 CFU/mL, detection time <50 min, linearity ranged from 10^1 to 10^8 CFU/mL, high specificity, and high reproducibility and regeneration efficiency. In a study conducted by (García-Aljaro et al., 2010) carbon nanotube (CNT) based immunosensors for the location of two sorts of microorganisms such as bacteria and viruses. A critical expansion in the opposition of the gadget was seen when the biosensor was presented to *E. coli O157:H7* entire cells or lysates with a detection limit of 10^5 and 10^3 CFU (colony forming units)/mL. On account of virus identification, a huge obstruction was recognized because of the connection of the bacteriophages with the Abs, with a detection limit of 10^3 plaque forming unit (PFU)/mL compared to 10^1 PFU/chip and excellent selectivity against MS2 bacteriophage.

García-Aljaro et al. (2010), Saini et al. (2019) functionalized strands of DNA (Deoxyribose Nucleic acid) to immobilize carbon walls coupled to AuNPs. The aftereffects of *S. enterica* got with common techniques for recognition in milk were compared on nanosensors. The sensor's explicitness was affirmed within the sight of other food microorganisms: the origin of *Klebsiella pneumoniae, E. coli, and Listeria monocytogenes* (Vaisocherová-Lísalová et al., 2016). Utilized a surface plasmon resonance shift biosensor, and discovered that a co-polymer brush-based coating demonstrated high fouling obstruction and biorecognition capacities for a variety of foods (such as milk, spinach, cucumber, burgers, and lettuce) for the identification of bacterial pollution, including *E. Coli O157:H7, E. Coli O145:H2, and salmonella*. The detection parameters were found to be within concentrations ranging from 1.5×10^2 to 1.5×10^7 CFU mL/l (colony forming unit per millilitre), 1.5×10^2 to 1.5×10^7 CFU mL/l, and 2.5×10^2 to 2.5×10^7 CFU mL/l, respectively.

Yang et al. (2008) developed a DNA biosensor using nano-membranes for detecting *Pseudomonas aeruginosa*. Nano-sized membranes have a surface area that is orders that are of magnitude heavier than continuous thin films. According to their research, using a nano-membrane increased response sensitivity by around three times and increased the amount of bound DNA by up to 3–5 times (Yang et al., 2008). The multi-striped nanowires were created for effective and precise biowarfare simulant multiplexing detection. Antibody-coated nanowires with some distinctive yet simply recognizable encode structures were used in a multiplexed identification of three non-pathological stimulant substances of *B. anthracis, Variola,* etc. with detection limits of 1.9×10^5 CFU/ml, 1.9×10^5 PFU/ml (Tok et al., 2006). Kim et al. (2007) developed a nanoparticle biosensor to identify *S. enteritidis* by monitoring the variances generated by a cell of the target which is attached to anti-Salmonella immobilized on gold electrodes. The nanoparticle-based biosensor detected 10^4 CFU S. enteritidis/ml in phosphate-buffered saline (PBS) in 3 minutes.

Eissa and Zourob (2020) portray that by using a biological enzyme, a bacterial peptide in an electrochemical biosensor, *Listeria monocytogenes* was found as

9 CFU mL/l as a limit of detection using the silver nanoparticles. Also, *Aspergillus niger* has been found with a limited detection range of 50 spores using silver nanoparticles in a colorimetric biosensor. With the usage of antibodies as a bio-recognition element, an electrochemical biosensor is used with a limited deduction level of 8.0 ng mL/l with a platinum nanoparticle and with the biological element, bilharzia, along with nano-strip, a limited detection level of 8.3887×10^{-2} ng mL/l was found (George Kerry et al., 2021). The detection limit of VP was 7.4×10^4 CFU/ml for the screen-printed electrode covered in agarose/nano-Au membrane and horseradish peroxidase (HRP) tagged with anti-*Vibrio parahaemolyticus* antibody (HRP-anti-VP) (Yang et al., 2008).

14.5.5 Detection of Mycotoxins

A novel amperometric biosensor utilizing antibodies was created to identify the mycotoxin class known as zearalenones. The biosensor made use of a glassy carbon working electrode that had been altered to improve mycotoxin collection and increase surface area using gold and platinum nanoparticles as well as polyethylenimine-functionalized multiwalled carbon nanotubes. Following surface modification, antibodies were immobilized to allow for the specific detection of zearalenone. With a low detection limit of 1.5 pg/mL, the biosensor operated at a steady 0.18 V. Additionally, even in the presence of other mycotoxins, the application of antibodies guaranteed strong selectivity for zearalenones (Liu et al., 2017).

Magnetic nanoparticles coated with antibody (anti-AFM1 or anti-OTA) were used to bind antigen (AFM1 or OTA) in milk and cereal contaminated with AFM1 and OTA. The results showed good repeatability and detection limits of 40 nM and 0.01μg/L, respectively (Selvaraj et al., 2015). Chuan et al. (2011) developed a Pt electrode biosensor modified with sol-gel aflatoxin-oxidase linked multi-walled carbon nanotubes (MWCNTs) that responds to the oxidation of AFB1 with a response time of 44 s, a high sensitivity of 0.33×10^{-2} A mol−1·L cm^{-2}, and a LOD of 1.6 nmol/L. Magnetic nanoparticles (MNPs) were employed in a study by (Kim et al., 2012) to collect and measure AFB1 and ZEN in feed. The nanoparticles' surface was modified by amino-organ silanes to aid in the conjugation of monoclonal antibodies (kj-AFB and kk-ZEN) against AFB1 and ZEN, respectively. The nanoparticles were subsequently coated in silica (Si,*O*-2) to lower the silica concentration. In 10 ppb AFB1 spiked corn and feed, the C-AFB1/MNP@Si, *O*-2. system recovered 90.28% and 83.7%, respectively, while the C-ZEN/MNP@Si, *O*-2. system recovered 99.9% and 86.0% of 50 ppb ZEN in spiked corn and feed. These findings showed that this technology outperformed more established Immuno Affinity Chromatography (IAC) techniques in its ability to collect mycotoxin from feed. The monitoring and control of feed contaminated with mycotoxin is made possible by this innovative collection method.

Mak et al. (2010) examined the application of a magnetic nanotag-based immunoassay for the simultaneous sub-picomolar concentration levels of mycotoxins

(AFB1, ZEN, and HT-2) detection. To quantify the change in resistance with mycotoxin binding, 10 nm MNPs coated with streptavidin-biotinylated antibodies (anti-AFB1, anti-ZEN, and anti-HT-2) are submitted to LOD at 50 pg/mL for AFB1, ZEN, and HT-2, and 333 pg/mL for AFB1 and AFB1, respectively… One possible point of use for this immune-MNP assay is mycotoxin testing. To detect the HT-2 toxin, (Eggenberger et al., 2019) devised innovative signal transduction by ion nano-gating (STING) technology. This method makes use of a functionalized quartz nanopipette as an electrochemical biosensor; a thio-ether bond forms between the maleimide group of the PLL and the anti-HT-2 monoclonal antibody. Poly-L-lysine (PLL) that has been physiosorbed on a negatively charged quartz surface immobilizes the nanopipette. This method makes use of a functionalized quartz nanopipette as an electrochemical biosensor; a thio-ether bond forms between the maleimide group of the PLL and the anti-HT-2 monoclonal antibody. The nanopipette is immobilised with poly-L-lysine (PLL) that has been physiosorbed on a negatively charged quartz surface. For HT-2 detection, the LOD for the STING platform was 100 fg/ml.

A multicolour immunochromatographic test strip (ICTS) biosensor has been developed for the fast and simultaneous visual detection of aflatoxin B1, fumonisin B1 (FB1), ochratoxin A (OTA), and zearalenone (ZEN) (AFB1). To target each analyte, hyperbranched Au plasmonic blackbodies (black), gold nanoflowers (blue), gold nanospheres (red) and gold nano cacti (purple), were created and functionalized with the matching antibodies. For FB1, OTA, ZEN, and AFB1, the developed multicolour platform enhanced visual detection capacity with cut-off values of 120, 2.5, 60and 5 ng/mL, respectively. It was determined that the limits of detection for OTA, ZEN, AFB1, and FB1 were 3.27, 0.10, 0.70, and 0.06 ng/mL, respectively.

14.5.6 Detection of Pesticides/Insecticides/Herbicides

Pesticides, commonly employed for pest management and crop enhancement, have become a concern due to their adverse effects on the environment and human health. However, traditional methods for analyzing pesticide residues are not feasible in resource-limited regions due to their high cost, complexity, and reliance on expensive instruments like GC/MS and LC/MS.

To find paraoxon-ethyl pesticide in spinach samples, an amperometric biosensor based on enzymes was created and it is disposable. Acetylcholinesterase (AChE) enzyme activity was inhibited by paraoxon-ethyl, which was the basis for the biosensor. The AChE enzyme, immobilized on the working electrode, catalyzed the hydrolysis of acetylthiocholine. The presence of pesticide molecules hindered this reaction, leading to a change in current, which was used to estimate pesticide concentration. The biosensor achieved a low detection limit of 0.03μg/L. Testing on spiked spinach samples showed excellent correlation with results obtained through HPLC analysis (Hua et al., 2019). In order to create an immunosensor for the detection of the pesticide atrazine, (Deep et al., 2014) examined the development of a printed carbon electrode doped with polyaniline and modified with styrene sulfuric acid. The anti-atrazine antibody that was mounted on the sensor surface interacted with atrazine as the mechanism of action. The amounts of atrazine detected ranged from 0.01 to 50 ngm L^{-1}, and the detection was highly sensitive and specific. Choi et al. (2001) created a fibre optic biosensor based on the suppression of the enzyme acetylcholinesterase that can identify organophosphate chemicals in contaminated water. Rapid detection in groundwater is crucial since these substances are utilised in agriculture to preserve crops. The reduction of the product o-nitrophenol in the presence of the substrate nitro phenylacetate, caused by the inhibition of acetylcholinesterase by organophosphorus chemicals, served as the basis for the detection. The biosensor has a response time of roughly 10 minutes and can detect organophosphate chemicals up to 2 ppm. A quartz crystal microbalance sensor was created by (Gupta et al., 2015) to detect atrazine residues in residual water. A film made of allyl-mercaptan and 2-hydroxyethyl methacrylate-phenol atrazine was imprinted onto the surface to cover it. The sensor's linear response ranged from 0.08 to 1.5 nM, with a detection limit of 0.028 nM. The outcomes demonstrated the sensor's effectiveness in addition to its high sensitivity, fast response, and affordable price. Cyclic voltammetry and electrochemical impedance experiments were used to examine the electrochemical activity of the newly developed graphene nanocomposite paste electrode and silver/copper alloy nanoparticles. After that, it was applied to isolate chlorpyrifos, the active component of organophosphate insecticides, from soil and water samples without the need for enzymes. The multicolour platform that was built enhanced the ability of bare-eye-based visual detection with cut-off values of OTA, AFB1, ZEN, FB1, and 2.5 ng/mL, respectively. The results showed that the criterion for OTA, AFB1, ZEN, and FB1 were 3.27, 0.70, 0.10, and 0.06 ng/mL, respectively. The nanocomposite's electrical conductivity and active site count were boosted by silver/copper nanoparticles, and it displayed great stability, sensitivity, strong selectivity, a wide linear range, low detection limits, and good repeatability and reproducibility (Sreedhar et al., 2015).

14.5.7 Detection of Allergens

Glinadin can currently be detected with a created and refined label-free impedimetric aptasensor. Because of their great sensitivity to low gluten concentrations, the sensor uses the poly (amidoamine) dendrimer of fourth generation (PAMAM G4) at 2 mg/ml and the Gli1 aptamer at 0.5μmol. At a 5μg/l (equal to 5 ppm of gluten), repeatability below 5%, and 2 months of storage stability at 4 °C, the aptasensor showed linearity in detecting gliadin concentrations ranging from 5–50 to 50–1000μg/l. The ELISA approach produced results that were consistent when testing gluten and gluten-free food products (Malvano et al., 2017). A DNA-based impedimetric biosensor was developed to detect Ara h1, the peanut allergen. The sensor utilized a stem-loop probe designed to switch from a closed to an open state in the presence of the target DNA. It achieved a limit of detection

(LOD) of 0.35 fM. Analysis of peanut milk samples yielded a concentration of 3.2×10^{-13} mol/L, indicating its outstanding pertinency for actual food samples (Sun et al., 2012).

14.5.8 Food Freshness

The food processing industry faces challenges in ensuring quality, safety, and authenticity due to limitations in traditional techniques such as chemical experiments and spectroscopy, which are costly and time-consuming. There is a need for cost-effective alternatives that provide objective and consistent measurements. Biosensors offer promise as they offer simplicity, real-time monitoring, selectivity, and affordability, addressing the industry's demand for effective solutions. Amperometric biosensors were employed to assess food freshness, targeting biogenic amines as indicators of spoilage. A dual-working electrode array was used to produce a dual enzymatic biosensor. Putrescine oxidase was functionalized on one electrode to detect putrescine, and histamine dehydrogenase on the other to detect histamine. Putrescine and histamine were found with LODs of 10μM and 8μM, respectively. Testing on red wine and fresh octopus samples showed a strong association with HPLC readings (Henao-Escobar et al., 2016). A different study achieved a substantially lower limit of detection (LOD) of 0.71μM for tyramine by developing an enzyme-based biosensor based on tyrosinase. With recovery rates surpassing 93%, this biosensor tested red wine, beer, yogurt, and Roquefort cheese with success. These findings highlight the efficacy of amperometric biosensors in monitoring food freshness (da Silva et al., 2019). Ghasemi-Varnamkhasti et al. (2012) developed enzymatic biosensors utilizing cobalt phthalocyanine for monitoring the ageing of beer during storage. Their study demonstrated that these biosensors were effective in tracking the ageing process of beer, indicating their potential for quality control in the brewing industry.

14.6 Challenges and Limitations

Probable electrochemical interferences and effective target analyte extraction from the complex food matrix are two pertinent challenges related to real sample analysis. Sample preparation and cleanup procedures must be carefully coordinated with surface chemistry and bioreceptor type to prevent electrochemical interferences. Furthermore, more thorough validation investigations are needed, and effective attention and testing should be given to the electrochemical biosensor's storage and operational stability under actual analysis settings. To assess how food processing affects these biosensors' ability to detect things, further work needs to be done using them to analyse samples that have been processed. The capacity of electrochemical biosensors to measure and distinguish between various analytes in a sample at the same time is crucial, and it should be a selling point for industrial use.

14.7 Future Trends of Biosensors in Non-destructive Technologies for Food Quality Analysis

Developments in biosensors seem to have a bright future in non-destructive technology for food quality analysis; they could potentially transform the industry. Based on a Markets and Markets study, the growing need for food safety and quality assurance measures is expected to propel the global biosensors market, which is expected to reach USD 36.7 billion by 2026 (Biosensors Market Size, Share, Trends and Revenue Forecast - *2030*, n.d.). Emerging trends indicate to a move toward mobility, miniaturizing, and integration with wireless communication technology, which makes it possible to monitor food quality criteria in real-time. Future research on innovative biosensors, like multifunctional nanocomposite biosensors, will meet the need for more affordable nanozyme-based techniques and improved food safety (Huang et al., 2019). Additionally, there is a lot of opportunity to improve detection sensitivity, accuracy, and specificity by integrating machine learning and artificial intelligence algorithms into biosensor platforms. Biosensors are predicted to become widely used in food production, processing, and distribution as they continue to advance, streamlining quality control procedures and reducing hazards related to pollutants and foodborne pathogens.

The majority of biosensors on the market today lack the ability to detect ultra-low concentrations of foodborne pathogens while also having good sensitivity and specificity. Therefore, studies aimed at improving biosensor performance ought to be conducted. Nanomaterials have not yet been found to have detection sensitivity greater than one or a few cells. However, none of those detection tests used sample enrichment. In terms of detection speed, real-time pathogen detection is possible with nanomaterials-based detection systems (Bobrinetskiy et al., 2021). These methods could be very helpful for determining the presence of pathogens in food samples before conducting more involved laboratory tests. While official validation and consumer adoption of nanomaterials-based detection tests may still be necessary, the potential of nanoparticles for rapid and sensitive pathogen detection appears promising.

The advancement of nanotechnology in biosensor development is revolutionary, and it could be accelerated further with strategic investments of scientific resources in smart technologies such as the Internet of Things (IoT) built with advanced telecommunication networks, big data analytics, and artificial intelligence (AI) with deep learning accessibility (Chugh et al., 2024; Mayer & Baeumner, 2019; Rodrigues et al., 2016). IoT refers to any physical objects, such as smartphones, that are connected to the internet and share data instantly via a unique identification, enabling safe data transfer without the need for human-to-human or human-to-computer interactions. Another growing area of technological hybridization is the application of AI for biomedical monitoring-based biosensors for point-of-care diagnostics. This combination might determine the significance of machine-learning algorithms for future nano-based biosensors, as well as IoT, computational approaches, and artificial intelligence.

14.8 Conclusion

The term "biosensor" refers to a powerful and novel analytical equipment that includes a biological sensing element. Biosensors come in several types based on its biological or transducing elements, each with unique advantages and disadvantages. Biosensors are known for their stability, low cost, and reproducibility. Some even offer on-site analyte detection capabilities. Biosensors have numerous applications, including detecting carbohydrates in food, detecting antibiotics and heavy metals in the environment, monitoring food allergies, and more. Despite decades of development, there are still numerous applications that require additional exploration. Biosensing has advanced significantly in all fields over the last decade, but there is still a long way to go before non-destructive biosensors can completely replace traditional approaches. Non-destructive biosensors have proven to be extremely effective in the biomedical diagnostic business, but their potential in the food sector remains to be fully assessed and validated. Despite extensive research and development efforts worldwide, commercially viable biosensors for the food industry remain rather limited. The food sector prioritizes quality control, necessitating quick monitoring systems. Conventional procedures are costly, time consuming, and labour-intensive. Developing efficient non-destructive biosensors will not only accelerate the process but will also save money.

References

Ahmad, O. S., Bedwell, T. S., Esen, C., Garcia-Cruz, A., & Piletsky, S. A. (2019). Molecularly imprinted polymers in electrochemical and optical sensors. *Trends in Biotechnology, 37*(3), 294–309. https://doi.org/10.1016/j.tibtech.2018.08.009

Akgönüllü, S., & Denizli, A. (2022). Recent advances in optical biosensing approaches for biomarkers detection. *Biosensors and Bioelectronics: X, 12*(August). https://doi.org/10.1016/j.biosx.2022.100269

Alaejos, M. S., & Montelongo, F. J. G. (2004). Application of amperometric biosensors to the determination of vitamins and α-amino acids. *Chemical Reviews, 104*(7), 3239–3265. https://doi.org/10.1021/cr0304471

Armstrong, R. D. (1991). Advances in electrochemical science and engineering. In *Electrochimica acta* (Vol. 36, Issue 14). https://doi.org/10.1016/0013-4686(91)85230-5

Arora, P., Sindhu, A., Dilbaghi, N., & Chaudhury, A. (2011). Biosensors as innovative tools for the detection of food borne pathogens. *Biosensors and Bioelectronics, 28*(1), 1–12. https://doi.org/10.1016/j.bios.2011.06.002

Artigues, M., Abellà, J., & Colominas, S. (2017). Analytical parameters of an amperometric glucose biosensor for fast analysis in food samples. *Sensors (Basel, Switzerland), 17*(11). https://doi.org/10.3390/s17112620

Balbinot, S., Srivastav, A. M., Vidic, J., Abdulhalim, I., & Manzano, M. (2021). Plasmonic biosensors for food control. *Trends in Food Science and Technology, 111*(December 2020), 128–140. https://doi.org/10.1016/j.tifs.2021.02.057

Becker, M. M., Ribeiro, E. B., de Oliveira Marques, P. R. B., Marty, J.-L., Nunes, G. S., & Catanante, G. (2019). Development of a highly sensitive xanthine oxidase-based biosensor for the determination of antioxidant capacity in Amazonian fruit samples. *Talanta, 204*, 626–632. https://doi.org/10.1016/j.talanta.2019.06.002

Besharati, M., Hamedi, J., Hosseinkhani, S., & Saber, R. (2019). A novel electrochemical biosensor based on TetX2 monooxygenase immobilized on a nano-porous glassy carbon electrode for tetracycline residue detection. *Bioelectrochemistry, 128*, 66–73. https://doi.org/10.1016/j.bioelechem.2019.02.010

Bhatia, D., Paul, S., Acharjee, T., & Ramachairy, S. S. (2024). Biosensors and their widespread impact on human health. *Sensors International, 5*(November 2023), 100257. https://doi.org/10.1016/j.sintl.2023.100257

Bhavadharini, B., Kavimughil, M., Malini, B., Vallath, A., Prajapati, H. K., & Sunil, C. K. (2022). Recent advances in biosensors for detection of chemical contaminants in food — A review. In *Food analytical methods* (Vol. 15, Issue 6, pp. 1545–1564). Springer. https://doi.org/10.1007/s12161-021-02213-y

Biosensors Market Size, Share, Trends and Revenue Forecast - 2030. (n.d.). Retrieved April 17, 2024, from https://www.marketsandmarkets.com/Market-Reports/biosensors-market-798.html.

Bobrinetskiy, I., Radovic, M., Rizzotto, F., Vizzini, P., Jaric, S., Pavlovic, Z., Radonic, V., Nikolic, M. V., & Vidic, J. (2021). Advances in nanomaterials-based electrochemical biosensors for foodborne pathogen detection. *Nanomaterials, 11*(10), 2700. https://doi.org/10.3390/NANO11102700

Bollella, P., & Gorton, L. (2018). Enzyme based amperometric biosensors. *Current Opinion in Electrochemistry, 10*, 157–173. https://doi.org/10.1016/j.coelec.2018.06.003

Bratov, A., Abramova, N., & Ipatov, A. (2010). Recent trends in potentiometric sensor arrays-A review. *Analytica Chimica Acta, 678*(2), 149–159. https://doi.org/10.1016/j.aca.2010.08.035

Calabrese, A., Battistoni, P., Ceylan, S., Zeni, L., Capo, A., Varriale, A., D'Auria, S., & Staiano, M. (2023). An impedimetric biosensor for detection of volatile organic compounds in food. *Biosensors, 13*(3). https://doi.org/10.3390/bios13030341

Chai, C., & Oh, S. W. (2020). Electrochemical impedimetric biosensors for food safety. *Food Science and Biotechnology, 29*(7), 879–887. https://doi.org/10.1007/s10068-020-00776-w

Chambers, J. P., Arulanandam, B. P., Matta, L. L., Weis, A., & Valdes, J. J. (2002). *Cimb-10-00001*. 1–12.

Chen, C., & Wang, J. (2020). Optical biosensors: An exhaustive and comprehensive review. *Analyst, 145*(5), 1605–1628. https://doi.org/10.1039/c9an01998g

Choe, W., Durgannavar, T., & Chung, S. (2016). Fc-binding ligands of immunoglobulin G: An overview of high affinity proteins and peptides. *Materials, 9*(12), 994. https://doi.org/10.3390/ma9120994

Choi, J.-W., Kim, Y.-K., Lee, I.-H., Min, J., & Lee, W. H. (2001). Optical organophosphorus biosensor consisting of acetylcholinesterase/viologen hetero Langmuir–Blodgett film. *Biosensors and Bioelectronics, 16*(9–12), 937–943. https://doi.org/10.1016/S0956-5663(01)00213-5

Chu, H.-W., Unnikrishnan, B., Anand, A., Lin, Y.-W., & Huang, C.-C. (2020). Carbon quantum dots for the detection of antibiotics and pesticides. *Journal of Food and Drug Analysis, 28*(4), 540–558. https://doi.org/10.38212/2224-6614.1269

Chuan Li, S., hua Chen, J., Cao, H., sheng Yao, D., & ling Liu, D. (2011). Amperometric biosensor for aflatoxin B1 based on aflatoxin-oxidase immobilized on multiwalled carbon nanotubes. *Food Control, 22*(1), 43–49.

Chugh, V., Basu, A., Kaushik, A., Manshu, Bhansali, S., & Basu, A. K. (2024). Employing nano-enabled artificial intelligence (AI)-based smart technologies for prediction, screening, and detection of cancer. *Nanoscale, 16*(11), 5458–5486. *Pubs.Rsc.Org*.

Chullasat, K., Kanatharana, P., Limbut, W., Numnuam, A., & Thavarungkul, P. (2011). Ultra trace analysis of small molecule by label-free impedimetric immunosensor using multilayer modified electrode. *Biosensors and Bioelectronics, 26*(11), 4571–4578. https://doi.org/10.1016/j.bios.2011.05.029

Ciucu, A. A., Negulescu, C., & Baldwin, R. P. (2002). Detection of pesticides using an amperometric biosensor based on ferophthalocyanine chemically modified carbon paste electrode and immobilized bienzymatic system. *Biosensors and Bioelectronics, 18*(2–3), 303–310. https://doi.org/10.1016/S0956-5663(02)00173-2

Clark, L. C., & Lyons, C. (1962). Electrode systems for continuous monitoring in cardiovascular surgery. *Annals of the New York Academy of Sciences, 102*(1), 29–45. https://doi.org/10.1111/j.1749-6632.1962.tb13623.x

Clemens, A. H., Chang, P. H., & Myers, R. W. (1977). The development of Biostator, a Glucose Controlled Insulin Infusion System (GCIIS). *Hormone and Metabolic Research = Hormon-Und Stoffwechselforschung = Hormones et Metabolisme*, (Suppl 7), 23–33.

Crivianu-Gaita, V., & Thompson, M. (2016). Aptamers, antibody scFv, and antibody Fab' fragments: An overview and comparison of three of the most versatile biosensor biorecognition elements. *Biosensors and Bioelectronics, 85*, 32–45. https://doi.org/10.1016/j.bios.2016.04.091

da Silva, W., Ghica, M. E., Ajayi, R. F., Iwuoha, E. I., & Brett, C. M. A. (2019). Tyrosinase based amperometric biosensor for determination of tyramine in fermented food and beverages with gold nanoparticle doped poly(8-anilino-1-naphthalene sulphonic acid) modified electrode. *Food Chemistry, 282*, 18–26. https://doi.org/10.1016/j.foodchem.2018.12.104

Damborský, P., Švitel, J., & Katrlík, J. (2016). Optical biosensors. *Essays in Biochemistry, 60*(1), 91–100. https://doi.org/10.1042/EBC20150010

Davis, G. (1985). Electrochemical techniques for the development of amperometric biosensors. *Biosensors, 1*(2), 161–178, Newcastle-upon-Tyne NE2 4HH (Great Britain).

Deep, A., Saraf, M., Neha, Bharadwaj, S. K., & Sharma, A. L. (2014). Styrene sulphonic acid doped polyaniline based immunosensor for highly sensitive impedimetric sensing of atrazine. *Electrochimica Acta, 146*, 301–306. https://doi.org/10.1016/j.electacta.2014.09.048

Dill, K., Stanker, L. H., & Young, C. R. (1999). Detection of salmonella in poultry using a silicon chip-based biosensor. *Journal of Biochemical and Biophysical Methods, 41*(1), 61–67. https://doi.org/10.1016/S0165-022X(99)00027-5

Dinçkaya, E., Akyilmaz, E., Sezgintürk, M. K., & Ertaş, F. N. (2010). Sensitive nitrate determination in water and meat samples by amperometric biosensor. *Preparative Biochemistry and Biotechnology, 40*(2), 119–128. https://doi.org/10.1080/10826060903558620

Dwivedi, H. P., Smiley, R. D., & Jaykus, L. A. (2010). Selection and characterization of DNA aptamers with binding selectivity to Campylobacter jejuni using whole-cell SELEX. *Applied Microbiology and Biotechnology, 87*(6), 2323–2334. https://doi.org/10.1007/S00253-010-2728-7/METRICS

Eggenberger, O. M., Ying, C., & Mayer, M. (2019). Surface coatings for solid-state nanopores. *Nanoscale, 11*(42), 19636–19657. https://doi.org/10.1039/C9NR05367K

Eissa, S., & Zourob, M. (2020). Ultrasensitive peptide-based multiplexed electrochemical biosensor for the simultaneous detection of Listeria monocytogenes and Staphylococcus aureus. *Microchimica Acta, 187*(9), 486. https://doi.org/10.1007/s00604-020-04423-3

Elyasi, M., Khalilzadeh, M. A., & Karimi-Maleh, H. (2013). High sensitive voltammetric sensor based on Pt/CNTs nanocomposite modified ionic liquid carbon paste electrode for determination of Sudan I in food samples. *Food Chemistry, 141*(4), 4311–4317. https://doi.org/10.1016/j.foodchem.2013.07.020

Foudeh, A. M., Fatanat Didar, T., Veres, T., & Tabrizian, M. (2012). Microfluidic designs and techniques using lab-on-a-chip devices for pathogen detection for point-of-care diagnostics. *Lab on a Chip, 12*(18), 3249. https://doi.org/10.1039/c2lc40630f

Gao, Y. (2006). Secondary structure effects on DNA hybridization kinetics: A solution versus surface comparison. *Nucleic Acids Research, 34*(11), 3370–3377. https://doi.org/10.1093/nar/gkl422

Gao, S., Guisán, J. M., & Rocha-Martin, J. (2022). Oriented immobilization of antibodies onto sensing platforms - A critical review. *Analytica Chimica Acta, 1189*, 338907. https://doi.org/10.1016/j.aca.2021.338907

García-Aljaro, C., Cella, L. N., Shirale, D. J., Park, M., Muñoz, F. J., Yates, M. V., & Mulchandani, A. (2010). Carbon nanotubes-based chemiresistive biosensors for detection of microorganisms. *Biosensors and Bioelectronics, 26*(4), 1437–1441. https://doi.org/10.1016/j.bios.2010.07.077

Gavrilaş, S., Ursachi, C. Ş., Perţa-Crişan, S., & Munteanu, F. D. (2022). Recent trends in biosensors for environmental quality monitoring. *Sensors, 22*(4). https://doi.org/10.3390/s22041513

Gehring, A. G., Patterson, D. L., & Tu, S. I. (1998). Use of a light-addressable potentiometric sensor for the detection of Escherichia coli O157:H7. *Analytical Biochemistry, 258*(2), 293–298. https://doi.org/10.1006/abio.1998.2597

George Kerry, R., Ukhurebor, K. E., Kumari, S., Maurya, G. K., Patra, S., Panigrahi, B., Majhi, S., Rout, J. R., Rodriguez-Torres, M. d. P., Das, G., Shin, H.-S., & Patra, J. K. (2021). A comprehensive review on the applications of nano-biosensor-based approaches for non-communicable and communicable disease detection. *Biomaterials Science, 9*(10), 3576–3602. https://doi.org/10.1039/D0BM02164D

Ghasemi-Varnamkhasti, M., Rodríguez-Méndez, M. L., Mohtasebi, S. S., Apetrei, C., Lozano, J., Ahmadi, H., Razavi, S. H., & Antonio de Saja, J. (2012). Monitoring the aging of beers using a bioelectronic tongue. *Food Control, 25*(1), 216–224. https://doi.org/10.1016/j.foodcont.2011.10.020

Giese, J. H. (2002). Food biosensors. *Food Technology, 56*(7), 72. https://doi.org/10.1039/9781782623908

Grieshaber, D., MacKenzie, R., Vörös, J., & Reimhult, E. (2008). Electrochemical biosensors - sensor principles and architectures. *Sensors, 8*(3), 1400–1458. https://doi.org/10.3390/s8031400

Guisan, J. M., Bolivar, J. M., López-Gallego, F., & Rocha-Martín, J. (Eds.). (2020). *Immobilization of enzymes and cells* (Vol. 2100). Humana New York. https://doi.org/10.1007/978-1-0716-0215-7

Gupta, V. K., Yola, M. L., Eren, T., & Atar, N. (2015). Selective QCM sensor based on atrazine imprinted polymer: Its application to wastewater sample. *Sensors and Actuators B: Chemical, 218*, 215–221. https://doi.org/10.1016/j.snb.2015.05.009

Janata, J. (1990). Potentiometric microsensors. *Chemical Reviews, 90*(5), 691–703

Hanefeld, U., Gardossi, L., & Magner, E. (2009). Understanding enzyme immobilisation. *Chemical Society Reviews, 38*(2), 453–468. https://doi.org/10.1039/B711564B

Heineman, W. R., & Jensen, W. B. (2006). Leland C. Clark Jr. (1918–2005). *Biosensors and Bioelectronics, 21*(8), 1403–1404. https://doi.org/10.1016/j.bios.2005.12.005

Henao-Escobar, W., del Torno-de Román, L., Domínguez-Renedo, O., Alonso-Lomillo, M. A., & Arcos-Martínez, M. J. (2016). Dual enzymatic biosensor for simultaneous amperometric determination of histamine and putrescine. *Food Chemistry, 190*, 818–823. https://doi.org/10.1016/j.foodchem.2015.06.035

Hu, Q., Wang, S., Duan, H., & Liu, Y. (2021). A fluorescent biosensor for sensitive detection of salmonella typhimurium using low-gradient magnetic field and deep learning via faster region-based convolutional neural network. *Biosensors, 11*(11), 447. https://doi.org/10.3390/bios11110447

Hua, Q. T., Ruecha, N., Hiruta, Y., & Citterio, D. (2019). Disposable electrochemical biosensor based on surface-modified screen-printed electrodes for organophosphorus pesticide analysis. *Analytical Methods, 11*(27), 3439–3445. https://doi.org/10.1039/C9AY00852G

Huang, L., Sun, D. W., Pu, H., & Wei, Q. (2019). Development of nanozymes for food quality and safety detection: Principles and recent applications. *Comprehensive Reviews in Food Science and Food Safety, 18*(5), 1496–1513. https://doi.org/10.1111/1541-4337.12485

Huang, S., Gao, Y., Hu, Y., Shen, F., Jin, Z., & Cho, Y. (2023). Recent development of piezoelectric biosensors for physiological signal detection and machine learning assisted cardiovascular disease diagnosis. *RSC Advances, 13*(42), 29174–29194. https://doi.org/10.1039/d3ra05932d

Hung Tzang, C., Yuan, R., & Yang, M. (2001). Voltammetric biosensors for the determination of formate and glucose-6-phosphate based on the measurement of dehydrogenase-generated NADH and NADPH. *Biosensors and Bioelectronics, 16*(3), 211–219. https://doi.org/10.1016/S0956-5663(00)00143-3

Kaur, A., Kumar, P., Gupta, A., & Sapra, G. (2023). Piezoelectric biosensors in healthcare. In *Enzyme-based biosensors: Recent advances and applications in healthcare* (pp. 255–271). Springer Nature Singapore. https://doi.org/10.1007/978-981-15-6982-1_11

Khansili, N., Rattu, G., & Krishna, P. M. (2018). Label-free optical biosensors for food and biological sensor applications. *Sensors and Actuators, B: Chemical, 265*, 35–49. https://doi.org/10.1016/j.snb.2018.03.004

Kim, G., Om, A. S., & Mun, J. H. (2007). Nano-particle enhanced impedimetric biosensor for detedtion of foodborne pathogens. *Journal of Physics: Conference Series, 61*, 555–559. https://doi.org/10.1088/1742-6596/61/1/112

Kim, H.-J., Kim, S.-H., Lee, J.-K., Choi, C.-U., Lee, H.-S., Kang, H.-G., & Cha, S.-H. (2012). A novel mycotoxin purification system using magnetic nanoparticles for the recovery of aflatoxin B1 and zearalenone from feed. *Journal of Veterinary Science, 13*(4), 363. https://doi.org/10.4142/jvs.2012.13.4.363

Kochana, J., Wapiennik, K., Knihnicki, P., Pollap, A., Janus, P., Oszajca, M., & Kuśtrowski, P. (2016). Mesoporous carbon-containing voltammetric biosensor for determination of tyramine in food products. *Analytical and Bioanalytical Chemistry, 408*(19), 5199–5210. https://doi.org/10.1007/s00216-016-9612-y

Koyun, A., Ahlatcıoğlu, E., & İpek, Y. K. (n.d.). *4 biosensors and their principles*. www.intechopen.com

Kumar, S., Dilbaghi, N., Barnela, M., Bhanjana, G., & Kumar, R. (2012). Biosensors as novel platforms for detection of food pathogens and allergens. *BioNanoScience, 2*(4), 196–217. https://doi.org/10.1007/s12668-012-0057-2

Liedberg, B., Nylander, C., & Lunström, I. (1983). Surface plasmon resonance for gas detection and biosensing. *Sensors and Actuators, 4*(C), 299–304. https://doi.org/10.1016/0250-6874(83)85036-7

Liu, N., Nie, D., Tan, Y., Zhao, Z., Liao, Y., Wang, H., Sun, C., & Wu, A. (2017). An ultrasensitive amperometric immunosensor for zearalenones based on oriented antibody immobilization on a glassy carbon electrode modified with MWCNTs and AuPt nanoparticles. *Microchimica Acta, 184*(1), 147–153. https://doi.org/10.1007/s00604-016-1996-z

Lu, Z., Li, C. M., Zhou, Q., Bao, Q.-L., & Cui, X. (2007). Covalently linked DNA/protein multilayered film for controlled DNA release. *Journal of Colloid and Interface Science, 314*(1), 80–88. https://doi.org/10.1016/j.jcis.2007.05.018

Luong, J. H. T., Bouvrette, P., & Male, K. B. (1997). Developments and applications of biosensors in food analysis. *Trends in Biotechnology, 15*(9), 369–377. https://doi.org/10.1016/S0167-7799(97)01071-8

Majdinasab, M., Mishra, R. K., Tang, X., & Marty, J. L. (2020). Detection of antibiotics in food: New achievements in the development of biosensors. In *TrAC - Trends in analytical chemistry* (Vol. 127). Elsevier B.V.. https://doi.org/10.1016/j.trac.2020.115883

Mak, A. C., Osterfeld, S. J., Yu, H., Wang, S. X., Davis, R. W., Jejelowo, O. A., & Pourmand, N. (2010). Sensitive giant magnetoresistive-based immunoassay for multiplex mycotoxin detection. *Biosensors and Bioelectronics, 25*(7), 1635–1639. https://doi.org/10.1016/j.bios.2009.11.028

Malvano, F., Albanese, D., Pilloton, R., & Di Matteo, M. (2017). A new label-free impedimetric aptasensor for gluten detection. *Food Control, 79*, 200–206. https://doi.org/10.1016/j.foodcont.2017.03.033

Malvano, F., Pilloton, R., & Albanese, D. (2020). Label-free impedimetric biosensors for the control of food safety–a review. *International Journal of Environmental Analytical Chemistry, 100*(4), 468–491. https://doi.org/10.1080/03067319.2019.1667096

Maraldo, D., & Mutharasan, R. (2007a). 10-Minute assay for detecting Escherichia coli O157:H7 in ground beef samples using piezoelectric-excited millimeter-size cantilever sensors. *Journal of Food Protection, 70*(7), 1670–1677. https://doi.org/10.4315/0362-028X-70.7.1670

Maraldo, D., & Mutharasan, R. (2007b). Preparation-free method for detecting Escherichia coli O157:H7 in the presence of spinach, spring lettuce mix, and ground beef particulates. *Journal of Food Protection, 70*(11), 2651–2655. https://doi.org/10.4315/0362-028X-70.11.2651

Mayer, M., & Baeumner, A. J. (2019). A megatrend challenging analytical chemistry: Biosensor and chemosensor concepts ready for the internet of things. *Chemical Reviews, 119*(13), 7996–8027. https://doi.org/10.1021/ACS.CHEMREV.8B00719

Mehrotra, P., Chatterjee, B., & Sen, S. (2019). EM-wave biosensors: A review of RF, microwave, mm-wave and optical sensing. *Sensors (Basel, Switzerland), 19*(5). https://doi.org/10.3390/s19051013

Mohtar, L. G., Aranda, P., Messina, G. A., Nazareno, M. A., Pereira, S. V., Raba, J., & Bertolino, F. A. (2019). Amperometric biosensor based on laccase immobilized onto a nanostructured screen-printed electrode for determination of polyphenols in propolis. *Microchemical Journal, 144*, 13–18. https://doi.org/10.1016/j.microc.2018.08.038

Morales, M. A., & Halpern, J. M. (2018). Guide to selecting a biorecognition element for biosensors. *Bioconjugate Chemistry, 29*(10), 3231–3239. https://doi.org/10.1021/acs.bioconjchem.8b00592

Narang, J., Chauhan, N., Singh, A., & Pundir, C. S. (2011). A nylon membrane based amperometric biosensor for polyphenol determination. *Journal of Molecular Catalysis B: Enzymatic, 72*(3–4), 276–281. https://doi.org/10.1016/j.molcatb.2011.06.016

Naresh, V., & Lee, N. (2021). A review on biosensors and recent development of nanostructured materials-enabled biosensors. *Sensors (Basel, Switzerland), 21*(4), 1–35. https://doi.org/10.3390/s21041109

Nguyen, H. H., Lee, S. H., Lee, U. J., Fermin, C. D., & Kim, M. (2019). Immobilized enzymes in biosensor applications. *Materials (Basel, Switzerland), 12*(1), 1–34. https://doi.org/10.3390/ma12010121

Nomngongo, P. N., Ngila, J. C., Nyamori, V. O., Songa, E. A., & Iwuoha, E. I. (2011). Determination of selected heavy metals using amperometric horseradish peroxidase (HRP) inhibition biosensor. *Analytical Letters, 44*(11), 2031–2046. https://doi.org/10.1080/00032719.2010.539738

Noy-Porat, T., Alcalay, R., Mechaly, A., Peretz, E., Makdasi, E., Rosenfeld, R., & Mazor, O. (2021). Characterization of antibody-antigen interactions using biolayer interferometry. *STAR Protocols, 2*(4), 100836. https://doi.org/10.1016/j.xpro.2021.100836

Oberhaus, F. V., Frense, D., & Beckmann, D. (2020). Immobilization techniques for aptamers on gold electrodes for the electrochemical detection of proteins: A review. *Biosensors, 10*(5). https://doi.org/10.3390/BIOS10050045

Pan, J., Chen, W., Ma, Y., & Pan, G. (2018). Molecularly imprinted polymers as receptor mimics for selective cell recognition. *Chemical Society Reviews, 47*(15), 5574–5587. https://doi.org/10.1039/C7CS00854F

Pejcic, B., & De Marco, R. (2006). Impedance spectroscopy: Over 35 years of electrochemical sensor optimization. *Electrochimica Acta, 51*(28), 6217–6229. https://doi.org/10.1016/j.electacta.2006.04.025

Perumal, V., & Hashim, U. (2014). Advances in biosensors: Principle, architecture and applications. *Journal of Applied Biomedicine, 12*(1), 1–15. https://doi.org/10.1016/j.jab.2013.02.001

Pividori, M. I., Merkoçi, A., & Alegret, S. (2000). Electrochemical genosensor design: Immobilisation of oligonucleotides onto transducer surfaces and detection methods. *Biosensors and Bioelectronics, 15*(5–6), 291–303. https://doi.org/10.1016/S0956-5663(00)00071-3

Pohanka, M. (2017). The piezoelectric biosensors: Principles and applications, a review. *International Journal of Electrochemical Science, 12*(1), 496–506. https://doi.org/10.20964/2017.01.44

Pohanka, M. (2018). Overview of piezoelectric biosensors, immunosensors and DNA sensors and their applications. *Materials (Basel, Switzerland), 11*(3). https://doi.org/10.3390/ma11030448

Pramanik, S., Pingguan-Murphy, B., & Osman, N. A. A. (2013). Developments of immobilized surface modified piezoelectric crystal biosensors for advanced applications. *International Journal of Electrochemical Science, 8*(6), 8863–8892. https://doi.org/10.1016/s1452-3981(23)12933-6

Ramanathan, K., & Danielsson, B. (2001). Principles and applications of thermal biosensors. *Biosensors and Bioelectronics, 16*(6), 417–423. https://doi.org/10.1016/S0956-5663(01)00124-5

Rashid, J. I. A., & Yusof, N. A. (2017). The strategies of DNA immobilization and hybridization detection mechanism in the construction of electrochemical DNA sensor: A review. *Sensing and Bio-Sensing Research, 16*, 19–31. https://doi.org/10.1016/J.SBSR.2017.09.001

Rodrigues, J. F., Paulovich, F. V., De Oliveira, M. C. F., & De Oliveira, O. N. (2016). On the convergence of nanotechnology and Big Data analysis for computer-aided diagnosis. *Nanomedicine, 11*(8), 959–982. https://doi.org/10.2217/NNM.16.35

Saini, K., Kaushal, A., Gupta, S., & Kumar, D. (2019). Rapid detection of Salmonella enterica in raw milk samples using Stn gene-based biosensor. *3 Biotech, 9*(11), 425. https://doi.org/10.1007/s13205-019-1957-4

Santini, A. O., Pezza, H. R., & Pezza, L. (2012). Development of a sensitive potentiometric sensor for determination of fumaric acid in powdered food products. *Food Chemistry, 134*(1), 483–487. https://doi.org/10.1016/j.foodchem.2012.02.104

Santos, S. B., Cunha, A. P., Macedo, M., Nogueira, C. L., Brandão, A., Costa, S. P., Melo, L. D. R., Azeredo, J., & Carvalho, C. M. (2020). Bacteriophage-receptor binding proteins for multiplex detection of Staphylococcus and Enterococcus in blood. *Biotechnology and Bioengineering, 117*(11), 3286–3298. https://doi.org/10.1002/BIT.27489

Schachinger, F., Chang, H., Scheiblbrandner, S., & Ludwig, R. (2021). Amperometric biosensors based on direct electron transfer enzymes. *Molecules, 26*(15). https://doi.org/10.3390/molecules26154525

Schasfoort, R. B. M. (2017). History and physics of surface plasmon resonance. In *Handbook of surface plasmon resonance* (pp. 27–59). The Royal Society of Chemistry. https://doi.org/10.1039/9781788010283-00027

Selvaraj, J. N., Zhou, L., Wang, Y., Zhao, Y., Xing, F., Dai, X., & Liu, Y. (2015). Mycotoxin detection — Recent trends at global level. *Journal of Integrative Agriculture, 14*(11), 2265–2281. https://doi.org/10.1016/S2095-3119(15)61120-0

Shahidi, S., & Branch, A. A. (2018). *Analytical & bioanalytical*. February.

Shan, D., Zhu, M., Han, E., Xue, H., & Cosnier, S. (2007). Calcium carbonate nanoparticles: A host matrix for the construction of highly sensitive amperometric phenol biosensor. *Biosensors and Bioelectronics, 23*(5), 648–654. https://doi.org/10.1016/j.bios.2007.07.012

Singh, R., Ashraf, J., Manzoor, S., & Dar, M. H. (2020a). Application of biosensors in food quality control. *Journal of Postharvest Technology, 2020*(1), 53–74.

Singh, S., Kumar, V., Dhanjal, D. S., Datta, S., Prasad, R., & Singh, J. (2020b). Biological biosensors for monitoring and diagnosis. In *Microbial biotechnology: Basic research and applications* (Environmental and microbial biotechnology) (pp. 317–335). https://doi.org/10.1007/978-981-15-2817-0_14

Skládal, P. (2016). Piezoelectric biosensors. *TrAC - Trends in Analytical Chemistry, 79*, 127–133. https://doi.org/10.1016/j.trac.2015.12.009

Sreedhar, N. Y., Sunil Kumar, M., & Krishnaveni, K. (2015). Sensitive determination of chlorpyrifos using Ag/Cu alloy nanoparticles and graphene composite paste electrode. *Sensors and Actuators B: Chemical, 210*, 475–482. https://doi.org/10.1016/j.snb.2015.01.016

Stobiecka, A., Radecka, H., & Radecki, J. (2007). Novel voltammetric biosensor for determining acrylamide in food samples. *Biosensors and Bioelectronics, 22*(9–10), 2165–2170. https://doi.org/10.1016/j.bios.2006.10.008

Subramanian, A, Kennel, S. J., Oden, P. I., Bruce Jacobson, K., Woodward, J., & Doktycz, M. J.. (1999). Comparison of techniques for enzyme immobilization on silicon supports. *Elsevier.* Retrieved May 8, 2024, from https://www.sciencedirect.com/science/article/pii/S0141022998000842.

Sun, X., Guan, L., Shan, X., Zhang, Y., & Li, Z. (2012). Electrochemical detection of peanut allergen Ara h 1 using a sensitive DNA biosensor based on stem–loop probe. *Journal of Agricultural and Food Chemistry, 60*(44), 10979–10984. https://doi.org/10.1021/jf3027233

Teles, F. R. R., & Fonseca, L. P. (2008). Trends in DNA biosensors. *Talanta, 77*(2), 606. https://doi.org/10.1016/J.TALANTA.2008.07.024

Thakur, M. S., & Ragavan, K. V. (2013). Biosensors in food processing. *Journal of Food Science and Technology, 50*(4), 625–641. https://doi.org/10.1007/s13197-012-0783-z

Tok, J. B.-H., Chuang, F. Y. S., Kao, M. C., Rose, K. A., Pannu, S. S., Sha, M. Y., Chakarova, G., Penn, S. G., & Dougherty, G. M. (2006). Metallic striped nanowires as multiplexed immunoassay platforms for pathogen detection. *Angewandte Chemie International Edition, 45*(41), 6900–6904. https://doi.org/10.1002/anie.200601104

Torrance, L., Ziegler, A., Pittman, H., Paterson, M., Toth, R., & Eggleston, I. (2006). Oriented immobilisation of engineered single-chain antibodies to develop biosensors for virus detection. *Journal of Virological Methods, 134*(1–2), 164–170. https://doi.org/10.1016/J.JVIROMET.2005.12.012

Trivedi, U. B., Lakshminarayana, D., Kothari, I. L., Patel, N. G., Kapse, H. N., Makhija, K. K., Patel, P. B., & Panchal, C. J. (2009). Potentiometric biosensor for urea determination in milk. *Sensors and Actuators, B: Chemical, 140*(1), 260–266. https://doi.org/10.1016/j.snb.2009.04.022

Tsai, W., & Hsieh, C. (2007). QCM-based immunosensor for the determination of ochratoxin A. *Analytical Letters, 40*(10), 1979–1991. https://doi.org/10.1080/00032710701484509

Turner, A. P. F. (2013). Biosensors: Sense and sensibility. *Chemical Society Reviews, 42*(8), 3184–3196. https://doi.org/10.1039/c3cs35528d

Updike, S. J., & Hicks, G. P. (1967). The enzyme electrode. *Nature, 214*(5092), 986–988. https://doi.org/10.1038/214986a0

Vaisocherová-Lísalová, H., Surman, F., Víšová, I., Vala, M., Špringer, T., Ermini, M. L., Šípová, H., Šedivák, P., Houska, M., Riedel, T., Pop-Georgievski, O., Brynda, E., & Homola, J. (2016). Copolymer brush-based ultralow-fouling biorecognition surface platform for food safety. *Analytical Chemistry, 88*(21), 10533–10539. https://doi.org/10.1021/acs.analchem.6b02617

Van Dorst, B., De Coen, W., Blust, R., & Robbens, J. (2010a). Phage display as a novel screening tool for primary toxicological targets. *Environmental Toxicology and Chemistry, 29*(2), 250–255. https://doi.org/10.1002/ETC.38

Van Dorst, B., Mehta, J., Bekaert, K., Rouah-Martin, E., De Coen, W., Dubruel, P., Blust, R., & Robbens, J. (2010b). Recent advances in recognition elements of food and environmental biosensors: A review. *Biosensors and Bioelectronics, 26*, 1178–1194. https://doi.org/10.1016/j.bios.2010.07.033

Vasconcelos, H., Coelho, L. C. C., Matias, A., Saraiva, C., Jorge, P. A. S., & de Almeida, J. M. M. M. (2021). Biosensors for biogenic amines: A review. *Biosensors, 11*(3). https://doi.org/10.3390/BIOS11030082

Wang, J., Jiang, L., Chu, Q., & Ye, J. (2010). Residue analysis of melamine in milk products by micellar electrokinetic capillary chromatography with amperometric detection. *Food Chemistry, 121*(1), 215–219. https://doi.org/10.1016/j.foodchem.2009.11.074

Wang, C. H., Wu, J. J., & Lee, G. B. (2019). Screening of highly-specific aptamers and their applications in paper-based microfluidic chips for rapid diagnosis of multiple bacteria. *Sensors and Actuators B: Chemical, 284*, 395–402. https://doi.org/10.1016/J.SNB.2018.12.112

Wu, H. Z., Lee, Y. C., Lin, T. K., Shih, H. C., Chang, F. L., & Lin, H. P. P. (2009). Development of an amperometric micro-biodetector for pesticide monitoring and detection. *Journal of the Taiwan Institute of Chemical Engineers, 40*(2), 113–122. https://doi.org/10.1016/j.jtice.2008.08.007

Xu, Y., Niu, X., Zhang, H., Xu, L., Zhao, S., Chen, H., & Chen, X. (2015). Switch-on fluorescence sensing of glutathione in food samples based on a graphitic carbon nitride quantum dot (g-CNQD)–Hg $^{2+}$ chemosensor. *Journal of Agricultural and Food Chemistry, 63*(6), 1747–1755. https://doi.org/10.1021/jf505759z

Xu, Q., Liu, Z., Fu, J., Zhao, W., Guo, Y., Sun, X., & Zhang, H. (2017). Ratiometric electrochemical aptasensor based on ferrocene and carbon nanofibers for highly specific detection of tetracycline residues. *Scientific Reports, 7*(1), 14729. https://doi.org/10.1038/s41598-017-15333-5

Yakovleva, M., Bhand, S., & Danielsson, B. (2013). The enzyme thermistor-A realistic biosensor concept. A critical review. *Analytica Chimica Acta, 766*, 1–12. https://doi.org/10.1016/j.aca.2012.12.004

Yang, H., Li, H., & Jiang, X. (2008). Detection of foodborne pathogens using bioconjugated nanomaterials. *Microfluidics and Nanofluidics, 5*(5), 571–583. https://doi.org/10.1007/s10404-008-0302-8

Yashini, M., Auddy, I., Shanmugasundaram, S., Vidyalakshmi, R., & Sunil, C. K. (2022). Characterization of antibody immobilization on chitosan/gelatin-modified electrode and its application to Bacillus cereus detection in cereal-based food. *Food Analytical Methods, 15*(9), 2382–2393. https://doi.org/10.1007/s12161-022-02299-y

Ye, Y., Guo, H., & Sun, X. (2019). Recent progress on cell-based biosensors for analysis of food safety and quality control. *Biosensors and Bioelectronics, 126*, 389–404. https://doi.org/10.1016/J.BIOS.2018.10.039

Zhou, J., Battig, M. R., & Wang, Y. (2010). Aptamer-based molecular recognition for biosensor development. *Analytical and Bioanalytical Chemistry, 398*(6), 2471–2480. https://doi.org/10.1007/s00216-010-3987-y

Ziegler, C., & Göpel, W. (1998). Biosensor development. *Current Opinion in Chemical Biology, 2*(5), 585–591. https://doi.org/10.1016/S1367-5931(98)80087-2

Chapter 15
Techniques Based on Electrical Properties of Food

Ashok Kumar Yadav and Sandeep Janghu

15.1 Introduction

The electrical characteristics of food is crucial in multiple domains such as agriculture and food science and technology. These characteristics offer important information on the make-up, consistency, and properties of food items. Factors like food composition, temperature, and moisture content affect electrical qualities including conductivity, permittivity, and impedance. Researchers and food technologists can ascertain features including moisture content, texture, and freshness all of which are critical for guaranteeing the safety and quality of food by examining these attributes (Kamonpatana & Sastry, 2022). Fundamentally, methods based on food's electrical characteristics are based on how electromagnetic fields interact with food ingredients. Different food ingredients have distinct electrical characteristics, which has led to the development of methods for efficiently analyzing and modifying characteristics. For example, food products' structural integrity and moisture content can be evaluated with the help of dielectric properties. Through the use of electromagnetic principles, scientists may develop complex equipment and techniques to monitor and regulate these attributes, enhancing procedures like food preservation, drying, and quality evaluation (Banti, 2020).

The study of electrical properties of food has become an essential component of current food science and technology. The electrical properties of food are the qualities that explain how food interacts with electrical fields. These characteristics are impacted by the composition, structure, and physical state of the food (Sarkar et al., 2022). Different food ingredients have distinct electrical properties, which can be used for a variety of non-destructive food quality analysis and applications.

A. K. Yadav · S. Janghu (✉)
Department of Food Technology, Rajiv Gandhi University, Rono Hills, Doimukh, Itanagar, Arunachal Pradesh, India

C. K. Sunil et al. (eds.), *Emerging Non-destructive Technologies for Food Quality Analysis*, https://doi.org/10.1007/978-3-031-95048-3_15

This chapter offers a variety of methods that take advantage of food's electrical characteristics for different objectives. Among these methods are electrical impedance tomography, dielectric spectroscopy, and impedance spectroscopy. By characterizing food components according to their dielectric characteristics at various frequencies, dielectric spectroscopy offers important insights into the composition and structure of the materials. Electrical impedance in food systems can be measured using impedance spectroscopy, which helps to evaluate characteristics including texture and moisture content. Using electrical impedance measurements, electrical impedance tomography is a non-destructive imaging method that maps the spatial characteristics of food, providing information on interior structures and quality (Binnur & Serap, 2011). The electrical characteristics of food are essential for improving our comprehension of food products.

15.2 Theories and Principles of Electrical Properties

Some of the main electrical properties of food are:

15.2.1 Dielectric Constant (Permittivity)

Permittivity is a measure of a material's ability to store electrical energy in an electric field. It is represented by the symbol ε (epsilon). Moisture content and composition influence a food material's dielectric constant (Ozturk et al., 2016). Water, a major component in many foods, has a high dielectric constant, and variations in moisture level can have a considerable impact on the food's overall dielectric constant (Guo et al., 2010; Curet et al., 2014).

15.2.2 Dielectric Loss Factor (Dissipation Factor)

Dielectric loss factor (tan (δ) or ε'') represents a material's ability to convert electrical energy into heat (Routray & Orsat, 2018). It describes the energy losses that occur within a material when exposed to an alternating electric field (Frabetti et al., 2023). The presence of ions and the mobility of polar molecules in the product both have an effect on the dielectric loss factor (Curet et al., 2014; Sosa-Morales et al., 2010).

15.2.3 Electrical Conductivity

Electrical conductivity (σ) measures a material's ability to conduct an electric current (Palaniappan & Sastry, 1991). Electrical conductivity in food is controlled by the presence of ions, such as those found in dissolved salts and acids. A food material's conductivity might indicate its ion content and its overall composition (Mitchell & de Alwis, 1989; de Alwis et al., 1989; Banti, 2020).

15.2.4 Impedance

Impedance (Z) measures a material's resistance to alternating current (AC). It is a multidimensional quantity that combines resistance (R) and reactance (X). Impedance can reveal information of the interior structure and composition of food. Impedance changes over a wide frequency range can be investigated using techniques such as dielectric spectroscopy (Zhao et al., 2000; Figura & Tiexeira, 2007).

15.2.5 Capacitance

Capacitance (C) indicates a material's ability to hold electrical charge. Capacitance is frequently related with the food's ability to store electrical energy in an electric field. It is proportional to the dielectric constant and is affected by variables such as moisture content (Curet et al., 2014).

These electrical properties are critical for understanding how food components respond to electrical fields, and they serve as the foundation for a variety of procedures used in non-destructive food quality analysis. The measurement and study of these qualities help to assess aspects like moisture content, composition, and structural changes in food products (de Alwis et al., 1989).

15.3 Measurement Techniques: Instrumentation and Equipment

To test the electrical properties of food, researchers and food scientists use a variety of methodologies, allowing them to get insights into features such as moisture content, composition, and structural changes (Frabetti et al., 2023). Here are some common techniques:

15.3.1 Dielectric Spectroscopy

This technique measures a material's dielectric characteristics over a wide range of frequencies. Dielectric spectroscopy can reveal details about the molecular dynamics and structure of food components (Jha et al., 2011). Dielectric spectroscopy usually uses instruments such as impedance analyzers and network analyzers (Curet et al., 2014; Jiao, 2019).

Dielectric spectroscopy measures a material's response to an electric field at different frequencies. Instruments such as impedance analyzers, network analyzers, and dielectric spectrometers measure the dielectric characteristics of food across several frequencies (Içier & Baysal, 2004b). This technique is ideal for examining the molecular dynamics and structure of dietary components (Dube et al., 1988; Jiao, 2019; Novas et al., 2023).

15.3.1.1 Principle

Dielectric Spectroscopy operates on the area of measuring food product's response to an applied electric field over a range of frequencies (Hlaváčová, 2003). When an external electric field is applied, it causes polarization in the material, and the material's dielectric characteristics impact how it responds to the electric field. These dielectric properties, which include the dielectric constant and dielectric loss factor, change with frequency and reflect the material's molecular movements and structure (Dube et al., 1988; Sosa-Morales et al., 2010).

As the frequency of the applied electric field varies, different chemical processes within the material influence the dielectric response (Frabetti et al., 2023). For example, molecules in the food relax at specific frequencies, such as dipole reorientation or ion movement. Dielectric spectroscopy, which analyzes the material's response at various frequencies (Fig. 15.1), gives useful information on its composition, molecular interactions, and structural properties (Figura & Tiexeira, 2007).

Dielectric Spectroscopy uses specialized equipment to measure the material's dielectric characteristics over a wide range of frequencies. Impedance analyzers assess the material's impedance, including resistance and reactance, at different frequencies (Jha et al., 2011). The impedance is then utilized to determine dielectric characteristics. Network Analyzers analyze a network's electrical behavior, including its reaction to different frequencies. They are frequently used to determine the complex impedance of a material (Dube et al., 1988; Novas et al., 2023).

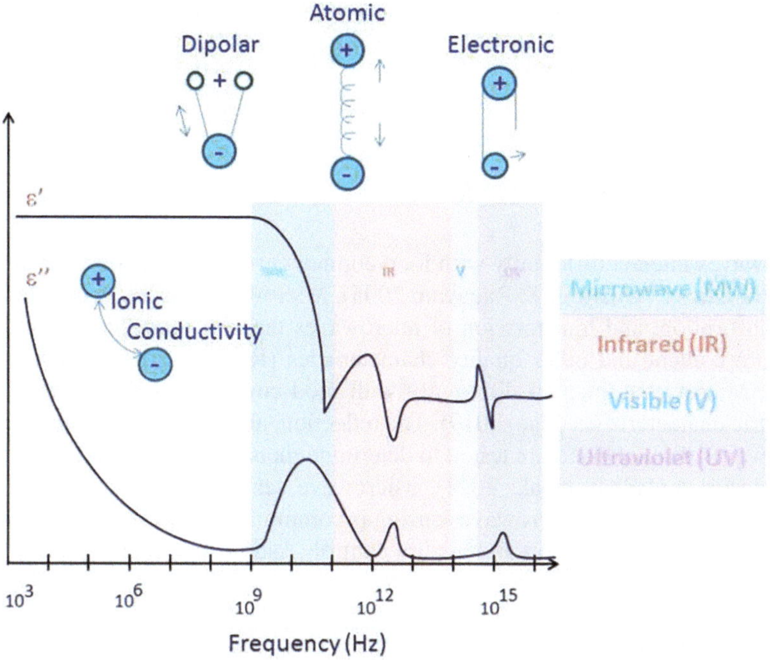

Fig. 15.1 Polarization at different frequency range (Khaled et al., 2016)

15.3.1.2 Application

Dielectric spectroscopy has many applications, but in the realm of food science, it is especially beneficial for examining the molecular dynamics and structure of food components (Jadhav et al., 2021). Some specific applications are:

Analyzing the dielectric response of dietary components at different frequencies might provide insights into their molecular dynamics, including water, proteins, and lipids.

Changes in dielectric characteristics can indicate changes in food quality (Knirsch et al., 2010; Pereira et al., 2024). Variations in moisture content, phase transitions during freezing or cooking, or the presence of pollutants can be detected (Curet et al., 2014; Leong & Oey, 2022).

Understanding the dielectric characteristics of food components can help optimize processing conditions. For example, in the food sector, understanding dielectric characteristics can help design effective drying or heating procedures. Dielectric Spectroscopy is a helpful tool for researching and developing innovative food products (Frabetti et al., 2023). It aids in understanding how various ingredients interact and affect the overall qualities of the cuisine (Jadhav et al., 2021). Dielectric Spectroscopy is an effective approach for analyzing the electrical response of food components, yielding useful information about their molecular dynamics, structure, and quality (Juansah et al., 2014). The apparatus utilized allows for exact measurements over a wide range of frequencies, making it a useful tool for researchers and food scientists (Banti, 2020).

15.3.2 *Microwave Sensing*

Microwaves interact differently with food components depending on their dielectric characteristics (Venkatesh & Ragavan, 2005). Microwave sensors detect the reflection, absorption, and transmission of microwaves through food, providing data on moisture content and other quality characteristics (Risman, 2009; Alkanan et al., 2021). Microwaves interact differently with food components depending on their dielectric characteristics (Jiao, 2019). The reflection, absorption, and transmission of microwaves through food are tested to determine moisture content and other quality characteristics (Jadhav et al., 2021). Microwave sensors measure the microwave characteristics of food. Microwave sensing is commonly utilized for non-destructive moisture content assessment and quality control (Astráin-Redín et al., 2024).

15.3.2.1 Principle

The theory behind microwave sensing is that microwaves interact differently with different food components depending on their dielectric characteristics. The dielectric constant and dielectric loss factor vary with the composition and structure of dietary ingredients (de Alwis et al., 1989; Sosa-Morales et al., 2010). Microwaves interact with food samples by reflection, absorption, and transmission, depending on the material's dielectric characteristics. Reflection refers to the fraction of microwave energy that bounces back off food surfaces (Risman, 2009). The amount of microwave energy absorbed by food depends on elements such as moisture, fat, and other ingredients (Juansah et al., 2014). Transmission is the amount of microwave energy that passes through the food sample (Feng et al., 2012). The interactions of microwaves with food components reveal information about the material's properties, particularly its moisture content and other quality parameters. Moisture, as a primary contributor to food's dielectric characteristics, has a considerable effect on microwave absorption and reflection (Astráin-Redín et al., 2024; Içier & Baysal, 2004a).

15.3.2.2 Instrumentation

Microwave sensors or meters are used to determine the microwave characteristics of food. Typically, the apparatus includes a microwave source (transmitter), a detector, and signal processing circuits (Duguay et al., 2017). Depending on the application, the system may run at a single frequency or a range of frequencies. The Microwave Source (transmitter) emits microwaves that interact with food samples. The detector measures microwave radiation that is reflected, absorbed, or transmitted via food (Zhao et al., 2000; Knirsch et al., 2010). Signal processing electronics analyzes received signals to determine food's dielectric characteristics, moisture content, and quality qualities (Içier & Baysal, 2004b). The design of microwave sensors varies depending on the application's specific requirements. Some sensors are designed for continuous monitoring in industrial operations, whereas others may be intended for laboratory or field use (Figura & Teixeira, 2023).

15.3.2.3 Application

Microwave sensing has a wide range of applications, the most common of which being non-destructive moisture content measuring and food quality monitoring. The link between water's dielectric characteristics and microwave absorption enables the accurate and speedy detection of moisture levels (Jadhav et al., 2021). Microwave sensing is used to evaluate the quality of food products, particularly during baking, drying, and thawing. Changes in dielectric characteristics can indicate compositional and structural differences (de Alwis et al., 1989). Microwave sensing enables real-time process control in industrial environments. Microwaves, for example, can be used in food manufacturing to ensure that items are heated or dried evenly. Dielectric characteristics of food and foreign bodies, such metal or glass, can be used to detect contamination in food products (Jadhav et al., 2021). Microwave sensing is a great technique for non-destructive study of food qualities, notably moisture content, and plays an important role in quality monitoring at various phases of food processing. Some specific applications of microwave sensing is shown in Table 15.1. The

Table 15.1 Applications of microwave sensing in food processing sector (Meng et al., 2017)

Food sample	Applications	Frequency
Pork	Determining added water in pork	0.13–18 GHz
Yoghurt	Determining sugar in yoghurt	1–20 GHz
Prawns	Composition analysis	0.2–12 GHz
Meat	Determination of water in meat	0.2–12 GHz
Ham	Determination of water activity, salt content and water content	2–6 GHz
Cereals and grains	Measurement of dielectric properties	5–17 GHz
Wheat	Measuring moisture content wheat	10–18 GHz
Fish products	Determination of prior treatment	0.4–6.2 GHz
Chicken	Determination of added water	0.2–12 GHz
Beef	Determination of fat and salt content	1–40 GHz
Potato	Dielectric properties in microwave and RF processing	1 M–1.8 GHz
Chicken breast	Quality measurement	0.01–1.8 GHz
Chicken meat	Effects of aging on dielectric properties	0.2–20 GHz

capacity to acquire real-time information makes it a useful and efficient tool in the food sector (Aurina & Sari, 2022).

15.3.3 Electrical Impedance Tomography (EIT)

EIT is an imaging technique that determines the impedance distribution of a food sample. It can be used to monitor changes in food's interior structure in real time during processing and storage (Venkatesh & Ragavan, 2005). EIT involves measuring the impedance distribution in food samples (Jiao, 2019). It provides a spatial representation of electrical properties and can be used in real time to monitor changes in food's interior structure. Impedance distribution is measured using specialized EIT equipment with electrode arrays (Knirsch et al., 2010). EIT is used to monitor and image operations in food manufacturing, including drying and baking (Curet et al., 2014).

15.3.3.1 Principle

Electrical Impedance Tomography (EIT) works on the principle of measuring impedance distribution within a substance, in this case a food sample. Impedance is the opposition to the flow of an alternating current (AC), which combines resistance (R) and reactance (X). EIT generates a spatial representation of electrical properties by measuring impedance at various places across the material. Impedance distribution changes reveal variations in the food's internal structure (Curet et al., 2014). EIT typically includes injecting a low-frequency electrical current into a food sample via an array of electrodes. The resulting voltage distribution is measured, and the impedance at different places is determined. The geographic distribution of these impedance values is then used to generate a representation of the food's interior structure (Içier & Baysal, 2004b).

15.3.3.2 Instrumentation

Specialized EIT systems are used to measure the impedance distribution within a food sample. The main components of EIT instrumentation are:

Electrodes are placed around food samples. These electrodes are used to inject electrical current and measure the resulting voltages. A regulated current source injects low-frequency electrical current into food samples via electrodes (Içier & Baysal, 2004b; Jiao, 2019). The voltage measurement system measures the voltage distribution across electrodes, allowing for accurate impedance calculations at various places. Then the Data Acquisition and Processing Unit collects and processes impedance data to create a spatial map of the electrical properties of food samples. Finally Image Reconstruction Software processes collected data to create an image

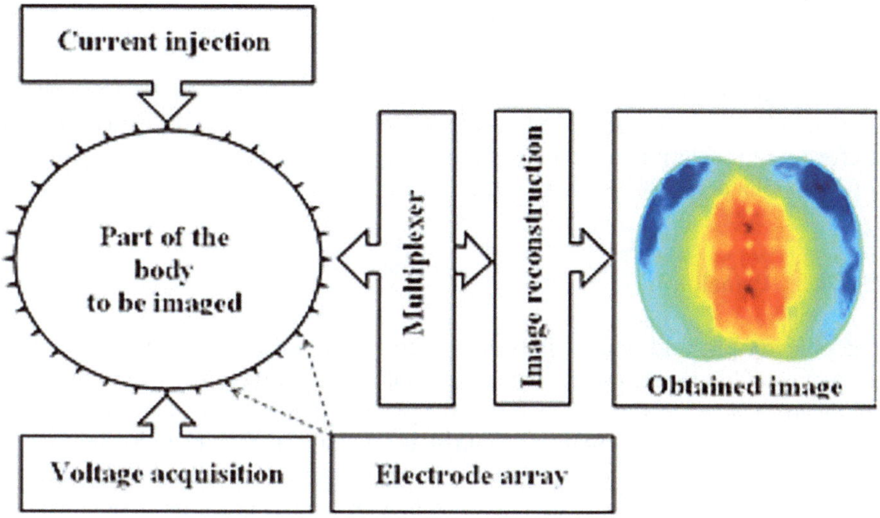

Fig. 15.2 Parts of EIT Imaging system

of the food's internal structure using impedance spatial distribution (Feng et al., 2012). Combined imaging system of Electrical Impedance Tomography has been shown in Fig. 15.2.

15.3.3.3 Application

Electrical Impedance Tomography (EIT) has a number of uses in food science and manufacturing operations. EIT enables real-time imaging and monitoring of processes in food manufacturing, including drying and baking. It offers information on changes in moisture content, temperature distribution, and other significant variables during these activities (Wang et al., 2013). EIT can identify abnormalities or non-uniformities in food product structures for quality control purposes. This is particularly useful for guaranteeing the consistency of processed meals. EIT enables non-destructive examination of food samples, revealing intrinsic features without affecting their structure (Jadhav et al., 2021). This technique can help to develop novel food products by analyzing the impact of different components and processing conditions on their internal structure. EIT techniques developed for medical imaging have been used for food quality research, providing a unique perspective on the interior structure of different food products (Binnur & Serap, 2011).

Electrical Impedance Tomography (EIT) is a useful tool for seeing and monitoring the interior structure of food samples in real time. Its uses in the food industry help with quality control, process optimization, and research and development, making it an important instrument in the field of food science and technology (Leong & Oey, 2022).

15.3.4 Conductivity Measurements

Electrical conductivity measures product's ability to conduct an electric current. Conductivity meters or probes measure food's electrical conductivity. This approach is effective for determining ion content and composition, particularly in liquids and solutions (Jha et al., 2011; Banti, 2020).

15.3.4.1 Principle

The objective behind conductivity measurements is to analyze how well a food product conducts an electric current. A material's electrical conductivity measures its ability to enable the flow of electric charge, which is often in the form of ions (Mitchell & de Alwis, 1989; de Halden et al., 2007). Conductivity in food is influenced by the presence of ions, such as those found in dissolved salts, acids, or other electrolytes (Duguay et al., 2017). When an electric potential is supplied to a material, the flow of charged particles (ions) through it increases its electrical conductivity (Curet et al., 2014).

15.3.4.2 Instrumentation

Conductivity measurements are performed with specialized instruments known as conductivity meters or probes. The main components of the instrumentation are:

The conductivity meter uses a probe that is immersed in the sample. The probe typically consists of electrodes spaced by a defined distance (Palaniappan & Sastry, 1991; Jiao, 2019). Electrical conductivity is measured as ions travel between these electrodes. The conductivity probe's electrodes allow electric current to flow through a sample (de Halden et al., 2007). They come into direct touch with the sample, making it possible to assess the conductivity (Fig. 15.3). The meter has a

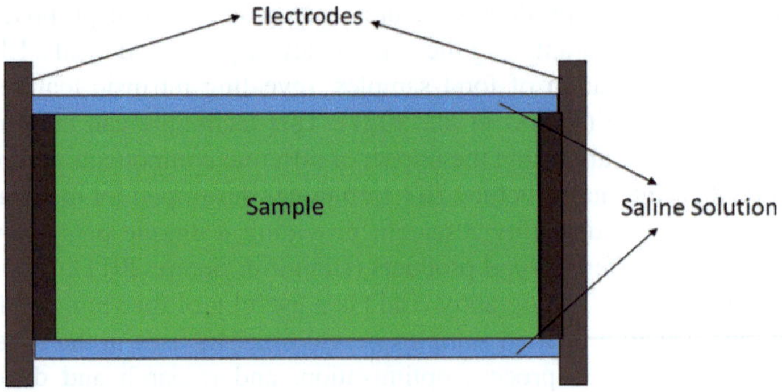

Fig. 15.3 Treatment chamber configurations: chamber covered on the sides with saline water (Astráin-Redín et al., 2024)

measurement circuit that applies an electric potential to the electrodes and measures the current flow. The sample's conductivity is then determined using the measured current and electrode geometry (Binnur & Serap, 2011).

15.3.4.3 Applications

Conductivity measurements are used in many parts of food analysis, particularly in liquids and solutions. Conductivity measurements help determine the ion concentration of liquids and solutions in food. This is especially important when determining the concentration of salts, acids, or other ionic species that may influence the taste, safety, or preservation of a food product (Palaniappan & Sastry, 1991; Duguay et al., 2017).

Quality Control for Beverages In the beverage business, conductivity tests are routinely used to determine liquid composition and quality. Changes in conductivity, for example, may reflect fluctuations in sugar content, mineral composition, or the presence of pollutants (Hlaváčová, 2003; Içier & Baysal, 2004a).

Monitoring Fermentation Processes Conductivity tests can track ion creation during food and beverage fermentation, including sugar to acid conversion. This is critical for managing and optimizing fermentation processes (Içier & Baysal, 2004a).

Conductivity tests can be used to detect adulteration in liquids. Deviations from expected conductivity levels could indicate the presence of foreign chemicals or dilution.

Conductivity measures can indicate water activity in products (Jiao, 2019). Water activity is an important characteristic for determining the stability and shelf life of different food products (Curet et al., 2014). Thus conductivity measurements provide a quick and easy way to determine the electrical conductivity of food samples. This approach is especially useful for determining ion content, overall composition, and detecting changes or deviations in liquid food products, which helps with quality control and ensures the integrity of various food and beverage items (Duguay et al., 2017).

15.3.5 Capacitance Measurements

Capacitance is used to assess a material's capacity to hold electrical charge. Capacitance meters or probes are used to measure dietary capacitance. Capacitance measurements are commonly used in conjunction with dielectric characteristics to determine moisture levels (Feng et al., 2012).

15.3.5.1 Principle

The principle behind capacitance measurements is to determine a product's ability to hold electrical charge. Capacitance is a property that measures a material's ability to store electrical energy when exposed to an electric field. Capacitance is very important in food since it indicates the moisture level of the product (Guo et al., 2010). Water's high dielectric constant has a considerable effect on the material's capacitance. Charges build up on a material's surfaces when an electric field is applied, and the capacitance of the material determines how much charge it can hold per unit voltage (Içier & Baysal, 2004b). Measurements of capacitance, which are useful in determining the moisture content of food because water is a polar molecule that increases capacitance, can be used to do this (Binnur & Serap, 2011).

15.3.5.2 Instrumentation

Capacitance meters or probes are used to measure capacitance in the context of food. Capacitance Meter/Probe: The food samples were brought in direct contact with a probe or sensor that makes up the capacitance meter. The purpose of the probe is to gauge the material's capacitance (Jha et al., 2011).

Measurement Circuit: Through the capacitance probe, the meter's measurement circuit applies an electric field to the substance. The capacity of the material to hold electrical charge is then determined by measuring the resulting capacitance. Complete measuring process of capacitance in food products is shown in Fig. 15.4 (Knirsch et al., 2010).

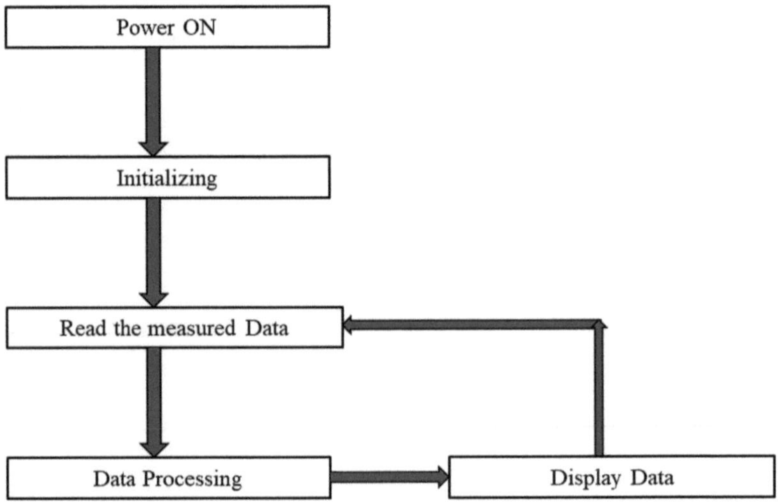

Fig. 15.4 Capacitance measuring process

15.3.5.3 Application

Measurements of capacitance, which are frequently combined with dielectric characteristics, are useful in many areas of food analysis, especially for determining the moisture content of food like bakery and confectionary. For a non-destructive evaluation of the moisture content of food products, capacitance measurements are frequently used. Changes in moisture levels can be linked to differences in capacitance since moisture content influences both capacitance and the dielectric characteristics (Figura & Teixeira, 2023).

Quality Control in Agriculture Soil moisture levels are monitored in agriculture using capacitance-based sensors. This information can be very important for the food business in terms of making the best irrigation decisions and guaranteeing healthy plant growth (Hlaváčová, 2003).

Process Control in Food Manufacturing Real-time monitoring and control of food manufacturing processes can be achieved by integrating capacitance measurements (Figura & Tiexeira, 2007).

Packaging Quality Evaluation The materials used in food packaging can be evaluated for quality using capacitance-based sensors. Leaks or changes in the packing material may be indicated by changes in the capacitance (Curet et al., 2014).

Predicting Shelf Life Food products' shelf lives can be determined and predicted by keeping an eye on variations in capacitance over time, especially in perishable commodities. This is essential to guaranteeing product quality and reducing food waste. Thus capacitance measurements offer a useful and non-destructive way to determine how much moisture is present in food items (Hlaváčová, 2003). The use of capacitance in food analysis aids in process optimization, quality assurance, and a general comprehension of the electrical properties of the material, especially with regard to its water content (Banti, 2020).

15.3.6 Time-Domain Reflectometry (TDR)

TDR's basic idea is to time how long it takes an electrical signal to move across a sample's transmission line. To evaluate the food's dielectric qualities, TDR devices which are frequently in the form of probes or sensors are employed. TDR is frequently employed to measure soil moisture, but it can also be modified for use in some food-related applications.

15.3.6.1 Principle

Time-Domain Reflectometry (TDR) is a method used to calculate the length of time an electrical signal takes to pass through a sample's transmission line. An electromagnetic pulse is propagated down a waveguide or transmission line, and the reflected signals are analyzed, as per the fundamental principle (Jha et al., 2011). The foundation of TDR is the idea that an electrical signal's speed as it passes through a substance is influenced by its dielectric characteristics (Ryynänen, 1995).

The TDR method is modified to evaluate the sample's dielectric characteristics in the context of food. Information on the composition and structure of the food product is obtained by measuring the reflections or changes in the electrical signal that is supplied through a probe or sensor. The working of TDR with the help of block diagram is shown in Fig. 15.5. Changes in the material, including fluctuations in moisture content, can be detected by TDR, which is especially sensitive to changes in the dielectric constant (Lin & Wang, 2020).

15.3.6.2 Instrumentation

Probes, sensors, and related instrumentation are usually included in TDR devices used to evaluate the dielectric characteristics of food. TDR probe or sensor is often a cable or rod that is put into the food sample and has a transmission line. This transmission line is used to send the electromagnetic pulse (Içier & Baysal, 2004b). The electromagnetic pulse for a short duration is produced by the pulse generator that is part of the TDR system. Then the receiver records the reflected signals or signal alterations brought on by interactions with the food sample (Ryynänen, 1995). The material's dielectric characteristics are then ascertained by analyzing the signals (Jiao, 2019).

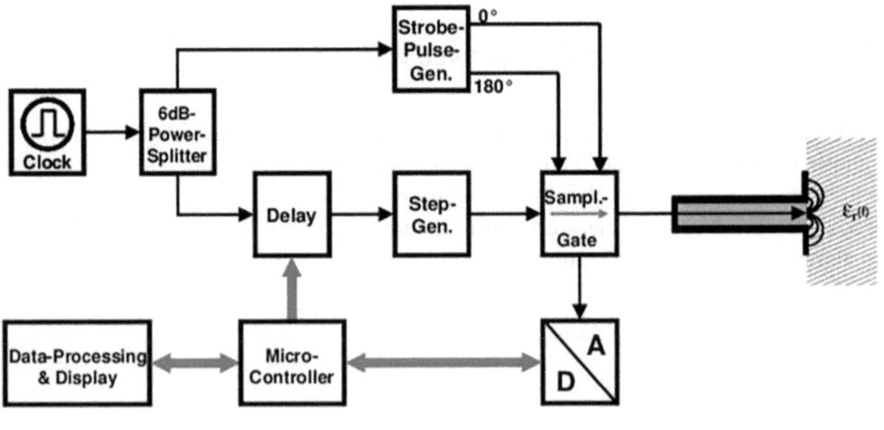

Fig. 15.5 Block diagram for time-domain reflectometry (Schimmer et al., 2004)

15.3.6.3 Applications

Although TDR is frequently linked to soil moisture measurements, it can also be modified for use in certain food-related contexts. Among the important applications are:

Moisture Content Measurement: TDR is used to determine how much moisture is present in food items. Variations in moisture content cause discernible changes in the TDR signal, and the dielectric characteristics of the material affect the electrical signal's speed (Curet et al., 2014).

Texture and Structural Analysis: TDR can be used to evaluate a food product's texture and structural qualities. The TDR reaction may alter in response to modifications in the material's structure or composition.

Quality Control in Processing: Real-time monitoring and control of food processing lines can be achieved with the integration of TDR. TDR measurements, for instance, can help optimize processing conditions and offer insights into moisture distribution in operations like baking and drying.

Packaging Evaluation: The quality and integrity of food packaging materials can be evaluated using TDR. It is possible to identify variations in the dielectric characteristics, such as those brought on by air presence or compositional changes in packing materials (Banti, 2020).

These methods offer important insights into the electrical characteristics of food and enable non-destructive evaluation and observation at different phases of food processing, manufacturing, and quality assurance (de Halden et al., 2007). The technique selected will rely on the particular properties being looked into as well as the features of the food item being examined.

15.4 Applications for Non-destructive Food Quality Analysis

In many applications related to food analysis, the electrical characteristics of food are vital. These qualities can be used to evaluate and keep an eye on the features, safety, and quality of food products. There are several non-destructive methods to evaluate food quality shown in Table 15.2 but overall, the electrical properties is preferable over others. The following are some significant uses of food's electrical characteristics in food analysis:

Table 15.2 Non-destructive methods for food quality evaluation

Non-destructive methods	Measured quality
Image processing	Size, shape, colour, outward defects
Reflectance, absorption, laser spectroscopy	Sugar, acidity, soluble solid content, colour, internal and external defects, stiffness
X ray imaging	The inner cavity structure, the degree of maturity
Vibration	Stiffness, viscoelasticity, the degree of maturity
Sound and ultrasound	Stiffness, viscoelasticity, internal cavity, density, sugar
NMR and MRI	Sugar, moisture content, the inner cavity
Artificial sensory evaluation	Acidity, sugar
Electrical properties	Moisture, quality, adulteration, ripeness, fermentation, texture, shelf life, freezing, thawing and real time monitoring

Table 15.3 The electrical conductivity (s/m) of food products measured at different temperatures (Zhang, 2007)

Food products	4 °C	22 °C	30 °C	40 °C	50 °C	60 °C
Beer	0.080	0.143	0.160	0.188	0.227	0.257
Coffee	0.138	0.182	0.207	0.237	0.275	0.312
Fruit juice	0.196	0.239	0.279	0.333	0.383	0.439
Milk	0.357	0.527	0.617	0.683	0.800	0.883
Vegetable juice	0.788	1.147	1.282	1.484	1.741	1.980

15.4.1 Determination of Moisture Content

Food's dielectric characteristics are very sensitive to its moisture content. Non-destructive assessment of food products' moisture content aid its quality assurance and standard compliance like in baking, dehydration, and drying (Frabetti et al., 2023). Food analysis relies heavily on the electrical properties of food, which provide important information on the attributes, safety, and quality of different food products (Kertesz et al., 2015). Measuring moisture content is one important application that is essential to guaranteeing food safety and quality (Jha et al., 2011). This is especially crucial in sectors where managing moisture levels is essential to product quality, like baking, dried, and dehydrated products. Food products' texture, shelf life, and general quality are all impacted by their moisture level (Ozturk et al., 2016; Banti, 2020).

The basic concept behind measuring moisture content using electrical properties is that water molecules have a higher dielectric constant than the majority of food ingredients. The dielectric characteristics of food vary with its moisture content and temperature which clearly depicts in Table 15.3 (Hlaváčová, 2003; Kertesz et al., 2015). This sensitivity makes it possible to quickly and non-destructively determine the food's moisture content without changing its composition. Actuality, food's dielectric characteristics are used by gadgets like moisture meters or sensors to determine how moist it is (Frabetti et al., 2023). These devices can give precise and up-to-date information on the sample's moisture content by monitoring the food's reaction to an applied electrical field (Kertesz et al., 2015). This application is especially useful in sectors where preserving product quality requires exact control over moisture content (Fryer, 1995). For instance, keeping an eye on moisture levels in the baking sector guarantees the proper texture and flavor of baked goods. Understanding and managing moisture content is crucial for food preservation and spoiling prevention throughout drying and dehydration operations (Jha et al., 2011). Food analyzers and manufacturers may use efficient quality control procedures, adhere to industry standards, and ultimately provide consumers with safer and higher-quality food items by using the electrical properties of food to detect moisture content (Jadhav et al., 2021).

15.4.2 Monitoring for Quality While Processing

Changes in electrical characteristics may be indicative of changes made to the product's structure and composition during preparation. Processing parameter optimization, homogeneity, and real-time deviation detection are made possible by the continuous monitoring of electrical properties that are generally applied during cooking, baking, and drying procedures (Jha et al., 2011). A key component of guaranteeing the uniformity, security, and general quality of food items is quality control during food processing. In this case, using the electrical characteristics of food offers a non-destructive, real-time way to evaluate how the composition and structure of the food change throughout processing or storage (Banti, 2020).

The presence of various components (e.g., sugars, fats, proteins) and variables like temperature and moisture content affect the electrical properties, including permittivity and conductivity (Wang et al., 2013; Pereira et al., 2024). Food's electrical characteristics alter throughout preparation along with changes in its content and structure. This application is frequently used in a variety of food processing methods, including frying, drying, and baking (Ozturk et al., 2016; Figura & Teixeira, 2023).

It's crucial to keep food processing quality constant. Food's electrical characteristics can be used as markers for internal product changes. Using sensors or probes to measure the electrical properties of the food at each stage of preparation is known as continuous monitoring. Processors are able to make timely adjustments thanks to this real-time feedback, guaranteeing that the finished product fulfills the intended quality criteria (Fryer, 1995). By comprehending the connection between electrical characteristics and food composition, processors can make the most of a number of parameters, including cooking time, moisture content, and temperature (Knirsch et al., 2010). For example, when baking, tracking the electrical characteristics of the dough during the heat treatment process can reveal information on the degree of gluten development, starch gelatinization, and moisture evaporation. Manufacturers are able to alter the production parameters in order to produce the desired final baked goods texture, color, and flavor by using these real-time measurements (Curet et al., 2014).

For foods to satisfy customers, uniformity is very essential. Disparities in a food batch's electrical characteristics may be a sign of uneven processing conditions or composition. Processors can detect and resolve problems that could result in inconsistent product quality by keeping an eye on these attributes (de Halden et al., 2007). This is especially significant in large-scale food manufacturing, where maintaining consistency is difficult but necessary to fulfill quality requirements. Rapid detection of deviations from the predicted processing conditions is made possible by the real-time monitoring of electrical properties (Feng et al., 2012). For instance, abrupt variations in electrical conductivity during frying could be a sign of problems with uneven heating or deteriorating oil. Early detection of these discrepancies allows operators to minimize waste and take corrective action, preventing the manufacture of subpar products (Fryer, 1995).

So, the use of electrical properties for quality monitoring in the food processing industry improves process control, aids in parameter optimization, guarantees consistency, and permits early deviation identification. This helps the various processing sectors produce food items that are safe, uniform, and of high quality (Leong & Oey, 2022).

15.4.3 Detection of Foreign Bodies

One of the most important aspects of food business quality control is the identification of foreign substances. It is vital to guarantee the end product is devoid of contaminants in order to comply with regulatory requirements, uphold consumer confidence, and avert significant health risks (Jha et al., 2011). Early identification of foreign objects not only ensures regulatory compliance but also minimizes product recalls and protects food makers' reputations. Automated techniques for detecting foreign bodies based on electrical characteristics are frequently incorporated into modern manufacturing lines (Weng et al., 2020). With the ability to rapidly scan and evaluate items in real-time, these devices can give the production process immediate feedback. Potential impurities are found and eliminated before the finished product is delivered to customers thanks to integration into production lines (Binnur & Serap, 2011).

The presence of impurities or foreign bodies can change an electrical property. Recognizing and identifying foreign objects or pollutants in food products is important to guarantee food safety and to monitor quality (Jha et al., 2011). Contaminants or other objects can change the electrical properties of food. When foreign components are added, the electrical field's relationship with the food's composition changes. The identification of foreign objects in food products is based on this alteration in electrical characteristics (Curet et al., 2014).

This application deals with locating and identifying foreign objects or pollutants in food items. This technique is widely applied to guarantee food safety and quality control at different phases of food processing and production. The idea that various materials have unique electrical properties forms the basis for the use of electrical properties in the detection of foreign substances. Variations in the electrical characteristics of the composite system might occur when a foreign material is included in a food product. These fluctuations, which are frequently quantifiable, may reveal information about the kind, location, or presence of foreign bodies (Leong & Oey, 2022). There are various electrical properties that can be used to find foreign objects. Commonly used terms are impedance, dielectric constant, and electrical conductivity. The selection of the property is contingent upon the particular attributes of both the foreign substance and the food item under examination. For example, the system's total conductivity will change if a foreign body has a different conductivity than the surrounding food matrix.

Depending on how foreign bodies affect electrical properties, different detection techniques are used to find them. Using metal detectors, which take advantage of variations in conductivity brought on by the presence of metallic foreign bodies, is one popular technique. This works especially well in sectors where metal pollution is a problem, like canning. For non-metallic foreign bodies dielectric characteristics are frequently used. Dielectric imaging methods, such as microwave sensing and capacitance imaging, can be used to detect changes in dielectric constants brought on by the presence of foreign elements. These techniques are useful for finding pollutants such as glass, plastic, and other non-metallic materials (Banti, 2020).

15.4.4 Evaluation of Fruit and Vegetable Ripeness

The ripening process is frequently linked to changes in electrical characteristics. Non-destructive evaluation of fruit and vegetable ripeness that aids in choosing the best time to harvest and guarantees product quality (Zhang, 2007). The ripening process of fruits and vegetables is frequently linked to changes in electrical characteristics (Kaur et al., 2016). Fruits and vegetables change as they ripen in terms of composition, water content, and other elements, all of which have an impact on electrical characteristics (Guo et al., 2010). The application entails the non-destructive evaluation of the ripeness of fruits and vegetables (Sarang et al., 2008). This provides important information for figuring out when to harvest the agricultural produce at its best and guaranteeing its overall quality (Binnur & Serap, 2011).

Fruits and vegetables undergo a variety of biochemical and physiological changes during the ripening process, such as changes in sugar content, acidity, water content, and cell structure (Hlaváčová, 2003). These changes can have an impact on the electrical properties of the produce; for instance, an increase in sugar content or modifications to the fruit's water distribution can result in variations in electrical conductivity and dielectric properties (Fryer, 1995; Ozturk et al., 2016). Conventional methods of determining ripeness frequently involve destructive sampling, which may have an adverse effect on the overall yield and quality of the harvested crop. The non-destructive alternative of using electrical properties for ripeness assessment is the measurement of the electrical characteristics of fruits and vegetables (Jha et al., 2011; Banti, 2020).

A variety of methods are used to measure electrical characteristics related to maturity without causing damage. Among the techniques employed are electrical impedance tomography, dielectric spectroscopy, and impedance spectroscopy (Kertesz et al., 2015). These methods give a thorough understanding of the electrical characteristics of the fruit or vegetable (Hlaváčová, 2003). The best time to harvest can be ascertained using the data gathered from electrical property measurements. It's essential to harvest fruits and vegetables at the ideal time of maturity to ensure the highest possible quality, flavor, and nutritional value (Kaur et al., 2016). In order to improve crop management generally, electrical property-based assessments give farmers a quantitative and objective way to determine when to harvest (Fryer, 1995; Juansah et al., 2014).

15.4.5 Quality Control in Storage and Distribution

The non-destructive evaluation of ripeness using electrical qualities is useful in operations related to storage and distribution even outside of the field. To ensure that only fruits and vegetables at the appropriate stage are delivered to market, it aids in the sorting and grading of food according to its levels of ripeness (Sarang et al., 2008). This helps the supply chain operate more efficiently and with less waste (Duguay et al., 2017). These agricultural practices can be made more sustainable by farmers who make well-informed decisions about crop management and resource optimization through the use of technology and data-driven techniques (Jadhav et al., 2021).

The farmers and other producers can have an accurate, non-destructive, and effective way to determine the freshness of fruits and vegetables by the use of electrical properties (Juansah et al., 2014). Better decision-making is facilitated by this at every stage of the agricultural supply chain, from harvest to distribution, guaranteeing that customers get ripe, high-quality fruit while reducing waste (Guo et al., 2010).

15.4.6 Monitoring Fermentation Processes

Electrical characteristics can be altered by fermentation processes. Tracking the development of fermentation in necessary to guarantee consistency and quality in food production, such as in the dairy and brewing sectors. The medium's electrical characteristics may alter as a result of fermentation activities (Leong & Oey, 2022). The conductivity, impedance, and other electrical properties of the substance fermenting are frequently influenced by the microbial processes that take place during fermentation, such as the conversion of carbohydrates into acids or alcohol.

In the production of food, particularly in the dairy and brewing sectors, the application entails the ongoing monitoring of the fermentation process. Maintaining consistency, quality, and ideal process control is ensured by this monitoring (Binnur & Serap, 2011).

Fermentation is a biological process in which various end products, like alcohol, acids, or gases, are produced by microorganisms like bacteria, yeast, or molds metabolizing substrates like carbohydrates. The fermentation medium's electrical characteristics may change as a result of these metabolic processes. For example, the conductivity, impedance, or capacitance of the fermenting material may be impacted by the generation of gasses or acids (Içier & Baysal, 2004a). In order to keep process control during fermentation, it is imperative that the electrical characteristics be continuously monitored. Producers can get real-time data during fermentation by installing electrical sensors and probes into the fermentation tanks. Fermentation conditions can also be improved with the help of the data obtained from electrical property monitoring. The electrical sensors provide real-time feedback that allows for the adjustment of many parameters, including pH, temperature, and nutrient levels (Içier & Baysal, 2004a). This increases the fermentation process's efficiency, which results in increased yields, quicker manufacturing, and better-quality of the final products.

In order to guarantee that the finished fermented foods and beverages have the appropriate flavor, fragrance, and texture profiles, consistency is essential. By spotting variations or inconsistencies in the fermentation process, electrical property monitoring helps to ensure uniformity across batches. This is especially crucial in sectors like brewing, where the fermentation process greatly influences the finished product's qualities (Binnur & Serap, 2011).

Similar to this, in the dairy sector, keeping an eye on modifications in the electrical characteristics of milk as it ferments into yogurt or cheese can help guarantee that the right kind of microbial activity is taking place. For real-time monitoring, modern fermentation methods frequently use cutting-edge technology. Tanks, bioreactors, and fermentation vessels are all equipped with carefully positioned sensors and instrumentation that can measure electrical qualities. These sensors' data can be incorporated into control systems to enable automated changes and provide operators a thorough understanding of the fermentation process. Using electrical properties to monitor fermentation processes helps with both regulatory compliance and quality control (Figura & Teixeira, 2023). To guarantee the security and quality of the finished food products, it is important to exhibit a consistent and regulated fermentation process.

15.4.7 Food Texture Testing

The structural alterations that affect the texture of food can be linked to electrical qualities. Non-destructive evaluation of food texture during processing, which enables modifications to preserve intended textural properties. Food's electrical characteristics are intimately related to the structural alterations that affect its texture. A food product's electrical properties vary along with changes in its structure and composition. There is a correlation between these electrical property changes and changes in food texture (de Alwis et al., 1989). Electrical properties are used in non-destructive food texture testing to continuously evaluate texture throughout processing. This enables modifications to be made in real-time to maintain the intended textural qualities of the food product.

Most important quality factor that affects consumer preferences is food texture. It is frequently linked to elements like chewiness, crispness, firmness, and general mouthfeel. Food's structural elements such as proteins, carbohydrates, and water content affect its texture. Alterations in these constituents throughout the processing phase may impact the food's electrical characteristics (Jadhav et al., 2021). Conventional techniques for evaluating food texture frequently were destructive testing, in which samples are broken or altered physically to assess characteristics like hardness or brittleness (Sarkar et al., 2022). An alternative technique is non-destructive testing, which uses electrical properties in which properties like conductivity, capacitance, or impedance are used obtain information about the structural changes influencing texture without compromising the food's physical integrity (Fryer, 1995). Electrical property changes during baking or frying may correspond with changes in moisture content, structural hardness, or starch gelatinization degree. Non-destructive testing provides real-time data that enables quick modifications to processing parameters. For instance, operators can adjust temperature, humidity, or processing time to obtain the appropriate textural characteristics if the electrical properties indicate that a product is becoming too soft or dry (Jha et al., 2011). This degree of control is especially useful in sectors where maintaining uniformity in texture is essential, like baking or snack foods. It ensures that cookies or crackers bake to the perfect crispness. It helps to keep snacks at the perfect crunch level during manufacture. When it comes to meat products, it can help to regulate things like juiciness and softness (Sarang et al., 2008). Manufacturers can apply efficient quality control procedures by combining non-destructive food texture testing (de Alwis et al., 1989). This aids to the general consistency and homogeneity of the items that reach the market in addition to guaranteeing that the finished product satisfies the required texture criteria.

15.4.8 Shelf Life Estimation

Any deviation or change in electrical characteristics may be associated with food degradation over time. Using electrical property measurements to forecast food product shelf life, which helps with inventory control and waste reduction. Alteration in the electrical properties of food may serve as a sign of deterioration over time (Jha et al., 2011). The food product's electrical properties may alter as a result of microbial growth, chemical interactions, and moisture changes, among other degrading processes. This reduces food waste, helps with inventory control, and guarantees that customers receive high-quality items (Banti, 2020). Chemical reactions, microbiological activity, and variations in moisture content are only a few of the variables that affect the process of food product deterioration. The food's qualities, composition, and structure may change as a result of these activities. These modifications have an impact on electrical characteristics including conductivity, impedance, and dielectric constant (Leong & Oey, 2022). Keeping an eye on these electrical characteristics over time yields important insights into the processes of continuous deterioration. By tracking a food product's electrical characteristics throughout time, a profile that corresponds to the rate of deterioration can be developed (Duguay et al., 2017). Analysis and modeling are done on the data that are gathered from the ongoing electrical property monitoring. Predictive models can be developed by creating connections between the electrical characteristics and recognized markers of food deterioration (such as bacteria counts, lipid oxidation, or variations in moisture content). These models aid in predicting the study of electrical properties over time, enabling the determination of the product's shelf life (Fryer, 1995; Varghese et al., 2014).

Utilizing electrical property measurements to forecast food product shelf life has important consequences for inventory control. Retailers and distributors can minimize the risk of selling products past their prime quality by optimizing their inventory with real-time estimates of each product's remaining useful life. Consequently, this reduces food waste and guarantees that customers receive goods at their peak freshness (Frabetti et al., 2023). Electrical property measurements for shelf life prediction help to guarantee that customers receive goods that live up to their high standards. It supports the overarching objective of giving customers a satisfying experience by delivering goods at the highest possible level of flavor and quality. Sensors and electronic components can now be integrated right into package materials thanks to developments in smart packaging technologies (Jha et al., 2011). These sensors are capable of monitoring various food quality indicators in addition to electrical qualities. Predicting shelf life more accurately and offering a more complete picture of product freshness are two benefits of smart packaging (Binnur & Serap, 2011).

15.4.9 Tracking the Processes of Freezing and Thawing

Phase transitions related to freezing and thawing affect electrical characteristics. Tracking the freezing and thawing procedures in the frozen food sector to guarantee product quality and maximize freezing conditions. Phase transitions related to freezing and thawing affect a material's electrical characteristics (Hlaváčová, 2003). Due to change in the material's composition and structure, water can experience changes in electrical properties during these transitions between its liquid and solid forms, including conductivity and dielectric constant. To maximize freezing conditions and guarantee the quality of frozen goods, electrical property measurements are used in the frozen food industry to monitor the freezing and thawing processes. Water goes through phase transitions throughout the freezing and thawing processes, changing from a liquid to a solid (freezing) and vice versa (thawing). The shape and arrangement of water molecules are altered during these phase transitions (de Alwis et al., 1989). Ice crystals form when water freezes, and when it thaws, the crystals melt back into liquid water. The electrical characteristics of the frozen substance may change as a result of these variations. Because it directly impacts the quality of frozen goods, the freezing process is crucial. Measurements of electrical properties, including dielectric constant or impedance, can be used to track the freezing process. The composition and molecular arrangement of water change as it turns into ice, causing a change in the product's electrical characteristics. Constant observation helps to optimize conditions for the highest-quality frozen products by providing real-time monitoring into the freezing process (Binnur & Serap, 2011).

Food business operators can maximize freezing conditions, such as temperature and freezing rate, by keeping an eye on electrical qualities during the freezing process (Wang et al., 2013). The purpose is to regulate ice crystal formation and size, which might affect the frozen food's texture and quality. By optimizing freezing conditions by electrical property measurements, frozen food product with desired qualities can be produced and problems like ice crystal formation and freezer burn can be reduced (Binnur & Serap, 2011).

In the frozen food sector, thawing is an essential stage, particularly for products that will be eaten raw. Measurements of electrical properties made during thawing shed light on how the liquid phase reforms. This can be very helpful in preserving the integrity of the product, avoiding unfavorable texture or flavor changes, and making sure that the thawing procedure doesn't degrade the quality of the product (Kamonpatana & Sastry, 2022). Electrical property measurements used to monitor the freezing and thawing processes is for quality assurance procedures. It aids in fulfilling legal requirements for the safety and quality of frozen food.

15.4.10 Food Product Authentication

Food product authentication, which helps to detect and stop fraud by guaranteeing that authentic items have consistent electrical characteristics. Because food differ in terms of composition, structure, and moisture content, their electrical characteristics can serve as distinctive markers for distinct food categories. As a unique signature, these intrinsic variations in electrical characteristics can be used to create a baseline or fingerprint for food product (Fryer, 1995). The process of verifying the authenticity of food products entails using their distinct electrical characteristics to spot and stop fraud, making sure that these characteristics are constant in real products, and looking for any differences that might point to tampered or fake products (Dube et al., 1988).

Food items differ in their contents, which include different proportions of water, carbohydrates, fats, proteins, and other ingredients. Foods' electrical characteristics, such as conductivity and dielectric constant, are influenced by these variations (de Halden et al., 2007). The idea is to realize that these electrical characteristics can serve as distinguishing indicators, giving each kind of food its own stamp. Baseline measurements of the electrical characteristics of real or authentic samples of each type of food are established in order to authenticate food products. Food's electrical characteristics can be measured using a variety of methods. The use of sensors, probes, or imaging technologies will help to determine the features of the food product. Electrical properties like impedance, dielectric, or electrical impedance tomography might be used (Banti, 2020).

A variety of food sectors can use electrical properties to authenticate food products. For example, electrical property measurements can assist in confirming the quality of the olive oil, where adulteration poses a serious threat to authenticity. Similar to this, authentication based on electrical qualities might help to ensure the product's legitimacy and purity in the case of honey, where several types may have unique electrical signatures (de Halden et al., 2007).

Thanks to technological advancements, automated systems that can quickly and accurately certify food items based on their electrical characteristics. These systems work well in production lines and distribution centers because they can handle big amounts of samples. Thus, electrical property-based food product authentication is a reliable way to guarantee the authenticity and reliability of food products (Binnur & Serap, 2011).

15.5 Future Trends

Developments in food science and technology appear to be highly promising when it comes to investigating and focussing the electrical properties of food. In the upcoming years, the following trends and areas of development might be expected:

15.5.1 Integration of Electrical Property Measurement Techniques with Other Advanced Technologies

To improve the overall capabilities of food analysis, the food business is embracing the integration of electrical property measurement techniques with other cutting-edge technology. The necessity for more thorough, precise, and effective means of quality control, monitoring, and guaranteeing the safety of food items could be fulfilled by integration techniques. The simultaneous investigation of chemical and electrical properties is made possible by integration with spectroscopic techniques like infrared spectroscopy or Raman spectroscopy (Lin & Wang, 2020; Novas et al., 2023). This makes it possible to characterize food samples more thoroughly by offering details on their molecular structure in addition to their electrical properties

(Leong & Oey, 2022). Properties within food products can be spatially mapped when electrical property measurements are combined with image technologies like computer vision or hyperspectral imaging. This can be useful for guaranteeing consistency in product quality and identifying variability in big batches. Adding nanoscale sensors to food analysis can improve accuracy and sensitivity. In addition to electrical property measurements, nanosensors may identify particular molecules or chemical changes, offering comprehensive information about the composition and quality of food products. Traceability and transparency in the food supply chain are improved by combining blockchain technology with electrical property measurements (Jha et al., 2011). By giving customers dependable information about the products' handling and origin, this integration contributes to the authenticity and quality of food products.

15.5.2 Combining Electrical Property Data with Artificial Intelligence (AI) and Machine Learning Algorithms

The future element pertains to the analysis and interpretation of electrical property data in food analysis through the utilization of artificial intelligence (AI) and machine learning (ML) algorithms. The combination of AI/ML and electrical property measurements has enormous potential to improve the precision, efficiency, and sophistication of food analysis procedures. Data analysis and pattern recognition are areas where AI and ML algorithms shine (Feng et al., 2012). These algorithms may understand intricate correlations between electrical property measurements and several quality indicators by being fed large datasets of observations. This makes it possible to create extremely precise prediction models for evaluating the safety, authenticity, and quality of food (Varghese et al., 2014). When AI and ML algorithms are combined with smart sensor networks, real-time monitoring and control in food production processes can be facilitated. Based on the constantly updated data from electrical property tests, these systems are capable of autonomously modifying processing parameters. Artificial intelligence and machine learning algorithms can be used to identify patterns linked to both normal and pathological fluctuations in electrical properties. This makes it possible to identify abnormalities or changes from predicted patterns to previous ones. Early detection in the food manufacturing process helps minimize waste and guarantee the product consistency (Feng et al., 2012).

15.5.3 Continued Efforts Towards the Development of Compact and Portable Instruments

Small and lightweight instruments are simpler to transport and use in a variety of settings, such as a field, a laboratory, or a production plant. This is especially useful for measurements that take place on-site, when it is not practicable to transfer large, heavy equipment. User-friendly interfaces are a common feature of miniature instruments, which increases their accessibility to a wider variety of users. This is especially important for use in places like farms or smaller food processing facilities where there may not be much access to specialist training. Quicker setup and measurements are made possible by smaller instruments, which speeds up the decision-making process. In terms of both the original outlay and continuing operating expenses, smaller devices are frequently more economical. These technologies are being adopted more widely in a variety of situations thanks in part to their affordability. The future trend of electrical property measurement device shrinking point to a desire to increase the accessibility, adaptability, and applicability of these technologies in a range of contexts. The miniaturization trend has the potential to change how we use and interact with electrical property measurement techniques in the food industry and beyond, from on-site applications in agriculture to integration into smaller food processing facilities and even possible consumer-level adoption.

15.5.4 Enhanced Analytics for Data

The need for advanced data analytics approaches will expand as the amount of data produced by electrical property tests rises. In the context of electrical property measurements, the future of advanced data analytics predicts a considerable rise in the number and complexity of data. Larger datasets will be produced by continuing technological advancements, therefore the focus will be on creating increasingly complex analytics methods to extract information (Jha et al., 2011). Big data analytics, pattern recognition, and sophisticated visualization techniques are used.

Continuous or high-frequency electrical property tests can produce enormous amounts of data. Large datasets can be handled and processed effectively with the help of big data analytics tools (Zhao et al., 2000). When data is created quickly or frequently, it might be difficult to keep up with the pace of data creation. Big data analytics can quickly process and evaluate data streams to deliver actionable result. Information gathered from electrical property tests might take many different forms, such as spectrum data, pictures, or numerical values. Big data analytics is capable of handling various data formats and identifying significant trends. Electrical property datasets can contain complex patterns that can be found using machine learning methods and pattern recognition techniques (Weng et al., 2020). Understanding the subtleties of how many elements contribute to variances in food quality is crucial.

One of the main purposes of data visualization is to make complex datasets easier to understand. Data visualization tools are essential for this purpose. Users such as academics, technicians, and decision-makers can gain a more intuitive understanding of trends, outliers, and patterns with the use of interactive visualizations. With a thorough grasp of the connections between electrical qualities and dietary attributes, it empowers stakeholders to make defensible judgments. The electrical property measurements in the food business is expected to undergo a radical change as a result of the current trend of using advanced data analytics combined with the future aspect of managing larger and more complex datasets (Feng et al., 2012). The amalgamation of big data analytics, pattern recognition, and sophisticated visualization techniques is poised to foster innovation in food production and analysis, expedite decision-making procedures, and enhance the food quality.

15.5.5 Sensor-Integrated Smart Packaging

The incorporation of sensors intended for the measurement of electrical properties is envisioned as a future feature of smart packaging. The next generation of smart packaging is expected to deliver real-time insights about the electrical properties of packaged food products. Conductivity, impedance, and dielectric constant are just a few examples of the electrical qualities that future smart packaging may contain sensors (Kamonpatana & Sastry, 2022). These measures can provide special information about the integrity, freshness, and makeup of the packaged food items (Guo et al., 2010). Throughout transportation, electrical properties are continuously monitored to make sure the products stay within predetermined quality standards. Any changes from the anticipated electrical characteristics can result in alarms, which enable quick action to preserve product quality. Smart packaging with electrical property sensors helps to reduce food waste by giving timely information on the state of the packaged goods. It makes it possible to evaluate product quality and freshness more precisely, reducing needless discards and improving inventory control. It also can reveal contamination or microbial activity (Binnur & Serap, 2011).

15.5.6 Precision Agriculture

The development and ongoing integration of electrical property measurements in soil and plant analysis is a crucial component of precision agriculture in the future. This includes creating increasingly advanced analytics and sensing systems to improve resource efficiency, further refine crop management techniques, and guarantee the production of premium agricultural goods (Duguay et al., 2017).

Soil moisture content can be analyzed from electrical conductivity measurements. More sophisticated sensors that can provide high-resolution, real-time data on soil moisture levels may be used in future applications. By using this information, one may precisely manage irrigation and avoid over- or under-irrigation. Soil composition and nutrient content affect electrical characteristics. In the future, sensors that measure pH, nutrient levels, and soil texture might be developed. These sensors would give farmers precise information for the best fertilization plans (Jha et al., 2011). To determine the salinity of soil, electrical conductivity measurements are frequently utilized. With the help of future technologies, farmers may be able to manage salinity more precisely by detecting variations in salinity with greater accuracy and sensitivity.

Plant's electrical characteristics can reveal information about its overall health. The creation of sensors that can evaluate physiological data in plants could be used in future applications to identify diseases, stress, and nutrient deficits early on. Plant tissues' electrical conductivity and nutrient uptake may be connected. Sensors could offer real-time data on plant nutrient levels, enabling accurate fertilization techniques to match the unique requirements of various crops (Juansah et al., 2014).

15.5.7 Environmental Monitoring in Food Production

Electrical property measurements will be integrated as a tool for environmental optimization in the future aspect of food production environmental monitoring. Using these analysis we could improve food processing's overall sustainability, reduce waste, and increase energy efficiency.

Processing parameters can be improved by using electrical property measurements, such as those pertaining to heat generation or electrical conductivity. Future uses could include the development of sensors that offer data on energy consumption in real time, enabling modifications to improve efficiency (Routray & Orsat, 2018). Improving the efficiency of heat transfer can be aided by knowing of the electrical characteristics of food items during heating procedures (Guo et al., 2010). This information can help to reduce the overall energy usage by energy-efficient heating and cooling systems. Maximizing processes for energy efficiency, reducing waste, and promoting overall sustainability are the main goals of using electrical property measurements in food production for environmental monitoring in the future. This comprehensive strategy is in line with the global trend of the food business adopting more accountable and ecologically friendly methods (Kamonpatana & Sastry, 2022).

15.5.8 Customized Food Production

The food business is currently seeing a rise in interest in and emphasis on customized nutrition. The understanding that every person has different dietary choices, nutritional needs, and health objectives is what is driving this trend (Routray & Orsat, 2018). Foods that meet their individual needs be the cause of dietary restrictions, medical issues, or lifestyle preferences. Customizing food items to fit each person's tastes and dietary requirements is known as personalized nutrition. Consumers have a wide range of dietary preferences, such as being gluten-free, vegetarian, vegan, or following other particular diet plans (Kamonpatana & Sastry, 2022). These preferences are taken into account, and solutions that fit certain lifestyles are provided through personalized nutrition. This approach takes into account each person's unique dietary requirements while accounting for age, gender, activity level, and medical concerns. The goal of this strategy is to offer specialized solutions to satisfy particular dietary needs (Binnur & Serap, 2011).

Customizing food manufacturing procedures to suit dietary requirements and personal preferences based on electrical property data. The development of technology for adaptive processing may be necessary for this. The future of customized food production involves incorporating electrical property data into the food manufacturing process so that items can be designed as per dietary requirements and personal preferences of consumers. In order to do this, electrical property measurements are used to optimize processing settings and develop adaptive technologies that allow for more precise customization of food products (Frabetti et al., 2023).

The production of customized food will likely move toward using electrical property data to develop customized and flexible processing systems. This strategy is in line with the trend of personalized nutrition, providing customers with food items that are designed to their unique tastes and dietary requirements.

These future elements, which emphasize innovation, sustainability, and increased precision in food analysis and manufacturing, reflect the continuous growth of electrical property assessment techniques in the food sector. These developments are probably going to have an impact on food science and help create more effective, sustainable, and consumer-focused food processing techniques.

15.6 Conclusion

The study of food's electrical properties combined with sophisticated measurement methods has greatly advanced the fields of food science and technology. Numerous applications in non-destructive food quality analysis have been made possible by thorough understanding of dielectric constants, conductivity, impedance, and other electrical properties. Methods including dielectric spectroscopy, microwave sensing, electrical impedance tomography, and conductivity tests have shown to be quite useful in determining the amount of moisture in food goods, tracking their processing, finding adulterated objects, and guaranteeing their general safety and quality. There are a lot of promising things that could happen to these electrical property measurement methods in the future. Anticipated trends include the development of smart packaging, sophisticated data analytics, instrument shrinking, and integration with artificial intelligence. These advancements seek to improve the technology's use, accuracy, and efficiency. They may find use in customized nutrition plans, precision agriculture, and ecologically friendly food production. A more comprehensive understanding of food qualities is anticipated as electrical property measurements are combined with other analytical techniques and technological advancements in future scenario. This all-encompassing technology has the potential to improve food manufacturing procedures, reduce waste, and provide consumers with higher-quality, nutritionally sound products.

References

Alkanan, Z. T., Altemimi, A. B., Al-Hilphy, A. R. S., Watson, D. G., & Singh, A. P. (2021). Ohmic heating in the food industry: Developments in concepts and applications during 2013–2020. *Applied Sciences, 11*, 2507. https://doi.org/10.3390/app11062507

Astráin-Redín, L., Ospina, S., Cebrián, G., & Álvarez-Lanzarote, I. (2024). Ohmic heating technology for food applications, from ohmic systems to moderate electric fields and pulsed electric fields. *Food Engineering Reviews*. https://doi.org/10.1007/s12393-024-09368-4

Aurina, K., & Sari, A. (2022). Ohmic heating: A review and application in food industry. *Advances in Biological Sciences Research, 19*, 107–113.

Banti, M. (2020). Review on electrical conductivity in food, the case in fruits and vegetables. *World Journal of Food Science and Technology, 4*(4), 80–89. https://doi.org/10.11648/J.WJFST.20200404.11

Binnur, K., & Serap, K. (2011). Using of electrical conductivity on food control and food process. *International Journal of Agriculture and Environmental Research, 2*(6), 1835–1846.

Curet, S., Rouaud, O., & Boillereaux, L. (2014). Estimation of dielectric properties of food materials during microwave tempering and heating. *Food and Bioprocess Technology, 7*, 371–384. https://doi.org/10.1007/s11947-013-1061-4

de Alwis, A., Halden, K., & Fryer, P. J. (1989). Shape and conductivity effects in the ohmic heating of foods. *Chemical Engineering Research and Design, 67*(2), 159–168.

de Halden, K., Alwis, A. A. P., & Fryer, P. J. (2007). Changes in the electrical conductivity of foods during ohmic heating. *International Journal of Food Science and Technology, 25*(1), 9–25. https://doi.org/10.1111/J.1365-2621.1990.TB01055.X

Dube, D. C., Lanagan, M. T., Kim, J. H., & Jang, S. J. (1988). Dielectric measurements on substrate materials at microwave frequencies using a cavity perturbation technique. *Journal of Applied Physics, 63*, 2466–2468. https://doi.org/10.1063/1.341024

Duguay, A. J., Ramaswamy, H. S., Zareifard, M. R., Zhu, S., Grabowski, S., & Marcotte, M. (2017). Electrical conductivity of cabbage and daikon radish as affected by electrical voltage, frequency, salt concentration and temperature. *Journal of Food Process Engineering, 40*(1). https://doi.org/10.1111/JFPE.12315

Feng, H., Yin, Y., & Tang, J. (2012). Microwave drying of food and agricultural materials: Basics and heat and mass transfer modeling. *Food Engineering Reviews, 4*, 89–106. https://doi.org/10.1007/s12393-012-9048-x

Figura, L. O., & Teixeira, A. A. (2023). Electrical properties. In *Food physics* (pp. 403–430). Springer. https://doi.org/10.1007/978-3-031-27398-8_9. isbn:978-3-031-27397-1.

Figura, L. O., & Tiexeira, A. A. (2007). Electrical properties. In *Food physics* (pp. 333–351). Springer. https://doi.org/10.1007/978-3-540-34194-9_8. isbn:978-3-540-34191-8.

Frabetti, A. C. C., Garnault, T., Curto, H., Thillier, A., Boillereaux, L., Rouaud, O., & Curet, S. (2023). Dielectric properties of low moisture foods measured by open-ended coaxial probe and cavity perturbation technique. *European Food Research and Technology, 249*, 2861–2873. https://doi.org/10.1007/s00217-023-04333-7

Fryer, P. (1995). Electrical resistance heating of foods. In G. W. Gould (Ed.), *New methods of food preservation* (pp. 205–235). Springer. https://doi.org/10.1007/978-1-4615-2105-1_10. isbn:978-1-4613-5876-3.

Guo, W., Wang, S., & Tiwari, G. (2010). Temperature and moisture dependent dielectric properties of legume flour associated with dielectric heating. *LWT - Food Science and Technology, 43*, 193–201. https://doi.org/10.1016/j.lwt.2009.07.008

Hlaváčová, Z. (2003). Low frequency electric properties utilization in agriculture and food treatment. *Research in Agricultural Engineering, 49*(4), 125–136.

Içier, F., & Baysal, T. (2004a). Dielectrical properties of food materials—1: Factors affecting and industrial uses. *Critical Reviews in Food Science and Nutrition, 44*(6), 465–471. https://doi.org/10.1080/10408690490886692

Içier, F., & Baysal, T. (2004b). Dielectrical properties of food materials—2: Measurement techniques. *Critical Reviews in Food Science and Nutrition, 44*(6), 473–478. https://doi.org/10.1080/10408690490892361

Jadhav, H. B., Annapure, U. S., & Deshmukh, R. R. (2021). Non-thermal technologies for food processing. *Frontiers in Nutrition, 8*, 657090. https://doi.org/10.3389/fnut.2021.657090

Jha, S. N., Narsaiah, K., Basediya, A. L., Sharma, R., Jaiswal, P., Kumar, R., & Bhardwaj, R. (2011). Measurement techniques and application of electrical properties for nondestructive quality evaluation of foods- A review. *Journal of Food Science and Technology, 48*(4), 387–411. https://doi.org/10.1007/s13197-011-0263-x

Jiao, Y. (2019). Measurement techniques of electrical properties for food quality evaluation. In J. Zhong & X. Wang (Eds.), *Evaluation technologies for food quality* (Woodhead publishing series in food science, technology and nutrition) (pp. 595–616). Woodhead Publishing. https://doi.org/10.1016/B978-0-12-814217-2.00023-8. isbn:978-0-128-14217-2.

Juansah, J., Budiastra, I. W., Dahlan, K., & Seminar, K. B. (2014). Electrical properties of Garut citrus fruits at low alternating current signal and its correlation with physicochemical properties during maturation. *International Journal of Food Properties, 17*(7), 1498–1517. https://doi.org/10.1080/10942912.2012.723233

Kamonpatana, P., & Sastry, S. K. (2022). Electrical conductivity of foods and food components: The influence of formulation processes. *Journal of Food Process Engineering*. https://doi.org/10.1111/jfpe.13992

Kaur, R., Gul, K., & Singh, A. K. (2016). Fatih Yildiz (Reviewing Editor) Nutritional impact of ohmic heating on fruits and vegetables— A review. *Cogent Food and Agriculture, 2*, 1. https://doi.org/10.1080/23311932.2016.1159000

Kertesz, A., Hlavacova, Z., Vozary, E., & Staronova, L. (2015). Relationship between moisture content and electrical impedance of carrot slices during drying. *International Agrophysics, 29*(1), 61–66. https://doi.org/10.1515/intag-2015-0013

Khaled, D. E., Castellano, N. N., Gázquez, J. A., Perea-Moreno, A. J., & Manzano-Agugliaro, F. (2016). Dielectric spectroscopy in biomaterials: Agrophysics. *Materials, 9*(5), 310–334. https://doi.org/10.3390/ma9050310

Knirsch, M. C., dos Santos, C. A., Vicente, A. A., & Penna, T. C. V. (2010). Ohmic heating - a review. *Trends in Food Science and Technology, 21*(9), 436–441. https://doi.org/10.1016/J.TIFS.2010.06.003

Leong, S. Y., & Oey, I. (2022). Application of novel thermal technology in foods processing. *Food, 11*, 125. https://doi.org/10.3390/foods11010125

Lin, B., & Wang, S. (2020). Dielectric properties, heating rate, and heating uniformity of wheat flour with added bran associated with radio frequency treatments. *Innovative Food Science & Emerging Technologies*. https://doi.org/10.1016/j.ifset.2020.102290

Meng, Z., Wu, Z., & Gray, J. (2017). Microwave sensor technologies for food evaluation and analysis-methods, challenges and solutions. In *Transactions of the institute of measurement and control*. https://doi.org/10.1177/0142331217721968

Mitchell, F. R. G., & de Alwis, A. A. P. (1989). Electrical conductivity meter for food samples. *Journal of Physics E: Scientific Instruments, 22*(8), 554–556. https://doi.org/10.1088/0022-3735/22/8/004

Novas, N., El-Khaled, D., Salvador, R. M. G., Portillo, F., Fernández-Ros, M., & Gazquez, J. A. (2023). *Dielectric spectroscopy for the non-destructive characterization of biomaterials:*

Fundamentals, techniques, and experimentations. IntechOpen. https://doi.org/10.5772/intechopen.1002493

Ozturk, S., Kong, F., Trabelsi, S., & Singh, R. K. (2016). Dielectric properties of dried vegetable powders and their temperature profile during radio frequency heating. *Journal of Food Engineering, 169*, 91–100. https://doi.org/10.1016/j.jfoodeng.2015.08.008

Palaniappan, S., & Sastry, S. K. (1991). Electrical conductivities of selected solid foods during ohmic heating. *Journal of Food Process Engineering, 14*(3), 221–236. https://doi.org/10.1111/J.1745-4530.1991.TB00093.X

Pereira, R. N., Rodrigues, R., Avelar, Z., Leite, A. C., Leal, R., Pereira, R. S., & Vicente, A. (2024). Electrical fields in the processing of protein-based foods. *Food, 13*, 577. https://doi.org/10.3390/foods13040577

Risman, P. (2009). Measurements of dielectric properties of foods and associated materials. In *Development of packaging and products for use in microwave ovens* (pp. 129–152). Elsevier.

Routray, W., & Orsat, V. (2018). Recent advances in dielectric properties–measurements and importance. *Current Opinion in Food Science, 23*, 120–126. https://doi.org/10.1016/j.cofs.2018.10.001

Ryynänen, S. (1995). The electromagnetic properties of food materials: A review of the basic principles. *Journal of Food Engineering, 26*(4), 409–429, ISSN 0260-87.. https://doi.org/10.1016/0260-8774(94)00063-F

Sarang, S., Sastry, S. K., & Knipe, L. (2008). Electrical conductivity of fruits and meats during ohmic heating. *Journal of Food Engineering, 87*(3), 351–356. https://doi.org/10.1016/J.JFOODENG.2007.12.012

Sarkar, T., Salauddin, M., Kirtonia, K., Pati, S., Rebezov, M., Khayrullin, M., Panasenko, S., Tretyak, L., Temerbayeva, M., Kapustina, N., Azimova, S., Gruzdeva, L., Makhmudov, F., Nikitin, I., Kassenov, A., Shariati, M. A., & Lorenzo, J. M. (2022). A review on the commonly used methods for analysis of physical properties of food materials. *Applied Sciences, 12*, 2004. https://doi.org/10.3390/app12042004

Schimmer, O., Daschner, F., Kent, M., & Knochel, R. (2004). Determination of quality and storage-time of dielectric mixturesusing time domain spectroscopy. *Advances in Radio Science, 2004*(2), 1–5.

Sosa-Morales, M. E., Valerio-Junco, L., López-Malo, A., & García, H. S. (2010). Dielectric properties of foods: Reported data in the 21st century and their potential applications. *LWT - Food Science and Technology, 43*(8), 1169–1179. https://doi.org/10.1016/j.lwt.2010.03.017. ISSN 0023-6438.

Varghese, K. S., Pandey, M. C., Radhakrishna, K., & Bawa, A. S. (2014). Technology, applications and modelling of ohmic heating: A review. *Journal of Food Science and Technology, 51*(10), 2304–2317. https://doi.org/10.1007/s13197-012-0710-3

Venkatesh, M. S., & Ragavan, G. S. V. (2005). An overview of dielectric properties measuring techniques. *Canadian Biosystems Engineering, 47*, 7.15–7.30.

Wang, Y., Zhang, L., & Gao, M. (2013). Temperature- and moisture-dependent dielectric properties of macadamia nut kernels. *Food and Bioprocess Technology, 6*, 2165–2176. https://doi.org/10.1007/s11947-012-0898-2

Weng, Y. K., Chen, J., Cheng, C. W., & Chen, C. (2020). Use of modern regression analysis in the dielectric properties of foods. *Food, 9*(10), 1472. https://doi.org/10.3390/foods9101472

Zhang, H. Q. (2007). Electrical properties of foods. In G. V. Barbosa-Canovas (Ed.), *Food engineering* (Encyclopedia of life support systems (EOLSS). Developed under the Auspices of the UNESCO) (Vol. 1, pp. 115–125). Eolss Publishers.

Zhao, Y., Flugstad, B., Kolbe, E., Park, J. A., & Wells, J. H. (2000). Using capacitive (RF) dielectric heating in food processing and preservation - A review. *Journal of Food Process Engineering, 23*(1), 25–55.

Chapter 16
Color and Texture Measurement Methods

Ravi Harini, K. Sandhya, Panoth Abhirami, M. Sivasathiya, and N. Venkatachalapathy

16.1 Introduction

Our five senses, vision, smell, hearing, touch, and taste, allow us to perceive the world around us. Of these, vision is typically the first to be used in identifying things that are happening in the visual world (Macdougall, 2010). Since the beginning of recorded human speculation and achievement, researchers have been fascinated by vision, one of the most investigated of all the senses. Several cooperative processes are involved in seeing, including the detection of movement and location of objects, the relationships between things and their environment, the light intensity and illuminance, and the color presence in the visual scene (Giese, 2003; Macdougall, 2010). Our eyes first detect these processes and subsequently interpret them in our brains (MacDougall, 2001). Numerous food items are measured using color measurement systems. Fruits and vegetables, both fresh and processed, cereals and grains, meat and dairy products, oils, spices, sugar, syrups, and drinks are among them (Giese, 2003). Color measurement is used in the inspection of food products for grading, identifying anomalies or damage, identifying particular contents, and assessing color changes. When buying a food product off the market, consumers consider visual superiority. Even though a product with superior quality costs a little bit more, consumers will still pick it. Conversely, producers who sell higher-quality products stand to gain from several factors, including faster sales at optimal pricing and increased profits (Siswantoro, 2019). A product's internal and external features make up its quality. The consumer typically evaluates external quality through visual means. Customers will determine the visual appearance by combining various

R. Harini · K. Sandhya · P. Abhirami · M. Sivasathiya · N. Venkatachalapathy (✉)
Department of Food Process Engineering, National Institute of Food Technology, Entrepreneurship and Management – Thanjavur, Ministry of Food Processing Industries, Government of India, Thanjavur, Tamil Nadu, India
e-mail: venkat@iifpt.edu.in

C. K. Sunil et al. (eds.), *Emerging Non-destructive Technologies for Food Quality Analysis*, https://doi.org/10.1007/978-3-031-95048-3_16

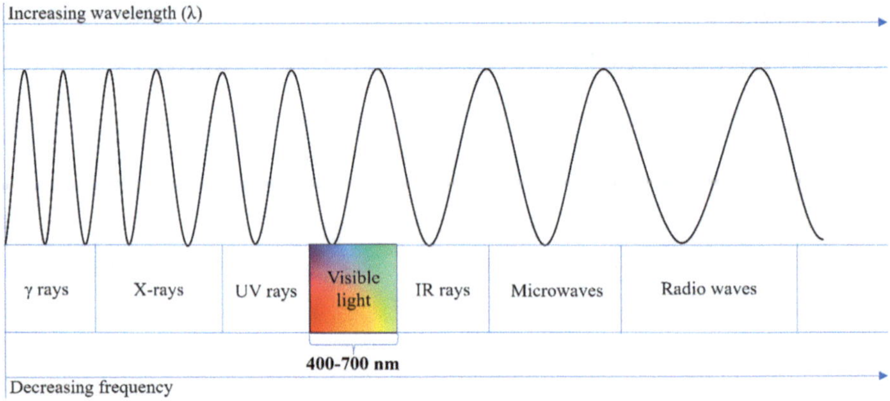

Fig. 16.1 Visible light ranges in the electromagnetic spectrum

external factors like color, dimensions, structure, and texture. Customers recognize a food product's surface color as its primary quality indicator. Customers frequently decide whether to take or disapprove of a product focusing only on the surface color (Siswantoro, 2019). Fruits are food whose price is affected by color. Color can be used to assess changes in product color during processing, identify particular matter in food products, identify the existence of any damage in the product, assess fruit maturity, and more.

Foods come in a vast array of visual characteristics. They can have glossy, diffuse, rough, porous, or flaky surfaces. Their colors can be consistent, sporadic, multilayered, or translucent. They can also be opaque, hazy, transparent, or translucent. Color and appearance can be greatly impacted by the way that light dispersion from food configuration interacts with pigment absorption (Macdougall, 2010). Within the electromagnetic spectrum, visible light falls between 380 and 780 nm. From below, it is surrounded by infrared light, and from above, by ultraviolet light, which is represented in Fig. 16.1. Upon impact, light can be absorbed, transmitted, or reflected by an object. Since the color of a substance is determined by reflected light, factors such as size, background variations, angle of view, light source, and amount of light can all affect how something appears (Giese, 2003). With the realization that the eye's lens had to portray perhaps a picture of the thing perceived onto the back of the eye, scientific investigations into the mechanisms underlying human perception of color and vision began in the seventeenth century (Macdougall, 2010; MacDougall, 2001).

All around the world, people eat a variety of foods. Individuals have experienced diverse food textures, and they have continued the custom of sharing meals over many decades (Kohyama, 2020). Individual subjects in diverse groups perceive the same food's texture in different ways. Humans perceive food's physical characteristics as its texture, which is defined as "all the mechanical, geometrical and surface attributes of a product perceptible by means of mechanical, tactile and, where appropriate, visual and auditory receptors." Since about the 1960s, texture analyses that incorporate both an instrumental evaluation and a sensory appraisal have been conducted extensively. Malcolm C. Bourne and Alina S. Szczesniak, who both died in 2016, were significant pioneers in the field of food texture research from the perspectives of physics and sensory assessment, respectively (Kohyama, 2020). At every stage of the oral process, there is a significant modification in the physical attributes of the food materials, and this modification is not insignificant (Kohyama, 2020; Lu, 2013). There are four unique habits of mouth behavior: cruncher, smoother, chewer, and sucker. The texture is less affected by the physical characteristics and structures of food in the early phases of processing, that is, when raw commodities are not consumed and in the post-swallowing stages of digestion. The majority of a food's textural data is detected by hand right before consumption and verbally during oral processing within the digestive system. While we eat, our tongue and teeth break up the bite-sized food before it is swallowed. After the meal is broken down and combined with saliva, a bolus is ready to be swallowed (Kohyama, 2020). Every person has a different perception of food texture depending on their abilities, physical activities, level of hunger or thirst, as well as other physiological responses like the period of day and surroundings (Jeltem et al., 2015; Kohyama, 2020). When creating a food's texture, food designers need to take consumer mouth habits into account. In healthy individuals, aging reduces oral adjustment to food textures, and oral impairments like tooth loss can also alter eating habits (Kohyama, 2020; Peyron et al., 2017). When serving well-homogenized foods, such as infant food and diets for people with dysphagia, oral processing can be omitted (Cichero, 2017; Kohyama, 2020). But in order to enjoy food as a necessary component of life and to perceive its flavor and texture, oral processing is required (Kohyama, 2015, 2020).

The sensory assessment of food quality highly depends on food texture, which is also crucial or significant for the marketing and quality evaluation of solid food. The term "food texture" is ill-defined because it refers to the collective human interpretation of a food product subjected to complex force or deformation that results in structural breakdown or changes. The fact that there is a wide range of foods with radically various textural qualities and that different people may have different expectations or explanations for various food types adds to the challenge of defining the texture of food. As a result, various terms have been used to describe the textural qualities; therefore, it is crucial to measure the textural attributes of food using objective, standard procedures. Because destructive or deformation methods frequently resemble or emulate human sensory evaluation, they are commonly employed for the objective assessment of the textural parameters of solid food. To ensure that the characteristics of food products will meet or surpass consumer expectations, objective evaluation of the textural characteristics at different stages of postharvest treatment is necessary. Due to their tendency to replicate or imitate the sensory assessment made by humans during evaluation, force-deformation techniques are frequently employed for the objective assessment of the textural characteristics of solid foods (Lu, 2013). In food and postharvest research, as well as industrial practice, characterization and modeling are crucial methods. Many techniques have been employed to assess instrumental results, which offer time-series information on product deformation and enable the calculation of a large number of texture features from force-displacement or force-time data. To evaluate the acceptability of raw and processed quality products, the food industries use texture as a critical quality attribute. One of the most crucial aspects of fruit and vegetable texture that is frequently used to gauge food freshness is hardness, also known as firmness (Chen & Opara, 2013a, Konopacka & Plocharski, 2004). Brittle, cellular, and crunchy food's essential characteristic is crispness (Zdunek et al., 2010). Adhesiveness, cohesiveness, springiness, and gumminess are important characteristics for the texture assessment of semisolid products. A significant amount of work has gone into developing the tools and measurement methods for the accurate evaluation of textural attributes (Chen & Opara, 2013; Oraguzie et al., 2009; Zdunek et al., 2010). Over time, a variety of instrumental evaluations have been employed in both industry and research to evaluate food texture.

16.2 Color Measurements in Foods

16.2.1 Color Vision by Human

Customers notice a food product's exterior color as its primary quality indicator. Customers frequently decide whether to accept or reject food based only on the surface color. Apples are one food whose price is affected by color. Color can be used to assess changes in products during processing, identify particular information in food products, identify the existence of any damage in the product, assess fruit maturity, and more (Siswantoro, 2019). To measure an object's color impression in a scene, three interacting factors must be considered. These include an awareness of how light affects objects in their surroundings, the nature of the resources witnessed, and the visual process used by humans (MacDougall, 2001). Just a portion of the total object recognition of the data that the eye and brain interpret is made up of the psychophysical condition known as color perception.

The field of vision in human eyes is nearly round, and they are made up of three membranes. The sclera, the outer layer, is continuous with the cornea anteriorly and the optic nerve sheath posteriorly. The iris as well as the capillary network-containing choroid, the middle layer, gives rise to the ciliary body, which hangs the lens. The retina, around inside the inner membrane that detects light, lines around inside the posterior lens. The rods and cones, which are named so due to their structures, are two types of light-sensing receptors found in the retina. Within the multi-folded disk-shaped formations in the outer part of the cones and rods, light energy is fixated on the retina and transformed into electrical signals by conformational changes in the photopigments. The so-called blue (B), green (G), and red (R) cone pigments, or the short-, mid-, and long-wave detectors, are most sensitive at roughly 440, 540, and 570 nm, respectively. Rhodopsin, the rod pigment, absorbs at a maximum of 505 nm (MacDougall, 2001).

16.2.2 Instrumental Color Measurement

To describe color, a range of color schemes or scales are employed. The Commission International de l'Eclariage (CIE) system, the Munsell color solid, and the Hunter L, a,b system are the ones that are most frequently utilized in the food industry. The most important system for describing color is the CIE system. Its foundation is the use of a standard observer and standard light source. The system converts the tristimulus qualities to the unreal primaries X, Y, and Z and obtains standard-observer CIE curves for the visible spectrum (Giese, 2003). The CIE color measurement system uses the spectral energy distribution of the light source and the color-matching mechanisms of the conventional observer to convert the object's reflection or transmission spectral range into three-dimensional space. The XYZ values were standardized by CIE as stimulus values. The XYZ space describes any color that the average human observer can interpret. The selection of this color space results in positive XYZ attributes describing every discernible visual stimulus. XYZ color space is commonly utilized as a reference color space and is not dependent on any particular device. X and Z are the two main virtual components; Y is the component of lightness (Siswantoro, 2019). The food industry frequently uses opponent-type devices, such as the HunterLab L*, a*, b*, and the modified CIE system known as CIELAB color scales (Giese, 2003). Lu* v* space and La* b* space are other CIE standard color spaces. These color spaces offer a perceptually equal space. This indicates a strong correlation between the Euclidean distance in these spaces between two colors and human visual interpretation. The two chromatic components are b∗(v∗) and a∗(u∗), and the color constituent L is lightness as well as luminance. The red-green axis defines a∗(u∗), while the yellow-blue axis defines b∗ (v∗) (Siswantoro, 2019).

Munsell color system: a means of defining colors and illustrating their relationships. Hue, chroma, and value are the three characteristics or qualities that make up each color. Munsell created numerical scales for all of these characteristics that have visually consistent steps. Any surface can be recognized by its color by comparing it to the chips when viewing conditions and lighting are appropriate. After that, the color is distinguished by its hue, chroma, and value. Spectrophotometers calculate the ratio of light transmitted or reflected from a food item to light from a recognized reference standard. Though spectrophotometers cost more, they are more precise than colorimeters. For transmission and liquids, spectrophotometers function best. The food particle and application type will determine which tool is best to use (Giese, 2003). One can also use an artificial neural network (ANN) or a mathematical model to perform a direct conversion from RGB space to La*b* space. Several parameters make up the mathematical model: the outputs are the estimated values of L, a*, and b*, and the inputs are the R, G, and B values derived from sample images. Minimizing the mean absolute error among estimated values and sample values out of a colorimeter would be how the model's parameters are estimated (Siswantoro, 2019).

16.2.3 Definition and Significance of Color in Food

Evaluating food quality based on its color has been a traditional practice since ancient times. The sensory analysis of food, which involves assessing its qualities through the senses, particularly sight, has been a prominent method of evaluation over the years. In which, color analysis plays a crucial role in determining the effectiveness of different treatments applied to post harvested treatment in foods. Recent advancements have highlighted the importance of measuring color not only as a quality parameter but also as an indicator of the internal composition of materials. In the broader context of assessing material appearance, attributes such as glossiness, matteness, transparency, and roughness are considered fundamental indicators of quality. The way consumers perceive fruits, vegetables, and processed foods is significantly influenced by their color (Gigilashvili, 2021; Pointer, 2003). Marketers often try to modify or enhance natural appearances to meet consumer expectations.

Color is a vital property for material analysis derived from light's spectral distribution. It depends on the illuminant and observer's eye. Various factors influence color perception: the light's spectral energy distribution, viewing conditions, the object's spectral characteristics, and the observer's eye sensitivity. Light, being the source of color, spans a narrow range from violet to red. Unequal distribution of light's energy creates color perception, with a white light indicating equal distribution. Material properties, lighting conditions, and light-object interactions affect an object's appearance. Additionally, light's interaction with the scene, including illumination type, spectral properties, and color rendering, plays a crucial role. Matte surfaces diffuse light differently, impacting color perception (Brainard & Maloney, 2011). Understanding these interactions is essential for accurate material assessment across different environments.

In agricultural and processed foods color is integral to consumer perception, with changes indicating flavor quality or maturity concerns. Understanding color perception is crucial given consumer expectations, though often irrational (Anderson, 2011). Color measurement is pivotal in correlating changes with process or sensory parameters, necessitating reliable quantification methods. Common models like Lab, RGB, and CMYK offer different approaches, but each has limitations, especially in replicating human perception that surpasses the capabilities of any color system (Schanda, 2007). Ongoing research is vital for enhancing color measurement methodologies and their application in food quality evaluation.

16.2.4 Commonly Used Color Models

Lab Model

- Recommended by the International Commission on Illumination (CIE), the Lab model provides a perceptually uniform color space, facilitating precise color measurement.

- Based on the XYZ color space, it offers a device-independent, absolute model, considering luminance (L*) and chromatic components (a* and b*).

RGB Model An additive color model utilizing red, green, and blue light to reproduce colors. - Device-dependent, it closely relates to human eye perception, with various proportions and intensities creating a broad array of colors.

CMYK Model A subtractive model based on cyan, magenta, yellow, and black, complementary to the RGB color model. Used in printing, it differs from RGB by considering white as the background color and black as the combination of colored inks (Hu et al., 2009).

The comparison and limitations of these models were that the Lab model offers a wide gamut encompassing colors from both RGB and CMYK models. RGB and CMYK models are device-dependent, limiting their general applicability. Human eye perception exceeds the gamut of any color model, indicating the need for cautious interpretation (Molada-Tebar et al., 2024).

16.3 Non-destructive Instrumental Analysis of Food Color

There exist multiple parameters for evaluating food, including its appearance (size, shape, color), texture, and aroma. Among these factors, food color is often regarded as the key indicator of food quality. Visual inspection is the most straightforward method for assessment. The human eye can discern nearly one million distinct colors. Yet, the color memory of the brain is often inadequate, making it challenging to recall the precise intensity of colors previously perceived. Visual color assessment is influenced by various intricate factors such as lighting conditions, individual perception, and angles of illumination. Consequently, instrumental techniques are deemed more reliable for accurate evaluation (Jain et al., 2021; Wrolstad & Smith, 2010).

Color can be described as the perception one undergoes upon exposure of the eye to radiation within the visible spectrum 400–800 nm. On the other hand, colorant is the pigment responsible for providing color to a product. Three essential components necessary for the manifestation of color are a colored object, a light source within the visible spectrum, and an observer. When white light interacts with the object being observed, it may be absorbed, reflected, or scattered. The specific absorption of particular wavelengths of light is the predominant factor that dictates the color of any food product (Berns, 2019). The color of any food product can be estimated by either visual or instrumental systems.

16.3.1 Visual System

The most renowned system for ordering colors visually was developed by A.H. Munsell in 1905 and is commonly referred to as the Munsell system. Within

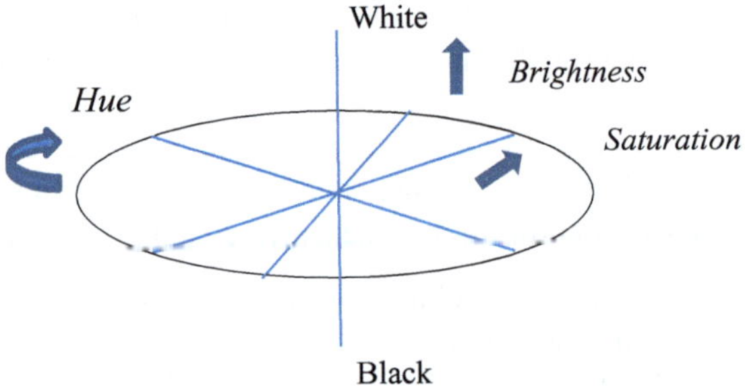

Fig. 16.2 Munsell color system

this framework, there are five primary colors: yellow, blue, purple, green and red, as well as five secondary pairs: red-purple, green-yellow, yellow-red, purple-blue and blue-green. These secondary colors are termed “Hue”. When the character of a color is described in terms of its brightness or darkness, ranging from white to gray to black, it is referred to as “Value.” Value is quantified on a scale from 0, representing absolute black, to 10, representing absolute white. Chroma denotes the intensity of each color, such as the addition of pure blue hue to gray until it matches the original blue. At this point, saturation is reached, and further changes in shade cease (Dutta & Nath, 2023) (Fig. 16.2).

16.3.2 Instrumental Methods

16.3.2.1 CIE Color System

The CIE color system is a widely recognized method for describing color. The CIE (Commission Internationale de l’Eclairage) is a significant international organization dedicated to the measurement of color. In many food industries, color measurement instruments operate on a color-space system: L*, a*, b*, as defined by CIE. This trichromatic system can reproduce any color by combining the three primary colors, which are red, green, and blue (RGB), represented by the X, Y, and Z axes, respectively. This CIE system employs the chromaticity diagram to pinpoint a specific color (Fig. 16.4). Uppercase X, Y, and Z denote the tristimulus values, whereas lowercase x, y, and z represent red, green, and blue, respectively. The value of x can be determined using the formula $x = X/(X + Y + Z)$, and the values of y and z can be obtained by substituting the value of x (Markovic et al., 2013).

16.3.2.1.1 Working Principle of Colorimeter

The principle of a colorimeter is grounded on the process wherein a beam of incoming light I_0 passes through a particular solution, where a portion of the light is reflected I_r, another part is absorbed I_a, and the rest is transmitted I_t. Incident light is the sum of reflected light, absorbed light and transmitted light.

$$I_0 = I_a + I_r + I_t$$

Where, I_0 is Monochromatic incident light

I_a is absorbed light
I_t is transmitted light
I_r is reflected light

Colorimeters operate primarily on the principles outlined in Beer-Lambert's Law, which asserts that the absorption of light as it passes through a medium is directly proportional to the concentration of that medium and the light path length through the solution.

A colorimeter serves to gauge the concentration of a colored solution or compound within the visible light spectrum, typically ranging from 400 to 800 nm. Its operational principle is rooted in the Beer-Lambert law. When a colored chemical compound is introduced into the cuvette as a sample, it absorbs a portion of the incident light, with the absorbed quantity correlating directly to the concentration of the compound in the sample. Within the colorimeter, a filament lamp serves as the light source, emitting light that traverses through the glass cuvette of fixed thickness. A fraction of this light is absorbed, while the remaining transmitted light reaches the photoelectric detectors. These detectors quantify the intensity of the incident light, facilitating the calculation of the absorbed light amount (Shrestha & Shrestha, 2023) (Fig. 16.3).

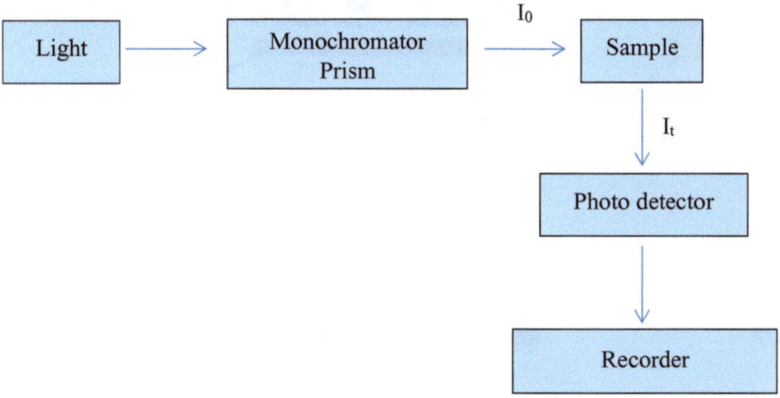

Fig. 16.3 Flow diagram representing basic components of colorimeter

16.3.2.2 Colorimeters

16.3.2.2.1 Tristimulus Colorimeters

Tristimulus values are derived optically rather than through mathematical computation, as the design of the apparatus is modeled after the cone cells found in the retina. The filters present in the instrument mimic the function of the three types of cone cells in the retina. The three main components of a tristimulus colorimeter consist of a light source, color filters capable of adjusting the energy distribution of incident or reflected light and a photoelectric detector that converts reflected light into electrical signals. Measurements taken with this colorimeter typically involve a comparative approach, highlighting the importance of calibrating standards using similar colors to those being measured (Aldughpassi et al., 2021) (Fig. 16.4).

16.3.2.3 Spectrophotometer

A spectrophotometer assesses the absorption of light as it traverses through a colored sample solution. Its fundamental principle closely resembles that of a colorimeter, with the key distinction being that a spectrophotometer can measure absorbance across various regions of the electromagnetic spectrum, including visible (400–800 nm), UV (200–400 nm), and infrared (700–1000 nm) (Akash et al., 2020).

There are two primary types of spectrophotometers: single-beam and double-beam.

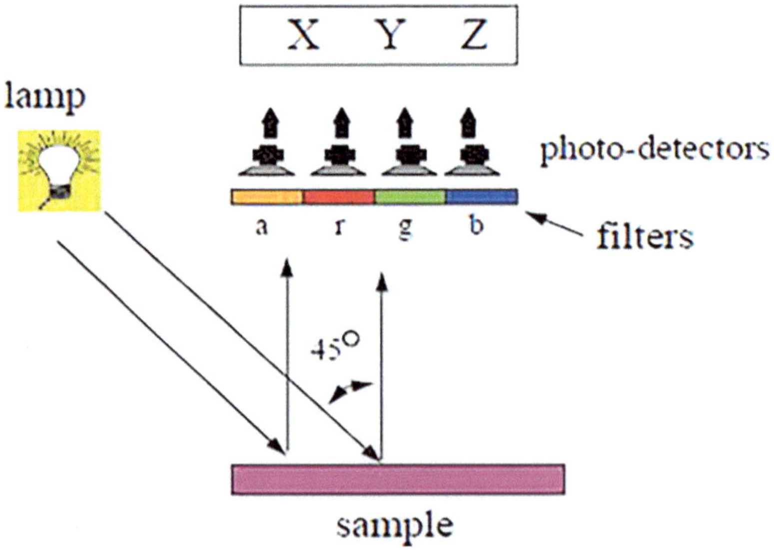

Fig. 16.4 Schematic diagram of tristimulus colorimeter. (Source: Dutta & Nath, 2023)

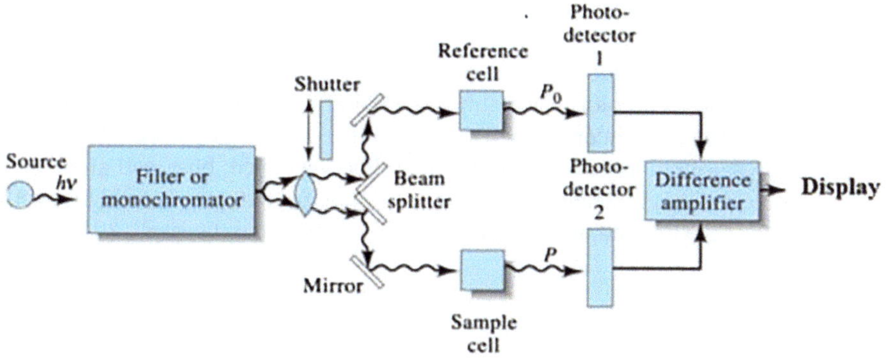

Fig. 16.5 Double beam spectrophotometer. (Source: Dutta & Nath, 2023)

In a single-beam spectrophotometer, the absorption of the blank sample (also known as the reference sample) is initially measured. Subsequently, the sample to be analyzed is inserted in place of the blank, and the absorption of light is measured again. To obtain the result, the absorption value of the blank sample is subtracted from the absorption value of the main sample. In a double-beam spectrophotometer configuration, the illumination source undergoes bifurcation into two beams subsequent to its passage through the lens (Fig. 16.5). One beam traverses through the blank sample, while the other passes through the original sample. Both transmitted rays are directed toward a photodetector before being recombined and directed toward the monochromator. The display section provides the reading by deducting the blank sample's measurement from that of the original sample. Figure 16.9 illustrates a schematic diagram of the double-beam spectrophotometer. The sample cuvettes are constructed from quartz, and the light sources employed consist of deuterium lamps for UV measurements and tungsten lamps for the visible range. In recent advancements, xenon arc lamps have emerged for use in both UV and visible spectrum analyses (Dutta & Nath, 2023).

16.3.2.4 Spectrocolorimeter

The fundamental principle of a Spectrocolorimeter closely resembles that of a spectrophotometer. Spectrocolorimeters, a specific type of colorimeter, are employed to evaluate color across the spectrum, capable of analyzing both the reflectance and transmittance of samples within the visible spectrum wavelengths. They can furnish colorimetric data in the form of X, Y, and Z tristimulus values, CIE L*, *a**, and b* values (Kim et al., 2023).

16.3.2.5 Densitometer

A densitometer gauges the color of a given food or substance by assessing its density. The perceived color of the substance being measured can vary depending on its density. The color intensity of a thin layer differs from that of a thicker film or layer. This type of colorimeter quantifies the reflection of red, blue, and green light from the object. Densitometers employ optical filters for red, green, and blue while measuring the transmission of cyan, magenta, and yellow (Salman et al., 2023; Dutta & Nath, 2023).

16.3.3 Evaluation Methods and Their Importance—Non-destructive Techniques

Non-destructive methods in food quality assessment offer notable advantages over traditional inspection techniques, primarily in their efficiency and speed, reducing time, resource consumption, and sample waste. These techniques, such as imaging-based approaches and sound-based methods, provide detailed information about surface and internal structures without the use of harsh chemicals, aligning with sustainability goals. They also advance automation and mechanization in the food industry, enabling facilities to achieve consistency and cost-effectiveness through real-time decision-making based on assessment data. However, successful implementation requires consideration of factors like product characteristics and desired quality attributes. Particularly beneficial in industries where destructive methods are impractical, non-destructive techniques enhance quality control and facilitate sorting based on predetermined criteria without compromising product integrity (El-Mesery et al., 2019). Covering mechanical, optical, electromagnetic, and dynamic approaches, these methods offer insights into physical, surface, and internal properties. As technology evolves, non-destructive methods are poised to play a crucial role in ensuring the safety and quality of agricultural products throughout the food supply chain.

16.3.4 Factors Influencing Color in Food

The color of food is a multifaceted aspect influenced by numerous factors, each playing a crucial role in determining its visual appeal and quality. These factors encompass various attributes intrinsic to the food itself, including the presence of pigments, the stage of maturity, and the specific variety of the food item. Additionally, external factors such as processing methods, packaging materials, and storage conditions further contribute to the overall color profile. Measuring the color of food serves as a valuable tool in quality control processes, allowing for the assessment of both the initial quality of the food and any changes that may occur over time. By monitoring color, food manufacturers can detect deviations from desired standards and take corrective actions to maintain consistency and freshness (Clydesdale, 1991, 1993). However, the impact of heat on natural color pigments presents a significant challenge in food processing. Thermal treatments can lead to the deterioration of these pigments, resulting in off-color development and a reduction in their biological viability. This degradation not only affects the aesthetic appeal of the food but also diminishes its nutritional value.

The stability of natural color pigments and flavor compounds is intricately tied to a multitude of factors. pH levels, exposure to light, temperature fluctuations, co-pigmentation processes, the presence of sulfites, antioxidants such as ascorbic acid, oxygen exposure, and enzymatic activity all play roles in determining the resilience of these essential components. So research suggests nonthermal emerging techniques like HPP, PL, UV, and Cold plasma have better retention of these factors than traditional methods (Pandiselvam et al., 2023).

16.3.5 Relationship Between Color and Food Quality

Color profoundly influences the perception and consumption of food, extending beyond mere visual cues to significantly impact gustatory experiences and consumer behavior. Recognizing this phenomenon is paramount for food manufacturers striving to meet consumer expectations and effectively market their products. Research has demonstrated that color plays a pivotal role in shaping various psychological aspects of the eating experience. Firstly, color influences flavor perception, as individuals often make assumptions about taste based on visual cues, leading to instances where the brain substitutes expected flavors for actual ones, potentially altering the overall taste experience (Francis, 1995). Additionally, color impacts the perception of fullness, potentially influencing the amount of food consumed, as studies suggest that the color of plates can subtly influence portion size selection. Moreover, color affects the enjoyment of food, with brighter hues often associated with better taste and superior nutrition, influencing consumers to select visually appealing options while avoiding those with off-putting colors signaling spoilage or inferior quality. Different colors evoke varying taste perceptions, with bright red hues often enhancing sweetness and diminishing bitterness. In contrast, unexpected colors like blue may deter consumption due to their rarity in natural foods. The challenge of off-coloring, whether natural or artificial, presents a significant hurdle for food manufacturers, necessitating careful selection of coloring agents to avoid negative effects on flavor perception and consumer acceptance. Experimental studies on the influence of color on taste perception primarily focus on artificially colored liquids, providing insights into color-taste associations. However, it's essential to acknowledge that the colors and flavors used often differ from those found in commercial products. Understanding the nuanced relationship between food/drink color and taste/flavor characteristics is crucial, as color-taste associations develop early in life and can significantly influence perception subconsciously. Cross-cultural differences in color-flavor associations underscore the importance of understanding individual color-taste perceptions, as they profoundly impact flavor judgments and food preferences globally (Spence, 2019). In summary, color's profound impact on the perception and consumption of food underscores its significance in food marketing and product development. As research in this area continues to evolve, it promises valuable insights into the intricate relationship between visual stimuli and eating behavior.

16.3.6 Imaging Techniques—Principle and Operation of Imaging Systems

Image analysis stands as a powerful, non-destructive method for assessing the color properties of food products. Unlike traditional colorimeters, this approach utilizes techniques like color imaging, where the entire surface of the food is meticulously examined pixel by pixel, providing comprehensive and descriptive data. One such tool is the Digieye imaging system, comprising an illumination box for consistent sample lighting and a digital camera linked to a computer for precise color measurement. Additionally, non-invasive methods like CT scans enable visualization of internal features without altering the integrity of the sample. Beyond color assessment, image processing techniques can be employed to analyze various other characteristics of food, such as size, shape, and texture (Pradeepika et al., 2019). Other imaging technologies, including cameras, ultrasound, MRI, and electrical tomography, further expand the scope of applications in food analysis, offering diverse insights into product quality and composition (Edwards et al., 2021).

16.3.6.1 RGB Imaging

RGB imaging plays a pivotal role in assessing food quality throughout the production chain, given its influence on consumer purchasing decisions. Parameters like size, shape, color, and the presence of visual defects are crucial indicators monitored during production. While human inspection has traditionally been relied upon for quality control, it is subjective, labor-intensive, expensive, and lacks scalability to meet high throughput demands. Moreover, human assessment is operator-dependent and not easily transferable between different settings. To address these challenges, automated systems leveraging computer vision techniques have emerged in the food industry, offering objective and reproducible evaluations. Image-based systems excel in characterizing food matrices due to their ability to capture spatial information, which is particularly important for heterogeneous compositions often found in food. These imaging systems are fundamentally built on the principle of replicating and presenting a scene to mimic human color perception. The human visual system, represented by the eye, detects color images through an array of three types of sensors known as cone cells, which respond to three distinct regions of the electromagnetic spectrum corresponding to red, green, and blue hues. Similarly, artificial color imaging systems like cameras, scanners, and webcams rely on sensor arrays, typically charge-coupled devices (CCDs), which capture intensity values for (R, at $\lambda \approx 630$ nm), green (G, at $\lambda \approx 545$ nm) and blue (B, at $\lambda \approx 435$ nm) spectral channels per pixel. This process emulates the spectral sensitivity of the human eye. The resultant RGB image is conceptualized as a three-dimensional data array comprised of numerous pixels arranged in rows (r) and columns (c), with each pixel characterized by its Red, Green, and Blue spectral variables. Artificial systems like cameras and scanners record intensity values for these channels, producing RGB images represented as three-dimensional data arrays. Leveraging this technology, image processing systems can analyze external attributes such as size, shape, color, texture, and defects in various food products. Additionally, research has explored the potential of color imaging to predict different food properties, further expanding the utility of RGB imaging in food analysis (Foca, 2016).

Color imaging techniques, utilizing representations like RGB and HSI systems, are integral in evaluating the visual appearance and color properties of food samples, with image acquisition systems comprising platforms for sample placement, digital cameras, light sources, and analysis software. By capturing scattered photons and converting them into electrical signals, these techniques enable comprehensive surface analysis and quantification of sample characteristics, which is crucial for various food quality assessments. Recent studies have demonstrated their efficacy in detecting adulteration in chicken burgers and differentiating ground meat samples, showcasing their potential for species identification and adulteration detection. However, future research should focus on expanding sample sizes, incorporating diverse meat types, and exploring transitions to online processing settings to enhance model performance and practical utility in real-time quality assurance processes within the food industry.

16.3.6.2 HSI Imaging

RGB imaging serves as the foundation for basic imaging systems, yet for certain applications, additional variables beyond the visible electromagnetic spectrum may be required. Hyperspectral imaging (HSI) systems, also known as chemical or spectroscopic imaging systems, extend the capabilities of RGB images, facilitating the exploration of chemical aspects within samples. HSI is gaining prominence in analytical chemistry for its ability to rapidly capture both spectral and spatial information simultaneously, representing it as image-like maps of chemical component distribution. Unlike RGB systems that record intensity values for red, green, and blue channels per pixel, HSI systems gather images across numerous spectral channels at varying wavelengths, typically encompassing visible and near-infrared ranges. Compared to traditional spectroscopic methods, HSI offers enhanced insights by combining digital imaging advantages with spectroscopic measurements.

The resulting data from HSI, often termed a hypercube, embodies a three-dimensional structure with spatial dimensions (y for pixel rows and x for pixel columns) and a spectral dimension (λ wavelengths). Thus, HSI not only aids in identifying chemical species and their concentrations but also reveals their spatial distribution, which is crucial for understanding the functionality of natural or artificial products, particularly in food analysis, and for screening and identifying contaminants in samples. Despite the higher cost associated with hyperspectral systems compared to RGB counterparts, their enhanced capabilities often yield significantly superior results. As a result, hyperspectral imaging applications have garnered increasing attention in the food industry recently (Foca, 2016).

Conventional NIR spectroscopy provides spectral information about a sample as a whole, lacking spatial details. Hyperspectral imaging (HSI) integrates spatial and spectral data, overcoming this limitation. Initially introduced by Goetz et al., this technology has since been adapted for various industries, including food, agriculture, and pharmaceuticals. HSI captures hyperspectral images in hypercubes, containing spectral data for each pixel akin to a chemical fingerprint. This facilitates the visualization and analysis of chemical components within samples, which is particularly useful for heterogeneous samples. NIR-HSI systems typically comprise a camera, spectrograph, detector, illuminator, and translation stage, offering three imaging configurations: whiskbroom, push-broom, and staredown. These configurations enable efficient acquisition of hyperspectral images with varying spatial and spectral resolutions suitable for different applications.

HSI leverages differences in how biological materials interact with light to create unique spectral signatures, aiding in sample identification and characterization. Spectral pre-processing eliminates biases before employing chemometric techniques to extract relevant information and discern differences among samples. This technique's appeal lies in its ability to rapidly and non-destructively analyze samples, crucial for quality control and process monitoring, particularly in online and at-line applications (Yu & MacGregor, 2003).

Studies by Kamruzzaman et al. explored NIR-HSI's potential for meat authentication, achieving high accuracies in discriminating between different meat types and detecting adulteration. These investigations emphasized wavelength selection optimization and demonstrated the method's viability for laboratory-scale applications. Further research is needed to validate its suitability for industrial-scale implementation. Additionally, studies on visible near-infrared (Vis-NIR) HSI for meat adulteration detection showcased promising results, suggesting its potential for rapid and non-destructive analysis in food authentication. Optimization of wavelength selection holds promise for developing cost-effective and reliable multispectral imaging devices tailored to specific applications, indicating a progressive shift towards industry-scale adoption of spectral imaging technologies (Kamruzzaman et al., 2012).

Food safety and quality are paramount in the food processing industry, yet conventional methods for analyzing them are often time-consuming. Emerging imaging techniques such as hyperspectral imaging, fluorescence imaging, backscattering imaging, magnetic resonance imaging (MRI), X-ray imaging, and ultrasound imaging offer promising solutions for enhancing food quality and safety.

Hyperspectral imaging (HIS) is an analytical spectroscopic imaging technique that combines traditional imaging with spectroscopy to capture both spectral and spatial information of an object. It generates hypercubes, 3-D cubes comprising 100 contiguous wavebands for every spatial position of the object. HIS has found wide applications in the food processing industry due to its speed, ease of use, reasonable start-up cost, and non-destructiveness. Various types of HIS, including Ultraviolet HIS, Visible HIS, and Vis/NIR HIS, have been utilized to differentiate between fresh and frozen-thawed fish or analyze parameters like soluble solids content and firmness in apples (Gowen et al., 2007).

16.3.7 Data Reduction Challenges in Image Analysis

Chemometrics, defined as the application of mathematical and statistical methods to extract relevant information from material systems, plays a vital role in imaging tasks such as image correction, segmentation, and mathematical modeling to determine sample properties directly from acquired images. However, a significant challenge, particularly with hyperspectral images, lies in data reduction. The immense computational load and time required for model computation, coupled with the extensive data conveyed by hyperspectral images, pose difficulties in handling large datasets, notably in the food industry, where various factors influence product quality. To address this limitation, numerous multivariate data reduction methods have been developed to extract useful information from images effectively.

16.3.7.1 Transformation from RGB Images to Colorgrams

In the realm of automated methods aimed at extracting pertinent information from RGB images, our research group has devised an algorithm for quantifying color-related data, leading to the creation of "colorgrams." These colorgrams offer a compressed representation of the color content of an image in a 1D signal, amalgamating various calculated parameters such as Red, Green, Blue, Intensity, ratios, Hue, Saturation, Lightness, and data from Principal Component Analysis (PCA). This compression, condensing millions of data points into a 4900-point signal, serves as a unique fingerprint of the image's color content, enabling streamlined analysis and facilitating the extraction of relevant information using multivariate analysis methods (Foca, 2016).

16.3.7.2 Transitioning from Hyperspectral Images to Hyperspectrograms

In dealing with hyperspectral imaging (HSI) data, a similar approach has been adopted to derive "hyperspectrograms." These 1D signals encapsulate essential information from each hypercube, representing both spatial and spectral aspects. By reducing each image to a vector of a few hundred points, hyperspectrograms allow for efficient analysis of multiple images simultaneously using common multivariate analysis techniques. Furthermore, through proper variable selection methods, hyperspectrograms can be refined to extract specific features relevant to the problem at hand. This approach of utilizing colorgrams and hyperspectrograms has been instrumental in various applications within the food industry, such as detecting defects in raw hams and identifying bruises in different apple varieties. Such techniques offer enhanced objectivity and transferability compared to traditional expert assessment methods, thereby addressing critical challenges in quality control and product evaluation (Foca, 2016).

16.3.8 Other Techniques

Another technique, backscattering imaging, involves the reflection of waves or radiation back to the direction from which they came. It is particularly useful for analyzing transparent or semi-transparent objects like food substances, allowing the study of texture and color. Backscattering imaging has been applied in food industries to analyze parameters like soluble solids content and firmness, aiding in identifying the ripening stage of fruits such as plums and apples (Pradeepika et al., 2019).

MRI, based on nuclear magnetic resonance technology, analyzes food by utilizing the water content and hydrogen molecules in the food, aligning protons in the direction of a magnetic field. It measures the magnetic properties of spins related to the physical or chemical properties of the object being analyzed. MRI is used in quality control of various foods such as meat products, cereals, fruits, and vegetables, with a focus on water content as a key indicator of food quality (Chen et al., 2013).

Ultrasound imaging, utilizing mechanical waves, is a cost-effective technique mainly used for food packaging and detecting foreign objects. It employs non-contact air-instability compensation ultrasound imaging to identify internal disorders and foreign objects in poultry and cheese.

X-ray imaging, employing electromagnetic waves, is widely used in the food processing industry to detect contaminants or foreign objects in food products. Soft X-ray imaging systems are designed based on the density of the food product, where contaminants appear as dark grey spots. This technique has been successful in detecting foreign objects in packed dry foods, particularly those with densities greater than water (Kwon et al., 2008).

These imaging techniques offer valuable tools for enhancing food safety and quality in the food processing industry. They provide rapid, nondestructive, and accurate analysis of various parameters critical for ensuring consumer health and satisfaction.

16.3.9 X-Ray Imaging and Computed Tomography

X-rays, spanning a range of 0.01–10 nm, offer a potent means of penetrating matter, forming the basis for X-ray imaging. This technique capitalizes on density disparities within materials, distinguishing between product and contaminant. Soft X-rays, falling within the 0.1–10 nm range, are favored for food inspection due to their reduced risk of contamination compared to hard X-rays. Soft X-ray imaging is rapid, typically producing radiographs in just 3–5 seconds, albeit at the expense of depth information. To address this limitation, computed tomography (CT) was developed, enabling the visualization of 3-D images by reconstructing a volume from multiple projections using algorithms like the Radon transform. CT scans offer superior image quality but are hindered by their costliness and prolonged image acquisition and processing times. X-ray micro-CT, employed to detect mechanically separated meat (MSM) in processed meat products, leverages bone fragments' high density for detection. While promising, studies like these underscore challenges such as operator dependency, cost, and time constraints, hindering widespread adoption in meat applications. Additionally, safety concerns regarding lead usage further complicate industrial implementation. Despite these limitations, X-ray imaging remains a valuable tool for non-destructive inspection, providing insights into sample composition and structure (Du et al., 2019).

X-ray CT technology, a pioneering advancement blending computer and X-ray technologies, was introduced in 1972 by British engineer Godfrey N. Hounsfield and American physicist Allan M. Cormack, earning them the title of "CT's father." Notably, Hounsfield was awarded the Nobel Prize in 1979 for his pivotal role in its development. Wilhelm Conrad Rontgen's discovery of X-ray radiation in 1895 marked the inception of its medical applications, eventually culminating in the groundbreaking use of CT for medical diagnoses. The exigencies of medicine catalyzed CT's theoretical and equipment advancements. In Fig. 16.6, X-ray CT scan process is given. In recent years, CT technology has seen diverse applications across various professional domains, including industry, engineering, safety testing, and agricultural science. Particularly within agriculture, CT technology has sparked interest among researchers seeking to visualize internal dynamic conditions. Its evolution has transitioned from external to internal quality inspection, enabling the examination and visualization of the two-dimensional (2D) and three-dimensional (3D) structures of agricultural products (Guelpa et al., 2016). With CT scanners, it's now possible to analyze various attributes such as pore topology, tiller numbers,

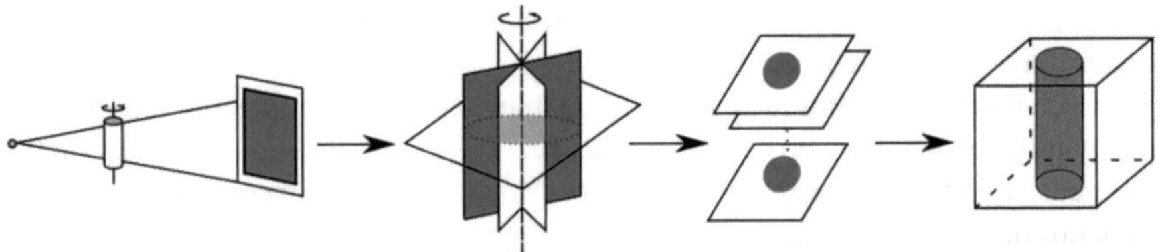

Fig. 16.6 X-Ray CT scan process. (Source: Wang & Zhou, 2022)

internal corruption, grain hardness, moisture, temperature, volumes and densities, and mechanical damage (Arendse et al., 2016). X-ray CT technology has become indispensable for non-destructive 3D characterization and tomographic imaging reconstruction. For instance, Takhar et al. employed micro-CT scanners to analyze moisture transport mechanisms in maize grains, aiding in predicting drying parameters with remarkable accuracy ($R^2 = 0.88 - 0.99$, CV = 3.5–9.5%) (Takhar et al., 2011). Additionally, CT technology has been instrumental in observing stress crack formations in maize grains due to temperature and moisture variations. Overall, X-ray CT facilitates the non-destructive inspection of agricultural product quality.

16.3.9.1 Basic Principles

CT scanning, with its inception at Atkinson Morley Hospital, comprises three main components: the scanning unit, computer system, and image display. The scanning unit, equipped with an X-ray source, detector, and gantry, collects data processed by the computer system and reconstructed into images displayed on a screen. The technology, rooted in Johann Radon's Radon transform principle, leverages differential mass attenuation properties of materials exposed to X-rays. By analyzing X-ray attenuation at various angles, CT gathers radiation attenuation data, utilizing unique decay characteristics of materials. Medical CT operates at 140 keV, while industrial CT spans 50 keV to 16 MeV, with energy selection in agricultural product inspection varying. Traditional CT rotates an X-ray tube around a sample, capturing radial projections forming 2D slice images, which are combined into 3D volumes. X-rays, produced when electrons interact with a sample, exhibit energy corresponding to electron kinetic energy. Planck's constant determines X-ray photon energy, the speed of light, and X-ray wavelength. X-rays with wavelengths from 0.1 mm to 0.01 nm demonstrate absorption, transmission, or scattering behaviors through materials with higher densities and atomic numbers, resulting in greater absorption. CT systems visualize differences in material density through changes in image intensity, quantified as CT numbers in Hounsfield units relative to water, spanning from −1000 to +4000 Hu (Du et al., 2019).

16.3.10 Image Processing and Analysis

Agricultural product analysis via CT involves extracting microstructural and physical, chemical, and mechanical properties from images. Image processing techniques facilitate accurate CT number extraction, enabling precise visualization and component identification. Segmentation techniques, such as threshold segmentation, are employed to distinguish between phase materials within images (Du et al., 2019). Textural features are extracted to identify regions of interest, with subsequent noise reduction techniques applied to address artifacts and defects. CT images, comprising grayscale pixels, reveal variations in material density, facilitating the visualization and analysis of agricultural product structures in 3D (Mao et al., 2016). Table 16.1 provides the applications of spectral imaging techniques in food quality assessment.

Quantitative and qualitative analysis of CT data is conducted using software like ImageJ, Avizo, Amira, and VGStudio Max. These tools enable a comprehensive examination of structural characteristics, aiding in the interpretation of agricultural product features. The image processing and analysis workflow involves reconstructing 3D models from 2D slice images, followed by noise reduction and segmentation based on gray value histograms (Mao et al., 2016). Finally, qualitative and quantitative analyses are performed on regions of interest, providing insights into the internal composition of agricultural products.

Table. 16.1 Applications of spectral imaging techniques in food quality assessment

Food type	Application	Method used	Inference	References
Apple	Quality assessment, fungal detection	Hyperspectral imaging	Hyperspectral imaging systems with UV emission can distinguish between normal fresh apples and those affected by fungus, aiding in food quality research.	Pradeepika et al. (2019)
Apple	Firmness and soluble solid content measurement	Multispectral imaging	Multispectral imaging using a neural network method predicts the firmness and soluble solid content of apples based on the scattering numerical range from 680 to 1060 nm.	Pradeepika et al. (2019)
Apple	Rotten tissue detection	Hyperspectral imaging	Hyperspectral imaging techniques, combined with spectral analysis, chemometric methods, and image processing tools, effectively distinguish between rotten and sound tissues in apples, enhancing automatic grading systems.	Zhang et al. (2015)

(continued)

Table. 16.1 (continued)

Food type	Application	Method used	Inference	References
Grape seeds	Harvest period estimation, soil type differentiation	Hyperspectral imaging	Hyperspectral imaging based on NIR spectra allows for the estimation of harvest period and differentiation between grape seed varieties and soil types, offering a non-destructive alternative to traditional analyses.	Rodríguez-Pulido et al. (2013)
Fish	Differentiating fresh and frozen-thawed varieties	VIS/NIR hyperspectral imaging	VIS/NIR hyperspectral imaging technique successfully distinguishes between fresh and frozen-thawed fish fillets with high accuracy using principal component analysis and machine learning classification.	Zhu et al. (2013)
Poultry	Contamination detection on carcasses and surface waste	Hyperspectral imaging	Hyperspectral imaging with visible or near-infrared spectroscopy effectively detects contamination on poultry carcasses and surface waste, achieving high accuracy through spectral processing methods.	Pradeepika et al. (2019)
Spinach leaves	Fecal contamination detection	Ultraviolet and violet spectral fluorescent imaging	Fluorescent imaging with ultraviolet and violet spectra achieves high accuracy in detecting fecal contamination in spinach leaves, offering a potential solution to reduce foodborne illnesses.	Everard et al. (2014)
Cereals	Impurity removal	NIR hyperspectral imaging	NIR hyperspectral imaging with chemometric tools effectively removes impurities from cereals, offering a time-saving alternative to manual methods.	Pierna et al. (2012)

16.3.11 Computer Vision

Computer vision technology has long been integral to the food industry, ranking among its top ten users of image processing techniques. Leveraging computer vision-based image analysis, the industry benefits from rapid, precise, and non-destructive evaluation of foods. This technology offers versatility, repeatability, and cost-effectiveness, facilitating real-time quality monitoring within processing plants and enhancing both production efficiency and packaging processes. A computer vision system (CVS) encompasses a camera, standard illumination, computer, and specialized image analysis software. Various types of cameras, including RGB, NIR, infrared, and X-ray, are employed as image acquisition devices within CVS setups. By using digital images as discrete representations of data with spatial and intensity information, color spaces such as RGB enable the representation of color in three-dimensional models. Each pixel in a digital image contains red, green, and blue components represented by integer values between 0 and 255. RGB cameras are predominantly utilized for image analysis and the extraction of visual features. The accompanying software comprises a range of algorithms, such as calibration, segmentation, and classification. Algorithms, serving as sequences of commands, are essential for completing tasks on a computer. These algorithms can be implemented using any software program equipped with image-processing capabilities (Mogol & Gökmen, 2014). A typical computer vision-based image analysis system comprises a camera for image acquisition, fluorescent lamps for object illumination, and a computer for data processing, with consistent image acquisition conditions being essential for reliable analysis. Various image processing approaches, including mean color information and color segmentation, are utilized for extracting specific quality-related data from digital images across different food products.

16.3.11.1 Mean Colour Information

Mean color information extracted from digital images is pivotal for evaluating food quality, leveraging color models such as XYZ, HSV, and, notably, CIE L*a*b*. The CIE L*a*b* space, preferred for its human perceptibility, provides a standardized framework for color measurement in foods. Within this space, luminance (L*) signifies lightness, while chromatic components (a* and b*) represent color variation along green-red and blue-yellow axes, respectively. Instruments like colorimeters, when adapted with the CIE L*a*b* space, enable surface color measurement of foods. However, challenges arise with non-homogeneous surfaces, commonly found in processed foods, due to chemical reactions like the Maillard reaction and caramelization during baking or frying. Computer vision-based image analysis addresses this challenge by allowing the measurement of color in specific regions of interest (ROIs), offering a more nuanced understanding of mean color information for non-uniform foods. By defining ROIs, such analysis captures variations in color across surfaces, ensuring a comprehensive and meaningful assessment of food quality (Mogol & Gökmen, 2014).

16.3.11.2 Featured Color Information

Utilizing pattern recognition techniques, such as image segmentation, is crucial for handling the non-uniform surface colors of processed foods, often characterized by surface browning resulting from thermal processing. This browning ratio serves as a distinctive feature that can be extracted from digital images through segmentation algorithms. Image segmentation partitions a digital image into multiple sets of pixels based on predefined reference color values, allowing the extraction of featured color information. Custom-designed Matlab® code facilitates the classification of pixels into subsets based on their Euclidean distance to these reference values, with additional consideration given to background pixels. By calculating the normalized area of specific pixel subsets, such as Set 2 pixels representing browning, distinctive features like the browning ratio can be determined. The flexibility to adjust the number and values of reference colors enables the calculation of various features tailored to specific food products, such as browning ratio for cookies and potato chips or porosity for breadcrumbs, enhancing the versatility of the segmentation algorithm for diverse food quality assessments (Ji et al., 2013; Mogol & Gökmen, 2014).

16.3.11.3 Correlation Between Color Information and Food Quality Characteristics

Extracting mean or featured color information from digital images via computer vision-based image analysis not only aids in assessing the visual quality of food products but also provides insights into their chemical and physical attributes. For example, the development of surface browning in thermally processed foods is indicative of the progress of the Maillard reaction, which occurs between carbonyl and amino compounds at elevated temperatures. This reaction not only leads to browning but also results in the formation of undesirable compounds, such as acrylamide and 5-hydroxymethylfurfural. Recent studies have shown a correlation between mean CIE a* values of food products and acrylamide concentration, demonstrating the potential of color information as an indicator of chemical composition. Furthermore, segmentation-oriented computer vision-based analysis allows for the extraction of specific features like the browning ratio, offering more precise and meaningful insights into food quality. The correlation between browning ratio and acrylamide concentration suggests the possibility of monitoring acrylamide formation during food processing using appropriate image analysis algorithms, potentially aiding in quality control. Additionally, digital image analysis techniques enable the objective evaluation of physical characteristics, such as breadcrumb grain quality, by measuring parameters like porosity index (Gökmen & Mogol, 2010). These techniques offer more accurate and reliable assessments than traditional methods, highlighting the value of computer vision-based image analysis in enhancing food quality control processes.

16.3.12 CNN—Computer Vision

Convolutional neural networks (CNNs) have become instrumental in this field, given their prowess in image classification tasks. CNNs are adept at recognizing patterns within images, making them well-suited for food recognition endeavors. To establish a food recognition system using CNNs, a dataset comprising images of diverse food items is amassed. This dataset serves as the foundation for training the CNN to classify images into distinct categories such as fruits, vegetables, meats, etc. Once trained, the CNN model can effectively classify new images of food, facilitating tasks such as nutritional analysis and color evaluation. By leveraging classification results, pertinent dietary information associated with each food type can be retrieved from a database. This capability proves invaluable for applications ranging from dietary tracking to offering recommendations for balanced meals.

In essence, computer vision-driven food recognition, particularly utilizing CNNs, revolutionizes color evaluation and nutrition analysis in the food industry. Its ability to automate these processes not only enhances efficiency but also empowers consumers with valuable insights into their dietary habits and choices (Wu & Sun, 2013).

Convolutional neural networks (CNNs) stand out as a crucial component in image and video analysis, particularly for tasks like image classification, object detection, segmentation, and medical image analysis. Their effectiveness lies in their ability to process and interpret visual data through a series of specialized layers, mimicking the hierarchical organization of the human visual system. At the core of CNNs are convolutional layers, which employ filters to detect specific patterns like edges and shapes within the input data. These patterns are progressively combined across layers to recognize increasingly complex features. Following the convolutional layers, pooling layers reduce the dimensionality of feature maps, enhancing the network's flexibility and computational efficiency.

Convolutional Neural Networks (CNNs) have diverse applications ranging from image classification to medical imaging, demonstrating several advantages in deep learning. Firstly, they autonomously learn pertinent features from raw data, eliminating the need for manual feature engineering and adapting well to various tasks. Secondly, CNNs excel in parallel processing, enabling efficient handling of large datasets and swift predictions. Moreover, they exhibit robust generalization to new data, ensuring reliable performance in real-world scenarios. Their versatility across domains makes them invaluable for tasks in image and video analysis. Overall, CNNs have revolutionized these fields, providing efficient solutions and extracting meaningful information from visual data (Shankaragowda & Latha, 2023). Table 16.2 provides the computer vision applications in food color assessment and quality evaluation.

16.3.13 AI & ML

Artificial intelligence (AI) encompasses the development of intelligent machines capable of emulating human-like thinking and behavior. Within this domain, machine learning (ML) plays a pivotal role, leveraging algorithms to enable machines to learn from data and make predictions without explicit programming. Supporting this is Artificial Intelligence Markup Language (AI & ML), a

Table 16.2 Computer vision applications in food color assessment and quality evaluation

Food type	Application	Method used	Inference	References
Beef	Predicting meat color	Computer vision	Computer vision techniques are effective for predicting meat color attributes such as lightness, redness, and chroma, offering potential applications in assessing meat freshness and quality.	Rahman et al. (2020)
Pork	Assessing fresh pork loin color	Computer vision	Computer vision systems can efficiently measure the sensory color of fresh pork, providing a practical tool for quality assessment.	Lu et al. (2000)
Fish	Evaluating salmon fillet color	Computer vision	Computer vision techniques offer a reliable method for quantifying color attributes in fish fillets, aiding in quality assessment and freshness determination.	Schubring (2009)
Wine	Measuring red wine color	Computer vision	Computer vision systems provide accurate measurements of red wine color attributes, facilitating quality control and assessment in wine production.	Condé et al. (2017)
Beer	Determining beer color	Computer vision	Computer vision technology enables precise determination of beer color, offering potential applications in brewery quality control and product development.	Viejo et al. (2016)
Potato chips	Assessing chip color	Computer vision	Computer vision systems offer a reliable and objective approach to assessing the color of potato chips, facilitating quality control in the potato processing industry.	Sanchez et al. (2020)
Wheat kernels	Measuring seed coat color	Computer vision	Computer vision techniques provide accurate measurements of wheat kernel seed coat color, offering potential applications in agricultural research and crop quality assessment.	Zapotoczny and Majewska (2010)
Bananas	Assessing banana ripeness and color changes	Computer vision	Computer vision systems offer a reliable method for assessing banana ripeness and color changes, providing insights into fruit quality and shelf-life monitoring.	Baglat et al. (2023)

programming language facilitating the creation of chatbots and other AI programs, enabling natural language interactions with humans through predefined rules and patterns. Machine learning encompasses various techniques, including supervised learning, unsupervised learning, semi-supervised learning, and reinforcement learning. These methods involve training models on labeled or unlabeled datasets to learn patterns and relationships, enabling them to make predictions or decisions based on new data.

Deep learning, a subset of machine learning, has profoundly impacted various industries, particularly in image and speech recognition, natural language processing, and predictive modeling. In computer vision, convolutional neural networks (CNNs) are pivotal, especially in food recognition and nutrient analysis tasks. CNNs excel in classifying images into food categories, detecting individual items, and estimating nutritional content, aiding in dietary analysis, weight management, and health tracking. Among object detection strategies, YOLOv2 (You Only Look Once version 2) stands out, offering efficient food recognition capabilities. By collecting, labeling, and training on food image datasets, YOLOv2 enhances accuracy, providing users with a convenient and accurate means to track their dietary intake and support various applications in health and nutrition management (Shankaragowda & Latha, 2023).

16.3.13.1 Process Flow

The creation of a computer vision-based food recognition system coupled with nutrition analysis involves a systematic process Fig. 16.7. Initially, a diverse dataset comprising images of various food items alongside their corresponding nutritional details is gathered through manual collection or web scraping techniques. Subsequently, both the images and nutritional data undergo preprocessing to ensure uniformity and readiness for further analysis, involving tasks like resizing images

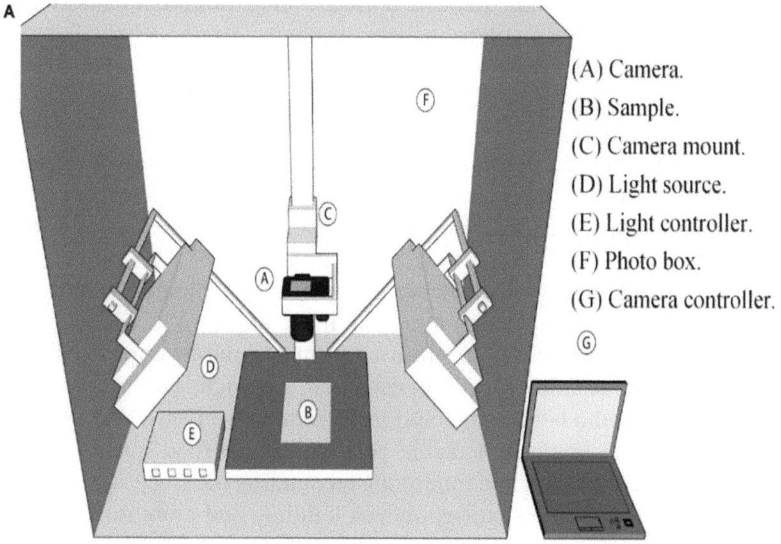

Fig. 16.7 Computer vision systems (Xiao et al., 2022)

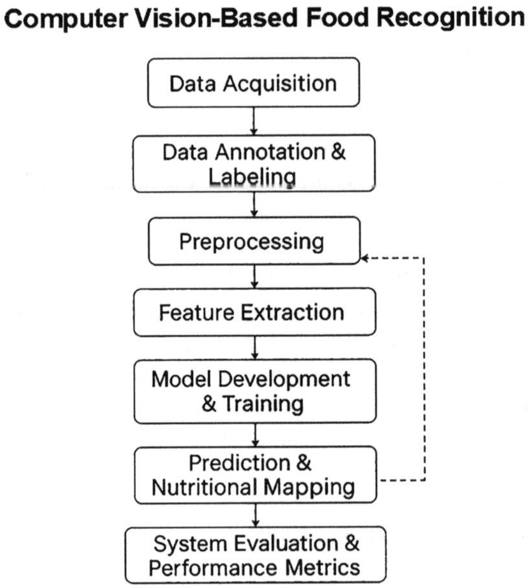

Fig. 16.8 Process flow of food recognition system in computer vision

and normalizing nutrition information. Feature extraction techniques, such as edge detection or color histograms, are then employed to represent images concisely. These features serve as inputs for training a machine learning model, utilizing techniques like supervised or unsupervised learning to distinguish between different food categories. Following model training, the system can accurately classify food items depicted in images and retrieve corresponding nutritional information for analysis. Finally, the system's performance is evaluated using metrics like accuracy, precision, and recall to ensure reliable food recognition and nutrition analysis across diverse food types (Shankaragowda & Latha, 2023). Figure 16.8 depicts the process flow of computer vision.

These steps constitute a typical procedure for erecting a computer vision-based food recognition system integrated with nutrition analysis. They are pivotal in ensuring the system's effectiveness and accuracy in recognizing and analyzing various food items.

Image processing involves enhancing image quality and addressing defects like geometric distortion and uneven illumination. It encompasses low-level processing, focusing on noise reduction and contrast adjustment; mid-level processing, which involves image segmentation for object detection; and high-level processing, utilizing deep learning methods like convolutional neural networks (CNNs). Deep learning techniques have shown remarkable performance in food quality assessment, automating feature extraction and improving recognition accuracy such as geometric distortion, focus irregularities, noise, uneven lighting, and motion blur induced by camera movement. Researchers are exploring various CNN models for tasks like defective carrot identification and peach variety detection, enhancing the intelligence and efficiency of food inspection systems. Despite its advantages, deep learning requires a large dataset for training and significant computational resources, posing challenges for widespread adoption (Xiao et al., 2022).

16.3.13.1.1 Low-Level Processing

The initial phase, known as low-level processing, involves the fundamental tasks of image acquisition and pre-processing. Since computer processing operates exclusively on digital data, information gathered from various collection devices needs transformation. This stage also addresses inherent flaws arising from environmental factors like lighting and humidity, as well as equipment-related issues such as device motion and transmission distance. Techniques such as noise reduction and contrast adjustment are vital here to enhance recognition accuracy. For instance, (Chen et al., 2021) performed low-order processing and reflection calibration to mitigate the impact of varying camera configurations and hyperspectral imaging systems, thereby improving subsequent recognition accuracy.

16.3.13.1.2 Mid-level Processing

Moving further, mid-level processing focuses on tasks like image segmentation, representation, and description. Image segmentation, a critical step, aims to delineate object regions from the background, thereby facilitating subsequent analysis. This intermediate phase plays a crucial role in machine vision applications. Researcher were done on enhanced watershed segmentation algorithm, integrating morphological filtering and gradient reconstruction, to segment rotten areas on apples swiftly. Similarly, Luo et al. utilized an improved watershed segmentation method for bruise detection in fruits using color variation, achieving an impressive detection accuracy of 99.5% (Luo et al., 2019).

16.3.13.1.3 High-Level Processing (Deep Learning)

The advent of deep learning, particularly convolutional neural networks (CNNs), has revolutionized image processing, especially in the domain of computer vision. Deep learning methods excel in automating feature extraction and pattern recognition, thereby enhancing the intelligence of food inspection systems. Notably, researchers have explored CNN models for diverse tasks such as identifying defective carrots, grading carrots, and detecting peach varieties. These efforts have significantly augmented the accuracy and efficiency of food inspection processes (Xiao et al., 2022).

16.4 Texture Measurements in Food

The texture of food is a multifaceted sensory attribute that significantly influences consumer perception, preference, and acceptance. It encompasses various qualities such as hardness, chewiness, crispiness, and viscosity, all of which contribute to the overall eating experience. Understanding and controlling food texture is of paramount concern in the food industry, as it directly impacts product quality, shelf-life, and consumer satisfaction. Consequently, the accurate measurement and analysis of food texture have become crucial subject of concern for researchers and food manufacturers alike (Bourne, 2002).

The texture of foods plays an important role in both processed and unprocessed fresh foods. In the case of processed foods, the crucial quality attribute is texture-dependent. Therefore, processing parameters such as frying, heating, and drying temperature and time are to be optimized to maintain the texture uniformly. In the case of unprocessed products such as fresh fruits and vegetables, texture is an important parameter determining the maturity of such foods for careful harvesting, storage and handling for future use (Chen & Opara, 2013a, b). Multidisciplinary collaboration in the field is essential to combining sensory and instrumental noninvasive textural measurement in food processing industries.

Traditionally, food texture analysis depended mostly on subjective sensory evaluation. With the advancement in research, destructive single-frame and dual-frame load cell-based destructive mechanical instruments were developed. However, these approaches are often time-consuming and labor-intensive. Moreover, destructive techniques alter the sample's integrity, rendering it unsuitable for further analysis or consumption (Chen & Opara, 2013a, b). In response to these limitations, non-destructive methods have emerged as promising alternatives, offering real-time monitoring capabilities without compromising sample integrity.

Non-destructive food texture measurement techniques utilize various principles and technologies to assess textural properties without altering the sample. These methods encompass a wide range of approaches, including acoustic, optical, rheological, and imaging techniques. Each technique exploits different physical properties of the food matrix to extract relevant textural information (Guillermic et al., 2021). For instance, acoustic methods measure sound propagation through the sample to infer its mechanical properties. At the same time, optical techniques analyze light scattering or absorption patterns to assess surface roughness or moisture content. Among the non-destructive techniques, imaging methods have garnered significant interest due to their versatility and potential for in-line monitoring in food processing plants. Imaging techniques encompass a diverse array of modalities, such as X-ray, ultrasound, infrared, and hyperspectral imaging, each offering unique insights into food structure and texture (Sun, 2009). These techniques can capture detailed spatial and spectral information, enabling comprehensive analysis of textural properties across various food products.

Research into imaging methods for food texture analysis has seen remarkable advancements in recent years. Studies have explored the application of X-ray imaging to assess the internal structural properties of foods such as bread, cheese, and meat. X-ray microtomography, in particular, allows for three-dimensional visualization of internal structures, facilitating the quantification of porosity, density, and distribution of components within the sample (Schoeman et al., 2015). Similarly, ultrasound imaging has been utilized to characterize the mechanical properties of food products non-invasively. By analyzing ultrasound wave propagation through the sample, the internal microstructure of the sample can bc cvaluated.

Infrared (IR) imaging has also emerged as a powerful tool for assessing food texture and quality. IR imaging exploits the interaction between infrared radiation and the chemical composition of the sample to generate spatial maps of various constituents, such as fat, protein, and moisture. This enables researchers to correlate textural attributes with chemical composition, providing insights into the underlying factors influencing food texture (Liu et al., 2017). Moreover, hyperspectral imaging extends the capabilities of traditional IR imaging by capturing spectral information across a wide range of wavelengths. By analyzing spectral signatures, hyperspectral imaging can discriminate between different food components and detect subtle changes in texture or composition.

The effectiveness of imaging methods in in-line monitoring within food processing plants depends on several factors, including speed, accuracy, and scalability. While imaging techniques offer unparalleled non-destructive capabilities, integrating them into industrial-scale production lines presents logistical and technical challenges. Researchers have focused on developing robust imaging systems capable of real-time monitoring and analysis tailored to the specific requirements of food processing environments. Additionally, advances in machine learning and data analytics have facilitated automated texture analysis, enabling rapid decision-making and quality control in manufacturing settings. A detailed review of emerging non-destructive techniques for the textural evaluation of food products is given elsewhere (Mishra et al., 2023).

16.4.1 Texture Measurements: Subjective and Objective

Firmness is a crucial texture characteristic of fresh produce, particularly, and it has long been used as a standard for classifying fruits and vegetables. Firmness can be understood in the context of the maximum penetration force, modulus of elasticity (E), or shear modulus in engineering terminology (Lu, 2013). Typically, non-destructive techniques like vibration, acoustics, impact, micro-deformation, and light absorption are used to measure firmness. The stiffness and commodities puncture force were correlated by the crunchiness index (CI), which was capable of determining the process treatments' intensity on the different products that were tested (Chen & Opara, 2013a, b).

Subjective sensory assessments grounded in psychophysics and objective instrumental measurement techniques grounded in physics and/or chemistry are used to evaluate texture. Instruments that measure acoustics, fracture, rheology, and, as well as microscopes and spectrometers that can characterize the structure of materials, can all be used to evaluate texture and associated physical characteristics. These sensory and instrumental results are frequently not consistent. Apart from the non-uniform properties of foods, possible explanations for this discrepancy could be the timing of the instrument use as opposed to the period of the sensory assessments, which usually take place subsequent to the start of oral processing. Conditions that are not well replicated by conventional models and can affect instrumental results include temperature, moisture from saliva, chewing motions, and deformation rates of food, all through oral processing (Kohyama, 2015).

16.4.2 Methods of Oral Texture Interpretation

There are various steps involved in sensory analysis, which take place both inside and outside the mouth. The very first bite and late phases of mastication, ingesting, and the aftertaste of food in the tongue and throat are instances of the latter stages. During the initial bite, certain mechanical attributes like fracturability, firmness, and viscosity are perceived. Meanwhile, the masticatory part evaluates chewiness, gumminess, and adhesiveness, while the remaining qualities are perceived after swallowing. A method for providing a quantitative description of the interpreted sensory properties is called quantitative descriptive analysis, or QDA. In sensory science, the primary method for gathering comprehensive, trustworthy, and repeatable data to characterize sensory profiles has been descriptive analysis (Kohyama, 2015).

Dynamic sensory evaluations have become a part of texture assessment during oral processing, and they are utilized to examine temporal variations in textural characteristics. A panelist describes the intensity of a chosen attribute in turn using the Time Intensity (TI) method. Using a list of attributes, a panelist continually selects the strongest aspect at each instant in the temporal dominance of sensations (TDS) method (Cheong et al., 2014; Fiszman & Tarrega, 2018).

16.4.3 Instruments-Based Texture Measurements

Generally, there are two types of mechanical food texture measurements: destructive and non-destructive (Lu, 2013). The destructive group, for instance, consists of the single-edge notched bend (SENB) test, the cutting "tooth method," which uses an incisor blade, the penetration and puncture tests, and the three-point bending test. Although these techniques mimic the mastication process and may be linked to molecular and microstructural mechanisms, they are destructive and their relationships to mouth feel are unclear. The quasi-static force-deformation, impact response, finger compression, as well as bioyield detection techniques are referred to as non-destructive measurement techniques because they can be used online and typically reveal no visible damage. Force is an important factor in both destructive and non-destructive assessments. For this reason, force versus time/distance /deformation are normal texture features (Chen & Opara, 2013a; Lu, 2013). The brittle fracture behavior of many crispy, hard, and crunchy solid food products is a typical feature. This behavior is typically escorted by a strident sound, also known as an acoustic vibration, and is closely related to the texture attributes of the food products. As a result, scientists have integrated acoustic monitoring with mechanical tests like penetration, compression, and three-point bending tests. The quantity and intensity of sound events reported are significant acoustic parameters; the amount of sound peak points was found to be the most effective parameter for differentiating between sample textures. During food texture assessments, texture profiles seem to be curves that track and document the spatial attribute events of samples. Food texture research includes an assessment of the features of both acoustic and mechanical measurements as a key component. Texture profile analysis (TPA) increases the predictability of food texture attributes by creating a "bridge" from objective assessment to subjective perception (Chen & Opara, 2013a).

16.4.3.1 Destructive Measurement Tests

There are many different approaches and/or strategies available for destructively measuring solid food. Destructive techniques can be basic or empirical. Empirical destructive techniques frequently resemble techniques used in sensory evaluations or food preparation. They involve a complicated type of loading where the physical stress and strain rates are far higher than the initial failure, and the measured quantity is frequently too small to be meaningfully understood in terms of fundamental engineering characteristics. Conversely, foundational destructive methods measure physical behavior like failure strength, yield strength, and others, along with the Poisson ratio, shear modulus, and Young modulus. However, these are frequently not well interpretable in terms of how humans perceive texture. Basic tests can be carried out straight on actual food samples with well-defined dimensions, but they frequently call for the use of specimens with particular dimensions that have been removed from food samples. As a result, the tests take longer than most of the empirical tests, and in tests like torsion and tension, mounting specimens on the testing apparatus and preparing them can occasionally be challenging. In general, empirical approaches have a stronger correlation with the sensory textural qualities of food. Still, fundamental approaches can provide us with a better understanding of a food's mechanical response, structural characteristics, and changes. Destructive techniques can include compression, puncture, shear, tension, twisting, bending, and more, depending on the loading pattern (Lu, 2013).

A food sample is 'punctured' to a predetermined depth, and the puncture test calculates the force needed to make that penetration. An empirical method that quite mimics the biting of the food product inside the mouth, the testing includes both pressure and shearing of the food. Measurements of the fundamental mechanical characteristics of a wide range of solid foods, such as fruits and vegetables, are frequently made using 'compression.' Using a universal testing machine, these tests are commonly performed on a cylindrical sample detached from food products under uniaxial applied stress. Additionally, unbroken food samples with very good geometry can have their simple mechanical characteristics measured using compression tests (Lu, 2013; Moiny et al., 2002). The term "shear" frequently describes the process of using force to split an element into two portions. This weak proper definition differs from the precise definition used in engineering, which is that pure shear cannot be experimentally determined except in cases of torsion and specific loading circumstances. Shear tests are particularly helpful in determining the texture of foods high in muscle. Depending on the shear characteristics of food, another method of assessing texture is 'torsion or twisting.' In such a torsion test, an object is subjected to force to rotate or twist one part in relation to another. Because applying torsional force to food samples is difficult and requires special preparation and mounting of the specimens, these tests are less frequently used for solid foods (Hamann et al., 2006; Lu, 2013). 'Tensile tests' calculate the amount of force needed to separate an unbroken food sample. Tensile tests help research the mechanical qualities of food tensile strength, especially with its structural failure attributes (Lu, 2013).

The only method that can still fully identify and characterize texture is through sensory evaluation, which makes use of a panel of individuals who have received training in measuring specific characteristics. However, for both research and industrial applications, instrumental assessments are desired over sensory tests because they are less cost-effective than sensory panels, easier to use, more broadly available, and likely to provide standard values even when used by inexperienced people (Lu, 2013).

16.4.3.1.1 Correlating Sensory and Instrumental Texture Examination

Examining the physical alterations that food undergoes during oral processing is essential to closing the information disparity between the subjective and objective approaches (Kohyama, 2015; Nishinari & Fang, 2018). The physicochemical characteristics recorded by equipment and the interpreted textures assessed by sensory panel members can be related through human physiological evaluation. Sensors that measure strain, vibration, load, displacement, and other variables are frequently connected to human participants using physiological methods in place of instruments. The oral processing method is the same as those of the sensory evaluations, and the sensor output values are objective, much like those of instrumental analyses (Kohyama, 2015). A single physical property cannot adequately describe food texture; multiple features might have to be examined in tandem to determine food texture. For instance, the TPA test in conjunction with the acoustic component may be a helpful method for assessing the texture properties. It is possible to analyze data from both instrumental measurement and sensory perception, but the difficulty lies in the absence of basic models for the wide range of foods. Consequently, the texture investigation of food commodities requires multidisciplinary cooperation between food engineers, material scientists, consumer scientists, and other experts (Chen & Opara, 2013a, b).

16.5 Non-destructive Food Texture Evaluation—Optical Techniques

The textural properties of foods are greatly influenced by the food microstructure, which plays an important role in its organoleptic and physical properties. Evaluation of food microstructure requires the need for sample preparation using destructive methods. However, the use of some novel techniques does not require extensive sample preparation or destruction of the sample to determine its textural properties. Imaging methods are one such technique where the internal structure can be enumerated non-destructively. This section explains imaging techniques such as X-ray imaging, X-ray microcomputed tomography, thermal imaging, fluorescence imaging, magnetic resonance imaging, and ultrasonic imaging.

16.5.1 X-Ray Microcomputed Tomography

This section deals with X-ray microcomputed tomography, which aids as a non-invasive, non-destructive alternative to determine the textural properties of food materials. Apart from determining non-destructively, it provides a three-dimensional structure characterization with changes in microstructure with a resolution up to the nanoscale level (Schoeman et al., 2015). The first utilization of computed tomography-based food evaluation was on green tomatoes to evaluate their maturity in 1991. The equipment scans the sample based on the radiographic imaging technique multiple times in order to evaluate the textural changes over a period of time. The experimental setup involves an X-ray source, sample stage and a detector where the evaluation of internal structure is done through image contrast, which is based on X-ray attenuation that combines absorption and scattering. This difference in attenuation brings about measurable values in density and compositional differences inside the sample (Lim & Barigou, 2004). The 2D images obtained by virtue of passing X-rays along different paths and directions yield images that exhibit density differences at thousands of points. These 2D images are stacked up to form 3D images of sample specimens. The various advantages of X-ray microcomputed tomography described by the authors Lim and Barigou (2004) are as follows: (a) the investigation of samples takes place in their natural states of atmospheric pressure and temperature without any sample preparatory steps, (b) provides a greater spatial resolution compared to other techniques, (c) X-rays can easily penetrate inside the food microstructure to evaluate its three-dimensional network, (d) the images obtained can be quantitatively determined based on image contract interpretation which gives the material density information from its microstructure. A detailed review of the food microstructure evaluation by X-ray microcomputed tomography is given by Olakanmi et al. (2023). The microstructure of various aerated products such as cake, bread, mousse, aerated chocolate, marshmallows, and muffins can be studied effectively using X-ray microcomputed

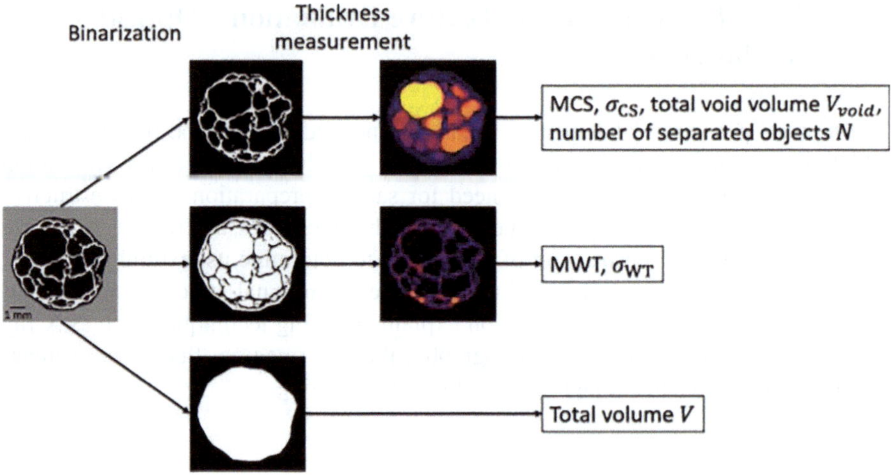

Fig. 16.9 Image analysis used to evaluate the microstructure of extrudates where MCS: mean cell size, MWT: mean wall thickness, σ value represents the standard deviation of the same. (Obtained with permission from Guillermic et al. (2021))

tomography. Studies also showed the impact of baking on the crumb thickness, which was evident in reconstructed 2D images followed by X-ray analysis.

An advancement in this technique is the use of X-ray grating-interferometry in foods, where the imaging modalities are obtained by calculating the absorption, refraction, and scattering characteristics of a food sample. In order to evaluate the soft tissues of pork containing lean and fat, this technique is best put into practice (Einarsdóttir et al., 2014). In a study conducted by Guillermic et al. (2021), microcomputed tomographical evaluation of red lentil based extruded snacks was assessed. The miniscule variation in the extrusion parameters greatly influences the mean cell size and cell wall thickness of extrudates which was easily observable in a binary image produced by the X-ray based on absorption and scattering. Figure 16.9 describes the binarization of the image to evaluate the different product parameters. Apart from assessing the textural properties, X-ray imaging also aids in the detection of internal injuries in fresh fruits. The internal rot in avocado fruits was identified using deep learning-based semantic segmentation of X-ray images by Matsui et al. (2023). The method developed was an in-line monitoring system with a high speed of data acquisition and image contrast that effortlessly provides a bright field of view in the presence of internal rot or injury. Such techniques help increase the efficiency of classification and can be used for the validation of products from seasonal or different geographical origins.

16.5.2 Thermal Imaging

This section deals with thermal imaging of food materials to infer the internal food microenvironment. The dielectric properties of foods have been studied for a long time now and are used in various heating and dryer designs. The dielectric properties of food that are considered in such imaging systems are moisture content and temperature, which play a crucial role in designing such instruments (Jha et al., 2011). Thermal imaging systems work in the principle that the invisible radiation from a food product is used to produce visible images through feature extraction and analysis. Thermal imaging utilizes the radiation that was emitted to produce a pseudo image of the thermal distribution of the object's surface. A large number of temperature data is generated about various points on the sample surface, which are processed to form a thermal map or a thermogram with color hue variations. Although thermal imaging is used as non-destructive internal texture evaluating method, it is of limited use owing to its poor sensitivity (Gowen et al., 2010). With the inclusion of computer vision-based systems, the application and efficacy of thermal imaging can be increased for in-line quality inspections in food processing plants. This technique uses a simple hand-held instrument that can be used to monitor the internal textural features using heat maps. A detailed review of the use of hand-held systems for the textural enumeration of foods such as fruits, vegetables, and meats is given elsewhere (Abasi et al., 2018). The incorporation of mathematical-based algorithms coupled with thermal imaging systems has the potential to provide more robust and accurate results in monitoring food samples. The exploration of new algorithms in the future can give promising results from thermal imaging (Mohd & Hashim, 2022).

16.5.3 Machine Vision System

A machine vision system or Computer vision system is a tool to obtain spatial information based on the principle of human vision using the color bands red, green, and blue. This technique works on the visible range spectrum and uses multiple cameras to capture images of the agricultural commodity at numerous angles to assess the internal and external features. The data obtained through this technique is a three-dimensional image cube, which consists of both spectral and spatial information related to the sample (Ropodi et al., 2016). With the advancement of computer engineering-based modeling, high-resolution and efficient real-time instruments can be developed for food processing plants. Computer vision techniques have emerged as powerful tools for the analysis of food texture and the detection of internal rot in food products (Adedeji et al., 2020). These techniques leverage image processing algorithms to extract meaningful information from visual data, enabling automated assessment of textural attributes and the identification of defects. For instance, convolutional neural networks (CNNs) have been employed to classify food textures based on visual features extracted from images, enabling rapid and accurate texture analysis. Additionally, computer vision algorithms can detect internal rot in fruits and vegetables by analyzing color, shape, and texture variations indicative of decay. This non-destructive approach offers advantages in terms of speed, accuracy, and scalability, facilitating quality control and waste reduction in food production and processing.

16.5.4 Acoustic Method

This section deals with the use of acoustic systems in determining the crunchiness and crispiness of food products. Samples that are evaluated using this technique have a filled portion such as a co-extruded cereal-based product with fillings such as toffee, milk, coconut, chocolate cream and fruit jelly (Jakubczyk et al., 2016). The intensity of sounds emitted for different fillings, along with the sensory scoring, results in controlling the texture parameters. The acoustic energy amplitude-time spectra are used to identify the texture parameters of the food sample. The deformation of the internal structure produces increased acoustic emission energy. This technique is easier to correlate with a sensory score in order to put forth a product with desired textural properties as perceived by the consumers. Figure 16.10 represents the acoustic vibration technique of pear samples wherein pear of different textures is taken for analysis. The acoustic vibration technique involves the use of a laser Doppler vibrometer (LDV) detector, which detects the signal emitting from the surface of the sample. Using parameters such as second resonance, amplitude and phase shifts, the vibration spectrum is obtained, and other parameters, such as elasticity index, can be calculated using the formula (Zhang et al., 2014).

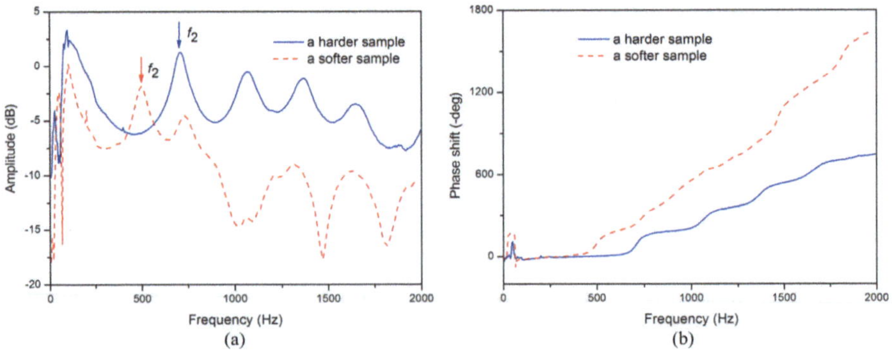

Fig. 16.10 (**a**) amplitude-frequency response and (**b**) phase-frequency response of pears of different textures. (Obtained with permission from Zhang et al. (2014))

16.5.5 Spectroscopic Methods

Moving on to the spectroscopic techniques in the measurement of food texture, methods such as Raman spectroscopy, NMR spectroscopy, and MRI spectroscopy prove to be effective, non-destructive alternative assessment methods. Apart from textural evaluation, spectroscopic methods are also used for non-destructive assessment of food fraud in the meat industry (Edwards et al., 2021). This technique is simple, non-invasive, fast and reliable data about adulteration in meat samples with online real-time monitoring. Figure 16.11 represents the non-destructive spectroscopic techniques, namely (a) NIR spectroscopy, (b) FTIR spectroscopy, (c) Hyperspectral imaging, and (d) Raman spectroscopy.

16.5.6 Magnetic Resonance Imaging Spectroscopy

This section throws light on the use of Magnetic Resonance Imaging (MRI) in the evaluation of food texture. The presence of a magnetic field results in the magnetization of the atomic nuclei, which gives the nature of the distribution of atoms in the

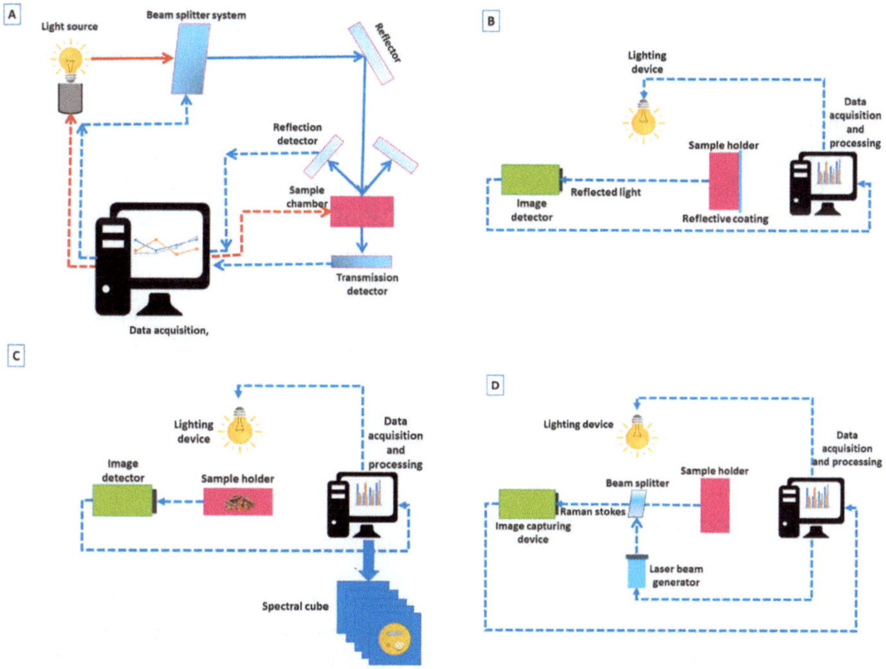

Fig. 16.11 Non-destructive spectroscopic techniques in food quality evaluation. (Obtained with permission from Modupalli et al. (2021))

sample (Sarkar et al., 2022). Magnetic resonance is generated by the interaction of atomic particles with the external magnetic field. If the interacting particle is a nucleus, it is termed nuclear magnetic resonance imaging. Monitoring of the proton distribution in the presence of an external magnetic field gives a 2D or 3D image where the quality of the image depends on the high signal-to-noise-ratio, high spatial resolution, and better image contrast (Cakmak, 2019). An advantage of the use of the MRI technique in the evaluation of texture is its ability to produce contrast images, implying a difference in internal densities. Therefore, such systems are useful in determining the internal texture assessment of meat slices where the layers of fat and the meat can be easily visualized using contrast images obtained through MRI (Silva et al., 2020).

16.5.7 Raman Spectroscopy

This section gives information about the use of Raman spectroscopy in the assessment of food texture. Raman spectroscopy has emerged as a promising technique for non-destructive food texture measurement, offering valuable insights into the chemical composition and structural properties of food matrices. Raman spectroscopy relies on the inelastic scattering of monochromatic light, wherein molecules undergo vibrational and rotational transitions, resulting in characteristic spectral fingerprints that can be used for qualitative and quantitative analysis (Wu et al., 2022). Several studies have demonstrated the utility of Raman spectroscopy for food texture analysis across various food categories.

Similarly, Raman spectroscopy has been utilized to assess the texture of meat products, such as beef and pork. Raman spectroscopy technique in the assessment of meat quality has garnered wide attention in food research (Silva et al., 2020). Spectral peaks are obtained where the intensity of peaks defines the composition of muscle fragments at that particular region, for example, tyrosine doublet at 826 and 853 cm^{-1} and α-helix at 930 cm^{-1} (Fowler et al., 2014). In some cases, the textural changes with respect to raw, cooked and thawed samples can be predicted using the Raman spectroscopic technique, which observes a spectral change in protein, amino acids and short peptides. Defects in meat samples, such as boar taint, can be easily observed in a Raman spectrum.

16.5.8 NIR Spectroscopy

This section focuses on the aid of UV vis/NIR spectroscopy to obtain spatial information on agricultural produce. A study focussed on the evaluation of mandarin oranges using NIR spectroscopy (Herna & He, 2006). NIR models are equipped using regression coefficient and partial least square tests to estimate the characteristics of the sample. The data is obtained in terms of absorbance spectra with absorbance as the y-axis and wavelength as the x-axis. Apart from determining the firmness of fresh produce, NIR spectroscopy finds its application in determining the superior variety of commodities from its various genotypes (Mancini et al., 2020). Thus, this multi-faceted instrument can help consumers evaluate fruits based on juiciness and firmness in the supply chain, thereby identifying a product that meets their sensory demands. Apart from firmness measurement, juiciness of pear using Vis-NIR spectroscopy combined with chemometrics is also studied (Wang et al., 2020). The data generated are fed to a linear regression correction combined with spectral ratio (LRCSR) with partial least square regression as a calibration model. The models are run using simulations such as Monte Carlo simulation to verify regression coefficient (R^2) and root mean square error (RMSE). The future of NIR spectroscopy depends on the combined use of artificial neural networks (ANN) in texture prediction. Studies prove that good qualitative models were obtained for hardness and chewiness, while no satisfactory model was obtained for cohesiveness in a NIR spectrum (Basile et al., 2022). Apart from fresh produce, the texture of frozen foods can also be evaluated using this technique. Various parameters, such as drip loss, hardness, chewiness, and gumminess. And gel strength can be predicted using principal component regression (PCR) and back propagation artificial neural network (BPANN) (Jiang et al., 2023).

16.5.9 Hyperspectral Imaging

Hyperspectral imaging (HSI) is an advanced feature of NIR spectroscopy wherein a combination of spectroscopy and imaging techniques are studied to collect information about the chemical properties of a material along with its spatial distribution (Caporaso et al., 2018). The progress in research can be witnessed from studying ground material to whole grains and further down to single grain kernels to predict damage and other properties such as protein content, α-amylase content and kernel hardness. The main advantages of HSI are individual assessment of grains, rapid data production and a non-contact system, which makes it widely acceptable in food industries. A review of the use of HSI in determining fruit maturity is reported elsewhere (Magwaza & Tesfay, 2015). The data obtained through hyperspectral imaging is termed a hypercube, which is a spectrum obtained through various wavelengths passed through the sample. It is helpful in determining the textural properties of meat products as multispectral images of the sample can be produced, denoting protein, fat, moisture, etc. (Kamruzzaman et al., 2013).

16.6 Future Trends

Significant advancements are anticipated in both hardware and software for non-destructive color evaluation in foods. Hardware breakthroughs will enable the creation of portable, on-field sensing devices, facilitating real-time monitoring of food quality and processing. These devices will deliver heightened accuracy and reliability for online evaluation, emphasizing the simplification of hardware and increased accessibility of wireless devices. Simultaneously, software enhancements will target improving the accuracy and reliability of non-destructive approaches, particularly for online applications. Rapid, cost-effective equipment development will expand the future applications of these techniques, bolstering food quality standards and offering promising solutions for the industry. Integration with hybrid technologies and the Internet of Things (IoT) is poised to enhance automated harvesting and grading operations, ensuring yield traceability and enhancing the efficiency of automated systems. Computer vision and imaging have emerged as highly efficient methods for color measurement in food, surpassing traditional approaches. Combining these techniques with spectroscopy will yield even more informative results for future innovations. High-end techniques such as AI, ML, and DNN, when integrated with existing hardware models, will enhance quality evaluations and decision-making processes. Utilizing advanced quality evaluation in real-time transportation assessments represents a significant achievement and innovation on the horizon.

In conclusion, non-destructive food texture measurement techniques signify a paradigm shift in food science and technology. Providing real-time, non-invasive analysis capabilities, these methods hold tremendous potential for improving product quality, consistency, and efficiency in the food industry. The integration of imaging techniques into food processing plants marks a new era of in-line monitoring and control, empowering manufacturers to optimize production processes and meet evolving consumer demands. Although data obtained from these techniques require modeling for correct correlation values, numerous data sets are modeled to enhance instrument efficiency. The future of non-invasive techniques should leverage artificial intelligence with artificial neural networks to monitor and sort defects during in-line processing. Various modeling and multivariate analyses of instruments, coupled with AI incorporation, can maximize the technology's potential for real-time and in-process texture measurement.

16.7 Conclusion

This chapter has extensively discussed the principles, applications, and advancements in non-destructive color & texture evaluation for food. In conclusion, non-destructive color & texture evaluation methods offer significant potential for food quality detection, providing rapid, cost-effective, and easy-to-operate alternatives to traditional methods. Despite challenges such as issues related to sample heterogeneity, sensitivity, and accuracy, these methods are poised to play a crucial role in the future of food quality assessment. By embracing advancements in hardware and software and addressing remaining challenges through ongoing research and development efforts, the food industry can revolutionize quality standards and ensure sustainable and impactful advancements in food analysis and inspection practices. Despite significant progress, challenges persist in designing computer vision, imaging and other advanced systems capable of handling the biological variations inherent in food products. Further research is needed to enhance system robustness, real-time capability, sample handling, and standardization. Challenges also arise from segmentation algorithms, which are crucial for successful computer vision-based quality measurement. Additionally, efficient and robust calibration methods are necessary to mitigate the influence of camera, illumination, and environmental changes.

Advancements in spectroscopy techniques, such as hyperspectral and multispectral imaging, NMR, IR, Raman, X-ray-based methods, fluorescence, and UV-visible, present a game-changing solution for food composition analysis. By providing valuable insights into food quality, chemical components, composition, structure-function relationships, and sensory attributes, spectroscopy eliminates the need for extensive sample preparation, saving time and resources. Coupled with appropriate chemometric multivariate methods, spectroscopy enables comprehensive and accurate analyses of various food materials.

To realize the full potential of non-destructive methods, further advancements are needed in hardware and software development. Portable and on-field sensing devices are desired to monitor the quality and processing of food products in real time. The accuracy and reliability of non-destructive approaches must be improved for online and real-time evaluation, with a focus on simplifying hardware and increasing the accessibility of wireless devices. Explorative studies and efforts to adopt rapid and low-cost equipment will contribute to the development of future applications of non-destructive techniques, ultimately improving food quality standards and providing promising solutions for the food industry.

References

Abasi, S., Minaei, S., Jamshidi, B., & Fathi, D. (2018). Dedicated non-destructive devices for food quality measurement: A review. *Trends in Food Science & Technology*. https://doi.org/10.1016/j.tifs.2018.05.009

Adedeji, A. A., Ekramirad, N., Rady, A., Hamidisepehr, A., Donohue, K. D., Villanueva, R. T., Parrish, C. A., & Li, M. (2020). Non-destructive technologies for detecting insect infestation in fruits and vegetables under postharvest conditions: A critical review. *Food, 9*(7), 1–28. https://doi.org/10.3390/foods9070927

Akash, M. S. H., Rehman, K., Akash, M. S. H., & Rehman, K. (2020). Ultraviolet-visible (UV-VIS) spectroscopy. In *Essentials of pharmaceutical analysis* (pp. 29–56). Springer.

Aldughpassi, A., Sidhu, J. S., Allafi, A., Al-Kandari, S., Al-Foudari, M., & Al-Othman, A. (2021). Tristimulus colorimetry for evaluating quality of high-fiber baked products made in a commercial bakery. *Annals of Biology, 37*(1), 104–111.

Anderson, B. L. (2011). Visual perception of materials and surfaces. *Current Biology, 21*(24), R978–R983.

Arendse, E., Fawole, O. A., Magwaza, L. S., & Opara, U. L. (2016). Non-destructive characterization and volume estimation of pomegranate fruit external and internal morphological fractions using X-ray computed tomography. *Journal of Food Engineering, 186*, 42–49.

Baglat, P., Hayat, A., Mendonça, F., Gupta, A., Mostafa, S. S., & Morgado-Dias, F. (2023). Non-destructive banana ripeness detection using shallow and deep learning: A systematic review. *Sensors, 23*(2), 738.

Basile, T., Marsico, A. D., & Perniola, R. (2022). Use of artificial neural networks and NIR spectroscopy for non-destructive grape texture prediction. *Food, 11*(3). https://doi.org/10.3390/foods11030281

Berns, R. S. (2019). *Billmeyer and Saltzman's principles of color technology*. Wiley.

Bourne, M. (2002). *Food texture and viscosity: Concept and measurement*. Elsevier.

Brainard, D. H., & Maloney, L. T. (2011). Surface color perception and equivalent illumination models. *Journal of Vision, 11*(5), 1–1.

Cakmak, H. (2019). Assessment of fresh fruit and vegetable quality with non-destructive methods. In *Food quality and shelf life*. Elsevier Inc. https://doi.org/10.1016/B978-0-12-817190-5.00010-0

Caporaso, N., Whitworth, M. B., & Fisk, I. D. (2018). Near-infrared spectroscopy and hyperspectral imaging for non-destructive quality assessment of cereal grains. *Applied Spectroscopy Reviews, 53*(8), 667–687. https://doi.org/10.1080/05704928.2018.1425214

Chen, L., & Opara, U. L. (2013a). Texture measurement approaches in fresh and processed foods—A review. *Food Research International, 51*(2), 823–835. https://doi.org/10.1016/j.foodres.2013.01.046

Chen, L., & Opara, U. L. (2013b). Approaches to analysis and modeling texture in fresh and processed foods - A review. *Journal of Food Engineering, 119*(3), 497–507. https://doi.org/10.1016/j.jfoodeng.2013.06.028

Chen, Q., Zhang, C., Zhao, J., & Ouyang, Q. (2013). Recent advances in emerging imaging techniques for non-destructive detection of food quality and safety. *TrAC Trends in Analytical Chemistry, 52*, 261–274.

Chen, G.-W., Chen, Y.-A., Chang, H.-Y., Huang, T.-C., & Chen, T.-Y. (2021). Combined impact of high-pressure processing and slightly acidic electrolysed water on Listeria monocytogenes proteomes. *Food Research International, 147*, 110494.

Cheong, J. N., Foster, K. D., Morgenstern, M. P., Grigor, J. M. V., Bronlund, J. E., Hutchings, S. C., & Hedderley, D. I. (2014). The application of temporal dominance of sensations (TDS) for oral processing studies: An initial investigation. *Journal of Texture Studies, 45*(6), 409–419. https://doi.org/10.1111/jtxs.12091

Cichero, J. A. Y. (2017). Unlocking opportunities in food design for infants, children, and the elderly: Understanding milestones in chewing and swallowing across the lifespan for new innovations. *Journal of Texture Studies, 48*(4), 271–279. https://doi.org/10.1111/jtxs.12236

Clydesdale, F. M. (1991). Color perception and food quality. *Journal of Food Quality, 14*(1), 61–74. https://doi.org/10.1111/j.1745-4557.1991.tb00047.x

Clydesdale, F. M. (1993). Color as a factor in food choice. *Critical Reviews in Food Science and Nutrition, 33*(1), 83–101. https://doi.org/10.1080/10408399309527614

Condé, B. C., Fuentes, S., Caron, M., Xiao, D., Collmann, R., & Howell, K. S. (2017). Development of a robotic and computer vision method to assess foam quality in sparkling wines. *Food Control, 71*, 383–392.

Du, Z., Hu, Y., Ali Buttar, N., & Mahmood, A. (2019). X-ray computed tomography for quality inspection of agricultural products: A review. *Food Science & Nutrition, 7*(10), 3146–3160. https://doi.org/10.1002/fsn3.1179

Dutta, K., & Nath, R. (2023). *Application of colorimetry in food industries*. IntechOpen.

Edwards, K., Manley, M., Hoffman, L. C., & Williams, P. J. (2021). Non-destructive spectroscopic and imaging techniques for the detection of processed meat fraud. *Food, 10*(2), 448.

Einarsdóttir, H., Nielsen, M. S., Miklos, R., Lametsch, R., Feidenhans'l, R., Larsen, R., & Ersbøll, B. K. (2014). Analysis of micro-structure in raw and heat treated meat emulsions from multimodal X-ray microtomography. *Innovative Food Science and Emerging Technologies, 24*, 88–96. https://doi.org/10.1016/j.ifset.2013.11.003

El-Mesery, H. S., Mao, H., & Abomohra, A. E.-F. (2019). Applications of non-destructive technologies for agricultural and food products quality inspection. *Sensors, 19*(4), 846.

Everard, C. D., Kim, M. S., & Lee, H. (2014). A comparison of hyperspectral reflectance and fluorescence imaging techniques for detection of contaminants on spinach leaves. *Journal of Food Engineering, 143*, 139–145.

Fiszman, S., & Tarrega, A. (2018). The dynamics of texture perception of hard solid food: A review of the contribution of the temporal dominance of sensations technique. *Journal of Texture Studies, 49*, 202–212. https://doi.org/10.1111/jtxs.12273

Foca, G. (2016). Imaging techniques: A rapid tool for food analysis. In *2nd IMEKOFOODS conference: Promoting objective and measurable food quality and safety 2016* (pp. 228–232) https://www.imeko.org/publications/tc23-2016/IMEKO-TC23-2016-046.pdf

Fowler, S. M., Schmidt, H., Van de Ven, R., Wynn, P., & Hopkins, D. L. (2014). Predicting tenderness of fresh ovine semimembranosus using Raman spectroscopy. *Meat Science, 97*(4), 597–601. https://doi.org/10.1016/j.meatsci.2014.02.018

Francis, F. J. (1995). Quality as influenced by color. *Food Quality and Preference, 6*(3), 149–155.

Giese, J. (2003). Color measurement in foods. *Food Technology, 57*(12), 48–49. +54.

Gigilashvili, D. (2021). *On the appearance of translucent objects: Perception and assessment by human observers*. https://ntnuopen.ntnu.no/ntnu-xmlui/handle/11250/2757506

Gökmen, V., & Mogol, B. A. (2010). Computer vision-based image analysis for rapid detection of acrylamide in heated foods: Rapid detection of acrylamide. *Quality Assurance and Safety of Crops & Foods, 2*(4), 203–207. https://doi.org/10.1111/j.1757-837X.2010.00072.x

Gowen, A. A., O'Donnell, C. P., Cullen, P. J., Downey, G., & Frias, J. M. (2007). Hyperspectral imaging–an emerging process analytical tool for food quality and safety control. *Trends in Food Science & Technology, 18*(12), 590–598.

Gowen, A. A., Tiwari, B. K., Cullen, P. J., McDonnell, K., & O'Donnell, C. P. (2010). Applications of thermal imaging in food quality and safety assessment. *Trends in Food Science and Technology, 21*(4), 190–200. https://doi.org/10.1016/j.tifs.2009.12.002

Guelpa, A., du Plessis, A., & Manley, M. (2016). A high-throughput X-ray micro-computed tomography (μCT) approach for measuring single kernel maize (Zea mays L.) volumes and densities. *Journal of Cereal Science, 69*, 321–328.

Guillermic, R. M., Aksoy, E. C., Aritan, S., Erkinbaev, C., Paliwal, J., & Koksel, F. (2021). X-Ray microtomography imaging of red lentil puffed snacks: Processing conditions, microstructure and texture. *Food Research International, 140*(September 2020), 109996. https://doi.org/10.1016/j.foodres.2020.109996

Hamann, D. D., Junhua, Z., Daubert, C. R., Foegedtng, E. A., & Diehl, K. C. (2006). Analysis of compression, tension and torsion for testing food gel fracture properties. *Journal of Texture Studies, 37*(6), 620–639. https://doi.org/10.1111/j.1745-4603.2006.00074.x

Herna, A., & He, Y. (2006). Non-destructive measurement of acidity, soluble solids and firmness of Satsuma mandarin using Vis / NIR-spectroscopy techniques. *Journal of Food Engineering, 77*, 313–319. https://doi.org/10.1016/j.jfoodeng.2005.06.036

Hu, J., Deng, J., & Sui, M. (2009). Color space conversion model from CMYK to LAB based on prism. In *2009 IEEE international conference on granular computing* (pp. 235–238) https://ieeexplore.ieee.org/abstract/document/5255123/

Jain, A., Pradhan, B. K., Mahapatra, P., Ray, S. S., Chakravarty, S., & Pal, K. (2021). Development of a low-cost food color monitoring system. *Color Research & Application, 46*(2), 430–445.

Jakubczyk, E., Gondek, E., & Tryzno, E. (2016). Application of novel acoustic measurement techniques for texture analysis of co-extruded snacks. *LWT - Food Science and Technology*. https://doi.org/10.1016/j.lwt.2016.10.013

Jeltem, M., Beckley, J., & Vahalik, J. (2015). Model for understanding consumer textural food choice. *Food Science and Nutrition, 3*(3), 202–212. https://doi.org/10.1002/fsn3.205

Jha, S. N., Narsaiah, K., Basediya, A. L., Sharma, R., Jaiswal, P., Kumar, R., & Bhardwaj, R. (2011). Measurement techniques and application of electrical properties for nondestructive quality evaluation of foods-a review. *Journal of Food Science and Technology, 48*(4), 387–411. https://doi.org/10.1007/s13197-011-0263-x

Ji, W., Koutsidis, G., Luo, R., Hutchings, J., Akhtar, M., Megias, F., & Butterworth, M. (2013). A digital imaging method for measuring banana ripeness. *Color Research & Application, 38*(5), 364–374. https://doi.org/10.1002/col.21741

Jiang, Q., Zhang, M., Mujumdar, A. S., & Wang, D. (2023). Non-destructive quality determination of frozen food using NIR spectroscopy-based machine learning and predictive modelling. *Journal of Food Engineering, 343*(December 2022), 111374. https://doi.org/10.1016/j.jfoodeng.2022.111374

Kamruzzaman, M., Barbin, D., ElMasry, G., Sun, D.-W., & Allen, P. (2012). Potential of hyperspectral imaging and pattern recognition for categorization and authentication of red meat. *Innovative Food Science & Emerging Technologies, 16*, 316–325.

Kamruzzaman, M., Elmasry, G., Sun, D., & Allen, P. (2013). Non-destructive assessment of instrumental and sensory tenderness of lamb meat using NIR hyperspectral imaging. *Food Chemistry, 141*(1), 389–396. https://doi.org/10.1016/j.foodchem.2013.02.094

Kim, D. H., Lee, J. G., Kwon, J. H., Kim, S. W., & Jun, H. B. (2023). Development of membrane filter colorimetry for determining tap water discoloration using a spectrocolorimeter. *Water, 15*(23), 4137.

Kohyama, K. (2015). Oral sensing of food properties. *Journal of Texture Studies, 46*(3), 138–151. https://doi.org/10.1111/jtxs.12099

Kohyama, K. (2020). Chapter 1: Food texture – Sensory evaluation and instrumental measurement. In *Textural characteristics of world foods*. https://doi.org/10.1002/9781119430902.ch1

Konopacka, D., & Plocharski, W. J. (2004). Effect of storage conditions on the relationship between apple firmness and texture acceptability. *Postharvest Biology and Technology, 32*(2), 205–211. https://doi.org/10.1016/j.postharvbio.2003.11.012

Kwon, J.-S., Lee, J.-M., & Kim, W.-Y. (2008). Real-time detection of foreign objects using X-ray imaging for dry food manufacturing line. In *2008 IEEE international symposium on consumer electronics* (pp. 1–4) https://ieeexplore.ieee.org/abstract/document/4559552/

Lim, K. S., & Barigou, M. (2004). X-ray micro-computed tomography of cellular food products. *Food Research International, 37*(10), 1001–1012. https://doi.org/10.1016/j.foodres.2004.06.010

Liu, Y., Pu, H., & Sun, D. W. (2017). Hyperspectral imaging technique for evaluating food quality and safety during various processes: A review of recent applications. *Trends in Food Science and Technology, 69*, 25–35. https://doi.org/10.1016/j.tifs.2017.08.013

Lu, R. (2013). Principles of solid food texture analysis. In *Instrumental assessment of food sensory quality: A practical guide* (pp. 103–128). https://doi.org/10.1533/9780857098856.1.103

Lu, J., Tan, J., Shatadal, P., & Gerrard, D. E. (2000). Evaluation of pork color by using computer vision. *Meat Science, 56*(1), 57–60.

Luo, W., Zhang, H., & Liu, X. (2019). Hyperspectral/multispectral reflectance imaging combining with watershed segmentation algorithm for detection of early bruises on apples with different peel colors. *Food Analytical Methods, 12*(5), 1218–1228. https://doi.org/10.1007/s12161-019-01456-0

MacDougall, D. B. (2001). Principles of colour measurement for food. In *Instrumentation and sensors for the food industry* (Second ed.). Woodhead Publishing Limited. https://doi.org/10.1533/9781855736481.1.63

Macdougall, D. B. (2010). Colour measurement of food: Principles and practice. In *Colour measurement: Principles, advances and industrial applications*. Woodhead Publishing Limited. https://doi.org/10.1533/9780857090195.2.312

Magwaza, L. S., & Tesfay, S. Z. (2015). A review of destructive and non-destructive methods for determining avocado fruit maturity. *Food and Bioprocess Technology, 8*(10), 1995–2011. https://doi.org/10.1007/s11947-015-1568-y

Mancini, M., Mazzoni, L., Gagliardi, F., Balducci, F., Duca, D., Toscano, G., Mezzetti, B., & Capocasa, F. (2020). Application of the non-destructive NIR technique quality parameters. *Food, 9*, 441.

Mao, H., Kumi, F., Li, Q., & Han, L. (2016). Combining X-ray computed tomography with relevant techniques for analyzing soil–root dynamics – An overview. *Acta Agriculturae Scandinavica, Section B - Soil & Plant Science, 66*(1), 1–19. https://doi.org/10.1080/09064710.2015.1067711

Markovic, I., Ilic, J., Markovic, D., Simonovic, V., & Kosanic, N. (2013). Color measurement of food products using CIE L* a* b* and RGB color space. *Journal of Hygienic Engineering and Design, 4*, 50–53.

Matsui, T., Sugimori, H., Koseki, S., & Koyama, K. (2023). Automated detection of internal fruit rot in Hass avocado via deep learning-based semantic segmentation of X-ray images. *Postharvest Biology and Technology, 203*(April), 112390. https://doi.org/10.1016/j.postharvbio.2023.112390

Mishra, G., Pandiselvam, R., Bhati, D., & Mahanti, N. K. (2023). Emerging nondestructive techniques to quantify the textural properties of food: A state-of-art review. *Journal of Texture Studies, 54*(2), 173–205. https://doi.org/10.1111/jtxs.12741

Modupalli, N., Naik, M., Sunil, C. K., & Natarajan, V. (2021). Emerging non-destructive methods for quality and safety monitoring of spices. *Trends in Food Science & Technology, 108*(June 2020), 133–147. https://doi.org/10.1016/j.tifs.2020.12.021

Mogol, B. A., & Gökmen, V. (2014). Computer vision-based analysis of foods: A non-destructive colour measurement tool to monitor quality and safety: Computer vision based analysis of foods. *Journal of the Science of Food and Agriculture, 94*(7), 1259–1263. https://doi.org/10.1002/jsfa.6500

Mohd, M., & Hashim, N. (2022). Non-destructive methods for detection of food quality. In *Future foods*. Elsevier Inc. https://doi.org/10.1016/B978-0-323-91001-9.00003-7

Moiny, V., Meullenet, J. F., & Xiong, R. (2002). Uniaxial compression of Cheddar cheese at various loading rates and its correlation to sensory texture profiles. *Journal of Texture Studies, 33*(3), 237–254. https://doi.org/10.1111/j.1745-4603.2002.tb01347.x

Molada-Tebar, A., Verhoeven, G. J., Hernández-López, D., & González-Aguilera, D. (2024). Practical RGB-to-XYZ color transformation matrix estimation under different lighting conditions for graffiti documentation. *Sensors, 24*(6), 1743.

Nishinari, K., & Fang, Y. (2018). Perception and measurement of food texture: Solid foods. *Journal of Texture Studies, 49*(2), 160–201. https://doi.org/10.1111/jtxs.12327

Olakanmi, S., Karunakaran, C., & Jayas, D. (2023). Applications of X-ray micro-computed tomography and small-angle X-ray scattering techniques in food systems: A concise review. *Journal of Food Engineering, 342*(August 2022), 111355. https://doi.org/10.1016/j.jfoodeng.2022.111355

Oraguzie, N., Alspach, P., Volz, R., Whitworth, C., Ranatunga, C., Weskett, R., & Harker, R. (2009). Postharvest assessment of fruit quality parameters in apple using both instruments and an expert panel. *Postharvest Biology and Technology, 52*(3), 279–287. https://doi.org/10.1016/j.postharvbio.2009.01.004

Pandiselvam, R., Mitharwal, S., Rani, P., Shanker, M. A., Kumar, A., Aslam, R., Barut, Y. T., Kothakota, A., Rustagi, S., & Bhati, D. (2023). The influence of non-thermal technologies on color pigments of food materials: An updated review. *Current Research in Food Science, 6*, 100529.

Peyron, M., Woda, A., Bourdiol, P., & Hennequin, M. (2017). Age-related changes in mastication. *Journal of Oral Rehabilitation, 44*, 299–312. https://doi.org/10.1111/joor.12478

Pierna, J. F., Vermeulen, P., Amand, O., Tossens, A., Dardenne, P., & Baeten, V. (2012). NIR hyperspectral imaging spectroscopy and chemometrics for the detection of undesirable substances in food and feed. *Chemometrics and Intelligent Laboratory Systems, 117*, 233–239.

Pointer, M. R. (2003). *Measuring visual appearance-a framework of the future. Project 2.3 measurement of appearance.* https://eprintspublications.npl.co.uk/2792/

Pradeepika, M., et al. (2019). Special issue of National Conference on innovations in bio, chemical and food technology (IBCFT -2019). *American International Journal of Research in Science, Technology, Engineering & Mathematics*, 388–394.

Rahman, M. F., Iqbal, A., Hashem, M. A., & Adedeji, A. A. (2020). Quality assessment of beef using computer vision technology. *Food Science of Animal Resources, 40*(6), 896.

Rodríguez-Pulido, F. J., Barbin, D. F., Sun, D.-W., Gordillo, B., González-Miret, M. L., & Heredia, F. J. (2013). Grape seed characterization by NIR hyperspectral imaging. *Postharvest Biology and Technology, 76*, 74–82.

Ropodi, A. I., Panagou, E. Z., & Nychas, G. E. (2016). Data mining derived from food analyses using non-invasive/non-destructive analytical techniques; determination of food authenticity, quality & safety in tandem with computer science disciplines. *Trends in Food Science & Technology*. https://doi.org/10.1016/j.tifs.2016.01.011

Salman, M., Farooq, U., Dar, A., Haq, I., Burhan, T., Anwar, J., et al. (2023). Determination of cyanide at trace levels by computational scanning densitometry. *Current Analytical Chemistry, 19*(6), 466–471.

Sanchez, P. D. C., Hashim, N., Shamsudin, R., & Nor, M. Z. M. (2020). Applications of imaging and spectroscopy techniques for non-destructive quality evaluation of potatoes and sweet potatoes: A review. *Trends in Food Science & Technology, 96*, 208–221.

Sarkar, T., Salauddin, M., Kirtonia, K., Pati, S., Rebezov, M., Khayrullin, M., Panasenko, S., Tretyak, L., Temerbayeva, M., Kapustina, N., Azimova, S., Gruzdeva, L., Makhmudov, F., Nikitin, I., Kassenov, A., Shariati, M. A., & Lorenzo, J. M. (2022). A review on the commonly used methods for analysis of physical properties of food materials. *Applied Sciences (Switzerland), 12*(4). https://doi.org/10.3390/app12042004

Schanda, J. (2007). *CIE Colorimetry: Understanding the CIE system* (Vol. 3, pp. 25–78). John Wiley & Sons.

Schoeman, L., Williams, P., & Anton du Plessis, M. M. (2015). X-ray micro-computed tomography (μCT) for non-destructive characterisation of food microstructure. https://doi.org/10.1016/j.tifs.2015.10.016.

Schubring, R. (2009). Quality assessment of fish and fishery products by color measurement. In *Handbook of seafood and seafood products analysis* (pp. 413–442). CRC Press.

Shankaragowda, B. B., & Latha, K. B. (2023). Computer vision based food recognition with nutrition analysis. *International Journal of Scientific Research in Engineering and Management. 7*(7)

Shrestha, Y. K., & Shrestha, S. K. (2023). *Fundamentals of colorimetry*. IntechOpen.

Silva, S., Guedes, C., Rodrigues, S., & Alimentos, A. T. (2020). Non-destructive imaging and spectroscopic techniques for assessment of carcass and meat quality in sheep and goats: A review. *Food, 9*(1074), 1–22. https://www.mdpi.com/791764

Siswantoro, J. (2019). Application of color and size measurement in food products inspection. *Indonesian Journal of Information Systems, 1*(2), 90–107. https://doi.org/10.24002/ijis.v1i2.1923

Spence, C. (2019). On the relationship(s) between color and taste/flavor. *Experimental Psychology, 66*(2), 99–111. https://doi.org/10.1027/1618-3169/a000439

Sun, D.-W. (2009). *Infrared spectroscopy for food quality analysis and control.* Academic.

Takhar, P. S., Maier, D. E., Campanella, O. H., & Chen, G. (2011). Hybrid mixture theory based moisture transport and stress development in corn kernels during drying: Validation and simulation results. *Journal of Food Engineering, 106*(4), 275–282.

Viejo, C. G., Fuentes, S., Li, G., Collmann, R., Condé, B., & Torrico, D. (2016). Development of a robotic pourer constructed with ubiquitous materials, open hardware and sensors to assess beer foam quality using computer vision and pattern recognition algorithms: RoboBEER. *Food Research International, 89*, 504–513.

Wang, Y., & Zhou, B. (2022). Recent progress in single and combined porosity-evaluation techniques for porous materials. *Materials, 15*(9), 2981.

Wang, F., Zhao, C., & Yang, G. (2020). Development of a non-destructive method for detection of the juiciness of pear via vis/nir spectroscopy combined with chemometric methods. *Food, 9*(12). https://doi.org/10.3390/foods9121778

Wrolstad, R. E., & Smith, D. E. (2010). Color analysis. *In Food analysis* (pp. 573-586). Boston, MA: Springer US.

Wu, D., & Sun, D. W. (2013). Colour measurements by computer vision for food quality control–a review. *Trends in Food Science & Technology, 29*(1), 5–20.

Wu, X., Liang, X., Wang, Y., Wu, B., & Sun, J. (2022). Non-destructive techniques for the analysis and evaluation of meat quality and safety: A review. *Foods, 11*(22), 3713,1-30.

Xiao, Z., Wang, J., Han, L., Guo, S., & Cui, Q. (2022). Application of machine vision system in food detection. *Frontiers in Nutrition, 9*, 888245.

Yu, H., & MacGregor, J. F. (2003). Multivariate image analysis and regression for prediction of coating content and distribution in the production of snack foods. *Chemometrics and Intelligent Laboratory Systems, 67*(2), 125–144.

Zapotoczny, P., & Majewska, K. (2010). A comparative analysis of colour measurements of the seed coat and endosperm of wheat kernels performed by various techniques. *International Journal of Food Properties, 13*(1), 75–89.

Zdunek, A., Konopacka, D., & Jesionkowska, K. (2010). Crispness and crunchiness judgment of apples based on contact acoustic emission. *Journal of texture studies, 41*(1), 75–91.

Zhang, W., Cui, D., & Ying, Y. (2014). Nondestructive measurement of pear texture by acoustic vibration method. *Postharvest Biology and Technology, 96*, 99–105. https://doi.org/10.1016/j.postharvbio.2014.05.006

Zhang, B., Fan, S., Li, J., Huang, W., Zhao, C., Qian, M., & Zheng, L. (2015). Detection of early rottenness on apples by using hyperspectral imaging combined with spectral analysis and image processing. *Food Analytical Methods, 8*(8), 2075–2086. https://doi.org/10.1007/s12161-015-0097-7

Zhu, F., Zhang, D., He, Y., Liu, F., & Sun, D.-W. (2013). Application of visible and near infrared hyperspectral imaging to differentiate between fresh and frozen–thawed fish fillets. *Food and Bioprocess Technology, 6*(10), 2931–2937. https://doi.org/10.1007/s11947-012-0825-6

Chapter 17
AI and Machine Learning Applications

Vinass Jamali, Vismaya K. Sachithanandhan, Eyarkai Nambi, Loganathan M., S. Shanmugasundaram, and V. Chandrasekar

17.1 Introduction to AI

John McCarthy, known as the father of artificial intelligence and the one who coined the term "Artificial Intelligence" in 1956, described it as the fusion of science and engineering aimed at creating intelligent devices for human benefit. AI encompasses the field of artificial intelligence science, utilizing computers to mimic human intelligence and training them to replicate human behaviours such as learning, judgment, and decision-making (Xu et al., 2021). It is a knowledge-centric endeavour that focuses on acquiring, analysing, and expressing knowledge to emulate human intellectual activities (Duan & Xu, 2012). AI integrates various disciplines including computer science, logic, biology, psychology, and philosophy, and has demonstrated significant achievements in areas like speech recognition, image processing, natural language processing, theorem proving, and intelligent robotics (Duan et al., 2009). AI plays a crucial role in societal advancement, offering revolutionary improvements in labour efficiency, cost reduction, human resource optimization, and the creation of new job opportunities (Duan et al., 2019).

Artificial General Intelligence (AGI) refers to a system capable of performing intellectual tasks just like humans, handling multiple processes simultaneously

V. Jamali · V. K. Sachithanandhan · E. Nambi (✉) · S. Shanmugasundaram
Department of Food Plant Operations, Incubation and Entrepreneurship, National Institute of Food Technology, Entrepreneurship and Management – Thanjavur, (NIFTEM-T), Thanjavur, Tamil Nadu, India
e-mail: eyarkainambi@iifpt.edu.in

Loganathan M.
Departmentof Food Packaging and Storage Technology, National Institute of Food Technology, Entrepreneurship and Management – Thanjavur, (NIFTEM-T), Thanjavur, Tamil Nadu, India

V. Chandrasekar
Department of Food Process Technology, National Institute of Food Technology, Entrepreneurship and Management – Thanjavur, (NIFTEM-T), Thanjavur, Tamil Nadu, India

C. K. Sunil et al. (eds.), *Emerging Non-destructive Technologies for Food Quality Analysis*, https://doi.org/10.1007/978-3-031-95048-3_17

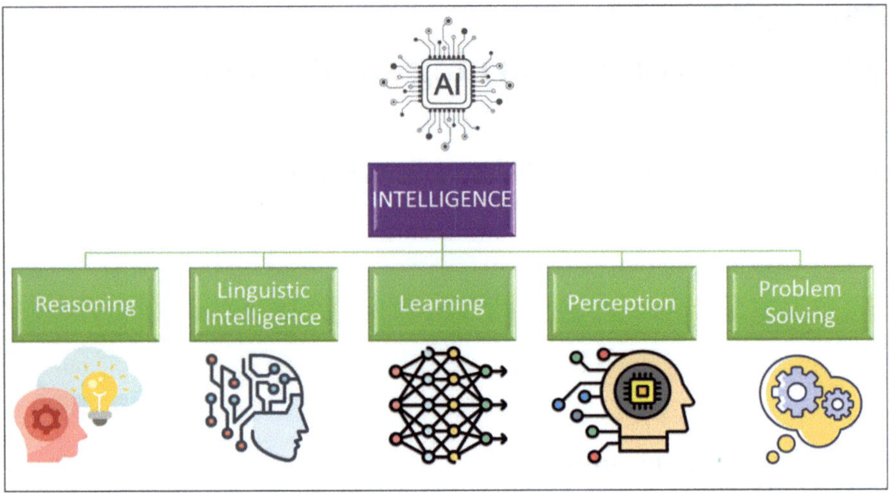

Fig. 17.1 Sectors in Artificial Intelligence (AI)

(Fig. 17.1). It also encompasses linguistic logic and reasoning (Russell & Norvig, 2016). Artificial intelligence can be categorized into two types:

1. Weak AI: The concept of Weak AI is that machines act as though they are intelligent. Weak AI demonstrates that virtual abilities such as thinking, speaking, and moving can be achieved by machines if they are programmed accordingly. For example, in a chess game, the computer can play and move pieces automatically. The computer doesn't actually think but is programmed to make the correct moves.
2. Strong AI: The concept of Strong AI is that machines will be able to perform calculations, think independently, and predict future outcomes. For instance, IBM's artificial intelligence supercomputer "WATSON" exemplifies Strong AI. In the future, we may see machines or humanoids capable of performing tasks and thinking more powerfully than humans.

Artificial intelligence is an interdisciplinary field involving information, logic, cognition, thinking, systems, and biology. It has applications in knowledge processing, pattern recognition, machine learning, and natural language processing. AI has been used in automatic programming, expert systems, knowledge systems, and intelligent robots. Besides logical thinking and imitation, emotion is also a crucial aspect of AI. The next breakthrough in AI could give computers emotional capabilities in addition to enhanced logical reasoning. Machine intelligence may soon surpass human intelligence (Zhang & Lu, 2021) .

17.2 Fundamental Differences Between Biological and Artificial Intelligence

We often conceptualize intelligence with an anthropocentric bias, using our own intelligence as the obvious and clear reference point. This framework is typically applied when reasoning about other, less familiar forms of intelligence, including various biological and artificial intelligences (Coley & Tanner, 2012). This perspective can lead to intriguing questions and ideas, such as discussions about when machines will achieve "intelligence at human level." For example, before attaining superintelligence, general AI will have cognitive capabilities equivalent to those of a human. Consequently, researchers extensively debate the timeline for reaching general AI (Goertzel, 2007). However, we argue that these questions may not be entirely accurate. There are potentially many different forms of general intelligence, of which human-like intelligence is just one. This suggests that AI development is governed by the constraints of physics and technology, rather than biological evolution. Just as a hypothetical extra-terrestrial visitor might possess a different inorganic structure with distinct characteristics compared to humans, artificial forms of general intelligence will also differ. Below, we summarize some fundamental differences between human and artificial intelligence (Bostrom, 2014).

- Basic structure: Biological (carbon-based) intelligence relies on neural "wetware," which is fundamentally different from artificial (silicon-based) intelligence. In silicon or digital systems, "hardware" and "software" are separate entities (Kosslyn & Koenig, 1992). When a biological system learns a new skill, it is intrinsic to that system. In contrast, an AI system's learned algorithms can be directly copied to all other similar digital systems.
- Speed: Signals in AI systems propagate at nearly the speed of light, whereas in humans, nerve conduction velocity is a maximum of 120 m/s, which is extremely slow compared to computer timescales (Siegel & Sapru, 2006).
- Connectivity and communication: Humans cannot communicate directly and rely on language and gestures, which have limited bandwidth and are slower and more challenging than the direct communication between AI systems. AI systems can be directly connected, enabling them to collaborate through integrated algorithms.
- Updatability and scalability: AI systems can be updated, scaled, and reconfigured with minimal constraints, ensuring they have the necessary algorithms and data processing/storage capacities for their tasks. This rapid structural expansion and improvement capability is not applicable to humans.
- Energy efficiency: Organic brains are significantly more energy-efficient than computers. The human brain consumes less energy than a light bulb, while a supercomputer with comparable computational performance requires enough electricity to power a village (Fischetti, 2011).

These differences in basic structure, speed, connectivity, updatability, scalability, and energy consumption lead to distinct qualities and limitations between human and artificial intelligence. For instance, human response speed to simple stimuli is thousands of times slower than that of artificial systems. AI systems can easily connect and function as an integrated system, reducing the risk of errors due to miscommunication (e.g., autonomous vehicles at an intersection), as they are parts of the same system and algorithm (Lee et al., 2016).

17.3 The Origin and Development of Artificial Intelligence

Artificial intelligence involves studying how to enable computers to perform tasks that previously required human intelligence (Huang et al., 2019c). In recent years, AI has advanced rapidly, significantly altering people's lifestyles (Huang et al., 2019a). The development of AI has become a crucial strategy for countries globally, enhancing national competitiveness and ensuring security (Rajakumar et al., 2018). To gain an edge in the new wave of international competition, many countries have introduced favourable policies and reinforced the focus on key technologies and talent (Xu et al., 2008). AI has emerged as a prominent research area in science and technology, with major companies like Google, Microsoft, and IBM dedicated to AI and increasingly applying it across various fields (Shi et al., 2007).

AI is a multidisciplinary technology capable of integrating cognition, machine learning, emotion recognition, human-computer interaction, data storage, and decision-making (Lu, 2019). First proposed by John McCarthy at the Dartmouth Conference in the mid-twentieth century, AI has achieved significant milestones since 1993. The rapid development of neural networks was driven by the wide application of the BP algorithm. In large-scale environments, the extensive use of expert systems has significantly reduced costs and improved efficiency across industries (Cai-Ming & Hao-Nan, 2020). For instance, the PROSPECTOR expert system successfully analysed mineral deposits worth hundreds of millions of dollars. Following this, efforts to develop general artificial intelligence programs faced serious obstacles and reached a deadlock, leading AI into a period of stagnation. However, in 1997, the success of "Deep Blue" revitalized AI development. Enhanced computing power broke AI's bottleneck, enabling the growth of deep learning and reinforcement learning based on big data. The continuous development of GPUs and the successful creation of custom processors have further improved computing power, laying the foundation for AI's explosive growth (Zhang, 2019).

Artificial intelligence has undergone a lengthy development process spanning over 70 years, divided into several stages. In 1943, the artificial neuron model was proposed, initiating the era of artificial neural network research (Duan & Xu, 2012). The early twenty-first century saw the mobile Internet bring more AI application scenarios (Zhang & Chen, 2020). In 2012, the proposal of deep learning led to significant breakthroughs in AI, particularly in speech and visual recognition.

17.4 Enabling Drivers and Technologies of Artificial Intelligence/Concepts

17.4.1 Big Data

Big data is essential for AI, serving as a key factor in enhancing AI's recognition rate and accuracy. With the development and widespread adoption of the Internet of Things, the volume of generated data has grown exponentially, exhibiting a significant annual growth rate. Alongside this increase in volume, the dimensionality of the data has also expanded (Zhang, 2020). This abundance of high-dimensional data makes the data more comprehensive and robust, thereby supporting AI development more effectively (Chen & Lin, 2014).

17.4.2 Algorithms

In traditional pattern recognition, researchers formulated laws and methods, but this abstraction approach had significant limitations and low accuracy. Observing that babies learn to recognize objects without being taught inspired researchers to propose a machine learning method to identify objects (Ling X Li, 1999a, b; Yang et al., 2001). For example, by inputting numerous photos of dogs into a computer, the computer can learn the characteristics of dogs through training models like neural networks and accurately recognize dogs in other photos based on these learned characteristics (Zhang and Fu 2021).

Machine learning allows computers to automatically learn from and analyse big data, enabling them to make decisions and predictions about real-world events. Algorithms are the foundation of AI (Chi-Hsien & Nagasawa, 2019). Beyond pattern recognition, AI-relevant algorithms have also yielded impressive results in fields such as speech recognition, search engines, semantic analysis, and recommendation systems (Zhang et al., 2019a). Each of these areas has made significant advancements driven by AI algorithms.

17.4.3 Machine Learning

The fundamental concept of machine learning involves using an algorithm that enhances its performance by learning from data (Nilsson, 1982). The four primary types of problems addressed by machine learning are prediction, clustering, classification, and dimensionality reduction (Erhan et al., 2010). In terms of learning method classification, machine learning can be categorized into four types: supervised learning, unsupervised learning, semi-supervised learning, and reinforcement learning (Bose, 2017). Table 17.1 lists out the Machine learning models and their training algorithms.

Table 17.1 Machine learning models and their training algorithms (Lawton, 2024)

Supervised learning	Unsupervised learning	Semi-supervised learning	Reinforcement learning
Data scientists contribute by providing input, output, and feedback to develop models. **EXAMPLE ALGORITHMS** **Linear regressions** Sales forecasting Risk assessment **Support vector machines** Image classification Financial performance comparison **Decision trees** Predictive analytics Pricing	Utilize deep learning to derive conclusions and identify patterns from unlabelled training data. **EXAMPLE ALGORITHMS** **Apriori** Sales functions. Word associations. Searcher. **K-means clustering** Performance monitoring Searcher intent **Artificial neural networks** Generate new, synthetic data. Data mining and pattern recognition.	It builds a model using a combination of labelled and unlabelled data, a set of categories, suggested labels, and example labels. **EXAMPLE ALGORITHMS** **Generative adversarial networks** Audio and video manipulation. Data creation. **Self-trained Naive Bayes classifier** Natural language processing	Self-interpreting yet based on a system of rewards and punishments learned through trial and error, seeking to maximize reward. **EXAMPLE ALGORITHMS** **Q-learning** Policy creation. Consumption reduction. **Model-based value estimation** Linear tasks Estimating parameters

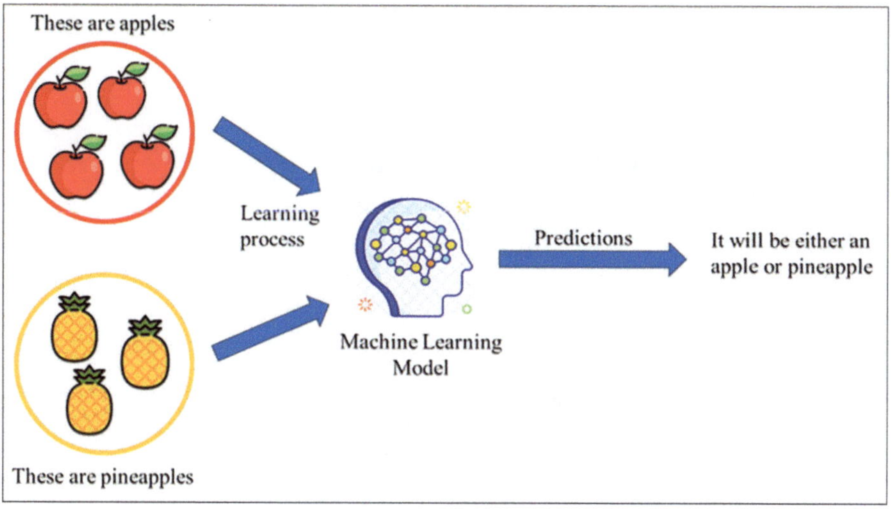

Fig. 17.2 Supervised learning

Supervised learning involves using labelled data to train models in order to predict the type or value of new data (Fig. 17.2). Based on the prediction results, supervised learning can be divided into two categories: classification and regression. Typical methods of supervised learning include SVM (Support Vector Machine) and linear discrimination (Najafabadi et al., 2015; Neyshabur et al., 2017).

Regression involves predicting continuous output values; for example, analysing housing price data, fitting it to sample data inputs, and obtaining a continuous curve to forecast housing prices. Classification, on the other hand, involves predicting discrete output values, such as determining whether a photo depicts a dog or a cat based on various features, with the output being 1 or 0 (Morocho-Cayamcela et al., 2019).

When data lacks labels, unsupervised learning becomes data mining. Unsupervised learning primarily involves clustering, where data is classified based on different characteristics without tags (Fig. 17.3). Typical methods of unsupervised learning include k-clustering and principal component analysis. K-clustering relies on the premise that differences between data points can be measured using Euclidean distance, and if not, they need to be converted to a usable Euclidean distance. Principal component analysis is a statistical technique that uses orthogonal transformation to convert correlated variables into uncorrelated variables, known as principal components. The main idea is to replace the original correlated indicators with a set of independent comprehensive indicators (Baryannis et al., 2019).

Semi-supervised learning, in its most literal sense, is a combination of supervised and unsupervised learning (Fig. 17.4). It involves mixing labelled and unlabelled data in the learning process. Typically, the amount of unlabelled data is much greater

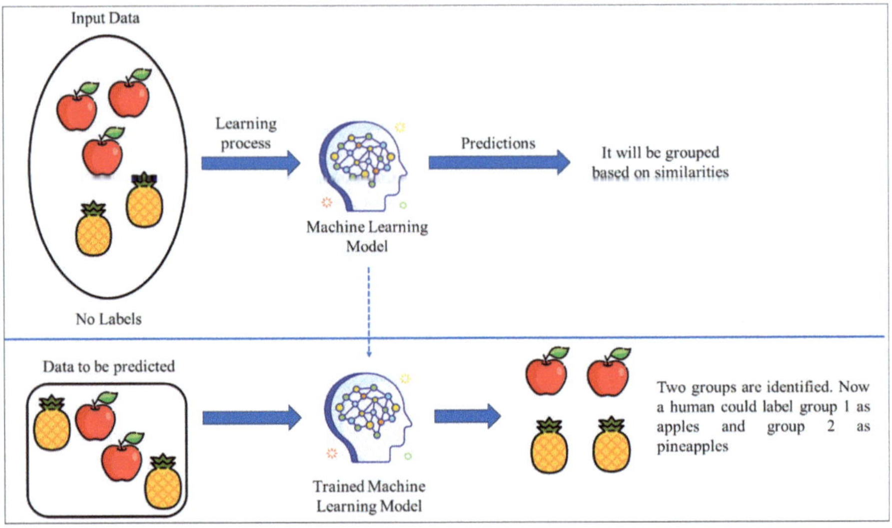

Fig. 17.3 Unsupervised learning

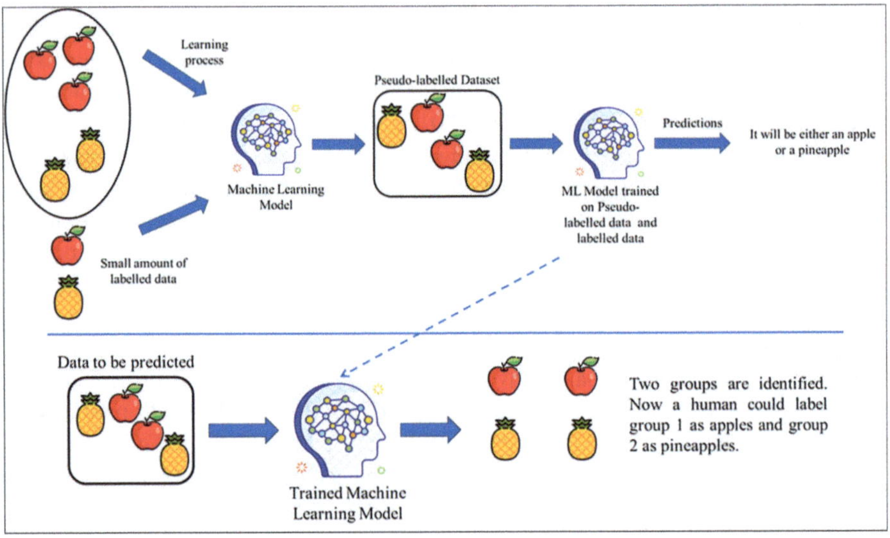

Fig. 17.4 Semi-supervised learning

than the labelled data. Although the concept of semi-supervised learning is appealing, it is not widely used in practical applications. Common semi-supervised learning algorithms include self-training, graph-based semi-supervised learning, and semi-supervised support vector machines (S3VM).

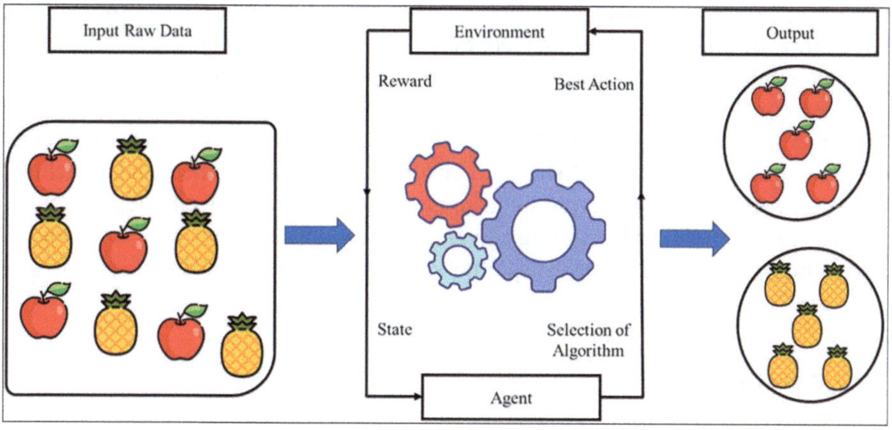

Fig. 17.5 Reinforcement learning

Reinforcement learning is a method where an agent interacts with the environment to receive rewards, assesses the quality of its actions based on these rewards, and then trains the model accordingly (Fig. 17.5). A key challenge in reinforcement learning is balancing exploration and exploitation: to achieve better rewards, one must choose actions that may yield the highest reward while also exploring unknown actions (Chollet, 2017). The principles of reinforcement learning are rooted in behavioral psychology. In 1911, Thorndike proposed the law of effect: actions leading to a comfortable state are likely to be repeated, while those causing discomfort are likely to be reduced (Collins & Moons, 2019). Essentially, reinforcement learning enhances rewarding behaviors and diminishes punishing behaviors. By using a trial-and-error mechanism, the model can be trained to find the optimal actions and behaviors to maximize returns. This approach mimics the learning processes of humans and animals, allowing agents to learn without specific guidance (Li et al., 2007; Qi et al., 2007; Wang et al., 2013).

17.4.4 National Language Processing (NLP)

Natural language processing (NLP) refers to the ability of computers to recognize and understand human text language, an interdisciplinary field that bridges computer science and human linguistics. A distinctive feature of natural language is that human thought is based on language, making NLP a significant goal of AI. NLP is divided into seven areas: grammatical and semantic analysis, information extraction, text mining, information retrieval, machine translation, question answering systems, and dialog systems (Zhang et al., 2019b).

Natural language understanding (NLU) is a technology that uses natural language to interact with computers. The key to processing natural language is enabling computers to "understand" it, which is why it is also known as computational linguistics, positioned at the intersection of language information processing and artificial intelligence. Initially, as a machine, it collects sound signals. NLP technology then converts these sound signals into text signals and interprets their meaning. The process involves converting sounds into words and words into meanings. Once these two processes are complete, the machine can hear and understand. Machines equipped with speech recognition and semantic understanding technologies continuously optimize algorithms through learning, allowing them not only to listen but also to understand—and even grasp emotions (Bostrom et al., 2014).

17.4.5 Hardware

In machine learning, "deep" neural network models are utilized to tackle complex problems. Machine learning is a method to achieve AI, and deep learning is a subset of machine learning (Makkar et al., 2021). The primary hardware platform for running deep learning is the GPU, primarily produced by NVIDIA. GPU acceleration employs massively parallel processors to enhance applications with parallel functions. While it previously took a CPU a month to obtain training results, a GPU can now achieve the same results in a day. The powerful parallel computing capabilities of GPUs alleviate the training bottleneck for deep learning algorithms, thereby unlocking the potential of artificial intelligence (Brynjolfsson & Mcafee, 2017).

17.4.6 Computing Vision

The aim of computer vision is to enable computers to perceive and comprehend the world through visual input, similar to human vision. This is primarily achieved through algorithms that identify and analyse images. The most common applications of computer vision are facial recognition and image recognition (Tan et al., 2019b).

Since 2015, image processing methods utilizing deep learning for classification have become widespread. Neural networks, constructed through interconnected neurons, employ activation functions to achieve a nonlinear fitting function (Dhingra et al., 2016). Once the input and output requirements for the model are established, the model can automatically learn the processes of feature extraction and classification training. Deep learning simplifies the most time-consuming and labour-intensive aspects of image classification, enhancing both its effectiveness and efficiency. The VGG convolutional neural network, ResNet (Residual Neural Network), and Startup are among the most commonly used structures in engineering. Projects must consider both efficiency and effectiveness, ensuring accuracy while maintaining speed. After training, models are fine-tuned and optimized. Faster R-CNN, Mask-RCNN, and YOLO (You Only Look Once) are popular network models known for their high accuracy and speed. In facial recognition, these models can detect and produce results in real time (Davenport & Ronanki, 2018).

Computer vision breaks down images into pixels and processes them. Semantic segmentation aims to understand the meaning of these segmented pixels, enabling the recognition of objects such as people, motorcycles, cars, and streetlights within an image. This requires distinguishing densely packed pixels. Convolutional neural networks have significantly advanced semantic segmentation algorithms (Zhao et al., 2019b). The basic approach in semantic segmentation involves classification prediction using a sliding window. The development of the Fully Convolutional Network (FCN) eliminated the need for the fully connected layer in networks. Building on FCN, the Encoder-Decoder architecture was created, where the decoder compensates for the encoder's inability to handle smaller spatial sizes by restoring spatial size and detailed information. Subsequently, the pooling operation was replaced by Dilated/Atrous convolution, which maintains spatial resolution. Additionally, the Conditional Random Field (CRF) method can further enhance segmentation outcomes (Zhao et al., 2019a).

17.5 The Research Fields of AI

The AI is applied in different research fields or areas, such as the expert system, machine learning, robotics, intelligent decision support system and pattern recognitions (Table 17.2). Application of AI in these fields is discussed in the below section.

17.5.1 The Expert System

An expert system is a knowledge-based system that relies on the expertise of human specialists. It was one of the earliest fields of AI research and is extensively used in areas such as medical diagnosis, geological surveying, and the petrochemical industry. Expert systems generally refer to various knowledge systems (Miotto et al., 2018). These intelligent computer programs utilize professional knowledge provided by human experts to simulate their thought processes, using knowledge and reasoning to solve complex problems typically handled by domain experts. An expert system contains a substantial amount of information and reasoning capabilities in a specific field, encompassing extensive professional knowledge and experience, and can simultaneously store, reason, and make judgments (Tung, 2019). Its core components are the knowledge base and the reasoning engine (Cai et al., 2018). The application of an expert system involves first storing the knowledge, experience, and research information of experts in a specific field within a database and knowledge base. These can then be accessed by the interpreter and reasoning engine and provided to experts as needed, facilitated through a human-computer interaction interface (Zappone et al., 2019). The advantage of using expert systems in education

Table 17.2 Research fields of AI

Domain	Justification
Expert system	An expert system is an intelligent computer software that possesses professional knowledge and experience. It mimics complex problems that are typically solved by experts by simulating human experts' problem-solving abilities through knowledge representation and knowledge reasoning approaches in artificial intelligence. It is capable of addressing problems to the same degree as a professional.
Machine learning	The field of machine learning encompasses various fields such as probability theory, statistics, approximation theory, convex analysis, algorithm complexity theory, and more. Machine learning is the study of how computers mimic or actualize human learning tendencies to pick up new abilities or knowledge, and how they rearrange the current knowledge structure to keep getting better at what they do.
Robotics	The science that deals with the development, implementation, and upkeep of robots is known as robotics. It primarily investigates the connection between object processing and robot control.
Decision support system	Decision support systems are closely related to "knowledge-intelligence" and are within the management science category.
Pattern recognition	The field of pattern recognition studies ways to provide robots perceptive abilities. It focuses mostly on the identification of auditory and visual patterns.

is that they are not constrained by time, space, environment, or emotional factors. They are particularly beneficial for distance education, where their widespread use and advantages are well recognized (Pouyanfar et al., 2018).

17.5.2 Machine Learning

For a computer to possess knowledge, it must either receive this knowledge as input in a form acceptable to the computer, or the computer must have the capability to acquire, continuously summarize, and improve knowledge through practice (Young et al., 2018). This approach is known as machine learning (Liu et al., 2020a). The research in machine learning focuses on (1) studying human learning mechanisms and brain thought processes, (2) exploring human learning methods, and (3) developing machine learning techniques to create learning systems for specific tasks (Liu et al., 2015). This research draws on various disciplines, including information science, brain science, neuropsychology, logic, and fuzzy mathematics (Hu et al., 2021).

The concept of deep learning originates from artificial neural networks. Common deep learning algorithms include the Restricted Boltzmann Machine (RBM), the deep belief network (DBN), the convolutional neural network (CNN), and the stacked auto-encoder (Litjens et al., 2017). In early artificial neural networks, issues arose when the network layer exceeded four layers. The traditional backpropagation algorithm is used to train convergence, and a multilayer perceptron with multiple hidden layers constitutes a deep learning structure. Deep learning identifies the distribution characteristics of data by combining low-level features to create high-level attribute categories or features (Shokri et al., 2015; Zhang et al., 2019c). Notable artificial neural network algorithms include the perceptron neural network (PNN), backpropagation (BP), the self-organizing network (SON), the self-organizing map (SOM), and learning vector quantization (LVQ) (Jamshidi et al., 2020; Yuan et al., 2010).

17.5.3 The Robotics

A robot is a machine capable of simulating human behavior, and research into robotics has gone through three generations of development (Liu et al., 2018).

The first generation involves program-controlled robots. These robots are programmed by designers, with the instructions stored in the robot to execute tasks under program control. Before the robot performs a task for the first time, a technician instructs it step by step (Xing Li & Li, 1999; Lu et al., 2018a; Yang et al., 2014). Each recorded operation is represented as an instruction. The second generation comprises adaptive robots (Zhang et al., 2019d). These robots are equipped with sensory sensors (such as vision, hearing, and tactile sensors) that allow them to acquire simple information about their environment and operating objects. This information is processed by a computer to control the robot's activities. The third generation consists of intelligent robots (Shone et al., 2018). These robots possess human-like intelligence and are equipped with high-sensitivity sensors (Sodhro et al., 2019). Their sensory abilities surpass those of ordinary people. Intelligent robots can analyse perceived information, control their behaviour, respond to environmental changes, and complete complex tasks (Sünderhauf et al., 2018; Tambe et al., 2019).

17.5.4 The Intelligent Decision Support System

The decision support system falls under the domain of management science and is closely linked to "knowledge-intelligence." During the 1980s, expert systems achieved notable success in various areas (Abduljabbar et al., 2019). The integration of AI, particularly the use of intelligence and knowledge processing technology in decision support systems, has broadened their application range and enhanced their problem-solving capabilities. This evolution has led to the development of intelligent decision support systems (Morocho-Cayamcela et al., 2019; Reichstein et al., 2019).

17.5.5 The Pattern Recognition

Pattern recognition focuses on developing machines with perception capabilities (Tan et al., 2010). It primarily deals with the recognition of visual and auditory patterns, such as identifying objects, terrain, images, and fonts. This field has numerous applications in both everyday life and the military. Recently, the application of fuzzy mathematical models and artificial neural network models has advanced rapidly, gradually replacing traditional statistical models and structural pattern recognition methods (Chen et al., 2020; Yang et al., 2019).

17.6 Domain Justification

Expert System An expert system is an intelligent computer software that possesses professional knowledge and experience. It mimics complex problems that are typically solved by experts by simulating human experts' problem-solving abilities through knowledge representation and knowledge reasoning approaches in artificial intelligence. It is capable of addressing problems to the same degree as a professional.

Machine Learning The field of machine learning encompasses various fields such as probability theory, statistics, approximation theory, convex analysis, algorithm complexity theory, and more. Machine learning is the study of how computers mimic or actualize human learning tendencies to pick up new abilities or knowledge, and how they rearrange the current knowledge structure to keep getting better at what they do.

Robotics The science that deals with the development, implementation, and upkeep of robots is known as robotics. It primarily investigates the connection between object processing and robot control.

Decision Support System Decision support systems are closely related to "knowledge-intelligence" and are within the management science category.

Pattern Recognition The field of pattern recognition studies ways to provide robots perceptive abilities. It focuses mostly on the identification of auditory and visual patterns.

17.7 Three Viewpoints of AI

Symbolism, connectionism, and behaviourism represent the three main conceptual viewpoints in AI research. These are the most significant theoretical frameworks in the development of AI and form the theoretical foundation for the discipline.

17.7.1 Symbolism

Also known as logicism, symbolism posits that symbols are fundamental to human cognition, and the cognitive process is one of symbolic calculation and reasoning. Symbolism expresses cognitive objects as symbols using mathematical logic and then employs the computer's symbol processing capabilities to simulate human cognitive processes. The core principles supporting symbolism include the physical symbol system hypothesis and the principle of limited rationality (Jha & Topol, 2016). Symbolism's primary research focus is on knowledge representation and reasoning technology, grounded in logic. Key figures in this field include Newell, Simon, McCarthy, Robinson, and Shotfield (Wang et al., 2007). Major contributions of symbolism include analytical reasoning methods, heuristic algorithms, expert systems, and knowledge engineering theory and technology. Symbolism has been notably successful in addressing logical problems, such as solving the "four-color conjecture" and the development and application of expert systems in the 1970s (Jin et al., 2014).

17.7.2 Connectionism

Also known as Bionics, connectionism posits that human intelligence is dependent on the physiological structure and functioning of the human brain. Connectionism asserts that the basic units of human cognition are the neurons in the brain and that the cognitive process involves the brain's information processing (Fu et al., 2018; Wang et al., 2020). Therefore, connectionism aims to mimic the brain's structure and functioning to truly simulate human intelligence in machines. The primary focus of connectionism research is on neural networks. Key figures in this field include McClelland and Rumelhart. Major contributions from connectionism include brain model research and the backpropagation (B-P) algorithm in multilayer networks (Dwivedi et al., 2021).

17.7.3 Behaviourism

Behaviourism is one of the evolution of cybernetics, behaviourism posits that intelligence depends on perception and action rather than knowledge, representation, or reasoning (Shi et al., 2020). Behaviourism emphasizes that fundamental human abilities include action, perception, sustaining life, and self-replication. Intelligent behaviour is manifested through interactions with the real-world environment, and AI should evolve gradually, akin to human intelligence (Haydari & Yılmaz, 2020). This approach focuses on simulating various human control behaviours. The primary representative of behaviourism is Brooke. Key developments in this field include the realization of intelligent control and intelligent robot systems. While behaviourism has not yet formed a complete theoretical system, its divergence from traditional AI perspectives has garnered significant attention within the AI community (Tokui et al., 2015; Wanigasekara et al., 2021). These three viewpoints—symbolism, connectionism, and behaviourism—have had a critical and far reaching influence on the development of artificial intelligence. It is under the guidance of these perspectives that various specific AI analysis models, implementation methods, and algorithms have emerged (Zhang & Tao, 2021). Even the latest advancements in data mining, knowledge discovery, and intelligent agent technologies are deeply influenced by these viewpoints.

17.8 AI in Food Industry

The application of AI in the food industry has been expanding for years due to various needs such as food sorting, parameter classification and prediction, quality control, and food safety. The role of AI in food industry is listed out in Fig. 17.6.

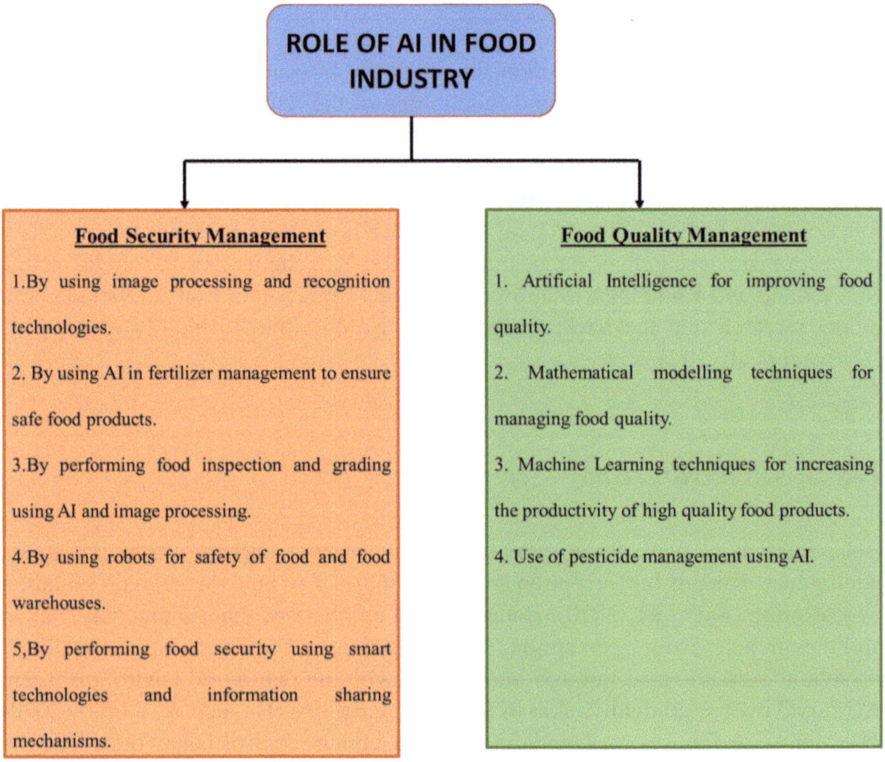

Fig. 17.6 Role of AI in food industry

Techniques like expert systems, fuzzy logic, artificial neural networks (ANN), adaptive neuro-fuzzy inference systems (ANFIS), and machine learning have become popular in the food industry. Prior to AI implementation, extensive studies on food have been conducted over the years to educate the public and improve outcomes related to food properties and production (Abd. Rahman et al., 2012; Belluco et al., 2013; Jayasooriya et al., 2004; Mozafari et al., 2008; Rahman et al., 2012; Saha & Bhattacharya, 2010). The use of AI methods offers numerous benefits and has been implemented in the food industry for decades, with a continuous increase to the present day (Corney, 2002; Doganis et al., 2006; Perrot et al., 2006). However, this chapter will focus on the application of AI in the food industry from 2015 onwards, highlighting the recent surge in innovation and implementation. It is important to note that certain methods such as partial least squares, gastrointestinal unified theoretical framework, in silico models, empirical models, sparse regression, successive projections algorithms, and competitive adaptive reweighted sampling used for prediction and enhancement in the food industry are not discussed here. Instead, the chapter narrows its focus to the broad application of AI in the food industry. And also, different algorithms and their applications are discussed. The advantages and limitations of important algorithms used in AI are listed in Table 17.3.

17.8.1 Knowledge Based Expert System in Food Industry

A knowledge-based system is a computer program created to address complex problems by utilizing knowledge from various sources. These systems are divided into expert systems, knowledge-based artificial intelligence, and knowledge-based engineering. Among these, the knowledge-based expert system (ES) is widely used in industries for its ability to emulate human decision-making by leveraging a knowledge base and an inference engine. ES has been employed in numerous sectors of the food industry to improve decision-making processes, food safety, quality control, and risk assessment. For example, it is utilized in white winemaking for supervision and intelligent control (Sipos, 2020), and in web-based applications to calculate nutritional values and aid small and medium-sized industries in obtaining food production certificates (Ardiansah et al., 2020). Additionally, ES models have been developed for food safety management, nutrition, and cost control, such as the digital learning tool MESTRAL for food processing (Suciu et al., 2021). A comprehensive review by Leo Kumar has highlighted the applications of ES in manufacturing planning and decision-making across various processes (Leo Kumar, 2019).

17.8.1.1 Applications of Expert Systems in the Food Industry

Expert systems are widely utilized in the food industry, enhancing processes like disease identification, quality control, and sustainability. For example, in the banana

Table 17.3 Advantages and limitations of AI algorithms

AI algorithm	Advantages	Limitations
Expert System (ES)	Reliable and understandable; Highly responsive; High performance; Error rate is lower than human errors; Useful in production capacities.	Construction and designing are expensive and rare as it requires expert engineers; Vocabulary utilized by the experts is limited and often difficult to be understood.
Fuzzy Logic (FL)	Imprecise, incomplete, and uncertain information can be solved; Simpler and direct results can be obtained; Accountable, noise tolerant, and robust to disturbances; Faster interpretation than ANN and support vector machine methods; Knowledge can be extended easily with the extension of the rules; Saves costs and time; Can improve the quality and safety of the products.	Generalization is not possible as it can only deduce the given rules; Sometimes requires the knowledge of an expert in creating the rules.
Artificial Neural Network (ANN)	Able to model complex functions or problems accurately and easily; Accurate, robust to disturbances, and noise tolerant; Has the ability to learn from patterns or examples; Has generalization ability; Affordable, noise tolerant, easy, and flexible method; Non-linear problems are more appropriate by using this method; Useful in prediction, classification, and control tool.	The performance of the model is hard to be explained compared to others as it appears as a black box model; Requires more time compared to the FL as the suitable number of layers should be determined; Need sufficient and reliable data.
Adaptive Neuro-Fuzzy Inference System (ANFIS)	Able to merge details from various resources; Noise tolerant, accurate, and effective method in solving complex problems; Higher performance compared to ANN and FL methods; Benefits from both ANN and FL methods; Classification and prediction tasks can be done more conveniently; Saves time and cost overall in general compared to manual methods.	Data available should be reliable to avoid any confusion or misinformation during the training process as it will affect the final results.

sector, expert systems diagnose banana diseases and suggest methods to combat them. As described by (Budiyanto et al., 2018), these systems can identify diseases in plants based on their stems, leaves, and roots, helping prevent disease spread.

In barley production, expert systems are employed to classify barley grains, supporting quality control. According to (Szturo & Szczypiński, 2017), the classification accuracy of the ES method exceeds 72%, which is considered satisfactory.

In the coffee industry, expert systems determine the appropriate process for the dry mill. (Hernández Vera et al., 2017) reported that the ES system could accurately select the correct production process for specific coffee types with an accuracy of 93%.

For grading coffee beans, expert systems are used to assess quality. Livio and Hodhod (2018) highlighted that the fuzzy ES, known as AI Cupper, could grade coffee with an accuracy of 95%.

In agriculture, expert systems help identify corn pests and diseases. Sumaryanti et al. (2020) found that these systems could detect pests and diseases with an accuracy of 76.6% and suggest control measures. Additionally, they provide detailed explanations of diagnosis results.

In terms of food safety, expert systems determine the halal safety rating for food additives. Zakaria et al. (2019) emphasized that the Halal Food Additive ES provides consumers with a safety rating based on their past food consumption experiences. It also helps consumers find nutritious foods that meet the Halalan-Toyyiban criteria.

For food products, expert systems monitor and predict product quality during production. Blagoveshchenskiy et al. (2020) discussed an intelligent ES capable of monitoring product quality indicators, making necessary changes to existing methods, and providing recommendations for the final product's quality and defect management.

Expert systems also optimize fresh food distribution. Bortolini et al. (2016) proposed the Food Distributor expert system, which can develop the most effective distribution method, reduce transportation carbon expenditure by 0.6%, increase operational efficiency by 1%, minimize waste during transportation, and optimize the distribution network.

From diagnosing banana diseases to optimizing fresh food distribution networks, expert systems play a crucial role in ensuring quality, safety, and sustainability within the industry.

17.8.2 Fuzzy Logic Technique in the Food Industry

Introduced by Zadeh in 1965, Fuzzy Logic (FL) employs human-like decision-making capabilities to manage imprecise, uncertain, and ambiguous data (Mahdi et al., 2019; Zadeh, 2015). Fuzzy set theory permits elements to belong to a fuzzy set with varying degrees of membership within the interval [0, 1] (Hannan et al., 2019). FL models consist of three main steps: fuzzification, inference, and defuzzification (Mutlag et al., 2016; Rajesh Mavani et al., 2021). Fuzzification transforms crisp values into degrees of membership, forming fuzzy input sets. Common membership functions include triangular, Z-shaped, S-shaped, trapezoidal, and Gaussian-shaped (Alsaqour et al., 2015).

The inference system converts fuzzy inputs into outputs using IF-THEN rules, where the IF part contains input parameters and the THEN part contains output parameters (Nafan et al., 2019). Inference can be conducted using Mamdani or Takagi–Sugeno Kang (TSK) methods. Defuzzification translates fuzzy outputs back into crisp values through methods such as the centre of gravity, mean of maximum, and centroid of area (Ali et al., 2016; Zareiforoush et al., 2016).

FL is extensively used in industry due to its simplicity and efficiency in quickly and accurately solving problems. In the food industry, FL is applied for modelling, control, classification, and managing food-related issues using human-like reasoning (Al Mahasneh et al., 2016). The implementation of FL in food manufacturing has reduced electricity losses by about 7% compared to traditional methods (Baliuta et al., 2020). FL is also crucial in sensory evaluation and provides quick solutions to problems through fuzzy rules (Cebi et al., 2019). Studies have demonstrated that FL effectively maintains food quality and serves as a prediction and control tool in food production processes.

17.8.2.1 Application of Fuzzy Logic in the Food Industry

In the context of aromatic foods, the fuzzy logic system was employed to rank the sensory attributes of aromatic foods packed in films made from corn starch, focusing on the overall ranking for tea and tastemaker as well as the aroma and taste of tea leaf and tastemaker, which were considered highly important sensory attributes (Chowdhury & Das, 2015).

For beetroot candy, fuzzy logic was used to rank the candy with different content ratios, optimizing and performing the ranking effectively (Fatma et al., 2016). In canned food, fuzzy logic was applied to control sterilization temperature, maintaining accuracy within ±0.5 °C, and enabling batch processing with reduced time, steam consumption, and over-sterilization risks (Chung et al., 2016).

In the coffee industry, a fuzzy logic-based expert system determined suitable processes on a dry mill based on customer requirements, enhancing decision-making accuracy and process validation with a 93% coefficient of determination (Hernández-Vera et al., 2017). For coffee beans, a control system for the roasting machine was introduced, ensuring consistent roasting levels (Harsawardana et al., 2020). The ranking of cupcakes according to their quality attributes was achieved using fuzzy logic, which determined the best conditions for sensory attributes (Singh et al., 2018). In dough making, fuzzy logic acted as a controller system, showing better performance in setting time and response compared to PID controllers (Yousefi-Darani et al., 2019).

The study also predicted physical parameters for fava beans with various moisture contents, showing high correlation and effectiveness in process management (Farzaneh et al., 2017a, b). For flaxweed seed, fuzzy logic ranked different extraction methods based on properties and preparation efficiency (Shahidi et al., 2020). In fresh mango juice, fuzzy logic studied the effects of high-pressure processes, demonstrating its benefits in product preservation (Kaushik et al., 2015).

The release of caffeine from hydrogel colloidosomes was estimated using a diffusion-fuzzy model, which improved uncertainty handling and generalization (Amiryousefi et al., 2017). The drying kinetics of onions were predicted with high accuracy using fuzzy logic (Jafari et al., 2016). Pineapple Rasgulla ranking based on parameters was successfully performed using fuzzy logic (Sarkar et al., 2020). In the pizza production industry, a fuzzy logic system identified the causes of defects and proposed improvements (Blasi, 2005).

Salt properties were estimated with high accuracy using fuzzy logic (Yulianto et al., 2017). The quality of sardine was assessed using biogenic amines with fuzzy logic, achieving high initial quality determination (Zare & Ghazali, 2017). For treated raw apple juice, sensory evaluation showed variations in customer acceptability using fuzzy logic (Basak, 2018). In wheat dough processing, fuzzy logic developed an effective sheeting method, optimizing the number of rolling steps (Mahadevappa et al., 2017). Lastly, the moisture ratio of white mulberry during drying was forecasted with high accuracy using fuzzy logic (Jahedi Rad et al., 2018).

17.8.3 ANN Technique in the Food Industry

Artificial Neural Networks (ANN) are widely utilized in the food industry to mimic the human brain's learning process through synaptic weights (Al-Waeli et al., 2019; Kumar, 2016). Designed for tasks such as data classification and pattern recognition, ANNs are highly adaptable and flexible for various problems and scenarios, as noted by Gonzalez-Fernandez (Gonzalez-Fernandez et al., 2019). They excel at modelling non-linear systems and can adapt to new situations, with their primary feature being non-linear regression (Gandhi et al., 2016). Common types of ANN include feedforward neural networks, radial basis function networks (RBFNN), Kohonen self-organizing neural networks, recurrent neural networks, convolutional neural networks, and modular neural networks (Maladkar, 2018).

Multilayer perceptron (MLP), RBFNN, and Kohonen self-organizing algorithms are among the most effective for addressing real-world problems (Gandhi et al., 2016). MLP is the most frequently used network for prediction and pattern recognition (Abdul Aziz et al., 2019; Correa et al., 2018; Khamis et al., 2018). ANN learning can be divided into supervised and unsupervised techniques (Funes et al., 2015). A typical ANN structure consists of an input layer, hidden layer, and output layer, with activation functions that can be either feed-forward or feedback (Ismail et al., 2018; Rashmi et al., 2019; Said et al., 2018; Trafialek et al., 2015). The backpropagation learning algorithm is commonly employed to minimize prediction error (Correa et al., 2018). Additionally, a bias input is incorporated into neurons to model phenomena with thresholds (Silva et al., 2015; Tanajura da Silva et al., 2015). ANN training encompasses supervised, unsupervised, and reinforcement learning algorithms (Shanmuganathan, 2016).

17.8.3.1 Applications of ANN in the Food Industry

The study on the application of artificial neural networks (ANN) in the food industry covers various applications and objectives. For cocoa powder, the model could predict the effect of process parameters on the properties of cocoa mixtures, achieving a high accuracy with an R^2 value of 0.934 and a low error of 0.053 (Benković et al., 2015). In dried vegetables, ANN was used to identify the quality of dried carrots and classify such products, with backpropagation being the most successful algorithm (Koszela et al., 2017).

In the context of eggplant, ANN described the mass transfer kinetics in osmotic dehydration, achieving the highest R^2 value of 0.9825 (Bahmani et al., 2016). For extra virgin olive oils, ANN evaluated the influence of light exposure conditions and packaging material on the stability of physicochemical characteristics, showing a high classification performance with greater than 90% accuracy for test data and 85% for training data (Silva et al., 2015; Tanajura da Silva et al., 2015).

The sensory quality of garlic was forecasted using ANN, achieving the best prediction with an R^2 value of 0.9866 (Liu et al., 2019). In honey, ANN predicted the stability of Indian honey crystallization with high accuracy, achieving an R^2 value of 0.9994 (Naik et al., 2019). For mangoes, ANN estimated the weight by applying different parameters in the model (Dang et al., 2019).

ANN also predicted the moisture content of mushrooms during the drying process with an R^2 value of 0.9817 (Omari et al., 2018). For temperature variety in mushroom growing halls, the prediction by using the Radial Basis Function (RBF) model had a higher accuracy compared to MLP, with R^2 values of 0.996 and 0.9612, respectively (Ardabili et al., 2020). In onions, ANN estimated the drying behaviour at different temperatures, achieving an R^2 value of 0.99956 (Jafari et al., 2016).

In potato cubes, ANN carried out analysis in a fluidized bed dryer under different conditions, achieving an R^2 value greater than 0.98 with an average of 0.99 (Azadbakht et al., 2017). For potato peels, ANN helped determine the optimal extraction conditions, achieving R2R^2R2 values greater than 0.93 (Anastácio et al., 2016). In quince fruit, ANN determined the moisture ratio during the drying process, achieving an R^2 value greater than 99% (Chasiotis et al., 2020).

For rice crops, ANN predicted rice production yield and investigated factors affecting the yield, achieving an accuracy of 97.54% with 96.33% sensitivity and 98.12% specificity (Gandhi et al., 2016). ANN also forecasted the benzopyrene content of smoked sausage, helping create a control system for smoking processes with a prediction accuracy greater than 0.90 (Chen et al., 2018). Lastly, ANN classified vegetable oils such as canola, sunflower, corn, and soybean with high accuracy, requiring minimal mathematical manipulations (Silva et al., 2015; Tanajura da Silva et al., 2015).

17.8.4 Machine Learning Techniques

Machine learning (ML), a branch of AI, consists of algorithms that improve automatically through experience (Butler et al., 2018). ML is categorized into supervised learning, unsupervised learning, and reinforcement learning (Erickson et al., 2017; Ramakrishna et al., 2020). Supervised learning focuses on predicting targets based on provided inputs (Mullainathan & Spiess, 2017), whereas unsupervised learning identifies patterns and classifies data without predefined outputs (Deo, 2015). Reinforcement learning involves the program interacting with its environment to achieve specific objectives (Carleo et al., 2019). Common ML models include artificial neural networks (ANN), decision trees (DT), support vector machines (SVM), regression analysis, Bayesian networks, genetic algorithms, kernel machines, and federated learning (Jordan & Mitchell, 2015; Mullainathan & Spiess, 2017).

ML excels at handling complex tasks and large datasets without needing predefined formulas, learning from examples instead of programmed rules (Rajkomar et al., 2019). In the food industry, ML methods such as ordinary least square regression (OLS-R), stepwise linear regression (SL-R), principal component regression (PC-R), partial least square regression (PLS-R), support vector regression (SVM-R), boosted logistic regression (BLR), random forest regression (RF-R), and k-nearest neighbors' regression (kNN-R) are used (Estelles-Lopez et al., 2017). ML helps reduce sensory evaluation costs, aids decision-making, and improves business strategies (Lu et al., 2020). Long short-term memory (LSTM) networks are utilized for pH detection in cheese fermentation (Li et al., 2021a), while genetic algorithms optimize food parameters, and neural networks predict fouling rates in food processing (Kim et al., 2020). Additionally, ML predicts food insecurity in the UK (Alaiz-Rodríguez & Parnell, 2020), forecasts sales trends in the food industry (Tsoumakas, 2019), and analyses food waste generation to provide insights into production systems (Garre et al., 2020).

17.8.4.1 Applications of ML in the Food Industry

The study on the application of machine learning (ML) in food industries includes several applications and objectives. For apples, ML methods such as linear discriminant analysis and adaptive boosting were able to classify apples accurately with a rate of 100% using collected acoustic emission signals (Li et al., 2018). In artichokes, ML models like MLP, RF, and BLR were used to analyse the characteristics of ion patterns for enzyme setup with a high prediction rate of 95% (Sabater et al., 2019).

For beer, ANN models classified the chemical components with a high overall accuracy of 95% (Gonzalez Viejo et al., 2018). In biscuits, a convolutional neural network classified and evaluated different types of biscuits with up to 99% accuracy (de Sousa Silva et al., 2020). The combination of mechanistic modelling with LSTM in cheese fermentation described changes in lactic acid, lactose, and biomass with high accuracy, achieving an R^2 value greater than 0.99, and predicted pH effectively (Li et al., 2021a).

For Citrus limetta (Mosambi peel), SVM-ANN and SVM-Gaussian process regression (SVM-GPR) predicted and classified the taste of lime powder treated with salt with an accuracy of 1.0, maintaining taste and polyphenol content (Younis et al. 2019). In fruits (Arbutus unedo L. fruits), RF, SVM, and ANN methods stabilized extracts in terms of aqueous and powder systems with overall coefficients ranging from 0.9128 to 0.9912 (Astray et al., 2020).

For lamb meat, SVM increased the classification accuracy of lamb meat fat from 89.70% to 93.89% (Alaiz-Rodríguez & Parnell, 2020). In mangoes, Naive Bayes and SVM detected maturity based on quality attributes (Pise & Upadhye, 2018). For meat, methods like OLS-R, SL-R, PC-R, PLS-R, SVM-R, RF-R, and kNN-R detected different kinds of microorganisms causing beef spoilage and predicted the spoilage accurately using regression ML, combining data from five different analytical methods (Estelles-Lopez et al., 2017).

For milk, SVM classifiers determined the presence and level of antibiotics in cow milk with high accuracy rates of over 83% and greater sensitivity compared to typical metrics (Gutiérrez et al., 2020). In salmon, TreeBagger classified normal and freezer burnt categories with high accuracy, achieving a correct classification rate of 0.914 for validation and 0.978 in cross-validation (Xu & Sun, 2017).

For wine, SVM, RF, and MLP algorithms evaluated wine quality, with the RF method achieving the best result with an average accuracy of 81.96%, while other methods had lower accuracy rates, indicating that the RF algorithm is effective for evaluating wine quality (Shaw et al., 2020).

17.8.5 Adaptive Neuro Fuzzy Inference System (ANFIS) Techniques

Adaptive Neuro-Fuzzy Inference System (ANFIS) combines fuzzy logic (FL) with artificial neural networks (ANN) to integrate human-like reasoning with computational and learning capabilities (Al-Mahasneh et al., 2016). In ANFIS, the learning procedure from the neural network is transferred to the FL system, developing a set of fuzzy rules with suitable membership functions from obtained data (Melin et al., 2018). ANFIS processes uncertain data with high accuracy and is recognized for its speed and robustness (Bouhoune et al., 2017; Mamat et al., 2019). Sharma et al. (2017) noted that ANFIS outperforms models like ANN and multiple regression in their study.

ANFIS typically uses the Takagi-Sugeno-Kang fuzzy inference system combined with a feedforward neural network and learning algorithms. Its structure comprises five layers: fuzzy layer, product layer, normalized layer, defuzzification layer, and total output layer. The backpropagation algorithm is commonly applied to avoid overfitting. A high correlation value (R^2) indicates the model's high accuracy and suitability for industrial applications (Viharos & Kis, 2015).

ANFIS is effective in solving complex problems in the food industry, particularly in prediction and classification tasks. It is used in various food processing technologies, covering five main categories: food property prediction, food drying, thermal process modelling, microbial growth, and quality control of food rheology (Al-Mahasneh et al., 2016).

17.8.5.1 Specific Applications of ANFIS in the Food Industry

In the context of fish oil, a model was developed to estimate oxidation parameters using ANFIS, multilayer perceptron, and multiple linear regression, with ANFIS showing the best accuracy in predicting the parameters (Asnaashari et al., 2016). For ice cream, ANFIS was used to investigate sensory attributes and predict the acceptability of taste with a minimum error of 5.11% and a high correlation value of 0.93 (Bahram-Parvar et al., 2017).

In Indian sweets (Pantoa), ANFIS predicted the heat transfer coefficient during frying, yielding a high R2R^2R2 value of 0.9984, important for designing process equipment and saving energy in commercial production (Neethu et al., 2016) For oranges, ANFIS predicted taste with higher performance compared to multiple regression models (Mokarram et al., 2019). In quince fruits, ANFIS accurately predicted moisture ratio, energy utilization, energy utilization ratio, exergy loss, and exergy efficiency during the drying process, with R^2 values of 0.9997, 0.9989, 0.9988, 0.9986, and 0.9978, respectively (Abbaspour-Gilandeh et al., 2020).

The ANFIS model was compared with the ANN model in various applications, showing higher accuracy in ANFIS, with higher R^2 values (Farzaneh et al., 2017a, b). For rapeseed oil, ANFIS predicted different outputs in oil extraction and cooking on an industrial scale, achieving a high correlation coefficient around 0.99 (Farzaneh et al., 2017b). In the prediction of *Salmonella enteritidis* inactivation by ultrasound, ANFIS showed high accuracy with a correlation coefficient of 0.974, significant for preventing food poisoning (Soleimanzadeh et al., 2015).

In taro, ANFIS optimized extraction conditions for antioxidants, validated prediction values, and provided consistent results (Kumar & Sharma, 2017). For vegetables (cantaloupe, garlic, potatoes), ANFIS predicted moisture diffusivity, specific energy consumption, and moisture ratio with high regression coefficients and minimal error values, showing higher efficiency compared to ANN models (Kaveh et al., 2018). In virgin olive oil, ANFIS predicted quality with high accuracy, visualizing the effects of temperature on total polyphenol, fatty acids profile, and tocopherol on oxidative stability (Arabameri et al., 2019).

Finally, in yams, ANFIS predicted moisture content and R^2 values of 0.9826 by using different drying methods, showing superior performance (Ojediran et al., 2020).

17.8.6 *Integrating AI with External Sensors for Real Time Detection in Food Industry*

The integration of Fuzzy Logic (FL) or Artificial Neural Networks (ANN) with various sensors has become increasingly prevalent for real-time detection, especially in the food industry. Sensors such as electronic nose (E-nose), electronic tongue (E-tongue), machine learning (ML), computer vision systems (CVS), and near-infrared spectroscopy (NIRS) are commonly used to enhance the accuracy and efficiency of detection processes. By combining these sensors and leveraging AI methods, it is possible to achieve more accurate and targeted results in a shorter time. This approach has seen significant adoption in the food industry in recent years, leading to improvements in quality control, safety, and overall product quality.

An electronic nose, or E-nose, is a device designed to detect odors or flavors similarly to the human nose. It comprises an array of electronic chemical sensors capable of recognizing both simple and complex odors (Yan et al., 2015). E-noses are utilized in gas sensing applications, where they analyze individual components or mixtures of gases and vapors. In the food industry, they are essential for quality control. Due to their capability to identify complex odors, E-noses are also used for environmental protection and the detection of explosive materials (Deshmukh et al., 2015). The primary hardware component of an E-nose is an array of nonspecific gas sensors that interact with various chemicals with different intensities, producing a unique response or "fingerprint" (Yan et al., 2015). The main software components include feature extraction and pattern recognition algorithms, which process the sensor responses to extract and select significant details. Therefore, the software significantly influences the E-nose's performance. Generally, an E-nose consists of three main parts: the sample delivery system, the detection system, and the computing system. Methodologies such as artificial neural networks (ANN), fuzzy logic (FL), and pattern recognition are employed in E-noses (Szulczyński et al., 2018).

The electronic nose (e-nose) has been extensively utilized for quality control and assurance in the food industry. It has been applied to wines, grains, cooking oils, eggs, dairy products, meat, fish products, fresh-cut and processed vegetables, tea, coffee, and juices for sample classification, detection, and quality control. E-nose has proven effective in classifying samples with different molecular compounds (Wojnowski et al., 2017). Moreover, Sanaeifar et al. have reviewed and confirmed the capability of e-nose to detect defects and contamination in foodstuffs (Sanaeifar et al., 2017). It has also been used to classify and differentiate various fruits (Baietto & Wilson, 2015). Additionally, a review has been conducted on the application of e-nose for monitoring food authenticity (Gliszczyńska-Świgło & Chmielewski, 2017). Furthermore, Mohamed et al. have performed a comprehensive review on using e-nose integrated with fuzzy logic (FL) and artificial neural network (ANN) methods for classifying food freshness (Mohamed et al., 2015).

17.8.6.1 Application of E-nose with AI in Food Industries

For beef, an Adaptive FL system (AFLS) and ANFIS were used to classify beef samples, achieving a high accuracy of 94.28% in overall correct classification, effectively tackling complex issues like meat spoilage (Kodogiannis & Alshejari, 2016). In cocoa, Decision Trees (DT), boosted tree Random Forest (RF), ANN, K-Nearest Neighbors (KNN), and Naive Bayes (NB) were employed to classify the fermentation time of cocoa beans. The ANN and boosted tree algorithms achieved acceptable classification rates, though the fermentation time was not well-determined by KNN and NB algorithms (Tan et al., 2019a).

For coffee beans, ANN predicted the acidity level with an accuracy of about 95% based on human sourness level scores (Thazin et al., 2018). In chicken meat, the Fuzzy K-nearest neighbors algorithm (FK-NN) classified fresh and freeze-thawed chicken meat with high performance, proving useful in identifying meat quality (Mirzaee-Ghaleh et al., 2020). ANN also detected adulteration in cow ghee and edible oil with high accuracy (Ayari et al., 2018; Karami et al., 2020). For fish, ANN and PCA classified fish spoilage, achieving an accuracy of 96.87% (Vajdi et al., 2019).

In fruits, Kernel Extreme Learning Machines (KELM) in E-nose recognized and perceived fruit odors with high accuracy and efficiency (Uçar & Özalp, 2017). ANN integrated with artificial bee colony algorithms classified data from the E-nose, outperforming backpropagation algorithms in odor classification (Adak & Yumusak, 2016). For fruit juice, RF, SVM, ELM, and PLSR were used to predict food additives like benzoic acid and chitosan, with RF showing the highest R^2 values (Qiu & Wang, 2017).

ANN and SVM classified Iranian ziziphus honey samples, showing high prediction rates in determining physicochemical properties (Faal et al., 2019). In lemons, ELM, RF, and SVM predicted quality, with ELM achieving the highest accuracy of 0.959 (Guo et al., 2018). For pears, FL optimized the drying process, enhancing food quality by retaining aroma and product quality (Li et al., 2021b).

In pork meat, BPANN and E-nose differentiated between fresh and frozen-thawed meat with high accuracy, distinguishing meat types with 85.1% sensitivity and 97.5% specificity (Górska-Horczyczak et al., 2017). FL ARTMAP, PCA, and MLR detected Sitophilus oryzae infestation in stored rice grains, effectively classifying infested, non-infested, and required treatment grains (Srivastava et al., 2019a). BPANN and E-nose also classified Rhyzopertha dominica infested rice grains, achieving the highest R^2 value (Srivastava et al., 2019b).

For shelled peanuts, FL assessed storage quality, predicting spoilage times and providing an eco-friendly alternative for quality analysis (Raigar et al., 2017). PCA, CNN, and SVM classified the freshness of salmon samples, achieving an accuracy greater than 90% (Feng et al., 2020). ANN detected saffron adulteration, showing high accuracy rates of 100% and 86.87% for different sample types (Heidarbeigi et al., 2015).

In identifying different aromas of Iranian saffron, E-nose coupled with ANN achieved 100% classification accuracy (Kiani et al., 2016). ANN detected postharvest freshness in spinach with 93.75% accuracy using the BPNN method (Huang et al., 2019b). ANN and ML classified and analysed strawberry juice, with ELM networks showing faster and more accurate classification (Qiu et al., 2015). FL assessed the frying disposal point of sunflower oil blend, showing potential for other fried product platforms (Upadhyay et al., 2017).

Lastly, FL determined granary weevil infestation in stored wheat grains, with E-nose sensors accurately identifying and monitoring infestations, proving useful for industry quality control (Mishra et al., 2018).

For ham, the Simplified Fuzzy ARTMAP neural network (SFAM) monitored the salt processing of hams salted differently, analysing and classifying data with high accuracy and determining optimal parameters for use in microcontroller devices (Gil-Sánchez et al., 2015). In honey, the Fuzzy ARTMAP neural network (FAM) differentiated types of honey based on antioxidant levels with a 100% classification success rate (Juan-Borrás et al., 2017).

For liquor, SVM and fuzzy evaluation classified different types of Chinese liquor flavours with 100% accuracy and aligned the output evaluation with human perception (Liu et al., 2020b). SVM detected adulteration in raw milk with high accuracy, identifying various types of adulterants with over 87% accuracy (Tohidi et al., 2018). In peanut meal, ANN and partial least squares (PLS) assessed taste attributes and compared predictive abilities, with RBFNN exhibiting better prediction ability than PLS (Wang et al., 2015a, b).

SVM, KNN, ANN, and RF classified pineapples based on sweetness levels, with KNN achieving an accuracy of 0.820 (Hasan et al., 2020). ANN distinguished different types of rice with 95% accuracy, and voltametric E-tongue provided qualitative analysis for rice (Wang et al., 2015a, b). ANN combined with similarity analysis (SA-ANN) predicted rice sensory quality more precisely than PCA-ANN, showing potential for rapid, high-accuracy predictions (Lu et al., 2015).

In sugarcane, ANN processed complex responses from the E-tongue for glucose analysis, proving suitable for hydrolyzed samples (de Sá et al., 2016). BPNN and ELM classified tangerine peel of different ages, with ELM showing the highest classification accuracy (Shi et al., 2018a, b). ANN and SVM distinguished different types of teas, identifying compositions effectively (Huang et al., 2019b, d).

ANN-PCA predicted freshness changes in tilapia fillets at different temperatures, distinguishing fillets stored at various days and temperatures (Shi et al., 2018a, b).

17.8.7 Computer Vision System

The computer vision system (CVS), a branch of AI, integrates image processing and pattern recognition techniques. It is a non-destructive approach that enables the analysis and extraction of image features to assist in designing classification patterns (Cárdenas-Pérez et al., 2017). CVS is also recognized for its effectiveness in measuring external features such as size, shape, colour, and defects. Typically, it consists of a digital camera, a lighting system, and software for image processing and analysis (de Oliveira et al., 2016). There are two main types of CVS: 2D and 3D. Its applications span various sectors, particularly in the food industry, where it is used to assess the ripeness of apples (Cárdenas-Pérez et al., 2017), predict the colour attributes of pork loin (Sun et al., 2016), determine the roasting degree of coffee (Leme et al., 2019), evaluate the quality of table grapes (Cavallo et al., 2019), and detect defects in pork (Chmiel & Słowiński, 2016). The integration of CVS with soft computing techniques is considered highly valuable in the food industry due to the combination's ability to provide accurate and rapid predictions.

17.8.7.1 Application of CVS with AI in Food Industries

For apples, CNN and SVM were used to sort defective and normal apples with a high accuracy rate of 96.5%, outperforming conventional methods which achieved an accuracy of 87.1% when combined with SVM (Fan et al., 2020). ANN tracked drying effects on apple slices, estimating quality changes during the drying process with an R^2 value greater than 0.92 for all analyses (Nadian ct al., 2015).

For bananas, ANN, SVM, KNN, and Decision Trees (DT) classified ripeness, with the ANN-based model showing the highest overall recognition rate of 97.75% (Mazen & Nashat, 2019). In barley flour, SVM, KNN, DT, and RF were used for prediction, achieving accuracy rates of 75% (KNN), 95% (SVM & RF), and 100% (DT) by using the spatial pyramid partition ensemble method (Lopes et al., 2019). ANN and ML forecasted beer acceptability based on sensory parameters, with the best model showing an R^2 value of 0.92 (Gonzalez Viejo et al., 2019).

ANN and FL described the ripeness level of bell peppers, with the ANN model achieving higher classification accuracy than FL (Villaseñor-Aguilar et al., 2020). For cape gooseberry, ANN, SVM, DT, and KNN classified ripeness, achieving accuracy greater than 86% using different colour spaces (Castro et al., 2019). ANN classified coffee beans, analysing and classifying green coffee beans with high accuracy (de Oliveira et al., 2016). ANN, KNN, DT, and SVM classified dry beans, with SVM achieving the highest overall classification rate of 93.13%, followed by DT, ANN, and KNN with rates of 92.52%, 91.73%, and 87.92%, respectively (Koklu & Ozkan, 2020).

ANN predicted the volume of eggs with a high linear coefficient of 0.9738 and a relative absolute error of 2.2078% (Siswantoro et al., 2017) Decision tree-FL classified figs based on visual features, with the REP decision tree showing the highest RRR value and lower RMSE values, enabling efficient classification (Banakar et al., 2017). ANN and PCA identified fish freshness with a success rate of 94.17% in training and 90.00% in prediction (Huang et al., 2016).

FL developed a quality control system for gluten-free cakes, studying texture with different ingredients and achieving optimal results (Rezagholi & Hesarinejad, 2017). ANFIS predicted the weight of Indian lime fruits with high accuracy, improving classification when combined with the fuzzy C-means clustering method (Phate et al., 2019). ANN estimated the fruit mass of mangoes, using CVS data as input parameters with an accuracy of 97% and an efficiency coefficient of 0.99 (Utai et al., 2019).

ANN and FL determined the appearance quality of mushrooms, with ANN achieving 95.6% accuracy using image processing systems (Nadim et al., 2019). Multi-class SVM (MCSVM) classified passion fruits based on ripeness, achieving 93.3% accuracy (Sidehabi et al., 2018). SVM assessed pork loin quality based on colour and quality attributes, meeting industry standards (Sun et al., 2018). FL developed a grading system for potatoes, combining CVS and fuzzy logic to allow faster grading and cost reduction (Bhagat & Markande, 2016).

FL also carried out qualitative grading of milled rice, achieving 89.83% total sensitivity and 97.45% specificity, proving useful for automatic grading in the industry (Zareiforoush et al., 2015). DT-FL classified Iranian green and black tea, with the REP decision tree proving more convenient than the J48 tree for developing a fuzzy classifier system (Bakhshipour et al., 2020). ANN detected tomato maturity, achieving 99.31% accuracy and 1.2% standard deviation (Wan et al., 2018). Lastly, FL classified vegetable seeds, distinguishing between cauliflower and Chinese cabbage seeds effectively (Garcia et al., 2020).

17.8.8 Near Infrared Spectroscopy

Near-infrared spectroscopy (NIRS) is widely utilized in the food industry due to its chemical-free nature, providing accurate and precise results within minutes or even continuously (Qu et al., 2015). This non-destructive, cost-effective, quick, and straightforward technique is a viable alternative to traditional, expensive, labour-intensive methods that consume significant time (Porep et al., 2015). NIRS's chemical-free approach makes it an environmentally friendly and health-safe option. It can perform a wide range of quantitative and qualitative analyses on gases, materials, slurries, powders, and solids. Samples need not be ground as the light passing through them reveals unique features or characteristics through their spectra. The spectra provide complex physical and chemical information about molecular bond vibrations, such as C–H, N–H, O–H, N–O, C–N, C–O, and C–C groups in organic materials, which can be recorded in reflection, interactance, or transmission modes (Qu et al., 2015).

Recently, NIRS has gained attention in the food industry for inspecting food quality, controlling study objectives, and evaluating food safety (Cortés et al., 2019; Cozzolino, 2016; Fu & Ying, 2016; Grassi & Alamprese, 2018; Porep et al., 2015; Qu et al., 2015). Researchers have applied NIRS to determine various properties of food, such as analysing the fatty acid profile in milk and goat milk fat groups (Núñez-Sánchez et al., 2016), predicting the composition of salted meat at different temperatures (Kartakoullis et al., 2019), and estimating sodium content in processed meat products (De Marchi et al., 2017). NIRS has also been used to detect and grade wooden breast syndrome in chicken fillets during processing (Wold et al., 2017), assess avocado maturity based on oil content (Olarewaju et al., 2016), predict acrylamide content in French-fried potatoes and potato flour (Adedipe et al., 2016), and determine fatty acid composition in lamb (Pullanagari et al., 2015). A review on the application of artificial neural networks (ANN) combined with NIRS for food detection and authenticity has been conducted (Liang et al., 2022). The integration of NIRS with soft computing techniques like ANN, fuzzy logic (FL), and machine learning (ML) enables rapid and accurate classification and prediction of samples.

17.8.8.1 Application of NIRS with AI in Food Industries

For cheese, PCA-ANN determined the characteristics of controlled-processing cheese, identifying sensory attributes with cost-effective and useful methods for quality control (Curto et al., 2020). In chicken meat, Decision Tree (DTree) classified meat based on quality grades, with DT models outperforming other methods in spectral analysis and classification (Barbon Jr. et al., 2018).

ANN was used to develop a quality assessment model for chocolate, fusing NIRS with ANN to predict chocolate quality based on chemical, physical, and sensory properties, achieving accurate predictions (Gunaratne et al., 2019). For civet coffee, ANN, SVM, and KNN detected alterations, with ANN showing better accuracy compared to other methods, achieving an accuracy range of 95–100% (Arboleda, 2018). In edible gelatin, ANN and SVM detected adulteration, with SVM showing the best recognition rate and ANN achieving a rate of 97.44% (Zhang et al., 2018).

RBFNN and PCA determined egg freshness with high accuracy, suitable for high-speed screening (Aboonajmi et al., 2016). SVM classified different types of food powders, achieving high accuracy for classifying wheat flour, organic wheat flour, tapioca starch, corn starch, and rice flour (Mohamed et al., 2019).

ANN, least square SVM (LSSVM), and Random Forest developed a model for discriminating different grades of Keemun black tea in China, with LSSVM showing higher performance (Ren et al., 2020). AFLS created a system for detecting meat spoilage, classifying meat into fresh, semi-fresh, and spoiled categories with a 95.94% correct classification rate (Alshejari & Kodogiannis, 2017). SVM detected adulteration in olive oils using NIRS and Raman spectral data, accurately predicting dopant contents (Xu et al., 2020).

ELM determined soluble solid contents in pears, with SPA-ELM outperforming conventional PCA-ELM methods (Lu et al., 2018b). ML techniques like RF, PCA, LDA, and PLS classified rice based on compositions and processing parameters, showing good performance (Rizwana & Hazarika, 2020). In wheat flour, multivariate SVM and RF predicted quality using NIRS, with RF overcoming performance issues (Barbon Junior et al., 2020). SVM predicted the origin of white asparagus, distinguishing German from imported products with high accuracy (Richter et al., 2019).

17.9 Future Trends

17.9.1 Future Trends in AI and Machine Learning Applications in Food Quality Analysis

Advanced quality control and food safety: AI powered technology involves in the early detection of food contaminants, pathogens in food products helps in ensuring the supply of safe products to the end customer. The intervention of machine learning techniques with analysis of the data from sensors, cameras, loggers and other sources where the Algorithms under transfer learning identifies the anomalies and indicates the presence of contaminants or harmful substances in the food with a signal. The prohibited items such as stones, metals, glass which are physical contaminants, and the biological contaminants like microbes and pathogens where the deep learning techniques would accurately detect and identify these contaminants, foreign materials.

Another key application of AI in food industry is the monitoring of production process in real time involves in real time data collection from the temperature sensors, humidity gauges, in-line cameras, and analysis of the collected data to check any deviations of the quality from the standards. The real time monitoring of production line would enhance traceability and ensures food safety and improves transparency in food supply chain. The combination of blockchain technology with AI helps in keep a digital record of the food chain from farm to plate. These information helps to track the origin of the raw materials, ingredients, to identify the points of potential contamination, and to verify the authenticity of the products used in food chain (Kumar & Saha, 2023). "Climate Foods Inc." employed AI in Deep intelligence platform for forecasting precision of the plant based food along with optimization of color, flavour, texture and nutritional value and sensory qualities.

17.9.2 Predictive Maintenance and Supply Chain Optimization

Predictive maintenance involves in real time prediction of failures of equipment before it occurs. The breakdown of the production process in the food industry can lead to the product losses due to the spoilage of the products and the production downtime, which can be overcome by predictive maintenance.

In terms of food equipment, the data collected from sensors in monitoring the temperatures, vibrations and other parameters of the machinery where the predictive maintenance involves the detection of wear and tear due to abnormality in temperature readings, increase in vibrations. This predictive maintenance involves in alerting the maintenance teams to schedule repairs. This would prevent in spoilage of products and avoid emergency repairs that adds up cost.

Supply Chain Optimization involves in simplification of raw materials and products flow from the suppliers to the customers. Efficient logistics is crucial in food industry to monitor the products where the food products are perishables that have short shelf life whose demands can rapidly fluctuate in food chain (Ramirez-Asis et al., 2022).

Supply chain optimization involves streamlining the flow of goods and materials from suppliers to customers. In the food industry, where products have a short shelf life and demand can fluctuate rapidly, efficient supply chain management is essential.

Supply chain optimization uses forecasting models in demand in predicting sales of food products. Based on the predictions made by models, the retailers can optimize inventory, reduces the risk of stocks and excess inventories. Retailers can optimize transportation with use of software to plan transportation routes, costs, ensures effective supply chain, predicting demands and reduces food wastes (Ejjami & Boussalham, 2024).

"Vivi nova" collaborated with ChatGPT for development of beverage with low sugar vegan beverage where the development time was reduced by 2 days with reduction in the production, streamline and cutting costs.

17.9.3 Personalized Nutrition and Dietary Recommendations

AI and machine learning algorithm are widely used in the field of nutrition and dietetics to provide dietary plans for individual based on their preferences and needs. The nutrition and diets are so important which plays a major role in maintaining a regular and health lifestyle for each and every individual. The effective dietary plan would reduce the risk of cardiovascular diseases, Diabetes (Type II), obesity, and other related health complications. The intervention of AI systems in dietetics involves in providing personalized meal plans for diabetics and other complications to reduce the consequences of the illness. Nonetheless, these AI systems encounter considerable obstacles arising from the complexities of prioritizing the genuine needs of people. The food plan suggested by the AI algorithm may not be suitable for all age groups, as an unbalanced diet can result in malnutrition (Zhu et al., 2023).

AI algorithm investigates the special needs of an individual based on the analysis of their data including portion of nutrients intake per day (Protein, fat requirement, carbohydrate intake, deficiencies in micronutrients) (Atwal, 2024). Based on the data analysis, AI algorithm recommends specific foods with calories to be consumed per day, meal plan, portion sizes of the meal. AI suggests meal plan by incorporating information about allergies and intolerances for individual.

AI powered application can track the dietary intake of individual and provide real time suggestions helping individual stay on track with their weights and diet consumption. Food manufacturers can develop personalized food products to meet the requirement of specific individual such as products with high protein, low carbohydrates, gluten free products (Roychowdhury, 2023).

17.9.4 Sustainable Food Production and Consumption

AI technology intervened drones and satellite monitors crops for the signs of diseases, stress, pest and insects attack, nutrition deficiencies help in farmers to understand and monitor these changes to fix resource wastage. AI algorithm analyse the previous history of yield and disease encounter information in order to predict the yield of crops, helps farmers for the optimization of decisions of sowing, planting and harvesting of crops. In terms of irrigation, AI sensors can be used to optimize the irrigation schedules based on the analysis of humidity of soil, weather patterns, water and nutrition requirements of crops using AI algorithm (Mavani et al., 2022).

AI algorithm analyse the consumer preferences, sales data to predict the demand and supply requirement of the particular food product, in order to reduce the waste and overproduction. It helps to optimize inventory levels to meet out demand without excess waste. It optimizes routes and schedules of transportation, reduces fuel consumption (Mavani et al., 2022).

17.9.5 Authenticity and Traceability

AI analyse data from RFID tags, sensors and other tracking devices to understand the view of supply chain. This would enable the retailer to identify and track any irregularities of the supply chain. AI analyse products attributes such as physical features, chemical properties, and digital signatures in order to verify their authenticity. This might avoid counterfeit products getting into the market, avoid tampering of products, disruptions of supply chain, ensure authenticity of products and enhance the loyalty and trust of consumers (González-Domínguez et al., 2022).

17.9.6 Virtual Tasting and Sensory Analysis

AI would create a virtual tasting experience using the advanced algorithm and sensory data by stimulating the aroma, taste, texture without physical consumption. The data from tasters, aroma preferences, evaluation of texture of the food product would help the AI algorithm to optimize the formulations of new products, develop new products. The AI algorithm could predict preferences and choices of consumers, market trend of the products and allows customer to select the products based on their diets and preferences (García-Esteban et al., 2018). Nestle utilizes virtual reality named "Web3 spaces" for analysis of trend in product purchase behaviour of consumers and products insights.

17.9.7 Intelligent Packaging Solutions

The packaging of food products is the crucial step in the food chain where the intelligent packaging, where the traditional packaging fused with technology creates revolution in the food supply chain. Intervening AI for the food packaging enhances the product quality. Food manufacturers could incorporate AI systems in packaging line to ensure the food safety, shelf life extension of the packed products and could provide customers with the specific information of the products in real time (Li et al., 2023).

Food manufacturers can incorporate AI systems in packaging line to ensure the food safety, in ensuring the changes in the products in response with the parameters like temperature, relative humidity, exposure in light can be tracked and monitored on real time. In food industries, sensors embedded in the packaging line would detect excessive temperature deviation from the standard set temperature to avoid the deterioration of the products in package and ensures the integrity of products. AI technology can check for any patterns in data of the product to verify the authenticity and genuineness of food product and ensure preventing the counterfeiting. Intervening of blockchain with Artificial Intelligence can create a record of transportation of the packed products from farm to plate. This ensures transparent and secured traceability with quick identification of the contamination of the products.

Use of QR tags, NFC tags can link to the information of the product, nutritional information that provide interactive experiences for the consumers. This ensures the loyalty of the brand and helps customers to make better choices (Nagaveni & Poosarla, 2024).

17.9.8 Food Allergen Detection and Prevention

AI algorithm analyse captured images of the food products in order to identify potential allergens on the visual inspection and detecting allergens in processed foods. Textural data of the products can be processed by applying Natural Language Processing (NLP) such as list of ingredients, customers review to find the allergens in food. This can prevent the mislabelling or allergic reactions caused by allergens in products. Chatbots and virtual assistants with intervention of AI technology can help customers to analyse the information regarding allergens in food and can adopt the food based on their food choices.

17.9.9 Consumer Engagement and Marketing

AI tools can monitor social media information, helps customers to find relevant keywords, and engage them in real-time for the purchase of the food products. Chatbots can provide personalized customer support, assists in recommendations of the food product that improves customer experience. AI technology can used to visualize Augmented reality and Virtual reality experiences for consumers before they buy and provides a better shopping experience. AI can analyze data from customers to recommend products based on their choices which helps in increase in size and satisfaction of customers.

17.10 Conclusion

The integration of AI and Machine learning techniques with non-destructive technologies has created the opportunities for analysis of food quality and safety. These technologies provide combination of data processing, recognition of pattern, and prediction. AI and ML algorithms can analyse the data from non-destructive sensors, identify and improves the efficiency and accuracy in the assessments of food quality, which leads to quick decision making and reduced recalls for the products. It leads to the monitoring of the food quality parameters through the food supply chain ensuring the freshness and quality of the product from farm to fork. These technologies work on low investment and operational costs in comparison with conventional destructive methods which is more accessible to food business entrepreneurs. These techniques enable early detection of quality issues of food products which can help in reduction of food wastage due to deterioration of the food products in storage and market level. AI intervening in agricultural sectors can contribute to the sustainability in the food production and supply chain with minimizing resources need and contributing less environmental impacts. AI involves in the enhancement of the confidence of the consumers on the purchase of the food products provided with information on quality and safety of the products transparently. This would improve the food business with stringent regulatory compliances and policies provided with accurate data on the food quality. Future advancements in the exploration of new applications with integration of advanced and less cost sensors, with improvised data analysis methodology were necessary in improving the efficiency in quality analysis in food industry.

References

Abbaspour-Gilandeh, Y., Jahanbakhshi, A., & Kaveh, M. (2020). Prediction kinetic, energy and exergy of quince under hot air dryer using ANNs and ANFIS. *Food Science & Nutrition, 8*(1), 594–611. https://doi.org/10.1002/fsn3.1347

Abd. Rahman, N., Hussain, M., Jahim, J., & Jahim. (2012). Production of fructose using recycle fixed-bed reactor and batch bioreactor. *Journal of Food, Agriculture & Environment, 1010*, 268–273.

Abdul Aziz, F. A. B., Abd. Rahman, N., & Mohd Ali, J. (2019). Tropospheric ozone formation estimation in Urban City, Bangi, using artificial neural network (ANN). *Computational Intelligence and Neuroscience, 2019*(1), 6252983. https://doi.org/10.1155/2019/6252983

Abduljabbar, R., Dia, H., Liyanage, S., & Bagloee, S. A. (2019). Applications of artificial intelligence in transport: An overview. *Sustainability*. https://www.mdpi.com/2071-1050/11/1/189

Aboonajmi, M., Saberi, A., Abbasian Najafabadi, T., & Kondo, N. (2016). Quality assessment of poultry egg based on visible–near infrared spectroscopy and radial basis function networks. *International Journal of Food Properties, 19*(5), 1163–1172. https://doi.org/10.1080/10942912.2015.1075215

Adak, M. F., & Yumusak, N. (2016). Classification of E-nose aroma data of four fruit types by ABC based neural network. *Sensors, 16*(3), Article 3. https://doi.org/10.3390/s16030304

Adedipe, O. E., Johanningsmeier, S. D., Truong, V.-D., & Yencho, G. C. (2016). Development and validation of a near-infrared spectroscopy method for the prediction of acrylamide content in French-fried potato. *Journal of Agricultural and Food Chemistry, 64*(8), 1850–1860. https://doi.org/10.1021/acs.jafc.5b04733

Alaiz-Rodríguez, R., & Parnell, A. C. (2020). A machine learning approach for lamb meat quality assessment using FTIR spectra. *IEEE Access, 8*, 52385–52394. IEEE Access. https://doi.org/10.1109/ACCESS.2020.2974623

Ali, J. A., Hannan, M. A., Mohamed, A., & Abdolrasol, M. G. M. (2016). Fuzzy logic speed controller optimization approach for induction motor drive using backtracking search algorithm. *Measurement, 78*, 49–62. https://doi.org/10.1016/j.measurement.2015.09.038

Al-Mahasneh, M., Aljarrah, M., Rababah, T., & Alu'datt, M. (2016). Application of hybrid neural fuzzy system (ANFIS) in food processing and technology. *Food Engineering Reviews, 8*(3), 351–366. https://doi.org/10.1007/s12393-016-9141-7

Alsaqour, R., Abdelhaq, M., Saeed, R., Uddin, M., Alsukour, O., Al-Hubaishi, M., & Alahdal, T. (2015). Dynamic packet beaconing for GPSR mobile ad hoc position-based routing protocol using fuzzy logic. *Journal of Network and Computer Applications, 47*, 32–46. https://doi.org/10.1016/j.jnca.2014.08.008

Alshejari, A., & Kodogiannis, V. S. (2017). An intelligent decision support system for the detection of meat spoilage using multispectral images. *Neural Computing and Applications, 28*(12), 3903–3920. https://doi.org/10.1007/s00521-016-2296-6

Al-Waeli, A. H. A., Sopian, K., Yousif, J. H., Kazem, H. A., Boland, J., & Chaichan, M. T. (2019). Artificial neural network modeling and analysis of photovoltaic/thermal system based on the experimental study. *Energy Conversion and Management, 186*, 368–379. https://doi.org/10.1016/j.enconman.2019.02.066

Amiryousefi, M. R., Mohebbi, M., Golmohammadzadeh, S., Koocheki, A., & Baghbani, F. (2017). Fuzzy logic application to model caffeine release from hydrogel colloidosomes. *Journal of Food Engineering, 212*, 181–189. https://doi.org/10.1016/j.jfoodeng.2017.05.031

Anastácio, A., Silva, R., & Carvalho, I. S. (2016). Phenolics extraction from sweet potato peels: Modelling and optimization by response surface modelling and artificial neural network. *Journal of Food Science and Technology, 53*(12), 4117–4125. https://doi.org/10.1007/s13197-016-2354-1

Arabameri, M., Nazari, R. R., Abdolshahi, A., Abdollahzadeh, M., Mirzamohammadi, S., Shariatifar, N., Barba, F. J., & Mousavi Khaneghah, A. (2019). Oxidative stability of virgin olive oil: Evaluation and prediction with an adaptive neuro-fuzzy inference system (ANFIS). *Journal of the Science of Food and Agriculture, 99*(12), 5358–5367. https://doi.org/10.1002/jsfa.9777

Arboleda, E. R. (2018). Discrimination of civet coffee using near infrared spectroscopy and artificial neural network. *International Journal of Advanced Computer Research, 8*(39), 324–334. https://doi.org/10.19101/IJACR.2018.839007

Ardabili, S., Mosavi, A., Mahmoudi, A., Gundoshmian, T. M., Nosratabadi, S., & Várkonyi-Kóczy, A. R. (2020). Modelling temperature variation of mushroom growing hall using artificial neural networks. In A. R. Várkonyi-Kóczy (Ed.), *Engineering for sustainable future* (pp. 33–45). Springer International Publishing. https://doi.org/10.1007/978-3-030-36841-8_3

Ardiansah, I., Efatmi, F., Mardawati, E., & Putri, S. H. (2020). Feasibility testing of a household industry food production certificate using an expert system with forward chaining method. *Jurnal Online Informatika*, 137–144. https://doi.org/10.15575/join.v5i2.579

Asnaashari, M., Farhoosh, R., & Farahmandfar, R. (2016). Prediction of oxidation parameters of purified Kilka fish oil including gallic acid and methyl gallate by adaptive neuro-fuzzy inference system (ANFIS) and artificial neural network. *Journal of the Science of Food and Agriculture, 96*(13), 4594–4602. https://doi.org/10.1002/jsfa.7677

Astray, G., Albuquerque, B. R., Prieto, M. A., Simal-Gandara, J., Ferreira, I. C. F. R., & Barros, L. (2020). Stability assessment of extracts obtained from Arbutus unedo L. fruits in powder and solution systems using machine-learning methodologies. *Food Chemistry, 333*, 127460. https://doi.org/10.1016/j.foodchem.2020.127460

Atwal, K. (2024). Artificial intelligence in clinical nutrition and dietetics: A brief overview of current evidence. *Nutrition in Clinical Practice, 39*(4), 736–742. https://doi.org/10.1002/ncp.11150

Ayari, F., Mirzaee- Ghaleh, E., Rabbani, H., & Heidarbeigi, K. (2018). Using an E-nose machine for detection the adulteration of margarine in cow ghee. *Journal of Food Process Engineering, 41*(6), e12806. https://doi.org/10.1111/jfpe.12806

Azadbakht, M., Aghili, H., Ziaratban, A., & Torshizi, M. V. (2017). Application of artificial neural network method to exergy and energy analyses of fluidized bed dryer for potato cubes. *Energy, 120*, 947–958. https://doi.org/10.1016/j.energy.2016.12.006

Bahmani, A., Jafari, S. M., Shahidi, S.-A., & Dehnad, D. (2016). Mass transfer kinetics of eggplant during osmotic dehydration by neural networks. *Journal of Food Processing and Preservation, 40*(5), 815–827. https://doi.org/10.1111/jfpp.12435

Bahram-Parvar, M., Salehi, F., & Razavi, S. M. A. (2017). Adaptive neuro-fuzzy inference system (ANFIS) simulation for predicting overall acceptability of ice cream. *Engineering in Agriculture, Environment and Food, 10*(2), 79–86. https://doi.org/10.1016/j.eaef.2016.11.001

Baietto, M., & Wilson, A. D. (2015). Electronic-nose applications for fruit identification, ripeness and quality grading. *Sensors, 15*(1), Article 1. https://doi.org/10.3390/s150100899

Bakhshipour, A., Zareiforoush, H., & Bagheri, I. (2020). Application of decision trees and fuzzy inference system for quality classification and modeling of black and green tea based on visual features. *Journal of Food Measurement and Characterization, 14*(3), 1402–1416. https://doi.org/10.1007/s11694-020-00390-8

Baliuta, S., Kopylova, L., Kuievda, Y., Kuyevda, V., & Kovalchuk, O. (2020). *Fuzzy logic energy management system of food manufacturing processes*. https://dspace.nuft.edu.ua/handle/123456789/37987

Banakar, A., Zareiforoush, H., Baigvand, M., Montazeri, M., Khodaei, J., & Behroozi-Khazaei, N. (2017). Combined application of decision tree and fuzzy logic techniques for intelligent grading of dried figs. *Journal of Food Process Engineering, 40*(3), e12456. https://doi.org/10.1111/jfpe.12456

Barbon, S., Jr., Costa Barbon, A. P. A. d., Mantovani, R. G., & Barbin, D. F. (2018). Machine learning applied to near-infrared spectra for chicken meat classification. *Journal of Spectroscopy, 2018*(1), 8949741. https://doi.org/10.1155/2018/8949741

Barbon Junior, S., Mastelini, S. M., Barbon, A. P. A. C., Barbin, D. F., Calvini, R., Lopes, J. F., & Ulrici, A. (2020). Multi-target prediction of wheat flour quality parameters with near infrared spectroscopy. *Information Processing in Agriculture, 7*(2), 342–354. https://doi.org/10.1016/j.inpa.2019.07.001

Baryannis, G., Validi, S., Dani, S., & Antoniou, G. (2019). Supply chain risk management and artificial intelligence: State of the art and future research directions. *International Journal of Production Research, 57*(7), 2179–2202. https://doi.org/10.1080/00207543.2018.1530476

Basak, S. (2018). The use of fuzzy logic to determine the concentration of betel leaf essential oil and its potency as a juice preservative. *Food Chemistry, 240*, 1113–1120. https://doi.org/10.1016/j.foodchem.2017.08.047

Belluco, S., Losasso, C., Maggioletti, M., Alonzi, C. C., Paoletti, M. G., & Ricci, A. (2013). Edible insects in a food safety and nutritional perspective: A critical review. *Comprehensive Reviews in Food Science and Food Safety, 12*(3), 296–313. https://doi.org/10.1111/1541-4337.12014

Benković, M., Tušek, A. J., Belščak-Cvitanović, A., Lenart, A., Domian, E., Komes, D., & Bauman, I. (2015). Artificial neural network modelling of changes in physical and chemical properties of cocoa powder mixtures during agglomeration. *LWT - Food Science and Technology, 64*(1), 140–148. https://doi.org/10.1016/j.lwt.2015.05.028

Bhagat, N. B., & Markande, S. D. (2016). Automatic grading of potatoes with fuzzy logic. In *2016 online international conference on green engineering and technologies (IC-GET)* (pp. 1–4). https://doi.org/10.1109/GET.2016.7916651

Blagoveshchenskiy, I. G., Blagoveshchenskiy, V. G., Besfamilnaya, E. M., & Sumerin, V. A. (2020). Development of databases of intelligent expert systems for automatic control of product quality indicators. *Journal of Physics: Conference Series, 1705*(1), 012019. https://doi.org/10.1088/1742-6596/1705/1/012019

Blasi, A. (2005). Scheduling food industry system using fuzzy logic. *Journal of Theoretical and Applied Information Technology, 96*(9), 6463–6473

Bortolini, M., Faccio, M., Ferrari, E., Gamberi, M., & Pilati, F. (2016). Fresh food sustainable distribution: Cost, delivery time and carbon footprint three-objective optimization. *Journal of Food Engineering, 174*, 56–67. https://doi.org/10.1016/j.jfoodeng.2015.11.014

Bose, B. K. (2017). Artificial intelligence techniques in smart grid and renewable energy systems—Some example applications. *Proceedings of the IEEE, 105*(11), 2262–2273. Proceedings of the IEEE. https://doi.org/10.1109/JPROC.2017.2756596

Bostrom, N. (2014). Superintelligence: Paths, dangers, strategies. *The Philosophical Quarterly*. Oxford Academic. https://academic.oup.com/pq/article-abstract/66/262/196/2460870

Bostrom, N., et al. (2014). Chapter 4: The ethics of artificial intelligence. In *Artificial intelligence safety and security*. Chapman and Hall/CRC. https://www.taylorfrancis.com/chapters/edit/10.1201/9781351251389-4/ethics-artificial-intelligence-nick-bostrom-eliezer-yudkowsky

Bouhoune, K., Yazid, K., Boucherit, M. S., & Chériti, A. (2017). Hybrid control of the three phase induction machine using artificial neural networks and fuzzy logic. *Applied Soft Computing, 55*, 289–301. https://doi.org/10.1016/j.asoc.2017.01.048

Brynjolfsson, E., & Mcafee, A. N. D. R. E. W. (2017). Artificial intelligence, for real. *Harvard business review, 1*(1), 1–31. https://starlab-alliance.com/wp-content/uploads/2017/09/AI-Article.pdf

Budiyanto, G., Ipnuwati, S., Gifari, S. A. A., Huda, M., Jalal, B., Abdul, A., Maseleno, A., & Hananto, A. L. (2018). Web based expert system for diagnosing disease pest on banana plant. *International Journal of Engineering and Technology, 7*(4), 4715–4721.

Butler, K. T., Davies, D. W., Cartwright, H., Isayev, O., & Walsh, A. (2018). Machine learning for molecular and materials science. *Nature, 559*(7715), 547–555. https://doi.org/10.1038/s41586-018-0337-2

Cai, H., Xu, L., Xu, B., Zhang, P., Guo, J., & Zhang, Y. (2018). A service governance mechanism based on process mining for cloud-based applications. *Enterprise Information Systems, 12*(10). https://www.tandfonline.com/doi/abs/10.1080/17517575.2018.1442933

Cai-Ming, Z., & Hao-Nan, C. (2020). Preprocessing method of structured big data in human resource archives database. In *2020 IEEE international conference on industrial application of artificial intelligence (IAAI)* (pp. 379–384). https://doi.org/10.1109/IAAI51705.2020.9332880

Cárdenas-Pérez, S., Chanona-Pérez, J., Méndez-Méndez, J. V., Calderón-Domínguez, G., López-Santiago, R., Perea-Flores, M. J., & Arzate-Vázquez, I. (2017). Evaluation of the ripening stages of apple (Golden Delicious) by means of computer vision system. *Biosystems Engineering, 159*, 46–58. https://doi.org/10.1016/j.biosystemseng.2017.04.009

Carleo, G., Cirac, I., Cranmer, K., Daudet, L., Schuld, M., Tishby, N., Vogt-Maranto, L., & Zdeborová, L. (2019). Machine learning and the physical sciences. *Reviews of Modern Physics, 91*(4), 045002. https://doi.org/10.1103/RevModPhys.91.045002

Castro, W., Oblitas, J., De-La-Torre, M., Cotrina, C., Bazán, K., & Avila-George, H. (2019). Classification of cape gooseberry fruit according to its level of ripeness using machine learning techniques and different color spaces. *IEEE Access, 7*, 27389–27400. IEEE Access. https://doi.org/10.1109/ACCESS.2019.2898223

Cavallo, D. P., Cefola, M., Pace, B., Logrieco, A. F., & Attolico, G. (2019). Non-destructive and contactless quality evaluation of table grapes by a computer vision system. *Computers and Electronics in Agriculture, 156*, 558–564. https://doi.org/10.1016/j.compag.2018.12.019

Cebi, N., Sagdic, O., Basahel, A. M., Balubaid, M. A., Taylan, O., Yaman, M., & Yilmaz, M. T. (2019). Modeling and optimization of ultrasound-assisted cinnamon extraction process using fuzzy and response surface models. *Journal of Food Process Engineering, 42*(2), e12978. https://doi.org/10.1111/jfpe.12978

Chasiotis, V. K., Tzempelikos, D. A., Filios, A. E., & Moustris, K. P. (2020). Artificial neural network modelling of moisture content evolution for convective drying of cylindrical quince slices. *Computers and Electronics in Agriculture, 172*, 105074. https://doi.org/10.1016/j.compag.2019.105074

Chen, X.-W., & Lin, X. (2014). Big data deep learning: Challenges and perspectives | IEEE Journals & Magazine | IEEE Xplore. https://ieeexplore.ieee.org/abstract/document/6817512

Chen, Y., Cai, K., Tu, Z., Nie, W., Ji, T., Hu, B., Chen, C., & Jiang, S. (2018). Prediction of benzo [a]pyrene content of smoked sausage using back-propagation artificial neural network. *Journal of the Science of Food and Agriculture, 98*(8), 3022–3030. https://doi.org/10.1002/jsfa.8801

Chen, H., Li, L., & Chen, Y. (2020). Explore success factors that impact artificial intelligence adoption on telecom industry in China. *Journal of Management Analytics, 8*(1). https://www.tandfonline.com/doi/abs/10.1080/23270012.2020.1852895

Chi-Hsien, K., & Nagasawa, S. (2019). Applying machine learning to market analysis: Knowing your luxury consumer. *Journal of Management Analytics, 6*(4) https://www.tandfonline.com/doi/abs/10.1080/23270012.2019.1692254

Chmiel, M., & Słowiński, M. (2016). The use of computer vision system to detect pork defects. *LWT, 73*, 473–480. https://doi.org/10.1016/j.lwt.2016.06.054

Chollet, F. (2017). *Xception: Deep learning with depthwise separable convolutions* (pp. 1251–1258) https://openaccess.thecvf.com/content_cvpr_2017/html/Chollet_Xception_Deep_Learning_CVPR_2017_paper.html

Chowdhury, T., & Das, M. (2015). Sensory evaluation of aromatic foods packed in developed starch based films using fuzzy logic. *International Journal of Food Studies, 4*(1), Article 1. https://doi.org/10.7455/ijfs/4.1.2015.a3

Chung, C.-C., Chen, H.-H., & Ting, C.-H. (2016). Fuzzy logic for accurate control of heating temperature and duration in canned food sterilisation. *Engineering in Agriculture, Environment and Food, 9*(2), 187–194. https://doi.org/10.1016/j.eaef.2015.11.003

Coley, J. D., & Tanner, K. D. (2012). Common origins of diverse misconceptions: Cognitive principles and the development of biology thinking. *CBE—Life Sciences Education, 11*(3), 209–215. https://doi.org/10.1187/cbe.12-06-0074

Collins, G. S., & Moons, K. G. M. (2019). Reporting of artificial intelligence prediction models. *The Lancet, 393*(10181), 1577–1579. https://doi.org/10.1016/S0140-6736(19)30037-6

Corney, D. (2002). Food bytes: Intelligent systems in the food industry. *British Food Journal, 104*(10), 787–805. https://doi.org/10.1108/00070700210448890

Correa, D. A., Montero Castillo, P. M., & Martelo, R. J. (2018). Neural networks in food industry. *Contemporary Engineering Sciences, 11*(37), 1807–1826. https://doi.org/10.12988/ces.2018.84141

Cortés, V., Blasco, J., Aleixos, N., Cubero, S., & Talens, P. (2019). Monitoring strategies for quality control of agricultural products using visible and near-infrared spectroscopy: A review. *Trends in Food Science & Technology, 85*, 138–148. https://doi.org/10.1016/j.tifs.2019.01.015

Cozzolino, D. (2016). Chapter 7: Near infrared spectroscopy and food authenticity. In M. Espiñeira & F. J. Santaclara (Eds.), *Advances in food traceability techniques and technologies* (pp. 119–136). Woodhead Publishing. https://doi.org/10.1016/B978-0-08-100310-7.00007-7

Curto, B., Moreno, V., García-Esteban, J. A., Blanco, F. J., González, I., Vivar, A., & Revilla, I. (2020). Accurate prediction of sensory attributes of cheese using near-infrared spectroscopy based on artificial neural network. *Sensors, 20*(12), Article 12. https://doi.org/10.3390/s20123566

Dang, N. T., Vo, M.-T., Nguyen, T.-D., & Dao, S. V. T. (2019). Analysis on mangoes weight estimation problem using neural network. In *2019 19th international symposium on communications and information technologies (ISCIT)* (pp. 559–562). https://doi.org/10.1109/ISCIT.2019.8905118

Davenport, T. H., & Ronanki, R. (2018). Artificial intelligence artificial intelligence for the real world. Don't start with moon shots. *Harvard Bus. Rev*, 1–10

De Marchi, M., Manuelian, C. L., Ton, S., Manfrin, D., Meneghesso, M., Cassandro, M., & Penasa, M. (2017). Prediction of sodium content in commercial processed meat products using near infrared spectroscopy. *Meat Science, 125*, 61–65. https://doi.org/10.1016/j.meatsci.2016.11.014

de Oliveira, E. M., Leme, D. S., Barbosa, B. H. G., Rodarte, M. P., & Pereira, R. G. F. A. (2016). A computer vision system for coffee beans classification based on computational intelligence techniques. *Journal of Food Engineering, 171*, 22–27. https://doi.org/10.1016/j.jfoodeng.2015.10.009

de Sá, A. C., Cipri, A., González-Calabuig, A., Stradiotto, N. R., & del Valle, M. (2016). Resolution of galactose, glucose, xylose and mannose in sugarcane bagasse employing a voltammetric electronic tongue formed by metals oxy-hydroxide/MWCNT modified electrodes. *Sensors and Actuators B: Chemical, 222*, 645–653. https://doi.org/10.1016/j.snb.2015.08.088

de Sousa Silva, M., Cruz, L. F., Bugatti, P. H., & Saito, P. T. M. (2020). Automatic visual quality assessment of biscuits using machine learning. In L. Rutkowski, R. Scherer, M. Korytkowski, W. Pedrycz, R. Tadeusiewicz, & J. M. Zurada (Eds.), *Artificial intelligence and soft computing* (pp. 59–70). Springer International Publishing. https://doi.org/10.1007/978-3-030-61534-5_6

Deo, R. C. (2015). Machine learning in medicine. *Circulation*. https://doi.org/10.1161/CIRCULATIONAHA.115.001593

Deshmukh, S., Bandyopadhyay, R., Bhattacharyya, N., Pandey, R. A., & Jana, A. (2015). Application of electronic nose for industrial odors and gaseous emissions measurement and monitoring – An overview. *Talanta, 144*, 329–340. https://doi.org/10.1016/j.talanta.2015.06.050

Dhingra, M., Jain, M., & Jadon, R. S. (2016). Role of artificial intelligence in enterprise information security: A review. In *2016 fourth international conference on parallel, distributed and grid computing (PDGC)* (pp. 188–191). https://doi.org/10.1109/PDGC.2016.7913142

Doganis, P., Alexandridis, A., Patrinos, P., & Sarimveis, H. (2006). Time series sales forecasting for short shelf-life food products based on artificial neural networks and evolutionary computing. *Journal of Food Engineering, 75*(2), 196–204. https://doi.org/10.1016/j.jfoodeng.2005.03.056

Duan, L., & Xu, L. D. (2012). Business intelligence for enterprise systems: A survey. *IEEE Transactions on Industrial Informatics, 8*(3), 679–687. IEEE Transactions on Industrial Informatics. https://doi.org/10.1109/TII.2012.2188804

Duan, L., Xu, L., Liu, Y., & Lee, J. (2009). Cluster-based outlier detection. *Annals of Operations Research, 168*(1), 151–168. https://doi.org/10.1007/s10479-008-0371-9

Duan, N., Liu, L.-Z., Yu, X.-J., Li, Q., & Yeh, S.-C. (2019). Classification of multichannel surface-electromyography signals based on convolutional neural networks. *Journal of Industrial Information Integration, 15*, 201–206. https://doi.org/10.1016/j.jii.2018.09.001

Dwivedi, Y. K., Hughes, L., Ismagilova, E., Aarts, G., Coombs, C., Crick, T., Duan, Y., Dwivedi, R., Edwards, J., Eirug, A., Galanos, V., Ilavarasan, P. V., Janssen, M., Jones, P., Kar, A. K., Kizgin, H., Kronemann, B., Lal, B., Lucini, B., et al. (2021). Artificial intelligence (AI): Multidisciplinary perspectives on emerging challenges, opportunities, and agenda for research, practice and policy. *International Journal of Information Management, 57*, 101994. https://doi.org/10.1016/j.ijinfomgt.2019.08.002

Ejjami, R., & Boussalham, K. (2024). Resilient supply chains in industry 5.0: Leveraging AI for predictive maintenance and risk mitigation. *International Journal For Multidisciplinary Research, 6*(4).

Erhan, D., Courville, A., Bengio, Y., & Vincent, P. (2010). Why does unsupervised pre-training help deep learning? In *Proceedings of the thirteenth international conference on artificial intelligence and statistics* (pp. 201–208) https://proceedings.mlr.press/v9/erhan10a.html

Erickson, B. J., Korfiatis, P., Akkus, Z., & Kline, T. L. (2017). Machine learning for medical imaging. *Radiographics, 37*(2), 505–515. https://doi.org/10.1148/rg.2017160130

Estelles-Lopez, L., Ropodi, A., Pavlidis, D., Fotopoulou, J., Gkousari, C., Peyrodie, A., Panagou, E., Nychas, G.-J., & Mohareb, F. (2017). An automated ranking platform for machine learning regression models for meat spoilage prediction using multi-spectral imaging and metabolic profiling. *Food Research International, 99*, 206–215. https://doi.org/10.1016/j.foodres.2017.05.013

Faal, S., Loghavi, M., & Kamgar, S. (2019). Physicochemical properties of Iranian ziziphus honey and emerging approach for predicting them using electronic nose. *Measurement, 148*, 106936. https://doi.org/10.1016/j.measurement.2019.106936

Fan, S., Li, J., Zhang, Y., Tian, X., Wang, Q., He, X., Zhang, C., & Huang, W. (2020). On line detection of defective apples using computer vision system combined with deep learning methods. *Journal of Food Engineering, 286*, 110102. https://doi.org/10.1016/j.jfoodeng.2020.110102

Farzaneh, V., Bakhshabadi, H., Gharekhani, M., Ganje, M., Farzaneh, F., Rashidzadeh, S., & Carvalho, I. S. (2017a). Application of an adaptive neuro_fuzzy inference system (ANFIS) in the modeling of rapeseeds' oil extraction. *Journal of Food Process Engineering, 40*(6), e12562. https://doi.org/10.1111/jfpe.12562

Farzaneh, V., Ghodsvali, A., Bakhshabadi, H., Ganje, M., Dolatabadi, Z., Carvalho, S., & I. (2017b). Modelling of the selected physical properties of the fava bean with various moisture contents using fuzzy logic design. *Journal of Food Process Engineering, 40*(2), e12366. https://doi.org/10.1111/jfpe.12366

Fatma, S., Sharma, N., Singh, S., Jha, A., & Kumar, A. (2016). Fuzzy analysis of sensory data for ranking of beetroot candy. *ETP International Journal of Food Engineering*. https://doi.org/10.18178/ijfe.2.1.26-30

Feng, H., Zhang, M., Liu, P., Liu, Y., & Zhang, X. (2020). Evaluation of IoT-enabled monitoring and electronic nose spoilage detection for Salmon freshness during cold storage. *Food, 9*(11), Article 11. https://doi.org/10.3390/foods9111579

Fischetti, M. (2011). *Computers versus brains*. Scientific American. https://www.scientificamerican.com/article/computers-vs-brains/

Fu, X., & Ying, Y. (2016). Food safety evaluation based on near infrared spectroscopy and imaging: A review. *Critical Reviews in Food Science and Nutrition, 56*(11), 1913–1924. https://doi.org/10.1080/10408398.2013.807418

Fu, Y., Wang, S., Wang, C. -X., Hong, X., & McLaughlin, S. (2018). Artificial intelligence to manage network traffic of 5G wireless networks | IEEE Journals & Magazine | IEEE Xplore. https://ieeexplore.ieee.org/abstract/document/8553655

Funes, E., Allouche, Y., Beltrán, G., & Jiménez, A. (2015). A review: Artificial neural networks as tool for control food industry process. *Journal of Sensor Technology, 05*(01), Article 01. https://doi.org/10.4236/jst.2015.51004

Gandhi, N., Petkar, O., & Armstrong, L. J. (2016). Rice crop yield prediction using artificial neural networks. In *2016 IEEE technological innovations in ICT for agriculture and rural development (TIAR)* (pp. 105–110). https://doi.org/10.1109/TIAR.2016.7801222

Garcia, J. A. A., Arboleda, E. R., & Galas, E. M. (2020). Identification of visually similar vegetable seeds using image processing and fuzzy logic. *International Journal of Scientific & Technology Research, 9*(02).

García-Esteban, J. A., Curto, B., Moreno, V., González-Martín, I., Revilla, I., & Vivar-Quintana, A. (2018). A digitalization strategy for quality control in food industry based on Artificial Intelligence techniques. In *2018 IEEE 16th international conference on industrial informatics (INDIN)* (pp. 221–226).

Garre, A., Ruiz, M. C., & Hontoria, E. (2020). Application of Machine Learning to support production planning of a food industry in the context of waste generation under uncertainty. *Operations Research Perspectives, 7*, 100147. https://doi.org/10.1016/j.orp.2020.100147

Lawton, G. (2024). *4 types of learning in machine learning explained*. https://www.techtarget.com/searchenterpriseai/tip/Types-of-learning-in-machine-learning-explained

Gil-Sánchez, L., Garrigues, J., Garcia-Breijo, E., Grau, R., Aliño, M., Baigts, D., & Barat, J. M. (2015). Artificial neural networks (Fuzzy ARTMAP) analysis of the data obtained with an electronic tongue applied to a ham-curing process with different salt formulations. *Applied Soft Computing, 30*, 421–429. https://doi.org/10.1016/j.asoc.2014.12.037

Gliszczyńska-Świgło, A., & Chmielewski, J. (2017). Electronic nose as a tool for monitoring the authenticity of food. A review. *Food Analytical Methods, 10*(6), 1800–1816. https://doi.org/10.1007/s12161-016-0739-4

Goertzel, B. (2007). Human-level artificial general intelligence and the possibility of a technological singularity: A reaction to Ray Kurzweil's The Singularity Is Near, and McDermott's critique of Kurzweil. *Artificial Intelligence, 171*(18), 1161–1173. https://doi.org/10.1016/j.artint.2007.10.011

Gonzalez Viejo, C., Fuentes, S., Torrico, D., Howell, K., & Dunshea, F. R. (2018). Assessment of beer quality based on foamability and chemical composition using computer vision algorithms, near infrared spectroscopy and machine learning algorithms. *Journal of the Science of Food and Agriculture, 98*(2), 618–627. https://doi.org/10.1002/jsfa.8506

Gonzalez Viejo, C., Torrico, D. D., Dunshea, F. R., & Fuentes, S. (2019). Development of artificial neural network models to assess beer acceptability based on sensory properties using a robotic pourer: A comparative model approach to achieve an artificial intelligence system. *Beverages, 5*(2), Article 2. https://doi.org/10.3390/beverages5020033

González-Domínguez, R., Sayago, A., & Fernández-Recamales, Á. (2022). An overview on the application of chemometrics tools in food authenticity and traceability. *Food, 11*(23), 3940.

Gonzalez-Fernandez, I., Iglesias-Otero, M. A., Esteki, M., Moldes, O. A., Mejuto, J. C., & Simal-Gandara, J. (2019). A critical review on the use of artificial neural networks in olive oil production, characterization and authentication. *Critical Reviews in Food Science and Nutrition, 59*(12), 1913–1926. https://doi.org/10.1080/10408398.2018.1433628

Górska-Horczyczak, E., Horczyczak, M., Guzek, D., Wojtasik-Kalinowska, I., & Wierzbicka, A. (2017). Chromatographic fingerprints supported by artificial neural network for differentiation of fresh and frozen pork. *Food Control, 73*, 237–244. https://doi.org/10.1016/j.foodcont.2016.08.010

Grassi, S., & Alamprese, C. (2018). Advances in NIR spectroscopy applied to process analytical technology in food industries. *Current Opinion in Food Science, 22*, 17–21. https://doi.org/10.1016/j.cofs.2017.12.008

Gunaratne, T. M., Gonzalez Viejo, C., Gunaratne, N. M., Torrico, D. D., Dunshea, F. R., & Fuentes, S. (2019). Chocolate quality assessment based on chemical fingerprinting using near infra-red and machine learning modeling. *Food, 8*(10), Article 10. https://doi.org/10.3390/foods8100426

Guo, T., Yin, T., Ma, Z., Wang, Z., Sun, X., & Yuan, W. (2018). Characterization of different processes lemon slice using electronic tongue. *IFAC-PapersOnLine, 51*(17), 683–688. https://doi.org/10.1016/j.ifacol.2018.08.117

Gutiérrez, P., Godoy, S. E., Torres, S., Oyarzún, P., Sanhueza, I., Díaz-García, V., Contreras-Trigo, B., & Coelho, P. (2020). Improved antibiotic detection in raw Milk using machine learning tools over the absorption spectra of a problem-specific nanobiosensor. *Sensors*. https://www.mdpi.com/1424-8220/20/16/4552

Hannan, M. A., Ghani, Z. A. B. D., Hoque, M. M., Ker, P. J., Hussain, A., & Mohamed, A. (2019). Fuzzy logic inverter controller in photovoltaic applications: Issues and recommendations. *IEEE Access, 7*, 24934–24955. IEEE Access. https://doi.org/10.1109/ACCESS.2019.2899610

Harsawardana, Samodro, B., Mahesworo, B., Suparyanto, T., Atmaja, D. B. S., & Pardamean, B. (2020). Maintaining the quality and aroma of coffee with fuzzy logic coffee roasting machine. *IOP Conference Series: Earth and Environmental Science, 426*(1), 012148. https://doi.org/10.1088/1755-1315/426/1/012148

Hasan, M., Sarno, R., & Sabilla, S. (2020). Optimizing machine learning parameters for classifying the sweetness of pineapple aroma using electronic nose. *International Journal of Intelligent Engineering and Systems, 13*(5), 122–132. https://doi.org/10.22266/ijies2020.1031.12

Haydari, A., & Yılmaz, Y. (2020). Deep reinforcement learning for intelligent transportation systems: A survey | IEEE Journals & Magazine | IEEE Xplore. https://ieeexplore.ieee.org/abstract/document/9146378

Heidarbeigi, K., Mohtasebi, S. S., Foroughirad, A., Ghasemi-Varnamkhasti, M., Rafiee, S., & Rezaei, K. (2015). Detection of adulteration in saffron samples using electronic nose. *International Journal of Food Properties, 18*(7), 1391–1401. https://doi.org/10.1080/10942912.2014.915850

Hernández-Vera, B., Aguilar Lasserre, A. A., Gastón Cedillo-Campos, M., Herrera-Franco, L. E., & Ochoa-Robles, J. (2017). Expert system based on fuzzy logic to define the production process in the coffee industry. *Journal of Food Process Engineering, 40*(2), e12389. https://doi.org/10.1111/jfpe.12389

Hu, S., Liang, Y.-C., Xiong, Z., & Niyato, D. (2021). Blockchain and artificial intelligence for dynamic resource sharing in 6G and beyond. *IEEE Wireless Communications, 28*(4), 145–151. IEEE Wireless Communications. https://doi.org/10.1109/MWC.001.2000409

Huang, X., Xu, H., Wu, L., Dai, H., Yao, L., & Han, F. (2016). A data fusion detection method for fish freshness based on computer vision and near-infrared spectroscopy. *Analytical Methods, 8*(14), 2929–2935. https://doi.org/10.1039/C5AY03005F

Huang, C., Cai, H., Xu, L., Xu, B., Gu, Y., & Jiang, L. (2019a). Data-driven ontology generation and evolution towards intelligent service in manufacturing systems. *Future Generation Computer Systems, 101*, 197–207. https://doi.org/10.1016/j.future.2019.05.075

Huang, D., Bian, Z., Qiu, Q., Wang, Y., Fan, D., & Wang, X. (2019b). Identification of similar Chinese congou black teas using an electronic tongue combined with pattern recognition. *Molecules, 24*(24), Article 24. https://doi.org/10.3390/molecules24244549

Huang, B., Huan, Y., Xu, L. D., Zheng, L., & Zou, Z. (2019c). Automated trading systems statistical and machine learning methods and hardware implementation: A survey. https://www.tandfonline.com/doi/full/10.1080/17517575.2018.1493145

Huang, X., Yu, S., Xu, H., Aheto, J. H., Bonah, E., Ma, M., Wu, M., & Zhang, X. (2019d). Rapid and nondestructive detection of freshness quality of postharvest spinaches based on machine vision and electronic nose. *Journal of Food Safety, 39*(6), e12708. https://doi.org/10.1111/jfs.12708

Ismail, M., Jubley, N. Z., & Ali, Z. M. (2018). Forecasting Malaysian foreign exchange rate using artificial neural network and ARIMA time series. *AIP Conference Proceedings, 2013*(1), 020022. https://doi.org/10.1063/1.5054221

Jafari, S. M., Ganje, M., Dehnad, D., & Ghanbari, V. (2016). Mathematical, fuzzy logic and artificial neural network modeling techniques to predict drying kinetics of onion. *Journal of Food Processing and Preservation, 40*(2), 329–339. https://doi.org/10.1111/jfpp.12610

Jahedi Rad, S., Kaveh, M., Sharabiani, V. R., & Taghinezhad, E. (2018). Fuzzy logic, artificial neural network and mathematical model for prediction of white mulberry drying kinetics. *Heat and Mass Transfer, 54*(11), 3361–3374. https://doi.org/10.1007/s00231-018-2377-4

Jamshidi, M., et al. (2020). Artificial intelligence and COVID-19: Deep learning approaches for diagnosis and treatment | IEEE Journals & Magazine | IEEE Xplore. https://ieeexplore.ieee.org/abstract/document/9115663

Jayasooriya, S. D., Bhandari, B. R., Torley, P., & D'Arcy, B. R. (2004). Effect of high power ultrasound waves on properties of meat: A review. *International Journal of Food Properties, 7*(2), 301–319. https://doi.org/10.1081/JFP-120030039

Jha, S., & Topol, E. J. (2016). Adapting to artificial intelligence: Radiologists and pathologists as information specialists | Artificial Intelligence | JAMA | JAMA Network. https://jamanetwork.com/journals/jama/article-abstract/2588764

Jin, C., Li, F., Wilamowska-Korsak, M., Li, L., & Fu, L. (2014). BSP-GA: A new genetic algorithm for system optimization and excellent schema selection. *Systems Research and Behavioral Science*. Wiley online library. https://onlinelibrary.wiley.com/doi/abs/10.1002/sres.2280

Jordan, M. I., & Mitchell, T. M. (2015). Machine learning: Trends, perspectives, and prospects. *Science, 349*(6245), 255–260. https://doi.org/10.1126/science.aaa8415

Juan-Borrás, M., Soto, J., Gil-Sánchez, L., Pascual-Maté, A., & Escriche, I. (2017). Antioxidant activity and physico-chemical parameters for the differentiation of honey using a potentiometric electronic tongue. *Journal of the Science of Food and Agriculture, 97*(7), 2215–2222. https://doi.org/10.1002/jsfa.8031

Karami, H., Rasekh, M., & Mirzaee-Ghaleh, E. (2020). Application of the E-nose machine system to detect adulterations in mixed edible oils using chemometrics methods. *Journal of Food Processing and Preservation, 44*(9), e14696. https://doi.org/10.1111/jfpp.14696

Kartakoullis, A., Comaposada, J., Cruz-Carrión, A., Serra, X., & Gou, P. (2019). Feasibility study of smartphone-based near infrared spectroscopy (NIRS) for salted minced meat composition diagnostics at different temperatures. *Food Chemistry, 278*, 314–321. https://doi.org/10.1016/j.foodchem.2018.11.054

Kaushik, N., Gondi, A. R., Rana, R., & Srinivasa Rao, P. (2015). Application of fuzzy logic technique for sensory evaluation of high pressure processed mango pulp and litchi juice and its comparison to thermal treatment. *Innovative Food Science & Emerging Technologies, 32*, 70–78. https://doi.org/10.1016/j.ifset.2015.08.007

Kaveh, M., Rasooli Sharabiani, V., Amiri Chayjan, R., Taghinezhad, E., Abbaspour-Gilandeh, Y., & Golpour, I. (2018). ANFIS and ANNs model for prediction of moisture diffusivity and specific energy consumption potato, garlic and cantaloupe drying under convective hot air dryer. *Information Processing in Agriculture, 5*(3), 372–387. https://doi.org/10.1016/j.inpa.2018.05.003

Khamis, N., Yazid, M. R. M., Hamim, A., Rosyidi, S. A. P., Yusoff, N. I. M., & Borhan, M. N. (2018). Predicting the rheological properties of bitumen-filler mastic using artificial neural network methods. *Jurnal Teknologi, 80*(1), Article 1. https://doi.org/10.11113/jt.v80.11097

Kiani, S., Minaei, S., & Ghasemi-Varnamkhasti, M. (2016). A portable electronic nose as an expert system for aroma-based classification of saffron. *Chemometrics and Intelligent Laboratory Systems, 156*, 148–156. https://doi.org/10.1016/j.chemolab.2016.05.013

Kim, D. H., Zohdi, T. I., & Singh, R. P. (2020). Modeling, simulation and machine learning for rapid process control of multiphase flowing foods. *Computer Methods in Applied Mechanics and Engineering, 371*, 113286. https://doi.org/10.1016/j.cma.2020.113286

Kodogiannis, V. S., & Alshejari, A. (2016). Neuro-fuzzy based identification of meat spoilage using an electronic nose. In *2016 IEEE 8th international conference on intelligent systems (IS)* (pp. 96–103). https://doi.org/10.1109/IS.2016.7737406

Koklu, M., & Ozkan, I. A. (2020). Multiclass classification of dry beans using computer vision and machine learning techniques. *Computers and Electronics in Agriculture, 174*, 105507. https://doi.org/10.1016/j.compag.2020.105507

Kosslyn, S. M., & Koenig, O. (1992). *Wet mind: The new cognitive neuroscience*. Simon and Schuster.

Koszela, K., Łukomski, M., Mueller, W., Górna, K., Okoń, P., Boniecki, P., Zaborowicz, M., & Wojcieszak, D. (2017). Classification of dried vegetables using computer image analysis and artificial neural networks. In *Ninth international conference on digital image processing (ICDIP 2017), 10420* (pp. 650–656). https://doi.org/10.1117/12.2281718

Kumar, K. (2016). Intrusion detection using soft computing techniques. *International Journal of Computer Science & Communication Networks, 6*(3), 153–169.

Kumar, P., & Saha, D. (2023, March). Revolutionizing the food industry: How artificial intelligence is transforming food processing. *Researchgate.Net*. https://www.researchgate.net/profile/Pankaj-Kumar-23/publication/369021766_Revolutionizing_the_Food_Industry_How_Artificial_Intelligence_is_Transforming_Food_Processing/links/6412931c92cfd54f8403c88b/Revolutionizing-the-Food-Industry-How-Artificial-Intell

Kumar, V., & Sharma, H. K. (2017). Process optimization for extraction of bioactive compounds from taro (Colocasia esculenta), using RSM and ANFIS modeling. *Journal of Food Measurement and Characterization, 11*(2), 704–718. https://doi.org/10.1007/s11694-016-9440-y

Lee, E. -K., Gerla, M., Pau, G., Lee, U., & Lim, J. -H. (2016). *Internet of Vehicles: From intelligent grid to autonomous cars and vehicular fogs*. https://journals.sagepub.com/doi/full/10.1177/1550147716665500

Leme, D. S., da Silva, S. A., Barbosa, B. H. G., Borém, F. M., & Pereira, R. G. F. A. (2019). Recognition of coffee roasting degree using a computer vision system. *Computers and Electronics in Agriculture, 156*, 312–317. https://doi.org/10.1016/j.compag.2018.11.029

Leo Kumar, S. P. (2019). Knowledge-based expert system in manufacturing planning: State-of-the-art review. *International Journal of Production Research, 57*(15–16), 4766–4790. https://doi.org/10.1080/00207543.2018.1424372

Li, et al. (2007). Introduction: Advances in intelligent information processing. *Information Systems, 32*(7) https://dl.acm.org/doi/abs/10.1016/j.is.2006.10.001

Li, M., Ekramirad, N., Rady, A., & Adedeji, A. (2018). *Application of acoustic emission and machine learning to detect codling moth infested apples*. https://doi.org/10.13031/trans.12548

Li, B., Lin, Y., Yu, W., Wilson, D. I., & Young, B. R. (2021a). Application of mechanistic modelling and machine learning for cream cheese fermentation pH prediction. *Journal of Chemical Technology & Biotechnology, 96*(1), 125–133. https://doi.org/10.1002/jctb.6517

Li, J., Li, Z., Li, L., Song, C., Raghavan, G. S. V., & He, F. (2021b). Microwave drying of balsam pear with online aroma detection and control. *Journal of Food Engineering, 288*, 110139. https://doi.org/10.1016/j.jfoodeng.2020.110139

Li, X., Liu, D., Pu, Y., & Zhong, Y. (2023). Recent Advance of Intelligent Packaging Aided by Artificial Intelligence for Monitoring Food Freshness. *Foods, 12*(15), 2976. https://doi.org/10.3390/foods12152976

Liang, N., Sun, S., Zhang, C., He, Y., & Qiu, Z. (2022). Advances in infrared spectroscopy combined with artificial neural network for the authentication and traceability of food. *Critical Reviews in Food Science and Nutrition, 62*(11), 2963–2984. https://doi.org/10.1080/10408398.2020.1862045

Ling X Li, L. (1999a). Proposing an architectural framework of a hybrid knowledge-based system for production rescheduling. *Expert Systems*. Wiley Online Library. https://onlinelibrary.wiley.com/doi/abs/10.1111/1468-0394.00119

Ling X Li, L. (1999b). Knowledge-based problem solving: An approach to health assessment[1]. *Expert Systems with Applications, 16*(1), 33–42. https://doi.org/10.1016/S0957-4174(98)00026-8

Litjens, G., Kooi, T., Bejnordi, B. E., Setio, A. A. A., Ciompi, F., Ghafoorian, M., van der Laak, J. A. W. M., van Ginneken, B., & Sánchez, C. I. (2017). A survey on deep learning in medical image analysis. *Medical Image Analysis, 42*, 60–88. https://doi.org/10.1016/j.media.2017.07.005

Liu, Z., et al. (2015). *Deep learning face attributes in the wild*. https://openaccess.thecvf.com/content_iccv_2015/html/Liu_Deep_Learning_Face_ICCV_2015_paper.html

Liu, J., Kong, X., Xia, F., Bai, X., Wang, L., Qing, Q., & Lee, I. (2018). Artificial intelligence in the 21st century. *IEEE Access, 6*, 34403–34421. IEEE Access. https://doi.org/10.1109/ACCESS.2018.2819688

Liu, J., Liu, L., Guo, W., Fu, M., Yang, M., Huang, S., Zhang, F., & Liu, Y. (2019). A new methodology for sensory quality assessment of garlic based on metabolomics and an artificial neural network. *RSC Advances, 9*(31), 17754–17765. https://doi.org/10.1039/C9RA01978B

Liu, L., et al. (2020a). Deep learning for generic object detection: A survey. *International Journal of Computer Vision*. https://link.springer.com/article/10.1007/s11263-019-01247-4

Liu, J., Zuo, M., Low, S. S., Xu, N., Chen, Z., Lv, C., Cui, Y., Shi, Y., & Men, H. (2020b). Fuzzy evaluation output of taste information for liquor using electronic tongue based on cloud model. *Sensors, 20*(3), Article 3. https://doi.org/10.3390/s20030686

Livio, J., & Hodhod, R. (2018). AI cupper: A fuzzy expert system for sensorial evaluation of coffee bean attributes to derive quality scoring. *IEEE Transactions on Fuzzy Systems, 26*(6), 3418–3427. IEEE Transactions on Fuzzy Systems. https://doi.org/10.1109/TFUZZ.2018.2832611

Lopes, J. F., Ludwig, L., Barbin, D. F., Grossmann, M. V. E., & Barbon, S. (2019). Computer vision classification of barley flour based on spatial pyramid partition ensemble. *Sensors, 19*(13), Article 13. https://doi.org/10.3390/s19132953

Lu, Y. (2019). Artificial intelligence: A survey on evolution, models, applications and future trends. *Journal of Management Analytics, 6*(1), 1–29. https://doi.org/10.1080/23270012.2019.1570365

Morocho-Cayamcela, M. E., Lee, H., & Lim, W. (2019). Machine learning for 5G/B5G Mobile and wireless communications: Potential, limitations, and future directions | IEEE Journals & Magazine | IEEE Xplore. https://ieeexplore.ieee.org/abstract/document/8844682

Xing Li, H., & Li, L. X. (1999). Representing diverse mathematical problems using neural networks in hybrid intelligent systems. *Expert Systems*. Wiley Online Library. https://onlinelibrary.wiley.com/doi/abs/10.1111/1468-0394.00118

Lu, L., Tian, S., Deng, S., Zhu, Z., & Hu, X. (2015). Determination of rice sensory quality with similarity analysis-artificial neural network method in electronic tongue system. *RSC Advances, 5*(59), 47900–47908. https://doi.org/10.1039/C5RA06310H

Lu, L., Xu, L., Xu, B., Li, G., & Cai, H. (2018a). Fog computing approach for music cognition system based on machine learning algorithm. *IEEE Transactions on Computational Social Systems, 5*(4), 1142–1151. IEEE Transactions on Computational Social Systems. https://doi.org/10.1109/TCSS.2018.2871694

Lu, M., Li, C., Li, L., Wu, Y., & Yang, Y. (2018b). Rapid detecting soluble solid content of pears based on near-infrared spectroscopy. In *2018 2nd IEEE advanced information management, communicates, electronic and automation control conference (IMCEC)* (pp. 819–823). https://doi.org/10.1109/IMCEC.2018.8469315

Lu, N.-V., Vuong, T.-N., & Dinh, D.-T. (2020). Combining correlation-based feature and machine learning for sensory evaluation of Saigon beer. *International Journal of Knowledge and Systems Science (IJKSS), 11*(2), 71–85. https://doi.org/10.4018/IJKSS.2020040104

Mahadevappa, J., Groß, F., & Delgado, A. (2017). Fuzzy logic based process control strategy for effective sheeting of wheat dough in small and medium-sized enterprises. *Journal of Food Engineering, 199*, 93–99. https://doi.org/10.1016/j.jfoodeng.2016.12.013

Mahdi, M., Ibrahim, M. F., Mahmood, S., Singam, P., & Baseri Huddin, A. (2019). Fuzzy logic system for diagnosing coronary heart disease. *International Journal of Engineering and Technology, 1*(7), 119–125.

Makkar, A., Garg, S., Kumar, N., Hossain, M. S., Ghoneim, A., & Alrashoud, M. (2021). An efficient spam detection technique for IoT devices using machine learning. *IEEE Transactions on Industrial Informatics, 17*(2), 903–912. IEEE Transactions on Industrial Informatics. https://doi.org/10.1109/TII.2020.2968927

Maladkar, K. (2018). *6 types of artificial neural networks currently being used in ML.* https://analyticsindiamag.com/ai-mysteries/6-types-of-artificial-neural-networks-currently-being-used-in-todays-technology/

Mamat, R. C., Kasa, A., Razali, S. F. M., Samad, A. M., Ramli, A., & Yazid, M. R. M. (2019). Application of artificial intelligence in predicting ground settlement on earth slope. *AIP Conference Proceedings, 2138*(1), 040015. https://doi.org/10.1063/1.5121094

Mavani, N. R., Ali, J. M., Othman, S., Hussain, M. A., Hashim, H., & Rahman, N. A. (2022). Application of artificial intelligence in food industry—A guideline. *Food Engineering Reviews, 14*(1), 134–175. https://doi.org/10.1007/s12393-021-09290-z

Mazen, F. M. A., & Nashat, A. A. (2019). Ripeness classification of bananas using an artificial neural network. *Arabian Journal for Science and Engineering, 44*(8), 6901–6910. https://doi.org/10.1007/s13369-018-03695-5

Melin, P., Miramontes, I., & Prado-Arechiga, G. (2018). A hybrid model based on modular neural networks and fuzzy systems for classification of blood pressure and hypertension risk diagnosis. *Expert Systems with Applications, 107*, 146–164. https://doi.org/10.1016/j.eswa.2018.04.023

Miotto, R., et al. (2018). Deep learning for healthcare: Review, opportunities and challenges. *Briefings in Bioinformatics*. Oxford Academic. https://academic.oup.com/bib/article/19/6/1236/3800524

Mirzaee-Ghaleh, E., Taheri-Garavand, A., Ayari, F., & Lozano, J. (2020). Identification of fresh-chilled and frozen-thawed chicken meat and estimation of their shelf life using an E-nose machine coupled fuzzy KNN. *Food Analytical Methods, 13*(3), 678–689. https://doi.org/10.1007/s12161-019-01682-6

Mishra, G., Srivastava, S., Panda, B. K., & Mishra, H. N. (2018). Prediction of Sitophilus granarius infestation in stored wheat grain using multivariate chemometrics & fuzzy logic-based electronic nose analysis. *Computers and Electronics in Agriculture, 152*, 324–332. https://doi.org/10.1016/j.compag.2018.07.022

Mohamed, et al. (2015). *Data mining techniques in food safety*. https://www.academia.edu/43020000/Data_Mining_Techniques_in_Food_Safety

Mohamed, M. Y., Solihin, M. I., Astuti, W., Ang, C. K., & Zailah, W. (2019). Food powders classification using handheld near-infrared spectroscopy and support vector machine. *Journal of Physics: Conference Series, 1367*(1), 012029. https://doi.org/10.1088/1742-6596/1367/1/012029

Mokarram, M., Amin, H., & Khosravi, M. R. (2019). Using adaptive neuro-fuzzy inference system and multiple linear regression to estimate orange taste. *Food Science & Nutrition, 7*(10), 3176–3184. https://doi.org/10.1002/fsn3.1149

Mozafari, M. R., Khosravi-Darani, K., Borazan, G. G., Cui, J., Pardakhty, A., & Yurdugul, S. (2008). Encapsulation of food ingredients using nanoliposome technology. *International Journal of Food Properties, 11*(4), 833–844. https://doi.org/10.1080/10942910701648115

Mullainathan, S., & Spiess, J. (2017). Machine learning: An applied econometric approach. *Journal of Economic Perspectives, 31*(2), 87–106. https://doi.org/10.1257/jep.31.2.87

Mutlag, A. H., Mohamed, A., & Shareef, H. (2016). A nature-inspired optimization-based optimum fuzzy logic photovoltaic inverter controller utilizing an eZdsp F28335 board. *Energies, 9*(3), Article 3. https://doi.org/10.3390/en9030120

Nadian, M. H., Rafiee, S., Aghbashlo, M., Hosseinpour, S., & Mohtasebi, S. S. (2015). Continuous real-time monitoring and neural network modeling of apple slices color changes during hot air drying. *Food and Bioproducts Processing, 94*, 263–274. https://doi.org/10.1016/j.fbp.2014.03.005

Nadim, M., Ahmadifar, H., Mashkinmojeh, M., & Yamaghani, M. R. (2019). Application of image processing techniques for quality control of mushroom. *Caspian Journal of Health Research, 4*(3), 72–75. https://doi.org/10.29252/cjhr.4.3.72

Nafan, E., Sulaiman, R., Mohamad Ali, N., Priambodo, B., Hamidy, A., Nseaf, A., Putra, Z., Fikri, H., & Handriani, I. (2019, December 1). *Improving the modelling of robot bunker with camera.*

Nagaveni, S., & Poosarla, V. G. (2024). Future trends in intelligent packaging solutions. In *Intelligent packaging* (pp. 383–411). Elsevier.

Naik, R. R., Gandhi, N. S., Thakur, M., & Nanda, V. (2019). Analysis of crystallization phenomenon in Indian honey using molecular dynamics simulations and artificial neural network. *Food Chemistry, 300*, 125182. https://doi.org/10.1016/j.foodchem.2019.125182

Najafabadi, M. M., et al. (2015). Deep learning applications and challenges in big data analytics. *Journal of Big Data*. https://link.springer.com/article/10.1186/s40537-014-0007-7

Neethu, K. C., Sharma, A. K., Pushpadass, H. A., Emerald, F. M. E., & Manjunatha, M. (2016). Prediction of convective heat transfer coefficient during deep-fat frying of pantoa using neurocomputing approaches. *Innovative Food Science & Emerging Technologies, 34*, 275–284. https://doi.org/10.1016/j.ifset.2016.02.012

Neyshabur, B., Bhojanapalli, S., Mcallester, D., & Srebro, N. (2017). Exploring generalization in deep learning. *Advances in Neural Information Processing Systems, 30*. https://proceedings.neurips.cc/paper/2017/hash/10ce03a1ed01077e3e289f3e53c72813-Abstract.html

Nilsson, N. J. (1982). *Principles of artificial intelligence*. Springer Science & Business Media.

Núñez-Sánchez, N., Martínez-Marín, A. L., Polvillo, O., Fernández-Cabanás, V. M., Carrizosa, J., Urrutia, B., & Serradilla, J. M. (2016). Near Infrared Spectroscopy (NIRS) for the determination of the milk fat fatty acid profile of goats. *Food Chemistry, 190*, 244–252. https://doi.org/10.1016/j.foodchem.2015.05.083

Ojediran, J. O., Okonkwo, C. E., Adeyi, A. J., Adeyi, O., Olaniran, A. F., George, N. E., & Olayanju, A. T. (2020). Drying characteristics of yam slices (Dioscorea rotundata) in a convective hot air dryer: Application of ANFIS in the prediction of drying kinetics. *Heliyon, 6*(3). https://doi.org/10.1016/j.heliyon.2020.e03555

Olarewaju, O. O., Bertling, I., & Magwaza, L. S. (2016). Non-destructive evaluation of avocado fruit maturity using near infrared spectroscopy and PLS regression models. *Scientia Horticulturae, 199*, 229–236. https://doi.org/10.1016/j.scienta.2015.12.047

Omari, A., Behroozi-Khazaei, N., & Sharifian, F. (2018). Drying kinetic and artificial neural network modeling of mushroom drying process in microwave-hot air dryer. *Journal of Food Process Engineering, 41*(7), e12849. https://doi.org/10.1111/jfpe.12849

Perrot, N., Ioannou, I., Allais, I., Curt, C., Hossenlopp, J., & Trystram, G. (2006). Fuzzy concepts applied to food product quality control: A review. *Fuzzy Sets and Systems, 157*(9), 1145–1154. https://doi.org/10.1016/j.fss.2005.12.013

Phate, V. R., Malmathanraj, R., & Palanisamy, P. (2019). Clustered ANFIS weighing models for sweet lime (Citrus limetta) using computer vision system. *Journal of Food Process Engineering, 42*(6), e13160. https://doi.org/10.1111/jfpe.13160

Pise, D., & Upadhye, G. D. (2018). Grading of harvested mangoes quality and maturity based on machine learning techniques. In *2018 international conference on Smart City and emerging technology (ICSCET)* (pp. 1–6). https://doi.org/10.1109/ICSCET.2018.8537342

Porep, J. U., Kammerer, D. R., & Carle, R. (2015). On-line application of near infrared (NIR) spectroscopy in food production. *Trends in Food Science & Technology, 46*(2, Part A), 211–230. https://doi.org/10.1016/j.tifs.2015.10.002

Pouyanfar, S., et al. (2018). A survey on deep learning: Algorithms, techniques, and applications. *ACM Computing Surveys, 51*(5). https://dl.acm.org/doi/abs/10.1145/3234150

Pullanagari, R. R., Yule, I. J., & Agnew, M. (2015). On-line prediction of lamb fatty acid composition by visible near infrared spectroscopy. *Meat Science, 100*, 156–163

Qi, J., Wu, F., Li, L., & Shu, H. (2007). Artificial intelligence applications in the telecommunications industry. *Expert Systems, 24*(4), 271–291. https://doi.org/10.1111/j.1468-0394.2007.00433.x

Qiu, S., Gao, L., & Wang, J. (2015). Classification and regression of ELM, LVQ and SVM for E-nose data of strawberry juice. *Journal of Food Engineering, 144*, 77–85. https://doi.org/10.1016/j.jfoodeng.2014.07.015

Qiu, S., & Wang, J. (2017). The prediction of food additives in the fruit juice based on electronic nose with chemometrics. *Food Chemistry, 230*, 208–214. https://doi.org/10.1016/j.foodchem.2017.03.011

Qu, J.-H., Liu, D., Cheng, J.-H., Sun, D.-W., Ma, J., Pu, H., & Zeng, X.-A. (2015). Applications of near-infrared spectroscopy in food safety evaluation and control: A review of recent research advances. *Critical Reviews in Food Science and Nutrition, 55*(13), 1939–1954. https://doi.org/10.1080/10408398.2013.871693

Rahman, M. S., Rashid, M. M., & Hussain, M. A. (2012). Thermal conductivity prediction of foods by Neural Network and Fuzzy (ANFIS) modeling techniques. *Food and Bioproducts Processing, 90*(2), 333–340. https://doi.org/10.1016/j.fbp.2011.07.001

Raigar, R. K., Upadhyay, R., & Mishra, H. N. (2017). Storage quality assessment of shelled peanuts using non-destructive electronic nose combined with fuzzy logic approach. *Postharvest Biology and Technology, 132*, 43–50. https://doi.org/10.1016/j.postharvbio.2017.05.016

Rajakumar, G., Ananth Kumar, T., Samuel, T. S. A., & Kumaran, E. (2018). IoT based milk monitoring system for detection of milk adulteration. *International Journal of Pure and Applied Mathematics, 118*, 21–32.

Rajesh Mavani, N., Lim, C. Y., Hashim, H., Abd. Rahman, N., & Mohd Ali, J. (2021). Fuzzy Mamdani based user-friendly interface for food preservatives determination. *Food and Bioproducts Processing, 126*, 282–292. https://doi.org/10.1016/j.fbp.2021.01.01

Rajkomar, A., Dean, J., & Kohane, I. (2019). Machine learning in medicine. *New England Journal of Medicine, 380*(14), 1347–1358. https://doi.org/10.1056/NEJMra1814259

Ramakrishna, R. R., Hamid, Z. A., Zaki, W. M. D. W., Huddin, A. B., & Mathialagan, R. (2020). Stem cell imaging through convolutional neural networks: Current issues and future directions in artificial intelligence technology. *PeerJ, 8*, e10346. https://doi.org/10.7717/peerj.10346

Roychowdhury, P. (2023). A review on the potencies of AI-driven nutritional assessment studies in enhancing public health. *The Indian Journal of Nutrition and Dietetics, 61*(1). https://doi.org/10.21048/IJND.2024.61.1.34685

Ramirez-Asis, E., Vilchez-Carcamo, J., Thakar, C. M., Phasinam, K., Kassanuk, T., & Naved, M. (2022). A review on role of artificial intelligence in food processing and manufacturing industry. *Materials Today: Proceedings, 51*, 2462–2465.

Zhu, Y., Koecher, K., Benoit, V., Normington, J., Menon, R., & Campbell, J. (2023). Chapter 20: Personalized nutrition: From science to consumer. In D. Ghosh, D. Bogueva, & R. B. T.-N. S. Smarta (Eds.), *Marketing nutrition, health claims, and public policy* (pp. 267–286). Academic. https://doi.org/10.1016/B978-0-323-85615-7.00017-3

Rashmi, W., Osama, M., Khalid, M., Rasheed, A., Bhaumik, S., Wong, W. Y., Datta, S., & Gupta, T. C. S. M. (2019). Tribological performance of nanographite-based metalworking fluid and parametric investigation using artificial neural network. *The International Journal of Advanced Manufacturing Technology, 104*(1), 359–374. https://doi.org/10.1007/s00170-019-03701-6

Reichstein, M., et al. (2019). Deep learning and process understanding for data-driven Earth system science. *Nature*. https://www.nature.com/articles/s41586-019-0912-1

Ren, G., Wang, Y., Ning, J., & Zhang, Z. (2020). Highly identification of keemun black tea rank based on cognitive spectroscopy: Near infrared spectroscopy combined with feature variable selection. *Spectrochimica Acta Part A: Molecular and Biomolecular Spectroscopy, 230*, 118079. https://doi.org/10.1016/j.saa.2020.118079

Rezagholi, F., & Hesarinejad, M. A. (2017). Integration of fuzzy logic and computer vision in intelligent quality control of celiac-friendly products. *Procedia Computer Science, 120*, 325–332. https://doi.org/10.1016/j.procs.2017.11.246

Richter, B., Rurik, M., Gurk, S., Kohlbacher, O., & Fischer, M. (2019). Food monitoring: Screening of the geographical origin of white asparagus using FT-NIR and machine learning. *Food Control, 104*, 318–325. https://doi.org/10.1016/j.foodcont.2019.04.032

Rizwana, S., & Hazarika, M. K. (2020). Application of near-infrared spectroscopy for rice characterization using machine learning. *Journal of The Institution of Engineers (India): Series A, 101*(4), 579–587. https://doi.org/10.1007/s40030-020-00459-z

Russell, S. J., & Norvig, P. (2016). *Artificial intelligence: A modern approach*. Pearson. https://thuvienso.hoasen.edu.vn/handle/123456789/8967

Sabater, C., Olano, A., Corzo, N., & Montilla, A. (2019). GC–MS characterisation of novel artichoke (Cynara scolymus) pectic-oligosaccharides mixtures by the application of machine learning algorithms and competitive fragmentation modelling. *Carbohydrate Polymers, 205*, 513–523. https://doi.org/10.1016/j.carbpol.2018.10.054

Saha, D., & Bhattacharya, S. (2010). Hydrocolloids as thickening and gelling agents in food: A critical review. *Journal of Food Science and Technology, 47*(6), 587–597. https://doi.org/10.1007/s13197-010-0162-6

Said, M., Ba-Abbad, M., Abdullah, S. R. S., & Mohammad, A. W. (2018). Artificial neural network (ANN) for optimization of palm oil mill effluent (POME) treatment using reverse osmosis membrane. *Journal of Physics: Conference Series, 1095*(1), 012021. https://doi.org/10.1088/1742-6596/1095/1/012021

Sanaeifar, A., ZakiDizaji, H., Jafari, A., & de la Guardia, M. (2017). Early detection of contamination and defect in foodstuffs by electronic nose: A review. *TrAC Trends in Analytical Chemistry, 97*, 257–271. https://doi.org/10.1016/j.trac.2017.09.014

Sarkar, T., Bhattacharjee, R., Salauddin, M., Giri, A., & Chakraborty, R. (2020). Application of fuzzy logic analysis on pineapple rasgulla. *Procedia Computer Science, 167*, 779–787. https://doi.org/10.1016/j.procs.2020.03.410

Shahidi, B., Sharifi, A., Roozbeh Nasiraie, L., Niakousari, M., & Ahmadi, M. (2020). Phenolic content and antioxidant activity of flixweed (Descurainia sophia) seeds extracts: Ranking extraction systems based on fuzzy logic method. *Sustainable Chemistry and Pharmacy, 16*, 100245. https://doi.org/10.1016/j.scp.2020.100245

Shanmuganathan, S. (2016). Artificial neural network modelling: An introduction. In S. Shanmuganathan & S. Samarasinghe (Eds.), *Artificial neural network modelling* (pp. 1–14). Springer International Publishing. https://doi.org/10.1007/978-3-319-28495-8_1

Sharma, L. K., Vishal, V., & Singh, T. N. (2017). Developing novel models using neural networks and fuzzy systems for the prediction of strength of rocks from key geomechanical properties. *Measurement, 102*, 158–169. https://doi.org/10.1016/j.measurement.2017.01.043

Shaw, B., Suman, A. K., & Chakraborty, B. (2020). Wine quality analysis using machine learning. In J. K. Mandal & D. Bhattacharya (Eds.), *Emerging technology in modelling and graphics* (pp. 239–247). Springer. https://doi.org/10.1007/978-981-13-7403-6_23

Shi, C., Yang, X., Han, S., Fan, B., Zhao, Z., Wu, X., & Qian, J. (2018a). Nondestructive prediction of tilapia fillet freshness during storage at different temperatures by integrating an electronic nose and tongue with radial basis function neural networks. *Food and Bioprocess Technology, 11*(10), 1840–1852. https://doi.org/10.1007/s11947-018-2148-8

Shi, F., et al. (2020). Review of artificial intelligence techniques in imaging data acquisition, segmentation, and diagnosis for COVID-19 | IEEE Journals & Magazine | IEEE Xplore. https://ieeexplore.ieee.org/abstract/document/9069255

Shi, Q., Guo, T., Yin, T., Wang, Z., Li, C., Sun, X., Guo, Y., & Yuan, W. (2018b). Classification of pericarpium citri reticulatae of different ages by using a voltammetric electronic tongue system. *International Journal of Electrochemical Science, 13*(12), 11359–11374. https://doi.org/10.20964/2018.12.45

Shi, Z., Huang, Y., He, Q., Xu, L., Liu, S., Qin, L., Jia, Z., Li, J., Huang, H., & Zhao, L. (2007). MSMiner—A developing platform for OLAP. *Decision Support Systems, 42*(4), 2016–2028. https://doi.org/10.1016/j.dss.2004.11.006

Shokri, R., et al. (2015). Privacy-preserving deep learning. In *Proceedings of the 22nd ACM SIGSAC conference on computer and communications security*. https://dl.acm.org/doi/abs/10.1145/2810103.2813687

Shone, N., Ngoc, T. N., Phai, V. D., & Shi, Q. (2018). A deep learning approach to network intrusion detection. *IEEE Transactions on Emerging Topics in Computational Intelligence, 2*(1), 41–50. IEEE Transactions on Emerging Topics in Computational Intelligence. https://doi.org/10.1109/TETCI.2017.2772792

Sidehabi, S. W., Suyuti, A., Areni, I. S., & Nurtanio, I. (2018). The development of machine vision system for sorting passion fruit using MultiClass support vector machine. *Journal of Engineering Science and Technology Review, 11*(5), 178–184. https://doi.org/10.25103/jestr.115.23

Siegel, A., & Sapru, H. N. (2006). *Essential neuroscience*. Lippincott Williams & Wilkins.

Silva, S. F., Anjos, C. A. R., Cavalcanti, R. N., & Celeghini, R. M. d. S. (2015). Evaluation of extra virgin olive oil stability by artificial neural network. *Food Chemistry, 179*, 35–43. https://doi.org/10.1016/j.foodchem.2015.01.100

Singh, V., Kumar, S., Singh, J., & Rai, A. K. (2018). Fuzzy logic sensory evaluation of cupcakes developed from the mahua flower (Madhuca longifolia), *5*(1).

Sipos, A. (2020). A knowledge-based system as a sustainable software application for the supervision and intelligent control of an alcoholic fermentation process. *Sustainability, 12*(23), Article 23. https://doi.org/10.3390/su122310205

Siswantoro, J., Hilman, M. Y., & Widiasri, M. (2017). Computer vision system for egg volume prediction using backpropagation neural network. *IOP Conference Series: Materials Science and Engineering, 273*(1), 012002. https://doi.org/10.1088/1757-899X/245/1/012002

Sodhro, A. H., Pirbhulal, S., & de Albuquerque, V. H. C. (2019). Artificial intelligence-driven mechanism for edge computing-based industrial applications. *IEEE Transactions on Industrial Informatics, 15*(7), 4235–4243. IEEE Transactions on Industrial Informatics. https://doi.org/10.1109/TII.2019.2902878

Soleimanzadeh, B., Hemati, L., Yolmeh, M., & Salehi, F. (2015). GA-ANN and ANFIS models and almonella enteritidis inactivation by ultrasound. *Journal of Food Safety, 35*(2), 220–226. https://doi.org/10.1111/jfs.12174

Srivastava, S., Mishra, G., & Mishra, H. N. (2019a). Fuzzy controller based E-nose classification of Sitophilus oryzae infestation in stored rice grain. *Food Chemistry, 283*, 604–610. https://doi.org/10.1016/j.foodchem.2019.01.076

Srivastava, S., Mishra, G., & Mishra, H. N. (2019b). Probabilistic artificial neural network and E-nose based classification of Rhyzopertha dominica infestation in stored rice grains. *Chemometrics and Intelligent Laboratory Systems, 186*, 12–22. https://doi.org/10.1016/j.chemolab.2019.01.007

Suciu, I., Ndiaye, A., Baudrit, C., Fernandez, C., Kondjoyan, A., Mirade, P. S., Sicard, J., Tournayre, P., Bohuon, P., Buche, P., Courtois, F., Guillard, V., Athes, V., Flick, D., Plana-Fattori, A., Trelea, C., Trystram, G., Delaplace, G., Curet, S., et al. (2021). A digital learning tool based on models and simulators for food engineering (MESTRAL). *Journal of Food Engineering, 293*, 110375. https://doi.org/10.1016/j.jfoodeng.2020.110375

Sumaryanti, L., Istanto, T., & Pare, S. (2020). Rule based method in expert system for detection pests and diseases of corn. *Journal of Physics: Conference Series, 1569*(2), 022023. https://doi.org/10.1088/1742-6596/1569/2/022023

Sun, X., Young, J., Liu, J. H., Bachmeier, L., Somers, R. M., Chen, K. J., & Newman, D. (2016). Prediction of pork color attributes using computer vision system. *Meat Science, 113*, 62–64. https://doi.org/10.1016/j.meatsci.2015.11.009

Sun, X., Young, J., Liu, J.-H., & Newman, D. (2018). Prediction of pork loin quality using online computer vision system and artificial intelligence model. *Meat Science, 140*, 72–77. https://doi.org/10.1016/j.meatsci.2018.03.005

Sünderhauf, N., Brock, O., Scheirer, W., Hadsell, R., Fox, D., Leitner, J., Upcroft, B., Abbeel, P., Burgard, W., Milford, M., & Corke, P. (2018). *The limits and potentials of deep learning for robotics*. https://journals.sagepub.com/doi/full/10.1177/0278364918770733

Szturo, K., & Szczypiński, P. M. (2017). Ontology based expert system for barley grain classification. In *2017 signal processing: Algorithms, architectures, arrangements, and applications (SPA)* (pp. 360–364). https://doi.org/10.23919/SPA.2017.8166893

Szulczyński, B., Gębicki, J., & Namieśnik, J. (2018). Application of fuzzy logic to determine the odour intensity of model gas mixtures using electronic nose. *E3S Web of Conferences, 28*, 01036. https://doi.org/10.1051/e3sconf/20182801036

Tambe, P., Cappelli, P., & Yakubovich, V. (2019). Artificial intelligence in human resources management: Challenges and a path forward. https://journals.sagepub.com/doi/abs/10.1177/0008125619867910

Tan, W., et al. (2010). A methodology toward manufacturing grid-based virtual enterprise operation platform. *Enterprise Information Systems, 4*(3). https://www.tandfonline.com/doi/abs/10.1080/17517575.2010.504888

Tan, J., Balasubramanian, B., Sukha, D., Ramkissoon, S., & Umaharan, P. (2019a). Sensing fermentation degree of cocoa (Theobroma cacao L.) beans by machine learning classification models based electronic nose system. *Journal of Food Process Engineering, 42*(6), e13175. https://doi.org/10.1111/jfpe.13175

Tan, L. T., Hu, R. Q., & Hanzo, L. (2019b). Twin-timescale artificial intelligence aided mobility-aware edge caching and computing in vehicular networks. *IEEE Transactions on Vehicular Technology, 68*(4), 3086–3099. IEEE Transactions on Vehicular Technology. https://doi.org/10.1109/TVT.2019.2893898

Tanajura da Silva, C. E., Filardi, V. L., Pepe, I. M., Chaves, M. A., & Santos, C. M. S. (2015). Classification of food vegetable oils by fluorimetry and artificial neural networks. *Food Control, 47*, 86–91. https://doi.org/10.1016/j.foodcont.2014.06.030

Thazin, Y., Pobkrut, T., & Kerdcharoen, T. (2018). Prediction of acidity levels of fresh roasted coffees using E-nose and artificial neural network. In *2018 10th international conference on knowledge and smart technology (KST)* (pp. 210–215). https://doi.org/10.1109/KST.2018.8426206

Tohidi, M., Ghasemi-Varnamkhasti, M., Ghafarinia, V., Bonyadian, M., & Mohtasebi, S. S. (2018). Development of a metal oxide semiconductor-based artificial nose as a fast, reliable and non-expensive analytical technique for aroma profiling of milk adulteration. *International Dairy Journal, 77*, 38–46. https://doi.org/10.1016/j.idairyj.2017.09.003

Tokui, S., Oono, K., Hido, S., & Clayton, J. (2015). *Chainer: A next-generation open source framework for deep learning.*

Trafialek, J., Laskowski, W., & Kolanowski, W. (2015). The use of Kohonen's artificial neural networks for analyzing the results of HACCP system declarative survey. *Food Control, 51*, 263–269. https://doi.org/10.1016/j.foodcont.2014.11.032

Tsoumakas, G. (2019). A survey of machine learning techniques for food sales prediction. *Artificial Intelligence Review, 52*(1), 441–447. https://doi.org/10.1007/s10462-018-9637-z

Tung, K. (2019). AI, the internet of legal things, and lawyers. *Journal of Management Analytics, 6*(4). https://www.tandfonline.com/doi/abs/10.1080/23270012.2019.1671242

Uçar, A., & Özalp, R. (2017). Efficient android electronic nose design for recognition and perception of fruit odors using Kernel Extreme Learning Machines. *Chemometrics and Intelligent Laboratory Systems, 166*, 69–80. https://doi.org/10.1016/j.chemolab.2017.05.013

Upadhyay, R., Sehwag, S., & Mishra, H. N. (2017). Electronic nose guided determination of frying disposal time of sunflower oil using fuzzy logic analysis. *Food Chemistry, 221*, 379–385. https://doi.org/10.1016/j.foodchem.2016.10.089

Utai, K., Nagle, M., Hämmerle, S., Spreer, W., Mahayothee, B., & Müller, J. (2019). Mass estimation of mango fruits (Mangifera indica L., cv. 'Nam Dokmai') by linking image processing and artificial neural network. *Engineering in Agriculture, Environment and Food, 12*(1), 103–110. https://doi.org/10.1016/j.eaef.2018.10.003

Vajdi, M., Varidi, M. J., Varidi, M., & Mohebbi, M. (2019). Using electronic nose to recognize fish spoilage with an optimum classifier. *Journal of Food Measurement and Characterization, 13*(2), 1205–1217. https://doi.org/10.1007/s11694-019-00036-4

Viharos, Z. J., & Kis, K. B. (2015). Survey on Neuro-Fuzzy systems and their applications in technical diagnostics and measurement. *Measurement, 67*, 126–136. https://doi.org/10.1016/j.measurement.2015.02.001

Villaseñor-Aguilar, M.-J., Bravo-Sánchez, M.-G., Padilla-Medina, J.-A., Vázquez-Vera, J. L., Guevara-González, R.-G., García-Rodríguez, F.-J., & Barranco-Gutiérrez, A.-I. (2020). A maturity estimation of bell pepper (Capsicum annuum L.) by artificial vision system for quality control. *Applied Sciences, 10*(15), Article 15. https://doi.org/10.3390/app10155097

Wan, P., Toudeshki, A., Tan, H., & Ehsani, R. (2018). A methodology for fresh tomato maturity detection using computer vision. *Computers and Electronics in Agriculture, 146*, 43–50. https://doi.org/10.1016/j.compag.2018.01.011

Wang, C., et al. (2007). *Conceptual design of remote monitoring and fault diagnosis systems.* https://www.sciencedirect.com/science/article/abs/pii/S0306437906000895

Wang, C. -X. et al. (2020). Artificial intelligence enabled wireless networking for 5G and beyond: Recent advances and future challenges | IEEE Journals & Magazine | IEEE Xplore. https://ieeexplore.ieee.org/abstract/document/9023918

Wang, L., Niu, Q., Hui, Y., & Jin, H. (2015a). Discrimination of rice with different pretreatment methods by using a voltammetric electronic tongue. *Sensors, 15*(7), Article 7. https://doi.org/10.3390/s150717767

Wang, L., Niu, Q., Hui, Y., Jin, H., & Chen, S. (2015b). Assessment of taste attributes of peanut meal enzymatic-hydrolysis hydrolysates using an electronic tongue. *Sensors, 15*(5), Article 5. https://doi.org/10.3390/s150511169

Wang, L., Zou, H., Su, J., Li, L., & Chaudhry, S. (2013). An ARIMA-ANN hybrid model for time series forecasting. *Systems Research and Behavioral Science, 30*(3), 244–259. https://doi.org/10.1002/sres.2179

Wanigasekara, C., Oromiehie, E., Swain, A., Prusty, B. G., & Nguang, S. K. (2021). Machine learning-based inverse predictive model for AFP based thermoplastic composites. *Journal of Industrial Information Integration, 22*, 100197. https://doi.org/10.1016/j.jii.2020.100197

Wojnowski, W., Majchrzak, T., Dymerski, T., Gębicki, J., & Namieśnik, J. (2017). Portable electronic nose based on electrochemical sensors for food quality assessment. *Sensors, 17*(12), Article 12. https://doi.org/10.3390/s17122715

Wold, J. P., Veiseth-Kent, E., Høst, V., & Løvland, A. (2017). Rapid on-line detection and grading of wooden breast myopathy in chicken fillets by near-infrared spectroscopy. *PLoS One, 12*(3), e0173384. https://doi.org/10.1371/journal.pone.0173384

Xu, L., et al. (2008). An approach to enterprise process dynamic modeling supporting enterprise process evolution. *Information Systems Frontiers*. https://link.springer.com/article/10.1007/s10796-008-9114-3

Xu, J.-L., & Sun, D.-W. (2017). Identification of freezer burn on frozen salmon surface using hyperspectral imaging and computer vision combined with machine learning algorithm. *International Journal of Refrigeration, 74*, 151–164. https://doi.org/10.1016/j.ijrefrig.2016.10.014

Xu, L. D., Lu, Y., & Li, L. (2021). Embedding blockchain technology into IoT for security: A survey. *IEEE Internet of Things Journal, 8*(13), 10452–10473. IEEE Internet of Things Journal. https://doi.org/10.1109/JIOT.2021.3060508

Xu, Q., Yang, X., Hu, Z., Li, Y., & Zheng, X. (2020). Identification of adulterated olive oil by fusion of near infrared and Raman spectroscopy. *Journal of Physics: Conference Series, 1592*(1), 012041. https://doi.org/10.1088/1742-6596/1592/1/012041

Yan, J., Guo, X., Duan, S., Jia, P., Wang, L., Peng, C., & Zhang, S. (2015). Electronic nose feature extraction methods: A review. *Sensors, 15*(11), Article 11. https://doi.org/10.3390/s151127804

Yang, B., Li, L. X., Ji, H., & Xu, J. (2001). An early warning system for loan risk assessment using artificial neural networks. *Knowledge-Based Systems, 14*(5), 303–306. https://doi.org/10.1016/S0950-7051(01)00110-1

Yang, G., et al. (2014). A health-IoT platform based on the integration of intelligent packaging, unobtrusive bio-sensor, and intelligent medicine box | IEEE Journals & Magazine | IEEE Xplore. https://ieeexplore.ieee.org/abstract/document/6747344

Yang, H., et al. (2019). An efficient edge artificial intelligence MultiPedestrian tracking method with rank constraint | IEEE Journals & Magazine | IEEE Xplore. https://ieeexplore.ieee.org/abstract/document/8633399

Young, T., et al. (2018). Recent trends in deep learning based natural language processing [Review Article] | IEEE Journals & Magazine | IEEE Xplore. https://ieeexplore.ieee.org/abstract/document/8416973

Younis, K., Ahmad, S., Osama, K., & Malik, MA (2019). Optimization of de-bittering process of mosambi (Citrus limetta) peel: Artificial neural network Gaussian process regression and support vector machine modeling approach. *Journal of Food Process Engineering, 42*(6). https://doi.org/10.3390/foods12152976

Yousefi-Darani, A., Paquet-Durand, O., & Hitzmann, B. (2019). Application of fuzzy logic control for the dough proofing process. *Food and Bioproducts Processing, 115*, 36–46. https://doi.org/10.1016/j.fbp.2019.02.006

Yuan, R., et al. (2010). An SVM-based machine learning method for accurate internet traffic classification. *Information Systems Frontiers*. https://link.springer.com/article/10.1007/s10796-008-9131-2

Yulianto, T., Komariyah, S., & Ulfaniyah, N. (2017). Application of fuzzy inference system by Sugeno method on estimating of salt production. *AIP Conference Proceedings, 1867*(1), 020039. https://doi.org/10.1063/1.4994442

Zadeh, L. A. (2015). Fuzzy logic—A personal perspective. *Fuzzy Sets and Systems, 281*, 4–20. https://doi.org/10.1016/j.fss.2015.05.009

Zakaria, Z., Nordin, N., Malik, A., Elias, S. J., & Shahuddin, A. (2019). Fuzzy expert systems (FES) for halal food additive. *Indonesian Journal of Electrical Engineering and Computer Science, 13*, 1073–1078. https://doi.org/10.11591/ijeecs.v13.i3.pp1073-1078

Zappone, A., Di Renzo, M., & Debbah, M. (2019). Wireless networks design in the era of deep learning: Model-based, AI-based, or both? *IEEE Transactions on Communications, 67*(10), 7331–7376. IEEE Transactions on Communications. https://doi.org/10.1109/TCOMM.2019.2924010

Zare, D., & Ghazali, H. M. (2017). Assessing the quality of sardine based on biogenic amines using a fuzzy logic model. *Food Chemistry, 221*, 936–943. https://doi.org/10.1016/j.foodchem.2016.11.071

Zareiforoush, H., Minaei, S., Alizadeh, M. R., & Banakar, A. (2015). A hybrid intelligent approach based on computer vision and fuzzy logic for quality measurement of milled rice. *Measurement, 66*, 26–34. https://doi.org/10.1016/j.measurement.2015.01.022

Zareiforoush, H., Minaei, S., Alizadeh, M. R., Banakar, A., & Samani, B. H. (2016). Design, development and performance evaluation of an automatic control system for rice whitening machine based on computer vision and fuzzy logic. *Computers and Electronics in Agriculture, 124*, 14–22. https://doi.org/10.1016/j.compag.2016.01.024

Zhang, C. (2019). Research on the fluctuation and factors of China TFP of IT industry. *Journal of Industrial Integration and Management, 04*(04), 1950013. https://doi.org/10.1142/S2424862219500131

Zhang, C. (2020). Research on the economical influence of the difference of regional logistics developing level in China. *Journal of Industrial Integration and Management, 05*(02), 205–223. https://doi.org/10.1142/S2424862220500049

Zhang, C., & Chen, Y. (2020). A review of research relevant to the emerging industry trends: Industry 4.0, IoT, blockchain, and business analytics. *Journal of Industrial Integration and Management, 05*(01), 165–180. https://doi.org/10.1142/S2424862219500192

Zhang, C., & Lu, Y. (2021). Study on artificial intelligence: The state of the art and future prospects. *Journal of Industrial Information Integration, 23*, 100224. https://doi.org/10.1016/j.jii.2021.100224

Zhang, C., Patras, P., & Haddadi, H. (2019a). Deep learning in mobile and wireless networking: A survey. *IEEE Communications Surveys & Tutorials, 21*(3), 2224–2287. IEEE Communications Surveys & Tutorials. https://doi.org/10.1109/COMST.2019.2904897

Zhang, C., Xu, X., & Chen, H. (2019b). Theoretical foundations and applications of cyber-physical systems: A literature review. *Library Hi Tech*, ahead-of-print. https://doi.org/10.1108/LHT-11-2017-0230

Zhang, S., et al. (2019c). Deep learning based recommender system: A survey and new perspectives. *ACM Computing Surveys, 52*(1). https://dl.acm.org/doi/abs/10.1145/3285029

Zhang, W., et al. (2019d). Domain ontology development of knowledge base in cardiovascular personalized health management. *Journal of Management Analytics, 6*(4). https://www.tandfonline.com/doi/abs/10.1080/23270012.2019.1694454

Zhang, H., Sun, H., Wang, L., Wang, S., Zhang, W., & Hu, J. (2018). Near infrared spectroscopy based on supervised pattern recognition methods for rapid identification of adulterated edible gelatin. *Journal of Spectroscopy, 2018*(1), 7652592. https://doi.org/10.1155/2018/7652592

Zhang, J., & Tao, D. (2021). Empowering things with intelligence: A survey of the progress, challenges, and opportunities in artificial intelligence of things. *IEEE Internet of Things Journal, 8*(10), 7789–7817. IEEE Internet of Things Journal. https://doi.org/10.1109/JIOT.2020.3039359

Zhang, C., & Fu, W. (2021). Optimal Model for Patrols of UAVs in Power Grid under Time Constraints *International Journal of Performability Engineering 17*(1), 103. https://doi.org/10.23940/ijpe.21.01.p10.103113

Zhao, Z. -Q., et al. (2019a). Object detection with deep learning: A review. *IEEE Journals & Magazine IEEE Xplore*. https://ieeexplore.ieee.org/abstract/document/8627998

Zhao, R., Yan, R., Chen, Z., Mao, K., Wang, P., & Gao, R. X. (2019b). Deep learning and its applications to machine health monitoring. *Mechanical Systems and Signal Processing, 115*, 213–237. https://doi.org/10.1016/j.ymssp.2018.05.050

Index

A
Allergen detection, 200, 535
Artificial intelligence (AI), 2, 11, 27–29, 44, 59, 60, 125, 140, 147, 161, 177, 188, 207, 208, 278, 299, 381, 406, 407, 442, 446, 475–480, 492, 501–536
Authentication, 40, 58, 77, 81, 82, 84, 86, 87, 155, 176, 189, 195, 198, 203–204, 207, 440–441, 466

B
Backscattering imaging, 215–232, 466, 468
Biosensors, 26, 314, 358, 379
Biospeckles, 112, 113, 116, 237–255

C
Chemical contaminants, 380
Color, 3, 50, 161, 187, 216, 242, 259, 302, 373, 433, 451, 532
Composition analysis, 69, 200–201, 423, 493
Computed Tomography (CT), 12, 22, 135–156, 464, 469–471
Computer vision system (CVS), 24, 100, 107–112, 229, 472, 473, 487, 525, 528, 529

D
Destructive analysis, 12–14

E
Electronic nose (e-nose), 18, 25, 196, 238, 351–356, 359, 360, 363, 364, 366, 372, 373, 525–528
Electronic tongue (e-tongue), 18, 25–26, 196, 351, 352, 356–358, 363–365, 367, 372, 373, 525, 528

F
Food, 1, 40, 65, 99, 135, 159, 188, 215, 239, 259, 287, 328, 351, 379, 417, 451, 516
Food characteristics, 3, 5, 11, 26, 72, 136, 155, 159, 196, 202, 203, 223, 380, 417, 418, 420–423, 430–432, 439, 445, 453, 454, 464, 484, 486
Food processing, 4, 5, 8, 13, 21, 23, 26, 100, 104, 107, 123, 135, 155, 160, 188, 189, 191, 193–209, 259–281, 325, 391, 395, 405, 423, 431, 433, 434, 443, 445, 446, 463, 466–469, 474, 480, 481, 487, 492, 517, 523, 524
Food quality, 1, 59, 99, 135, 159, 189, 222, 239, 270–281, 311, 341, 379, 421, 453, 520
Food quality and safety, 2, 10, 11, 15, 19, 29, 40, 162, 189, 232, 255, 287, 302, 316, 393, 467, 536
Food quality control, 1, 2, 29, 143, 268, 474
Food quality evaluations, 136, 163, 165, 167, 288, 432, 456, 489
Food safety, 1, 106, 135, 188, 238, 259, 287, 380, 432, 466, 516

C. K. Sunil et al. (eds.), *Emerging Non-destructive Technologies for Food Quality Analysis*, https://doi.org/10.1007/978-3-031-95048-3

Food wastes, 28, 99, 255, 395, 429, 438, 439, 444, 523, 533
Fourier transform infrared (FTIR) spectroscopy, 14, 15, 19–20, 41, 42, 44, 45, 53, 60, 135, 290, 489

H
Honey, 50, 69, 77–80, 306, 312, 365, 373, 399, 441, 522, 527, 528
Hyperspectral imaging (HSI), 13, 49, 103, 159, 187, 238, 264, 299, 442, 465

I
Imaging, 20, 48, 88, 102, 139, 159, 187, 219, 238, 261, 288, 326, 418, 462
Imaging techniques, 21, 112–118, 125, 136, 139, 152, 161, 187–209, 228, 237, 246, 272, 299–302, 315, 317, 342, 424, 464–467, 469, 471, 472, 480, 481, 485, 491, 492
Image processing, 21, 100, 105–107, 110, 119, 120, 123, 125, 142–143, 150–151, 165, 190–192, 202, 222–226, 263, 264, 271, 272, 432, 464, 465, 471–473, 478, 479, 487, 501, 510, 528, 529
Internet of Things (IoT), 27, 28, 59, 111, 406, 492, 505

K
Kovats Retention Index, 362

L
Light propagation, 217, 219, 220, 227

M
Machine learning (ML), 2, 59, 119, 160, 192, 223, 278, 299, 364, 381, 442, 472, 502
Magnetic resonance imaging (MRI), 12, 16, 21, 72, 87–89, 91, 101, 112, 113, 115–116, 136, 151–156, 330, 432, 464, 467, 468, 485, 489–490
Multispectral imaging (MSI), 17, 20–21, 100, 104, 161, 167, 187–209, 466, 471, 493

N
Nanotechnology, 380, 381, 406
Natural language processing (NLP), 477, 501, 502, 506, 509–510, 535
Near-infrared (NIR), 13–20, 40–60, 100, 103, 108, 109, 115, 124, 159–161, 163, 164, 171, 173, 175, 177, 189–191, 195, 196, 199, 203, 206, 208, 219, 221, 238, 244, 261, 288, 291, 306, 343, 346, 465–467, 472, 473, 489–491, 525, 530
NMR in food component analysis, 73
Non-destructive, 13, 41, 91, 116, 135, 159, 189, 228, 237, 259, 287, 325, 351, 382, 417, 462, 528
Non-destructive analysis, 12, 200, 255, 331–347, 466
Non-destructive techniques (NDT), 12–15, 21, 23, 263, 287, 341, 462, 480, 481, 493
Non-destructive testing (NDT), 326–330, 343, 347, 438
Nuclear magnetic resonance (NMR), 16, 21, 24, 41, 65–91, 115, 116, 155, 288, 432, 468, 489, 490, 493

P
Packaging inspections, 189, 195, 204–205, 209, 268
Pathogen detection, 28, 406

Q
Quality, 1, 40, 77, 99, 135, 159, 188, 221, 237, 259, 287, 329, 352, 380, 417, 451, 509
Quality analysis, 2, 11–28, 60, 89, 135, 151, 159, 167–176, 248, 303, 304, 341, 346, 351, 379, 380, 382, 393, 395, 406–407, 417, 419, 431–441, 446, 527, 531–532, 536
Quality assessments, 7, 21, 25, 28, 29, 59, 60, 90, 107, 114, 117, 118, 136, 143, 144, 154, 196–197, 203, 204, 209, 227, 231, 238, 248, 260, 272, 288, 289, 314, 316, 462, 465, 471, 474, 476, 478, 493, 531
Quality control, 2, 20, 27, 56, 59, 91, 125, 135, 136, 152, 159, 160, 167, 170, 171, 177, 188, 189, 193, 199, 200, 207, 223, 230, 255, 262, 263, 266–269, 273–274, 278, 279, 281, 289, 291, 343–347, 391, 400, 405–407, 422, 425, 427, 429, 431, 433, 434, 436–438, 441, 462, 464, 466, 468, 474, 476, 481, 488, 516–518, 524, 526, 527, 529, 531
Quality testing, 177, 302, 381

R
Real time analysis, 59, 266

S
Safety, 1, 57, 79, 117, 136, 159, 188, 221, 237, 259, 287, 330, 379, 417, 462, 519
Sensory analysis, 25, 89, 274, 351, 373, 456, 482, 534
Spatial analysis, 240–243
Spectroscopy, 13, 40, 65, 100, 154, 159, 196, 221, 238, 288, 396, 418, 466, 530

T
TD-NMR, 69, 83, 91
Temperature monitoring, 260, 266, 268
Texture, 2, 89, 99, 152, 161, 188, 216, 237, 273, 302, 342, 351, 417, 452, 529
Thermal imaging, 17, 22–23, 113, 116–117, 259, 260, 262–266, 268, 270–281, 485, 487
Time domain, 41, 69, 155, 238, 289–293, 299, 300, 304–307, 312, 340
Time history speckle pattern (THSP), 239, 241–243, 253

U
Ultrasonic analysis, 341, 343–345
Ultrasound, 13, 18, 59, 100, 112, 114, 117–118, 326–346, 432, 464, 467, 468, 480, 481, 525
UV-visible, 51–60, 493

V
Vibrations, 14, 23, 43, 46, 50, 99, 100, 117, 263, 290, 291, 312, 316, 328, 432, 481, 483, 484, 488, 530, 532

X
X-ray, 21, 100, 136, 189, 261, 288, 467